THE WORLD OF THE AMERICAN WEST

The World of the American West is an innovative collection of original essays that brings the world of the American West to life and conveys the distinctiveness of this diverse, constantly changing region. Twenty scholars incorporate the freshest research in the field to take the history of the American West out of its timeworn "Cowboys and Indians" stereotype right up to the major issues being discussed today, from water rights to the presence of the defense industry. Other topics covered in this heavily illustrated, highly accessible volume include the effects of leisure and tourism, Western women, politics and politicians, Native Americans in the twentieth century, and, of course, oil.

With insight both informative and unexpected, *The World of the American West* offers perspectives on the latest developments affecting the modern American West, providing essential reading for all scholars and students of the field so that they may better understand the vibrant history of this globally significant, ever-evolving region of North America.

Contributors: Gordon Morris Bakken, Peter J. Blodgett, Robert Carriker, Alicia M. Dewey, Lisa E. Emmerich, Randal Fulkerson, Neil Greenwood, Elizabeth Kuebler-Wolf, Sheila McManus, K. D. Motes, Charles R. Porter, Andrea G. Radke-Moss, Brent M. Rogers, Kurt Russo, Shannon D. Smith, Gregory E. Smoak, Robert L. Spude, W. R. Swagerty, Susan Tschabrun, and Aaron Steven Wilson.

Gordon Morris Bakken is Professor of History at California State University, Fullerton. He is the author of many books, his most recent being *Invitation to an Execution: A History of the Death Penalty in the United States.*

THE ROUTLEDGE WORLDS

THE ELIZABETHAN WORLD
Edited by Susan Doran and Norman Jones

THE BYZANTINE WORLD
Edited by Paul Stephenson

THE VIKING WORLD
Edited by Stefan Brink in collaboration with Neil Price

THE BABYLONIAN WORLD
Edited by Gwendolyn Leick

THE EGYPTIAN WORLD
Edited by Toby Wilkinson

THE ISLAMIC WORLD
Edited by Andrew Rippin

THE WORLD OF POMPEII
Edited by Pedar W. Foss and John J. Dobbins

THE RENAISSANCE WORLD
Edited by John Jeffries Martin

THE EARLY CHRISTIAN WORLD
Edited by Philip F. Esler

THE GREEK WORLD
Edited by Anton Powell

THE ROMAN WORLD
Edited by John Wacher

THE HINDU WORLD
Edited by Sushil Mittal and Gene Thursby

Forthcoming:

THE OTTOMAN WORLD
Edited by Christine Woodhead

THE ATLANTIC WORLD
Edited by William O'Reilly

THE VICTORIAN WORLD
Edited by Martin Hewitt

THE WORLD OF THE INDIGENOUS AMERICAS
Edited by Robert Warrior

THE WORLD OF THE AMERICAN WEST

Edited by

Gordon Morris Bakken

Routledge
Taylor & Francis Group
NEW YORK AND LONDON

First published 2011 by Routledge
605 Third Avenue, New York, NY 10017
2 Park Square, Milton Park, Abingdon, Oxon OX14 4RN

Routledge is an imprint of the Taylor & Francis Group, an informa business

Typeset in Garamond by EvS Communication Networx, Inc.

Library of Congress Cataloging in Publication Data
The world of the American West / edited by Gordon Morris Bakken.
p. cm.—(The Routledge worlds)
"Simultaneously published in the UK"—T. p. verso.
Includes bibliographical references and index.
1. West (U.S.)—History. 2. West (U.S.)—Geography.
3. Regionalism—West (U.S.) 4. West (U.S.)—Social conditions.
I. Bakken, Gordon Morris.
F591. W8744 2010
978—dc22
2010008269

ISBN 13: 978-0-415-98995-4 (hbk)
ISBN 13: 978-1-03-240260-4 (pbk)
ISBN 13: 978-0-203-84605-6 (ebk)

DOI: 10.4324/9780203846056

For

Bob and Donia Bovee, Big Timber, Montana,

and Mike and Mona Lovely, McLeod, Montana

couples living the world of the American West

CONTENTS

ILLUSTRATIONS

FIGURES

TABLES

CONTRIBUTORS

Gordon Morris Bakken is Professor of History at California State University, Fullerton. He is the author of many books, his most recent being *Invitation to an Execution: A History of the Death Penalty in the United States.*

Peter J. Blodgett is the H. Russell Smith Foundation Curator of Western Historical Manuscripts at the Huntington Library and Executive Director of the American Historical Association's Pacific Coast Branch. He received his bachelor's degree from Bowdoin College and his master's and doctoral degrees in Western history at Yale University. Since joining the Huntington Library's Manuscripts Department in 1985, he has been responsible for the acquisition, processing, exhibition, and reference service of the library's rare original documents concerning the history of the trans-Mississippi West across the past two centuries. Active in a variety of professional organizations, he has spoken and written widely on national parks, tourism, and recreation as well as on the management of manuscripts and archives.

Robert Carriker is the Alphonse and Geraldine Arnold Distinguished Professor of the College of Arts and Sciences at Gonzaga University in Spokane, Washington. Among his books are several on the Lewis and Clark Expedition, including *Ocian in View! Lewis and Clark on the Columbia River* (2005). He is a co-author of *America Looks West: Lewis and Clark on the Missouri* (2002). Dr. Carriker has directed eight National Endowment for the Humanities seminars on Lewis and Clark, and he has been a scholar-leader for Smithsonian Journeys and Lindblad–National Geographic expeditions.

Alicia M. Dewey received her PhD from Southern Methodist University in 2007 and is Associate Professor of History at Biola University in La Mirada, California. Her current research focuses on Anglo and Tejano entrepreneurs in South Texas in the early twentieth century, including their roles in transforming local economies and the ways in which they navigated success and failure.

Lisa E. Emmerich is a professor of history and the Coordinator of the American Indian Studies Program at California State University, Chico. Her fields of research and publication include American Indian women, American Indian health care in the nineteenth and twentieth centuries, and American Indian gaming. Her current research project focuses on representations of the American Indian past in roadside historical markers.

Randal Fulkerson earned his PhD in social-cultural history from the University of Nebraska. Randal worked as a movie and television extra while earning his MA in American history from California State University, Fullerton. He is Adjunct Professor at Southeast Community College in Lincoln, Nebraska, teaching U.S., World, and Hollywood history.

Neil Greenwood earned a BA in history at California State University, Fullerton, in 1983 and a MA and PhD in history at University of California, Los Angeles, in 1985 and 1992, respectively. He has been teaching history at Cleveland State Community College in Cleveland, Tennessee, since the fall semester of 1992.

Elizabeth Kuebler-Wolf is Assistant Professor and the Program Director for Art History at the University of St. Francis in Fort Wayne, Indiana.

Sheila McManus is Associate Professor of History at the University of Lethbridge in Lethbridge, Alberta. She is the author of *The Line which Separates: Race, Gender and the Making of the Alberta–Montana Borderlands in the Late Nineteenth Century* (2005), *Choices and Chances: A History of Women in the United States West* (2010), and is co-editor with Elizabeth Jameson of *One Step over the Line: Toward a History of Women in the North American Wests* (2008).

K. D. Motes is Assistant Professor of History at Rockhurst University in Kansas City, Missouri. A tribal member and native of the Choctaw Nation of Oklahoma, his research focuses on Choctaw culture in the nineteenth century. Dr. Motes holds a PhD from the University of California, Riverside.

Charles R. Porter, Jr. is a real estate and construction expert and Adjunct Professor of History at St. Edward's University in Austin, Texas. He researches, writes, and teaches the history of Spanish Catholics in the New World and their influence on Texas law, specifically law relating to water rights, community formation, and real-estate transfer. He was the winner of the Texas Catholic Historical Society's Robert S. Weddle Award in 2009 and St. Edward's University Presidential Award in 2008. Certified as an instructor by the Texas Real Estate Commission, he writes and teaches core- and continuing-education courses, including, among others, "Mediate, Arbitrate, Litigate" (winner of the Best Education Program in Texas in 2005) and "Water Rights for Texas Agents" (winner of the Best Ethics Program in Texas in 2008).

Andrea G. Radke-Moss received her PhD in history from the University of Nebraska, Lincoln, in 2002 and is Assistant Professor of History at Brigham Young University, Idaho. Her book, *Bright Epoch: Women and Coeducation in the American West* (2008) was a finalist for the 2009 Great Plains History Distinguished Book Prize. Radke-Moss's research interests include women's experiences at the Chicago World's Columbian Exposition of 1893, Mormon women, and women and education in the American West. She anticipates the publication of book chapters on Northern Great Plains women and Mormon women's experiences at the 1893 World's Congress of Representative Women.

Brent M. Rogers earned a bachelor's degree in history from San Diego State University in 2005 and a master's degree in history with an emphasis in public history from California State University, Sacramento, in 2007. A PhD candidate in history at the University of Nebraska, Lincoln, Brent specializes in the nineteenth-century American West and digital history. He has authored essays on the place of Utah Territory in national debates over popular sovereignty during the mid-nineteenth century and on history in the digital age.

Kurt Russo received his PhD in history from the University of California, Riverside, an MS in Forestry from the University of Washington, and a BS in Forestry from the University of Montana. He has worked with indigenous communities throughout the American West and Alaska, as well as in Canada, Mexico, Guatemala, the Caribbean, Brazil, and Chile. He is Research Fellow at the California Center for Native Nations (University of California, Riverside) and the Executive Director of the Native American Land Conservancy based in southern California. He lives in Bellingham, Washington, where he works with the Lummi Indians on poverty-reduction programs.

Shannon D. Smith is Fellow at the EDUCAUSE Center for Applied Research and teaches online courses in history and social justice at Oglala Lakota College on the Pine Ridge Indian Reservation. She is a PhD student of Western history at the University of Nebraska, Lincoln, and the author of *Give Me Eighty Men: Women and the Myth of the Fetterman Fight,* which was awarded the 2009 Wyoming State Historical Society Award for Nonfiction.

Gregory E. Smoak, Associate Professor of History at the University of Utah, specializes in American Indian, American Western, and public history. He completed an MA at Northern Arizona University and a PhD at the University of Utah in 1999. He has taught at the University of Minnesota and Colorado State University. He also served as the Director of Historical Research at the American West Center of the University of Utah where he worked with numerous American Indian tribes. His book *Ghost Dances and Identity: American Indian Ethnicity, Racial Identity, and Prophetic Religion* was published in 2006.

Robert L. Spude received his bachelor's and master's degrees from Arizona State University and a PhD in history from the University of Illinois at Urbana, Champaign. Since 1977 he has worked at various posts for the National Park Service, including Alaska parks, Anchorage regional office, and Denver regional office, and is presently Regional Historian, Intermountain Region, based in Santa Fe. His publications include essays on the parks, the history of the American mining West, and a forthcoming co-authored history of New Mexico.

W. R. Swagerty is Professor of History and Director of the John Muir Center at University of the Pacific in Stockton. Raised in New Mexico, he holds degrees in history from the Colorado College (1973) and the University of California, Santa Barbara (1981). His specialties are the exploration and fur-trade eras of the North American West, American Indian history, and the history of the North American environment.

Susan Tschabrun holds a Ph.D in History from the University of Wisconsin, Madison and a Masters in Library Science from the University of California, Los Angeles. She has published in the fields of history of the American West, African history, and archival studies. She is a member of the library faculty at the Pollak Library at California State University, Fullerton, where she serves as Department Chair.

Aaron Steven Wilson is Resident Associate Professor at Creighton University. He received his PhD from the University of Nebraska, Lincoln. His major interests include imperialism in the American West, territorial history, and comparative U.S. history. He is working on a manuscript that explores the relationship between Hawaii, American colonialism in the Pacific, and Western expansion.

INTRODUCTION

When I agreed to edit this volume on the American West, my mind reflected upon the task of identifying the twenty chapter topics and convincing authors to go on this scholarly journey. In contemplating the topics, Irena Pratis's poem "Cairns" roused my imagination.

> I find them when I need them most—
> When faced with an arroyo
> That's washed the trail away,
> When coming onto a mesa
> Of featureless stone, when trees are woven
> With so many paths I don't know which to take—
> At each impasse, I stop and think,
> "I can't go on." I consider retracing my steps:
> "If I can't get where I'm going,
> I can at least return to where I started."
> And then I see it, a cairn, a pile of stones
> That marks the way—and I pick the trail
> Up again. I've read of cairns as burial markers,
> And those that rest on sacred places,
> And I wonder at how they hold
> To memory and meaning, offering guides,
> Memorials, a road to follow when all seems lost.
> They touch the past and gesture
> Toward the future. They speak
> To me: Someone I will doubtless never meet
> Came this way before, and seeing
> The importance of the place,
> Or knowing the risks of losing the trail,
> Stopped and set stones on stones
> As carefully as placing words near words,
> And through attention given in that moment,
> Many since have found their way.[1]

The history of the American West has a beginning and no end, yet a road to the future is marked by cairns and guides. Without debating the geographic limits of the American West, the first two chapters set out geomorphology as well as the plants and animals of the West. American Indians followed hunting and gathering in this terrain. Explorers entered the region as did the fur trappers and traders. The gold rush to California moved the American center to the West. Some of these topics have been marked well by Frederick Jackson Turner, Ray Allen Billington, and Martin Ridge. Yet when I hit the first arroyo, it was clear that few books took a serious look at the physical geography of the West. Robert L. Spude did so, and we start our journey with scientists of the early twentieth century puzzling over the West. An even less worn path was the flora and fauna of the American West. Kurt Russo moves the scholarly journey off of a train and into bioregions now familiar to some environmental historians. Lewis and Clark and numerous other explorers trekked through the terrain and trampled sacred places of many tribes and bands. The Corps of Discovery, in particular, created a distinct cairn and a memory most recently revisited by thousands of tourists. W. R. Swagerty, following fur traders and trappers throughout the West, explores clear markers left by historians of the early nineteenth century but creates cairns distinctly Swagerty in interpretation. Sheila McManus takes us out of the rendezvous and onto the borderlands of the American West with all of its peoples. The West is a place of dynamic social, economic, and cultural contact. Throughout this volume you will encounter women in a variety of roles. Their place in the American West is evolving with the work of historians, and I trust you will find these pages adding to that knowledge.

Placing words by words in crafting the topics of this volume, I thought that urbanization, irrigation, and recreation deserved scholarly attention ranging from the nineteenth to the twenty-first century. Cities came to dominate the West and brought educated people together to form an economic critical mass of talent capable of sustaining computer, airframe, and defense industries. Water enabled this growth, and recreation followed, now sustaining regional economies no longer able to maintain mining, forestry, or manufacturing sectors. Each topic has a chapter, and many themes course well throughout the second half of this volume.

Another topic that transcends chapters is the role of women in the West. Whether in education, science, or politics, women in the West made a difference and molded the trajectory of the region.

The American West is a region of extremes. Beyond terrain, economic disparity clearly marks the successes and failures of politics and economics. Yet the West was the focus of imaginative artists who won the American art world and philanthropists who funded rich collections of their art. These artists often represented American Indians in their works. In the twenty-first century, the image of the American Indian is far different than the extremes of the nineteenth century. Tribal success in gaming and economic development characterizes some tribes and bands. Yet poverty dogs others.

Finally, the West, particularly California and Texas, grew into an economic powerhouse giving the region world-ranked economic status. When thinking about who should write this final chapter, my memory recalled Joan Voss Greenwood's haiku written in 1991:

Twenty-seven years
ago we put away my
baby bassinet
Neil and Mary used;

now we dust it off for Mike's
and Mary's first child.[2]

History is a journey marked off by many words, many cairns, many bassinets of knowledge, but it was only fitting that the first historian to answer my call to scholarship was Neil Greenwood, and it is only fitting that he should have the last word.

In addition to the authors, I must thank my editorial board of Mary Murphy, Montana State University, Bozeman; Matthew Whittaker, Arizona State University; and Vanessa Gunther, Chapman University. Several colleagues were significant in suggesting authors: Brian Dippe, University of Victoria; John Wunder, University of Nebraska; Laura Woodworth-Ney, Idaho State University; Paula Mitchell Marks, St. Edwards University; Leo Lyman, Victor Valley College; and Clifford Trafzer, University of California, Riverside. These historians filled many of history's bassinets with scholarship and were most kind to read chapters, suggest authors, and encourage the scholarship in this volume. These chapters are now the cairns on the trail of the history of the American West. The paths they mark out are broader and far longer than the ones I hiked even a decade ago.

Gordon Morris Bakken
Fullerton, California

NOTES

1. *Touch* (Georgetown, Ky.: Finishing Line Press, 2004), p. 17.
2. Joan Voss Greenwood, *Stirring Dawn* (Indian Trail, NC: D-N Publishing, 2009), p. 35.

CHAPTER ONE

THE PHYSICAL GEOGRAPHY OF THE WEST

Robert L. Spude

THE "GEOGRAPHERS' SPECIAL" TRAIN; OR, THE VIEW OF THE GEOMORPHOLOGY OF THE WEST FROM A RAILROAD OBSERVATION CAR WINDOW

On August 22, 1912, at 8:46 A.M., the aging Professor William Morris Davis, the "father of American Geography," could relax as the nine-car "geographers' special" train left Grand Central Station, New York City, for its long-planned cross-country journey. Onboard were forty-three foreign geographers from thirteen different countries and a host of their American colleagues. Davis had successfully convinced several railroad barons to support a two-month traveling symposium that not only would study and publicize the West, but that would also alter the way in which we define its varied landscapes today.

This would not be a rigorous expedition, thanks to the help of the Pullman Railroad Car Company and the polished varnish observation cars "Huelma" and "Oronso," a dining car, and staff. The Poland Springs Water Company of Maine supplied water, and the Anheuser-Busch Brewing Company provided their special refreshment for the travelers. The private cars of railroad officials joined the group as they journeyed along each line, on the Northern Pacific system through the Northwest to Seattle and Portland, with the Southern Pacific down the coast to San Francisco and over the Sierra and across the Great Basin to Salt Lake City. From the Mormon capital to Denver, the Denver & Rio Grande system hauled the special, and south into New Mexico and Arizona, the great Santa Fe line catered to the savants. A list of those tagging along reads like a who's who of early twentieth-century railroad men. They ensured the success of the rail tour.

Davis brought with him a group of his Harvard students and disciples. Mark Jefferson of Michigan State University, Ypsilanti, Richard E. Dodge of Columbia, and Isaiah Bowman of Yale were marshals to help usher the group along. Davis appointed Nevin M. Fenneman, University of Cincinnati, "journalist," to chronicle the tour as well as deal with reporters. Nevin M. Fenneman was born 1865 and raised in Ohio, earned his doctorate in the physical sciences from the University of Chicago in 1901. In 1907, he joined the geology and geography faculty of the University of Cincinnati, where he spent the rest of his career. He also worked mapping the St. Louis area for the US Geological Survey (USGS), which brought him to the attention of his colleagues, and to Davis who invited him on the railroad tour.

Lawrence Martin, University of Wisconsin, worked with Fenneman to publish a "Daily Bulletin" on the train. Albert Perry Brigham of Colgate would be trip historian. All would be among the leading figures in American geography during the ensuing decades. Davis had come to organize the event as part of the sixtieth anniversary of the American Geographical Society, as well as to get his colleagues in the field. He was a major advocate of distant expeditions to evaluate and observe the land as well as the exchange of ideas with colleagues from abroad.

During the first days, the train made several stops along the way West, such as at Niagara Falls and Chicago, but, in the words of Alfred Ruhl from Berlin, "when we get outside of the big cities we hope to see something of the real American People."[1] As the train rolled through Minnesota, through the "bonanza wheat farms" of Fargo, a look through the observation car windows revealed a landscape that became more barren and arid. In 1912, the farms dotting the trackside became fewer and far between after the train crossed the Missouri River at Bismarck.

At their first stop west of the 98th meridian, Medora, North Dakota, the assembled small-town boosters, business leaders, politicians, and reporters met the well-publicized tour party—an act that would be repeated in other towns, large and small, for the tour's duration. Horses and wagons took the geographers into the Dakota badlands, where Teddy Roosevelt had once ranched and been a cowboy. Roosevelt was in the news in that summer of 1912 as he returned for another bid for the Presidency. His Bull Moose Progressives party had supporters in the Bad Lands, but politics was not to be a topic of discussion for the geographers.

William Davis, author of treatises on the "cycles of erosion," was in his element in the

Figure 1.1 Badlands developed on sedimentary rocks, Theodore Roosevelt National Park, North Dakota.
Source: Courtesy of Wayne Engstrom.

Figure 1.2 Mammoth Hot Springs with active travertine precipitation, Yellowstone National Park. Source: Courtesy of Wayne Engstrom.

badland's grayish-brown broken landscape. His studies of geomorphology had helped better understand landscapes and lead in classifying and defining landscape changes. The German scientist Baron Alexander von Humboldt had first developed a classification system for physical geography, and American John Wesley Powell, Director of the USGS, in 1880 had used it to prepare the first crude draft on physiographic divisions. With Professor Davis on the train was geographer Nevin M. Fenneman, who had already considered improving that work, and his meticulous reporting of the journey led to his proposal at the next meeting of the American Geographical Society to prepare the definitive classification and division system. Over the next dozen years, the physical division or geomorphology published by Fenneman became recognized even by those who may have disagreed in specific instances with his boundaries, and they are convenient for more particular comment on the landforms. As the train rolled into Montana, the geomorphology of the West unfolded before them.

According to Fenneman, the Great Plains Province extended into the eastern parts of New Mexico, Colorado, Wyoming, and Montana. The province's tilted plateau was created from the alluvium of streams flowing east out of the Rocky Mountains. Davis would explain this to the train party in the badlands, and as they rolled west up the Yellowstone River that the rivers picked up rock, sand, and dirt deposited on the broad descent to the Mississippi. The continual action of mountain uplift and erosion left a maze of riverbeds, extensive breaks, and badlands from West Texas to Montana.

As the train steamed by the Missouri River, then up the Yellowstone River, the railroad land offices had hoped the geographers' press releases would be a boon for land sales. Geographer Howard Simpson of the University of North Dakota and Daniel E. Willard, formerly with that college but in 1912 with the Northern Pacific's publicity department,

Figure 1.3 Old Faithful Geyser, Yellowstone National Park, Wyoming. Source: Courtesy of Wayne Engstrom.

helped promote settlement under the promotion of "Dry Land" farming. Both were passengers on the train. The railroads concurred that scientific farming, their term for dry farming, could enhance the value of their lands, which the Granger railroads from the Rock Island line in West Texas and New Mexico to the Burlington and Northern Pacific lines in the Dakotas and Montana promoted new homesteads and settlement. Unfortunately, this Dry Land farming boom of the 1910s, one of the last great migrations to the West, would be defeated by the inevitable droughts and the resultant dust bowl, which by the 1930s left a string of ghost towns across the Great Plains Province.[2]

Debates on soils, climate, and glacial activity fed the group's discussion, especially as the train sped toward Yellowstone National Park, in the heart of the Rocky Mountain Province. On September 3, the party arrived at Gardiner, Montana, for six days of exploring the nation's first national park. Yellowstone National Park was created by Congress March 1, 1872, "as a public park or pleasuring ground for the benefit and enjoyment of the people."[3] The Northern Pacific Railroad had promoted the park as a tourist destination from the date of its founding, if not before. It had built grand lodges, catering facilities, and other concessions while the government debated how to manage the park—the National Park

Service (NPS) would not be created until 1916, so the US Army, which patrolled the park against poachers, guided the geographers around.

They toured Old Faithful geyser and geyser basin, Mammoth Hot Springs and boiling mud pools, Yellowstone Falls, and Yellowstone Lake. Exchanges of ideas were highlighted by evening talks in front of the roaring fireplace at the grand Old Faithful Inn. Formal papers were given that evening: Gunner Anderson of Stockholm on Yellowstone's flora with reference to the evergreen mountains of Scandinavia; Emmanuel Martonne of Paris on the park's geomorphology; and Emile Chaix of Geneva on the geysers.

Late on September 9, the special pulled out of the Livingston, Montana, station bound for a stop at the bustling mining town of Butte. Founded as a placer gold camp in 1864, Butte in 1912 was the largest industrial center in the Rocky Mountain province. The city's chamber of commerce sponsored an electric trolley tour to copper mines, mills, and the technical school. After two hours, the train steamed uphill through the broad valleys of small rail towns, high country farms, and ranches which dotted the green landscape into the Hell Gate crossing of the northern Rockies. In the spectacular mountain valleys and forests, it rumbled past the straggling mining centers of northern Idaho into Spokane, for a late evening, electric-light tour of the capital of the inland empire.

The next major stop was a visit to the Grand Coulee region of eastern Washington. From Almira, Washington, the group took an auto tour of the orchards, the wheat fields, and then the volcanic landscape of the Columbia Plateau province at the Grand Coulee. The geographers were interested in how the great Columbia River could have had a waterfall grander than Niagara here, and then by force of volcanic activity changed its course, leaving its former falls high and dry. Fenneman's Columbia Basin province was one of three major physiographic divisions separating the Rockies and the Pacific Coast ranges. The foreign geographers commented most on the layers of ash and lava caused by volcanic activity of the mountain ranges

Figure 1.4 Umatilla Rock at Dry Falls, Grand Coulee, Channeled Scablands, Columbia Plateau, Washington.
Source: Courtesy of Wayne Engstrom.

to the west and on the expanding wheat farms a result of irrigation, which was changing the appearance of the sage-brush-covered basin.

At Missoula, Montana, Fred G. Plummer, the chief geographer of the US Forest Service, had boarded the train for the ride to Seattle. Plummer, an early conservationist, had written about earthquakes and volcanic activity, but his best-known publications were focused on the ecology of forest fires. The Forest Service had evolved from the thinking of earlier conservationists such as George Perkins Marsh who had posited that denuded forests had caused the fall of earlier civilizations. In 1891, Congress recognized the need to wisely manage the nation's forests by passing the Forest Reserve Act followed by the creation of the US Forest Service in 1897. Plummer had first been hired to assist in assessing the reserves, but he quickly became an advocate for conservation. On the train he was able to show his fellow geographers the vast extent of Pacific Northwest's evergreen forests, the nation's largest. He also led discussions on the volcanic origins of the peaks of the coast range. Plummer's notes and correspondence, and those of other forest rangers, helped Fenneman develop the physical division of the Pacific Coast province. Unfortunately, Plummer died shortly after the tour without contributing to its ongoing academic debates and publications.

Once over the Coast Range, the geographers convened at the University of Washington and shared their findings about delineating the great Northwest. Seattle, like other communities along the route, feted the participants with dinners, tours, and publicity moments. A respite was taken from the train by a boat ride to Tacoma and discourses on the evolution of the volcanic landscape before them, dominated by the view of Mount Rainier. After a short stop in Portland, the train pulled into Medford, in southern Oregon, where another local boosters' greeting party led the group on an 80-mile auto tour to Crater Lake

Figure 1.5 Crater Lake caldera formed during an eruption about 7,600 years ago and younger cinder cone forming Wizard Island, Crater Lake National Park, Oregon.
Source: Courtesy of Wayne Engstrom.

Figure 1.6 Yosemite Valley, a classic example of a glacial trough, Yosemite National Park. Source: Courtesy of Wayne Engstrom.

National Park. That trip led to a healthy discussion of volcanism by the geographers. Over a campfire in the frosty air, the geographers had a memorable view of the blue waters of Crater Lake; they spent the night in a tent camp in the forest.

On September 18, Southern Pacific railroad engines pulled the special on an express run from Medford to San Francisco, where Andrew C. Lawson and Ruliff S. Holway of the University of California Geology and Geography Departments met the group. Both men had been on the State Earthquake Commission, and they retold of the rebuilding after the fire and earthquake six years before. The geographers were more eager to tour the "rift valley" of the San Andreas fault and to discuss recent quakes and volcanic action. Holway, in fact, would be the first to the top of Lassen volcano after its eruption in 1915. The valley of the Sacramento, where farmers were reclaiming the wetlands for their orchards and fields, sparked a debate about irrigation's expanding role in California. The Central Valley back of the San Francisco Bay was one of the two great valleys being opened for farming between the Sierra Nevada and the coast ranges. In California, the party met with horticulturist Luther Burbank on the evolving variety of fruits, vegetables, and other agricultural products that propelled the state's expanding economy.

A side trip north of the bay brought the party to Muir Woods National Monument, a remnant of the once vast redwood forests along the Marin Peninsula. John Muir met with the party and shared his philosophy and his concerns. In his final years, the philosopher-conservationist had done much for the protection of the scenic areas of the West, particularly in the Sierras. The party noted the contrast between Muir Woods' lush forest and grand redwoods and the denuded hills surrounding it; nowhere was the effect of overcutting more stark. At the time, Muir was in the middle of his last battle, the fight over preserving the Hetch Hetchy valley, a part of his beloved Yosemite National Park, a fight that would be lost the next year.

On September 21, the party reboarded the train in the San Francisco area, chugged on to Salt Lake City, and then to Denver, where it arrived on September 27. The trip, to judge by press reports, was a whirl of greetings by town boosters, testimonial dinners, and quick side tours over hill and dale to flourishing fields, prosperous mines, and bustling towns. For the geographers, this six-day stretch was an opportunity to view a cross-section of the US West's geography. One can image Nevin M. Fenneman sitting in the observation car staring out the window watching the landscape pass by, noting its characteristics and keeping tab in his daily reports. On the railroad cars gliding through the West he had experts at hand to engage in argument, critique, and debate. These included Ruliff Holway of Berkeley over the Sierra Nevada; Edwin S. Hinckley of Brigham Young University, Provo, Utah, across the Great Basin; Robert Marshall, Chief Geographer of the USGS over the Rocky Mountains; and dozens of European colleagues. The tableau of defining characteristics of the provinces—Pacific Coast, Great Basin, Rocky Mountain—would pass outside the window before them.

Based on what he saw, Fenneman would soon sketch a rough landmark map of the physical divisions or geomorphology of the USA. The cross-section of landscape of the Pacific Coast Province lay out his window, the Coast Range with its cliffs dropping into the Pacific Ocean, then the farmlands of the low Central Valley, which give way to the Sierra Nevada. The train passed glacier-carved canyons and evergreen forests, where some of the highest annual rainfalls fell, from the Sierra Heights to the Cascades. As the train dropped to the eastern foot of the Sierra, the sage-brush landscape of the Great Basin province unfolded before him. Following the Humboldt River from its sink, the geographer noted the closed basin, where the waters of the rivers had no exit to the ocean. The Great Salt Lake was

Figure 1.7 Aeolian cross-bedding in Navajo Sandstone, Zion National Park, Utah. Source: Courtesy of Wayne Engstrom.

the catch basin for the rivers of the eastern portion of the Great Basin province. They could denote the Lahontan beaches, the ancient Lake Bonneville shore line above its shore. North and south of the train, the distinctive basin and range landforms gave the province its distinctive variety of vegetation, from range top to valley floor.

On the eastern edge of the Great Salt Lake the geographers' special passed numerous Mormon farm villages where worthy Saints had built their homes and farms around cooperatively constructed irrigation systems. This island of green gave way to the steep Wasatch range, where the train passed over into the Colorado Plateau province, with its bare cliffs, reddish rocks, and dramatic canyons. Fenneman noted the arid landscape had only patches of green, where Colorado river water was diverted for irrigation projects, such as at Grand Junction, where the Colorado and Grand rivers met and where the steep canyon entry into the Rocky Mountains began.

Fenneman and his followers debated the exact boundary between the Colorado Plateau and Rocky Mountain provinces, but those on the train knew when the train left Basalt, Colorado, they were entering the granite blocks that uplifted as the Rocky Mountains. The train's route traveled along chiseled cliffs, beneath 14,000 foot peaks, and near timber line at Hagerman Pass. The increase in moisture, change in vegetation, and the broad mountain parks along the Arkansas River was also evidence to the party of the changed geomorphology. Then, abruptly, at the mouth of the Royal Gorge of the Arkansas, the train steamed out of the mountains and onto the Great Plains. Blocked by a direct route over the mountains into Denver, the railroad had to turn north, around the foot of Pikes Peak, to reach the Queen City of the Plains.

At Denver, the usual greetings by the governor and entourage of boosters holding welcoming receptions were followed by testimonials on the campus at Boulder and then short tours into the mountains. The party went above tree line at Corona Pass and hiked across 12,000 foot Hagerman Pass on foot rather than ride the train through the railroad tunnel. The explorer-geographer Gottfried Merzbacher of Munich commented on the ease of hiking above tree line in the Rockies as compared to the Alps' steeper slopes above tree line, which allowed for observation of mountain formation. On September 29, the party departed for Raton, New Mexico, and a view of its volcanic district. A side trip on the 20-mile long, ambitiously named St. Louis, Rocky Mountain, and Pacific Railroad to Capulin Volcano and back added to the debate on the formation of the geographic province along the southern Rockies.

Into Santa Fe on September 30, the group met anthropologist Edgar Lee Hewett and American Indian musicians and dancers. Hewett had been a significant participant in the passage of the 1906 Antiquities Act, which provided for the creation of national monuments in order to preserve prehistoric and other sites. A series of archeological sites across the Southwest had been immediately set aside for protection as national monuments by Teddy Roosevelt. Earlier, at the University of Utah, the group had heard Professor Byron Cummings speak about recent discoveries in the cliff dwellings of the Four Corners region of the Colorado Plateau province, the richest area for archeological investigations in the USA. Hewett took the party on an auto tour to Tusuque Pueblo and more interaction with the Pueblo Indians and discussions about the emerging field of cultural geography.

An evening reception at the Palace of the Governors showcased the new history and archeology museum, the Museum of New Mexico, headed by Hewett. The Hispanic community, led by former mayor Jose Sena, hosted the group with foods and scenes of the ancient capital. Isaiah Bowman of Yale, just beginning his thoughts on environmental

Figure 1.8 Coconino Plateau developed on flat-lying rocks, South Rim, Grand Canyon National Park.
Source: Courtesy of Wayne Engstrom.

controls and human responses, appreciated the opportunity to visit the ancient city, the "old Indian and Spanish realm," which had survived the centuries.[4] Eugen Oberhummer of Vienna included in his formal paper a discussion of "the typical Spanish city," represented by Santa Fe.[5]

Discussions of *acequias* and water systems and talk of desert peoples carried the geographers into the evening as the train hurried west for an early morning stop at Petrified Forest, then Meteor Crater, and, finally, the Grand Canyon. Two nights at El Tovar hotel allowed for travels along the rim and mule trips to the bottom of the canyon and back. Ellsworth and Emery Kolb gave an evening lecture with photographs of a trip down the canyon while some of the visitors bought pottery and jewelry from the Navajo and Hopi selling their handcrafted art at the train station. All agreed the Grand Canyon was "nature's masterpiece on the continent."[6]

From the canyon, the Santa Fe railroad took the geographers' special to Phoenix. Passing from the magnificent ponderosa forest around Williams, off the Colorado Plateau through Prescott, and then a drop to the saguaro cactus-covered Sonoran desert on the approach to Phoenix allowed the party to experience the rapid changes in vegetation. Besides the usual reception by the governor and an auto tour with boosters of the irrigated farms and orchards, the group had a chance to speak with Forrest Shreve, who arrived from the Carnegie Institute's research center in Tucson. Twenty-five-year-old Alan Grant Ogilvie of Oxford, later Edinburgh, was captivated by the desert biota; he would present and then publish his observations on American vegetation.

From Phoenix, an expedition along the Apache Trail, a dirt road built by the government as an example of the best mode of travel in the desert (and agreed to in a paper by Ole Olufsen of Copenhagen) took the party to the recently completed Roosevelt Dam. The dam was an early crowning achievement of the nascent Bureau of Reclamation. Organized after the passage of the Newlands Act in 1902, the Bureau had supported the construction of irrigation projects in seventeen Western states by the time of the geographers' trip. The dam across the Salt River brought large-scale irrigation to the Salt River Valley. The project impressed the geographers, where irrigation was making it, in the words of Anton Vacher of Lille, France, "a garden spot of the nation."[7] Expecting only to make an overnight visit at Roosevelt, a seasonal monsoon drenched the party, a flash flood causing the party to extend its stay. Uncontrolled water, in the desert of all places, readily showed the necessity of such dams to capture wasted runoff and to make the desert flower as the rose. The road wash-out also brought the weary geographers that unstated delightful moment called "down time."

Back in Phoenix, the tour organizers needed to shorten the tour. The group boarded the train for a hurried return to the East—but not without a stop near Gallup, New Mexico, to see the Nutria Monocline and an observation of the ancient lava flows of El Malpais. Their two-month expedition was near its end, which occurred October 12, at the University of Virginia. The guests participated in their last symposium, this on geography education. The party disbanded at Charlottesville, many going on to Washington, DC, as guests of the National Geographic Society, visiting colleagues in Federal agencies, or departing for home.

Upon their return to New York City, William Morris Davis and the remaining participants celebrated the opening of the new headquarters building of the American Geographical Society. Toasts and testimonials followed. Edouard-Alfred Martel of Paris said of the trip, "perfect organization and information—the height of comfort—explanations given by the most competent gentlemen on the spot—charm and interest of scenery and geological features—is such a journey a dream or a 'conte de fee'?"[8]

THE PHYSICAL PROVINCE OF THE WESTERN USA

After the journey, the participants had agreed to share their findings and observations through scholarly and popular publications, idea-sharing, and continued correspondence. But after a spate of publications, World War I intervened. Edouard-Alfred Martin would not meet such a diverse group of colleagues until after the war, during the negotiations of the Treaty of Versailles and the advisory board of geographers, some of whom had rolled through the American West on the "geographers' special" train.

Probably the most significant publication was the proposal of Nevin M. Fenneman to prepare the definitive map and guide on the physical divisions of the USA. His notes and reports on the train ride were a sound basis to begin the work. Shortly after the 1912 tour, Fenneman proposed to prepare a map of the physical divisions of the USA, classifying the distinct regions to better help Americans understand the land in which they lived.

The American Geographical Society supported the effort, and Fenneman chaired the preparation of a map, which was accepted by the USGS in 1916 (which also published a series of geological guidebooks to the rail route followed by the party). Working with Fenneman on the map was Chief Geographer of the USGS, Robert Marshall, a fellow passenger on the train ride out West. Though revised and improved, especially the graphic qualities through the use of computer graphics, the maps of Fenneman and his colleagues are still used. They exemplify the substantial accomplishments of the first generation of American geographers. The provinces in the Far West are as follows.

Figure 1.9 Eroded uplands in Rocky Mountain National Park, Colorado.
Source: Courtesy of Wayne Engstrom.

Great Plains Province

The distinguishing feature of the Plains is its apparent vast, level landscape, though there is a slight easterly tilt. There are few trees, mostly along river beds and water holes; vegetation is grasslands, with short grass prairies dominating the further west one travels. Past the 98th meridian there is insufficient rain for farming, at least as experienced by Americans east of the Mississippi River. Except where irrigation has been introduced or where drills tap aquifers for farmlands, the region is best known for its cattle ranches, some larger than states in the Northeast. The Black Hills are a distinct volcanic intrusion in the Great Plains. The Wyoming Basin province is an extension of the Great Plains in between the northern and southern Rocky Mountains. Settlement patterns are dictated by water, small farm communities where aquifers or irrigation is tapped, while large portions of the area contains abandoned homesteads and towns, a result of the dust bowl bust of the 1930s and after.

Rocky Mountains Provinces

The region is divided into the northern and southern Rocky Mountains provinces. The high mountains with peaks over 14,000 feet abruptly rise from the Great Plains, a distinct break between the two geologic zones. From the range flow the headwaters of the great rivers of the West, the Columbia, the Missouri, the Colorado, the Arkansas, and the Rio Grande. The mountains have higher rainfall but poorer soils for farming. Ranching is found in the broad valleys and parks, while mining and lumber camps dot the mountain landscapes, most of them ghost towns today. Mining and timber harvesting continues, but at a conflict with recreational users and new residents concerned about environmental degradation.

The southern end of the Rocky Mountains is home to centuries-old Hispanic communities that developed separate systems, imitative of Iberian models and Pueblo Indian practices, to subsist on small, irrigated farm plots and timberlands. Denver is the largest city and regional hub serving the province, though it is located in the Great Plains at the base of the Rocky Mountains.

Columbia Plateau Province

West of the northern Rocky Mountains, the Columbia Plateau was created by the layering of ash and lava over the ions. The plateau is carved by the Columbia River and its tributaries, primarily the Snake River. This is part of the sage-brush desert, a relatively level expanse that has been brought under cultivation through monumental irrigation systems. Regional, mid-size cities—Spokane, Boise—serve the area.

Great Basin Province

The distinctive Basin and Range landscape of the Great Basin extends from the Columbia Plateau south into the deserts of the Southwest, including the Mojave, Sonora, and Chihuahua. With less than 10 inches of annual rain fall, this is the driest and dreariest landscape to the Easterner. Its expanses of sage brush, in Nevada and Utah, and then creosote further south in California, Arizona, and New Mexico, define the landscape, as do the appearance of saguaro cactus define the Sonora desert, and the lechugia plant define the Chihuahuan. Formerly, this was the landscape of the cowboy, buckaroo and vaquero, the desert prospector, and miner. In the Great Basin, Death Valley, with the nation's lowest point, is towered over by Mount Whitney, the highest point in the lower forty-eight states.

Figure 1.10 Youthful fault scarp of Amargosa Range rising above Death Valley floor which is marked by salt-rich playa and alluvial fans, Death Valley National Park, Mojave Desert, California. Source: Courtesy of Wayne Engstrom.

The twentieth-century introduction of irrigation systems along the Colorado and its tributaries and the Rio Grande has created extensive farmlands that are now being overtaken by urban growth in Salt Lake City, Phoenix, Tucson, and El Paso. Las Vegas is an anomaly in relation to farmlands, though its growth is fed by the abundant waters of Lake Mead.

Colorado Plateau Province

Abutting the Rocky Mountains on the east and the Great Basin on the west, the high Colorado Plateau is distinctive for its grand carved canyons. The Grand Canyon of the Colorado is best known, but its incised side canyons isolate the region from east–west traffic, except at its southern and northern reaches. Aridity defines the slick rock country of Utah, while uplands contain beautiful forested hillsides. As in the Great Basin, Mormon pioneers carved out small, irrigated homelands, a stark contrast to the surrounding red-rock canyon walls along the stream beds. Many tribes, especially the Navajo Nation, the largest American Indian population, call this landscape home, where their basic agricultural activities have been surrounded by a series of extractive booms, for uranium, then oil and gas. Without the diversion of Colorado River water, Los Angeles' phenomenal twentieth-century growth would not have happened. Without water, one finds the driest of areas, such as the Colorado River sand dunes west of Yuma.

Pacific Coast Province

The Sierra Nevada and Cascade ranges border and de-mark the boundary with the Great Basin and Columbia Plateau provinces to the east. From these high barriers cascade streams that water the world-class farmlands of the Great Central Valley of the Sacramento and San Joaquin rivers and the Willamette valley. Beyond these valleys run the Coast Range from Canada to Mexico, probably the most active volcanic and earthquake-prone region in the lower forty-eight states. The timberlands of the Northwest are the most extensive and valuable in the USA. And the gold mines of California were world-changing. Minerals and timber are still extracted from the province. The great cities of the West lay along the coast: San Diego, Los Angeles, San Francisco, Portland, and Seattle the greatest concentrations of population in the region.

Debates will continue on the boundaries of these provinces. Yet, for the basic geographic outline of the West, these help understand the different regions. Fenneman's 1928 final delineation of physical regions quickly became the accepted standard. It continues in use to the present. They remain valid, in part, because of the 1912 field expedition and data gathered while seated in an observation railroad car viewing the landscape on a train rolling back and forth, north, south, east, and west through these provinces.

On the 1912 train through the West was forty-five-year-old Harlan H. Barrows, head of the Geography Department of the University of Chicago, another disciple of William Morris Davis. But with Fenneman's work in hand, Barrow broke with the traditional study of geomorphology and geographic classifications to become one of the pioneers in the study of historical geography. Geographers and historians are not just interested in the description of a landscape but also in how a landscape shaped human history. Barrows (1877–1960) was not a proponent of environmental determinism, the belief that we are the result of our environment. Although he saw the environment as important, the central player was human choice. People can adjust to their environment in active not passive ways.

Barrows' studies of the Illinois River Valley for the Illinois Geological Survey more thoroughly investigated the history of peoples and their actions with the environment. By the time of his 1922 presidential address to the American Geographical Society, he had coined the term "human ecology" as the study of people in their environment. His popular lectures on historical geography inspired students to do their own studies. Best known is the work of Texan Walter Prescott Webb, whose *Great Plains* is a classic of such literature.

In the arid Great Plains Province, in that region once known as the "Great American Desert," settlers adapted to climate and geography. Webb credits the windmill, six-shooter, and barbed wire, among other factors, as the reason people could move into the Plains and prosper. In a similar approach, Ross Calvin, in his *Sky Determines,* another classic, describes how climate, particularly lack of rain, influenced the Southwest from the time of the Pueblo Indians to the present. The most influential modern historical geographer is David W. Meinig, whose initial study, the classic *The Great Columbia Plain: A Historical Geography, 1805–1910,* has led to other works on the Southwest and Texas and most recently to the four-volume *Shaping of America,* his magnum opus.

THE PUBLIC DOMAIN, THE WEST'S BIGGEST LAND OWNER

Onboard the 1912 geographer's special were a dozen or more government employees. All were on the train for only parts of the trip, but they provided the geographers with information they might use in the discussions since most worked for the US Weather Service and the USGS. Fred G. Plummer of the Forest Service, US Department of Agriculture, and L. F. Schmeckebier, custodian of Crater Lake National Park with the US Department of the Interior, were the exceptions. The USGS proved the most useful to the scientists, especially with Robert Marshal, Chief Geographer of the agency, along for half the trip. Marshall also had been essential in the mapping of several national parks, was an advocate for creating new parks (especially Rocky Mountain National Park in Colorado), and would be general superintendent of all the parks, with an office in Washington, DC, at the time of the creation of the NPS. The government employees and the interaction with the geographers on the train represented the ongoing interchange between those in the academy and those responsible for managing the public domain.

These public lands have over a century and a half since acquisition been divided up among several different government departments. The land-managing agencies of the federal government are a major presence in the West today. Nearly 50 percent of the land in the eleven far Western states is federal land—from a low of 28 percent of Washington to a high of 83 percent in Nevada. The largest land manager, especially in Nevada, is the Bureau of Land Management (BLM), the late-twentieth-century name of the former General Land Office.

The BLM is under the Department of the Interior, a department established in 1849 during the presidency of Millard Fillmore, shortly after the signing of the Treaty of Guadalupe Hidalgo and the Treaty with Britain for the Oregon Country. The agency managed almost all of the present Far West including the sliver of land called the Gadsden Purchase encompassing southern Arizona and the boot heel of New Mexico, acquired in 1854. Some military and local government lands were excluded, but not the Spanish and Mexican grants which it needed to adjudicate validity before acceptance of prior title as valid. With Mexican war debts facing Congress, the Far West was seen as a land to exploit for the benefit of the nation. The effort to sell lands was hastened with the rising cost of the Civil War. By the Gilded Age, otherwise known as the Great Barbecue, the nation was divesting itself of forests, minerals, waters, and lands at an accelerating pace. The General Land Office was, well, doing a land-office business.

The surprising denudation of forests, obvious in the East by mid-century, the destruction of wild life, especially birds, and the fear of the loss of valuable historic or scientifically significant places brought the first efforts of conservationists to save the land. In the Forest Reserve Act of 1891, Congress provided for the setting aside of forest lands in order to ensure that if they were harvested they would be replanted, allowing for new growth for use by later generations. President Benjamin Harrison set aside the first national forest in Wyoming, now known as the Shoshone National Forest. The act was also used by President Harrison in 1892 to set aside the Casa Grande ruins, an archeological site in Arizona, the first land thus reserved from acquisition by homesteaders.

The Progressive movement of the late nineteenth and early twentieth century, especially its conservation efforts, spawned the growth of new Department of Interior agencies: the Forest Service, Bureau of Reclamation, Fish and Wildlife Service (FWS), and NPS. The Forest Service was transferred to the US Department of Agriculture but retained its primary mission: to protect the watersheds of the forest lands and to provide for the continued tree growth in areas harvested for commercial use. Gifford Pinchot, new head of the Forest Service and a personal friend of President Theodore Roosevelt, expanded the forest reserves much to the consternation of Western congressmen. In 1907, an Act limiting the use of the presidential powers to set aside forests caused Pinchot and Roosevelt to declare 16 million acres within the Western states before the new Bill would go into effect. The Forest Service settled in as long-term multiple-use manager of federal lands, sometimes criticized as handmaiden of the timber industry or for its policies. But its acreage in the West, mostly in the Pacific Northwest, has little changed since the selection of those 1907 midnight reserves.

Similarly, the FWS's system of wildlife refuges were begun after outcries against the decline in water fowl, a result as much from over-hunting as the use of feathers in the hats used for women's fashions of the day. In 1903, President Roosevelt established the first refuge in Florida's Pelican Island. Some date the agency's founding to the 1871 Commission on Fish and Fisheries, an agency which aimed to study the declining fisheries. The bringing together of the wildlife refuge system (584 refuges with 150 million acres today) and the research arm of numerous agencies did not occur until 1940, when the FWS began its existence. The FWS is today known for its research activities as much as its land management.

Popular at its founding, the President Roosevelt-backed Bureau of Reclamation brought engineering skill, organizational abilities, and funds from the federal government to build dams for local irrigation projects. Boosters saw the agency as a panacea for all the ills of the dry lands. Established in 1902 by the Newlands Act, not surprisingly its first major project was irrigation projects that would benefit the lands owned by Senator Newlands in Nevada. The first projects were small scale in comparison to the mammoth construction projects of the 1930s, from the Grand Coulee on the Columbia River to the Hoover Dam on the Colorado River. These projects not only managed the flooding and downstream flows but also helped create vast new agricultural communities, and with them powerful political bases pushing for more projects by the Reclamation Service. The number of dams

Figure 1.11 Lower Yosemite Falls is an example of a hanging valley, Yosemite National Park. Source: Courtesy of Wayne Engstrom.

built in the West, their scale, and environmental impact was truly phenomenal. Many asked were the dams necessary, and by the 1970s the answer was no—especially once the Grand Canyon itself was under assault. The New Conservation born in the 1960s put an end to the big dam building of the Reclamation Service.

Like the Forest Service, FWS, and Bureau of Reclamation, the NPS was born during the Progressive era. Though President Roosevelt had desired a national park service and had established through the 1906 Antiquities Act a series of national monuments that would become part of the agency, Congress did not pass the enabling legislation until August 25, 1916, signed into law by President Woodrow Wilson. The law gave the Park Service the dual mission to protect the parks for future generations while providing for the public enjoyment of those lands. When the park service took over Yellowstone, Yosemite, Crater Lake, Grand Canyon, and other parks, the "public enjoyment" in many instances had the upper hand. Bear feedings at the dump, fire falls of logs off the cliff, railroad corporate influence, and other issues needed to be changed. The development of a core of superintendents and park rangers took time, especially with the exigencies of World War I, but by the 1920s the first director, Stephen Mather, had issued policy statements, hired key staff, and created a park system that had not existed before.

In the Southwest, Mather's subordinate Frank "Boss" Pinckley managed the national monuments, primarily archeological sites such as the Casa Grande ruins. A philosophy for preservation and stabilization was developed and evolved and continues to be reshaped. The biggest reshaping of the NPS came in 1933, with the reorganization of agencies by Franklin Roosevelt. The NPS acquired the monuments and Civil War battlefields then managed by the Forest Service and the US Army. During the 1960s and 1970s new units were created by Congress almost monthly until December 1980, when President Jimmie Carter doubled the acreage of the system by designating sixteen new Alaska parklands. Today the agency manages 391 units, from Acadia National Park, Maine, to War in the Pacific National Historical Park in Guam, a total of 84.4 million acres.

Since World War II and the military buildup, an expanding land-holder in the West has become the Department of Defense. The military presence in the West has dated from the time of acquisition, 1846, but the posts were relatively small and military reserves a few miles square. Since World War II, the need for US Army tank maneuver bases, US Air Force bombing and missile ranges, naval stations, and nuclear test sites has gobbled up Western acreage. The Yuma Proving Grounds/Barry Goldwater bombing range in Arizona is bigger than Rhode Island; the White Sands Missile range is larger than Connecticut; the Fort Carson tank maneuver base is as large; while the Nevada and Utah testing and proving grounds are huge (exact acreage is unavailable). The military presence throughout the West, especially the Department of Defense dollars, fuels the local economies. As an agency, the Department of Defense, once oblivious to local concerns, has begun to be a better neighbor through environmental programs, historic preservation activities, and public outreach.

The largest land manager in the West, BLM lacked an enabling legislation until the passage of the Federal Land Policy and Management Act of 1976. Since then, the agency has endured the trials of supporting the mining and grazing industries of the rural West while appeasing the new users who seek recreation, environmental protection, and wilderness experiences. The Wilderness Act of 1964 tasked the NPS, FWS, and Forest Service with studying and recommending for designation lands that should remain undeveloped, without vehicular access, where people would be but visitors. The BLM land has since been added. The Wilderness Act allowed the NPS and the Forest Service to designate areas they had already been advocating for further protection. Indeed, the Forest Service, following the recommendations of ranger Aldo Leopold, created the Gila Wilderness in Southwest, New Mexico, in 1924. After leaving the Forests Service, Leopold conducted forest research in Wisconsin and wrote the landmark *A Sand County Almanac,* the bible for a land ethic. Since the passage of the Wilderness Act, some 107 million acres, or 16 percent of these agencies' lands, have been proposed as part of the wilderness system.

NOTES

1. See Alfred Ruhl quoted in *New Orleans Times Picayune* of August 28, 1912.
2. A dust bowl is the area plagued by wind and soil erosion most graphically experienced in the 1930s in Kansas and Oklahoma.
3. Emmanuel de Martonne, "Le Parc du Yellowstone: Esquisse Morphologique," in *Memorial Volume of the Transcontinental Excursion of 1912 of the American Geographical Society of New York* (New York: American Geographical Society, 1915), p. 231.

4. Bowman, quoted in *Albuquerque Journal* October 10, 1912.
5. See E. Oberhummer, "Amerikanische und europaische Stadte," *Memorial Volume of the Transcontinental Excursion of 1912 of the American Geographical Society of New York* (New York: American Geographical Society, 1915), p. 167.
6. See Albert Perry Brigham, "History of the Excursion," in *Memorial Volume of the Transcontinental Excursion of 1912 of the American Geographical Society of New York* (New York: American Geographical Society, 1915), pp. 22–3.
7. Vacher, quoted in Phoenix *Arizona Republican* October 12, 1912.
8. E.-A. Martel, "Appreciation," in *Memorial Volume of the Transcontinental Excursion of 1912 of the American Geographical Society of New York* (New York: American Geographical Society, 1915), p. 26.

CHAPTER TWO

THE FLORA AND THE FAUNA

Kurt Russo

INTRODUCTION

The numbers startle the imagination. Prior to the arrival of the Euro-Americans, up to 60 million buffalo, 30 million whitetail deer, and 15 million pronghorn antelope roamed the Western portion of the American continent; 400 million blacktailed prairie dogs populated one "dog town" covering virtually the entire Texas Panhandle; up to 100 million beaver could be found in the continent's waterways; and 100,000 grizzly bears inhabiting every corner of the American West, including habitat in the Pacific Northwest in the world's largest temperate rainforest. In general, "a larger proportion of North America was covered by a greater number of species historically than it is now."[1] The biological richness and abundance of the American West prior to the arrival of the Euro-Americans was described by Dan Flores:

> There were so many passenger pigeons in the annual migration flights up from Texas that some scholars believe the Cross Timbers, a four-hundred-mile-long strip of oak woods reaching almost to Kansas, was planted by their droppings. [...] And old-growth giant trees, firs and redwoods and sequoias in the Pacific Northwest and the Sierra, parklands of fat, yellow-bellied ponderosas from the Mogollon rim of Arizona to the Flathead Valley in Montana. And prairie, of course, the only apt metaphor for which is still today, *oceans.*[2]

A great deal has changed since the time of the first European arrivals. According to the US Fish and Wildlife Service, there are now 612 threatened and endangered animal species in North America, including eighty-four mammal species, ninety species of birds, and 139 species of fish, as well as 746 threatened and endangered plants.[3] Of course, the disappearance of species is not new to North America. Approximately two-thirds of big mammals in North America, or thirty-nine genera, are believed to have gone extinct between 12,000 to 10,000 years B.P., possibly due to some combination of climatic factors and anthropogenic influences.

While there is an ongoing debate over what caused the extinction of these prehistoric animals, it is widely agreed that America's "primeval" landscape had been impacted by indigenous peoples for thousands of years. By one estimate, the cumulative indigenous population north of Mexico in Canada and the USA over the five centuries prior to European contact numbered between 50 million and 200 million people.[4] Their footprint on the landscape was unmistakable throughout North America, though largely underestimated and just as often overgeneralized.

What was the condition of the flora and fauna of the American West prior to and immediately after contact with the Europeans? How has the human footprint changed the natural world of the American West in the ensuing centuries? This chapter will provide some answers to these questions, as well as some insight into how the nature and direction of environmental change influenced the origins, growth, and narrative arc of American environmentalism.

THE ABORIGINAL LANDSCAPE OF THE AMERICAN WEST

The forty-eight contiguous states of the USA encompass 3.6 million square miles, 2.3 billion acres, and extend from the 24th parallel to the 49th parallel and between 66 and 125 degrees longitude. A major feature of the North American continent is the Mississippi River that drains more than half of the American landmass. Immediately west of the Mississippi River is the Great Plains, an area approximately 500 miles east to west, that extends from North Dakota south to Texas and Oklahoma. Further to the west are the mountain ranges of the Rockies, Sierra Nevadas, and Cascades, interrupted by the intermontane Great Basin between 4,000 and 6,000 feet above sea level. This is also the province of the major desert systems of North America, including the Mojave, Sonorran, Great Basin, and Chihuahuan that shade into grasslands, mesic shrub, and montane forests. To the west and northward, parallel strips of mountains (the Coast Range and the Cascade Range) protect the inland regions from moisture from the Pacific.

When the Wisconsian ice sheet was at its maximum (18,000 B.P.), the North American continent was a land of ecotones not monotones, "of unending edges between treeland and grassland, wetland and dry, where life was most exuberant."[5] A veritable ark of a world, North America was landscape where saiga antelopes, moose, and lions from Eurasia mixed with indigenous short-faced bears and camels, bison (including now extinct forms of large bison), mammoths, mastodons, ground sloths, giant beaver, saber-tooth tigers, and horses.[6] Beginning in the early Holocene (10,000 years B.P.), North American forests expanded and "covered almost half of the country's surface."[7]

The Archaic cultures (8,000 years B.P.), and particularly those that evolved around 2, 500 years B.P., were actively engaged as purveyors of environmental change. By 1492, the activity of America's First Peoples had "modified vegetation and wildlife [and] created earthworks, roads, and settlements throughout the Americas."[8] As a result of anthropogenic changes from the earliest peoples up to the present day, the physical environment of North America has been "transformed more rapidly at the hands of people than any other large area of the world."[9]

Prior to the Euro-American era, the Great Plains west of the 100th meridian was an expanse of grasslands, a unique ecology divided into tall-grass and short-grass prairies (see Figure 2.1). Native peoples had been hunting bison on the Great Plains for more than 8,000 years, constituting the "longest sustained human lifeway in North American history."[10] Some historians place the number of bison at over 60 million animals at the time of European contact.[11] A single herd could number as many as 12 million animals covering 50 square miles.[12] The abundance of game on the Great Plains was such that, "[o]nly on the

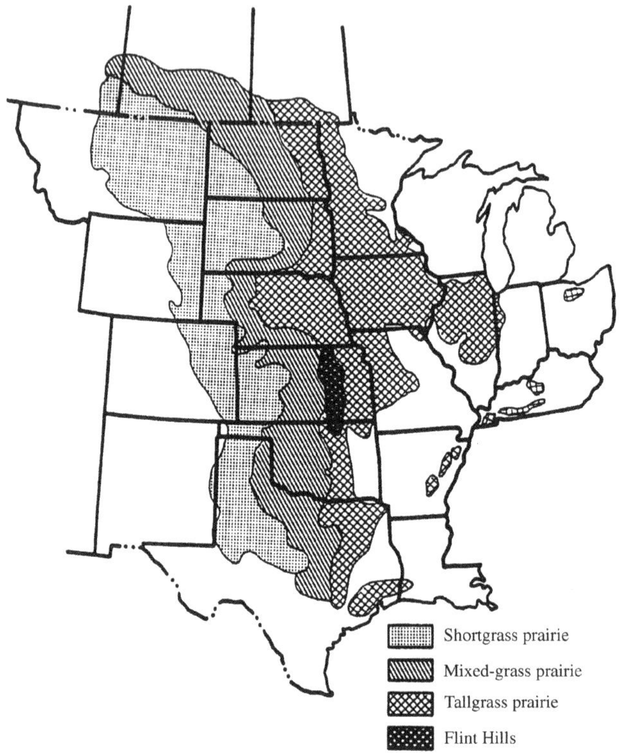

Figure 2.1 Prairie distribution at the time of European settlement prairie conservation. Source: Fred B. Samson and Fritz L. Knopf, *Preserving North America's Most Endangered Ecosystem*, 1996.

African veldt in the pioneering days [...] has there been anything comparable since man began recording history."[13]

It was primarily anthropogenic fire that played a major role in the spread of the grasslands, if not their ultimate origins.[14] As a consequence, the bison greatly increased in numbers and "extended their range into the anthropogenic grasslands of the Alabama prairie belt long before the Europeans arrived on the scene," and were even found in southern refugia in Louisiana, as well as in New Jersey, Florida, and in portions of the Southwest.[15] Nor was the use of fire limited to the Great Plains and the Southwest. The fires set by indigenous peoples across western North America produced a greenhouse effect so widespread that it "may have helped end the Ice Age as well as bringing the next cold cycle (the Little Ice Age of 1550–1850) to a speedy, early end."[16]

At around 2,500 years B.P., big game on the Great Plains largely consisted of bison, mule deer, pronghorn antelope, and whitetail deer, as well as bighorn sheep in the southern Rocky Mountains. As big-game hunting gave way to more frugal hunting and gathering, aboriginal communities began the intensive cultivation of food crops such as maize, potatoes, squash, watermelon, and kidney beans (see Figure 2.2). In some portions of the Southwest, indigenous agriculture was so extensive that it "led to fundamental biotic change and frequently to soil erosion."[17]

The availability of water was the limiting factor on human settlement, with much of the western two-thirds of the USA receiving less than 10 inches of precipitation per year. By 10,000 B.P., aridity came to be the rule in the Southwest region, characterized by grass vegetation that included tall grass, short grass, and mesquite grass, along with desert vegetation of sage brush and creosote bush, and forest vegetation dominated by woodland species. The mounting evidence of the degree to which indigenous communities in the Southwest and elsewhere altered the landscape for agricultural landforms led one historian to observe that "an apparent mania for earth moving, landscape engineering on a grand scale runs as a thread through much of New World prehistory."[18]

In the Rocky Mountains, the landscape supported an almost unimaginable richness and abundance of fauna. Frederic Wagner estimated that 20–30 million animals aggregated in the Rockies, including up to 10 million bison, 15 million pronghorns, 2 million bighorn sheep, 5 million mule deer, and 2 million elk.[19] The flora of Montana west of the Continental Divide consisted of grassland and forested valleys surrounded by heavily forested mountains of ponderosa pine, Douglas fir, western larch, and lodgepole pine.[20] The Palouse bioregion in west-central Idaho, southwestern Washington, and northeastern Oregon was dominated by bunch grass that grew in clumps of about 8 inches in diameter and often over 3 feet in height.[21] In the Pacific raincoast of western Oregon and western Washington, the largely coniferous montage of spruce, cedar, hemlock, pine, and fir constituted the world's largest temperate rainforest "covering 90% of the land in the late eighteenth century."[22]

Further south is the vast Floristic Province, an area that covers 70 percent of California, portions of southwestern Oregon, a small part of western Nevada, and northern Baja California. The native vegetation of the province, one of only four areas in the world with a Mediterranean-type climate, included chaparral, foothill, woodland, coastal scrub, and montane evergreen forest. The indigenous peoples of California burned portions of the landscape, including the coastal range chaparral, to maintain a diverse assemblage of habitat patches in a state of intermediate disturbance.[23] Looking over the aboriginal use of

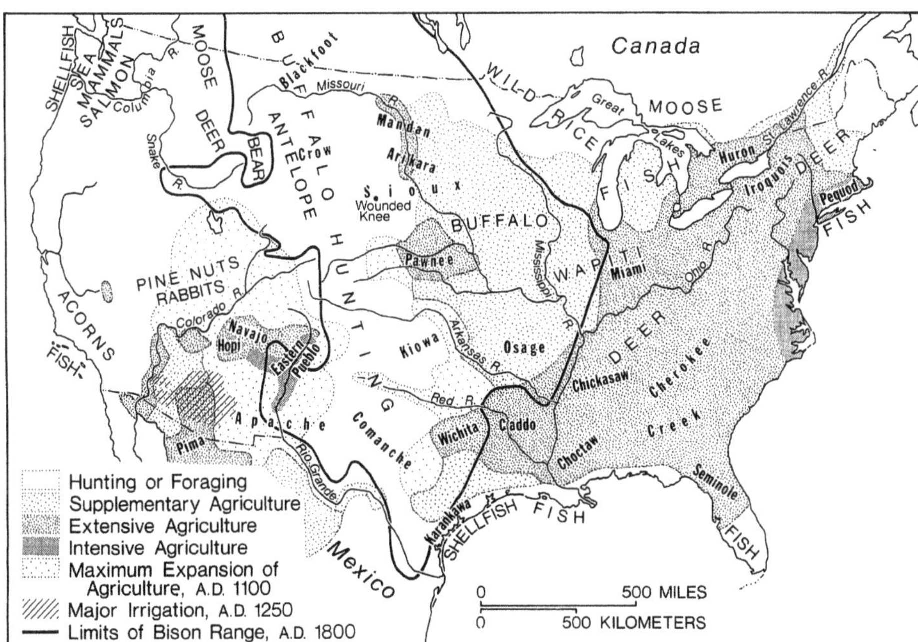

Figure 2.2 Patterns of Indian subsistence in Euro-American contact times, sixteenth to nineteenth centuries.
Source: Michael P. Conzen (ed.), *The Making of the American Landscape*, 1990.

California landscapes, Theodora Kroeber and Robert Heizer concluded that "the white invaders wrested from [the California Indians] a garden, not the wilderness it salved their conscience to call it."[24]

EARLY EURO-AMERICAN ARRIVALS

By the mid-sixteenth century, the Spanish were moving north from Mexico City up the coast of California from where they would later dispatch their Manila galleons bound for the Philippines laden with New World silver. Further to the north, Russian fur traders made their way down the west coast from Alaska to northern California. These early explorers and traders brought with them plants and animals that formed part of the "grunting, lowing, neighing, crowing, chirping, snarling, buzzing, self-replicating and world-altering avalanche" of the Columbian Exchange.[25] According to the ecologist Harold Heady, the impact of these early European arrivals in the coastal areas could easily have spread inland:

> A brief land exploration, especially one with horses, from the first sailing vessel to reach the California shores probably left new plants. [...] Once a plant species produced seeds [...] birds could have carried them to other locations. California lies in the path of many bird migrations, especially north–south routes with one terminus in Mexico.[26]

The early fur trade caused additional changes in the marine ecosystems, with the sea-otter population suffering some of the worst declines. Ray Dasmann estimated that "[p]erhaps 300,000 or more [sea otters] swam in the offshore waters" at the beginning of the Mission period (ca. 1769).[27] Their numbers were greatly reduced with the establishment of a trading outpost near Bodega Bay by the Russian-American Fur Company in 1811, followed by the American trappers and the Hudson's Bay Company working in the rivers, lakes, sloughs, and marshes of California.

By the time the English sailed for James Bay, the Spanish were already moving north from their mines in Santa Barbara, Mexico, into the Rio Grande region above El Paso.[28] The settlement corridor extended through the Rio Grande Valley from the semi-arid and montane headwaters in Colorado to the Gulf of Mexico, passing through the northern extension of the Chihuahuan Desert.[29] The pattern of Spanish settlement of the Rio Grande Valley and other portions of the Southwest was in large part determined by the availability of water. The institution of Spanish *acequia madre* irrigation systems provided for the cultivation of new crops and coincided with the initiation of what would by the mid-nineteenth century become the "kingdom of cattle."

The native plant communities in the Southwest region had been subject to intermittent grazing by large ungulates such as bison prior to the Spanish period but now were experiencing the transformative effect of intensified, year-round grazing of these newly introduced hooved animals. The impact of grazing included not only the cropping of native vegetation but also soil compaction and the attendant problems of erosion, as well as the transportation of invasive species. In addition, efforts to protect livestock from predators contributed to the contraction of habitat and the depletion of the historic richness and abundance of large carnivore species such as gray wolves, grizzly bears, black bears, and mountain lions which were endemic to large portions of the desert and semi-arid biomes of New Mexico and Arizona during the late Holocene (4500–450 years B.P.).[30]

INTO THE WEST: EARLY EURO-AMERICAN SETTLERS

The settlement of the American Southwest provides an example of dramatic environmental as well as cultural change. By the seventeenth century, the lifeways of the Native peoples of the Southwest and the Great Plains had been changed by domestic livestock, new techniques of warfare, Christianity, and alphanumeric literacy. The Europeans introduced into the aboriginal landscape three agents of change that transformed the cultures and biology of the American West: horses, guns, and disease.

By some estimates, horses that were reintroduced into the Americas by the Europeans had already begun to disperse north out of Mexico by 1550. By 1700, "all of the Plains tribes south of the Platte River and north of the Spanish settlements in Mexico were familiar with the horse."[31] In the Great Plains during the period 1680–1880, three dozen Native groups adopted the horse as a mainstay of their hunting cultures.

In some instances, a tribal community of 1,000 individuals could have as many as 20,000 horses and mules. By the second quarter of the nineteenth century, the horse herds of the Southern Plains alone may have numbered up to half a million animals in addition to wild mustang herds.

The dispersion of the horse had a significant impact on bison ecology. The horse has similar water requirements as bison and has an 80 percent dietary overlap with bovines. Along with newly introduced sheep, goats, and mules, these new arrivals not only grazed the grasslands but also climbed slopes where the potential for erosion was greatest and had a deleterious effect on the fragile network of native flora. The concentration of large numbers of horses in the river valleys during the winter encampment season brought about addition ecological changes.

The horse also helped provide a surplus of animal products—particularly hides and furs—the tribes could now trade for firearms, gunpowder, and other items. In his analysis of the impact of European trade among the eastern Algonquins, John Richards observed that to obtain trade goods, "the Indians hunted harder and killed more [moose, lynx, otter, marten, badger, and muskrats] than they ever had before."[32] The story played out in the Lake States was repeated in portions of the grasslands west of the Mississippi with the spread of the hide trade.

Accompanying the new herds of horses were livestock, introducing diseases such as anthrax that began to infect the bison populations. All of the surviving bison in captivity in the late nineteenth century had high rates of infection of this disease, along with tuberculosis and brucellosis brought to the Great Plains by cattle. More ominous for the Native populations, the horse helped to diffuse smallpox by the late eighteenth century. William Denevan placed the precolonial indigenous population of North America at 3.8 million and estimated that there were no more than 1 million American Indians in North America by 1800.[33]

As a consequence of this rapid and widespread depopulation of the Southwest and Great Plains, terraced hillsides and raised fields were laid to waste. Forests advanced and grasslands and savannas retreated. When the prairie fires ceased by the 1860s, "the forest was renewed and advanced more rapidly than the original," leaving "much more forest primeval in 1850 than in 1650."[34] The genealogy of what William Denevan described as the "pristine myth" traces to the early settlers "being unaware of human impacts that may be obvious to scholars today, particularly in wildlife and vegetation."[35] The encounter with a "virgin land" was in reality an encounter with a widowed land in which Native populations plummeted as diseases swept through their communities.

In the interior midland areas, fur trade drove much of the Euro-American exploration beginning in the eighteenth century. Hunting and trapping are often the vanguard of settlement, and the area west of the Mississippi River was no exception. The fur trade led to a quickening of exploration and settlement west of the 100th meridian, with far-reaching implications for the biology of the aboriginal landscape.

One of the species that experienced the full force of the fur trade was the beaver. It has been estimated that there were 60–100 million beaver in North America prior to European settlement.[36] Beaver pelts fetched a good price in Europe after King Charles I decreed in 1638 that all hats manufactured in England had to be made of beaver fur. As a consequence, the early frontier east and west of the Mississippi River turned into an ongoing battleground between the British and the French for control of the lucrative fur trade.

The British voyager James Cook brought the imperial era of European expansion to the Douglas-fir region of the Pacific Northwest in the late eighteenth century. These journeys provided a new avenue for exploiting the fur trade, with a single voyage providing up to $25,000 in profits. But it was the 1792 journey of Robert Gray to the Pacific Northwest that captured the imagination of the Americans. The news that he had located and explored the "Great River of the West," which he named after his ship *Columbia,* aroused the interest of fur traders and settlers alike.

Few of the early Northwest settlers tried to establish homesteads in the coastal forests, which was described as "impregnable [as the] great Chinese wall."[37] In his journal from 1811, Alexander Ross characterized the landscape as "so thick" with forest "and so close the trees together, that in its fall it would often rest its ponderous top on some other friendly tree."[38] In 1833, William Tolmie, traveling on the Columbia River from Fort George to Fort Vancouver, commented on the "sombre gloominess of the wilderness" of unbroken forest of pines that covered the surrounding hills.[39] The trees were so massive that it sometimes required two days or more to fell one tree in these coastal forests.

Timber harvesting in these primeval forests was not significant until the late nineteenth century, with timber from the Midwest keeping interest away from the forests of the Pacific Northwest. The bioregion of western Oregon and western Washington was a montage of spruce, cedar, hemlock, pine, and fir and would remain the world's largest temperate rainforest through the dawn of the twentieth century. Over time, this Douglas-fir region would occupy the center of the American experience in the arena of international empires as furs financed international trade from Asia. The frenzy for fur would also stir the debate over expansionism from Jefferson through the 1840s, by which time:

> most of the land west of the Mississippi had been explored, and wagon trains of settlers were following the Oregon Trail westward to Oregon's Willamette Valley, just south of the Columbia River. President James K. Polk envisioned the USA as extending across the entire breadth of the continent, and his election to the presidency in 1844 triggered the acquisition of additional lands west of the Mississippi.[40]

During the Spanish Mission era in California (ca. 1769–1823), the most discernible environmental impacts were restricted to the marine-influenced strip west of the inner coastal ranges. Within this area, however, the impacts were extensive. The coastal prairies, oak savannas, prairie patches in coastal redwood forests, and riparian habitats kept open and fertile through Indian burning, "became grazing land for vast herds of cattle, sheep, goats, hogs, and horses owned by the Spanish missions and rancheros."[41] By 1832, there were 320,000 sheep, goats, and hogs and 60,000 horses and mules associated with the Spanish missions and rancheros. In addition, agricultural fields replaced native ecosystems that soon "overwhelmed the native species, markedly changing the character and diversity of grasslands and other habitats west of the inner Coastal ranges."[42]

IMAGINED RICHES AND INESCAPABLE ECOLOGIES

The Great Plains

The Great Plains experienced some of the most dramatic, and immediate, impacts of the westward migration in the nineteenth century. The fur trade had already injected thousands of Europeans onto the Great Plains and led to the creation of fur-trading sites. These trading sites opened the door to a westward expansion, with settlements rising across the Great Plains.

When the region was acquired as part of the Louisiana Purchase in 1803, President Jefferson characterized it as an "immense and trackless desert" unfit for agriculture or human habitation.[43] Meriwether Lewis had a somewhat different impression of the Plains during his expedition of 1803–1805, describing its velvety seas of switchgrass, bluebunch, mid-height grams, bluestems, and fescues as resembling a "bowling green in fine order."[44] In 1823, Edwin James sided with Jefferson, concluding that this region was "almost wholly unfit for cultivation, and of course, uninhabitable by a people depending upon agriculture for their subsistence."[45]

This "uninhabitable area," teeming with wildlife, would soon be transformed with the arrival of the first wave of American settlers in the 1840s. In the Great Plains, the symbol of change was the settler moving east to west out of the Mississippi River Valley in the 1840s. These settlers were "primarily a forest people [and] when they met the grassland of the interior of the continent they misunderstood it in many ways."[46] As noted by Walter Prescott Webb, the Great Plains, unlike the area east of the Mississippi, was "level, timberless, and semi-arid." As a consequence "every institution that was carried across it was either broken and remade or else greatly altered." He goes on to say that: "east of the Mississippi civilization stood on three legs—land, water, and timber; west of the Mississippi not one but two of these legs were withdrawn,—water and timber,—and civilization was left on one leg—land. It is small wonder that it toppled over in temporary failure."[47]

This early phase of settlement was more subsistence-oriented farming "characterized by diversified husbandry that brought in the new without banishing the old."[48] In the second half of the nineteenth century, however, the settlement of the Great Plains was increasingly interconnected with new technologies such as windmills, high-grade steel ploughs, barbed wire, harvesters that allowed the production of monocultures, the completion of the transcontinental railroad in Utah in 1869, and the railroad system that reached the Colorado plateau in 1884. Government agencies also served as motors of transformation, expanding the farmer's frontier with the passage of the Homestead Act in 1862, allowing people to claim title to 160 acres of land.

The unintentional exchange of plants was a prominent feature of settlement of the Great Plains, complemented by "systematic, commercially and scientifically informed botanic importation."[49] Settlers introduced and cultivated in the west "the tame grasses he had learned to depend upon: timothy, orchard grass, bluegrass, etc, and the clovers." Between 1860 and 1920, North American land converted to crop production increased from 15,000 square miles to 960,000 square miles, the vast majority of it west of the Mississippi River.[50] As a consequence, species losses have been significant for the Great Plains grassland, savanna, and shrubland biome, an area that has undergone substantial land conversion from native vegetation to agricultural lands. By 1994, only 4 percent of the presettlement tallgrass prairie remained.[51] Overall, the North American prairies, one of the most highly altered grasslands in the world, have declined by an average of 79 percent since the early 1800s.[52]

One keystone species, the bison, was among the first to feel the impact of this wave of immigrants to the Great Plains. Their population briefly increased with Indian depopulation of the Great Plains, numbering up to 30 million animals. With the settlement of the Great Plains in the mid-nineteenth century, and the increase in livestock (goats, horses, sheep, and cattle), their numbers plummeted due to drought, widespread loss of habitat (including the loss of their southern refugia), disease, and hunting, including new pressures from some of the Great Plains tribes. During the nineteenth century, Native horsemen with rifles became more technologically capable of pressuring the resources. The shift among some of these Great Plains tribes from "subsistence/ecosystem hunters" to hunters linked to the European market system "may have added catalyst stress to a bison hard already being eaten away on a variety of fronts."[53]

By some estimates, the buffalo had lost 99 percent of its habitat and numbered only 800 animals in the wild and another 1,000 in parks or ranches in 1895 when the US government passed a law protecting the herd (see Figures 2.3, 2.4, and 2.5).[54] In contrast, by the final decade of the nineteenth century there were 40 million sheep in West. By 1920, the number of cattle exceeded 15 million animals. Overall, the number of domestic animals on ranges

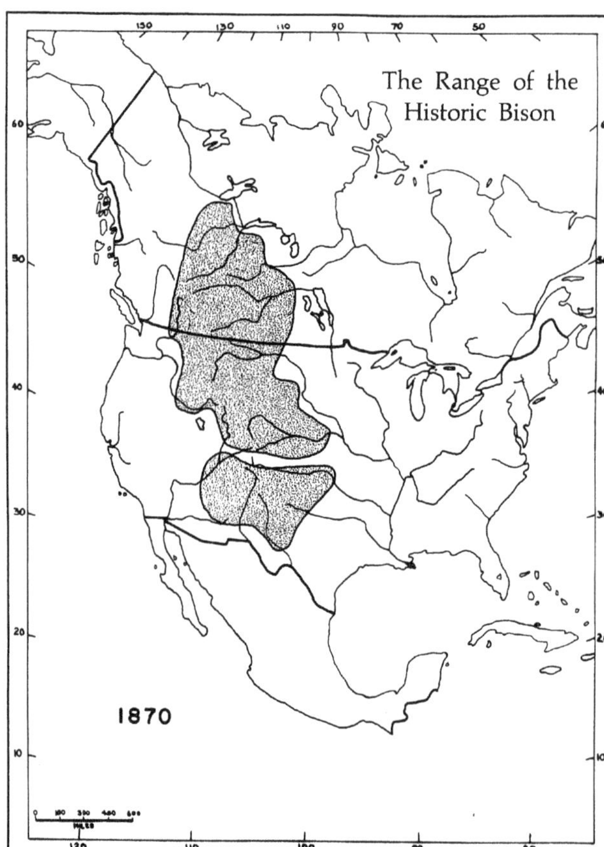

Figure 2.3 The range of the historic bison (ca. 1870). Source: J. Albert Rorabacher, *The American Buffalo in Transition: A Historical and Economic Survey of the Bison in America*, 1970.

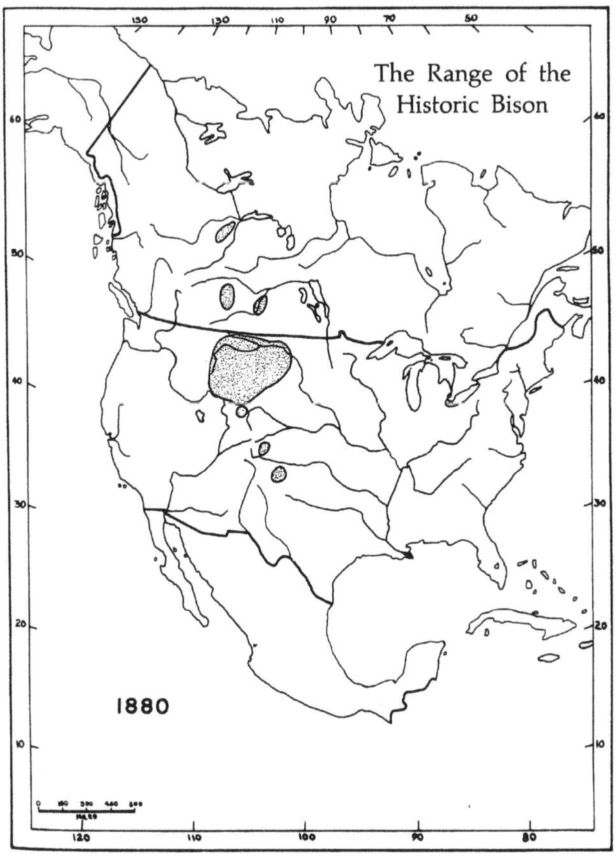

Figure 2.4 The range of the historic bison (ca. 1880). Source: J. Albert Rorabacher, *The American Buffalo in Transition: A Historical and Economic Survey of the Bison in America*, 1970.

in the Great Plains between 1890 and 1940 were "twice the values for wild ungulates in presettlement times while pressure on the habitat was half again those of wild animals."[55]

The settlement of the American West was accompanied by another agent of transformation: fire suppression. Prior to the nineteenth century, frequent low-intensity surface fires were dominant ecological processes in most pine and mixed-conifer forests in the Southwest. Fire played an important role in the character of these distributions and was "intimately linked with the presence or absence of certain species, as well as the age, structure, composition, and spatial arrangement of extant communities."[56] The decline in the frequency and intensity of fires during the mid- to late nineteenth century allowed forests to come back in Kansas, Nebraska, and elsewhere. In the Southwest, fire suppression combined with grazing "triggered wholesale encroachment of trees and shrubs into rangelands."[57] This encroachment was not an invasion but a "recovery from past disturbances, including prehistoric humans."[58]

Agriculture, the growth of urban population centers, and resource extraction have been the most prominent change agents in the American West in the modern era. By 1920, the centers of agricultural lands had shifted from east to west of the Mississippi into the Great Plains region and down through Texas, the Palouse bioregion, coastal California and the California Valley. A new wave of large farms rolled westward by 1950, supported by large irrigation projects in the arid and semi-arid Western regions. The number of hectares of croplands in the Great Plains increased from 30 million in 1900 to 50 million by 1920

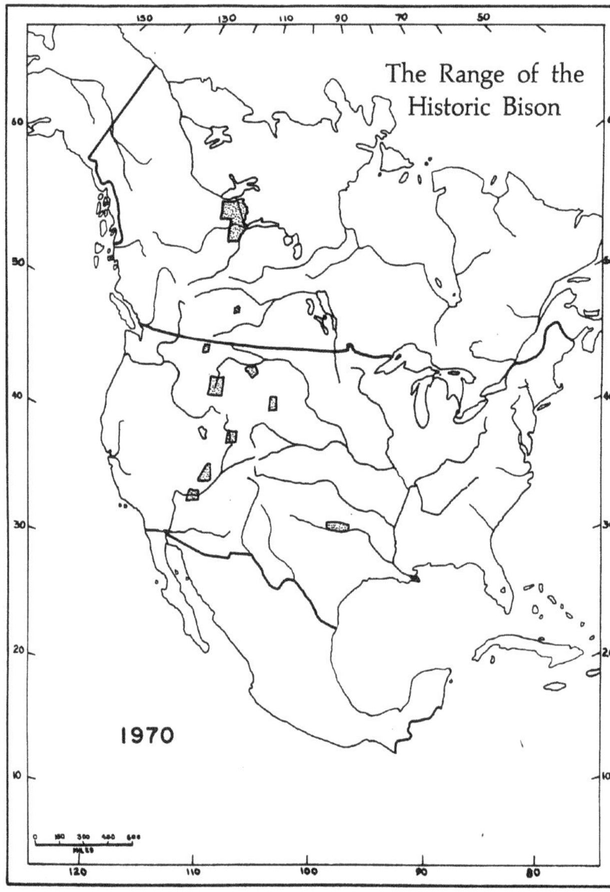

Figure 2.5 The range of the historic bison (ca. 1970). Source: J. Albert Rorabacher, *The American Buffalo in Transition: A Historical and Economic Survey of the Bison in America*, 1970.

and over 60 million by 1980, while the percentage of the American population engaged in farming nationwide declined from 27 percent to 3.4 percent of the labor force during the same time period.[59]

At the same time, the average farm size in the Great Plains tripled between 1935 and 1980 to 5.4 square kilometers. These large, capital-intensive farms "are the basic social institution through which human transformation of the region has occurred" in the modern era.[60] Increasingly, the trend in the Great Plains has been extensive and intensive dryland as well as irrigated farming to produce small grains for global markets. Demands on the Ogallala aquifer that underlies much of the southern and central Great Plains resulted in a ten-foot drop through a fifth of its area in the 1980s. A detailed study of the aquifer published in 1982 concluded that, due to its degraded condition, the Plains would require water imports early in the twenty-first century or face serious social disruption. This conclusion was derived from the other major conclusion of the study, "that irrigated farmers could not easily revert to dryland agriculture because of altered soil conditions and economics."[61]

Several keystone species convey the magnitude of the environmental change in the Great Plains. For example, whitetail deer numbered at least 30 million animals before the mid-nineteenth century, roughly twice the present population.[62] The blacktailed prairie dog that historically occupied up to 100 million acres now has less than 2 million acres of

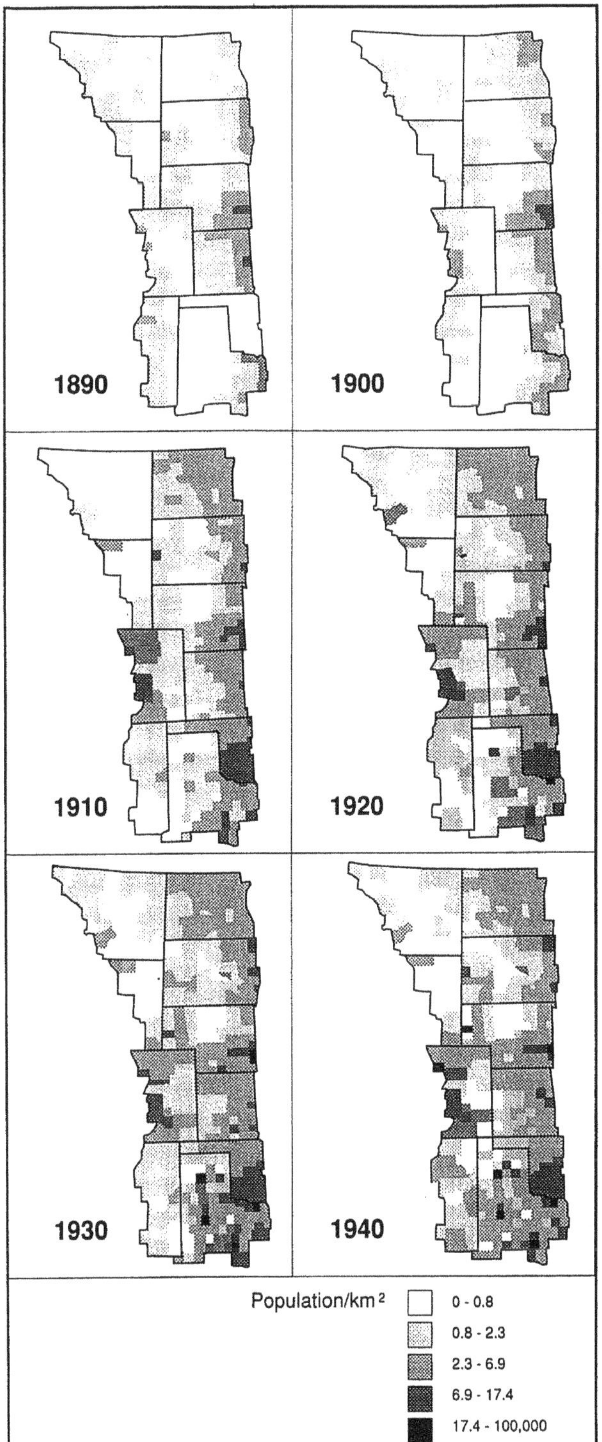

Figure 2.6 Great Plains population density (1890, 1900, 1910, 1920, 1930, 1940). Source: B. L. Turner II et al., *The Earth as Transformed by Human Action: Global and Regional Changes in the Biosphere over the Past 300 Years*, Cambridge: Cambridge University Press, 1990.

habitat and is being considered for listing under the Endangered Species Act. The grizzly bear, widely distributed throughout the Great Plains prior to the arrival of the settlers, is now largely extirpated from the region (see Figure 2.6).

Looking over the pauperization of the Great Plains in particular, Barry Lopez observed that "If you count the buffalo for hides and the antelope for backstraps and the passenger pigeons for target practice and the Indian ponies ([killed] by whites, to keep the Indian poor), it is conceivable that 500 million creatures died."[63]

The historic range lost by prominent Great Plains species is presented in Table 2.1.

The Floristic Province

Along the California coast the fur trade and grazing by livestock had already expanded inland into the Central Valley and the Coast Ranges by the 1830s. The fur trade greatly depleted the beaver populations by the 1840s, while in the mountains the wolverine, marten, mink, and river otter populations also declined precipitously. Native plants such as California bromes, blue wild rye, and clover were eaten by cattle and sheep before they could set seed. But it was the rush for the glittering metal that was the signature event of the nineteenth century, ushering in a new era of environmental transformation.

From 1852 until it was banned in 1884, hydraulic mining for gold, which required a vast system of canals and ditches, destroyed the salmon runs in many of the rivers and streams. Miners changed the course of whole rivers, with devastating effects on aquatic life and streamside vegetation. In June of 1860, C. A. Kirkpatrick wrote, "Salmon fish are fast disappearing from our waters—that is, upon all the streams upon which mining is carried on to any extent."[64] By 1882, there were "6,000 miles of ditches, sluices, and flumes in the Sierras."[65]

The era of hard-rock mining that began in 1884 brought new impacts to the California landscape. Hard-rock mining required the use of mercury that poisoned the rivers for fish and made the water unhealthy for human use. Human activities greatly reduced the original cover of oaks and other woodland species which were cut and used for cabins, sluice boxes, or were cut to make new lands for grazing. The use and occupancy of the Sierras also had a significant impact on the wildlife. The grizzly bear, native to the western slope of the Sierras, was extinct by 1924. The populations of mule deer, cougar, small mammals, and birds were also effected by mining, logging, fire, and grazing.

The latter half of the nineteenth century was a period of increased activity in the coastal

Table 2.1 Range contraction by species

Species	*% of historic range lost*
Black bear	87
Elk	90
Fisher	77
Gray wolf	85
Grizzly bear	91
River otter	78
Wolverine	94

Source: Adapted from Laliberte and Ripple, "Range Contractions of North American Carnivores and Ungulates," *Bioscience* 54(2) (2004):123–138.

and inland forested areas, particularly in northern California. In 1860, timber production was concentrated in the redwood forests, where logging began earlier and occurred more extensively than in other Westside forests.[66] Timber harvesting quickly spread, and by 1870 the value of production along the California coast had quadrupled as valley and blue oaks were harvested for steam-powered locomotives, canyon live oaks were used for ships' knees, wagon wheels and axles, and interior and coast live oaks fueled the smelters.[67]

By 1880, timber production spread throughout much of northern California. Agriculture also moved to a place of prominence in the latter half of the nineteenth century. By 1870, the economy of farming and those living in the broader geography exceeded the number of those employed in all the mines of the Sierra foothills. More than 11 million acres were in cultivation in California by 1870 and over 28 million by 1900, by which time "40 percent of California's 31 million acres of old-growth forest had been logged."[68]

Some of the most dramatic alterations anywhere in the continental USA occurred in the California grasslands. The original interior grasslands blanketed much of California's Central Valley and low elevations along the central and southern coast, representing one-fourth of California's landmass. The ecosystem supported large herds of pronghorn antelope and tule elk, which no longer exist outside nature preserves. Agricultural production in California increased after 1920, particularly along coastal areas and in California's Central Valley where "marshes were drained, underground water was tapped by artesian wells, streams and rivers were dammed and diverted for irrigation, and lands were fenced. In the process, huge tracts of former native grasslands, riparian corridors, and vernal pools were converted to artificial, human-managed agricultural systems."[69]

The overall impact on the landscape of sixty years of settlement was such that:

> In the short span of sixty years, from 1850–1910, relatively few Europeans and Asians [...] caused major declines in dozens of bird and mammal populations, denuded entire landscapes, caused major changes in the state's hydrological processes, vastly accelerated rates of erosion, destroyed countless acres of productive wildlife and plant habitats, radically altered much of the state's vegetation, and decimated Indian populations and destroyed their cultures. [...] [T]he historian Alfred Crosby Jr. calls this period "the greatest biological revolution in the Americas since the end of the Pleistocene era."[70]

One measure of the nature and extent of change in the ecology of the west coast is the presence of introduced plant species. In the southern two-thirds of California, exotic species took root and flourished, replacing native grasses first on the south and central coasts and then into the San Joaquin Valley and the Sacramento Valley. These species would "eventually Europeanize some 50 to 90 percent of California's grasslands."[71] Today, about 600 plants (10 percent of the native species) are declining in numbers and are in danger of extinction, and approximately 17 percent of all plants in California are non-native. With over 1,000 plant species brought in to California, there is "hardly a plant community or habitat left in California unaffected by some introduced plant species."[72]

With the exception of the transformation of the California grasslands, the botanical adjustments triggered by the colonial process up to the mid-nineteenth century were largely imperceptible. The situation was quite different with the native fauna of California. The Old World animals (sheep, pigs, goats, poultry, mules, cattle, and horses) brought ecological modifications as a result of grazing pressure, the dissemination of exotic plant species at the expense of native species, and the radical alteration of Native land-use practices. Because of the disruptions unleashed by Spanish settlement, most notably the decimation of the indigenous peoples, the population of elk, antelope, and deer reached "numbers and concentrations not seen since the Pleistocene Epoch."[73]

A significant amount of native habitat and farmland around metropolitan areas has also been lost to expanding West Coast population centers. By 1920, Los Angeles, which had population of 11,000 in 1880, had grown to over 577,000 residents. During the same period, Portland increased from a population of 18,000 to over 258,000 residents, and San Francisco grew from 234,000 to 507,000. In the space of seventy years (1850–1920), the population of California grew from 93,000 to over 3 million residents.[74]

By 1992, farmland in the West Coast urban areas surrounding Los Angeles, San Francisco, Portland, and Seattle had already begun to decline with the spread of the suburban culture and other urban influences. This was part of a broader nation trend as the total number of prime farmland acres declined from 137 million acres in 1982 to 133 million acres in 1992.[75]

Along the coast of southern California, "urban residential development is currently the principal land-use that fragments the native shrub habitats, coastal sage scrub sand chaparral."[76] The anthropogenic disturbances associated with fragmentation are likely to be "significant influences on abundance and extinction rates [for] predators, parasites, competitors, resource species, or mutualists [that] in turn affect species with which these interact."[77] As an example, a study of the avifauna of coastal southern California noted a strongly or moderately negative effect of habitat fragmentation on two-thirds of the native bird species.[78]

As a result of logging, grazing, agriculture, mining, oil extraction, and urbanization, less than one-quarter of California's Floristic Province remains in a more or less original condition. Native grasslands have been reduced to about 1 percent of their original extent due to agriculture, livestock, and urban development. The redwood forests, which once covered over 2 million acres in northern California, have been reduced by 95 percent.[79] Three-fourths of the original acreage of California's mixed-conifer forest has been select-cut or clear-cut at least once; riparian forests in the Province have been reduced by 90 percent, and southern maritime sage scrub have been reduced to 10 percent or less of their original area. As a result of the loss of 90 percent of the coastal wetlands and 98 percent of the interior wetlands, 25 percent of all plant and 55 percent of all animal species designated by state agencies as either threatened or endangered have wetlands as their essential habitat.[80] This pauperization of California's Floristic Province due to human pressures has rendered California one of the four most ecologically degraded states in the country.

The Pacific Northwest

A sea change was visited upon the ecology of the Pacific Northwest with the arrival of fur traders followed by a wave of settlers. In the 1820s, the Hudson's Bay Company ordered field representatives to exterminate valued fur-bearing animals to forestall competition with the USA and to ensure company control over the Pacific Northwest fur trade. They hoped to create a "fur desert around the boundaries of the Pacific Northwest."[81] At one point, beaver pelts were exported to Europe "at the rate of 200,000 per annum, and the animals were rapidly being exterminated."[82] The intense hunting of beaver continued until the 1840s, when the demand for beaver fur rapidly declined and the market demand in Europe and America for other animal skins increased.

Cattle were brought to Vancouver Island in 1789, but by 1825 were only found in isolated settlements in Puget Sound and the lower Columbia River region. The transformative effects of domesticated animals on the bioregion began in earnest after 1870 with the exponential growth in the number of sheep. Their influences on the ecosystem included eating native vegetation, damage to soil due to compaction, concentration of nutrients caused by defecation, the dispersal of exotic plant species via animal fur, fecal deposition, and harm to native wildlife. During this period, bighorn sheep were locally extirpated because of diseases spread through contact with domestic sheep.[83]

In the 1860s, cattlemen moved into the Palouse bioregion. Initially, the settlers used the hills as pasture, but they were soon farming these slopes as well. By 1869, the first grain crops were grown in this bioregion, and within ten years of the first plowing water tables in low-lying meadows dropped below the level to allow tilling. The increase in the conversion of native prairie vegetation to field or pasture resulted in the draining of wetlands. Equipment enabled farmers to utilize the steep slopes, further shrinking the refugia for native flora and fauna and imposing a new disturbance regime on the landscape. By the early 1880s, the Palouse Prairie had been badly overgrazed, and large-scale effects of non-native herbivores were found throughout much of the rangeland of the interior Columbia Basin as cheat grass replaced native plant species in the steppe region. Between 1940 and 1992, the number of cattle more than doubled, with cattle becoming "the dominant grazing animal in the region."[84]

Since 1900, 94 percent of grasslands and 97 percent of wetlands in the Palouse have been converted to pasture lands, crop, and hay production.[85] Native flora such as the Idaho fescue, rough fescue, smallheat goldenweed, and Jessica's aster have been replaced by wheat, barley, and lentils, as well as yellow star thistle and twenty-nine other species of noxious weeds. The once common fauna and flora of the region exist only in small areas of grassland, shrub, and forest, and these are threatened by herbicide drift and introduced species. The disturbance of the ecosystem has also resulted in the severe reduction in the richness and abundance of native fauna such as the mule deer, spotted frog, and sharp-tailed grouse, while the whitetail jack rabbit and the ferruginous hawk have been nearly extirpated as breeding populations. Of the once-continuous native prairie dominated by mid-length perennial grasses, little more than 1 percent remains, making the Palouse Prairie "one of the most endangered ecosystems in the United States."[86]

The nineteenth century brought enduring changes to the ecologies of other portions of the Pacific Northwest. When the American settlers first arrived in the region in the early 1800s, "the Douglas-fir forest occupied a broad belt extending from southern British Columbia to below Cape Mendocino in northern California."[87] The forested lands in Oregon and Washington included over 23 million acres of trees greater than 125 years old, including vast stands of Douglas fir, western hemlock, mountain hemlock, western red cedar, and sitka spruce.[88] This forest biome supported one of the most densely populated nonagricultural populations of American Indians in the pre-contact world of North America.

More than any other single event, it was the rush for gold that spurred the development of the Pacific Northwest timber industry. In a pattern characteristic of the period 1850–1900, "mining served to accelerate the commercial logging of northwestern forests."[89] Prior to this time, the West Coast timber industry was limited and local in nature. The gold rush of 1849 "upset that relatively simple local system; the resultant demand for building and mining material far outstripped the few meager shipments that came from the East Coast and the South of the United States."[90]

In 1852, the first timber mill was established on Puget Sound. Within ten years, the mill was shipping 9 million board feet annually to foreign and domestic markets. The exploitation of the forest accelerated with the introduction of the long-handled, double-bitted ax in 1878, the first steam donkey, mechanized yarding, the skid road, and new, high-capacity mills. New developments such as the completion of the Northern Pacific Railroad in 1883 and the Union Pacific Railroad to Portland in 1884 brought in more people and greater access to inland markets. As a result, timber production increased from 600 million board feet in 1879 to 2.6 billion board feet in 1889.[91] By 1910, two-thirds of Washington State's wage earners depended on the lumber industry for their livelihood.[92]

The fledgling industry was becoming larger, more complex, and more concentrated, as well as a more potent force in local, regional, and national politics. The position of this commanding industry was reflected in the remarks of E. T. Allen before the National Conservation Congress in 1914. Fearing that the designation of forest reserves would reduce the amount of available timber, he argued that, "there is entirely too much forgetting that forests are useless unless used; that not forests, but forest industry, is what we really seek to perpetuate."[93]

Despite the rapid growth of the timber industry, at the dawn of the twentieth century only a relatively small portion of the forested lands west of the Cascades in Oregon and Washington had been cut or settled for agriculture. In 1903, 11 percent of the 32,000 square miles of merchantable timberlands in Washington and Oregon west of the Cascades were designated as cutover lands (see Figure 2.7).[94] East of the Cascades in Oregon and Washington, slightly more than 5 percent of the merchantable timberlands had been harvested. As recently as 1936, when the Pacific Northwest was the nation's leading lumber-producing region, old-growth forests of all types still made up "73% of all commercial forests in eastern Oregon and Washington."[95]

By 1980, 838 billion board feet of timber had been harvested in the forests of western Oregon and Washington, by which time old-growth forests accounted for 13 percent of forest land cover on public and private forest land in this bioregion.[96] According to Robert Bunting, less than twenty years later, 6 percent of western Oregon and western Washington's old-growth forests remained, primarily in fragmented islands of habitat on public lands.[97] The rapid and radical transformation of the Douglas-fir region has severely disrupted the ecology for 1,500 species of invertebrates associated with old-growth forests and 118 vertebrate species whose primary habitat is in mature and old-growth forest systems.[98]

The rivers and wetlands of the Pacific raincoast have also been transformed. By 1865, the first series of dikes were introduced into the Puget Sound region in Washington State. Between the 1780s and the 1980s, wetland acreage in Washington and Oregon declined by 31 percent and 38 percent, respectively. In the Columbia River estuary, the impacts were particularly pronounced, recording a loss of 65 percent of its swamps and marshlands since 1870, due largely to dikes.[99]

The Columbia River is arguably the most significant environmental force in the Pacific Northwest region. The Columbia River drains a 259,000-square-mile basin including areas in seven states (Oregon, Washington, Idaho, Montana, Nevada, Wyoming, and Utah). By the 1830s it was the site of the first commercial fishing operations, with Hudson Bay shipping salted salmon from the Columbia River to the East Coast. But it was the development of the salmon cannery industry in 1866 that triggered the "salmon rush." By 1883, forty canneries processed a total of 43 million pounds of salmon and reached its maximum harvest in 1911 when 50 million pounds of salmon were packed.[100]

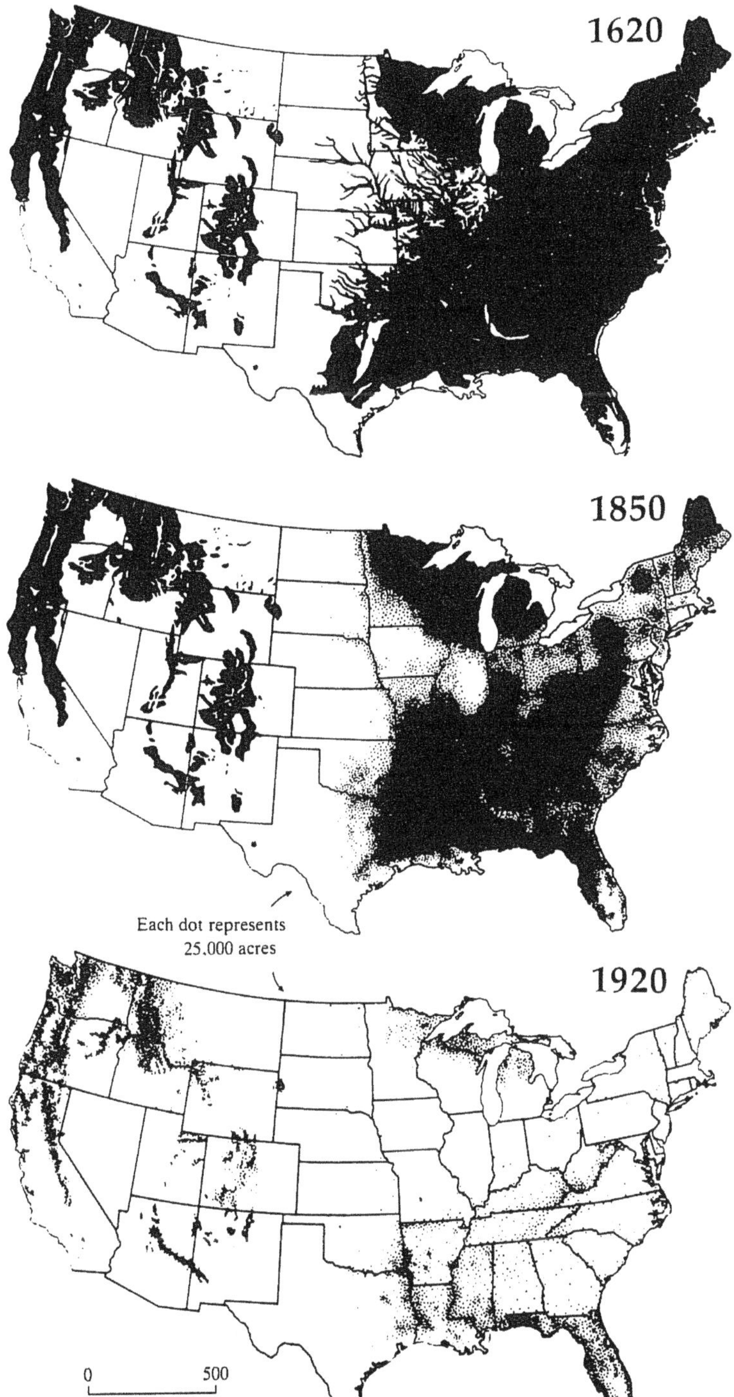

Figure 2.7 Area of "virgin" forest in 1620, 1850, and 1920.
Source: Michael P. Conzen (ed.), *The Making of the American Landscape*, 1990.

By 1975, the salmon harvest dropped to less than the harvest of 1867, the second year of cannery operations on the Columbia River. One of the major factors in the decline of the salmon was the engineering of the Columbia River through a network of hydroelectric projects. In their 1976 report "Columbia Basin Salmon and Steelhead Analysis," Ed Chaney and L. Edward Perry concluded that habitat loss caused by all other land uses is "relatively insignificant" in comparison with that caused by the damming of the region's rivers.[101]

The damming of the Columbia River began in 1888 with the Cascade Locks and was followed by the Dalles–Celilo project in 1915. The engineering activities reached a new level of environmental transformation in 1930s when work began on the Bonneville Dam on the lower river, the Rick Island Dam, and the Grand Coulee Dam on the upper river. By 1975, eleven dams stood on the mainstem with dramatic consequences for the premier species of the river, the six species of anadromous salmon that spawn within the Columbia River basin. The cumulative effect of these dams has been to reduce salmon spawning habitat from 163,000 to 73,000 square miles, with the salmon disappearing from 40 percent of its historic range on the Columbia River (see Figure 2.8).[102]

The fate of the salmon on the Columbia River was repeated throughout the Pacific Northwest in the years following the settlement era. According to a 1993 Wilderness Society study, all but one (pink salmon) of the ten anadromous fish species are "extinct" or "at risk of extinction" throughout "the majority" of their historic ranges in the Pacific Northwest and California.[103] Of the 214 stocks of anadromous fish in the Pacific Northwest, 47 percent are at risk of extinction due to dams as well as the introduction of exotic fish species, fish harvesting, and the degradation of freshwater habitat (see Figure 2.9).[104] Public concern for the old-growth forests of the Pacific Northwest, as well as the condition of the region's rivers, streams, and salmon populations, played an important role from

Figure 2.8 Celilo Falls (Washington), with Indian fishermen on platforms, "two weeks before being drowned by dammed waters."
Source: Oregon Historical Society, #OrHi89454. Photo by Thomas E. Reynolds.

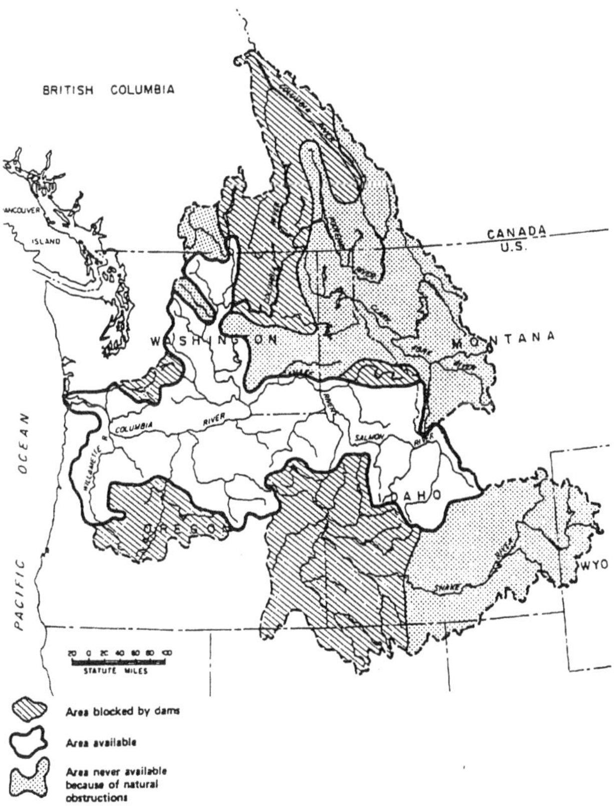

Figure 2.9 The Columbia River basin, showing spawning areas blocked by dams. Source: Bonneville Power Administration.

the early days of the conservationist movement and throughout the evolution of modern environmentalism.

THE ROOTS AND BRANCHES OF AMERICAN ENVIRONMENTALISM

The arc of the American environmental narrative closely resembles what Alan Taylor described as "a tripartite structure" that characterizes the classic environmental histories: "initial abundance, transforming settlers, and a legacy of diminished nature."[105] The foreknowledge of the threat of a diminished natural world was already evident in the call for wildlife reserves as early as 1844, when George Bird Grinnell campaigned against market hunting and on behalf of realistic game laws. In the American West, hunters noted the scarcity of buffalo and other big game up and down the Great Plains. In 1871, Colonel Richard Irving Lodge, a military man and enthusiastic hunter, told how the air was "foul with the sickening stench, and the vast plain [{. . .}] was a dead, solitary, putrid desert."[106] In the beginning, it was primarily, though by no means exclusively, this concern for vanishing wildlife that sparked the conservation movement.

In 1872, President Ulysses S. Grant signed an Act designating over 2 million acres in Wyoming as Yellowstone National Park. Yellowstone, the first wildlife refuge in the country and the world, would launch the spirit of natural conservation across the country. Over time it would become "a patriotic symbol of the nation's paradisiacal origins, a lightning rod for the unending American debates between economic development and environmental conservation."[107] Two years later, thanks in large part to the efforts of John Burroughs and John Muir, Yosemite National Park was created, suggesting that a "transformation was gradually taking hold in the United States in the people's relationship to wildlife."[108]

Equally important for this emerging conservation ethic was the work of the indomitable John Wesley Powell and his "Report on the Lands of the Arid Region of the United States, with a More Detailed Account of the Lands of Utah." The report, published in 1878, pointed out the severe water problems on land west of the 100th meridian and argued for a new pattern of settlement for the region. In 1890, Powell presented a congressional committee with a map of the arid West, mapping out twenty-four major natural provinces and further subdividing the region into some 140 candidates for commonwealth status based on drainage and topographical cohesion. Though Congress did not act on his recommendations, his work would eventually bear fruit as the ethic of conservation found an acceptable place in the American environmental narrative.

Concern about the future of the forests of the American West increased in the late nineteenth century. In 1891, Congress passed as part of the General Revision Act that included a little-debated section that empowered the President to withdraw from public entry any federally owned forestland. This was followed two years later by the creation of the Cascade Range Forest Reserve, the first of many forest reserves in western Oregon and western Washington. By 1893, there were sixteen western reserves with more than 17 million acres. In 1897, President Grover Cleveland added another 21 million acres of forest reserves, with no mention of utilitarian management objectives.

That same year, under pressure from the forestry and mining industry, Congress passed the Organic Act to make clear that these reserves were not preserves and that their purpose was to simply "protect water flow and to ensure the continuous supply of timber."[109] The struggle behind passage of the Organic Act embodied the contest between doctrinaire utilitarianism, where all value is homogenized in terms of utility of marketplace consequences and the moral imperative of the aesthetic conservation. The two streams of utilitarianism and preservationism fed into this conservation river and "expressed the ambivalence Americans were experiencing as they moved toward an urban, industrial, bureaucratic society."[110] While only a trickle at the time, this tributary of aesthetic conservation would eventually "swell into the main channel of modern environmentalism."[111]

By the time Congress passed the Lacy Act in 1900 to promote wildlife conservation, all of the Western states had passed some laws to conserve wildlife. In 1901, John Muir worked with President Theodore Roosevelt to set aside another 148 million acres of forest reserves. In 1905, Roosevelt, an avid hunter and outdoorsman, appointed Gifford Pinchot Chief Forester of the US Forest Service. Though Pinchot was not a proponent of wilderness preservation, he was a devotee of scientific management and promoted the application of rudimentary ecological science to the management of the national forests.

The 1930s proved to be a watershed in the evolution of the environmental movement. No other decade or administration did so much to save wildlife. President Franklin Roosevelt's Civilian Conservation Corps brought attention to the conservation effort, planting more than 3 billion trees and expending more than 7 million man-days to enhance or restore wildlife habitat. There were new laws to manage and protect vulnerable species, including the addition of 8.7 million acres in the Mississippi River Valley, set aside as critical flyway for migrating ducks and geese. In 1934, Congress passed the Taylor Grazing Act which for the first time established a significant measure of control over unappropriated public domain. In that same year, Robert Marshall, Director of the Forestry Division of the US Office of Indian Affairs, issued a memorandum to Secretary of the Interior Harold L. Ickes, advocating the preservation of undeveloped public lands. Although his recommendation for the creation of a wilderness planning board was ultimately rejected, it served as a harbinger of things to come.

It was Aldo Leopold who, as much as anyone, helped shape the modern environmental movement. According to the historian David Worster, "Leopold has been one of the two or three most important figures in twentieth century American conservation thinking" whose "biocentricism" became "the dominant theme" in the advancing environmental movement.[112] Throughout the 1930s and up until his death in 1947, Leopold served as a "crucial link between turn-of-the-century natural resource conservation and modern environmentalism."[113]

In 1935, Aldo Leopold was invited to serve as the first president of the newly formed Wilderness Society. Though he did not accept the appointment, he did serve in an advisory capacity as the organization worked to promote what Robert Yard, the President of the organization, referred to as "the gospel of wilderness."[114] The organization was soon working the Halls of Congress on issues ranging from Florida's Everglades to Washington's Olympic Peninsula. Just as importantly, it began to influence the public discourse through its periodical *Living Wilderness,* advocating that "wilderness is a natural resource having the same basic relation to man's ultimate thought and culture as coal, timber, and other physical resources have to his material needs."[115]

In 1949, the publication of Aldo Leopold's *A Sand County Almanac and Sketches Here and There* proved to be a landmark event in the American environmental movement. In this paean to the natural world, Leopold brought together the modern impulse to embrace wild nature, the complex relationship between the material forces in nature, the role of technological and scientific progress, and the importance of economies of abundance. Leopold linked wild nature with the well-being of human communities and cultures, describing wilderness as the bedrock "to which man returns again and again to organize yet another search for a durable scale of values."[116]

The intervening years between the posthumous publication of Leopold's book in 1949 and Rachael Carson's influential book *Silent Spring* in 1962 witnessed a number of historical conjunctures. It was an era of unprecedented economic growth, providing America's growing middle class with the means and the leisure time to explore and appreciate America's public lands. Accompanying this era of prosperity were the environmental hazards of urban sprawl, accelerated harvest and fragmentation of public forestlands, the despoliation of streams, rivers, and lakes, the spread of industrial pollution, and the prodigious use of synthetic chemicals with insecticidal properties—an offspring of World War II—that found their way onto the nation's agricultural lands and into the food chain.

Silent Spring was not the first exposé of the dangers of industrial pollution. In her classic text, *Industrial Poisons in the United States,* published in the 1920s, Alice Hamilton exposed the environmental consequences of industrial activity and was a "core figure in the urban and industrial roots of environmentalism."[117] But Carson's book was more effective in linking traditional wildlife-protection issues with urban-industrial ones. Described by Linda Nash as the "seminal environmental work of the mid-twentieth century," *Silent Spring* popularized the new science of ecology and linked human health to environmental health, revitalizing the understanding of medical topography that "linked human health to the surrounding landscape."[118]

Carson's book also anticipated the revitalized women's movement and ecological feminism that would emerge in the 1960s. Mainstream women's groups such as the League of Women's Voters took an early lead in studying and pressing for clean air and water legislation. But the women's movement injected into the environmental debate a gender dimension in which "modern science came across as a hard male onslaught with imperialistic designs on a softer feminine nature."[119] At the same time, the conjunction of the women's movement with the ecology and social-justice movements brought into focus the issue of liberation and the need to "reform a capitalist system that creates profits at the expense of nature and working people."[120]

The events of the 1960s and 1970s brought into play other forces, faces, and factors that would continue to shape, and be shaped by, the environmental movement. The growth of the environmental movement during this period coincided with the rise of civil disobedience, a resurgent democratic populism and more inclusive grass-roots community organizing, American Indian environmental activism and the iconic image of Iron Eyes Cody, and the lunar "Earthrise" photograph of our lovely and fragile planet taken by Apollo astronaut Bill Anders. It was a time of exponential growth of mass media, particularly television, that could reach a broad spectrum of the American public. Even before 1970, one-third of all American households were two-television families. It was also the era of the Vietnam war-protest movement, which would play an important role in environmental legislation.

The Wilderness Act in 1964 and the Endangered Species Act of 1973 are testimonies to the historical origins, the ecological vision, and the expanded base of modern environmentalism. The names and ideas of Henry David Thoreau, John Muir, and Robert Marshall appeared time and again during the congressional hearings. Secretary of Interior Stewart L. Udall discussed ecology and a land ethic, referring to Aldo Leopold as the "instigator of the modern wilderness movement."[121] Even more impressive from the legislator's point of view for passage of the Wilderness Act was the impact of the movement's grass-roots base. Thousands of citizens "with no greater commitment to wilderness than having enjoyed a pack or canoe trip" took their opinions to Congress.[122] By the end of the 1960s, the environmental movement had come of age and was a political force hard to ignore.

The political currency of the environmental movement was evident throughout the years of President Richard Nixon's Administration. Embroiled in the increasingly controversial war in Vietnam, he recognized that the environmental movement was strong in key electoral-college states, particularly among the younger voters. Hundreds of administration staffers fanned out around the country on April 12, 1970, showing their support for the newly inaugurated Earth Day celebration. The use of environmentalism as a wedge issue dated back to Nixon's support for the first Endangered Species Conservation Act in 1969, the National Environmental Policy Act of 1969, the creation of the Environmental Protection Agency in 1970, and, in 1973, the passage of a more comprehensive Endangered Species Act (ESA).

The ESA was an indication that the environmental movement had reached a new level of influence. The ESA basically stated that no species should be allowed to go extinct by human hands. In advocating the passage of the ESA, the environmental movement continued to shape the contours of the long-standing debate between the strict economic or recreational value of a species versus their intrinsic value as part of a fragile and complex ecology. Despite its many shortcomings, the ESA represented an important turning point in policy, committing the government to the protection of endangered species "not just for their commercial and material value, but also because they represent irreplaceable ecological, scientific, recreational, aesthetic, and ethical values as well."[123]

The passage of the ESA promoted Leopold's position of the intrinsic, biotic right of species to exist regardless of what they offer us in the way of economic advantage. Unlike the earlier conservationists who sought "greater control over natural resources in order to use them better," environmentalists now pursued a revolution in how we think about nature. E. O. Wilson, an iconic figure in the modern-day environmental movement, invoked the spirit of both humility and egalitarianism in his argument for the value of the natural world: "The more we know of other forms of life, the more we enjoy and respect ourselves. Humanity is exalted not because we are so far above other living creatures, but because knowing them well elevates the very concept of life."[124]

In 1972, the Club of Rome published its report, *The Limits to Growth,* calling for new measures to address pollution and resource depletion. Later that year, UNESCO founded the World Heritage Fund with the Convention Concerning the Protection of World Cultural and Natural Heritage, which has since been ratified by 185 nation-states. The stage was set for the expansion of American international environmentalism that took part in global environmental forums such as the Earth Summit in 1992, the Kyoto Protocol on Global Climate Change in 1997, and the Johannesburg World Summit on Sustainable Development in 2002. On a broader scale, the Subalternist School, led by writers such as Dipesh Chakrabarty, argued for revisiting the genealogy of Western thought that had marginalized and provincialized non-Western concepts of nature during the European imperial era.[125]

As the orthodoxy of multiculturalism gained public support throughout the 1980s and into the twenty-first century, the environmental movement formed biracial social alliances similar to those of the civil-rights movement of the 1960s, expanding its environmental social-justice initiatives in urban areas. Throughout the final two decades of the twentieth century, the growing awareness of the global nature of the environmental crisis—from the accelerating rate of extinctions to the fate of the rainforests and the oceans, and climate change—joined to World Wide Web-enabled infrastructures ushered in an era of coordinated global environmental activism.

AT PLAY IN THE FIELD OF VALUES

The origins, emergence, growth, and maturity of the environmental movement is part of the American narrative. The form as well as the content of this narrative discourse draws on storied realities associated with the lost landscapes of the American "Eden," the settling and closing of the "frontier," the central saga of "progress," the denomination of "naturalness," "sustainable" development and, ultimately, the meaning of the natural world vis-à-vis the human enterprise.[126]

As in the past, the environmental movement is faced with one of the major narrative plot lines deeply embedded in American environmental history: capitalism, and the secular religion of economic growth, that "has been the decisive factor in this nation's use of nature."[127] Though capitalism has been an enduring feature of Western economies for at least the past 200 to 300 years, it is now part of a larger postmodern discourse. Described by Keith Jenkins as the "last great gambit of capitalism," postmodernism is the lodestar of Western ways of knowing.[128] One important consequence of the postmodern enterprise for the environmental movement is the "collapse of definitional value gradients" which is now "the major fact of our intellectual and social circumstances."[129] The postmodern "rejection of innateness" is evident not only in the valorization of the marketplace and the commodification of nature but, more broadly, in the quantitative, mechanistic approach to nature and the schematization of biotic species and landscapes—and persons—as objects or resources largely devoid of intrinsic value.[130] In this "postist" world, as Barbara Ehrenreich and Deirdre English point out, "the marketplace, which long ago expanded to include the relations of production, has now expanded to include all relationships," including our relationship with nature as a strictly commodified "resource."[131]

One of the important challenges facing the environmental movement is whether it will be able to emplot the intrinsic (nature-in-itself) value of nature in this postmodernist performance model of discourse embraced by an increasingly minority–majority society. This certainly requires what Louis Warren describes as an environmentalism of "equity and social justice, an environmentalism of linked natural and human environments, an environmentalism of transformation."[132] It also requires a pragmatic understanding of postmodernism's death of the subject, the modern dissolution of external order, and the persisting opposition of the abstract and concrete self to the natural world. Environmentalism of the twenty-first century needs to engage this complex, multilayered discourse on self and the environment.

The tragedy of extinction—the greatest since the end of the Mesozoic Era 65 million years ago—is not simply the handiwork of fate. Rather, it is the result of a field of culturally complex representations produced by what Hayden White described as the "apparatus" of discourse.[133] This discursive field, legitimated by a nexus of power and knowledge, is acted out through the relationship between ecology, production, and ideology.

It is a propitious time for the environmental movement, one filled with possibilities to influence the discourse and the terms of the argument on the contested field of values. But stepping onto the field will require going beyond a concern for the growing list of endangered species and threatened habitat. It will also require moving adeptly within the governmentalities of power and thinking deeply about positions of power and concepts of knowledge, tropological space, epistemology, ontology, and ethics. Environmentalism is positioned to help reinvent the environmental imaginary, to redress the covenant with nature, and to participate in the reconstruction of the environmental paradigm. This goes beyond debates over "naturalness" or the dynamics of ecosystems, each of which contain their own ideologics. Instead, environmentalism can help modern individuals reimagine the magnitude and magnificence of nature in order to "re-enter" their relationship with it.

Speaking of his vision atop the Crazy Mountains where he witnessed in his mind's eye the end of the last of the buffalo, the nineteenth-century Crow leader Plenty Coups spoke his lament for a changed universe with the cryptic phrase: "After this, nothing happened."[134] More than a century after he spoke those words, refrains of Plenty Coups, as well Leopold's call for a durable scale of values, can be heard in the forewarnings of E. O. Wilson about what a failure of imagination might mean for the future of modern societies if they persist in the vision of human exceptionalism:

> To the extent that we depend on prosthetic devices to keep ourselves and the biosphere alive, we will render everything fragile. To the extent that we banish the rest of life, we will impoverish our own species for all time. And if we should surrender our genetic nature to machine-aided ratiocination, and our ethics and art and our very meaning to a habit of careless discursion in the name of progress, imagining ourselves godlike and absolved from our ancient heritage, we will become nothing.[135]

NOTES

1. Andrea S. Laliberte and William J. Ripple, "Range Contractions of North American Carnivores and Ungulates," *BioScience,* 54 (2) (2004): 123–138.
2. Dan Flores, *The Natural West: Environmental History in the Great Plains and Rocky Mountains* (Norman, Okla.: University of Oklahoma Press, 2001), p. 188.
3. Code of Federal Regulations-Title 50.17.11, "Endangered and Threatened Wildlife" (December 2005).
4. Flores, *Natural West,* p. 189.
5. William H. MacLeish, *The Day before America: Changing the Nature of a Continent* (Boston, Mass.: Houghton Mifflin, 1994), p. 52.
6. MacLeish, *The Day before America,* p. 52.
7. Joseph M. Petulla, *American Environmental History* (Columbus, OH: Merrill, 1988), p. 7.
8. William M. Denevan, "The Pristine Myth: The Landscape of the Americas in 1492," in Louis S. Warren (ed.), *American Environmental History* (Malden, Mass.: Blackwell Pub., 2003), pp. 5–48.
9. Stanley W. Trimble, "Nature's Continent," in Michael P. Conzen (ed.), *The Making of the American Landscape* (London and New York: Routledge, 1994), pp. 9–26.
10. MacLeish, *The Day before America,* p. 161.
11. Petulla, *American Environmental History,* p. 8.
12. Carolyn Merchant, *The Columbia Guide to American Environmental History* (New York: Columbia University Press, 2002), p. 15.
13. Donald Worster, *An Unsettled Country: Changing Landscapes of the American West* (Albuquerque, N. Mex.: University of New Mexico Pres, 1994), p. 61.
14. D. I. Axelrod, "Rise of the Grassland Biome, Central North America," *Botanical Review,* 51 (2) (1985): 163–201.
15. Joseph M. McCann, "Before 1492: The Making of the Pre-Columbian Landscape. Part II: The Vegetation, and Implications for Restoration for 2000 and Beyond," *Ecological Restoration,* 17 (1) (1999): 15–30. Erhard Rostlund, "The Geographic Range of the Historic Bison in the Southeast," *Annals of the Association of American Geographers,* 50 (1960): 395–397.
16. McCann, "Before 1492, The Making of the Pre-Columbian Landscape (Part I: The Environment)," p. 23.
17. Karl W. Butzer, "The Americas before and after 1492: An Introduction to Current Geographical Research," *Annals of the Association of American Geographers,* 82 (2) (1992): 345–368.
18. J. J. Parsons, "Raised Field Farmers as Pre-Columbian Landscape Engineers: Looking North from San Jorge (Colombia)," in I. S. Farmington (ed.), *Prehistoric Intensive Agriculture in the Tropics* (Oxford: British Archaeological Reports, 1985), pp. 156–174.
19. Frederic Wagner, "Livestock Grazing and the Livestock Industry," in Howard Brokaw (ed.), *Wildlife in America: Contributions to an Understanding of American Wildlife and its Conservation* (Washington, DC: Council on Environmental Quality, 1978), pp. 121–145.
20. Stephen W. Barrett and Stephen F. Arno, "Indian Fires as an Ecological Influence in the Northern Rockies," *Journal of Forestry,* 80 (10) (1982): 647–651.
21. Donald W. Meinig, "Environment and Settlement in the Palouse, 1868–1910," Master's thesis, University of Washington, 1950, pp. 15–16.
22. Robert Bunting, *The Pacific Raincoast: Environment and Culture in an American Eden, 1778–1900* (Lawrence, Kans.: University Press of Kansas, 1997), p. 3.
23. McCann, "Before 1492," p. 110.
24. Theodora Kroeber and Robert F. Heizer, *Almost Ancestors: The First Californians* (San Francisco, Calif.: Sierra Club, 1968), p. 24.
25. Alfred W. Crosby, *Ecological Imperialism: The Biological Expansion of Europe, 900–1900* (Cambridge: Cambridge University Press, 1986), p. 194.
26. Harold F. Heady, "Valley Grassland," in M. G. Barbour and J. Major (eds.), *Terrestrial Vegetation of California* (Sacramento, Calif.: California Native Plant Society, 1988), pp. 491–514.

27. Raymond F. Dasmann, "Environmental Changes before and after the Gold Rush," in James J. Rawls and Richard. J. Orsi (eds.), *A Golden State: Mining and Economic Development in Gold Rush California* (Berkeley, Calif.: University of California Press, 1999), pp. 105–122.
28. Petulla, *American Environmental History,* p. 135.
29. James L. Wescoat, Jr., "Challenging the Desert," in Michael P. Conzen (ed.), *The Making of the American Landscape* (London and New York: Routledge, 1994), pp. 186–203.
30. Laliberte and Ripple, "Range Contractions," p. 127.
31. Alfred W. Crosby, *The Columbian Exchange: Biological and Cultural Consequences of 1492* (Westport, Conn.: Greenwood Press, 1972), p. 104.
32. John F. Richards, *The Unending Frontier: An Environmental History of the Early Modern World* (Berkeley, Calif.: University of California Press, 2005), p. 473.
33. Denevan, "The Pristine Myth," p. 6.
34. Henry Allan Gleason, "The Vegetation of the Middle West," *Annals of the Association of American Geographers,* 12 (1922): 39–85; E. Rostlund, "The Myth of a Natural Prairie Belt in Alabama: An Interpretation of Historical Records," *Annals of the Association of American Geographers,* 47 (4) (1957): 392–411.
35. William M. Denevan, "The Pristine Myth: The Landscape of the Americas in 1492," in Louis S. Warren (ed.), *American Environmental History* (Malden, Mass.: Blackwell, 2003), pp. 5–48.
36. Richards, *Unending Frontier,* p. 470.
37. Bunting, *Pacific Raincoast,* p. 75.
38. Alexander Ross, *Adventures of the First Settlers on the Oregon or Columbia River, 1810–1813* (Lincoln, Nebr.: University of Nebraska Press, 1986), pp. 90–92.
39. William Fraser Tolmie, *The Journals of William Fraser Tolmie, Physician and Fur Trader* (Vancouver: Mitchell Press, 1963), pp. 166–167.
40. Merchant, *American Environmental History,* pp. 80–81.
41. M. Kat Anderson, Micheal Barbour, and Valerie Whitworth, "A World of Balance and Plenty: Land, Plants, Animals, and Humans in Pre-European California," *California History,* 76 (summer and fall) (1997); p. 76.
42. M. Kat Anderson, *Tending the Wild: Native American Knowledge and the Management of California's Natural Resources* (Berkeley, Calif.: University of California Press, 2005), p. 77.
43. D. W. Meinig, *The Shaping of America: A Geographical Perspective on 500 Years of History,* Vol. II: *Continental America, 1800–1867* (New Haven, Conn.: Yale University Press, 1993), p. 76.
44. Flores, *Natural West,* p. 188.
45. Meinig, *Shaping of America,* p. 76.
46. James C. Malin, *The Grasslands of North America: Prolegomena to Its History* (Gloucester, Mass.: Smith, 1967), p. 461.
47. Walter Prescott Webb, "Great Plains Ecology," in Carolyn Merchant, *Major Problems in American Environmental History* (Boston, Mass.: Houghton Mifflin, 2005), p. 289.
48. William Beinart and Peter Coates, *Environment and History: The Taming of Nature in the USA and South Africa* (London and New York: Routledge, 1995), p. 54.
49. Beinart and Coates, *Environment and History,* p. 40.
50. Laliberte and Ripple, "Range Contractions," pp. 127–128.
51. Fred Samson and Fritz Knopf (eds.), *Prairie Conservation: Preserving North America's Most Endangered Ecosystem* (Washington, DC: Island Press), p. 41.
52. Laliberte and Ripple, "Range Contractions," p. 136.
53. Flores, *Natural West,* p. 67.
54. D. P. Boyd, "Conservation of North American Bison: Status and Recommendations," Master's thesis, University of Calgary, Alberta, 2003.
55. Wagner, "Livestock Grazing," p. 124.
56. Thomas Swetnam, "Applied Historical Ecology: Using the Past to Manage for the Future," *Applied Historical Ecology,* 9 (4) (1999): 1–19.

57. Swetnam, "Applied Historical Ecology."
58. Swetnam, "Applied Historical Ecology."
59. B. L. Turner I et al., *The Earth as Transformed by Human Action: Global and Regional Changes in the Biosphere over the Past 300 Years* (Cambridge: Cambridge University Press, 1990), p. 563.
60. Turner et al., *The Earth as Transformed by Human Action,* p. 562.
61. Turner et al., *The Earth as Transformed by Human Action,* p. 567.
62. MacLeish, *The Day before America,* p. 102.
63. Barry Holstun Lopez, *Of Wolves and Men* (New York: Scribner & Sons, 1978), p. 180.
64. C. A. Kirkpatrick, "Salmon Fishery on the Sacramento River," in R. R. Olmsted (ed.), *Scenes of Wonder and Curiosity from Hutchings' California Magazine, 1856–1861* (Berkeley, Calif.: Howell-North), pp. 47–66.
65. Merchant, *American Environmental History,* p. 86.
66. T. Luke George and L. Arriana Brand, "The Effects of Habitat Fragmentation on Birds in Coast Redwood Forests," in T. Luke George and David S. Dobkin (eds.), *Effects of Habitat Fragmentation on Birds in Western Landscapes: Contrasts with Paradigms from the Eastern United States* (Lawrence, Kans.: Cooper Ornithological Society, 2002), pp. 92–102.
67. Anderson, *Tending the Wild,* p. 100.
68. Anderson, *Tending the Wild,* p. 100.
69. Anderson, *Tending the Wild,* p. 96.
70. Anderson, *Tending the Wild,* p. 120.
71. William Preston, "Serpent in the Garden," in Ramón A. Gutiérrez and Richard J. Orsi (eds.), *Contested Eden: California before the Gold Rush* (Berkeley, Calif.: University of California Press, 1998), pp. 260–298.
72. M. Kat Anderson, Michael G. Barbour, and Valerie Whitworth, "A World of Balance and Plenty: Land, Plants, Animals, and Humans in a Pre-European California," in Ramón A. Gutiérrez and Richard J. Orsi (eds.), *Contested Eden: California before the Gold Rush* (Berkeley, Calif.: University of California Press, 1998), pp. 12–47.
73. Preston, "Serpent in the Garden," p. 275.
74. Jeanne Kay Guelke and David Hornbeck, "The Far West, 1840–1920," in Thomas F. McIlwraith and Edward K. Muller (eds.), *North America: The Historical Geography of a Changing Continent* (New York: Rowman & Littlefield, 2001), pp. 261–283.
75. Margaret Maizel, R. Denis White, Ralph Root, Stuart Gage, Susan Stitt, Leon Osborne, George Muehlbach, "Historical Interrelationships between Population Settlement and Farmland in the Conterminous United States, 1790–1992," in Tom D. Sisk (ed.), *Perspectives on the Land Use History of North America: A Context for Understanding Our Changing Environment,* available online at http://purl.access.GPO.gov/GPO/LPS13770 (accessed May 18, 2010).
76. Douglas T. Bolger, "Habitat Fragmentation Effects on Birds in Southern California: Contrast to the 'Top-Down' Paradigm," in T. Luke George and David S. Dobkin (eds.), *Effects of Habitat Fragmentation on Birds in Western Landscapes: Contrasts with Paradigms from the Eastern United States* (Lawrence, Kans.: Cooper Ornithological Society, 2002), pp. 141–157.
77. Bolger, "Habitat Fragmentation Effects," p. 144.
78. Bolger, "Habitat Fragmentation Effects," p. 149.
79. Anderson, "A World of Balance and Plenty," p. 24.
80. Anderson, "A World of Balance and Plenty," p. 22.
81. Bunting, *Pacific Raincoast,* p. 32.
82. E. W. Gilbert, *The Exploration of Western America, 1800–1850* (Cambridge: Cambridge University Press, 1933), p. 79.
83. William L. Robinson and Eric G. Bolen, *Wildlife Ecology and Management* (New York: Macmillan, 1989).
84. Kathleen A. Dwire, Bruce A. McIntosh, and J. Boone Kauffman, "Ecological Influences of the Introduction on Livestock on Pacific Northwest Ecosystems," in Dale Goble and Paul Hirt (eds.),

Northwest Lands, Northwest Peoples: Readings in Environmental History (Seattle, Wash.: University of Washington Press, 1999), pp. 313–335.

85. Anne E. Black, J. Michael Scott, Eva Strand, R. Gerald Wright, Penelope Morgan, and Cortney Waatson, "Biodiversity and Land-Use History of the Palouse Bioregion: Pre-European to Present," in Tom D. Sisk (ed.), *Perspectives on the Land Use History of North America: A Context for Understanding Our Changing Environment,* available online at http://purl.access.GPO.gov/GPO/LPS13770 (accessed May 18, 2010).
86. Black et al., "Biodiversity and Land-Use History of North America."
87. Thomas M. Bonnicksen, *America's Ancient Forests: From the Ice Age to the Age of Discovery* (New York: Wiley & Sons, 2000), p. 388.
88. Douglas E. Booth, *Valuing Nature: The Decline and Preservation of Old-Growth Forests* (London: Rowman & Littlefield, 1994), p. 110.
89. Bunting, *Pacific Raincoast,* p. 124.
90. Michael Williams, *Americans and Their Forests: A Historical Geography* (Cambridge: Cambridge University Press, 1989), p. 292.
91. Williams, *Americans and Their Forests,* p. 304.
92. Bunting, *Pacific Raincoast,* p. 157.
93. Bunting, *Pacific Raincoast,* p. 156.
94. Williams, *Americans and Their Forests,* p. 305.
95. Committee on Environmental Issues in Pacific Northwest Forest Management (Norman Christensen, Jr., Chair), *Environmental Issues in Pacific Northwest Forest Management* (Washington, DC: National Academy Press, 2000), p. 70.
96. Booth, *Valuing Nature,* p. 96.
97. Douglas Booth estimates the percentage of remaining old-growth forests in Western Oregon and Western Washington is 16.9 percent and 18.2 percent, respectively, though he agrees with Robert Bunting that these forests are found almost exclusively on public lands (Booth, *Valuing Nature,* p. 117; Bunting, *Pacific Raincoast,* p. 1).
98. Booth, *Valuing Nature,* p. 109.
99. Bunting, *Pacific Raincoast,* p. 79.
100. Dale E. Goble, "Salmon in the Columbia Basin: From Abundance to Extinction," in Dale Goble and Paul Hirt (eds.), *Northwest Lands, Northwest Peoples: Readings in Environmental History* (Seattle, Wash.: University of Washington Press, 1999), pp. 229–263.
101. Ed Chaney and L. Edward Perry, *Columbia Basin Salmon and Steelhead Analysis* (Portland, Oreg.: Pacific Northwest Regional Commission, 1976), p. 3.
102. Joseph C. Dupris, Kathleen S. Hill, and William H. Rodgers, Jr., *The Si'lailo Way: Indians, Salmon and Law on the Columbia River* (Durham, NC: Carolina Academic Press, 2006), p. xxiv.
103. Dupris et al., *The Si'lailo Way,* p. xxiv.
104. Dwire et al., "Ecological Influences," p. 330.
105. Alan Taylor, "'Wasty Ways': Stories of American Settlement," in Louis S. Warren (ed.), *American Environmental History* (Oxford: Blackwell, 2003), p. 103.
106. Richard Irving Dodge, *The Plains of North America and Their Inhabitants,* ed. Wayne R. Kime (Newark, Del.: University of Delaware Press, 1989), p. 140.
107. Peter Nabokov and Lawrence Loendorf, *Restoring a Presence: American Indians and Yellowstone National Park* (Norman, Okla.: University of Oklahoma Press, 2002), p. 9.
108. Petulla, *American Environmental History,* p. 243.
109. Nancy Langston, *Forest Dreams, Forest Nightmares: The Paradox of Old Growth in the Inland West* (Seattle, Wash.: University of Washington Press, 1995), p. 95.
110. Bunting, *Pacific Raincoast,* p. 152.
111. Beinart and Coates, *Environment and History,* p. 46.
112. Worster, *An Unsettled Country,* pp. 82–83.
113. Beinart and Coates, *Environment and History,* p. 82.

114. Roderick Nash, *Wilderness and the American Mind* (New Haven, Conn.: Yale University Press, 1967), p. 207.
115. Nash, *Wilderness and the American Mind,* pp. 207–208.
116. Nash, *Wilderness and the American Mind,* p. 199.
117. Robert Gottlieb, "Reconstructing Environmentalism: Complex Movements, Diverse Roots," in Louis Warren (ed.), *American Environmental History* (Oxford: Blackwell, 2003), pp. 245–270.
118. Linda Nash, "Finishing Nature: Harmonizing Bodies and Environments," *Environmental History,* 8(1) (2003): 25–53.
119. Beinart and Coates, *Environment and History,* p. 95.
120. Carolyn Merchant, *The Death of Nature: Women, Ecology, and the Scientific Revolution* (San Francisco, Calif.: Harper & Row, 1980), p. 294.
121. Nash, *Wilderness and the American Mind,* p. 224.
122. Nash, *Wilderness and the American Mind,* p. 225.
123. Stephen R. Kellert, *The Value of Life: Biological Diversity and Human Society* (Washington, DC: Island Press, 1996), p. 158.
124. Louis Warren, "Environmental Protection and the Environmental Movement," in Louis Warren (ed.), *American Environmental History* (Oxford: Blackwell, 2003), pp. 262–289. Edward. O. Wilson, *Biophilia* (Cambridge, Mass.: Harvard University Press, 1984), p. 22.
125. Dipesh Chakrabarty, *Provincializing Europe: Postcolonial Thought and Historical Difference* (Princeton, NJ: Princeton University Press, 2000).
126. An intriguing theory advanced by Margaret Somers on the construction of the metanarrative of the public and private spheres and the conceptual space of progress can be found in her article "The Privatization of Citizenship: How to Unthink a Knowledge of Culture," in Victoria E. Bonnell and Lynn Hunt (eds.), *Beyond the Cultural Turn* (Berkeley, Calif.: University of California Press, 1999), pp. 121–161.
127. Donald Worster, *Dust Bowl: The Southern Plains in the 1930s* (New York: Houghton Mifflin, 1979), p. 27.
128. Keith Jenkins, *Re-Thinking History* (London and New York: Routledge, 1991), p. 80.
129. Jenkins, *Re-Thinking History,* p. 76.
130. Charles Taylor, *Sources of the Self: The Making of the Modern Identity* (Cambridge, Mass.: Harvard University Press, 1989), p. 165. Jenkins, *Re-Thinking History,* p. 80.
131. Barbara Ehrenreich and Deirdre English, *For Her Own Good* (New York: Knopf, 1979), p. 276.
132. Gottlieb, "Reconstructing Environmentalism," p. 255.
133. Hayden White, *The Content of the Form: Narrative Discourse and Historical Representation* (Baltimore, Md.: Johns Hopkins University Press, 1987), p. 42.
134. Plenty Coups, "Vision in the Crazy Mountains," in Frank Bergon (ed.), *The Wilderness Reader* (New York: Mentor, 1980), pp. 223–229.
135. Edward O. Wilson, *Consilience: The Unity of Knowledge* (New York: Vintage, 1999), p. 326.

CHAPTER THREE

THE NATIVE WEST BEFORE 1700

Gregory E. Smoak

A Plains Indian warrior sits astride his pony, face painted, wearing beaded buckskin finery and a feathered war bonnet, with lance and shield in hand. It is a romantic and iconic image familiar to nearly every American, and for some it is all they know of the continent's Native peoples or their histories. In the popular imagination the Plains Indian has become the "real" American Indian. Yet in 1492 no Native person in the American West looked anything like that Plains warrior. By 1700, only a handful of Native peoples had begun to embrace an equestrian life, and the golden age of Plains culture that followed would last slightly more than a century and a half. And so, the Plains Indian stereotype, as noble as it may be, must be set aside in order to grasp the historical processes that shaped the Native West before 1700 and, by extension, the American West as a region.

First, we must acknowledge that the "American West" as defined in this volume had no meaning for the region's aboriginal inhabitants. Defining the boundaries of the American West has produced years of spirited and sometimes acrimonious debate among historians. Yet, no matter which set of borders one accepts, none have meaning outside the context of European colonization, and even then only in relation to histories of the USA and Canada (for Mexico, the West is "El Norte"). The region called the "West," then, is largely an historical creation, and to fully understand that history we must begin with its first human inhabitants. While this essay focuses on the North American continent stretching from the tall grass prairies to the Pacific coast, the narrative ranges east to the Mississippi valley, where the great city of Cahokia peaked around 1300 and served as a nexus for far-reaching trade networks, to the south into central Mexico where the domestication of corn transformed indigenous lifeways beginning perhaps 9,000 years ago, and to the far north, from where waves of human migrants likely first entered the American continents. Notwithstanding the human migrations and historical changes described in the following pages, American Indians were, and remain, essentially placed-based peoples with strong cultural and spiritual ties to their homelands. It should always be remembered that their lands were not their "West," but rather their center place, their home.

Counter to any static or monolithic image, the Native West was a diverse and dynamic place, home to hundreds of distinct peoples who spoke different languages, practiced different religions, and employed countless ways to make a living from the land. The story of the West's Native peoples was always about more than horses and buffalo. In a very real sense, America's Native peoples have been colonized by history. Rather than being treated as real people who faced choices and made decisions that shaped the course of history, they have more often appeared as tropes—literary devices that alternately critique or justify conquest. Images of the "noble savage" (simple, honorable, close to nature, yet destined for extinction in the face of superior civilization) or the "ignoble savage" (cruel, bloodthirsty, barely human) have filled the pages of countless volumes. In neither case were Native cultures accorded their own logic or their own histories. To the contrary, the Native West was the product of thousands of years of human decisions and actions before any European ever set foot there. Simply put, Native peoples were not sitting around in the Americas waiting for Europeans to arrive in order for history to "start." This chapter explores the histories of the West's Native peoples, how they shaped the region before contact, and how they understood and survived the early years of conquest.[1]

THE FIRST AMERICANS

The Native peoples of the West understand their origins and that of the Earth in culturally meaningful ways. For the Zuni people, life began in the undermost world within the Earth Mother where all the people and animals lived in darkness and confusion. The Sun Father created twin sons and instructed them to bring all of his children into the daylight world. The twins travelled west to the Grand Canyon and then, using lightning arrows, they opened a path into the Earth Mother and descended to First World. The twins made fire and planted a pine tree from the north and led the people upward to the second, "Water Moss World." A spruce tree from the west provided a path to the Mud World above, while an aspen from the south allowed them to climb to the fourth "Wing World." There, the twins instructed the people in sacred knowledge before leading them up a silver spruce to the daylight or outer world. In each world along the way the people acquired the knowledge necessary for life in the next. Once on the surface they eventually set off to find the "middle of the world." After an arduous journey, the people came to that middle place, where the Zuni Pueblo still stands.[2] Many other Native groups, especially in the Southwest, similarly understand their origins in terms of emergence from chaos into light.

In other cultures, creation is explained as a process of making the Earth itself. Among the Shoshone-Bannock people of Idaho, it is explained that originally the world was covered with water. Apa, the creator, had three water children: the Beaver, the Otter, and the Muskrat. One day, Apa decided that if there was dry land there could be many other creatures, and so he asked his children to help him find out what was under all that water. While Beaver and Otter were at first doubtful, the Muskrat suggested they all try to find the bottom. Beaver dove first and remained under for a very long time but returned to the surface gasping for breath without finding the bottom. Otter went next but also failed. At last, Muskrat dove straight and deep and ignored the fish that called out his name. In short order he returned to the surface with a ball of mud upon his nose. (In most versions of the "Earth Diver" story, the smallest and weakest creature eventually succeeds.) Apa took the mud and breathed on it. He rolled it around in his hands, and it grew until it became the Earth that all live upon today.[3]

Another common Shoshonean tale explains the creation of human beings and the origins of ethnic difference. The central figure in this story is the Trickster, Coyote. Tricksters are often comic figures that embody the ambiguities of life. They often bring great things but are also regularly responsible for misery and despair. Shoshone people in the Great

Basin tell how Coyote followed two women to their home on an island in a large body of water. He spent many days with the women, and both began to give birth at an astounding rate. They put all the babies in a woven water jug and told Coyote he must go home, carrying the basket with him. The women also warned Coyote to be careful to keep the lid in place, lest any of the babies escape. But the jug was heavy, and whenever Coyote put it down to rest he let his curiosity get the better of him and lifted the lid to look inside. Each time some of the babies escaped. Some became the Shoshones, while others became the Utes, Paiutes, Bannocks and Gosiutes. They all spoke different languages and were of the land where they lived. In Skull Valley, Utah, it is said that the last baby to emerge was "covered with dust. He was the Gosiute. He is tougher than other people."[4] Like the Zuni story that told of a voyage to the center place of the Earth, the Shoshone Coyote story explains the origins of a people and how they came to occupy their rightful place in the world.

The scientific and scholarly community explains the origin of American Indian peoples in a very different manner. For some Native people, the widely accepted theory that their ancestors migrated to the Americas from Asia is offensive. It suggests that their origin stories are mistaken or that in some way their ancestors were little different than European conquerors. Other Native people and groups, however, accept, or even embrace the migration theory. Visitors to the Mashantucket Pequot museum in Connecticut, for instance, begin their tour by descending an escalator through a faux glacier, suggesting the Ice Age passage of Native peoples to the Americas. Nor is the scholarly and scientific community of a single mind concerning the process and the implications of the human settlement of the Americas, although the premise that the migration from Asia was the final stage in the human colonization of the planet is not disputed. Still, in the past quarter century new discoveries, theories, and methodologies have produced a far more complicated picture of the early human habitation of the American continents.

For generations, schoolchildren were taught that the peopling of the Americas took place in a fairly uniform and rapid fashion at the close of the last Ice Age. As early as 1590, the Jesuit Joseph de Acosta theorized that American Indians had migrated from Asia. Two centuries later, when Thomas Jefferson wrote that the "resemblance between the Indians of America and the Eastern inhabitants of Asia, would induce us to conjecture, that the former are the descendants of the latter" he was expressing a fairly common belief.[5] It was not until archeological discoveries in the 1920s and 1930s, however, that the modern "Bering Strait theory" gained prominence. In its basic form, the theory posits that the first Americans arrived some 12,000 to 14,000 years B.P. as falling sea levels exposed a 1,000-mile-wide land bridge between Siberia and Alaska. The migrants then followed the McKenzie Basin corridor, an ice-free passage between the ice sheets that paralleled the Rocky Mountains. Once clear of the ice, these "Paleoindians" rapidly populated both North and South America.[6]

The theory meshed well with mid-twentieth-century understandings of North American archeology that were based, in large part, on discoveries made at Folsom and Clovis, New Mexico. In 1926, archeologists working near Folsom uncovered a distinctive set of fluted stone projectile points in close proximity to the remains of an extinct species of bison. Other sites with Folsom points were discovered later, and all have been dated between 9,000 and 10, 500 B.P. In the mid-1930s, slightly over 200 miles south of Folsom, an even older stone tool tradition, which eventually took its name from the nearby town of Clovis, was identified. The bifacial, fluted "Clovis points" were not as finely made as the Folsom points, but generally dated to 13,000 B.P. For decades, Clovis points stood as the oldest evidence of human habitation in the Western Hemisphere, and the phrase "Clovis First" came to neatly summarize archeological orthodoxy. Taken together, the Clovis and Folsom discoveries were interpreted as evidence that Paleoindians migrated to the Americas rapidly at the end of the Ice Age, bringing with them a fairly uniform and specialized big game hunting culture.[7]

The rapid extinction of some thirty-five genera of North American megafauna—including mammoths, mastodons, camels, ground sloths, dire wolves, saber-toothed tigers, and the earliest horses—approximately 11,000 years ago has been proposed as both confirmation of the Bering Strait/Clovis First orthodoxy and the first great consequence of human presence in the Americas. Paul S. Martin, the original proponent of the "Pleistocene Overkill" theory, argued that the changing climate at the end of the Ice Age would not account for the mass extinctions. Instead, he proposed that nothing less than a "blitzkrieg" of Clovis hunters emerged from the ice sheets to make easy prey of the lumbering beasts.[8] Critics of the overkill theory have pointed out that many species that were not subject to human predation also went extinct while others that were, such as the bison, survived. The absence of kill sites for all but two of the vanished species raises further questions. Critics have also argued that the overkill thesis has become a "faith-based" creed, kept alive by popular authors who use the theory as a cautionary tale of the ecological consequences of human rapaciousness. Current consensus in the scientific community is that human predation may well have been a contributing factor but was not the principal reason for the mass extinctions.[9]

Since the 1970s, new archeological discoveries have proven a far more serious challenge to the "Clovis First" orthodoxy than the eclipse of the Pleistocene overkill theory. In all, some twenty-six sites in North America and nine in South America have yielded pre-Clovis dates.[10] In 2008, researchers announced that they had recovered the oldest strain of human DNA yet found in the Americas at Paisley Caves in southeast Oregon. At 14,300 years old, it is at least 1,000 years older than the Clovis artifacts.[11] By far the most famous, controversial, and consequential find has proven to be Monte Verde in southern Chile. The site contained a pole frame structure, bone and stone tools, bits of rope, a child's footprint, and other evidence of human occupation dated to at least 12,800 years B.P. The initial dating of the Monte Verde artifacts, reported in the late 1970s, met great skepticism within the archeological community. Clearly, a pre-Clovis site so old and so far to the south raised a serious challenge to the traditional Bering Strait theory. As more evidence came to light, however, the antiquity of Monte Verde became widely accepted, although a few dissenting opinions can still be heard.[12]

The presence of so many pre-Clovis sites not only pushed back the timeline for human arrival in the Americas, it also led some scholars to conclude that the first Americans had followed a very different route, one along the Pacific coast. Water travel would have been easier and faster than the arduous continental route. Moreover, groups such as Australia's Aborigines made substantial open-water crossings in at least 40,000 B.P. Several important archeological finds as well as circumstantial evidence also lend weight to the coastal migration theory in the Americas. Paleo-ecological evidence indicates that glaciers retreated from the Northwest Pacific coast some 16,000 years ago, while an abundance of coastal plant and animal life could have supported human residents. And, while rising sea levels have submerged the vast majority of the archeological evidence for human presence, there is hard evidence that some Paleoindian peoples possessed a developed maritime culture. The partial skeleton of "Arlington Springs Man" (anthropologists have debated the gender of the remains for decades) recovered on Santa Rosa Island off the Southern California coast is at least 11,000 and perhaps 13,000 years old. The presence of these remains, which are contemporary with Clovis, on an island that could only be reached by boat even during the period of lowest sea levels indicates that the earliest Californians possessed watercraft and the skills to navigate coastal waters. Thousands of miles to the south, two fishing and sea-bird-hunting campsites of similar antiquity have been excavated in Peru, and in 2008 researchers at Monte Verde reported that ancient seaweed recovered there confirmed human occupation more than 14,000 years ago and indicated that residents had much stronger ties to the Pacific Coast (some 35 miles away) than previously suspected. For many, the weight of the evidence suggests that the human migration to the Americas began at least 16,000 years ago, well before the retreat of the continental glaciers, and most likely followed a Pacific coastal route.[13]

Coastal-migration theories, however, are not limited to the Pacific. Based on an analysis of lithic technology, Dennis Stanford of the Smithsonian Institution and Bruce Bradley of Exeter University have proposed a second Ice Age migration to North America, this one of European peoples along the Atlantic coast. Arguing that no clear antecedents for the Clovis tradition have been found in Asia, and the many similarities between Clovis points and the projectile points produced by the Solutrean Culture of southwestern Europe (ca. 15,000–20,000 B.P.) they believe that at least some of the earliest migrants to the Americas arrived via a treacherous passage along the Atlantic ice shelf. They contend that the transitional stone points found in pre-Clovis sites such as Meadowcroft and Cactus Hill illustrate the evolution of the Solutrean tradition into the Clovis. The "Solutrean hypothesis" remains highly controversial. Critics point out that, in addition to the overwhelming physical difficulties of the proposed journey, there are actually vast differences between Clovis and Solutrean traditions. The latest Solutrean points are thousands of years older than the earliest Clovis points. Moreover, there is no evidence that Solutrean peoples possessed watercraft or the fishing technology that would have made the voyage even possible.[14]

Since the 1980s, the study of genetics has also entered the debate. Using mitochondrial DNA (mtDNA), which is only passed down through one's mother, molecular geneticists and anthropologists have sought to trace human lineages. Genetic researchers have found consistent links between modern American Indian and Asian populations through four principal genetic lineages that they have labeled A through D. Based on this data, linguist Joseph Greenberg and others theorized that there were three principal migrations from Asia. Each migration corresponded to an over-simplified classification of all indigenous American languages into three large groups: Amerind, Na-Dene, and Eskimo-Aleut. Later, a fifth lineage common in western Eurasia, marked X, caused a furor when it was discovered among some North American Native populations. Some theorized that this was evidence that European peoples had indeed migrated to the Americas in ancient times.[15] Later studies have dismissed this suggestion. One recent report concluded that all five lineages were present in a single pre-Clovis population and interpreted this finding as evidence for a rapid settlement of the Americas along a Pacific coastal route.[16]

The discovery of some two dozen sets of ancient human remains have further complicated the story and in some cases have had real political implications for modern American Indian communities. The 1996 discovery of "Kennewick Man" along the Columbia River in Washington state sparked the greatest controversy. The remains were over 9,000 years old but, according to some physical anthropologists, exhibited "Caucasoid" features. Kennewick Man quickly became a media event as the remains, combined with the recent discovery of the genetic lineage X, became evidence in the popular press for a prehistoric European colonization of the Americas. The case turned political when an alliance of five tribes sued for custody and reburial of the "Ancient One" under the 1990 Native American Graves Protection and Repatriation Act (NAGPRA). The Act was meant to redress the centuries of shameful treatment American Indian peoples had suffered at the hands of Western science. Several prominent anthropologists won a counter suit, arguing that the remains could not be linked to any modern Native population and therefore could not be subject to NAGPRA. Most scientists, who view Kennewick Man as a key to understanding the early human settlement of the Americas, hailed the decision. For many Native people, however, the treatment of the "Ancient One" (who today lies in court-ordered storage at the Burke Museum in Seattle) was simply another in a long line of injustices.[17]

What then is to be made of the conflicting theories concerning human migration to the Americas? It is clear that the traditional Bering Strait theory—a single, rapid migration of peoples through an ice-free corridor in the center of the continent—is no longer defensible. While scholars may interpret the evidence in differing and sometimes contradictory ways, the preponderance of that evidence suggests a far more complicated story. It is likely that the First Americans began their journeys before the end of the last Ice Age, 16,000 or perhaps more than 20,000 years ago; that more than one wave of peoples came to the American continent; that many followed Pacific coastal routes to the South; and that, in addition to a big-game hunting tradition, they brought with them and developed diverse ways of life, including a maritime culture.

PEOPLES AND CIVILIZATIONS IN MOTION

From the end of the Paleoindian period to the time of the Spanish *entrada* (an official Spanish term denoting formal entrance into a new land), the natural and human landscape of the Native West was repeatedly transformed. Hunting, fishing, and gathering peoples developed new technologies and methods as they adapted to changing environmental conditions and exploited new opportunities. The rise of agriculture was even more consequential, and peoples formed powerful and populous societies. Finally, Native Peoples in the pre-contact West did not sit still but participated in great migrations, some spanning half a continent. At times, the result was conflict, but just as often these migrations revealed new patterns of trade and cultural exchange.

Archeologists refer to the time between 10,000 B.P. and 3,000 B.P. as the Archaic period. Both the diversification of hunting cultures and the growing reliance on plant foods differentiate the Archaic from the earlier Paleoindian period, yet there is not a single defined sequence of these events throughout North America. Instead, scholars have defined several regional traditions. As the climate warmed on the Great Plains, bison herds filled the niche left by extinct Ice Age mammals and became the principal subsistence of numerous small bands known collectively as the Plano culture. Plano hunters used atlatls, spear throwers, to make their hunts more deadly. In later periods, Plains hunters used various methods to take buffalo including impounds and buffalo jumps. Farther to the west, the Desert Archaic tradition emerged in the Great Basin, where smaller game, including waterfowl, was just as important. Desert Archaic peoples also used seed-grinding tools to process wild grass seeds and pine nuts. Likewise, adaptations for big-game hunting were not particularly evident along the coast. In the Pacific Northwest, for instance, a coble tool tradition (simple stone hand tools modified mostly through use) and a focus on maritime resources distinguished life during the Archaic period. Late in the Archaic, Native peoples in the southwest began to produce pottery and to cultivate plants, developments that foreshadowed great changes to come.[18]

The rise of agriculture provided the engine for even greater social and cultural change in Archaic Native America. Wild-plant management was the earliest form of agriculture. Weeding out unwanted species facilitated the growth of favored plants, while broadcast-sowing wild seeds improved harvests. True agriculture—the dedicated planting, tending, and harvesting of a crop—began perhaps in 9,000 B.P. in the Tehuacán Valley of central Mexico when farmers began raising bottle gourds both as a food source and as storage containers. Similar developments occurred independently in other locations in Mesoamerica. The domestication of the most important crop of all, corn or maize, took place perhaps as early as 8,700 B.P. in the Balsas River Valley in central Mexico.[19] Between 7,000 and 5,000 B.P., agriculture became more elaborate as farmers cleared lands and grew a wide assortment of crops including chili peppers, beans, avocados, and squashes. While corn became the cornerstone of farming, Native peoples never practiced a monocrop system. Throughout much of North America, people planted the "three sisters"—corn, beans, and squash—together. The three plants enjoy a symbiotic relationship. Beans fix nitrogen, the principal nutrient for the other two crops, in the soil. Corn stalks provide the beans with a structure to climb, while the squash vines fill the garden floor, monopolizing the sunlight and keeping weeds to a minimum. The system was highly effective but appeared chaotic to early European observers accustomed to straight plowed furrows and monocrop fields. Moreover, in most Native societies women carried out the daily agricultural labor, the opposite of the European sexual division of labor. Regardless of the cultural biases that allowed Europeans to dismiss the ingenuity and productivity of Indian farming, agriculture transformed Native North America in the millennia before contact. In some cases, agriculture resulted in soil erosion and the forced abandonment of village sites. The social and cultural impacts of agriculture included greater population densities, longer-term residence in a single location, specialization of labor, and eventually, in certain cultures, the emergence of a stratified social order.[20]

Agriculture fueled the rise of great population centers in the Mississippi and Ohio river valleys as well as in the desert southwest. Around 1,900 years ago, or A.D. 100, the Hopewell culture arose in the Ohio River country and flourished for the next four centuries. Hopewell peoples (the cultural complex was not limited to a single ethnic or linguistic group) practiced a mixed economy of agriculture, hunting, and trade. Corn was introduced into the region in approximately A.D. 200, but sunflowers, squash, and various native grasses were more important. Earthen enclosures built in geometric patterns, surrounding burial and ceremonial mounds characterize Hopewell sites. Burials contained everyday items as well as exotic trade goods such as shells, copper, obsidian, and grizzly-bear teeth, indicating that Hopewell peoples lived at the nexus of trade networks stretching from the Atlantic coast to the Rocky Mountains and from the Great Lakes to the Gulf of Mexico. In about A.D. 400 the culture began a decline that ended with the abandonment of all the towns by A.D. 550. Some scholars theorize that more intensive localized cultivation of corn undercut the focus on trade. It is also possible that ecological changes and new hunting technologies, such as the bow and arrow, contributed to the eclipse of the Hopewell tradition.[21]

Within two centuries of the Hopewell collapse, a far more widespread and long-lasting agricultural society began its rise. The Mississippian cultural complex eventually influenced most of what is today the USA south of the Great Lakes from the Mississippi valley to the Atlantic Ocean. Several theories suggest that Mississippian culture drew some traditions from the Hopewell culture. Mississippian villages first appeared around A.D. 700 and, within a century and a half, the culture was evident throughout the region. The culture peaked between 1100 and 1300. Large platform mounds sat at the center of Mississippian towns and were often surrounded by densely populated residential areas. Corn was the principal crop, and long-distance trade networks were central to Mississippian life. The largest and most elaborate Mississippian sites were constructed in the middle Mississippi and Ohio river valleys. Further to the south, towns appeared to be more ceremonial than residential. Later Mississippian settlements that developed southeast of the Appalachians were generally smaller and less elaborate than those found in the Mississippi and Ohio valleys. While the largest towns of the Mississippian core were abandoned by 1500, some chiefdoms in the southeast survived into the post-contact era. Evidence suggests that the Mississippian social order was highly stratified with religious and political power resting in the hands of an elite chiefly class. Mississippians also produced a distinctive iconography which archeologists labeled the Southeastern Ceremonial Complex, or more simply, the "Southern Cult."[22]

The pinnacle of Mississippian culture was the great city of Cahokia situated near the confluence of the Illinois and Mississippi rivers. The fertile lands surrounding the site, known as the American Bottom, had been settled perhaps as early as A.D. 600. Small farming villages existed in the area for the next four centuries. Around 1050, environmental conditions changed, allowing full-scale corn agriculture to begin, and the population swelled. The construction of massive mounds began, and, eventually, the Mississippians built over 100 major earthworks on the 6-square-mile site. The largest earthen structure in pre-contact North America, Monks Mounds, was the centerpiece of a ceremonial plaza completed around 1300. Densely populated residential districts surrounded the city center. The population of Cahokia probably peaked at between 30,000 and 40,000, making it in 1300 equal in size to London and the most populous city north of Mexico at any given time until Philadelphia grew larger in 1800. While corn agriculture allowed such population densities, far-reaching trade networks, extending from the Atlantic coast into the Native West, brought wealth and exotic trade goods to Cahokia. Some of the obsidian uncovered in Cahokia, for instance, had been quarried in what is today Yellowstone National Park. Elaborate burials that included human sacrifice, and the apparent segregation of residential districts indicate a stratified social order. Around 1300, Cahokia began to decline in size and importance. Some scholars argue that resource exhaustion was the principal reason for the decline, but the increasing number of stockades built during the era also suggests growing conflict. Regardless of the reasons for Cahokia's decline, it was not the end for the Mississippians. After 1300, smaller Mississippian towns appeared in the southeast, and the cultural complex influenced Caddoan peoples farther to the west.[23]

During the same centuries that saw the arc of Hopewellian and Mississippian cultures, four distinctive agricultural societies arose in the desert Southwest: the Patayan, the Hohokam, the Mogollon, and the Ancestral Puebloans popularly known as the Anasazi. All emerged from the hunting and gathering southwestern Archaic tradition, and all depended upon agriculture to varying degrees. Yet each was distinctive in material culture and agricultural adaptations. The Patayan culture, sometimes referred to as Hakataya, covered a vast area of modern western Arizona, southern California, and northern Baja California from 700 to 1500. It is the least studied and least well understood of the four Southwestern cultures. Patayan culture has been termed "rock-oriented" due to the emphasis on stone construction techniques and rock shrines and trail markers found throughout the area. The use of inundation agriculture (the reliance on seasonal flooding) instead of irrigation was another characteristic of Patayan life.[24]

To the east of the Patayans, the Hohokam people wrested a living from the Sonoran desert of Arizona. They practiced ingenious methods of dry farming but are best known as the architects of the largest irrigation system in pre-contact North America. By the 500s the Hohokams began building canals to deliver water from the Gila and Salt rivers to their fields. Between 800 and 1300 they built ever-more-elaborate irrigation systems. Two-thirds of the canals were built along the Salt River, where more resistant bedrock meant shallower channel cuts and the easier diversion of water. Along the Salt, the Hohokams constructed 350 miles of main canals and over 1,000 miles of feeder canals to water perhaps 10,000 acres. Hohokam crops included corn, drought-resistant beans, squashes, tobacco, cotton, and agave. Irrigated agriculture probably supported 50,000 people in the Salt River Valley and a regional population of 130,000. Although earlier theories that the Hohokams themselves were immigrants from the south are today generally dismissed, cultural connections with Mesoamerica are evident in Hohokam sites. Ball courts, on which a pre-contact ritual sport was played throughout Mesoamerica, first appeared in Hohokam villages around A.D. 700, and by 1000 they were present in nearly every village. In later phases the Hohokams also built large rectangular platform mounds and central plazas in the largest towns.[25]

By 1450, however, the Hohokam had abandoned their towns and canals and "disappeared." Controlling water in an arid environment was the basis of Hohokam success, but water was also a likely cause of the decline. A massive flood had struck around A.D. 900 but the people rebuilt and expanded their canal network. A second devastating deluge in the 1350s, however, destroyed much of the canal network, and this time there was no recovery. Although it cannot be proven, increased salinity in the soil as a result of inadequate drainage was probably also a contributing factor to their decline. After centuries of intense irrigation, Hohokam fields simply could no longer support the people. For some, the story of the Hohokam provides evidence that, contrary to popular beliefs, the First Americans were not the "first environmentalists." Drawing such a conclusion is both an example of presentism and the power of popular stereotypes. No human society in 1400 understood ecology in modern scientific terms, and no human being alive could be considered an "environmentalist." Nor did the Hohokam vanish mysteriously. The abandonment of towns and whole regional systems was a repeated phenomenon in the pre-contact Southwest as peoples adapted to changing social and environmental conditions. It meant the end of one possible adaptation, not the end of an entire people. The Hohokams were likely some of the ancestors of the modern Akimel O'odham and Tohono O'odham peoples.[26]

South and east of the Hohokams, the Mogollon people covered a vast area that straddles the modern US–Mexican border. The earliest Mogollon sites date to around A.D. 200 and were characteristically small pit-house communities located on ridges and bluffs above good agricultural land. Floodwaters provided irrigation for Mogollon fields. Between 950 and 1150 major changes came to the Mogollon world. The population increased substantially and the people began to build above-ground masonry Pueblos and produce more refined black-and-white pottery similar to the Ancestral Puebloans to the north. Indeed, there were many connections between the two cultural traditions. Between 1275 and 1400, for instance, the Mogollon population increased again, and it is likely that the increase was partially due to the upheavals and abandonments that took place to the north during this period. Mogollon people abandoned their villages in Arizona and New Mexico by 1450 and migrated in search of new farming lands. Some undoubtedly were ancestors of the modern Pueblo peoples.[27]

North and east of the Mogollon homeland was the fourth, and best known, cultural tradition in the pre-contact Southwest. For decades these peoples were known as "Anasazi," with the term popularly translated as "ancient ones." In fact, Anasazi is a Navajo word that literally means "ancestors of our enemies." Today, many scholars refer to these peoples with the less colorful but more accurate and respectful term "Ancestral Puebloans." (The Mogollon culture was also ancestral to some modern Pueblos, but for the purposes of this essay the term is applied more narrowly.) About 2,000 years ago, Native peoples began inhabiting caves and rock shelters in southwestern Colorado and adjacent areas of New Mexico and Arizona. In around A.D. 400 the area's residents began to construct pit houses, often on mesa tops, and to produce a simple gray pottery. Hunting and gathering supplemented agriculture and sustained the people. The next several centuries saw the intensification of agriculture, a shift toward above-ground masonry structures, and more elaborate forms of pottery. Around A.D. 800, as populations increased in both upland and canyon settlements, shared "community houses" increasingly replaced pit houses. What many scholars deem the zenith of Ancestral Puebloan culture came next: the Chaco Canyon period.[28]

Chaco culture thrived from A.D. 850 to 1150. Dozens of "great houses," or major towns, were constructed early in the period. The Chacoans oriented the great houses with solar and lunar observations as well as with the cardinal directions. With 700 or more rooms and over three dozen kivas (circular underground ceremonial rooms), the magnificent Pueblo Bonito sat at the center of the Chacoan world. Four hundred miles of engineered roads connected Pueblo Bonito and the other great houses to other towns throughout the region as well as with the San Juan River, Mount Taylor, and the Chuska Mountains. The roads ran for great distances in nearly straight lines, leading many archeologists to believe they held great ceremonial and symbolic meaning. They also likely served a practical purpose as much of the corn consumed at Pueblo Bonito as well as the vast majority of the over 1 million stones used in the town's construction were brought from upwards of 50 miles away. Like the Hohokam, the Chacoans built water-control structures that enabled agriculture in the arid canyon. Chaco also sat at the nexus of a regional trade network, and the great houses of the canyon produced vast quantities of finished turquoise jewelry. By 1050, Chaco Canyon was a ceremonial and political center with a population of between 5,000 and 15,000 that dominated a vast area, yet within a single century the great houses would be abandoned. Environmental changes were likely one root cause of the Chaco decline. Around 1100, a half-century-long drought struck the region, and within two decades construction in Chaco Canyon had ceased. The depletion of local resources surely magnified the drought's impact, but some scholars argue that the collapse of a discredited political and religious elite was also a contributing factor.[29]

In the popular imagination, the "Anasazi" world was a peaceful and egalitarian place. Many archeologists argue instead that the evidence suggests a highly stratified and centralized sociopolitical structure characterized by planned settlement patterns, substantial labor investment in the construction of the great houses and water-control structures, mortuary customs, and differing access to material goods. More controversial is the theory proposed by a smaller number of scholars that chronic violence, including incidents of cannibalism, enforced this social order. Supporting evidence for the theory—burned structures, haphazardly buried or unburied bodies, and telltale damage to human remains—has been found in dozens of sites either spatially or temporally associated with Chaco. Anthropologists

Christy Turner and Jacqueline Turner have even suggested that the "rapid development of things Chacoan," was directly linked to the arrival of peoples from central Mexico who "used cannibalism not only for ritual purposes but also, in their homeland, as an element of intimidation and social control."[30] Their theory is highly speculative, and no concrete evidence for a physical migration of peoples from the south exists. It does present an intriguing, if unsettling, theory for the rise and the fall of Chaco culture, as environmental conditions could have intersected with the violent excesses attributed to the ruling elite.

What is certain is that after 1150 the population shifted from Chaco Canyon and other low basins toward higher country. The lands north of the San Juan River including Mesa Verde saw substantial population growth. Great houses generally gave way to smaller structures, although after a half-century hiatus, magnificent buildings such as Yellow Jacket and Aztec were once again built. Continued evidence of violence suggests that regional conflict had not ended with the Chaco collapse. Large or small, community houses were no longer built in exposed open country but rather in more easily defended sites. Mesa Verde's Cliff Palace and Balcony House were built within seemingly inaccessible cliff overhangs. Like Chaco, the upland sites experienced a rapid depopulation. Chronic violence may have been a factor, but changing environmental conditions certainly contributed to the decline. Between 1200 and 1250, ample rainfall led to more reliable corn harvests, expanding trade networks, and a growing population. When the rains and crops once again failed after 1260, however, the Puebloans quickly felt the impact of over-hunting and resource depletion. The drier, west-facing, and fortified mesa-top sites failed first. By 1300, only the best-watered eastern-facing villages remained in the region. Soon, they too would be abandoned as the people moved southeast to Pueblos along the Rio Grande, or south and west to the Zuni or Hopi mesas. The Pueblo world that the Spanish would encounter in the sixteenth century was beginning to take shape.[31]

Just as the impact of Mississippian societies was felt beyond the Mississippi valley, the agricultural traditions of the Southwest also influenced peoples throughout a larger region. The Fremont culture, centered in the modern state of Utah, was very likely a local adaptation of resident hunting-and-gathering peoples to the northward spread of agriculture around 500. Fremont agriculture depended on one variety of corn, known as Fremont dent. Hunting and gathering also remained relatively more important among the Fremont than other groups. Villages were generally small collections of pit houses along with associated storehouses and granaries. The ultimate fate of the Fremont people is poorly understood, although the migration of Numic peoples into the region may have led to the amalgamation of some groups and the migration of others to the South.[32]

Great human migrations, then, such as the movement of Numic-speaking (Shoshone, Paiute, and Ute) peoples across the Great Basin and into the Rocky Mountains and Great Plains, were another factor that shaped the pre-contact Native West. Archeology, linguistics, and oral traditions all provide evidence for the Numic migration, but scholars disagree on its timing. Beginning in the 1930s, some archeologists began documenting a Numic expansion that began some time between A.D. 800 and 1000. Later anthropologists used linguistics to build upon these findings and to argue that the "Numic spread" took place very rapidly and fairly late in the pre-contact period. By the time of sustained European contact, however, Numic languages, with relatively few dialectical differences, were spoken across a vast arc stretching from the Western Basin across the present states of Nevada, Oregon, Idaho, Utah, Colorado, and Wyoming. Advocates of the "Numic spread" theory argued that this fact is indicative of a migration that took place so rapidly that numerous dialectical differences simply had no time to develop. They propose that the migration began sometime after A.D. 800 and that Shoshonean peoples may not have reached the Snake River region until after the fourteenth century, but more recent archeological work in the Snake River area suggests that Shoshonean peoples have resided there for at least 3,500 to 4,000 years. Whatever the case, Numic peoples moved into areas once occupied by groups of the Fremont and Mountain traditions, either absorbing or replacing those peoples. Moreover, the great Numic migration did not end with contact; rather, the new opportunities presented by an equestrian life spurred further expansion. Some Shoshones would move as far north as the plains of Alberta and Saskatchewan by the early 1700s while others moved onto the Southern Plains where they became known as Comanches.[33]

Sometime between the Puebloan abandonment of the upland sites and the Spanish *entrada,* Athapaskan-speaking ancestors of the Apaches and Navajos moved into the Rio Grande and Colorado Plateau region after a long migration south along the flanks of the Rocky Mountains. Their migration likely began sometime around A.D. 200 in what is today northwestern Canada and Alaska. Bringing with them a hunting technology that included the bow and arrow and the distinctive Avonlea projectile point, Athapaskans may have arrived in modern Montana and Wyoming soon thereafter. Some scholars believe Athapaskans were in northwestern New Mexico as early as A.D. 700, while others place their arrival far later, within a century of Spanish contact. Some of the Athapaskans became the Lipan, Mescalero, Jicarilla, Chiricahua, and Western Apaches. Into the historic period they maintained an economy based upon hunting, gathering, and raiding. Others became the Navajos, or Diné, who may have resided in the Four Corners region by the mid-1400s and who developed a complex relationship with the Pueblos marked by warfare and raiding as well as intermarriage and cultural borrowing.[34]

THE NATIVE WEST IN 1492

Anthropologists have employed the "culture area" concept to understand and compare Native cultures in North America. For the most part, culture areas are defined by environment and economics. Diverse environments posed unique challenges and presented distinct opportunities, and within a given area Native cultures exhibited general similarities in regard to economic strategies and expressions of material culture.[35] On the northwest coast, for example, salmon fisheries and other marine resources were central to the economic lives of all Native peoples, and this fact was evident in their rich material cultures. One must be aware, however, of the concept's limitations. First, a rigid application of the concept obscures the great diversity—religious, linguistic, and even economic—that could occur within a single culture area, particularly on its geographic margins. Second, there is a danger that over-reliance on the culture-area concept can create a static and environmentally deterministic view of Native cultures. Because one group adapted to a given environment at a specific time did not make that adaptation inevitable or unchanging. One final note: the group names used below generally reference linguistic divisions and are included to give the reader a point of reference. In nearly every case, the modern tribal peoples who emerged from pre-contact groups had yet to develop distinct sociopolitical identities and would not have referred to themselves by these names before contact. With those caveats in mind, a survey of culture areas in 1492 provides vivid illustration of the diversity of Native North America at the time of contact.

California was the most populous and linguistically diverse of the pre-contact culture areas. Approximately 300,000 people lived in permanent settlements concentrated along the coasts and rivers. They spoke at least eighty mutually unintelligible languages, further divided into hundreds of dialects. Along the northern coast were Pomos, Yukis, and Sinkyones. Nisenan, Miwok, and Yokuts speakers lived in the Sacramento and San Joaquin valleys and the Sierra foothills. Chumash people occupied a large portion of the coast in the Santa Barbara area as well as the Channel Islands. Yet with all this linguistic diversity, California Indians practiced similar subsistence strategies made possible by the richness of their environment. Men fished and hunted small and large game. Women collected and processed acorns from the oaks that covered the hillsides to make gruels and breads. California Indians actively husbanded many other varieties of "wild" medicinal and food plants. They used fire to clear underbrush, promote fruit production, and create the edge habitat favored by deer, while broadcast-scattering of seeds ensured sustained growth of grasses and other semi-domesticated plants. There were, of course, regional variations. Coastal people such as the Chumash could harvest shellfish and marine mammals not available to inland peoples. The intricate trade ties that rounded out the California Indian subsistence strategies, however, brought products not locally available to peoples throughout the region.[36]

Like California, the Pacific Northwest coast provided an incredible natural bounty, and the peoples who lived from northern California to southeastern Alaska developed perhaps the richest material culture in pre-contact North America. As in most other culture areas, linguistic diversity was the rule. Along the Oregon coast, tongues from at least nine language families could be heard. Numerous dialects of Coast Salish were spoken around Puget Sound and north into modern British Columbia. Farther north were Haida, Tsimshian, and Tlingit speakers. Autonomous villages were the essential political unit on the northwest coast, but intricate kinship ties bound numerous villages together in trade and social networks. While hunting and gathering of berries and other wild-plant foods were important, marine resources were the heart of the economy. Coast Indians possessed an elaborate material culture to take and process fish, shellfish, and marine mammals. Some peoples, such as the Makah, went to sea in large dugout canoes to hunt whales. Most important of all, however, were the salmon that ascended the region's rivers by the millions in every season but winter. The fish were speared, netted, and trapped in weirs at stations usually placed at falls. Salmon was eaten fresh, and millions of pounds were dried on racks for later consumption.[37]

The region's wealth was reflected in the peoples' material culture as well as a highly stratified and hierarchal social order that included an indigenous form of slavery. Northwest coast peoples generally lived in large multi-family log and cedar plank homes. From Puget Sound to the north, elaborate carved totem poles recounting clan lineages and group history stood at the entrance of the houses. A stratified social order and the cultural importance of wealth were unique features of northwest-coast life. At the top of the social order was a wealthy elite made up of the chiefs and their lineage members. Below the elites were the "commoners," and at the bottom of the order were slaves. Slaves could either be purchased or captured and were subject wholly to their master's will. Social status came through birth, but the acquisition and redistribution of wealth was also a cultural imperative that gave rise to the region's most distinctive ritual: the potlatch. Kin-groups' leaders hosted these elaborate give-away ceremonies, where property that was accumulated over months was gifted to invited guests. Anthropologists have long debated the myriad cultural meanings applied to the potlatch, but most agree that its central purpose was to raise the status of the hosts and their children. In some very rare cases it was possible for commoners to overcome their birth status and to become a member of the elite through careful potlatching.[38]

Salmon was also central to the economic lives of the peoples who lived east of the Cascade Mountains on the great plateau drained by the Columbia River and its tributaries. The northern reaches of the plateau were home to Interior Salish speakers such as the Shuswap, Lillooet, Coeur d'Alene, and Okanagan. In the more arid lands on the southern plateau, Salish gave way to Sahaptin peoples such as the Yakama, Nez Perce, Walla Walla, and Umatilla. (Other peoples such as the Cayuse and the Klamaths spoke unrelated languages.) Plateau peoples lived in a diverse landscape that ranged from high, forested mountains to vast rolling plains, to the rugged volcanic "scablands." Despite this physiographic diversity, rivers were the lifeblood of the people. Plateau peoples generally wintered in semi-permanent villages. Large mat-covered lodges housed multiple families. At other times of the year the people were on the move, either to the fisheries during the great salmon runs or in search of terrestrial game and wild-plant foods. Fisheries were situated at the base of falls, and plateau groups possessed an impressive array of fishing technologies that included weirs and traps as well as spears with detachable barbed tips. Over thirty varieties of root vegetables, including camas, wild onions, bitter-root, and biscuit-root, were of particular importance in the plateau diet. People traveled long distances to access critical resources at specific times of the year. Compared with the salmon-fishing peoples of the coast, plateau peoples held far fewer material possessions and placed far less cultural emphasis on the accumulation of goods. In the post-contact era, the Nez Perce and other plateau groups expanded their travels even farther, incorporating equestrian bison hunting east of the Rocky Mountains as yet another part of their subsistence cycle.[39]

South of the Columbia plateau, the vast inland Great Basin between the Rocky Mountains and the Sierra Nevada was unique among culture areas as the majority of its residents spoke dialects of one or more related Numic languages: Shoshone, Ute, and Paiute. Paiute groups lived throughout what is today western Nevada and into eastern Oregon and southwestern Idaho. Shoshones lived across the northern half of Nevada, southern Idaho, and northern Utah, and some were already pressing out onto the plains of Wyoming. Ute speakers occupied the majority of modern Utah as well as the western half of Colorado. The Great Basin has been the most misunderstood and stereotyped of the culture areas. Basin peoples faced comparatively harsh environments and developed numerous strategies to wrest a living from the arid landscape. Many mid-twentieth-century anthropologists minimized this diversity and posited a direct connection between the harsh environment and an overly simplified vision of basin life marked by small kin-based social units, low population densities, and a simple subsistence culture focused on sheer survival. They treated the Great Basin as an ethnological laboratory to study the most "primitive" and "simple" of American Indian peoples and, as a result, created a static, ahistorical vision of its peoples.[40] In their works, they privileged the piñon nut harvest, the collection of grass seeds and other wild-plant foods, and the taking of small game as indicative of basin subsistence and consequently obscured the great variations that existed throughout the region. Fishing was of critical importance to some groups, most notably the Paiutes of western Nevada and the Shoshone and Paiute speakers of the Snake River plain. The collection of camas and other root vegetables took place on the high prairies at the foot of the mountains, while some Shoshones, known as Tukudika or "Sheep-eaters," relied upon big-game hunting at even higher elevations north of the Snake River. Like the peoples of the plateau, many basin groups practiced a complex annual subsistence cycle that only became more expansive with the acquisition of horses in the post-contact period.[41]

The culture area most associated with horses is, of course, the Great Plains. But in 1492 the most numerous and prosperous Plains peoples were horticulturalists who lived in earth-lodge villages along the river courses. Caddoan speakers, including the ancestors of the modern Caddo, Wichita, Pawnee, and Arikara, moved north and west from their homelands in the lower Mississippi valley, beginning around A.D. 900. They brought with them corn agriculture and elements of the Mississippian cultural complex. They settled first along the well-watered and more humid tributaries of the Missouri and Mississippi, including the Arkansas River where the Caddos made their home. The Pawnees' and Arikaras' migrations continued further, out onto the more arid plains of modern Kansas and the Dakotas. Another population center was along the middle Missouri River in present-day North and South Dakota, where Siouan-speaking Mandans and Hidatsas established permanent villages in around 1100. Later Siouan arrivals to the eastern margins of the Plains included the ancestral Otoes, Iowas, Omahas, and Poncas. Earlier attempts to grow corn on the arid Western prairies ended with the sustained drought around the beginning of the thirteenth century, but, as conditions got better, people again moved west and resumed agriculture. By the 1400s, the Pawnees were farming along the Loup and Republican rivers in Nebraska and Kansas.[42] With regional variations, all of these peoples came to practice mixed economies that combined a central focus on agriculture with hunting and gathering of wild food crops. Gardens were situated in alluvial river bottoms. Women carried out the majority of the agricultural labor, and, in many groups, such as the Arikara, they owned the fields and the crops. Men helped at harvest time, hunted, and fished. Bison presented an attractive source of protein on the hoof, but hunting on foot was an arduous task restricted to times that the herds came near the villages. The acquisition of horses, beginning in the later half of the seventeenth century, allowed a growing focus on buffalo hunting among the Pawnees and other Plains earth-lodge peoples.[43]

Away from the rivers and farther west, more nomadic groups already depended upon the bison herds. Kiowa oral traditions, for instance, indicate they migrated east across the Plains toward the Black Hills in the centuries before contact. The Crows also began to move, splitting off from the Hidatsas sometime after 1450 and making their way toward the Bighorn Mountains. Pre-horse Plains peoples pursued the buffalo herds on foot and, like, the river people, used buffalo jumps and impounds. Like the later peoples of the region, they lived in skin lodges. Their principal beasts of burden, however, were dogs, which limited the loads that could be carried and the distances that could be covered. Conspicuously absent from the Great Plains in 1492 were many of the iconic peoples who came to define the equestrian Plains culture, including the Comanches, Cheyennes, Lakotas, and Blackfeet. The people who later would be known as Comanches were part of the great Numic migration across the Intermountain and Rocky Mountain West and had yet to develop a distinct social and political identity. Both the Cheyennes and the Lakotas were peoples of the Great Lakes woodlands and prairie margins. Their migrations to the west were one of the great consequences of contact as the expansion of the fur trade and of peoples farther east pushed them west at the same time as new opportunities presented by the equestrian culture pulled them toward the Plains. Similar forces motivated the Blackfeet migration out of the woodlands and onto the plains of Alberta sometime before 1700.[44]

The Southwest was perhaps the most diverse region of the Native West. To the north and east along the Rio Grande River and on the Colorado plateau, sedentary Pueblo horticulturalists and nomadic Athapaskans had developed a complex set of social and economic relations. By 1492, the ancestors of the Apaches and Navajos had resided in the Southwest for around two centuries. They had adapted to the land and the opportunities and limitations it presented. For the most part, they subsisted through hunting and gathering, supplemented by a cycle of trading with and raiding the Pueblos. This pattern became more pronounced during the historical period and incorporated intermarriage and cultural borrowing. Eventually, a limited number of Athapaskans adapted some Pueblo horticultural methods, and, in the post-contact era, the Spanish distinguished them from other Athapaskan speakers as the *Apaches de Navajo:* the "Strangers of the Cultivated Fields."[45]

While the Spanish deemed the sedentary peoples they found in New Mexico "Pueblo" (or village) Indians, they were a diverse group defying simple generalizations. Upwards of 60,000 people lived in over ninety permanent towns at the end of the fifteenth century. At the time of contact, Pueblo peoples spoke languages from four language groups. Most of the Rio Grande Pueblos spoke Tanoan languages (Tiwa, Tewa, and Towa). A few others, as well as the central Pueblos of Acoma and Laguna, spoke Keresean tongues. Hopi is a Uto-Aztecan language, while Zuni has no known cognates. Socially, politically, and religiously, the Pueblos also exhibited great diversity. Along the Rio Grande, most groups were patrilineal and lived in bilateral family households. Summer and winter moieties (dualistic social and kinship subdivisions within a community) were responsible for the ceremonial needs of the community, and politics tended to be more centralized in the hands of a cacique (chief) and council. Farther west, in the more isolated mesa villages, such as Hopi, Zuni, and Acoma, people lived in matrilineal households. Politically, the western Pueblos operated more as a theocracy, with public opinion enforcing the consensus of leaders.[46] Yet, with all these differences, the Pueblo peoples had more in common with each other than with the Athapaskan-speaking Apaches and Navajos who migrated into the southwest in the centuries before the Spanish arrival.

West of the Pueblos lived many other peoples who developed their own unique ways of living in a desert country. In the lands where the Hohokam culture had flourished several centuries before, the Piman-speaking ancestors of the modern Pimas and Tohono O'Odhams made their homes. Living in the valleys of the Gila and Salt rivers, the Pimas relied heavily on agriculture, supplemented by hunting and gathering. Corn and cotton were their principal crops, and the Pimas were renowned for their basketry and cotton blankets. The Tohono O'Odhams (Papagos) lived to the south and east, in some of the driest reaches of the Sonoran desert. They knew the desert country intimately and collected a wide range of edible plants, including the fruit of the saguaro cactus, and took game large and small. North and west of the Piman homelands ranging to the Colorado River and beyond were numerous Yuman-speaking groups ancestral to the modern Havasupais, Yavapais, Hualapais, Mohaves, Quechans (Yumas), and Maricopas. With regional variations, these peoples farmed maize in the bottomlands of watercourses and supplemented that diet through hunting and gathering. Their lands ranged from low deserts to high mountains and included the Havasupai and Hualapai homelands, the spectacular country along the south rim of the Grand Canyon.[47]

For the past century, scholars have hotly debated pre-contact population numbers for the Americas. Estimating a population without extensive historical documentation is an inexact science based on a range of methodologies. The earliest estimates failed to account fully for the impact of the epidemics and therefore were far too low. In 1939, for instance, A. L. Kroeber, using older estimates and assumptions about environmental "carrying capacity" drawn largely from North America, proposed a total pre-contact population of 8.4 million for the entire Western Hemisphere. The number was ridiculously low considering the massive populations documented in Central and South America. Henry Dobyns was the first to seriously consider the impact of post-contact epidemics. Using known depopulation figures from Central America as well as carrying capacities, he estimated 112 million for the same area. Estimates for North America range from under 1 million to over 10 million. Currently accepted figures tend to fall somewhere in the middle with Cherokee demographer Russell Thornton estimating a population of 72 million hemisphere-wide, 7 million north of Mexico and slightly over 5 million within the modern borders of the coterminous USA in 1492.[48] Narrowing the demographic focus to the Native West is even more problematic. Sixteenth-century population estimates are rudimentary and limited to the Pueblos of New Mexico. All later population estimates came after a relentless cycle of depopulation had already begun. While exact numbers may elude demographers, the most densely populated regions of the Native West were certainly those with the richest resources bases: California, the Northwest Coast, the Southern Plains, and the Rio Grande Valley of New Mexico. California alone may have been home to nearly one-third of the 1 million or so residents of the pre-contact Native West.[49]

THE IMPACT OF COLONIZATION AND CONQUEST

In 1700, the vast majority of Native peoples in the American West had yet to meet a European, but their lives had already seen great changes as a result of the colonizers' presence in the Americas. The effects of conquest radiated outward from the first European settlements like shockwaves radiating from the epicenter of an earthquake. Most American Indian peoples first felt those waves through interactions with other indigenous groups. In some cases, factors largely beyond Native peoples' control, such as epidemic diseases and the resulting demographic collapse, were the catalysts of change. Yet Native peoples were not simply the passive victims of colonization. They remained agents in their own history and in many instances shaped or even initiated change in order to exploit new opportunities or to meet new challenges. The adoption of the equestrian lifestyle epitomizes this second course. The America West in 1700, then, was not a pristine aboriginal landscape, but it was one still largely of Native making. The two centuries that followed Columbus's voyages were a critical period that revealed historical patterns and processes that would shape the future American West.

The beginning of the European conquest marked the beginning of the greatest biological exchange in human history. For millennia, separate flora and fauna had developed in the New World and the Old. In the years following Columbus's arrival, plants, animals, and pathogens made their way from one hemisphere to the other, in the process transforming life and ecosystems on a global scale. American food crops proved particularly advantageous for the Europeans. Avocados, chili peppers, cocoa, peanuts, pineapples, pumpkins, strawberries, and tomatoes all enhanced Old World diets. Potatoes and the most important of all American crops, corn, quickly became global staples that improved standards of living throughout Eurasia and Africa. Less advantageous was another American import: tobacco. The Europeans brought with them apple, peach, and pear trees; a range of grains including wheat, barley, millet, and rye; lettuce, cabbage, carrots, and other vegetables. Cotton and sugarcane were two Old World plants that became the staple crops of colonial plantation economies throughout the Americas. Before contact, the Native peoples of the

Americas had domesticated very few animals: dogs and turkeys in North America; llamas, alpacas, and guinea pigs in South America. By contrast, the colonizers brought with them a whole host of domesticated animals—including cattle, horses, sheep, goats, swine, and fowl—in addition to unwelcome pests such as brown and black rats. Old World flora and fauna aggressively colonized the American landscape alongside their human benefactors, making life as the Europeans knew it easier and Native ways more difficult. Environmental historian Alfred Crosby deemed this process "ecological imperialism."[50]

The most deadly of all European imports could not be seen: the microbes that caused smallpox, measles, plague, yellow fever, and a host of other epidemic diseases for which America's Native peoples possessed relatively weak immunity. "Virgin-soil epidemics," so called because they strike populations without previous exposure, have occurred throughout human history and across the globe. The most famous of all virgin-soil epidemics, the "Black Death" (most likely bubonic plague), carried away approximately one-third of the European population between 1347 and 1352. In the four centuries following Columbus's first voyage, a relentless succession of pandemics ravaged the Native population of the Americas. They were a primary factor in a catastrophic hemispheric population decline. The Arawak population of the Caribbean, by many estimates nearly 4 million people before contact, all but disappeared within a century. Declines of 80 to 90 percent among many Native groups became commonplace. Any understanding of the impact of the European conquest of the Americas must then begin with a consideration of the epidemics.[51]

The Americas in 1492 were not a disease-free paradise. Analysis of pre-contact skeletal remains suggests that a virulent form of syphilis may have been present, as was chicken pox and some other viral maladies. Yet the worst killers—smallpox, measles, influenza, and plague—arrived with the Europeans. Why then were there apparently so few destructive pathogens present in the Americas before the beginning of sustained European contact? One theory is that the strenuous migration to the Americas through a frozen unforgiving landscape functioned as a "disease filter," with only the healthy surviving to populate the Americas. More important was the relative absence of domesticated animals in the Americas. New diseases emerge most readily where large human populations are in close daily contact with large numbers of domesticated animals. In such situations, the chances of a viral or bacterial mutation spreading to a human population are exponentially greater. Europe, Asia, and Africa were, in a sense, one large disease community. Epidemics arose, killed many, and then subsided, leaving behind a population with developed antibodies better able to survive the next appearance of that particular disease. By the time of European contact, however, the Americas had seen thousands of years of relative isolation from the Old World disease community. Circumpolar contact had been constant but was limited to small groups of people (including the Norse), who apparently did not introduce devastating epidemics. While American Indians were not "biologically defenseless," as many historians have suggested, their lack of exposure to European diseases left them more vulnerable.[52]

Historians must then move beyond biology alone to understand and explain the devastation that the epidemics wrought. Whereas a single virgin-soil epidemic might kill between 30 and 40 percent of an affected population, many American Indian groups experienced losses at double those rates or more. Certainly the sheer number of new diseases that struck in relentless succession was a factor, but even that cannot explain the human toll. American Indian peoples experienced the epidemics against a backdrop of warfare, slave raids, and social dislocation. Native healing methods such as sweat baths were not suited for treating the infections, and some historians believe they may have actually increased mortality rates. Historian and physician David S. Jones has argued that a range of contingent factors including poverty, malnutrition, and environmental stresses were just as important as genetic or developmental susceptibility to infectious disease for understanding the course of the epidemics.[53] The impact of the epidemics transformed Native life beyond simple demographics. So many deaths meant the loss of knowledge and economic productivity, and, while the epidemics did not cause dependency on European manufactured goods, they very likely increased the pace of the process. The first scholars to emphasize the impact of the epidemics also theorized that the inability of Native healers to cope with the new illnesses led to a wholesale decline in Native spiritual practices. But here again, the process was complex, and at least among some southeastern Indians, innovative spiritual and medical practices helped keep mortality rates down and illustrated cultural resilience.[54]

What is not debated is that the epidemics spread from one group to another so rapidly that they reshaped the human landscape of the Americas long before most Native peoples ever laid eyes upon a European. Archeologist Anne Ramenofsky used an analysis of the number of settlements, total settlement area, and the roofed area of earth lodges in villages on the middle Missouri River to estimate pre-contact population sizes and decline. She concluded that the first epidemics brought a "catastrophic loss of population" sometime between 1612 and 1680, coinciding with the "appearance of European trade goods in the region" but preceding the arrival of Europeans themselves by decades.[55] Because the epidemics nearly always preceded sustained contact, the first Europeans to enter any given area usually found a Native population that had been radically reduced from its pre-contact numbers, yet they assumed these populations to be the historic norm. Such assumptions helped to shape the invaders' attitudes about Native peoples and have had political implications to this very day. The first Puritan colonists of New England in the 1620s and 1630s generally viewed the epidemics as a sign of "providential destruction" whereby God had cleared the land for their taking. The comparatively sparse (by European standards) Native populations played into Puritan justifications of conquest. John Winthrop, the first governor of Massachusetts Bay colony, employed a doctrine he called *vacuum domicillium*—literally "empty house"—to argue that the Indians' small numbers and "uncivilized" land-use patterns did not warrant their vast estate. "Civilized" Europeans, on the other hand, put the land to higher use and were therefore justified in their actions. Ideas such as Winthrop's have echoed through American history: from Thomas Jefferson to Frederick Jackson Turner to the western films of John Wayne, the North American continent was a "virgin land" open to settlement and a guarantee of the nation's future greatness. Native peoples were but an anachronistic and minor obstacle to that destiny. But, as the historian Francis Jennings reminds us, America was not an "empty continent," it was a "widowed land."[56]

Yet, not all things that the Europeans brought were detrimental, and, more than any other, horses represented the opportunities that Native peoples seized in the post-contact era. The earliest horses evolved in the Americas but had gone extinct at the end of the Pleistocene. Horses returned to the American mainland with Hernán Cortéz's 1519 invasion of Mexico. The Spanish also brought with them a highly developed equestrian culture. Early theories explaining the acquisition of horses by peoples in North America focused on strays that had escaped from the De Soto and Coronado expeditions. Yet the herds in question were quite small, suffered few reported losses, and were made up largely of geldings. Furthermore, the "stray" theory could not account for the transfer of the knowledge necessary to care for and use horses. Native peoples acquired horses, then, not from far-flung expeditions but from Spanish settlements and ranches, first in northern Mexico and later in New Mexico. Although official Spanish policy made it illegal to teach an Indian to ride, in practice the conquerors had little choice but to teach their workers the skill. The process began as a trickle as Indian laborers escaped on horseback. That trickle became a flood with the Pueblo Revolt of 1680. And as more peoples acquired horses, the Spanish herds in New Mexico, Texas, and northern Mexico became the target of constant raids.[57]

Once out of Spanish hands, horses entered a distinctly Native trade network. Following pre-existing routes that paralleled the Rocky Mountains, Utes, Comanches, and others traded horses to the north. The diffusion of the animals as well as the knowledge for their care and use took place rapidly. Annual trade fairs, part of what John C. Ewers called the "aboriginal intertribal trade pattern," became the nexus of the horse trade. The Mandan-Hidatsa Arikara villages on the Missouri River became an important conduit of horses for peoples to the north and east.[58] West of the Rockies, a particularly important trade fair took place each summer at a place that the Newe, or Shoshone, called *sehewooki'*: "willows standing in rows like running water." Today, it is known as Idaho's Boise valley. Utes most likely brought the first horses to the region as early as 1690. Many Newe quickly embraced the equestrian lifestyle as well as the horse trade. At *sehewooki'* they provided the Nez Perce and other Columbia plateau tribes with their first horses. From that beginning the Nez Perce built massive herds and practiced selective breeding that resulted in a distinctively American breed, the Appaloosa.[59]

European ideas would impact Native peoples and cultures, particularly in regard to land tenure, economic exchange, and gender relations. In very general terms, European notions of property emphasized individual, exclusive ownership as opposed to the often communal and usufruct systems of American Indians. As Native economies engaged and were eventually absorbed by the European market economy, corresponding shifts took place within Indian societies. One result was economic dependency—the state where customary Native economies were no longer viable. American Indians would increasingly be forced to conform to European concepts of economic exchange. In many matrilineal Native groups, for instance, women saw a relative decline in their status and power. On the Great Plains, the demands upon women's labor increased exponentially as their peoples became enmeshed in the trade. Whereas a man hunting from horseback could kill many more bison with a minimal increase in effort, the greater number of hides still had to be tanned in a laborious process that took from three to ten days. In some hunting groups, polygamy also increased along with the demands of the market. Yet the apparent decline in the status of women did not happen quickly nor did it affect all Native peoples evenly.[60]

The imperial policies of three principal European powers—Spain, France, and Great Britain—also shaped the Native West (Russian imperial policy had a more limited influence among Alaskan Native peoples). The conversion of American Indians to Catholicism was the official goal of the Spanish Empire, but military and economic conquest more often took precedent. The Spanish sought to incorporate Native peoples at the bottom of their social order as the labor necessary in the fields and mines. To this end, the Spanish employed a series of institutions to manage and distribute Indian labor: first the *encomienda* (a hereditary grants of Indian labor), and later the *repartimiento* and the *hacienda*. The Spanish also created an elaborate racial/social hierarchy that incorporated Indian peoples as well as the mixed-race peoples (mestizos, mulattos, etc.) that made up a growing segment of the colonial population. The French colonial effort was never as well financed or manned and, as a result, relied heavily upon far-reaching Native alliances established via the fur trade for its survival. As in the Spanish Empire, mixed-race peoples and Catholic missionaries played major, albeit quite different, roles. And while the French are usually portrayed as more sympathetic toward Native peoples, their strategies reflected their relative weakness rather than a greater respect for Native cultures. British Indian policy, on the other hand, was not based on the incorporation of Native peoples into their society. Indians could be allies in war and trading partners in peace but essentially had no official place within the borders of established colonies or the colonial social order unless they rejected their own cultures and adopted British ways. For instance, Puritan "praying towns" such as Natick Massachusetts were founded as a means to culturally reeducate Native peoples. In practice, however, Indians continued to live widely dispersed throughout the British settlements, working as laborers and often acting as intermediaries between their peoples and the English. By the eighteenth century, the imperial policies of Britain and France began to impact the lives of Native peoples in the West, but before 1700 it was the expansion of the Spanish Empire that brought the most consequential changes to the Native West.[61]

SPAIN AND THE NATIVE WEST, 1500–1700

The first Spaniard to see and report his observations of the Native West was truly an accidental tourist. Álvar Núñez Cabeza de Vaca accompanied Pánfilo de Narváez's ill-fated invasion of Florida in 1528. When the expected resupply ships failed to arrive, the Spaniards built boats and set off for Mexico. They made it as far as the Texas coast near Galveston Bay. Narváez and most of the party set off again to an unknown fate. Cabeza de Vaca and a handful of others remained among the Karankawa Indians and other Native groups for the next six and a half years in various forms of captivity. In the summer of 1535, Cabeza de Vaca and three others, including a Moorish slave named Estavanico, made good their escape and struck out, hoping to reach the Spanish settlements overland.[62] Cabeza de Vaca and his companions were the first non-Natives to encounter the American Indian peoples living on the southern margins of what we have defined as the American West. They also gained an understanding of the vastness of North America's bison herds and their importance to Native groups. Cabeza de Vaca reported that the "cows" had short horns like "Moorish cows," but long shaggy fur. He found their meat better tasting and more plentiful than that of Spanish cattle. Vast herds extended hundreds of miles along their route, and de Vaca wrote, "the people who inhabit [the valleys] come down and sustain themselves on them, and they supply the land with a great quantity of hides." As the group made its way northwest, the Native dependence on bison became ever-more apparent. At the confluence of the Rio Conchos and Rio Grande, near present-day Presidio, Texas, the travelers first met the "People of the Cows," who may have been the Jumanos. They lived in permanent earthen houses and gave their visitors beans, squash, and corn. By the time they approached the area around modern-day El Paso, the Native people were dependent almost entirely upon meat. There had not been consistent rain for two years and so they had not planted. They told the travelers of two roads, one to the west where they would find corn, and one to the north into the heart of the buffalo herds. Cabeza de Vaca and his men chose the western road, and their grand voyage came to a conclusion in April 1536 when they reached the Spanish settlements in Sinaloa.[63]

Tales of Cabeza de Vaca's journey swept through the Spanish colony and set plans in motion to exploit the imagined wealth of the north. Some Spanish believed that the fabled

"Seven Cities of Cíbola," supposedly founded by seven Catholic bishops who had fled the Moors, must lay just to the north of where Cabeza de Vaca changed course. Reports obtained through Native trade networks of the Rio Grande Pueblos further fueled Spanish desires. Fray Marcos de Niza led a reconnaissance from New Spain in 1539, with Estavanico acting as the party's guide. As Marcos journeyed north he heard more tales of Cíbola, a city of many streets and plazas with some houses reaching ten stories into the sky. Estavanico reached "Cíbola"—in reality the small Zuni town of Hawikuh—in advance of the friar and the others. His overbearing demands for women and turquoise, or perhaps the Zuni perception that he was a spy, got him killed. Two survivors of his party brought word of Estavanico's death back to the main party. Marcos continued on, he claimed, until Cíbola was within view. "It has the appearance of a very beautiful town," he wrote, "the best I have seen in these parts." The friar also estimated that the Pueblo was "bigger than the city of Mexico," and reported that several Indians accompanying him swore that it was the "least of the seven cities." Back in Mexico City his account provided all the confirmation necessary that large, wealthy cities lay to the north.[64]

Francisco Vásquez de Coronado led the "multi-ethnic invasion force" that returned to conquer "Cíbola" in 1540. Like Hernando De Soto who had launched an invasion of the southeast a year earlier, Coronado believed he could replicate the success of Cortez and Pizzaro by conquering massive native empires. Over 300 Europeans accompanied the expedition, but Mestizos and Mexican Indians, numbering between 800 and 1,000, made up the bulk of Coronado's army. On July 7, 1540, the expedition arrived at "Cíbola" and began to curse Fray Marcos, for they found not a city of gold but what expedition chronicler Pedro de Castañeda de Nájera called a "small, rocky Pueblo, all crumpled up." The Spanish had arrived in the midst of the Zunis' summer-solstice ceremonies. Zuni warriors blocked the road, and Bow Priests spread cornmeal across the Spaniards' path as a symbolic warning not the enter the Pueblo. Instead, Coronado ordered an attack, and, after a brief, desperate battle, the Spanish took possession of Hawikuh. Despite the disappointment, the Spaniards' dreams of wealth were stoked by new rumors of wealthy native provinces. Before leaving Zuni in November, Coronado dispatched a small contingent led by Pedro de Tovar and the Franciscan, Juan de Padilla, to explore the northwestern province of "Tusayan." The group sacked one of the Hopi Pueblos and became the first Europeans to peer into the Grand Canyon but found no great wealth.[65]

Coronado's interests then turned to the east. A delegation from Pecos Pueblo, led by a chief the Spanish called Bigotes, offered to guide them to the Rio Grande and beyond. Hernando de Alvarado and Fray Padilla led a small detachment east to survey the route. When they reached Pecos, Alvarado purchased a Plains Indian slave, likely a Pawnee or Wichita, as a guide and interpreter. The man was dubbed El Turco, "The Turk," due to his appearance, but historian James Brooks argues that the moniker also reflected Mediterranean systems of captivity familiar to the Spanish. His presence at Pecos, among the easternmost of the Pueblos, also illustrated the far-reaching trade and social ties between the Pueblo world and the Native peoples of the Great Plains. The entire army followed by December and occupied Tiguex Pueblo, near modern-day Albuquerque, as the Spanish headquarters. Spanish–Pueblo relations quickly deteriorated. Coronado began to distribute Indian labor and tribute among his loyal Spanish followers. Alvarado returned to Pecos to recover gold that the Turk said had been taken from him and hidden from the Spanish. As a result, Bigotes, the Pecos leader who had befriended the Spanish, was taken prisoner, tortured, and held captive for six months before being allowed to return to his people. Alvarado's actions resulted in widespread mistrust and growing displays of hostility toward the Spanish. A Spaniard's rape of a Pueblo woman proved to be the final straw, and outright war between the invaders and the twelve Tiwa-speaking Pueblos of Tiquex followed. The war dragged on into the spring of 1541. Rather than surrender, the Tiquex people fled into the mountains and refused to return until the Spaniards had left their country.[66]

The Turk's fanciful tales of his homeland of "Quivira" finally induced the Spanish to leave. Far across the Plains to the east, he claimed, were large towns along a great river. The nobles of Quivira were said to travel in massive sail canoes manned by twenty oarsmen, and gold and silver was everywhere for the taking. Driven by the insatiable thirst for gold and glory, the Spanish suspended their doubts and plunged east into the heart of Great Plains. Travel across the vast and seemingly featureless Llano Estacado ("staked plain") of the Texas panhandle terrified the Spanish. About two weeks' travel from Pecos the expedition encountered a *rancheria* (village) of "Querechos," most likely Plains Apaches. The foot-going buffalo hunters possessed many dogs, and Castañeda reported that they "live in tents made of dressed skins of the cattle [bison]." "When they move from place to place," Coronado wrote, "they load the dogs with their tents, poles, and other things." The Coronado expedition thus provided some of the earliest descriptions of Plains Indian life before the acquisition of horses. Farther out on the Llano Estacado were the "Teyas," who may have been another Apache group or, just as likely, Caddos to Jumanos. Like the Querechos, they traveled the Plains using dogs as pack animals. It was among the Teyas that the Spanish first heard inklings that the Turk's stories and directions were not accurate. After wandering northeast for over two months, the Spanish arrived at Quivira, near what is today the Kansas–Nebraska border. They found no cities of gold, only the earth-lodge villages of Caddoan speakers. An enraged Coronado demanded answers. The Turk confessed that it was a plan hatched at Pecos to lead the Spanish astray on the Plains and to weaken them, "believing that they would not know how to hunt or survive without maize," so when they returned to the Pueblos they could be easily destroyed. The Turk paid for the deception with his life, and the expedition began retracing its steps. Unrest among the Native peoples of northern Mexico demanded a precipitous withdrawal from the region. Coronado's army met constant harassment as it passed through Pueblo lands. By June 1542, Coronado was back in Mexico and facing trial for mismanaging the expedition. From the Spanish perspective, the expedition had been an abject failure. Intermittent trade and contact with the Pueblos continued, and various missionaries made their way north, yet over half a century would pass before the Spanish returned to the upper Rio Grande valley in force.[67]

Juan de Oñate led the Spanish return to New Mexico in 1598. Unlike Coronado's expedition, which one historian characterized as a pack of bandits with no intention of creating a "permanent Kingdom," Oñate brought with him the means to establish a permanent colony among the Pueblos. He also held a real commitment to Spain's imperial goal of Christianizing America's Native peoples. Oñate organized his *entrada* at Zacatecas, the heart of Mexico's silver-mining region where his family had grown wealthy. He recruited 129 soldier-settlers (the only individuals recorded on the official muster roll) and their families. Along with Indian servants and ten Franciscan missionaries, the company numbered over 500. The column that rumbled north in the spring of 1598 was nearly 2 miles long and included eighty wagons and carts, three small artillery pieces, and 7,000 head of livestock.[68]

If Oñate's goals were different, his tactics proved to be as heavy-handed and violent as his predecessors. Adhering to the colonial policies and expectations established during the conquest of Mexico and Peru, Oñate expected Native peoples to provide the colonists with tribute and sustenance. On the march north he dispatched advance parties to forcibly extract foodstuffs from the Pueblos. Once he had arrived in the upper Rio Grande valley, Oñate occupied the Tewa-speaking Pueblo of Yunge, which he renamed San Gabriel, as the Spanish capital. It lay directly across the river from its larger sister Pueblo of Okhay Owingeh, commonly known as San Juan until its residents reasserted its indigenous name in 2005. Erasing Native placenames was a more subtle, yet effective and persistent form of colonialism that all European powers employed to some extent. Oñate's other actions were far more overt. He distributed *encomiendas* to his followers and demanded tribute from neighboring Pueblos. He demanded that Pueblo people accept Christianity and admit the Franciscans to their villages. And he made war upon those who resisted. His treatment of the Acoma Pueblo was particularly ruthless. In December 1598, a small detachment arrived at the Pueblo and demanded supplies for a larger expedition heading toward Zuni. The Spaniards became abusive and were taken captive. When a relief party arrived, things turned violent, and eleven Spaniards were killed, including the expedition's commander. Oñate's retaliation was swift. He demanded the surrender of Acoma's leaders and that the people abandon the fortress-like Pueblo and accept missionaries. On January 22, 1599, after the Acoma people refused the demands, the Spanish stormed the Pueblo. The Indians fought with "determination and fury," and the battle stretched into the next day. Perhaps 800 died while nearly 600 were taken captive and an unknown number of others escaped. Oñate convened a court to try the survivors and to make an example of Acoma. Twenty-four men over the age of twenty-five were condemned to have one foot cut off and to spend twenty years in personal servitude. The males between twelve and twenty-five, as well as all females over twelve were also made servants for twenty years. Children under twelve were turned over to the Church for a "Christian upbringing," and as many as sixty young girls were sent off permanently to the convents of Mexico City. Finally, two Hopi men captured among the Acomas each had one hand chopped off and were returned to their people as a warning. Within a year, nearly all of the Acoma prisoners escaped servitude and returned to rebuild the Pueblo where their descendants still live. Oñate remains a controversial historical figure, celebrated in some segments of the Hispanic community as the founder of New Mexico and reviled in the Pueblos as an invading butcher.[69]

Despite Oñate's hopes and extravagant claims that great mines would make New Mexico a thriving colony, the province faltered, and only the intercession of the Franciscans prevented its abandonment. With no source of wealth beyond the production of the Pueblos, the Spanish burden on the Native population grew steadily worse. Demands for tribute and food pushed some Pueblos to the brink of starvation and disrupted established trade patterns between the Pueblos and their Athapaskan and Numic neighbors. Following the pattern set at Acoma, Pueblo resistance was met with brutal retaliation. Mineral wealth remained elusive, and, while Oñate explored neighboring areas, the Spanish population began to desert New Mexico. In 1606, the Viceroy recalled Oñate to Mexico City to face charges of misconduct and cruelty. In 1608, he was found guilty and forced to resign as governor. The Council of the Indies then recommended that the colony be abandoned. Fray Lázaro Ximénez, pleading the Franciscans' case before the Viceroy, claimed that the order had converted over 7,000 Indians and thousands more awaited baptism. The friar's report swayed King Philip III, who transformed New Mexico into a royal mission colony in 1609. A new governor, Pedro de Peralta, arrived and established a new Spanish capital at Santa Fe, which remained the only Spanish community in New Mexico for the rest of the seventeenth century.[70]

The Pueblos suffered most intensely from colonization during the seventeenth century. From a pre-contact population of perhaps 60,000, only around 17,000 survived by the 1670s. Aside from the thousands who died or fled as a direct consequence of the invasion, epidemics carried away many more. Spanish slave raids against neighboring peoples led to retaliation and greater stress on the Pueblos' already-diminished ability to sustain themselves. The Spanish also moved to impose their own social, political, and legal systems as well as their religion. Secular Spanish officials appointed governors to handle civil affairs for each Pueblo while the Franciscans appointed *fiscales* from among their converts to ensure that Christian discipline was enforced. Unlike some Catholic orders that attempted to use Native languages and cultures to convert Indians (most famously the Jesuits in New France), the Franciscans approach was heavy-handed and rigid. The most zealous of the order believed that Christianity could only be renewed through severe discipline, obedience, and poverty. They also viewed martyrdom as a supreme act of obedience to God akin to Christ's crucifixion. Often with the support of soldiers, the friars forced Indians to build churches and to attend services. Native ceremonies were banned and sacred objects destroyed. The friars also tried to remake Pueblo gender roles and attitudes about sexual relations that they deemed sinful. In some cases, such as at Zuni in 1631 and Taos in 1639, the people rose up and killed the missionaries. Still, thousands accepted baptism because the Church offered protection and food. Moreover, thousands of Pueblo converts created a syncretic form of Catholicism that blended the power of the invaders with elements of Pueblo religion. Meanwhile, traditional beliefs survived by remaining underground in the kivas. While some Pueblo people resisted alien ways, others adopted and adapted them, and the cultural differences between Spaniard and Indian in New Mexico began to blur.[71]

The impact of the Spanish conquest, however, was not limited to the Pueblo world. It radiated outward from New Mexico, creating a political economy of violence that rearranged relations between Native groups and eventually shaped a hybrid society in New Mexico. Spanish slaving was at the heart of the process. In Iberian culture as well as among some Native groups, captivity held both social and economic meanings and offered a gendered path to social status based upon the control of women and children. While historians disagree over the nature of pre-contact Native captivity and its impact on the evolving institution, women and children were clearly the focus of Spanish raids, and acts of sexual violence against Native women became commonplace throughout the region. Moreover, the scale of Spanish slavery was unlike anything seen before. In 1637, Governor Luis de Rosas, ignoring Franciscan complaints that slaving hindered the missionary effort, launched an "unjust war against the 'Utaca' nation," bringing back eighty captives. Kinspeople of these Ute prisoners soon launched retaliatory raids, and in the decades that followed the Utes adapted to Spanish presence by becoming slavers in their own right and displacing the violence onto the Numic peoples of the Great Basin. Some of the enslaved were sent south to labor in the mines, but many remained as servants in New Mexico. This burgeoning population of detribalized Indian slaves became known as *genizaros,* a term derived from the Ottoman word "janissary," meaning "new troops." *Genizaros* served as military auxiliaries, carrying out slave raids and defending the province, and by the eighteenth century *genizaro* communities such as Abiquiu served as buffers between New Mexico's Hispanic towns and the Utes, Comanches, Navajos, and Apaches.[72]

By the 1660s and 1670s, drought, crop failures, and the growing weight of Spanish colonialism drove the Pueblos toward rebellion. The increasingly oppressive actions of the Franciscans as well as a growing public rift between civil and religious authorities eroded Pueblo respect and obedience. Governor López de Mendizábal was openly hostile to the Franciscans and even allowed the public performance of some Pueblo ceremonies. In response, in 1661 the friars prohibited Katchina dances, raided the kivas, and destroyed thousands of sacred objects. The next year, their complaints led to Mendizábal's removal as governor. The dispute was indicative of the rapacious competition between the secular and religious segments of Spanish society to extract labor and tribute from the Pueblos. As a result, the Pueblos had little or no surplus to trade with the Navajos, Apaches, Utes, and Plains groups, and raiding rather than trade increasingly characterized intertribal relationships. The power of the Franciscans was further diminished by their personal behavior. Contrary to their vows of chastity, many carried on sexual relations with Native women and fathered numerous children. In the words of one historian, they were no longer viewed as "supermen." Nor were the secular Spanish, whose drab villages and dress appeared more Indian and less powerful as the years passed. As cultural differences in New Mexico blurred, Span ish authority declined. By 1680, this complex series of factors reached a breaking point, and the Pueblos rose in rebellion and drove the Spanish from New Mexico.[73]

The Pueblo Revolt of 1680 was the result of a remarkable inter-Pueblo alliance. Popé, a spiritual leader from Okhay Owingeh who had been arrested and whipped as part of a 1675 crackdown on Pueblo religion, was a principal leader of the rebellion. Luis Tupatú of Picuris and El Jaca of Taos, along with leaders from Santo Domingo, Jemez, San Lazaro, and Keres, joined Popé at Taos to plan the revolt. Runners carrying a knotted cord were sent to each of the Pueblos with instructions to untie one knot each day. When last knot was undone, the attack would commence. After two young men were captured and revealed the plot to the Spanish, Popé ordered the plan set in motion one day early, August 10, 1680. Two dozen Pueblos rose as one. In some cases, such as at Tanos, the people revolted even though their leaders opposed the plan. As far away as Hopi and Zuni, the Spanish were killed or driven out. Many of the marginalized mixed-blood people known as *castas* also joined in the rebellion. Of the approximately 2,500 Spanish residents of New Mexico, over 400 died in the attack, including nineteen priests and two lay brothers. The survivors gathered at Santa Fe and Isleta Pueblo, one of the few Pueblos that had not revolted. Thousands of Indians besieged the capital, forcing Governor Antonio de Otermín and his followers to fight their way out before fleeing south to the vicinity of El Paso. While the Pueblo Revolt is rightly celebrated as the first successful Native anti-colonial movement in American history, it was in fact part of an "epidemic of rebellions" against Spanish colonization mounted by Native peoples in the northern reaches of the empire in the seventeenth century.[74]

A new Pueblo golden age did not follow the Spanish departure. Popé and other revolutionary leaders set out to erase all evidence of the Spanish presence. Churches were burned, all crosses and rosaries were to be destroyed, and the Spanish language was forbidden. The official Spanish archives were piled up in the Santa Fe plaza and burned. Christian marriages were to be dissolved, and European goods and tools were to be shunned. But the Pueblo world in 1680 was radically different to that of 1598. Social and cultural changes meant that many Pueblo people had internalized Spanish ways and were loath to follow Popé's dictates. The unity that had made the revolt successful was short-lived. Civil war between the Pueblos followed hard times and increased raiding by Numic and Athapaskan groups. A political struggle between Popé and Luis Tupatú increased discord by the mid-1680s. Meanwhile, the Spanish launched expeditions to retake the province in 1681, 1688, and 1689. All failed, but they destroyed numerous Pueblos, cost hundreds of lives, and further reduced the Pueblo ability to resist the eventual reconquest.[75]

What Diego de Vargas intended to be only a preliminary reconnaissance in 1692 has become known as the "bloodless" reconquest of New Mexico. The small military expedition met only limited resistance, and Vargas's promises of peace and a pardon for all who lay down their arms gradually won over the suspicious, including Luis Tupatú. In return, he promised to help the Spanish re-establish peace and Christianity. Vargas then journeyed west to Zuni and Hopi before returning to El Paso to mount a much larger colonizing expedition in the fall of 1693. It was then that the reconquest turned bloody. Indians at Santa Fe resisted the Spaniards, and Vargas resorted to typically ruthless tactics. Once the resistance was broken he ordered the execution of seventy young men and the distribution of 400 women and children to his followers as slaves. By September 1694, he had succeeded in reconquering all of the Pueblos with the exceptions of Taos and Picurís in the north and Zuni and Hopi in the west.[76]

Vargas's peace was only temporary, and the Pueblos launched one more attempt to drive out the Spanish. There were already rumors of a second revolt in the summer of 1695. A severe winter increased hardships on both the Indians and the Spanish, and by the spring of 1696 the missionaries suspected that a revolt was imminent. The attack came on June 4. Five Franciscans and twenty-one other Spaniards were killed. But, unlike 1680, the rebellion was poorly coordinated, and divisions between the Pueblos turned to civil war. Warriors from Zia and other Pueblos joined the Spanish. Vargas prosecuted a vigorous campaign to put down the rebellion while promising pardons to those who surrendered. In July, Lucas Naranjo, whom the Spanish viewed as the principal rebel leader, died in battle. After laying siege to Acoma in August, Vargas marched north against Taos and Picurís. The people abandoned the villages and fled into the mountains. With the onset of winter there was no choice but to surrender or to flee west, where some found refuge among the Hopis. In the aftermath of the Pueblo Revolt of 1696, the Spanish not only strengthened their military presence but also relied more upon diplomacy. The missionaries also adopted a less intrusive stance toward their Pueblo charges. After 1700, Pueblo–Spanish relations became more stable and never again was the presence of the colonizers challenged so directly.[77]

Ten years after the second Pueblo uprising Spanish officials noted the presence of a new and expanding people to the north. In their native Shoshonean tongue they were the *Numunu,* the "people," but the Spanish called them *komántcia,* a Ute word for foreigner or enemy. Their lives had been shaped by the actions and decisions of countless generations before a European ever set foot on the American continents, as well as by European colonization. Their journey from the Great Basin had begun centuries earlier as part of the great Numic migration. They acquired horses around the time of the Pueblo Revolts and became an equestrian people. In 1706, they were new arrivals on the Southern Plains, and for roughly the next century and a half they, and many other peoples, would use the power of the horse to carve out their own lives and empires on the Great Plains.[78]

The emergence of the Plains equestrian culture was not, however, the inevitable culmination of the historical processes explored in this essay. It was just one possible outcome. In the decades and centuries that followed 1700, the Native West remained a diverse and dynamic place. Horse ownership was a single factor that transformed the lives of many peoples but affected others in far more limited ways. Many peoples strived to make a living from the earth in ways they always had: farming, fishing, gathering, and trading. But ultimately none could escape the myriad effects of the European conquest. In the eighteenth and nineteenth centuries the pace and gravity of changes increased. Still, Native nations and individuals would endure and leave their imprint on the region and its future course. The Native West, what would become the American West, was never a pristine land inhabited by timeless peoples but the product of a dynamic and diverse human history of migration, adaptation, and colonization. Just as the region's history did not begin with European contact, the importance of Native peoples to its history would not end with conquest.

NOTES

1. The best survey of topics covered in this essay is Colin G. Calloway's award-winning *One Vast Winter Count: The Native West before Lewis and Clark* (Lincoln, Nebr.: University of Nebraska Press, 2003). See also Alvin Josephy Jr. (ed.), *America in 1492: The World of American Indian Peoples before the Arrival of Columbus* (New York: Vintage, 1991). An essential source on American Indian cultures and history is William C. Sturtevant (gen. ed.), *Handbook of North American Indians,* planned at 20 vols. (Washington, DC: Smithsonian Institution, 1978–2008), hereafter cited as *HNAI.*
2. Matilda Coxe Stevenson, *The Zuñi Indians: Their Mythology, Esoteric Fraternities, and Ceremonies* (Washington, DC: Government Printing Office, 1904), pp. 23–26; *The Zunis: Self-Portrayals,* trans. Alvina Quam (Albuquerque, N. Mex.: University of New Mexico Press, 1972), pp. 129–137.
3. Eleanor B. Heady, *Sage Smoke: Tales of the Shoshoni-Bannock Indians* (Chicago, Ill.: Follett Publishing Company, 1973), pp. 15–19.
4. Julian Steward, *Some Western Shoshoni Myths* (Washington DC: Government Printing Office, 1943), p. 267; Anne M. Smith, *Shoshone Tales* (Salt Lake City, UT: University of Utah Press, 1993), pp. 12–14.
5. Thomas Jefferson, *Notes of the State of Virginia* (1782; reprint edn, New York: Library Classics of the United States, 1984), p. 226; Joseph F. Powell, *The First Americans: Race, Evolution, and the Origin of Native Americans* (Cambridge: Cambridge University Press, 2005), pp. 18–21.
6. Powell, *The First Americans,* pp. 118–120; E. C. Pielou, *After the Ice Age: The Return of Life to Glaciated North America* (Chicago, Ill.: University of Chicago Press, 1991).
7. E. James Dixson, *Bones, Boats, and Bison: Archaeology and the First Colonization of Western North America* (Albuquerque, N. Mex.: University of New Mexico Press, 1999), pp. 3–17.
8. See Paul S. Martin, "Prehistoric Overkill," in Paul S. Martin and H. E. Wright Jr. (eds.), *Pleistocene Extinctions: The Search for a Cause* (New Haven, Conn.: Yale University Press, 1967), pp. 75–120; Paul S. Martin, *Twilight of the Mammoths: Ice Age Extinctions and the Rewilding of America* (Berkeley, Calif.: University of California Press, 2005).
9. See Donald K. Grayson and David J. Meltzer, "A Requiem for North American Overkill," *Journal of Archaeological Science,* 30 (2003): 585–593; Stuart Fiedel and Gary Haynes, "A Premature Burial: Comments on Grayson and Meltzer's 'Requiem for Overkill,'" *Journal of Archaeological Science* 31 (2004): 121–131; Donald K. Grayson and David J. Meltzer, "North American Overkill Continued," *Journal of Archaeological Science,* 31 (2004): 133–136.
10. Thomas D. Dillehay, *The Settlement of the Americas: A New Prehistory* (New York: Basic Books, 2000).
11. "Evidence Supports Earlier Date for People in North America," *The New York Times,* April 4, 2008.
12. Dillehay, *The Settlement of the Americas.*
13. K. R. Fladmark, "Alternate Migration Corridors for Early Man in North America," *American Antiquity,* 44 (January) (1979): 55–69; Dixon, *Bones, Boats, and Bison,* pp. 129–130; Robert N. Zeitlin and Judith Francis Zeitlin, "The Paleoindian and Archaic Culture of Mesoamerica," in R. E. W. Adams and Murdo Macleod (eds.), *The Cambridge History of the Native Peoples of the Americas,* Vol. II: *Mesoamerica,* Part 1, pp. 51–53; Powell, *The First Americans,* pp. 120–122, 154; "Ancient Seaweed Chews Confirm Age of Chilean Site," *Reuters,* May 8, 2008.
14. Dennis Stanford and Bruce Bradley, "Ocean Trails and Prairie Paths? Thoughts about Clovis Origins," in Nina G. Jablonski (ed.), *The First Americans: The Pleistocene Colonization of the New World*

(San Francisco, Calif.: California Academy of the Sciences, 2002), pp. 255–271. For a summary of the theory's critics, see Powell, *The First Americans,* pp. 124–126.

15. Dixon, *Bones, Boats, and Bison,* p. 21. For summaries of the genetic data, see Powell, *The First Americans,* pp. 58–84, 188–190; Michael H. Crawford, *The Origins of Native Americans: Evidence from Anthropological Genetics* (Cambridge: Cambridge University Press, 2001).
16. Nelson J. R. Fagundes, Ricardo Kanitz, Roberta Eckert, Ana C. S. Valls, Mauricio R. Bogo, Francisco M. Salzano, David Glenn Smith, Wilson A. Silva, Jr., Marco A. Zago, Andrea K. Ribeiro-dos-Santos, Sidney E. B. Santos, Maria Luiza Petzl-Erler, and Sandro L. Bonatto, "Mitochondrial Population Genomics Supports a Single Pre-Clovis Origin with a Coastal Route for the Peopling of the Americas," *The American Journal of Human Genetics,* 82 (March) (2008): 583–592.
17. David Hurst Thomas, *Skull Wars: Kennewick Man, Archaeology, and the Battle for Native American Identity* (New York: Basic Books, 2000); James C. Chatters, *Ancient Encounters: Kennewick Man and the First Americans* (New York: Simon & Schuster, 2001); Native American Graves Protection and Repatriation Act, Public Law 101–601, 25 USC 3001, 1990; Kathleen S. Fine-Dare, *Grave Injustice: The American Indian Repatriation Movement and NAGPRA* (Lincoln, Nebr.: University of Nebraska Press, 2002).
18. Dean R. Snow, "The First Americans and the Differentiation of Hunter-Gatherer Cultures," in Bruce G. Trigger and Wilcomb E. Washburn (eds.), *The Cambridge History of Native Peoples of the Americas,* Vol. I: *North America, Part 1* (Cambridge: Cambridge University Press, 1996), pp. 125–200.
19. Anthony Ranere, Dolores R. Piperno, Anthony J. Ranere, Irene Holstb, Jose Iriarte, and Ruth Dickau, "Starch Grain and Phytolith Evidence for Early Ninth Millennium B.P. Maize from the Central Balsas River Valley, Mexico," *Proceedings of the National Academy of Sciences,* March 24, 2009.
20. R. Douglas Hurt, *Indian Agriculture in America: Prehistory to the Present* (Lawrence, Kans.: University Press of Kansas, 1987), pp. 1–26; Calloway, *One Vast Winter Count,* pp. 68–71.
21. James E. Fitting, "Regional Cultural Development, 300 B.C. to A.D. 1000," *HNAI 15: Northeast,* pp. 44–51; Hurt, *Indian Agriculture,* p. 13; Calloway, *One Vast Winter Count,* pp. 96–97.
22. Timothy R. Pauketat, *Ancient Cahokia and the Mississippians* (Cambridge: Cambridge University Press, 2004), pp. 1–25; Calloway, *One Vast Winter Count,* pp. 98–99; F. Kent Reilly and James F. Garber (eds.), *Ancient Objects and Sacred Realms: Interpretations of Mississippian Iconography* (Austin, Tex.: University of Texas Press, 2007).
23. Pauketat, *Ancient Cahokia and the Mississippians;* Calloway, *One Vast Winter Count,* pp. 98–104.
24. Linda S. Cordell, *Prehistory of the Southwest* (New York: Academic Press, 1984), p. 77; Linda S. Cordell, *Archaeology of the Southwest* (New York: Academic Press, 1997), pp. 210–212; Albert H. Schroeder, "Prehistory: Hakataya," in *HNAI 9: Southwest,* pp. 100–107.
25. Cordell, *Archaeology in the Southwest,* pp. 199–202; Patricia L. Crown and W. James Judge (eds.), *Chaco and Hohokam: Prehistoric Regional Systems in the American Southwest* (Santa Fe, N. Mex.: School of American Research Press, 1991); George J. Gumuerman and Emil W. Haury, "Prehistory: Hohokam," in *HNAI 9: Southwest,* pp. 75–90; Shepard Krech III, *The Ecological Indian: Myth and History* (New York: W. W. Norton, 1999), pp. 45–72; Calloway, *One Vast Winter Count,* pp. 77–80.
26. Krech, *The Ecological Indian.*
27. Cordell, *Archaeology in the Southwest,* pp. 202–210; Cordell, *Prehistory in the Southwest,* pp. 70–75; Paul S. Martin, "Prehistory: Mogollon," in *HNAI 9: Southwest,* pp. 61–74; Stephen Plog, *Ancient Peoples of the American Southwest* (London: Thames & Hudson, 1997).
28. David E. Stuart, *Anasazi America* (Albuquerque, N. Mex.: University of New Mexico Press, 2000); Michael S. Berry, *Time, Space, and Transition in Anasazi Prehistory* (Salt Lake City, UT: University of Utah Press, 1982).
29. Kendrick Frazier, *People of Chaco: A Canyon and Its Culture* (New York: W. W. Norton, 1986); Stuart, *Anasazi America,* pp. 51–124; Calloway, *One Vast Winter Count,* pp. 83–88.
30. Lynne Sebastian, "Sociopolitical Complexity and the Chaco System," in Patricia L. Crown and W. James Judge (eds.), *Chaco and Hohokam: Prehistoric Regional Systems in the American Southwest*

(Santa Fe, N. Mex.: School of American Research Press, 1991), pp. 109–134; Christy G. Turner and Jacqueline A. Turner, *Man Corn: Cannibalism and Violence in the Prehistoric American Southwest* (Salt Lake City, UT: University of Utah Press, 1999), pp. 470, 484; Stephen A. LeBlanc, *Prehistoric Warfare in the American Southwest* (Salt Lake City, UT: University of Utah Press, 1999).

31. Stuart, *Anasazi America,* pp. 125–145.
32. Cordell, *Prehistory of the Southwest,* pp. 96–98.
33. David B. Madsen and David Rhode (eds.), *Across the West: Human Population Movement and the Expansion of the Numa* (Salt Lake City, UT: University of Utah Press, 1994); Richard N. Holmer, "Prehistory of the Northern Shoshone," *Rendezvous: Idaho State University Journal of Arts and Letters,* 26 (1) (1990): 41–59.
34. Richard J. Perry, "The Apachean Transition from the Subarctic to the Southwest," *Plains Anthropologist,* 25 (November) (1980): 279–296; Calloway, *One Vast Winter Count,* pp. 57–58; *HNAI 10: Southwest.*
35. Here I have adopted the culture areas as defined in the *HNAI* that are specific to the American West: northwest coast, California, southwest, Great Basin, Plains, and plateau.
36. *HNAI 8: California;* Calloway, *One Vast Winter Count,* pp. 50–51.
37. *HNAI 7: Northwest Coast.*
38. *HNAI 7: Northwest Coast;* Richard D. Daugherty, "People of the Salmon," in Alvin Josephy Jr. (ed.), *America in 1492: The World of American Indian Peoples before the Arrival of Columbus* (New York: Vintage, 1991), pp. 49–70.
39. *HNAI 12: Plateau.*
40. Gregory E. Smoak, *Ghost Dances and Identity: Prophetic Religion and American Indian Ethnogenesis in the Nineteenth Century* (Berkeley, Calif.: University of California Press, 2006), pp. 15–20.
41. *HNAI 11: Great Basin.*
42. See articles on "Plains Village Tradition," in *HNAI 13: Great Plains, Part 1;* William R. Swagerty, "History of the United States Plains until 1850," in *HNAI 13: Great Plains, Part 1,* pp. 256–257; Calloway, *One Vast Winter Count,* pp. 60–61, 104–113.
43. Peter Iverson, "Taking Care of Earth and Sky," in Alvin Josephy Jr. (ed.), *America in 1492: The World of American Indian Peoples before the Arrival of Columbus* (New York: Vintage, 1991), pp. 87–98.
44. *HNAI 13: Great Plains, Part 1.*
45. *HNAI 10: Southwest.* Southwest peoples are treated in two volumes, with Vol. X devoted to Athapaskan- and Piman-speaking peoples and Vol. IX focused on the Pueblos.
46. *HNAI 9: Southwest.*
47. *HNAI 10: Southwest.*
48. A. L. Kroeber, "Cultural and Natural Areas of Native North America," *University of California Publications in American Archaeology and Ethnology,* 38 (Berkeley: University of California Press, 1939); Henry F. Dobyns, "Estimating Aboriginal American Population: An Appraisal of Techniques with a New Hemisphere Estimate," *Current Anthropology,* 7 (1966): 395–416; Russell Thornton, *American Indian Holocaust and Survival: A Population History since 1492* (Norman, Okla.: University of Oklahoma Press, 1987).
49. Thornton, *American Indian Holocaust and Survival,* pp. 28–32.
50. Alfred W. Crosby, *The Columbian Exchange: Biological and Cultural Consequences of 1492* (Westport, Conn.: Greenwood Press, 1972); Alfred W. Crosby, *Ecological Imperialism: The Biological Expansion of Europe, 900–1900* (Cambridge: Cambridge University Press, 1986).
51. Alfred W. Crosby, "Virgin Soil Epidemics as a Factor in the Aboriginal Depopulation in America," *William and Mary Quarterly,* 33 (April 1976): 289–299; David S. Jones, "Virgin Soils Revisited," *William and Mary Quarterly,* 60 (October) (2003): 703–742; David S. Jones, *Rationalizing Epidemics: Meanings and Uses of Indian Mortality since 1600* (Cambridge, Mass.: Harvard University Press, 2004).
52. Crosby, "Virgin Soil Epidemics"; Jones, "Virgin Soils Revisited"; Thornton, *American Indian Holocaust and Survival,* pp. 40–41.

53. Jones, "Virgin Soils Revisited."
54. Crosby, "Virgin Soil Epidemics"; Colin G. Calloway, *New Worlds for All: Indians, Europeans, and the Remaking of Early America* (Baltimore, Md.: Johns Hopkins University Press, 1997), pp. 70–74; Paul Kelton, "Avoiding the Smallpox Spirits: Colonial Epidemics and Southeastern Indian Survival," *Ethnohistory,* 51 (winter) (2004): 45–71.
55. Anne Ramenofsky, *Vectors of Death: The Archaeology of European Contact* (Albuquerque, N. Mex.: University of New Mexico Press, 1987), pp. 123–124.
56. David S. Jones, *Rationalizing Epidemics: Meanings and Uses of American Indian Mortality since 1600* (Cambridge, Mass.: Harvard University Press, 2004), pp. 21–67; William Cronon, *Changes in the Land: Indians, Colonists and the Ecology of New England* (New York: Hill & Wang, 1983), pp. 56–57; Francis Jennings, *The Invasion of America: Indians, Colonialism, and the Cant of Conquest* (Chapel Hill, NC: University of North Carolina Press, 1975).
57. Calloway, *One Vast Winter Count,* pp. 268–276.
58. John C. Ewers, *The Horse in Blackfoot Indian Culture,* Bureau of American Ethnology Bulletin 159, 1955 (reprint edn, Washington, DC: Smithsonian Institution Press, 1980), pp. 3–15; John C. Ewers, *Indian Life on the Upper Missouri* (Norman, Okla.: University of Oklahoma Press, 1968), pp. 14–30.
59. Smoak, *Ghost Dances and Identity,* pp. 22–23; Alvin Josephy Jr., *The Nez Perce Indians and the Opening of the Northwest* (New Haven, Conn.: Yale University Press, 1965), pp. 28–29.
60. Richard White, *The Roots of Dependency: Subsistence, Environment, and Social Change among the Choctaws, Pawnees, and Navajos* (Lincoln, Nebr.: University of Nebraska Press, 1984); Patricia Albers and Beatrice Medecine (eds.), *The Hidden Half: Studies of Plains Indian Women* (Lanham, Md.: University Press of America, 1983).
61. See chapters on "National Policies" in *HNAI 4: History of Indian–White Relations.*
62. Rolena Adorno and Patrick Charles Pautz, *Álvar Núñez Cabeza de Vaca: His Account, His Life, and the Expedition of Pánfilo de Narváez,* 3 vols. (Lincoln, Nebr.: University of Nebraska Press, 1999); Andrés Reséndez, *A Land So Strange: The Epic Journey of Cabeza de Vaca* (New York: Basic Books, 2007); Calloway, *One Vast Winter Count,* pp. 122–126.
63. Adorno and Pautz, *Álvar Núñez Cabeza de Vaca,* Vol. I, pp. 147, 223–231, 247–249; Calloway, *One Vast Winter Count,* pp. 121–126.
64. Cleve Hallenbeck, *The Journey of Fray Marcos de Niza* (Dallas, Tex.: Southern Methodist University Press, 1987), pp. 26, 28–35, 76–79; Calloway, *One Vast Winter Count,* p. 133.
65. George P. Hammond and Agapito Rey (eds.), *Narratives of the Coronado Expedition* (Albuquerque, N. Mex.: University of New Mexico Press, 1940), pp. 208–209, 213–217; Edmund J. Ladd, "Zuni on the Day the Men in Metal Arrived," in Richard Flint and Shirley Cushing Flint (eds.), *The Coronado Expedition to Tierra Nueva: The 1540–42 Route across the Southwest* (Niwot, Col.: University Press of Colorado, 1997), pp. 229–233; Jack D. Forbes, *Apache, Navaho, and Spaniard* (Norman, Okla.: University of Oklahoma Press, 1960), pp. 8–24.
66. Hammond and Rey, *Narratives of the Coronado Expedition,* pp. 220–231; Forbes, *Apache, Navaho, and Spaniard,* pp. 12–14; James Brooks, *Captives and Cousins: Slavery, Kinship, and Community in the Southwest Borderlands* (Chapel Hill, NC: University of North Carolina Press, 2002), pp. 45–48.
67. Hammond and Rey, *Narratives of the Coronado Expedition,* pp. 235–238, 241–242; Forbes, *Apache, Navaho, and Spaniard,* pp. 15–24. For the period between Coronado and Onate, see George P. Hammond and Agapito Rey, *The Rediscovery of New Mexico, 1580–1594* (Albuquerque, N. Mex.: University of New Mexico Press, 1966).
68. George P. Hammond and Agapito Rey (eds.), *Don Juan de Oñate: Colonizer of New Mexico, 1595–1628* (Albuquerque, N. Mex.: University of New Mexico Press, 1953); Marc Simmons, *The Last Conquistador: Juan de Oñate and the Settling of the Far Southwest* (Norman, Okla.: University of Oklahoma Press, 1991), pp. 91–97.

69. "Trial of the Indians of Acoma," in George P. Hammond and Agapito Rey (eds.), *Don Juan de Oñate: Colonizer of New Mexico, 1595–1628* (Albuquerque, N. Mex.: University of New Mexico Press, 1953), pp. 428–479; Simmons, *The Last Conquistador,* pp. 133–146.
70. Simmons, *The Last Conquistador,* pp. 178–185; Ramón Gutiérrez, *When Jesus Came the Corn Mothers Went Away: Marriage, Sexuality, and Power in New Mexico, 1500–1846* (Palo Alto, Calif.: Stanford University Press, 1991), pp. 54–55.
71. Edward H. Spicer, *Cycles of Conquest: The Impact of Spain, Mexico and the United States on the Indians of the Southwest, 1533–1960* (Tucson, Ariz.: University of Arizona Press, 1962); Gutiérrez, *When Jesus Came the Corn Mothers Went Away;* Andrew L. Knaut, *The Pueblo Revolt of 1680: Conquest and Resistance in Seventeenth Century New Mexico* (Norman, Okla.: University of Oklahoma Press, 1995).
72. Brooks, *Captives and Cousins,* pp. 121–138; Ned Blackhawk, *Violence Over the Land: Indians and Empires in the Early American West* (Cambridge, Mass.: Harvard University Press, 2006), pp. 24–27; Gutiérrez, *When Jesus Came the Corn Mothers Went Away,* pp. 149–156.
73. David J. Weber (ed.), *What Caused the Pueblo Revolt of 1680?* (New York: Bedford St. Martins, 1999); Knaut, *The Pueblo Revolt of 1680;* Gutiérrez, *When Jesus Came the Corn Mothers Went Away.*
74. Charles W. Hackett (ed.), *Revolt of the Pueblo Indians of New Mexico and Otermín's Attempted Reconquest,* 2 vols. (Albuquerque, N. Mex.: University of New Mexico Press, 1942); Weber, *What Caused the Pueblo Revolt of 1680?;* Calloway, *One Vast Winter Count,* pp. 177–185.
75. J. Manuel Espinosa (ed.), *The Pueblo Indian Revolt of 1696 and the Franciscan Missions in New Mexico* (Norman, Okla.: University of Oklahoma Press, 1988).
76. Espinosa, *The Pueblo Indian Revolt of 1696.*
77. Espinosa, *The Pueblo Indian Revolt of 1696;* Calloway, *One Vast Winter Count,* pp. 188–196.
78. Pekka Hämäläinen, *The Comanche Empire* (New Haven, Conn.: Yale University Press, 2008); Thomas J. Kavanaugh, "Comanche," in *HNAI 13: Great Plains,* pp. 886–906.

CHAPTER FOUR

HOSTS AND HORSES
Envirocultural Effects of Contact in the Amerindian West

K. D. Motes

> There are low, broken walls on the tabletops and smoke-blackened caves in the cliffs, where still there are metates and broken bowls and ancient ears of corn, as if the prehistoric civilization had gone out among the hills for a little while and would return; and then everything would be restored to an older age, and time would have returned upon itself and a bad dream of invasion and change would have been dissolved in an hour before the dawn.[1]

> A mounted buffalo chase was one of the most exciting things a man could do. There was always the risk of injury and death, but having once hunted buffalo from the back of a horse, no man could resist trying it over and over again. It was one thing to wait in ambush along a trail for blacktailed or whitetailed deer or at a waterhole for the white-bellied pronghorn, but chasing buffalo was the pursuit of life itself, testing one's nerves and skills to the limit as fast as a horse could run.[2]

Culture is the artistic, ideological, religious, political, and emotional expression of a people's relationship to their environment—the contested ground within which members of a group attempt to define their physical and metaphysical place in the world. A culture exists within a sort of envirocultural feedback loop in which environment shapes culture and culture alters environment. Cultures typically are not stone monoliths that rigidly restrict the thoughts and actions of all members of the group. Instead, they are something like playing fields where individuals and subunits of the group act somewhat independently from one another, while all agreeing to play the same game. Constant contestations regarding meaning, values, acceptability, behavior, lifestyle, social structure, and a host of other elements within a culture are the means by which a culture changes.

Though inherently conservative, cultures change when innovations cross the boundary from novel to common because increasing numbers of the group find those innovations useful, beneficial, or necessary. Even in circumstances in which external forces exert enormous pressures on a people to change, in all but the most extreme and limited cases people maintain the ability to direct that change and to make it mesh with their existing culture. In this way, cultural change is rarely a process of wholesale replacement but rather one of adaptation, selective addition and subtraction, and continuity. A culture can change dramatically over a relatively short period of time in certain contexts, but rapid and even fundamental change does not erase or invalidate what came before. Furthermore, because no culture is static it is preposterous to label one era of a people's culture "authentic" and a subsequent era "inauthentic." These issues are especially important in considering American Indian history, since certain well-known and powerfully influential stereotypes continue to dominate outsider views of Native peoples and their cultures.

For decades, histories of the American West focused almost exclusively on the stuff of pulp fiction and Hollywood westerns. Just as in those manifestations of American popular culture, gunfights and Indian raids dominated the narrative. Savage Indians appeared as threats to the spread of civilization and obstacles to be overcome by hearty frontier farmers and the US military. Within this context, the primary events that many observers presented as defining the history of the West were battles, massacres, and European and American manipulations that led to forced assimilation or armed resistance. In this history, Indian peoples always reacted to events beyond their control—often with violent irrationality or piteous deference—as their world shifted beneath their feet. This is the oft-proffered history of outmatched Pueblos falling under Spanish rule, French schemes to pit one naive Native people against another, and the eventual American domination of the Great Plains after defeating mounted "war-like" Lakotas, Comanches, Cheyennes, and numerous other buffalo-hunting peoples. This history claimed to detail the replacement of the cultures of Native peoples by a more sophisticated and rationalized European civilization in a process of historical progression that finally led to the rise of the "exceptional" American nation.

That story is culturally biased, oversimplified, and grossly incomplete. This old narrative may produce exciting films that occasionally allow viewers to pity the poor Natives while celebrating the birth of the American nation, but it hardly captures the true complexity of the West. The history of any region is complicated, but the West may be seen as especially so. The American West of the seventeenth through to the nineteenth centuries was home to hundreds of Native peoples who pursued their own interests in their interactions with other Native groups and agents of Spanish, French, British, and American expansion. Native peoples, Europeans, and Americans were all rational actors who sought to leverage whatever resources they possessed to their own best advantage, with varying degrees of success. Just as the Spanish, the French, and the Anglo-Americans all viewed each other as competitors, so too did various Indian nations view each other and these empires as rivals to oppose or potential allies to court. Each collection of historical agents was at times aided and at other times constrained by the environment it inhabited, the resources at its disposal, and the culture that shaped its motivations and outlook on the world of the West. In this intricate web of trade relations, shifting alliances, changing ecosystems, and social adaptations, the history recent scholars have sought to explicate is necessarily nuanced, often perplexing, and nearly always complicated.

Perhaps the two most important recent trends in the study of the West are the recognition of agency on all sides and the vitally important role of geography and the natural environment in the unfolding of historical events. Understanding how peoples shaped their own histories or the nature of the relationship between any two groups requires comprehending how the landscape set parameters on relations. In contrast to studies written in the early twentieth century that described landscapes only to dramatize the settings of battles between groups locked into unalterable historical trajectories, more recent scholarly work has emphasized the interplay of agency and environment.

Rather than mere pawns of processes scripted before they occurred on scenic but uninvolved landscapes, historical actors now appear as agents capable of radical departures from expected courses who were both propelled and restrained by their environs, their cultures, and their numbers. Due to the widespread aridity and concomitant resource limitations that characterized much of the West, as well as the extremes of temperature, elevation, and biomass, attentiveness to the interplay of environment, culture, and rational decisions is especially appropriate. Regardless of the region of the West or the international relationships under discussion, individual agency and environmental influence both played important roles.

Features of the natural environment of the West served to magnify the effects of novel European introductions on Native lifeways, just as they did innovations developed internally. Environment also shaped the courses Europeans and Americans pursued in the West, offering incredible opportunities but also establishing limitations. Geography and the environment shaped European perceptions of the value of a particular area and the timeframe in which Europeans invaded a given location. All societies are enmeshed in envirocultural feedback loops in which environment shapes culture and culture alters environment. Native societies, like all human societies, changed over time in response to environmental factors long before the arrival of Europeans. Change was not new; the specific changes that Europeans facilitated, however, were radical in their transformative potential.

Of primary importance among novel European introductions to the Native West were two biological entities, pathogens and horses, both of which stimulated significant alterations in Native societies as Indian peoples sought to adapt to population loss and maximize the potential horses offered. The scale of the devastation wrought by microorganisms on Amerindian peoples is difficult to comprehend, and it now appears incontrovertible that the history of European imperialism in the West is the story of the aftermath of apocalyptic epidemics. Following those initial epidemics, horses offered some of the survivors opportunities that were unimaginable in previous generations. Though the arrival of European guns has been cited often as a revolutionary process in Native history, horses were certainly the more transformational resource. Firearms could alter the balance of power in a given locale, but horses had already contributed to a Native reshaping of the West before guns became widely available. The trade networks through which guns traveled and the evolving relationships that the movement of guns highlighted are more important to understanding the history of the Native West than the use of guns in battle. It is likely that into the eighteenth century the desire to obtain guns as gifts or in trade motivated more change than any tactical advantage offered by possession of them.

The possession of horses, however, dramatically altered the geographic distribution of peoples, the strength of some relative to others, the immediacy of peoples' trade options, and the lifestyles they pursued. Horses offered Native groups the opportunity to make their existing lifestyles more efficient or to enter into significantly modified relationships with their environments, in essence fundamentally revising many peoples' envirocultural matrices, or the structures through which their cultures shaped and were shaped by their environments. Even peoples who did not dramatically alter their lifestyles in response to this new opportunity were forced to adapt to the increased ranges and strengths of those who did.

The aftermath of European and then American entry into the West experienced by Native populations defies generalized explication, as the results varied from region to region and from people to people. Indeed one of the most common mistakes in the interpretation of Amerindian history is the tendency to present "Indian" actions, responses, or cultural changes as if "Indians" were a single entity rather than hundreds of separate nations. Various peoples were differently situated to deal with contact and reacted to contact in divergent ways. Some tribes and bands benefited from the creation of new relationships with Europeans, Americans, or other Native peoples. Others suffered losses of population and land to the point of near extinction, surviving if at all only by joining larger tribes or bands. Some subgroups within certain tribes gained influence over others. With any given population, some individuals sought to capitalize upon the opportunities offered by contact, while others steadfastly pursued what they believed to be in the best interests of their people. Countless peoples allied with Europeans as a means to gain advantage over other Native groups through trade or military strength. At various times and places, other Native peoples banded together to resist European invasion and colonization. Perhaps the only generalization that one can employ to describe accurately the impacts of contact in the Native West is that change was ubiquitous as diseases, animals, and manufactured goods stimulated the recasting of trade networks, local and regional power structures, and patterns of geographic distribution of peoples and resources, affecting all Western peoples.

EPIDEMICS

More than any other factor—more than European colonial violence, more than the introduction of horses, firearms, steel, and other trade items that allowed Native peoples to transform their lives dramatically and rapidly within one or two generations—European pathogens forever altered the lives of the indigenous population of North America and thereby the fate of the continent. So-called "virgin-soil epidemics" spread throughout the Americas from Columbus's first arrival, resulting in mortality rates of perhaps 90 percent among some populations. Because Native populations had no contact with the "Old World" of Europe, Africa, and Asia during the long history of repeated outbreaks of diseases on those continents, Amerindians had no acquired immunity to these pathogens. Born into a world teeming with the microbes that caused epidemics, Europeans, Africans, and Asians—especially those born into crowded and filthy European cities—were exposed to them almost from birth, and repeatedly throughout their lives. If they survived childhood, most residents of the Old World cities were in effect immunized. They could still become sick, but they were likely to survive. Europeans who came to the "New World" were the survivors of centuries of disease.[3]

When they came to the Americas, Europeans brought their diseases with them. Native peoples with no immunities died in tragic and incomprehensible numbers. Epidemic outbreaks of such Old World diseases as smallpox, scarlet fever, cholera, measles, whooping cough, and influenza often reached Native populations long before direct contact with Europeans by spreading from Native to Native through trade networks. The events surrounding mass death by pathogen were, more than any other, the shared experience of nearly all Native peoples after the arrival of Europeans in the Americas. Extant evidence indicates that all cases of "first contact" spread of pathogens led to enormous population losses.[4]

Initial outbreaks of disease transferred by Spanish explorations of the mid-sixteenth century killed untold numbers of Native people in the Americas, first in the Caribbean, then in mainland Mexico. From there it is likely that disease spread both north and south, though it is difficult to determine the extent or precise path of this spread in the first decades of the sixteenth century. The Soto expedition through the Gulf South and the Coronado expedition through the Southwest, both around 1540, likely spread disease to the Native populations of those regions, though it is difficult to assess the outcomes with precision. Atlantic Coast populations experienced repeated waves of epidemic diseases spread by European explorers and fishermen from various empires by the early seventeenth century. By the eighteenth century, entire tribal nations had disappeared from what had become British America and surviving peoples faced enormous social and environmental challenges as thousands perished and thousands more fled west to escape both disease and European aggression by joining other groups and assimilating into them.[5]

In the West, records indicate that epidemic disease struck the Pueblos of New Mexico around 1640, though it is unlikely that this was the first European epidemic to spread through the area. Following the establishment of a permanent Spanish presence, the Pueblos suffered repeated epidemics: smallpox in 1719, measles in 1728, smallpox again in 1733 and 1737, unknown contagions in 1748 and 1770, and smallpox in 1781, 1799, 1816, 1840, 1853, and 1898. These likely resulted from commerce and colonization from the heart of the Spanish colonial population in Mexico, as well as from periodic infection from contact with the growing and frequently supplemented Spanish livestock herds. Old World diseases were devastating to Native populations both initially and cumulatively. During the period from 1530 to 1630, virgin-soil death rates likely exceeded 90 percent in many locations in New Mexico. Following Spanish colonization and the establishment of missions, documented disease episodes became recurrent. Even when mortality rates for a specific outbreak were lower, recurrence of the same disease or another disease in an already weakened community resulted in cumulative death rates that were often cataclysmic. This was especially the case when hunger resulting from a lack of healthy food producers, European-generated violence, land loss, and resource depletion accompanied repeated infection.[6]

The Spanish passed disease to the Amerindian peoples of California in the eighteenth and nineteenth centuries via the mission system. As they established missions along the coast of California from San Diego to San Francisco, priests and the forces that accompanied them used coercive violence to gather Native people in the missions. There the priests put Natives to work for the missions while attempting to convert them to Christianity and to remake them to fit Spanish cultural expectations. Spanish treatment of mission Indians no doubt increased the magnitude of disease mortality by weakening the Native peoples through forced labor, malnutrition, and overt violence. Native populations in California experienced tragic losses, particularly those who lived in closest proximity to the missions. As the majority of villagers close to the missions died, Spanish priests pushed farther inland in search of potential converts, spreading disease to inland populations in the process. For some, population decline topped 95 percent in the mission period.[7]

Disease led to calamitous population losses on the Northern Plains as well, beginning in the seventeenth century. Some diseases spread north from the Spanish settlements, as was the case with a smallpox epidemic that began in the valley of Mexico in 1778 and spread through New Mexico and all the way to the Northern Plains. Other outbreaks likely spread from French fur trappers and traders in the Great Lakes and Mississippi valley regions. Epidemics struck Native populations on the Northern Plains in 1781 and 1837 with approximately 70 percent mortality. Northern Plains peoples' winter counts—historical records in the form of pictographs on hides or paper—attest to approximately thirty-six significant disease epidemics among Lakotas, Nakotas, Blackfoots, Mandans, and Kiowas between 1714 and 1919, or, on average, a serious outbreak at least once every five to ten years. This rate of occurrence suggests that the societies non-Natives encountered on the Northern Plains were significantly altered by disease prior to those encounters and throughout the early history of interaction between Natives and Europeans and Americans. The fighting strength a people could muster against foes, the geographic locations of peoples, and even the very survival of some groups depended on the recurrence and severity of disease outbreaks.[8]

The same was true for the Pacific Northwest, where disease was also a significant factor in Native history following the arrival of Europeans. By the late eighteenth century, smallpox epidemics had reached Pacific Northwest populations, bringing catastrophic population decline. When the Lewis and Clark Expedition arrived in the first years of the nineteenth century, they came across Natives scarred by smallpox and societies still trying to recover. The oral traditions of numerous peoples of the Northwest indicate that epidemic diseases struck their peoples repeatedly. These waves of sickness took advantage of Native lifestyles, which included travel by canoe, densely settled populations, and frequent visits to members of family and clan networks, all of which served as effective catalysts of contagion. When Americans took over political control of the Oregon Country from the British and began colonization in earnest in the 1840s, they found Native populations that had been severely affected by disease and the social turmoil that followed in its wake.[9]

Though scholars of various stripes continue to debate the pre-epidemic Native population size of North America, the mortality rates inflicted by various diseases, the frequency of epidemics, and the means by which various pathogens spread, a consensus has arisen among most scholars of American Indian history that epidemic diseases of Old World origin dramatically altered the human ecology of North America in the early contact period, rapidly reconfiguring many aspects of life and presenting to European explorers and colonizers a reality that was far different from that of centuries prior to Columbus's voyage. The history of interactions between Amerindian peoples and Europeans and Americans that followed must always be considered within the context of epidemic aftermath. Many Native peoples suffered population losses that required generations—even centuries—before recovery was possible. An accident of history played a leading role in the success of European imperialism. The fact that Amerindians had not been exposed to virulent Old World pathogens was the Europeans' most deadly apparatus of conquest. Without the initial massive depopulation that followed contact, European conquest of the Americas likely would have turned out very differently. Perhaps it would not have occurred at all.

Population losses not only made European colonization possible, they also shifted the balance of Native power from certain groups to others. Due to their ability to amass surplus resources, sedentary agricultural peoples who lived in concentrated permanent settlements on fertile lands and traded excess foodstuffs along well-traveled waterways tended to consist of significant numbers and therefore to possess substantial strength relative to nomadic or semi-nomadic peoples in a given area prior to the microbial onslaught Europeans unleashed. When epidemics such as smallpox hit, however, the settled agriculturalist traders tended to contract disease earlier and to suffer more ruinous mortality rates than the nomadic peoples who lived in more widely dispersed configurations that often fled from disease when it became apparent. In the process, over the three centuries following the beginning of contact between Europeans and Native peoples of the West, many agricultural peoples experienced ruinous declines in numbers and power while neighboring nomadic or highly mobile groups' relative strength rose. Disease, therefore, was a vector of expansion not only for European colonizers of North America but also for some Native groups who benefited from their neighbors' disastrous losses.[10]

PUEBLOS

Spanish influences were not the first to arrive at the Pueblos from the south. Long before the viceroy of New Spain sent Coronado to the region in 1540, cultural elements that originated far to the south among the Native peoples of what is now Mexico had contributed substantially to the lifeways of the Amerindians of the Southwest. Maize cultivation and cultural forms that grew up around corn agriculture arrived from the south centuries before, shaping the societies that would first meet Spaniards in the Southwest. Innovation, incorporation of new ideas, and cultural adaptation were elements of the Pueblos' history long before the arrival of the Spanish.[11]

The Spanish nevertheless altered the Pueblos' world in dramatic ways. Spanish colonialism in the seventeenth century wreaked havoc on Pueblo society, as disease, oppression, and violence brought devastating consequences. As Spaniards pushed north from New Spain in what is now Mexico, they instituted imperial systems of control over the various Pueblo peoples. Initially, some Native leaders believed that the best way to deal with the Spanish was to form trade alliances with the newcomers and thereby gain access to the exotic weapons and other goods they carried. Native leaders with this mindset hoped that such trade would make it possible for their people to dominate the region in which they lived, or more benignly, to protect it from invasion. The Spaniards, however, did not enter the Southwest to initiate trade but rather to locate and plunder fabled cities filled with silver and gold.[12]

Marching into Pueblo lands with an army of hundreds that included a thousand Native Tlaxcalan allies, over 200 mounted soldiers and dozens of Spanish infantry accompanied by ferocious war dogs, Coronado and his army overcame Pueblo resistance by force in Pueblos such as the Zuni town of Hawikuh and by intimidation in others. They helped themselves to food stores and other supplies, raped Pueblo women, and treated others harshly as they continued to search for towns filled with precious metals, enraging Pueblos and leaving them vulnerable to winter. Pushed by ill treatment, Tiwa people along the Rio Grande fought back. Those who had welcomed the newcomers as potential trade partners joined with those who never had and attacked the Spaniards and their horse herds, waging a war of resistance for three months. The Spaniards responded with total war in which they burned captives at the stake and often killed Indians indiscriminately.[13]

When it became clear that the Spanish forces would prevail, many Pueblos fled from their villages as the Spanish approached. Some took up defensive positions in the mountains, while others gathered in a couple of the larger fortified Pueblos where they were eventually overrun and either killed or enslaved. Within two years, following a failed venture to find one of the mythical seven cities of Cíbola among the Wichitas across the vast plains in Kansas, an injured Coronado and his men returned to Mexico no richer than they had been before their rampage through the lands of the Pueblos.[14]

Over the next six decades, the Pueblos faced only small parties of Spaniards coming and going. During this time, Pueblo peoples returned to their towns, rebuilt their structures, restocked their food stores, reestablished their trade networks, and got on with their lives. The close of this period began in 1598, when the Don Juan de Oñate and a large expeditionary force of several hundred arrived from New Spain with the crown mission of subjugating the Pueblos and establishing permanent Spanish colonies among them. As Oñate's forces arrived at each town, Pueblo delegations proceeded carefully but pursued amicable relations. Oñate remained at each long enough to commandeer food supplies, claim possession of the Pueblo for the Spanish crown, and persuade Pueblo leaders to submit to the rule of Spain in hurried and untranslated ceremonies they surely did not understand. Oñate's men stole from, tortured, raped, and murdered Pueblo people along the way. At one town, Oñate ordered the Indians to vacate the premises, which he renamed San Juan and established as the Spanish capital of New Mexico. From there, his friars spread out to the various Pueblos to begin their conversion efforts.[15]

At the Native town of Acoma, the peace broke down. The inhabitants of Acoma had disagreed with one another regarding how to react to the Spanish reentry into New Mexico when Oñate arrived, but those who wanted to avoid conflict won majority support and the official delegation at Acoma greeted Oñate's expedition calmly. A month later, those who favored repelling the Spanish gained sway when Oñate's nephew arrived with a group of reinforcements and demanded food and supplies. When sixteen Spaniards scaled the walls of Acoma and entered without permission, Acoma warriors killed twelve. The remaining four escaped to the force below and upon telling what had happened, the entire expedition fled to San Juan to report the occurrence to Oñate. Within a month, Oñate sent a punitive expedition to Acoma. Dozens of Spaniards and a cannon proved a formidable offensive force, and over the course of a three-day battle, 800 residents of Acoma died. The Spaniards rounded up close to 600 survivors and took them to Oñate. After a show trial, the women and children became slaves and Oñate had the men punished by public mutilation: Spaniards amputated one foot from each man.[16]

Over the next several decades, Spaniards established the trappings of empire among the Pueblos: missions, political officers, law enforcement, and various systems of physical and psychological control. Some Pueblos openly resisted the imposition of Spanish hegemony, but many came to accept the Spanish presence while maintaining their own separate Native social structures. Most Pueblos had little choice but to deal with the Spanish presence as best they could, as Spanish regulation of their foodstuffs and trade goods had created a constant threat of attacks by Apache and Navajo raiders who could no longer obtain supplies through trade for Pueblo surpluses. Disease, including a smallpox epidemic in 1640, left the Pueblos' population numerically thinned and physically weakened, as did Spanish control of the region's food. In the face of Spanish military strength and the apparent spiritual power wielded by the bearded men who continued to survive waves of sickness that killed massive numbers of Natives, some Pueblos sought accommodation with the invaders; others fled their towns at the approach of the Spanish; still others fought Spanish hegemony from the beginning or rebelled shortly after the Spaniards attempted to assert political control. As the Spanish established a permanent presence in "New Mexico," they crushed rebellions with unflinching violence.[17]

The Spanish government awarded some Spanish colonists in New Mexico *encomiendas,* or royal grants that gave the grantees, or *encomenderos,* rights under Spanish law to force local Indians to contribute labor and goods in the form of food and cotton blankets to them. The Pueblos trapped in this system essentially became slaves on feudal estates. This system proved devastating to the Pueblo population, especially when Spaniards demanded higher tributes during harsh winters when the Pueblos could least afford to meet those demands. Not only did these demands deprive the Native people of the resources they needed to survive, they also prevented the Pueblos from carrying on trade with Apache and Navajo groups, leading to raids by these former trade partners.[18]

Prior to the arrival of the Spanish, the peoples of the Pueblos maintained extensive ties with semi-nomadic peoples through trade. Pueblos supplied buffalo hunters with agricultural produce, vessels, and other goods in exchange for buffalo meat, hides, and other products. This trade extended for thousands of miles, from the Southwest to the Northern Plains, from the Rocky Mountains to the far side of the Mississippi River, connecting far-flung peoples in an expansive network. The arrival of Spanish colonizers upset this trade in a number of ways. When the Spanish demanded that Pueblos provide trade goods to them, they effectively reduced the availability of those goods to the buffalo hunters of the Basin and Plains. When semi-nomadic peoples could not obtain the Pueblo goods on which they had come to depend, many turned to raiding to fill the void. Here too, the Spanish impact was profound. It was the Spanish introduction of horses to the Southwest that enabled mobile peoples to become effective raiders. Prior to the arrival of horses, Pueblo architecture and population strength had served to defend against such raiding effectively. Spanish diseases undermined Pueblo demographic strength and the possession of Spanish horses neutralized architectural defenses, making raiding much more attractive and profitable to groups such as Apaches and Navajos.[19]

As the Spaniards' systems of control became increasingly burdensome and Apache and Navajo raids continued largely unabated by relatively ineffective Spanish forces, Pueblos' discontent with the Spanish presence grew. Pueblos continued to die in large numbers as a result of European pathogens, violence, and scarcity, undermining Spanish priests' claims to spiritual power and suggesting to some Pueblos that a cleansing return to tradition was necessary to regain balance. To this end, some Pueblos began ignoring Spanish strictures against their traditions and defiantly practicing Native rites in public. In response, Spanish missionaries clamped down on expressions of Pueblo tradition, banning religious ceremonies and driving non-Catholic Pueblo spirituality underground in efforts to Hispanicize the Pueblos.[20]

These combined pressures, along with drought in the 1660s and 1670s, pushed the Pueblos to revolt in August 1680. At that point, a majority of the Pueblo people across two dozen communities, aided by a common understanding of the Spanish language and allied Apache bands, rose in unison to destroy missions and take control of horses, guns, and other supplies. The rebels drove Spanish officials and settlers from New Mexico entirely, forcing them hundreds of miles south to El Paso. Pueblos killed perhaps a fifth of the colonizers and half of the priests in New Mexico, inflicting a severe blow against Spanish imperialism.[21]

After the Pueblo Revolt of 1680, a series of Native revolts against the Spanish throughout New Mexico and other parts of the Spanish Empire in what is now Mexico continued over the next two decades. The most significant of these in New Mexico occurred in 1696, when Pueblos killed five priests and twenty-one other colonizers. By the time these revolts ended around 1698, Native peoples had demonstrated to the Spanish that they could not maintain political control of the northern borderland of their empire solely through force. They would need to pursue alliances and accommodation with at least some local Native peoples. With the power of mounted Apaches and other raiding groups rising as they continued to increase their own herds and obtain weapons and goods through trade, the Spaniards needed Pueblo warriors to supplement Spanish forces to defend New Mexico. To maintain order and reduce the potential for rebellion, Spanish priests moderated the vigilance with which they monitored Native religious activities, realizing that they could not obviate Pueblo religious beliefs and practices entirely. Over time, some Pueblos in turn acknowledged some Christian and Spanish cultural elements by incorporating them into Pueblo culture. For some, these additions were rather superficial accommodations aimed at easing Spanish fears. Other Pueblos, however, genuinely drew from these cultural constituents and combined them with their own traditions to create entirely new cultural manifestations.[22]

Over time, the interaction and combination of Native cultural traditions with those brought by the Spanish produced the unique culture of the American Southwest that continues to this day. It is important to note, however, that some Native communities changed little in response to Spanish cultural introductions, while others actively participated in shaping the broader culture of the Southwest. Native responses to the presence of Europeans were as varied as the peoples themselves. Peoples such as the Pueblos were not a passive audience that simply watched as historical events transpired all around them. Nor were they feeble victims of mighty Europeans, entirely unable to defend themselves against Spanish aggression and steel. The cultures, environment, geography, and social structures in which the various Pueblo peoples lived shaped their decisions. Yet they also innovated and adapted in efforts to respond to their present and prepare for their future.

HORSES

Though the Pueblos had earlier direct contact with Spanish horses than did Plains peoples, it seems horses offered less in the way of advantageous opportunities for change in their settled agricultural lives. On the Great Plains, however, horses allowed Native peoples to alter their lifestyles radically. It is difficult to overstate the impact of Spanish horses on the history of the American Plains, and thus the West. Horses allowed peoples who had long hunted buffalo on foot to become much more efficient in this pursuit. Perhaps more importantly, however, the availability of horses lured many new peoples to the Great Plains where they adopted an entirely new way of life, modified their cultures both conceptually and materially, and became some of the most powerful of the equestrian tribes.

Many peoples became horsemen long before they established direct contact with Europeans as horses filtered out from New Mexico north and east across the Plains. Native horsemanship began with the Pueblos, some of whom became the earliest "cowboys" in America when Spanish *encomenderos* employed their Native laborers as mounted herdsmen and caretakers of horses. From New Mexico, Native horses proliferated through trade and raids. As early as the late sixteenth century, Indians in the El Paso area attempted to acquire Spanish horses through raiding. By 1598, these Native peoples likely had access to some mounts. Pueblos fled from Spanish rule in large numbers beginning in the first decade of the seventeenth century, joining bands of Apaches and other mobile groups and taking Spanish horses with them—thus introducing the animals and their care to these groups. By 1606 the Spanish had provoked open hostility on the part of Navajo and Apache allies of Pueblos, resulting in repeated instances of raiding in which these groups obtained Spanish horse herds. Education in horsemanship may not have been necessary, as Apaches traded with the Pueblos throughout the early decades of the seventeenth century, camping entire bands near enough to trade locations to see Spanish equestrian methods firsthand. By the late seventeenth century, Navajos and Apaches to the northwest regularly raided New Mexican ranches, procuring significant numbers of horses.[23]

A period of relative peace ended when Spanish slave raids around 1630 and 1640 again provoked Plains Apache and Navajo attacks and raids, resulting in more horses among these groups. The Spanish also attacked the Utes during this period, instilling anti-Spanish sentiment and pushing Utes to spread that feeling, along with horses and horsemanship, to northern peoples. Again in the 1650s, Spanish slave raids led to Apache and Navajo responses that resulted in the capture of large numbers of horses. By the 1660s, Apaches and Navajos had so many horses that the Spanish no longer saw a need to prevent them from gaining more and began trading horses to them for slaves. From 1667 to 1680, vigorous Apache war against the Spaniards led to such a depletion of Spanish herds that more were required from Mexico. Large numbers of horses thus fell into Indian possession during this period.[24]

During the Pueblo Revolt of 1680, hundreds of horses made their way into Native hands, adding substantially to herds among the peoples who alternately traded with and raided the Pueblos, primarily Apaches, Navajos, Utes, and Comanches. Mounted Pueblo warriors also delayed the reassertion of Spanish dominance in the Pueblos. By the final decade of the seventeenth century, all or nearly all of the Plains groups of Texas had obtained horses, and by 1717 Caddos herded horses across the Plains to the Illinois country to trade regularly. Their obvious utility contributed to the rapid diffusion of horses from New Mexico over the next several decades. In 1719, Pawnees in Oklahoma probably constituted the "horse frontier." But just five years later, Pawnees, Kansas, and Comanches had a well-established trade in horses. Within a decade or two, horses were numerous and allowing Native peoples to transform their economies as far north as the Mandans of the Missouri River region and the Blackfoot lands of southern Canada. By the middle of the eighteenth century, the "horse frontier" had spread northeast all the way to Minnesota, where Lakota and Dakota bands were becoming expert horsemen.[25]

Horses and horsemanship were likely transmitted to the Plains by Apaches, Comanches, Utes, and Jumanos, who likely learned from the Pueblos who first had intimate contact with Spanish herds. Horses and knowledge of them passed through the trade networks to Pawnees and Osages across the Plains, and from those peoples to points beyond. They rapidly became a most prized commodity among peoples far and wide, commanding rich sums in trade. Native peoples added to and adapted the knowledge they gained along with the horses they obtained and made themselves into some of the greatest horsemen known to history.[26]

The spread of horses was a process predominantly directed by Native peoples rather than by the Spanish, who actively sought to prevent Amerindians from obtaining horses. Presented with a resource that offered obvious economic, military, and subsistence advantages, Native peoples pursued various strategies to obtain it. The extent to which peoples altered their societies in response to obtaining horses differed dramatically from group to group. Many settled agricultural peoples incorporated horses into their lives without necessarily altering their lifestyles in a fundamental way. Horses allowed agricultural groups such as the Mandans, Arikaras, and Caddos who lived near the buffalo Plains to increase their hunting efficiency and thus to obtain more meat, hides, and other products, without giving up their agricultural practices. Village-based tribes could not fully realize the potential that mounted hunting offered, however, because they could not follow the buffalo herds too far from home lest they leave their villages undefended.[27]

For other peoples, horses provided entrance into a radically altered lifestyle of long-distance seasonal rounds of buffalo hunting. Once mounted, groups such as the Lakotas and the Blackfoots migrated onto the Plains to adopt a lifestyle based almost entirely on the products of the hunt. Horses allowed buffalo hunters to travel great distances in less time and reduced the number of hunters needed to pursue a successful hunt. While still dangerous, hunting from the back of a carefully chosen horse trained to respond quickly to subtle gestures and a range of buffalo and rider behaviors was significantly less hazardous than hunting on foot. The increased efficiency offered by mounted buffalo hunting also made it possible for mobile bands of hunting peoples to live and travel in larger camps. Because they were far more effective draft animals than the dogs on which Native peoples had previously relied, horses also allowed Native families to construct larger tipis and to transport them, along with larger supplies of food and hides, more rapidly and over greater distances.[28]

TRADE

As the Spanish colonized the Southwest with imperial systems of control and permanent settlements, the French took a different approach. By the late seventeenth century, competition with the English and the desire to save souls led French traders and missionaries to establish contacts of some sort with Native peoples across a large portion of North America. Shortly after establishing Québec in 1608, Samuel de Champlain formed alliances with the Algonquin and Montagnais peoples in the area and thus established the French as an enemy of their enemies, the Iroquois. Champlain also befriended the Hurons and in 1609 undertook joint campaigns with them against the Iroquois. Upon securing firearms and ammunition from the Dutch years later, Iroquois warriors began a string of attacks against those peoples who aligned with the French. Though the muskets of the day were inherently limited, the Iroquois used them to great effect and successfully routed enemy forces repeatedly. Once the Iroquois engaged in efforts to control the burgeoning trade in beaver pelts from the Northeast to the Great Lakes in the middle of the seventeenth century, the Native peoples against whom they fought increasingly turned to the French for assistance. The French in turn armed the Montagnais and Abenakis of eastern Canada by the late 1640s, as well as the Ojibwes and Potawatomies in the following decades. Those peoples then spread French guns farther west through trade.[29]

Following the Dutch cession of New Netherlands to England in 1665 and the subsequent English establishment of the Hudson's Bay Company in 1670, France sent licensed fur traders into the western hinterlands of the Great Lakes, the Mississippi River, and beyond as a countermeasure. By 1680, some 800 French traders were in the field obtaining furs from Amerindians of various affiliations, often in exchange for firearms. Traveling with Native guides in Native canoes along waterways that served as the Amerindian highways of the day, Frenchmen established trade relationships with peoples up and down the Mississippi River on the prairies at the eastern edge of the Great Plains, and especially in the Great Lakes region. Not to be outmaneuvered, the English armed increasing numbers of Native groups from the eastern seaboard to the Mississippi River to secure alliances against the French. Facing an English population in North America that was growing at a much faster rate than their own, as well as highly competitive English trade prices, the French increasingly turned to Native allies for aid. This meant extending and intensifying their trade relationships and gift giving, which included offering firearms. Though it did not spark radical change on anything near the scale that the diffusion of horses did, this arms race would prove important in the history of the Great Plains because French-armed Ojibwes and Illinois were pressuring the Lakotas and Nakotas, and Dakotas to their west and south by the late seventeenth century.[30]

Where the Spanish sought to subject Native peoples to their colonial regime while maintaining racial and caste separation from Spaniards, French traders tended to try to insert themselves into the Native world. The apparent French respect for Native protocols, their acceptance of the importance of annual reciprocity gifts in the form of firearms, gunpowder, and ammunition, among other goods, and their convention of providing gunsmith services to Native allies all enhanced French efforts to court Amerindian trade partners along the Mississippi and to the west. The French no doubt would have preferred to incorporate Amerindian peoples into a world controlled by Frenchmen for the benefit of France, but several factors hindered their ability to pursue the kind of imperialism Spain was attempting to implement in the Southwest. The French faced direct competition from the nearby English, and Native political realities, as well as the relatively small French population in North America, required the French to approach their Native contacts as something more akin to partners and neighbors than subjects. The French did not possess the coercive power to impose their will on the Native peoples of the region. Of course, by befriending some peoples, they became de-facto enemies of others. The French were no less brutal than the Spaniards or English in fighting their foes; it was their treatment of Native allies that differed.[31]

French traders did not just befriend Native allies; many married Native women and often became incorporated into Amerindian societies. By doing so, they solidified profitable trade relationships for themselves, but they also submitted themselves to Native expectations and social constraints in many cases. This differed substantially from Spanish relationships with Native women, which nearly always existed within a Spanish-dominated context. French trade also differed from the methods of their imperial rivals. Because they offered trade goods—including firearms—without trying to subjugate Native peoples as the Spanish or pressure them for their lands as the English, the French were relatively adept at gaining Amerindian allies. In the Great Lakes region, French willingness to submit to the Native social order contributed to a conceptual and physical "middle ground," a space in between the two cultures where Amerindians and Frenchmen together created new cultural forms, despite their lack of understanding of the other's culture. Intermarriage also created a new people, the métis, who developed a culture that was and remains something different from and more than a mere combination of French and Amerindian.[32]

Competition with the English motivated many French actions; competition with the English and with indigenous enemies elicited Native responses as well. For many Amerindian peoples, the French were extremely attractive potential allies not only because they were less interested in colonizing Native lands and more willing to enter into "middle grounds" with them, but also because they offered military support and firearms that could be used against their enemies. Trade relationships between Native peoples and the French became necessary for both sides to protect their own interests in various locations between the Appalachian Mountains and the Great Plains, from the Great Lakes to the Gulf of Mexico. By the middle of the eighteenth century, many Native peoples depended on French weapons for defense, and the survival of New France depended on the cooperation of many Native peoples.[33]

French firearms found their way onto the Northern Plains by the late seventeenth century, via Ojibwe and Cree traders. Ojibwes and Crees also pressured the Lakotas, Nakotas, and Dakotas—peoples who did not yet have dependable sources of firearms—to migrate from the northern Great Lakes toward the Plains. The extent to which that migration was a product of the push of enemies versus the pull of opportunities offered by the horse culture is difficult to assess, but pressure from the east surely played a part. Both English and French traders commonly offered guns, ammunition, and powder to one Native group to persuade them to attack another, for reasons ranging from geopolitics to personal revenge. Lakotas, Nakotas, and Dakotas were more than targeted in this manner by such appeals to their enemies.[34]

In their constant efforts to expand their share of the North American market for trade goods and to shore up their military position, French traders pushed into the West to trade with Osages and Pawnees on the prairies and eastern Plains by the second decade of the eighteenth century. Gifts of weapons and ammunition won many allies for France in the area. Further west, Hasinais and Comanches of the Southern Plains also obtained guns from the French either directly or indirectly through Pawnee and other Native middlemen, which they employed against Apache enemies. Though still a limited commodity in the West in the late seventeenth and early eighteenth century, firearms offered significant advantages against enemies who did not possess them, especially in the context of Native confrontations that often involved small numbers of combatants. During this time, peoples up and down the Plains began to acquire firearms, and the French aspired to curb the influence of the Hudson's Bay Company by cornering as much of the trade as possible.[35]

The appearance of firearms on the Great Plains escalated existing tensions when Native groups sought to block their enemies from obtaining weapons and ammunition, especially by the late eighteenth century. Sometimes they attempted to do this by absorbing the traders' goods before they could reach their enemies. On other occasions, some groups used force to preclude commerce between arms suppliers and their foes. Gros Ventres, for example, often employed force against English traders to prevent them from trading guns to their Cree and Assiniboine enemies. When this failed, they demolished multiple English forts to punish them after the fact. This was serious business, since a few guns could sometimes change the balance of power in a given location. Armed Gros Ventres and Blackfoots used their firearms advantage to enlarge their Northern Plains hunting grounds by pushing some of their enemies out. When those enemies acquired guns from British and then American traders, Blackfoot warriors attacked the traders in retribution and to discourage further trades of the sort. Militarily, they had good reason. Once the Salish and Kootenai peoples of the Northwest obtained firearms, they were more formidable opponents in battle. When armed Salishes attacked a Blackfoot village in 1810, Blackfoot villagers blamed the traders and sought peace with the Salishes. In the period that immediately followed, Salishes were able to hunt buffalo on Blackfoot land and repel any Blackfoot attacks against them.[36]

The firearms trade on the Northern Plains allowed village peoples such as the Arikaras, Mandans, and Hidatsas to defend themselves while increasing their draw as trade centers. From these villages on the upper Missouri River, guns finally reached the peoples of the Columbia Plateau of present Washington, Oregon, and western Idaho such as the Walla Wallas, Paluses, and Yakamas. The upper Missouri villages also spread firearms to Nez Perces, Bannocks, Salishes, and Kutenais through the Crows, and to Lakotas, Utes, and other peoples to the south via the Araphos and Cheyennes.[37]

CHEYENNES

In a "masterpiece of adaptive fluency," peoples such as the Cheyennes, Comanches, Kiowas, and Lakotas entered the Plains to pursue a route to economic and military gain by capitalizing on the seemingly endless buffalo herds. The Cheyennes provide an excellent example of a people who actively shaped their own history through innovation and adaptation on an impressive scale. They both migrated a great distance from their original homeland and utilized horses to alter their way of life in the face of pressures that made their older lifestyle increasing difficult to maintain.[38]

Cheyenne oral history acknowledges that the Cheyenne people once lived to the east of the Great Plains as a western group of Algonquins who migrated in a southwesterly direction from a location north of the Great Lakes over an uncertain length of time. Crees armed with guns procured from Hudson's Bay Company traders around 1660 may have driven the Cheyennes southwest. Around 1680, Cheyenne bands lived near Dakotas in the forested woodlands of the upper Mississippi River. By the 1760s, Ojibwes armed with French guns and intentions to control the fur trade in the region drove both the Dakotas and the Cheyennes south and west to the prairies of the Minnesota River area in what is now southern Minnesota and to the James River in southeastern North Dakota. Here

Cheyenne bands lived in earthen lodges similar to those of neighboring Mandans and Hidatsas and pursued agriculture, seasonal buffalo hunts, and trade. But by the 1780s further Ojibwe raids in service of monopolizing the western Great Lakes and Eastern Plains fur trade coupled with a series of epidemics led the Cheyennes to move farther west onto the Great Plains. In 1800, Cheyennes occupied the High Plains near the Black Hills of South Dakota. By 1830, the Cheyenne Nation split into the Northern Cheyennes, located west of the Black Hills in present Wyoming, and the Southern Cheyennes in eastern Colorado and western Kansas.[39]

The Cheyenne transition from settled agriculturalists to dedicated buffalo hunters was not entirely a reaction to pressures from outsiders. Most Eastern Native peoples faced external pressures between the arrival of Europeans and the nineteenth century, but few became nomadic buffalo hunters. Cheyennes chose this route both as a defense against those external pressures and as a proactive strategy for bringing about a better future for themselves and their descendants. The opportunities provided by the buffalo herds and Plains trade motivated this move. Because they had acquired horses by this time, Cheyenne bands would be able to exploit these opportunities to great effect. It was not necessary for the Cheyennes to create their new lifeway entirely from scratch, since they had models to follow in the Crows and Kiowas, among others. The specific Cheyenne version of this life, however, was uniquely their own, embodied in the story of the prophet Sweet Medicine receiving the sacred arrows and establishing the Cheyenne social organization, ceremonies, and military societies of their horse-based life. Sweet Medicine's encounter encapsulated the historical experience Cheyennes were shaping: they were transforming into a new people who would need new tools and new knowledge.[40]

The transition was not a thing to be undertaken lightly, but the rewards could be enormous. Cheyennes could overcome previous limitations on hunting because hunters would no longer be tied down by the necessity to defend permanent villages. Further, if the entire band traveled the Plains together, both hunting and the processing of carcasses could be done with greater efficiency. All of the hunting-age males could hunt together, horses could transport large loads of buffalo remains, and all of the band's women were nearby to quickly harvest, clean, tan, and preserve the various materials the buffalo offered. And the buffalo likely seemed an inexhaustible resource, with herds even covering hundreds of square miles during the mating season. Mobility also multiplied opportunities for Cheyenne hunters to take other game animals such as deer, antelopes, elk, and bears.[41]

Being more mobile made it easier for the Cheyennes to avoid devastating raids of the sort the Ojibwes had inflicted on them in previous decades. Once their herds had grown large enough to afford every warrior a mount, the horse lifestyle also made Cheyennes a more formidable force when they were the aggressors. Because they could procure sufficient buffalo meat throughout the year, Cheyenne bands could support larger populations and thus greater numbers of warriors. It became much easier to amass surplus than in past centuries, and Cheyennes and other successful equestrian peoples faced a less precarious subsistence environment than many agricultural peoples pushed by disease, encroachment, raiding, or drought to the limits of their capacity to produce food.

The move to the Plains offered more than increased food supplies and military prowess; it also offered economic opportunities. The Great Plains region was a crossroads for long-distance trade, as it had been well before the arrival of Europeans. By the eighteenth century, however, goods of European origin were some of the most valuable commodities changing hands. Native agriculturalists and semi-nomadic peoples living along the rivers of the Eastern Plains offered vegetable foods and other traditional trade items along with manufactured goods—including guns and other metal implements—obtained from the French and the British. Plains peoples from the south and west traded buffalo products and Spanish horses and goods. Cheyennes consciously positioned themselves in the middle of the primary trade routes and established themselves as middlemen in the movement of both French and British guns from east to west and Spanish horses from west to east. In the process, they amassed substantial herds and collections of arms of their own, solidifying their position as one of the major economic and military powers of the Great Plains.

After driving the Kiowas south and then being driven south themselves by the Lakotas, the Cheyennes became the dominant power of the Central Plains for decades. Over time the bulk of Cheyenne trade shifted from interactions with other Native groups to increased contact with American traders as Cheyennes adapted to fluid circumstances. The very adaptability to trade that made the Cheyennes powerful led to the split between the Northern and Southern Cheyennes in the 1830s, when the nation divided to take advantage of trade at two trading posts separated by great distance.[42]

Cheyenne dominance of the Central Plains did not go uncontested, however, as other mounted peoples who had pursued similar courses of hunting and trade in the Southern Plains fought the Cheyennes and allied Arapahos to maintain their own dominance. An alliance of Kiowas, Comanches, and Plains Apaches battled Cheyennes repeatedly over decades from the early nineteenth century until the Great Peace of 1840, when Cheyennes, Arapahos, Lakotas, Comanches, Kiowas, and Plains Apaches formed an international friendship. This did not mean that Cheyennes' fighting days were over, though. Peoples from the Eastern Plains—Omahas, Osages, Pawnees, and Potawatomies—remained bitter enemies and provided ample opportunities for young warriors to prove their mettle.[43]

Violent encounters on the Great Plains were more frequent and larger in scope than most Native battles between forest peoples in the eastern woodlands or the Pacific Northwest. By the nineteenth century, Cheyennes and other Plains peoples had become skilled horsemen and had developed deadly cavalry tactics. Plains warriors carried specialized short but powerful bows that allowed them to fire arrows accurately from horseback, and many were crack shots with firearms as well. The rapidity of attack and withdrawal made raiding attractive, bringing violence to peoples who had something worth the trouble of taking. Raids and counter-raids by Plains groups against other Plains groups resulted in an enormous human toll. Nevertheless, this would pale in comparison to the devastation Americans were just beginning to bring to the Plains by the middle of the century in their attempt to fulfill their "manifest destiny." Horses helped the Plains peoples remain more independent than their village counterparts and made them more capable combatants against their enemies, including American settlers and the US military, than were smaller sedentary groups, especially after firearms became available on the Plains. Both of these factors would contribute to American willingness to revert to violence against Plains peoples, however.

Horses were not a panacea. In addition to increased violence and fracture over trade, horses brought limitations. Although they had initially allowed populations to grow larger, past a point they restricted population. If a group possessed ten horses per person—not an uncommon situation among Cheyennes, Lakotas, and other Plains peoples—a band of fifty people would have 500 horses in need of pasturage and care. Providing for enormous numbers of horses, gathering and herding them from one location to another, finding locations with ample grass and water, and protecting them from theft by individuals or raiding parties were all drains on time and energy. If bands grew too large, the horses themselves would undermine the very efficiency they allowed. The largest gatherings, therefore, were only possible in summers adequately watered by rains, when grasses were plentiful. From late fall to early spring, equestrian peoples traveled and lived in smaller bands.[44]

Horses had helped the Cheyennes to enter into a new environment and allowed them to develop an extremely profitable envirocultural matrix, but this extraordinary resource also pulled them into an array of consequences that were difficult to predict. Before the Americans brought massive military violence against Plains peoples, they brought violence of other sorts. As hundreds of thousands of Americans crossed the Plains in the mid-nineteenth century on their way to the West Coast with dreams of bountiful land or instant mineral wealth, waves of disease struck the Plains peoples with whom they traded along the journey.[45]

Another biological invader also ravaged the Native West: millions of livestock accompanying American interlopers ate tons of grass and trampled giant swaths, reducing the amount available for the buffalo and the herds of Cheyennes and other Plains peoples. American livestock also littered the landscape with their waste and often their carcasses, both of which drove buffalo from the area. Additionally, American interlopers heedlessly felled trees and harvested animals that Native peoples needed when winter came. Over years, the mass of Americans traveling along a few paths created areas that buffalo herds avoided. The peoples who lived near the trails thus were forced to travel farther to find the buffalo, sometimes in the hunting ranges of other groups, leading to increased confrontation that only escalated as the buffalo herds declined due in large part to American activities. When droughts struck in the 1840s, 1850s, and 1860s, they exacerbated these other factors and led to dramatic consequences for Plains peoples.[46]

COMPLEXITY

The arrival of Europeans in the West led to many changes for the Native peoples who lived there. Old World epidemics left many populations disoriented and faced with the difficult task of charting a course to recovery. This course was all the more difficult to plot because many of the people upon whom others would normally rely in trying times—the elders—were among the dead. Because young people were the other group most likely to have perished, it took a substantial amount of time before populations could recover, if they were able to at all. With so many elders gone in such a short time span, their specialized religious, cultural, and material knowledge, as well as their wisdom, often died with them. This left physically weakened populations culturally weakened as well, making the process of recovery even more daunting. For populations that suffered repeated waves of disease, it became difficult to hold their world together.

Disease and concomitant population losses led to dramatic reorientations of Native geographies. Some populations, finding traditional adversaries debilitated, moved to increase their own resources by absorbing those of their enemies. Others found once invincible enemies' strength significantly reduced. Still others moved into emptied lands where there were no previous occupants left alive to resist. The appalling consequences of epidemics also led to cooperation and ethnogenesis, as diminished population fragments joined together and formed new confederations based on cultural amalgamations and radical innovations. Some such peoples grew from scattered remnants of old populations to become some of the most powerful tribes of the eighteenth and nineteenth centuries.[47]

Often still recovering from the devastation and anomie wrought by disease, most Native peoples first met Europeans and later Americans in eras of great fluidity and relatively rapid social change. Native peoples could not know Europeans' intentions initially, nor could Europeans comprehend the full complexity or meaning behind Native cultures. At first contact, it was only natural for each to view the other as exotic and alien. Misunderstandings and competing interpretations of events often led to violence. It must be remembered, however, that Native violence, even when aggressive, was essentially defensive. Europeans were, after all, invaders. Amerindian peoples who invited groups of Europeans to stay in their territories, or permitted them to do so, expected certain behaviors in return. When Europeans failed to live up to those obligations, consequences ensued. Spanish imperialism overcame such events with brute force, but the French and English, having entered the New World at locations that were more densely populated by peoples who possessed a deeper reservoir of material resources, had little alternative but to engage in negotiations of various sorts. As long as the Native people of a given area were powerful enough to check European aggression, they could and did force Europeans to acknowledge Native envirocultural matrices and cultural parameters.[48]

When Frenchmen or Englishmen first arrived in a new area where they did not possess the power to dictate the terms of interaction with Amerindian peoples, they were forced to enter into Native social frameworks if they wished to get results. At places and times in which neither side possessed the power to coerce the other, a "middle ground" that was more than accommodation, created through the conceptual cooperation of both sides, could arise—though this did not always occur. Once Europeans held the upper hand, they typically dispensed with negotiations and made demands upon Native peoples, compelling them with coercive force.[49]

Regardless of who had the upper hand or the precise nature of the terms of interaction, in any given time and at any given place, Native peoples usually sought to better their lives through trade and personal relationships with those who were not their enemies. This often resulted in cultural adaptations motivated by rational judgments that specific changes would prove beneficial. Cultural adaptations could range from replacing stone knives with steel to transforming an entire people's lifestyle from settled agriculturalists to nomadic buffalo hunters. Obvious factors such as a people's population size and resource stability affected their ability to dictate the pace and extent of change they were willing to incorporate into their culture. Once horses and guns arrived, a people's ability to obtain them also influenced their options. Large populations with ready access to food, horses, firearms, and ammunition were best situated to resist Europeans while working out plans for the future. In the West, the equestrian peoples of the Plains, beyond the imperial reach of any European power, armed, powerful, and mobile, occupied such a position.

Most California Native peoples were not. Even large populations such as the Chumashes lived in many smaller autonomous villages spread over thousands of square miles—a situation that was not conducive to organized widescale military action. Though they had extensive trade networks, they were not tapped into the French and English trade in firearms as the Plains peoples were. Though some learned to ride Spanish horses and eventually raided to obtain them, it was at a later date than the Pueblos and Plains peoples, and their environment did not offer the same advantages to the possession of horses that the Plains did. Native peoples in the eastern portion of what is now the USA had been dealing with the European presence for over two centuries by the time California's Amerindian population experienced sustained contact with Europeans. Most Plains peoples who did not have such a long history with Europeans still had substantially more time to prepare for invasion than did the peoples of California, and until the nineteenth century their contacts were with Europeans who did not yet have the ability to overwhelm them. Though Europeans began exploring the coast of California as early as the 1540s, contacts between Europeans and California Native peoples remained scarce and cursory until the Spanish invasion began in 1769. By then, New Spain was well established and had the ability to exert coercive force.[50]

In California, the Native peoples' environmental circumstances, social organization, and late entry into European imperialism made the Spanish and American conquests possible. Chumashes rebelled against the missions in 1824, but the rebellions did not last long, at least not openly. Two short decades later, American gold miners arrived and California peoples suffered what has rightly been termed both genocide and holocaust. Armed groups of miners and then the government of the state of California were able to overwhelm Native peoples with sheer violence. Although on a less thorough scale, the government of the USA would overwhelm the holdouts on the Plains with shocking violence shortly after. The Native peoples of California were no less well adapted to their environment, no less proud of their heritage, no less willing to fight to defend it than were the Plains peoples depicted in countless films and other products of American popular culture. But they did not encounter the same opportunities. A combination of their environment, their adaptations to it, and historical sequences that originated far beyond their world shaped their experiences after sustained contact with Spaniards and Americans and prevented them from mounting the impressive military exploits or creating the lasting impression in the American imagination generated by the Plains peoples of the horse.[51]

In the same manner, countless other Native peoples of the West are often left out of the region's history, at least in popular accounts. The stories of peoples who lived in settled villages on the prairies where they grew corn and carried on a bustling trade with both the Plains and the East are as much a part of the history of the West as any other, for the goods and relationships flowing through their trade networks shaped countless other peoples. The histories of the fishing peoples of the Pacific Northwest, the diverse Pueblo peoples of the Southwest, and the multitudinous Native peoples of California are all vitally important constituents of the broader history of the West, as well as significant individually. After contact with Europeans and Americans, all of these peoples faced an altered world that they proceeded to further alter to varying degrees and in countless ways. Most sought to protect their lands and their lives in the manners that seemed most appropriate and most likely to succeed. All Native populations were, like any society, collections of disparate individuals who possessed varying talents, skills, knowledge, goals, and dreams. In each population, there were disagreements concerning the best course of action to pursue in the wake of epidemics, attacks, migrations, and the arrival of strange men with exotic goods.

All Native peoples were active historical agents who participated in the construction of the world in which they lived. They adapted and responded to events beyond their control, but they also shaped those events that were not. Even when they suffered unthinkable losses to disease, were enslaved or outnumbered, they acted. Amerindian peoples were every bit as creative, rational, and complex as the Europeans and Americans who colonized North America. As the example of the rise of horse cultures demonstrates, Native peoples were among the most adaptable and innovative in the world. European conquest was not a simple story of strong versus weak or sophisticated versus primitive. It was, rather, the outcome of an incredibly complicated tangle of historical processes, geographic oddities, and biological accidents.

The rise of the frequently stereotyped horse-culture peoples was largely the result of the order of introductions. Disease had already favored mobile or semi-sedentary peoples over settled agriculturalists. The introduction of horses favored the rise of horsemen more than it did the agriculturalists. The ascendancy was made possible by diseases and horses then nurtured by trade, competition between European powers, and Native peoples' decisions, creating the situation Americans entered into after the Louisiana Purchase. The different responses and actions of various peoples led to differing results: the Pueblos survived where they were; Cheyennes became powerful through migration and sweeping change; and some peoples disappeared entirely.

The history of the West following contact with Europeans was not a centuries-long Hollywood western. It was not merely a series of wars and raids, though both did occur. When Europeans arrived in North America, they entered into a world constructed by Native peoples. Through trade, violence, biological exchange, misunderstanding, marriage, friendship, hatred, competition, sex, hope, fear, and a host of other channels, Amerindians and Europeans then created a "New World" together.

NOTES

1. N. Scott Momaday, *House Made of Dawn* (New York: HarperCollins, 1999), p. 52.
2. Joseph M. Marshall III, *The Journey of Crazy Horse: A Lakota History* (New York: Penguin, 2004), p. 104.
3. See Alfred W. Crosby Jr., "Virgin Soil Epidemics as a Factor in the Aboriginal Depopulation in America," *William and Mary Quarterly,* 33 (2) (1976): 289–299; Russell Thornton, *American Indian Holocaust and Survival: A Population History since 1492* (Norman, Okla.: University of Oklahoma Press, 1987); Henry F. Dobyns, "Brief Perspective on a Scholarly Transformation: Widowing the 'Virgin' Land," *Ethnohistory,* 23 (2) (1976): 95–104; Henry F. Dobyns, *Their Number Became Thinned: Native American Population Dynamics in Eastern North America* (Knoxville, Tenn.: University of Tennessee Press, 1983).
4. Henry F. Dobyns, "Disease Transfer at Contact," *Annual Review of Anthropology,* 22 (1993): 273–291.
5. Ann F. Ramenofsky, *Vectors of Death: The Archaeology of European Contact* (Albuquerque, N. Mex.: University of New Mexico Press, 1988), pp. 42–71; Charles H. Fairbanks, "From Exploration to Settlement: Spanish Strategies for Colonization," in R. Reid Badger and Lawrence A. Clayton (eds.), *Alabama and the Borderlands from Prehistory to Statehood* (Tuscaloosa, Ala.: University of Alabama Press, 1985), pp. 128–139; pp. 134–136; Patricia Galloway, *Choctaw Genesis, 1500–1700* (Lincoln, Nebr.: University of Nebraska Press, 1995), pp. 131–143; Alfred W. Crosby (ed.), *Ecological Imperialism: The Biological Expansion of Europe, 900–1900* (Cambridge: Cambridge University Press, 1986), pp. 209–215. For difficulty of assessing population numbers from the sixteenth century, see Ann M. Palkovich, "Historic Population of the Eastern Pueblos: 1540–1910," *Journal of Anthropological Research,* 41 (4) (1985): 401–426.
6. Dobyns, "Disease Transfer at Contact," pp. 282–283, 285; Ann F. Ramenofsky, "The Problem of Introduced Infectious Diseases in New Mexico: A.D. 1540–1680," *Journal of Anthropological Research,* 52 (2) (1996): 161–184; Marc Simmons, "Spanish Attempts to Open a New Mexico-Sonora Road," *Arizona and the West,* 17 (1) (1975): 5–20.
7. Robert H. Jackson, "Grain Supply, Congregation, and Demographic Patterns in the Missions of Northwestern New Spain: Case Studies from Baja and Alta California," *Journal of the West,* 36 (1) (1997): 19–25; Robert H. Jackson, "Patterns of Demographic Change in the Missions of Southern Baja California," *Journal of California and Great Basin Anthropology,* 8 (2) (1986): 273–279; Robert H. Jackson, "Epidemic Disease and Population Decline in the Baja California Missions, 1697–1834," *Southern California Quarterly,* 63 (4) (1981): 308–346; Steven W. Hackel, "The Competing Legacies

of Junipero Serra: Pioneer, Saint, Villain," *Common-Place: The Interactive Journal of Early American Life*, 5 (2) (2004); Steven W. Hackel, *Children of Coyote, Missionaries of Saint Francis: Indian-Spanish Relations in Colonial California, 1769–1850* (Chapel Hill, NC: University of North Carolina Press, 2005); James A. Sandos, *Converting California: Indians and Franciscans in the Missions* (New Haven, Conn.: Yale University Press, 2004), pp. 8, 56, 111; David E. Stannard, "The Consequences of Contact: Toward an Interdisciplinary Theory of Native Responses to Biological and Cultural Invasion," in David Hurtz Thomas (ed.), *Columbian Consequences*, Vol. III: *The Spanish Borderlands in Pan-American Perspective* (Washington, DC: Smithsonian Institution, 1991), pp. 519–539.

8. Ramenofsky, "Problem of Introduced Infectious Diseases," pp. 169–170; Ramenofsky, *Vectors of Death;* Linea Sundstrom, "Smallpox Used Them Up: References to Epidemic Disease in Northern Plains Winter Counts, 1714–1920," *Ethnohistory*, 44 (2) (1997): 305–343; pp. 308–310, 324–325.
9. Robert Boyd, "Commentary on Early Contact-Era Smallpox in the Pacific Northwest," *Ethnohistory*, 43 (2) (1996): 307–316; Clifford E. Trafzer, *Death Stalks the Yakama: Epidemiological Transitions and Mortality on the Yakama Indian Reservation, 1888–1964* (East Lansing, Mich.: Michigan State University Press, 1997), pp. 2–3; D. W. Meinig, "American Wests: Preface to a Geographical Interpretation," *Annals of the Association of American Geographers*, 62 (2) (1972): 159–184; p. 165; Clifford E. Trafzer, "Smohalla, Washani, and Religion as a Factor in Northwestern Indian History," *American Indian Quarterly*, 9 (3) (1985): 309–324; pp. 311–312.
10. Ernest L. Schusky, "Evolution of Indian Leadership on the Great Plains," *American Indian Quarterly*, 10 (1) (1986): 72.
11. Carroll L. Riley, *Rio del Norte: People of the Upper Rio Grande from Earliest Times to the Pueblo Revolt* (Salt Lake City, UT: University of Utah Press, 1995), pp. 57–132; David Roberts, *The Pueblo Revolt: The Secret Rebellion that Drove the Spaniards Out of the Southwest* (New York: Simon & Schuster, 2004), pp. 30–31.
12. David J. Weber, *The Spanish Frontier in North America* (New Haven, Conn.: Yale University Press, 1992), pp. 14–29; Richard White, *"It's Your Misfortune and None of My Own": A New History of the American West* (Norman, Okla.: University of Oklahoma Press, 1993), p. 6.
13. John L. Kessell, *Spain in the Southwest: A Narrative History of Colonial New Mexico, Arizona, Texas, and California* (Norman, Okla.: University of Oklahoma Press, 2002), pp. 24–41; Colin G. Calloway, *One Vast Winter Count: The Native American West before Lewis and Clark* (Lincoln, Nebr.: University of Nebraska Press, 2003), pp. 134–136, 138–139; Weber, *The Spanish Frontier in North America*, p. 49.
14. White, *"It's Your Misfortune,"* p. 6; Calloway, *One Vast Winter Count*, pp. 140–142; Alvin M. Josephy Jr., *500 Nations: An Illustrated History of North American Indians* (New York: Alfred A. Knopf, 1994), pp. 160–164.
15. Weber, *The Spanish Frontier in North America*, pp. 77–78, 85; White, *"It's Your Misfortune,"* pp. 6–8; Josephy, *500 Nations*, pp. 165–167.
16. Weber, *The Spanish Frontier in North America*, pp. 85–86; White, *"It's Your Misfortune,"* p. 8; Josephy, *500 Nations*, pp. 167–172.
17. Robert H. Jackson, *From Savages to Subjects: Missions in the History of the American Southwest* (Armonk, NY: M. E. Sharpe, 2000), pp. 3–5; George P. Hammond and Agapito Rey (eds.), *Narratives of the Coronado Expedition, 1540–1542* (Albuquerque, N. Mex.: University of New Mexico Press, 1940), pp. 223–227, 333–334, 355–358; Calloway, *One Vast Winter Count*, pp. 147–151; Weber, *The Spanish Frontier in North America*, pp. 85–86, 99, 119; Elinore M. Barrett, "The Geography of the Rio Grande Pueblos in the Seventeenth Century," *Ethnohistory*, 49 (1) (2002): 123–169; pp. 123–124; George P. Hammond and Agapito Rey (eds.), *Don Juan de Onate: Colonizer of New Mexico, 1595–1628*, 2 vols. (Albuquerque, N. Mex.: University of New Mexico Press, 1953), Vol. I: pp. 337–362.
18. H. Allen Anderson, "The Encomienda in New Mexico, 1598–1680," *New Mexico Historical Review*, 60 (4) (1985): 353–377; White, *"It's Your Misfortune,"* pp. 10–12.
19. White, *"It's Your Misfortune,"* pp. 10–11; Calloway, *One Vast Winter Count*, pp. 167–170, 279–281.
20. Edward H. Spicer, *Cycles of Conquest: The Impact of Spain, Mexico, and the United States on the Indians of the Southwest, 1533–1960* (Tucson, Ariz.: University of Arizona Press, 1962), pp. 157, 303–304,

388–392; Carroll L. Riley, *The Kachina and the Cross: Indians and Spaniards in the Early Southwest* (Salt Lake City, UT: University of Utah Press, 1995), p. 93; Anderson, "The Encomienda in New Mexico"; Andrew L. Knaut, *The Pueblo Revolt of 1680: Conquest and Resistance in Seventeenth-Century New Mexico* (Norman, Okla.: University of Oklahoma Press, 1995), p. 54.
21. Alan Taylor, *American Colonies: The Settling of North America* (New York: Penguin Books, 2002), p. 89.
22. Cynthia Radding, *Wandering Peoples: Colonialism, Ethnic Spaces, and Ecological Frontiers in Northwestern New Mexico, 1700–1850* (Durham, NC: Duke University Press, 1997), pp. 9, 32–40; Roberto Mario Salmon, *Indian Revolts in Northern New Spain: A Synthesis of Resistance (1680–1786)* (Lanham, Md.: University Press of America, 1991), John L. Kessel, *Kiva, Cross, and Crown: The Pecos Indians and New Mexico, 1540–1840* (Albuquerque, N. Mex.: University of New Mexico Press, 1987), p. 293; Jack D. Forbes, *Apache, Navaho, and Spaniard* (Norman, Okla.: University of Oklahoma Press, 1960), p. 279; Calloway, *One Vast Winter Count,* pp. 174–196.
23. Peter Iverson, "When Indians Became Cowboys," *Montana: The Magazine of Western History,* 45 (1) (1995): 16–31; pp. 17–18; Jack D. Forbes, "The Appearance of the Mounted Indian in Northern Mexico and the Southwest, to 1680," *Southwestern Journal of Anthropology,* 15 (2) (1959): 189–212; pp. 195, 197–200; Sandra L. Myers, "The Ranching Frontier: Spanish Institutional Backgrounds of the Plains Cattle Industry," in David J. Weber (ed.), *New Spain's Northern Frontier: Essays on Spain in the American West* (Dallas, Tex.: Southern Methodist University Press, 1988), pp. 79–93; D. E. Worcester, "The Spread of Spanish Horses among the Plains Indian Tribes," *Pacific Historical Review,* 14 (4) (1945): 409–417; Calloway, *One Vast Winter Count,* pp. 268–269.
24. Forbes, "The Appearance of the Mounted Indian," pp. 200–201; Ned Blackhawk, "The Displacement of Violence: Ute Diplomacy and the Making of New Mexico's Eighteenth-Century Northern Borderlands," *Ethnohistory,* 54 (4) (2007): 725–734.
25. Francis Haines, "The Northward Spread of Horses among the Plains Indians," *American Anthropologist,* New Series, 40 (3) (1938): 429–434; Worcester, "Spanish Horses among the Plains Tribes"; Forbes, "The Appearance of the Mounted Indian," pp. 201–202, 204. French expeditions traded with Indians for Spanish horses on the eastern edge of the Great Plains in the 1680s. See Forbes, "The Appearance of the Mounted Indian," p. 207.
26. Calloway, *One Vast Winter Count,* p. 268; Worcester, "Spanish Horses among the Plains Tribes," p. 416; White, *"It's Your Misfortune,"* pp. 20–24.
27. Forbes, "The Appearance of the Mounted Indian," p. 208; Worcester, "Spanish Horses among the Plains Tribes," p. 412; White, *"It's Your Misfortune,"* p. 21.
28. Worcester, "Spanish Horses among the Plains Tribes," p. 417; Joseph M. Marshall III, *The Journey of Crazy Horse: A Lakota History* (New York: Viking, 2004).
29. Richard White, *The Middle Ground: Indians, Empires, and Republics in the Great Lakes Region, 1650–1815* (Cambridge: Cambridge University Press, 1991), pp. 1–49; Daniel K. Richter, *The Ordeal of the Longhouse: The Peoples of the Iroquois League in the Era of European Colonization* (Chapel Hill, NC: University of North Carolina Press, 1992); Donald E. Worcester and Thomas F. Schilz, "The Spread of Firearms among the Indians on the Anglo-French Frontiers," *American Indian Quarterly,* 8 (2) (1984): 104–105. While I do not agree with all points of their interpretation, Worcester and Schilz offer a succinct encapsulation of the chronological spread of firearms.
30. Denys Delâge, *Bitter Feast: Amerindians and Europeans in Northeast North America, 1600–1664* (Vancouver: University of British Columbia Press, 1993); White, *The Middle Ground;* Worcester and Schilz, "Spread of Firearms," pp. 104–106, 108.
31. Worcester and Schilz, "Spread of Firearms," p. 108; White, *The Middle Ground.*
32. Susan Sleeper-Smith, *Indian Women and French Men: Rethinking Cultural Encounter in the Western Great Lakes* (Amherst, Mass.: University of Massachusetts Press, 2001); White, *The Middle Ground,* pp. 60–75; Michael Witgen, "The Rituals of Possession: Native Identity and the Invention of Empire in Seventeenth-Century Western North America," *Ethnohistory,* 54 (4) (2007): 639–668.
33. Delâge, *Bitter Feast;* White, *The Middle Ground.*
34. Worcester and Schilz, "Spread of Firearms," pp. 110–111.

35. White, *The Middle Ground,* pp. 104, 113, 131–137; Worcester and Schilz, "Spread of Firearms," p. 110; Robert Ross McCoy, *Chief Joseph, Yellow Wolf, and the Creation of Nez Perce History in the Pacific Northwest* (London and New York: Routledge, 2004), pp. 47–50.
36. Worcester and Schilz, "Spread of Firearms," pp. 111–112.
37. McCoy, *Chief Joseph,* pp. 47–50; Worcester and Schilz, "Spread of Firearms," pp. 111–113.
38. Elliott West, "Called Out People: The Cheyennes and the Central Plains," *Montana: The Magazine of Western History,* 48 (2) (1998): 2–15; p. 3.
39. West, "Called Out People," pp. 4–5; George Bird Grinnell, *The Cheyenne Indians: Their History and Ways of Life* (Lincoln, Nebr.: University of Nebraska Press, 1972), p. 4; John H. Moore, *The Cheyenne* (Malden, Mass.: Blackwell Publishing, 1999), pp. 14–15, 18–19, 25–26, 129–138; George E. Hyde, *Indians of the High Plains: From the Prehistoric Period to the Coming of the Europeans* (Norman, Okla.: University of Oklahoma Press, 1959), p. 47.
40. West, "Called Out People," p. 8.
41. Moore, *Cheyenne Nation,* pp. 30–33; West, "Called Out People," p. 7.
42. West, "Called Out People," pp. 5–6, 10–12.
43. George Bird Grinnell, *The Fighting Cheyennes* (Norman, Okla.: University of Oklahoma Press, 1955), pp. 63–69; Donald J. Berthrong, *The Southern Cheyennes* (Norman, Okla.: University of Oklahoma Press, 1963), pp. 82–84; West, "Called Out People," p. 8–9; John H. Moore, "Cheyenne Political History, 1820–1894," *Ethnohistory,* 21 (4) (1974): 329–359.
44. West, "Called Out People," pp. 12, 14.
45. West, "Called Out People," p. 14.
46. West, "Called Out People," p. 14.
47. For an illuminating example of the formation of a historic tribal nation from the Southeast, see Patricia Galloway, *Choctaw Genesis, 1500–1700* (Lincoln, Nebr.: University of Nebraska Press, 1998).
48. Examples of social fluidity on the Northwest Plains can be found in Theodore Binnema, *Common and Contested Ground: A Human and Environmental History of the Northwestern Plains* (Norman, Okla.: University of Oklahoma Press, 2001). There are many detailed descriptions of views of the "Other" from both sides in the historical literature on contact between Amerindians and Europeans. For some explanations of the big picture, see Francis Jennings, *The Invasion of America: Indians, Colonialism, and the Cant of Conquest* (New York: W. W. Norton & Company, 1976); Weber, *Spanish Frontier in North America,* pp. 9–29; Daniel K. Richter, *Facing East from Indian Country: A Native History of Early America* (Cambridge, Mass.: Harvard University Press, 2003); Calloway, *One Vast Winter Count;* William Cronon, *Changes in the Land: Indians, Colonists, and the Ecology of New England* (New York: Hill & Wang, 1983). For an example of French difficulties understanding Native societies, see Witgen, "The Rituals of Possession. "
49. Philip J. Deloria, "What is the Middle Ground, Anyway?" *The William and Mary Quarterly,* 63 (1) (2006): 15–22; White, *The Middle Ground.*
50. Ramón A. Gutiérrez and Richard J. Orsi (eds.), *Contested Eden: California before the Gold Rush* (Berkeley, Calif.: University of California Press, 1998), pp. 1–195.
51. Clifford E. Trafzer and Joel Hyer have documented the holocaust suffered by Native peoples of California during the California Gold Rush in heartbreaking detail. See Clifford E. Trafzer and Joel Hyer, *Exterminate Them! Written Accounts of the Murder, Rape, and Slavery of Native Americans during the California Gold Rush, 1848–1868* (East Lansing, Mich.: Michigan State University Press, 1999). See also Joel R. Hyer, *"We Are Not Savages": Native Americans in Southern California and the Pala Reservation, 1840–1920* (East Lansing, Mich.: Michigan State University Press, 2001). For the Spanish treatment of Native peoples in California, see Rupert Costo and Jeannette Henry Costo, *The Missions of California: A Legacy of Genocide* (San Francisco, Calif. The Indian Historian Press, 1987); Gutiérrez and Orsi, *Contested Eden.*

CHAPTER FIVE

THE LEWIS AND CLARK EXPEDITION

Robert Carriker

When Meriwether Lewis, William Clark, and an accompanying unit of men, launched their three watercraft into the Missouri River at the frontier town of St. Charles, in the newly transferred Louisiana territory, the journey of the Corps of Discovery began in earnest. Behind them on that day in May 1804 was the memory of a winter spent alongside the Wood River in the Illinois country; ahead of them stretched the tantalizing prospect of a western horizon that ended at the Pacific Ocean. In some ways it was a clumsy group of maybe forty-five men; we do not know the exact number. Probably twenty-five men already belonged to the "permanent party," the group scheduled to go the entire distance with Lewis and Clark. A separate detachment of seven men under Corporal Richard Warfington was assigned to go only to some yet undetermined point on the upper Missouri River and then return to St. Louis.[1] Nine French boatmen were also attached to the "temporary party." It is unclear if Clark counted his African-American manservant, York, among the expedition personnel.

President Thomas Jefferson himself had authorized the Corps of Discovery. During the planning stage he so appreciated the enthusiasm and assistance of Lewis, his private secretary, that in February of 1803 the president named him the leader of the expedition.[2] Jefferson reflected on his choice in a letter to Benjamin Rush: "Capt. Lewis is brave, prudent, habituated to the woods, & familiar with Indian manners & character."[3] The part about his familiarity with frontier Indians is hyperbole, but the rest is accurate.

About that same time, spring 1803, Jefferson discovered that Albert Gallatin, his secretary of the Treasury, was a secret collector of maps and that he possessed a vast knowledge of western geography. Anxious to help, Gallatin commissioned cartographer Nicholas King to prepare a travel-map for Lewis that would contain the very latest geographic information. For example, Captain George Vancouver's 1792 charts of the "Coast of N. W. America," had been published in London. And fur trader Alexander Mckenzie's 1793 map of his overland routes to the Pacific Ocean across Canada had appeared in 1801 as part of his book *Voyages from Montreal.* Cartographer Aaron Arrowsmith's *A Map Exhibiting All the New Discoveries in the Interior Parts of North America ... with Additions to 1802,* had used these sources, plus others, and it was Gallatin's task to see that King's map included everything available. Reliable longitude and latitude was of utmost importance to Gallatin, as it was to the President.

Anxious to be a contributor, the President selected several maps and books from his private collection of 6,000 volumes and placed them at the disposal of Lewis. Some of them became part of the expeditions "traveling library."[4] Among the books Jefferson especially valued in his personal library, and with which Lewis likely became familiar, was one by Pierre Francois Xavier de Charlevoix, SJ published in the 1740s. It urged readers to envision a "pyramidal height of land" in the American West from which one could easily move between the sources of western rivers such as the Mississippi, the Missouri, and the Minnesota. Another book in Jefferson's library promoted the theory that rivers west of the Mississippi were symmetrical with those east of that river and that if one laid out a map of the USA, plus the adjacent western lands, and then inked in the river system east of the Mississippi before folding it over, the river system west of the Mississippi would suddenly appear on the previously blank part of the map.

Jefferson imposed on the patience of five of his colleagues in the American Philosophical Society by asking them to please instruct Lewis in the fields of astronomy, botany, zoology, and medicine. Would they also assist Lewis in learning the mechanics of the sextant, the chronometer, and the printed tables necessary to determine longitude and latitude? Accordingly, for several weeks in spring 1803, Lewis learned new skills and upgraded old ones in both Philadelphia and Lancaster, Pennsylvania.

While in Philadelphia, Lewis laid in supplies for the journey. The merchants of the city eagerly sold him $2,324 worth of tobacco, ink powder, mosquito netting, shirts, cooking utensils, scientific instruments, oiled linen, and so forth, the money coming from a congressional appropriation. Among the purchases was $289.50 for a bland concoction called "portable soup," an amount nearly three times what Lewis paid for medicine. Indian presents included beads, awls, needles, fishing hooks, and pocket mirrors. Together it all weighed in the vicinity of 2,700 pounds. Later, at the army-supply depot at Harper's Ferry, Lewis requisitioned 800 pounds of rifles, pistols, flints, lead, powder horns, and knives. He added also an iron contraption that could be folded shut for transport but when open and fitted with leather skins became a watercraft. Lewis's final stop in the preliminary phase of the expedition took place at a boatyard outside Pittsburgh. Here he supervised the finishing touches on the construction of a stout keelboat. Lewis required that the vessel be sturdy enough to resist the treacherous snags, current, and sandbars of the Missouri River. During the wait he purchased a Newfoundland dog which he named Seaman.[5]

Meantime, on June 20, 1803, President Jefferson wrote a set of instructions for the expedition and sent them to Lewis. At the time, the President did not know that the purchase of Louisiana, something that he had authorized in January, but whose negotiations took place in distant Paris, had been completed on May 2, 1803. He would not find out the specifics of the purchase until July.[6] Nevertheless, Jefferson confidently penned his instructions in June, placing special emphasis on western geography:

> The object of your mission is to explore the Missouri river, & such principal stream of it, as, by it's course and communication with the waters of the Pacific ocean, whether the Columbia, Oregan, Colorado or any other river may offer the most direct & practicable water communication across this continent for purposes of commerce.[7]

By the time America received news of the Louisiana Purchase Lewis was already in Pittsburgh, ready to shove off into the Ohio River on August 31, 1803. During more than 1,000 miles on the Ohio River, Lewis continued his preparations for the journey ahead. At Clarksville, Indiana Territory, Lewis halted long enough to welcome on board William

Clark, the man he had earlier asked to share the expedition command with him.[8] Clark, an acquaintance of Lewis's from earlier days and now a civilian engaged in family business in Kentucky and Indiana territory, possessed a wealth of frontier skills. Some of it, doubtlessly, he had learned from his older brother, George Rogers Clark, a Revolutionary War patriot widely known for his exploits in the Illinois country.[9] William Clark also possessed an innate skill for map making. Not the least of Clark's talents was his ability to select good companions, something he set about doing as soon as Lewis contacted him with the offer to share command. Thus, Clark had with him seven young men who he thought would work out quite well as expedition members, the lot of them chosen after conducting more than 100 interviews.[10] When Secretary of War Henry Dearborn formally authorized the Corps of Northwest Discovery on July 2, 1803, he limited the size of the party to a dozen men. Lewis and Clark, however, never felt constrained by this requirement and gathered thirty-six likely prospects from frontier military posts along the Ohio and Mississippi rivers, anticipating that there would be a steep attrition rate once the daily workload enveloped the corps.[11]

Once they set up their winter camp at River Dubois in the Illinois part of Indiana territory, choosing a site where they could look across the Mississippi River and see the mouth of the Missouri River, the intensity level of preparation accelerated.[12] The recruits received traditional military training; Lewis and Clark, meantime, spent much of their time between December 1803 and May 1804 making inquiries about Missouri River travel with talkative frontiersmen in St. Louis and Cahokia, two towns on different sides of the Mississippi River. Thus the co-captains of the expedition had an inkling as to where they were going and how long it would take them to get there. Well, at least as far as the Mandan Indian villages on the middle Missouri River.

Jean Pierre and Rene Auguste Chouteau, prominent St. Louis merchants, provided Lewis with significant practical information gleaned from clients who had previously ascended the river. In addition, John Hay, the postmaster at Cahokia, reminisced about his twenty years exploring the river, also taking the time to translate several French and Spanish documents for Lewis. President Jefferson sent Lewis handwritten extracts copied from a trade journal written by Jean Baptiste Truteau, a St. Louis schoolteacher who changed careers between 1794 and 1796 and became a fur trader. With eight other men, Truteau paddled more than 1,300 miles upriver to the villages of the Arikara Indians. His goal to reach the Mandan Indian villages unrealized, Truteau resumed his career in education. Among the most important opportunities for the captains to gain knowledge of the Missouri River took place when they read notes taken by John Evans and viewed maps drawn by his partner, James Mackay. In both 1795 and 1797 these two men led thirty men up the Missouri River to the Mandan villages and then returned.[13] On January 10, 1804, Clark interviewed Mackay at Camp Dubois. Two additional maps came into the possession of Lewis and Clark through the generosity of William Henry Harrison, the Governor of Indiana Territory. One was an "Indian Office Map," and the other was a chart of landmarks on the Missouri as far as the Knife River. When the assembled maps were used in coordination with a map of Upper Louisiana drawn by Antoine Soulard in the 1790s, it gave the captains confidence that they had a handle on the principal geographic features they would encounter beyond St. Louis. Secondarily, the interviews and the written materials referenced the American Indian tribes of the Missouri River. This proved to be immensely valuable inasmuch as President Jefferson's second emphasis, when he wrote his instructions to Lewis and Clark, related to Indians.

For several years, Jefferson had quizzed his friends in the American Philosophical Society about the characteristics of American Indians. Thus, when he wrote his instructions to Lewis, Jefferson used his backlog of unanswered questions. Historian James Ronda's research into the interests of Jefferson and his fellow Enlightenment intellectuals demonstrates the depth and breadth of American scholarly curiosity about American Indians at this time. Jefferson urged his captains to become "acquainted, as far as a diligent pursuit of your journey shall admit" with the names and numbers of the Indian nations, their tribal limits, language, traditions, monuments, economy, domestic accommodations, diseases and remedies, distinguishing physical traits, and peculiarities in laws, custom, religion and disposition.[14]

THE FIRST YEAR ON THE TRAIL: SPRING 1804–SPRING 1805

The process of transferring possession of the Louisiana territory from Spain to France took place in December of 1803. Three months later, Lewis and Clark witnessed the formal ceremony in St. Louis that consigned these same lands to the USA. Thus, when the expedition departed their Wood River encampment in two groups on May 14, 1804, the captains were confident that they traveled in American-owned territory.

The expedition keelboat, built to Lewis's design and specifications, was 55 feet long and 8 feet wide with a draft of only 3 feet, a necessary attribute for sailing on the sandbar-shallow Missouri River. The craft could carry several tons of cargo. If sailing proved difficult, as it often did, then twenty-two men pulling at the oars could move the craft upstream against a current that Clark calculated at 2.36 miles an hour but which was probably twice that. "Climbing the pole" furnished another source of power, but only if the depth of the river allowed the men to reach bottom with their poles. Cordelling, that is, pulling the vessel by an elkskin towrope, was another option, but the most unpleasant of choices because the brush and timber on the shoreline hampered the men. Rocks, brush, roots, hornets, snakes, cold water, and blistered hands made towing unpopular with the men. Two accompanying pirogues, one painted red and the other white, sat low in the water on flat bottoms, each approximately 25 feet long and 8 feet wide. They too could sail but usually required either six or seven strong oarsmen. Clearly, there was plenty of work to share between the permanent and temporary parties of the expedition just to make 12 to 17 miles each day on the river.

Calculations made by Clark over the winter caused him to believe that it would take the expedition 125 days to reach the Mandan Indian villages traveling at the rate of 12 miles per day.[15] Based on information from St. Louis fur traders Clark believed that the Mandan village was 1,500 miles upstream. In actuality, the expedition used 164 days to reach the Mandan village after forcing their flotilla 1,641 miles up the Missouri River.

Snags, sandbars, and rocks made the river dangerous; fog made it disorienting and eerie. On May 24, the current of the Missouri River turned treacherous and for the first time the watermen in the expedition had to fight for their lives:

> passed a Verry bad part of the River Called the Deavels race ground, this is where the Current Sets against Some projecting rocks for half a mile [...] The Swiftness of the Current wheeled the boat, Broke our *Toe* rope, and was nearly over Setting the boat, all hand Jumped out on the upper Side and bore on that Side untill the Sand washed

> from under the boat and wheeled on the next bank by the time She wheeled a 3rd Time got a rope fast to her Stern and by the means of Swimmers was Carred to Shore.[16]

On the following day, not yet 75 miles beyond the Mississippi River, the expedition observed a small French village of seven houses and as many families that called itself La Charrette. Clark, Sergeants John Ordway, Charles Floyd, and Patrick Gass, plus Private Joseph Whitehouse all noted in their journals that, "This is the Last Settlement of Whites..."[17]

Religiously the captains wrote day in and day out in their journals, an obligation requested by President Jefferson. The President only had to suggest it once because both Lewis and Clark were eager to write journals. In addition, Lewis wrote to Jefferson, "We have encouraged our men to keep journals, and seven of them do so."[18] Only six are known to exist, however. Lewis began writing in his journal on the day he left Pittsburgh in August 1803, but he could be inconsistent, resulting in several writing gaps.[19] Only one man wrote an entry for each day of the expedition: Sergeant John Ordway. William Clark wrote with an appealing literary style, though he is sometimes criticized for his unique spelling, capitalization, and punctuation. It is estimated that when the extant journals of the enlisted men are added together the total number of words penned on the expedition exceeds 1 million.

The two objectives of the expedition during their time in the present-day state of Missouri were to work as a team and to hone expedition skills. Hunters, for example, adapted to the environment and, between May and July 1804, killed seventy deer, a dozen or more black bears, three wild turkeys, a rabbit, a woodchuck, and a goose. Quizzing fur traders on their way to St. Louis about conditions upstream was another important activity. A highlight took place when Pierre Dorion, Sr., agreed to abandon his downstream course in a canoe and to join the expedition for several months as an interpreter. The expedition carpenter, meantime, had his moment to demonstrate his value to the expedition when he repaired a broken mast on the keelboat. On several occasions the captains brought out the medical kit to heal wounds. They also convened courts martial to discipline the men. And all the while the river and the weather kept the captains guessing. On July 14, 1804:

> The Storm which passd over an open Plain from the N. E. Struck our boat on the Starbd. quarter, and would have thrown her up on the Sand Island dashed to peces in an Instant, had not the party leeped out on the Leward Side and kept her off with the assistance of the ancker & Cable, untill the *Storm* was over.[20]

By mid-July the expedition had entered the Great Plains. Clark notes that on July 19 he enjoyed a breakfast of roasted deer ribs washed down with coffee and then took a walk on shore when he "Came Suddenly into an open and bound less Prarie, I Say bound less because I could not See the extent of the plain in any Derection, [. . .] this Prarie was Covered with grass about 18 Inches or 2 feat high and contained little of any thing else."[21] Clark called the moment "Sudden & entertaining," but it was not the last time the Great Plains landscape would surprise him.[22] For fourteen months during 1804 and 1805 the expedition participants lived on the Great Plains—roughly described as the Missouri River Valley between the Missouri–Kansas border and Three Forks, Montana—and during that time they collected specimens of plants, later found to represent at least seventeen new species, and they also discovered, or at the very least scientifically described for the first time, seven Great Plains species of mammals. Birds, fish, and reptiles extended their list.[23] The higher the expedition went up the Missouri the fewer the trees they found in the tallgrass prairie. The campsite of July 21 was the first of many without timber for either shade or firewood.

Ten miles north of the confluence of the Platte River with the Missouri, after 642 miles of travel, the expedition paused from July 22–27, 1804, to catch its breath and to "refresh" the men at a location they called Camp White Catfish. The pause also allowed the captains to calculate and to recheck their course and distance measurements by sighting the position of the stars and the moon over several evenings. After reviewing their situation, Lewis and Clark decided against sending the temporary party back to St. Louis from this location. The training of the permanent party was going well, but there were still too many unknowns to face farther upstream. One of those unknowns manifest itself a few days later.

Expedition scouts had noticed signs of Indian presence for several days before July 28 when expedition member George Drouillard made the first encounter. At Drouillard's invitation the man visited the expedition camp where Lewis persuaded the man to take one of the expedition linguists with him back to the Oto village where he lived. Meantime, the expedition would continue upriver and wait for the Indians to join them, an event that took place on August 2 at sunset. The expedition's first official conference with Great Plains tribesmen took place the following day at Council Bluffs. The meeting was cordial in part because the Otos and Missourias brought with them a Euro-American fur trader to interpret for them. If there was any awkwardness it was because Lewis and Clark were learning a new skill, how to conduct an Indian parley. Sheltered from the sun by the sail from the keelboat and with a stars-and-stripes flag flying from a makeshift flagstaff, gifts were given, military drills performed, and Lewis delivered an eloquent summary of the federal Indian policy of President Thomas Jefferson. Mostly, Lewis emphasized the change in leadership from the Spanish and French "fathers" to the "great chief of the Seventeen nations," or states.

Additional councils with Indians followed in August and September as long days on the river continued upriver. Several Omaha tribal headmen accepted an invitation, for example, to visit the expedition camp between August 17 and 20, but misunderstandings about the amount of gifts the Americans were prepared to offer put a negative spin on a session. Diplomacy became even more tedious after the tribes of the Lakota alliance got involved.[24] The Yankton Lakota on August 31, for example, needed confirmation that they would have a principal role in any new trade system introduced by the Americans on the Missouri River. Half Moon, a Yankton chief, suggested to Lewis and Clark that the Tetons, another Lakota tribe, would be even less cooperative when he predicted that, "I fear those nations above will not open their ears, and you cannot I fear open them."[25]

For a decade or more the Teton had exploited their well-known reputation for aggressive behavior toward Euro-American traders from a village at the mouth of the Bad river near present-day Pierre, South Dakota. Even President Jefferson had heard of the tribe, which is why he included an appeal for their special treatment in his June 1803 instructions to Lewis. Even if they had known of the President's personal interest, the Teton would have been unmoved for the stakes for them were high. The James River paralleled the Missouri River in this region and down it came goods from Northwest Company fur-trade posts in Canada. If the Teton acquiesced in allowing Americans from St. Louis to enter the middle Missouri trade it would upset their control of the manufactured trade goods in the region. A first meeting by the Corps with the Tetons grew tense, partly because of the absence of an interpreter. Also, neither Lewis nor Clark was able to correctly discern which chief, among many present, was, in fact, the dominant one attending the meeting. A chorus of complaints and demands by the Indians ultimately led three young Indian men to attempt to seize one of the expedition pirogues. Jostling, insolent gestures, and, finally, the readying of weapons on both sides pushed the conference toward a confrontation. At the critical moment, however, Black Buffalo took control of the Indians and defused the situation. During the next two days the expedition remained in their camp near the Teton village trying to repair the damage done, but each day a new situation with the Indians only confirmed Clark's original assessment, written in his journal, that the captains, "Suspect they are treacherous."[26]

Happily, a five-day visit with the Arikara Indians in early October was more productive for the Americans. The presence of two Euro-American traders facilitated communication, and the tribe even agreed to make peace with their rivals, the Mandan. As a sign of good faith, an Arikara chief accompanied the expedition north to the Mandan villages on the Knife River where more than 4,400 Mandan and Hidatsa Indians lived in five villages in a rectangle about 8 miles by 2 miles. The Mandan–Hidatsa compound was one of the larger cities in America at the time, with a population greater than Washington, DC.[27]

After constructing living quarters for the Corps at Fort Mandan in November 1804, the expedition leaders and members found much to occupy their time until they resumed their journey in the following spring. Keeping warm became a full-time occupation as the temperature dipped to 45 degrees below zero before Christmas and hovered in that vicinity for several months. Other activities included doctoring both the Natives and the expedition members, holding councils with the Indians, making and repairing equipment, conducting science experiments, and hunting. Not the least of accomplishments was expanding the permanent party by adding Toussaint Charbonneau, his Shoshone wife Sacagawea, and their infant son, Jean Baptiste.

Figure 5.1 C. M. Russell, *Indians and Scouts Talking*, ca. 1890, watercolor. Source: Courtesy of the Montana Historical Society.

THE SECOND YEAR ON THE TRAIL: SPRING 1805–SPRING 1806

Preparations for the expedition's departure from Fort Mandan began in March 1805 and continued into April. The members of the permanent party had one set of duties; the return party going back to St. Louis had other obligations. Both units left Fort Mandan on April 7, 1805. From the outset of the expedition the plan had been for a temporary party of men to escort the keelboat part way up the Missouri River and then to return with it to St. Louis. The return party, in addition to the leader, Sergeant Richard Warfington, consisted of seventeen men, including two who had been expelled from the permanent party. At various times on the journey downstream Warfington allowed Indians and traders to hitch a ride on the keelboat either part way or all the way to St. Louis. Precious cargo carried by the keelboat was addressed directly to President Jefferson. Four boxes, one trunk and three cages held maps, letters, and journals, plus sixty-eight items of special interest to the President: a Mandan buffalo robe, Indian corn, tobacco, tools and weapons, elk horns, soil samples, plants, four living magpies, a box of insects, and a live prairie dog.

The permanent party repacked their stores on April 3 and two days later loaded bundles into two large pirogues plus six newly constructed dugout canoes. They pushed off from shore on April 7 after 163 eventful days at Fort Mandan. Besides the captains, the permanent party included twenty-nine men, Sacagawea and her 55-day-old infant, plus "our dog." Lewis confided to his journal that, "The party are in excellent health and sperits, zealously attached to the enterprise, and anxious to proceed; not a whisper of murmur or discontent to be heard among them, but all act in unison, and with the most perfect harmony."[28] He adds, "we were now about to penetrate a country at least two thousand miles in width, on which the foot of civilized man had never trodden."[29] Neither Lewis nor Clark knew much about the French-Canadian exploration that had taken place west of the Great Lakes so they could not know that in all likelihood the Corps of Discovery would not break virgin trails until they were west of the Yellowstone River, a location they would reach in nineteen days after about 250 miles of paddling.[30]

Contact with Indian tribes ceased when Fort Mandan faded into view, but a whole new series of adventures kept the men of the expedition alert. On twenty occasions grizzly bears resisted the intrusion of mankind into their habitat. Another time a bison swam across the river and stampeded through the expedition camp.[31] High winds several times nearly swamped the boats. The spring runoff bloated the river causing towlines to snap. Still, there were glorious moments such as May 26, 1805, when Lewis wrote about

> arriving to the summit [of] one of the highest points in the neighbourhood I thought myself well repaid for any labour; as from this point I beheld the Rocky Mountains for the first time, [...] while I viewed these mountains I felt a secret pleasure in finding myself so near the head of the heretofore conceived boundless Missouri.[32]

As early as January 16, 1805, Lewis and Clark studied maps of the Missouri River country drawn in the dirt by Hidatsa cartographers. Hidatsa warriors knew from experience the landmarks and tributaries of the Missouri River going west. Their instructions, however, lost something in the translation—how to identify, for example, the "river which scolds at all others"—so when the expedition reached the confluence of the Marias River on June 2, some 750 miles beyond Fort Mandan, they questioned which branch remained the true Missouri. It took them seven days to reconnoiter and consider their options. The leaders sought the opinions of the men, made a decision, and three days later were proved correct when an advance party of the expedition became the first Americans to view the cascading Great Falls of the Missouri.

Making their way to the Mandan villages had been easy for the expedition. For the most part, Lewis and Clark, fortified with maps and advice from earlier fur traders, knew where they were going. The spring and summer of 1805, however, began a period when the captains needed to use all their intelligence in order to make correct choices. Their first opportunity had come at the Marias River. And the second came only a few days later when the expedition found it necessary to portage their equipment around the five waterfalls that make up the Great Falls of the Missouri. The strenuous endeavor occupied a full month, from June 13 to July 14. First, scouts laid out an 18.25-mile-long trail around the falls. Next, an advance party laid out two camps, one at each end of the pathway. Simultaneously, the men hid the two pirogues and buried some of their equipment in a cache until their return. Carts had to be built and fitted with sails. Then came the backbreaking and dangerous work of moving several tons of equipment up rock-strewn creek beds and across undulating plains. The sun seared the exposed skin of the men and a hailstorm pelted them with baseball-sized ice balls. And there were other odd happenings, too. Meriwether Lewis mused that, "I have scarcely experienced a day since my arrival in this quarter, without experiencing some novel ocurance among the party or witnessing the appearance of some uncommon object."[33] He referred not only to the times when the men were stalked by grizzly bears, or challenged by a wolverine, or charged by a buffalo, but also to moments such as when the men stumbled upon a gushing natural spring or they heard at sunset what seemed to them like "mountain artillery." There were two big disappointments, as well: the sinking of the iron-frame boat that they had carried with them all the way from Harper's Ferry, Virginia, and the distribution of the last gill from the 120 gallons of whiskey they had lugged all the way from St. Louis.

After portaging around the Great Falls, the expedition resumed travel on the Missouri River. In his instructions to Lewis, President Jefferson had specifically asked them to "explore the Missouri river," meaning its entire course. Where the Missouri merged with the Mississippi River had been identified as early as 1673, but neither the source of the great river nor its length was known. Until July 27, 1805, that is, when the entire Lewis and Clark Expedition assembled at Three Forks of the Missouri in present-day Montana. At that spot, three streams combine to form the longest flowing river in North America at 2,464 miles. Once again, information received from the Hidatsa Indians at Fort Mandan had proved helpful to the expedition. Indian informants had informed Lewis about the Three Forks, of which, he was told, the most northern stream was navigable to the foot of a chain of high mountains. The northernmost stream did, in fact, seem to head toward distant mountains. Lewis waited until August 9 to name the three watercourses, choosing the Madison, the Gallatin, and the Jefferson, all names associated with the current presidential administration.

The Hidatsa Indians encouraged Lewis and Clark to believe that the Shoshone Indians would assist them in crossing the Rocky Mountains.[34] Partly because of this advice, Lewis and Clark had hired Toussaint Charbonneau as an interpreter, knowing that he brought to the agreement his Shoshone wife, Sacagawea. But once past Three Forks of the Missouri the best-laid plans were put on hold because after the expedition transferred to the Jefferson River there were no Shoshone Indians to be seen. In fact, the expedition members had viewed no Indians of any tribe for over 120 days and counting. After five days, the

Jefferson River split into three branches, then divided again. At the same time that the watercourse diminished the mountains loomed closer. Lewis assessed the situation candidly: "We begin to feel considerable anxiety with rispect to the Snake Indians. If we do not find them or some other nation who have horses I fear the successful issue of our voyage will be very doubtfull."[35]

Sacagawea several times, beginning on July 22, recognized landmarks of her native land. Nevertheless, it took until August 11, 1805, for an advance party of the expedition to make contact with a Shoshone Indian. Lewis observed a lone horseman from a distance at a place later called Shoshone Cove. As Lewis moved toward him, the rider spooked, wheeled his horse around, and vanished into the brush. A sudden rainsquall wiped out his tracks. But at least the Indian showed the direction toward a trail spiraling up the mountain.

The next day, Lewis and his small party of three others became the first non-Indian Americans to cross the Continental Divide, achieving that distinction at the summit of present-day Lemhi Pass between Montana and Idaho. After pausing to look far into the distance at "immence ranges of high mountains still to the West of us with their tops partially covered with snow," and to imagine that beyond those peaks covered with snow was the Pacific Ocean, Lewis's contingent continued down the mountain slope.[36] The initial encounter, with a group of startled women, was awkward and fearful until Chief Cameahwait, surrounded by sixty warriors, arrived on the scene. He looked at the first Euro-American men he had ever seen, he listened, and then, to the surprise of Lewis, he hugged the American, this being a signal for a celebration to commence. Lewis writes that "we wer all carresed and besmeared with their grease and paint till I was heartily tired of the national hug."[37] A council with speeches and gifts took place before a triumphant march into the Indian camp along the Lemhi River. Lewis and his men ate roasted salmon that night for dinner. It was, Lewis wrote in his journal, "the first salmon I had seen and perfectly convinced me that we were on the waters of the Pacific Ocean."[38] He did not mention that camp that evening took place outside the boundary of the Louisiana Purchase territory. After 456 days in that territory, Lewis's small unit now rested in the Oregon country, an unmapped land of indeterminate boundaries, portions of which were claimed by Spain, Russia, Great Britain, and the USA.

While Lewis waited for Clark and the main body to catch up he occupied himself with several tasks. Quizzing Cameahwait about the best possible route west was important to Lewis. The Shoshone chief offered two routes: one by water and one by land. The water route—the Salmon River—sounded dangerous. And the overland route would require a guide. That decision would have to wait for input from Clark. Of more immediate concern was the necessity to secure horses from the Shoshone. They had about 400 of the animals, so that should be easy. But persuading Indians to assist the expedition as porters would be more difficult. Some warriors of the tribe viewed Lewis and his men as an enemy reconnaissance team, most likely one sent by the Blackfeet nation. Another complication was that the Shoshone were just now packing up to relocate to their fall buffalo-hunting grounds at the Three Forks. The Indians were not interested in crossing and recrossing Lemhi Pass with the American's main party when it arrived; that waste of time would not only delay the buffalo hunt, it would also extend their days at Three Forks and possibly antagonize the Blackfeet, who considered Three Forks their territory.

Lewis used the force of his personality to extract a promise of help from Cameahwait. Together, the Indians and Americans would cross Lemhi Pass on August 15 and wait for Clark and the main party. Upon Clark's arrival, the Shoshone would assist the explorers in

Figure 5.2 E. S. Paxson, *Lewis and Clark at Three Forks*, Mural at the Montana State Capitol 1912, oil on canvas.
Source: Courtesy of the Montana Historical Society.

bringing their supplies across the pass to the Lemhi River. But when Cameahwait assembled his people on the designated morning, a portion of his tribe resisted leaving camp for they feared Lewis was leading them into an ambush. This forced Lewis to now threaten the Indians with the loss of future US trade. He also feigned great disappointment in having accepted the promise of Cameahwait. In the end, Lewis got his way—the Indians crossed the pass and waited for Clark at Shoshone Cove—but the Shoshone lost valuable time in their trek toward Three Forks.

Clark and the main party rendezvoused with Lewis and the Shoshone on August 17, but not before Lewis, to allay fears by Cameahwait, gave the chief his own gun and told him to shoot him if he felt the tribe was going into a trap. On another occasion, one that he said "set a little awkward" with him, Lewis retrieved a note that he had left on a stick for Clark a week earlier and lied to Cameahwait that the note was from Clark, who would be along shortly. Increasingly, Cameahwait warriors, a large contingent of them still fearing an ambush, slept not around the campfire, but concealed in the dark, tangled underbrush.

Clark and the main party finally arrived at the Shoshone camp before noon on August 17. The reunion of the expedition members was made even more joyous when Sacagawea was reunited with one of her girlhood chums. And then, in an almost impossible coincidence, she recognized Cameahwait as her brother! Henceforth, there was no more talk of possible entrapment; instead there was a parlay about how the Shoshone could assist the Americans transport their baggage across Lemhi Pass. That monumental task was completed with only a few minor setbacks, and, by August 31, after nineteen agreeable days with Sacagawea's people, the expedition moved on, taking with them twenty-nine recently purchased horses and a mule.

For several days prior to leaving the Lemhi River camp, Lewis and Clark consulted again with Shoshone headmen about which route they should use to continue going west. Clark made it a point to field-test the Salmon River inasmuch as Cameahwait had early on told Lewis that it was a possibility, though not a good choice. After due consideration Clark concurred with the judgment of Sergeant Patrick Gass: "The water is so rapid and the bed of the river so rocky, that going by water appeared impracticable; and the mountains so amazingly high, steep and rocky, that it seemed impossible to go along the river by land."[39] Now what? Neither Lewis nor Clark completely understood what Shoshone geographers explained to them, but it seemed that they could go south—the details are vague—into Spanish territory, something the captains did not wish to do. The only real choice seemed to be to go overland to another major river, subsequently to be known as the Snake River, which would take them to an even greater river (meaning the Columbia River) and the salty water of the Pacific Ocean. This is what they decided to do.

Clark's hired Shoshone guide, Old Toby, led the way on a rugged trail assisted by his son. Soon Gass believed the expedition was cutting a road—"the worst road (if road it can be called) that was ever traveled"—more than it was following a marked trail.[40] Horses are in perpetual danger, he wrote, as the expedition climbed up and up and then, in the words of Private Joseph Whitehouse, "we descended a Mountain nearly as Steep as the roof of a house."[41] Suffering from both exasperation and exhaustion the expedition crossed the Continental Divide at either Lost Trail Pass or Chief Joseph Pass (historians are not sure!) and found themselves on September 4 entering the land of the Flathead Indians. The Flathead village of twenty-nine lodges and 400 people had likely never seen a Euro-American before, but they welcomed the visitors anyway. Flathead oral history says that a warrior observed the Corps as they struggled down the mountainside, and he reported back to camp that the black face of Clark's servant, York, might be a sign of war, though their casual manner said otherwise. The villagers took a risk and met the visitors in peace, even offering berries and roots. The tribesmen had no meat because they, like the Shoshone, were en route to the Three Forks of the Missouri to hunt buffalo.

The Salish language of the Indians puzzled the Americans. Sergeant John Ordway thought he detected a lisp. Clark described the speech pattern as a throaty, gurgling sound, difficult to understand. As a result, when Clark called a formal council with the Flatheads, a complex translation system that used five interpreters came into play. Clark spoke in English to François Labieche, who then relayed it to Charbonneau in French. Charbonneau, in turn, communicated the message in Hidatsa to his wife, Sacagawea, so that she could repeat it in her native language to a Shoshone boy who was a prisoner of the Flatheads and fluent in their language.

Continuing north along the Bitterroot River in present-day Montana, the men of the Lewis and Clark Expedition felt uneasy. It wasn't the rain and hail, or even the spotty success of the hunters. For a time there was an unfounded concern that the eleven horses and three colts purchased from the Flathead Indians might bolt from the expedition horse herd during the night and instinctively return to their home. What really bothered the Corps of Discovery was the glistening snow on the mountains to the west of them. If they hoped to place their feet on the shores of the Pacific Ocean before winter they would have to ultimately cross those white peaks. Would Old Toby, their Shoshone guide, recognize the trail leading to the pass across that jumble of mountains? Hopefully he would. But when Lewis and Clark asked questions about the watercourses, most times Old Toby, "could not inform us."[42]

On September 9, Clark records that "our guide informes that we should leave the river at this place," the present-day junction of US Highways 12 and 93, south of Missoula, Montana.[43] They halted to rest the horses and to take some celestial observations at a place they called Travelers Rest. In front of them lay the approach to Khusahna Ishkit, or, as it has subsequently come to be known in history, the Lolo Trail. This 100-mile-long passageway had been used since the 1730s by the Nez Perce Indians to cross the Bitterroot Mountains. In the evening, one of the hunters, John Colter, returned to camp with three Indians. Clark believed them to be Flathead Indians, but their knowledge of the trail across the mountains, and the statement by one of the men that he lived on the "plains below the mountains," suggest they may have been Nez Perce tribesmen. The Indians said they were seeking the recovery of twenty-three horses stolen by raiders. Two of the Indians departed the camp of the explorers after sunset, but the third volunteered to remain with the Corps as an auxiliary trail guide. But he only stayed one night and mysteriously moved on.

Rough terrain on the route to Lolo Pass extended some days deep into the night. On the third day out from Traveler's Rest, evening camp was made in an extensive meadow on the west side of Lolo Pass, bringing the Corps once again across the Continental Divide. On September 14 the Corps marched "9 miles over a high mountain steep & almost inaxcessible [with] much falling timber which fatigues our men & horses exceedingly."[44] It need not have been so strenuous. Unfortunately, Old Toby misread a junction in the trail, and he followed the wrong path coming off Lolo Pass. Instead of following the main trail going to the ridge of the mountains, he led the expedition down a footpath to a fishing creek at the headwaters of today's Lochsa River. That night, Sergeant Patrick Gass lamented that "none of the hunters killed any thing except 2 or 3 pheasants; on which, without a miracle it was impossible to feed 30 hungry men and upwards, besides some Indians."[45] For the first time, Lewis passed out portable soup to the men, an unappealing concoction of flour, cornmeal, sugar, coffee, dried apples, and lard, of which he had purchased 193 pounds in Philadelphia just for emergencies such as this. Gass asserts that "Some of the men did not relish this soup, and agreed to kill a colt; which they immediately did, and set about roasting it."[46] After dinner, Sergeant John Ordway took refuge in his tent from the thunder, hail, and rain. Even so, he could still see fearsome "high Mountains covered with Snow and timber."[47]

Eventually, Lewis and Clark would come to understand that they had strayed from the traditional trail and it would be necessary for them to climb from the river to the ridge of the mountains just to regain it. Fortunately, one of the men located a seldom-used Indian trail that zigzagged its way back to the ridge, and the main trail, above them. The narrow, ill-defined trail showed few marks of traffic, and for good reason. It rose about 3,500 feet in elevation in 10 miles "winding in every direction." Two horses gave out after they "Sliped and roled down Steep hills which hurt them verry much" but even so when the men made camp that evening they had triumphantly returned to the main trail.[48] They also slept amid snow patches.

Proceeding along the Lolo Trail tested the mettle of the Lewis and Clark Expedition as never before. The road crawled along a hog-back ridge that separated the Lochsa drainage from that of the North Fork of the Clearwater River. At an elevation approaching 7,000 feet, snow accumulated to 10 inches, frequently obscuring the trail but always exhausting the packhorses, sometimes spooking deer and elk and constantly freezing the hands and feet of the men. Clark admitted, "I have been wet and as cold in every part as I ever was in my life."[49] Several years after the event, Clark remembered that, "The want of provisions together with the dificuely of passing those emence mountains dampened the Spirits of the party."[50]

In an attempt to uplift the mood of the men, the captains split the party on September 18, Clark taking with him six hunters whose objective was to bag game and to leave it for the slower and noisier main party. That afternoon Clark ascended a ridge and could see salvation in the distance. The next day Lewis saw it, too: "we to our inexpressable joy discovered a large tract of Prairie country lying to the S. W."[51] He added, "this plain appeared to be about 60 Miles distant, but our guide assured us that we should reach it's borders tomorrow. the appearance of this country, our only hope for subsistance greately revived the sperits of the party already reduced and much weakened for the want of food."[52] Sergeant Gass spoke for the enlisted men when he wrote in his journal, "there was much joy and rejoicing among the corps, as happens among passengers at sea, who have experienced a dangerous and protracted voyage, when they first discover land on the long looked for coast."[53]

On September 20, Clark's cadre broke out of the mountains and entered Weippe Prairie. Three miles into this plain he encountered members of the NeMeePoo or Nez Perce tribe. His arrival, he said, "raised great Confusion" among a people who may have heard rumors about Euro-Americans, but had not yet been visited by one.[54] "Those people treated us well," writes Clark, as they sheltered and fed the men, mostly roots and salmon.[55] He found it all "tolerably good" until he became "verry unwell all the evening from eateing the fish & roots too freely."[56] The next afternoon his hosts led Clark into the canyon of the Clearwater River to meet Chief Twisted Hair. At the end of the strenuous journey Clark confided to his journal: "I am verry Sick to day and puke which relive me."[57]

Meantime, Lewis and the lagging main party still struggled in the mountains for two extra days before finally walking into the same Nez Perce camp of eighteen lodges on Weippie Prairie that had given comfort to William Clark and his six hunters. When Clark encountered his reassembled exploring party he was shocked at their weakened condition. He cautioned the men about "the Consequences of eateing too much &c."[58] But his warning came too late. The men of the Corps of Discovery, whose normal daily diet consisted almost exclusively of red meat, had already eaten heartily at the Nez Perce buffet of berries, roots, and salmon. Clark saw Lewis become so weak that he was "Scercely able to ride on a jentle horse," and several men were "Compelled to lie on the Side of the road for Some time."[59] Clark rummaged through the expedition medicine chest, but no pills, salts, or tonic brought from Philadelphia could relieve the men's distress. Private Joseph Whitehouse believed that the condition was "caused by a suddin change of diet and water as well as the climate."[60] Even a week later Clark lamented that, "Our men nearly all Complaining of ther bowels, a heaviness at the Stomack & Lax."[61]

In spite of the illness in camp, there was work to be done. First on Clark's to-do list was convening a cartography conference with the Nez Perce. There being no interpreter when he interviewed Twisted Hair, "we were Compelled to converse alltogether by Signs."[62] Drawing on a white elk skin with charcoal, Twisted Hair traced the westward flow of the Clearwater River, conveying as best he could the disposition of Indians ahead and noting two major tributaries whose distances he recorded as "2 Sleeps" and "5 Sleeps" beyond.[63] Next there was the matter of constructing canoes. Aided by Indians, Clark scouted along the Clearwater River in a hunt for "trees Calculated to build Canoes."[64] He established his so-called canoe camp on the north fork of the Clearwater River where there grew several large, straight ponderosa pine trees, each with a minimum of knots. Best of all, they could be felled in close proximity to the river.[65]

Shaping dugout canoes—essentially hollowed-out logs—with an adz and ax is labor intensive, a fact Clark had learned earlier in the year. At both Fort Mandan and Great Falls of the Missouri River members of the expedition had built a total of eight dugout canoes. From this experience Clark knew that, "our axes are Small & badly Calculated to build Canoes of the large Pine."[66] Happily, Nez Perce craftsmen offered good advice to the carpenters of the Corps. Instead of chopping deep into the log with axes, Indians urged the explorers to flatten the topside of the tree and then blanket it with hot coals. When the wood charred sufficiently it could be easily chipped out. The recommended process took more time but less work, an important consideration since many of the explorers felt ill.

The first two of five planned canoes were finished and launched on October 5. With the canoes coming along, it was time to provide for the expedition horses. Lewis and Clark planned to retrace their steps back to St. Louis from the Pacific Ocean in the spring of 1806, so they would need horses again for the eastward journey across the Lolo Trail and Lemhi Pass. On the same day as the first canoes were completed, collecting the thirty-eight horses belonging to the Corps of Discovery began. For purposes of future identification each animal had its mane cropped before it received a brand on the near fore shoulder: "US Capt. M. Lewis."[67] With those tasks completed, Clark delivered the horses to Nez Perce men whom they had come to trust. Clark received their promise "to be attentive to our horses untill we Should return."[68] Next on the agenda was to secretly bury the packsaddles, plus whatever supplies were being left behind.

During all of these work assignments the Nez Perce tribesmen proved to be excellent partners with the Corps of Discovery. Lewis and Clark referred to them as Chopunnish, which they understood to be the tribal word for "Pierced Noses." Later, French-Canadian fur traders translated the "Pierced Noses" into Nez Perce. The villages of the nation—probably the most numerous and powerful of all Pacific Northwest tribes at this time—followed a seasonal cycle for roots, wild vegetables, berries, salmon, and big game across the present-day states of Washington, Oregon, Idaho, and Montana.

The final three dugout canoes constructed by the expedition carpenters entered the Clearwater River on the evening of October 6, 1805, and they were loaded and launched the next morning. Thus far in the expedition—going on eighteen months at this point—the Corps of Discovery had not yet paddled with the current of a river. Previously they had to push, pull or pole against a current. The men were a little unsure of themselves in the dugout canoes at first but generally they avoided upsets. When a "Sad axident" did occur they forced the men who could not swim to hang on to their canoe until the rest of the flotilla could reach shore, unload a rescue canoe, and retrieve the men.[69] There was a silver lining, however, to one of the time-consuming accidents. While drying soaked equipment in the sun, Twisted Hair and Tetoharsky, two Nez Perce who had earlier promised to join the expedition, caught up with the Corps and declared their intention to remain as guides. Almost as if on cue, the following day Old Toby and his son, the helpful Shoshone guides who had been with the Lewis and Clark Expedition since late in August, left the expedition without a good-bye. Lewis and Clark were disappointed and urged Twisted Hair to send a horseman after Old Toby to return and receive some pay and presents, but the chief declined, explaining frankly that other Nez Perce "would take his things from him before he passed their camps."[70]

On October 10, as the canoeists approached their sixtieth mile on the Clearwater River, a large southerly fork about 250 yards wide came into view. Nez Perce informants had previously spoken of this merge, and Lewis and Clark now speculated that since the "water of the South fork is a greenish blue, [and] the north as clear as cristial," the new fork must be the one that Clark had tested, and rejected, back at the Shoshone camp.[71] It was an educated guess and was very close to being accurate. In truth, the Clearwater had just joined the Snake River at present-day Lewiston, Idaho, and Clarkston, Washington. The Snake is a twisting stream of more than 1,000 miles with scores of tributaries, including the one Clark had investigated and today is known as the Salmon River. So, this southerly fork was not the Salmon River as Clark thought, though the Snake River at this point did include the flow of the Salmon River. The Nez Perce guides confirmed that two rivers joined together several days away. Lewis and Clark would not begin to truly understand the complexity of the Snake–Salmon hydrographic system until they returned to the land of the Nez Perce in 1806.

The first camp of the Lewis and Clark Expedition in present-day Washington state was located about a mile from the confluence of the Clearwater and Snake rivers. In the morning the attention of the expedition would be focused on the turbulent Snake River. In 30 miles on October 11, for example, the canoes had to run nine rapids. On October 12, the explorers duplicated that feat. The unrestricted Snake River flowed with such velocity through present-day Washington state that by October 14, 1805, Sergeant John Ordway, the third in command with the Corps of Discovery, wished he could portage the canoes around the worst rapids. Sergeant Patrick Gass agreed: "This river in general is very handsome, except at the rapids, where it is risking both life and property to pass."[72] But Lewis and Clark decided against portages because the autumn was already "So far advanced and time precious with us."[73] In a race against winter weather, the Corps of Discovery needed to quickly locate with the Columbia River and after that continue on to the Pacific Ocean.

When the Nez Perce guides indicated to Lewis and Clark that the Snake would soon merge with an even greater river, the captains urged Twisted Hair and Tetoharsky to move ahead and announce their approach to the resident tribesmen. The Nez Perce accomplished their assignment, with the result that on October 16 several Indians came forward to meet the expedition and more or less escort them to the "Point" or "the big forks," the spot where the Snake River enters the Columbia River. Both American Indians and Euro-Americans celebrated heartily that evening. The Corps of Discovery marveled at their own accomplishment of having reached the Columbia River after 195 days and 3,714 miles on a route that had taken them around the Great Falls of the Missouri, over three mountain passes, and now a wild canoe ride of half a hundred rapids on the Clearwater and Snake rivers. The Indians exhibited their joy by forming a half-circle around the visitors, "Singing and beeting on their drum Stick and keeping time to the musik."[74]

President Thomas Jefferson had specifically requested that Meriwether Lewis collect information on the Columbia River. Therefore, on October 17 and 18, Lewis and Clark took elaborate longitude and latitude readings and estimated the width of the rivers, and Clark made a brief side trip upstream on the Columbia. The river was "Crouded with Salmon," and so clear, he said, that a fish "may be Seen at the deabth of 15 or 20 feet."[75] Clark estimated he "must have seen 3 or 400 dead and maney living."[76] Neither Lewis nor Clark understood that Pacific salmon are born in freshwater streams, migrate to the ocean in the spring and spend most of their adult life there before returning to fresh water to spawn and die.

Late on the afternoon of October 18 the expedition resumed travel, now for the first time on the Columbia River. Every man in the expedition knew the Pacific Ocean lay in front of him, to the west, but how far away was it? The men got their first hint later that afternoon. From a height of land Clark viewed in the distance a snow-covered mountain. The next day he saw another far-off mountain. Clark knew that in 1792 Lieutenant William Broughton of the British Navy had named both Mount Hood and Mount St. Helens from a point on the Columbia River estimated to be about 125 miles upstream from the Pacific Ocean. Clark surmised that if he could see what Broughton had named then the ocean could not be far away. The reality is that the two mountains that Clark viewed were Mount Hood on the first day and Mount Adams on the second, not Mount St. Helens. But it does not matter. What is important is that at this moment Clark knew that the expedition was back on the map for the first time since leaving Fort Mandan.

Between Clark's viewpoint and Broughton's naming point there were two long and very different stretches of swift water on the Columbia River. The first section of approximately 120 miles was fast but manageable. The second section of approximately 65 miles, however, was crowded with major obstacles to navigation. The canoes of the Corps of Discovery were halted by the first of these obstructions at Wyam, or Celilo Falls. Lewis and Clark had missed the fall salmon run on the Columbia, an event that made this location a magnet for several thousand peoples from a variety of regional tribes, so they never fully realized its significance until their return journey in 1806. But what they did comprehend immediately was the danger and labor that would be required to pass the falls.

The carrying of goods around the falls commenced on October 22, the heaviest articles being carried on Indian horses contracted to do the job. The canoes had to be taken through narrow channels in the river. Sergeant Gass described the scene:

> for three miles down, the river is so confined by rocks (being not more than 70 yards wide) that it cannot discharge the water, as fast as it comes over the falls, until what is deficient in breadth is made up in depth. About the great pitch the appearance of the place is terrifying, with vast rocks, and the river below the pitch, foaming through different channels.[77]

That evening, Lewis exchanged the expedition's smallest canoe for a native one which he found "neeter made than any I have ever Seen and Calculated to ride the waves, and carry emence burthens."[78]

Just 3 miles beyond Celilo Falls the Short Narrows of The Dalles halted the progress of the Lewis and Clark Expedition for the second time. Clark estimated the width of the channel at 45 yards. Two other journalists in the party thought it closer to 20 yards wide. High basalt walls made a portage impossible. "I deturmined to pass through this place notwithstanding the horrid appearance of this agitated gut swelling, boiling & whorling in every direction," proclaimed Clark.[79] When the men of the expedition did exactly that, Clark sheepishly added an aside to his journal entry: "from the top of the rock [it] did not appear as bad as when I was in it [...] we passed Safe to the astonishment of all the Inds."[80]

The next day, the Corps confronted yet a third barrier in the Columbia: the Long Narrows of The Dalles. But by now the men of the expedition had a routine for passing such dangerous places so there was no damage. Clark "felt my Self extreamly gratified and pleased."[81] But also sad, because a day earlier Twisted Hair and Tetoharsky, the Nez Perce guides, asked permission of Lewis and Clark to return to their homelands admitting that they could be of no further service. They knew nothing of the river below Celilo Falls, and, besides, the Nez Perce had sometimes been at war with some of the nations beyond. After the falls, Twisted Hair and Tetoharsky purchased horses from the Indians, turned them to the east, and began their journey home.

Navigating the Columbia River between October 22 and 26, 1805, had been exceedingly difficult for the members of the Lewis and Clark Expedition. Happily, the conquest of three obstructions earned everyone a day of rest on October 27. Hunters filled the larder

Figure 5.3 Harold Von Schmidt, *The Rapids*, 1954, oil on canvas.
Source: Courtesy of the Montana Historical Society.

with four deer, a grouse, and a squirrel. Meriwether Lewis jotted down some vocabulary words uttered by Indians. William Clark, meantime, wondered why "all the Bands flatten the heads of the female Children, and maney of the male children also."[82] Clark had first noticed the Indian practice of head-flattening ten days earlier when the expedition spent two days at the confluence of the Snake and Columbia rivers. The desire, it seemed to him, was to compress the head "so as to form a Streight line from the nose to the Crown of the head. [...] This amongst those people is considered as a great mark of buty."[83] He even made a drawing of the technique.[84]

Finally, the expedition navigated the river's fourth natural obstruction, the "Great Shute," now called the Cascades of the Columbia. The men managed the feat by laboriously walking the "baggage over the Portage of 940 yards, after which we got the 4 large Canoes over by Slipping them over the rocks on poles placed across from one rock to another."[85] At the end of the Cascades rose "a remarkable high detached rock [...] about 800 feet high and 400 paces around," a conspicuous landmark protected today in Beacon Rock State Park near Skamania, Washington.[86]

Now on the western side of the Cascade Mountains, the environment of the Columbia River influenced the Lewis and Clark Expedition in new ways. For example, the Cascades form a climatic barrier wherein the temperature moderates but rainfall increases. The annual average rainfall at The Dalles, Oregon, is 12 inches. At the Cascades of the Columbia, just 40 miles away, it exceeds 70 inches annually. Consequently, forests grow tall and close, allowing wild creatures to find protection and multiply. More importantly, in disbelief, Lewis and Clark watched the river on which they had just logged 185 turbulent miles suddenly turn docile. Clark was quick to conclude that the water "had everry appearance of being effected by the tide."[87] It is true. While the tide of the Pacific Ocean flows only 20 miles deep into the estuary of the Columbia River, the effects of tidewater can be felt at Beacon Rock which is 140 miles up river.

Even as the environment changed, so too did the attitude of the Indians. For some time Lewis and Clark had noticed the Indians were wearing machine-made clothing, an indication that they had entered the trade zone of "Boston men," seafaring traders who visited the mouth of the Columbia River. Lewis and Clark observed that while the Indians continued to barter with the expedition, offering nuts, berries, dried fish, baskets, and rain hats, the prices had gone "high with what they have to Sell, and Say the white people below Give them great Prices for what they Sell to them."[88]

All indications pointed to the nearness of the Pacific Ocean, an idea that gained support on the evening of November 3, 1805, when, according to Private Joseph Whitehouse, "towards evening we met Several Indians in a canoe who were going up the River. they Signed to us that in two Sleeps we Should See the Ocean vessels and white people &c. &c."[89] Each succeeding day brought additional evidence to the Corps of Discovery that the Pacific Ocean lay just ahead. Indians dressed in sailor jackets; some carried muskets and used tin flasks to hold their powder; and at least one native could curse in English. Most of the Indians seemed unafraid of the Americans, not assigning to their visitors any special status, as had the more remote tribes of the Rocky Mountains. At last, on November 7, 1805, after 34 miles on the water, Clark could proudly write in his journal: "Great joy in camp we are in View of the Ocian, this great Pacific Octean which we been So long anxious to See. and the roreing or noise made by the waves brakeing on the rockey Shores (as I Suppose) May be heard distictly."[90] The expedition was, at this point, approximately 28 miles from the breaking surf of the Pacific Ocean.

Leaning into their paddles the next morning, the men of the Corps of Discovery forced their dugout canoes toward the ocean by paddling against both tide and wind. But the Columbia estuary is 8 miles wide at that point and the river's ocean-like swells soon overwhelmed the vessels. Several in the party became sea sick. "We thought it imprudent to proceed," writes Clark.[91] The refuge that Clark chose on the beach, however, proved to be quite uncomfortable. As anxious as the members of the party were to abandon that spot the next morning, they had no choice but to hunker down in the stormy weather and stay put for another day and night. A bad situation got worse at 2 P.M. when rain, driven sideways by wind, pounded the camp and the floodtide pushed driftwood ashore almost on top of the canoes and makeshift camp. It was only "with every exertion and the Strictest attention by every individual of the party [that it] was Scercely Sufficient to Save our Canoes from being crushed by those monsterous trees maney of them nearly 200 feet long and from 4 to 7 feet through," reports Clark.[92] In spite of everything—for these days at the estuary of the Columbia River are, in fact, the most dangerous and trying of the entire expedition—Clark believed that the members of his command remained "chearful and anxious to See further into the Ocian."[93]

Fortunately, on November 10, the wind and waves subsided, so the expedition members loaded their canoes and resumed their journey. By hugging the Washington shore, paddling from cove to cove, the Corps of Discovery managed to log hard-won miles in salty water and then camp. Hunters returned each evening to say that the hills were "So high &

Steep, & thick with undergroth and fallen Timber" that foot travel was as impossible as canoe travel.[94] One afternoon "at a time when the wind was verry high and waves tremendous five Indians Came down in a Canoe loaded with fish."[95] The natives sold their supply of salmon and roots, and then they departed. Their route, writes a disbelieving Clark, took them straight across the river, "which is about 5 miles wide through the highest Sees I ever Saw a Small vestle ride."[96] It is certain, he added in admiration, "they are the best canoe navigators I ever Saw."[97]

Clark assessed the condition of the men, their rotting clothes, the lack of food, and the frightening thunder and lightning. "Our Situation is dangerous," he concluded.[98] Eight days of struggle against a river current that clashed with tidewater amidst wind-driven rain yielded the Corps of Discovery a net advance of only 13 miles. But it was enough distance to take them to a sandy beach, roughly 15 miles from Pacific Ocean breakers and in clear sight of Cape Disappointment at the mouth of the Columbia River. "This I could plainly See," Clark scribbled excitedly in his journal, "would be the extent of our journey by water."[99] Sergeant Gass sounds even more excited:

> We are now at the end of our voyage, which has been completely accomplished according to the intention of the expedition, the object of which was to discover a passage by the way of the Missouri and Columbia rivers to the Pacific ocean; notwithstanding the difficulties, privations and dangers, which we had to encounter, endure and surmount.[100]

Goal accomplished. Destination reached.

Lewis and Clark felt compelled to verify their accomplishment. To that end, Clark busied himself taking astronomical calculations and drawing a preliminary map. The sergeants saw to the task of establishing a camp that would be the home of the expedition for the next ten days. Clark ultimately named the place "Station Camp." Lewis, meantime, made a reconnaissance to Cape Disappointment, a mariner's landmark from the 1780s whose name reflects the frustration of attempting to pilot a ship across the bar of the Columbia River. Clark led eleven other men to the same location the following day.

The most poignant episode to take place at Station Camp involved Sacagawea. She was, in all respects, a valuable and uncomplaining auxiliary member of the otherwise all-male Corps of Discovery. One afternoon—the date is either November 20, or 22 or 23 because, oddly, four expedition journalists who reported the event can not agree when it took place—some Chinook Indians visited Station Camp, and Meriwether Lewis took a strong liking to an Indian robe made from sea-otter skins. Lewis offered a blanket, even a coat, in exchange, but the Chinook man held out for blue beads, of which Lewis, unfortunately, had none. Lewis had purchased nearly 50 pounds of multicolored trade beads in Philadelphia, but he had no way of knowing at the time that the Indians would exhaust his supply of prized blue beads first. Clark judged the robe "more butifull than any fur I had ever Seen" and he upped the ante to two blankets, but the owner said no, he would not take even five.[101] Finally, according to Clark, "at length we precured it for a belt of blue beeds which the Squar—wife of our interpreter Shabono wore around her waste."[102] Did the captains ask Sacagawea for her belt or did they order it taken off? Did she volunteer? No one says. All we know is that on another day Clark reports, "we gave the Squar a Coate of Blue Cloth for the belt of Blue Beeds we gave for the Sea otter Skins."[103] Was a cloth coat of equal value with a personal belt of blue beads?

On the tenth day of their stay at Station Camp, Lewis and Clark asked themselves, what happens now? They had fulfilled President Jefferson's major requirements for their expedition. Now what? Return to St. Louis and Washington, DC, of course, but not until 1806 for almost certainly snowstorms had already choked closed the high elevation passes at Lolo and Lemhi. "Being now determined to go into Winter quarters as Soon as possible," Clark writes, the captains focused on their newest task.[104]

In June 1805, when at the confluence of the Missouri and Marias rivers, Lewis and Clark evaluated the two possible routes and then asked the men for their opinion. Now on November 24, 1805, the captains convened another meeting of expedition members to seek advice. "At night," recounts Sergeant Patrick Gass, "the party were consulted by the Commanding Officers as to the place most proper for winter quarters."[105] Private Joseph Whitehouse similarly used the word "consult" in his journal entry.[106] Sergeant John Ordway characterized the event only slightly different, writing that the officers drew conclusions from "the oppinion of the party."[107] Currently it is fashionable to use the word "vote" for what took place that evening, but casting votes, as we understand it today, is not supported by any journal entry. Three sergeants were the first to answer the captains. Privates, interpreters, and York followed. Toussaint Charbonneau declined to offer an opinion. Sacagawea made known her preference for "a place where there is plenty of Potas."[108] It is not clear from Clark's tally if Sacagawea—on this occasion referred to as "Janey"—was asked her opinion or if she just spoke up spontaneously.[109]

The sergeants all held the same view: cross the Columbia to the south bank and seek a place to build a fort, all the while monitoring the disposition of the local Indians. It was essential that there be herds of elk to supply both food and clothing. Among the rest of the men, only one disagreed with the sergeants. Clark added more good reasons to stay near the coast. First, an American trading vessel just might enter the Columbia River.[110] Second, salt, a necessary item for preserving meat, can be boiled from ocean water. Lastly, the coast range of mountains on the southern shore would provide protection from the constantly blowing wind.

It having been agreed that the southern (Oregon) shore should be explored, the captain's follow-up question asked the men to indicate their secondary preference should the first choice prove to be undesirable. This time, three possibilities emerged from the meeting, but it never became an option once a suitable site was located, and Fort Clatsop, a structure of seven rooms surrounded by a palisade, took shape during December.

As the headquarters for their second winter in the field, Fort Clatsop witnessed a daily schedule of activity. Much of it took place indoors because of the inclement weather: making a list of American traders who came to the mouth of the Columbia; rewriting 179 pages of the journals that had been hastily written during the past several months; and examining more than three dozen local plant species plus numerous land and marine mammals, birds, reptiles, and fishes. Clark began work on his comprehensive map of the expedition. The men busied themselves by renewing the expedition equipment, handmaking moccasins for the return journey, and recovering from illness, boredom or injury. The men also took their turns manning the fires beneath five large kettles boiling ocean water at a salt-making camp on the coast. And Lewis wrote numerous ethnographic notes about the local Indian tribes who frequently came to the fort to trade or just out of curiosity.

The principal reason to be out of doors—consider that it rained all but a dozen days in three and one-half months at Fort Clatsop—was to hunt. Each day, three or four hunters went out to find fresh meat because wet wood and high humidity made it difficult to cure meat for storage. Sergeant Gass reports that all together the hunters bagged 131 elk and twenty deer. Private Whitehouse counted 155 elk—the leather of which furnished 338 pair of moccasins for the return journey.[111] Once, Clark led a party of curiosity seekers to the coast upon hearing that a whale had beached itself and the Indians were stripping it down. Clark purchased about 300 pounds of blubber plus a few gallons of oil. Sergeant John Ordway reflected that while at Fort Clatsop the expedition lived "as well as we had any right to expect, and we can Say that we were never one day without 3 meals of Some kind."[112]

THE THIRD YEAR ON THE TRAIL: SPRING 1806–FALL 1806

Meriwether Lewis, usually the more expressive of the two men who led the Corps of Discovery, noted a momentous event on March 23, 1806 with an uncharacteristic economy of words: "at 1 P. M. we bid a final adieu to Fort Clatsop."[113] The initial date of departure had been set for April 1, but everyone was anxious to begin the journey home so the captains acquiesced believing it would be good for the men's morale. Another consideration was that the fort was running out of food.

The good idea had a downside, however. The spring runoff in the Columbia River made it extremely tiring to paddle five canoes against the current, sometimes also against a stiff wind. For the first 140 miles the rhythmic tides added to their labor, sapping the strength of the expedition men. Inasmuch as they had spent the winter ashore and mostly indoors the men need not have felt embarrassed at being a little out of shape. But another episode probably did shame them. On March 17, as the captains made final preparations for the return journey, Lewis bought a canoe from a Clatsop Indian at a price that required him to give up his uniform coat plus a quantity of tobacco. He confided to his journal that "we yet want another canoe, and as the Clatsops will not sell us one at a price which we can afford to give we will take one from them in lue of the six Elk which they [reportedly] stole from us in the winter."[114] The following day, Sergeant Ordway reports, "4 men went over to the prarie near the coast to take a canoe which belonged to the Clatsop Indians, as we are in want of it."[115] The canoe remained concealed near the fort until it was time to depart. But on the first afternoon out, when the expedition got lost among some islands in the estuary, an Indian paddled over to assist, and he recognized his lost canoe in the flotilla! Embarrassed, Lewis writes that the man consented very willingly to take an elk's skin for the vessel and then immediately departed.[116] Is receiving an elk skin a fair trade for a handcrafted canoe? Probably not, unless a man stands alone before thirty-one others, each holding a weapon.

Retracing their route on the Columbia River had an advantage in that it allowed the journalists in the expedition to confirm some of their earlier observations. Sergeant Gass, for example, had only to spend a few days with the clumsy dugout canoes plodding against the current on the surging Columbia River to remember that "The natives of this country ought to have the credit of making the finest canoes, perhaps in the world, both as to service and beauty."[117] Lewis also appreciated the native canoes, and on April 1 he purchased one for 36 feet of strung colored beads. Twice the expedition located tributaries of the Columbia that they had missed on the journey going west, causing Clark to hurriedly include the Lewis River and the Willamette River on his master map.

The most noticeable change in attitude by the expedition members on the return journey was their stance toward Indians. They are, complained Lewis, "very unfriendly, and seemed illy disposed."[118] Late one night, "the centinel detected an old indian man attempting to creep into camp in order to pilfer" a spoon.[119] According to Lewis, the sentry "gave the fellow a few stripes with a switch and sent him off."[120] Lewis characterized the inhabitants of one village as "great rogues" and obliged his men "to keep them at a proper distance from our baggage."[121] Portaging the canoes around the rapids and falls of the Columbia River, however, only brought the two cultures into even closer contact. Near the cascades of the Columbia two Indians attempted to rob John Shields, the expedition blacksmith, after which they "pushed him out of the road."[122] Shields, Lewis continues,

> had nothing to defend himself with except a large knife which he drew with an intention of putting one or both of them to death before they could get themselves in readiness to use their arrows, but discovering his design they declined the combat and instantly fled through the woods.[123]

That same afternoon, "villains" of the same tribe stole Seaman, Lewis's Newfoundland dog. As soon as he heard the news Lewis dispatched "three men in pursuit of the thieves with orders if they made the least resistence or difficulty in surrendering the dog to fire on them."[124] After a chase of 2 miles the expedition patrol came within sight of the bandits whereupon the Indians abandoned the dog and fled.[125]

Fed up, Clark informed the Indians by signs that if they "insulted our men or Stold our property we Should Certainly put them to death."[126] Lewis spoke bluntly as well, saying that next time he

> would shoot the first of them that attempted to steal an article from us. that we were not affraid to fight them, that I had it in my power at that moment to kill them all and set fire to their houses, but it was not my wish to treat them with severity provided they would let my property alone.[127]

In an earlier aside, Lewis wrote, "I am convinced that no other consideration but our number at this moment protects us."[128]

Canoes were both necessary for transportation and an extreme liability when the expedition reached that part of the swollen Columbia River where the Cascades, The Dalles, and Celilo Falls block the river in succession. Spring runoff pushed the river 20 feet higher than the previous fall. Bulky canoes had to be lined through passages, with the added pressure that if the vessel became damaged a replacement would have to be purchased from the Indians at an extravagant price. As a result, the expedition sought to replace their canoes with horses. Lewis mistakenly assumed that the natives would not place a high value on their few horses because the country was too thickly timbered to use them to run game. He reasoned wrong. From the first the Indians dismissed his offers as being inadequate, but at this point the Corps of Discovery had been in the field for one month shy of two years so their supply of trade items was woefully low. Negotiating for horses, especially when one party has little to offer, can be a tedious exercise. And so it was for William Clark, the expedition's designated bargainer.

The Indians offered only poor stock. An impatient Lewis urged Clark to double any price he was offering because delay is expensive, too. Halting their progress only forced the expedition to buy expensive fuel and food from the Indians. Horse-trading became even more contentious at The Dalles. The Indians at this location—which the explorers considered "the Great Mart of all this Country"—loved the challenge of a good barter. A twice-yearly salmon-fishing event at The Dalles brought dozens of tribes, some from hundreds of miles distant, to come to take fish and also to trade wolf hides, cedar bark baskets, elk-teeth jewelry, and a hundred other items. Frustrated by their independence, Lewis characterized the Indians as "poor, dirty, proud, haughty, inhospitable, parsimonious and faithless in every rispect."[129]

East of Celilo Falls the landscape of the Columbia River Valley flattened out and opened up. More and more Indian villages held horses. Soon the expedition had enough animals to pack all the supplies so the canoes were broken up or sold. As Lewis and Clark acquired additional horses the captains divided the men into two alternating groups that rode horses for one day and walked the next. For a time, a Nez Perce man, returning to his home in the Clearwater valley, served as a volunteer guide. The expedition also placed their trust in a Walla Walla guide who offered to lead the Corps on a "passable road" to the confluence of the Snake and Columbia rivers. From there the expedition returned to the homeland of the Nez Perce. Sergeant Gass felt relieved: "All the Indians from the Rocky Mountains to the [Celilo] falls of Columbia are an honest, ingenuous, and well-disposed people; but from the falls to the sea-coast, and along it, they are a rascally, thieving set."[130]

Any euphoric feeling at that moment, however, was tempered by bad news from Nez Perce leaders. The Indians, Lewis reported, believe "that the snow is yet so deep on the mountains that we shall not be able to pass them untill the next full moon or about the first of June; others set the time at still a more distant period."[131] Clark considered this "unwelcom intelligence" to men confined to a diet of horsebeef and roots and all of them anxious to return to St. Louis and their homes.[132] But there were no options, so the Lewis and Clark Expedition waited from May 14 to June 10, 1806. During this downtime of twenty-eight days, Lewis and Clark used the opportunity to hold a formal peace council with the Nez Perce. Clark also developed a medical practice, at one point involving forty people.

The Corps of Discovery chose their extended campsite for its advantage of pasture not only for the horses but also for its potential for hunting, fishing, and roots. Sacagawea had, on several occasions, pointed out edible roots to Lewis and Clark, and she did so again at the Long Camp. Lewis considered her choices not only agreeable tasting but nourishing as well, noting that "The men who were complaining of the head ake and Cholicks yesterday and last night are much better to day."[133] He added in his journal that, "we would make the men collect these roots themselves but there are several speceis of hemlock [. . .] that it is difficult to discriminate [. . .] and we are affraid that they might poison themselves."[134]

When food supplies dwindled, the captains became proactive and ordered Sergeant Ordway to take a few men and explore a river where, they heard from the Indians, one could procure some salmon. Ordway and his companions were absent for a week but returned with seventeen salmon. Lewis thought the "fish were as fat as any I ever saw; sufficiently so to cook themselves without the addition of grease; those which were sound were extreemly delicious; their flesh is of a fine rose colour."[135]

For weeks Lewis could only look longingly at "that icy barier which seperates me from my friends and Country."[136] But in early June the water in the Clearwater River rose, a good sign that the snows were melting in the mountains. According to the Indians, when the water in the river slowed enough for horses to swim across then the spring melt would be complete and the mountain trails would become passable.

Anxious to get started after twenty-seven days in camp, the expedition repositioned themselves to a staging camp directly in front of the Lolo Trail. "Every body seems anxious to be in motion," Lewis noted, "convinced that we have not now any time to delay if the calculation is to reach the United States this season."[137] Clark confessed his concerns about returning to the Lolo Trail when he scribbled in his journal that he could not help but "Shudder with the expectation," knowing from past experience the "great dificuelties in passing those Mountains."[138]

A hard rain fell during most of the first day on the Lolo Trail. The single-file line of men on horseback forded creeks heavy with water then pushed upward through broken country where fallen timber and slippery roads continued to slow their progress. On the second day out, the line of sixty-six horses encountered 3 feet of snow. Gass writes that when the snow became 12 feet deep landmarks were obscured so the captains called a halt "to determine what was best to be done, as it appeared not only imprudent but highly dangerous to proceed without a guide of any kind."[139] It took two hours to reach the decision no one wanted to hear: they must retrace their steps back to the staging camp and hire a guide. Lewis realized that

> if we proceeded and should get bewildered in these mountains the certainty was that we should loose all our horses and consequently our baggage instruments perhaps our papers and thus eminently wrisk the loss of the discoveries which we had already made if we should be so fortunate as to escape with life.[140]

To do otherwise, he concluded, was madness.

Lewis and Clark sent two men back to the Nez Perce village to offer a rifle to anyone who would serve as a guide. Meantime, the rest of the expedition sullenly retraced their steps down the mountain and camped only 2 miles from where they had started. It took several days for the Corps of Discovery to add five Nez Perce guides. But, as a result, the reconfigured expedition's second attempt at crossing the Bitterroot Range of mountains went much better. Day after day the travelers hugged a trail along the steep side of a mountain, literally walking themselves out of the snow pack. Looking somewhat like a triumphant procession, the Corps crossed the summit of the mountains and began the descent. Their first stop on the east side of the mountains was at Lolo Hot Springs where "both the Men and the indians amused themselves with the use of a bath this evening."[141] Clark's journal entry for June 30, 1806, gets right to the point. After 156 miles on the Lolo Trail, the expedition, he said, "Descended the mountain to Travellers rest leaveing those tremendious mountanes behind us—in passing of which we have experiensed Cold and hunger of which I shall ever remember."[142] Lewis's journal entry for the day looks forward to resting before we "make our final arrangements for Seperation."[143] The important word in his statement is "Seperation. "

So confident had Lewis and Clark become in their judgment, and the abilities of the men they led, that in February, while at Fort Clatsop, they devised a travel plan for the return trip that divided the expedition into two groups. One unit would explore an Indian path reputed to link the Lolo Trail with the Great Falls of the Missouri. Later it would investigate the Marias River as a possible fur-trade route into Canada. The other unit, also led by a captain, would seek out the previously unexplored Yellowstone River tributary of the Missouri. The finalized plan, on July 1, at Traveler's Rest, involved three additional units, each commanded by a sergeant. All units would rendezvous at the junction of the Missouri and Yellowstone rivers, approximately the location of today's border between Montana and North Dakota, and roughly 850 miles distant from where they stood. It was a bold plan, the purpose of which was to include in their final report to President Thomas Jefferson additional details about Louisiana territory geography. Maybe the "Separation" seems foolhardy today, since it put small units of men at risk from hostile tribesmen and unpredictable natural elements. But at the time no such fears existed among the men who eagerly volunteered for one or the other of the patrols. And so, on July 3, 1806, the Lewis and Clark Expedition separated at Traveler's Rest. "We saddled our horses and set out," Lewis recorded in his journal, "I took leave of my worthy friend and companion Capt. Clark and the party that accompanyed him. I could not avoid feeling much concern on this occasion."[144]

Clark's unit consisted of twenty-three members of the expedition riding or leading fifty horses. The route down the Bitterroot Valley was familiar—the reverse of the way the expedition had proceeded to Traveler's Rest in September of 1805—so they made 36 miles the first day and 29 the second. As he passed through the campsite where he had visited the Flatheads in August 1805, Clark remembered some information imparted to him by the tribe with the result that instead of retracing his steps to the Lemhi River by way of Lost Trail Pass, he led his party across the mountains at present-day Gibbons Pass and into the previously unexplored Big Hole River Valley. Sacagawea seemed familiar with the area. From the Big Hole River it was an easy transition to the Beaverhead River, and, by July 8, after 164 miles on the trail, Clark's branch of the expedition rested at Camp Fortunate, the location where Lewis had secreted their canoes in August 1805 after meeting up with Clark. Six of the seven canoes were serviceable. Most of the unit now took to the water, paddling with the current all the way to Three Forks. Meantime, Sergeant Nathaniel Pryor and six men herded the horses to the same destination, both parties arriving within an hour of each other. It had taken them four days to travel in 1806 what took them three weeks in summer 1805.

Wasting little time at Three Forks, Clark's unit divided again. Sergeant John Ordway and nine men resumed paddling on the Missouri River in the direction of the Great Falls. Here he planned to meet up with a secondary unit from Lewis's group and together the two of them would portage the canoes around Great Falls and then join Lewis, who would be exploring the Marias River. Clark, with ten men plus Sacagawea and her child, Pomp, would in the meantime take the horses and make their way east. Happily, a well-trod buffalo path soon presented itself, and Clark's unit crossed Bozeman Pass (5,800 feet in elevation) and soon intercepted the Yellowstone River. All of this country was new to the men of the expedition.

Lewis's unit, meantime, consisted of nine members of the expedition either riding or leading seventeen horses north from Traveler's Rest. For a short period they benefited from the company of five Nez Perce guides. In the vicinity of present-day Missoula, Montana, they transferred from the Bitterroot River to the Clark Fork River and then to the Blackfoot River, the latter of which was known to the Nez Perce as the "River of Road to Buffalo," meaning Great Falls of the Missouri. The road proved to be obscure and rugged, especially where it crossed the Continental Divide at 6,284 feet elevation. But once across it the Corps of Discovery had reentered the Louisiana Purchase territory of the USA and easily followed the downward slope of a vast plain watered by the Sun River, which Lewis took directly into the Great Falls. Lewis estimated there were 10,000 bison grazing on the plains, and this time, unlike 1805, the expedition had horses with which to hunt them! How efficient was this route? It had taken the expedition fifty-seven days in 1805 to travel between the Great Falls and Traveler's Rest, but in 1806 Lewis achieved the same result in just eight days! Little by little the geographic mysteries of the Louisiana Purchase territory were being unlayered by the Lewis and Clark Expedition.

Wasting little time, Lewis's unit divided at Great Falls. Sergeant Gass received orders to wait for the arrival of Sergeant Ordway, who was bringing the canoes from Three Forks, and their combined units would transport the canoes and baggage around the falls and then proceed to the confluence of the Marias River with the Missouri and wait for Lewis and a small contingent who would be testing the usefulness of the Marias River. Lewis left orders with Gass to,

> wait at the mouth of Maria's river to the 1st of Sept. at which time, should he not arrive, we were to proceed on and join Capt. Clarke at the mouth of the Yellow-stone river, and then to return home: but informed us, that should his life and health be preserved he would meet us at the mouth of Maria's river on the 5th of August.[145]

Ordway and his men joined Gass and his cadre on July 19, four days after Lewis departed for the Marias. In spite of harassment by mosquitoes and grizzly bears—one bear attacked a man on horseback, knocking both to the ground; the horse recovered and ran away and the man climbed a tree and waited out the bear for three hours—the two sergeants accomplished their task in eight days. In 1805, it had taken eleven days, but then the sergeants had no horses to assist them. Five canoes made it through the portage unharmed, and the white pirogue was recovered from its cache before the combined unit moved on to the Marias River rendezvous location. Two days later, on July 28, Lewis and his men reunited with the sergeants. They arrived disoriented and exhausted after having ridden approximately 120 miles since dawn of the day before.

The story Lewis told at the Marias held his men spellbound. Going up the Marias, from the beginning Lewis was on the lookout, knowing that

> the Blackfeet indians rove through this quarter of the country and as they are a vicious lawless and reather an abandoned set of wretches I wish to avoid an interview with them if possible. I have no doubt but they would steel our horses if they have it in their power and finding us weak should they happen to be numerous wil most probably attempt to rob us of our arms and baggage.[146]

When the stream began to peter out Lewis realized it had not the importance which he had once given it. He camped in order to take some scientific readings by starlight, for this was as far north as the expedition would ever get, and he called the location Camp Disappointment. The food began to run out. Then a blustery weather system blew across the plains that made astrological observations impossible. When Lewis's watch failed, it convinced him that "the fates were against me."[147] Indeed they were.

On the morning his unit abandoned Camp Disappointment, Lewis spied in the distance "a very unpleasant sight" of eight Indians with horses about a mile away.[148] Were they a war party or a hunting party? Lewis approached them directly, everyone shaking hands. Lewis distributed some presents to the Piegan Indians, a division of the Blackfeet tribes. The Indians talked of trading at a post in Canada where they "obtain arm amunition sperituous liquor blankets &c in exchange for wolves and some beaver skins."[149] Lewis talked about having "been to the great waters where the sun sets and had seen a great many nations all of whom I had invited to come and trade with me on the rivers on this side of the mountains."[150] Everyone camped together that evening, Lewis and George Drouillard even sleeping inside the Indian tipi. The other two expedition members stood watch outside. A little after daybreak, Lewis awoke to angry voices. One of the Blackfeet, the one wearing a peace medal given to him by Lewis the night before, had grabbed the guns of Lewis's dozing sentries, and another Indian seized the rifles belonging to Lewis and Drouillard. The Americans chased the Indians, wrestling one to the ground, but the other suffered a stab wound to the heart. In the confusion, the remaining Indians stampeded the explorers' horses. Lewis shot one of the Indians in the stomach with his recovered rifle. The wounded man fired back. Six Indians hurried away, quickly followed, going in the opposite direction by Lewis and his men on four of their horses, plus several of the Indian's mounts. Fearing the Blackfeet had friends nearby, the explorers rode as far as they could before resting, then continued on after dark. When the sun rose the next morning, they estimated that in twenty-four hours they had traveled 112 miles.[151] Later that morning of July 28, from a vantage point above the Missouri River, the riders saw and heard the expedition canoes under command of the sergeants and hailed their canoes to stop. The reunion brought "unspeakable satisfaction" to Lewis and his men.[152]

With no time to lose, all the horses were set free, and all hands turned out to excavate the red pirogue, previously cached at the confluence of the Marias River. But it suffered too much decay to be recovered, though some of the supplies were still useable. Lewis acknowledged that, "I was so soar from my ride yesterday that I could scarcely stand."[153] Nevertheless, into the boats went everyone and everything, and they did not stop for the night until they had put 15 watery miles between themselves and the Marias, and maybe the Blackfeet.

Each day for the next two weeks the outfit under Lewis's command paddled with the current of the Missouri River toward the junction with the Yellowstone River. One day they logged 73 miles on the bloated, free-flowing river. The hunting was good, too. That same day hunters killed twenty-nine deer. Snags in the river several times upset the vessels, even throwing men into the water. On August 7, after 83 miles of travel, Lewis's party reached the Yellowstone. Here they found a "few words written or traced in the sand" from Captain Clark saying he would be waiting a few miles below.[154] There had also been a note posted on a rack of elk horns, but that was gone, taken by the leader of the second party to reach this location, Sergeant Nathaniel Pryor.

Clark's speed through the Bozeman Pass had been facilitated by both buffalo tracks worn deep into the dirt and the memory of Sacagawea: "The indian woman who has been of Great service to me as a pilot through this Country recommends a gap in the mountain more south which I shall cross."[155] His unit's fast pace on the Yellowstone River was made possible not only by the fact that his men carried only personal baggage but also that they were able to locate suitable cottonwood trees and to shape them into two dugout canoes. The horses became less important upon completion of the canoes, and that fact was underscored when two dozen of the horses disappeared overnight, likely stolen by Crow Indians. Thus, Sergeant Pryor and three privates of his choosing had only to look after twenty-six horses when Clark ordered them to herd the remaining mounts to the Mandan villages. Ultimately, Pryor suffered the rest of the horses to be stolen as well, and, rather than walk, he and his men also took to the water. Relying on skills they had learned from the Mandans in 1804, Pryor's men killed and skinned buffalo and fashioned two 7-foot-diameter circular "bull boats" reinforced by sticks that were roped at right angles. Their ingenuity paid off, and, after 150 miles on the Yellowstone they had nearly caught up to Clark, who by that time had abandoned the prearranged rendezvous point in the face of marauding mosquitoes and waited on a sandbar farther down the Missouri River. It was Pryor who discovered Clark's note to Lewis at the mouth of the Yellowstone, and, thinking that Lewis was ahead of him, he just brought it with him. In spite of the disappointment of learning that the horses had been stolen, Clark was much relieved to see Pryor. Several times, while canoeing down the Yellowstone, Clark had angered grizzly bears on the shore, at least two of them taking to the water to attack his canoes. For Pryor, in the bull boats, this could have been disastrous.

Even before Lewis and his unit caught up to the rest of the expedition Clark met two fur traders, the first of that profession he had seen since April 13, 1805. During the next several weeks he would meet many more trappers. St. Louis and Illinois fur traders, it appeared, were not waiting for Lewis and Clark to return. Actually, one of the traders informed the explorers that people long since given up on them and they were "almost forgotton." None of that mattered when, on August 12, Lewis caught up to the rest of the expedition, and the five units that had acted independently for six weeks were fully reunited.

The Mandan Indians must have been surprised when all thirty-three of the expedition members who had set out in the spring of 1805 returned to their villages on August 14, 1806. While no member of the expedition had died, Lewis had been recently injured. Pierre Crussat, who was nearsighted in one eye and blind in the other, had shot Lewis in the buttocks while hunting. Lewis's wound was not dangerous, and he would recover in enough time to say his heartfelt farewells to the Mandans, to Sacagawea and her son, Pomp, and to John Colter who wished to leave the Corps early and received permission to do so. Charbonneau, who received $500.33 for his services plus the use of his horse and tipi, was similarly detached from the expedition. It seems likely that Lewis and Clark were, on one hand, sad to leave their companions, but, on the other hand, home beckoned. Besides, for five days at the Mandan village, the captains were dragged back into the uneasy world of geopolitics. The Chief of the Minitari complained about the loss of his son, killed by the Blackfeet; the Sioux had attacked some Mandan hunters; the Arikaras were blamed for stealing from the Mandans. And so on.

The Missouri River between the Mandan villages and St. Louis ran fast, as expected. The river had changed some since 1804, however. Willows several feet high covered once-naked sandbars, and the entrance to several tributaries and creeks were now clogged shut by mud. Indian villages where the expedition had paused in 1804 now whisked by as the canoes covered 70 miles or more on some days. At one point the men floated the same distance in fifteen days that had taken them fifty-one days to traverse on the upstream route two years earlier. When they got near the Teton Sioux, Sergeant Gass reports, the men let the Indians know through an interpreter "that if they troubled us, we would kill every one of them."[156] Clark agreed: "I told this man to inform his nation that we had not forgot their treatment to us as we passed up this river &c. that they had treated all the white people who had visited them vry badly."[157] Their scorn was not reserved for only the Tetons, either. With regard to other tribes, the attitude was, according to Elliott Coues, one of the editors of the expedition journals, to hold themselves "in readiness to fire upon any Indian who should offer us the slightest indignity; as we no longer needed their friendship."[158] Hunters killed not only for food but also for specimens that might please President Jefferson. Sergeant Ordway, perhaps anticipating a business opportunity after the expedition, decided that the horns of buffalo could be taken back to the states where they would make excellent knife and fork handles. It was fun for the expedition to camp in some of the same locations as they had on the way upriver in 1804 and to reminisce. Sometimes the fall weather did not cooperate, as days and nights alternated between windy and foggy, but usually very wet.

The first sign of their return to civilization came on September 20 when the men saw cows grazing on the hillside of a Missouri farm. That was soon followed by the sight of the village of La Charette and then a joyous reception at St. Charles where "the people of the Town gathered on the bank and could hardly believe that it was us for they had heard and had believed that we were all dead and were forgotten."[159] Many of the men spent the night in homes of the happy villagers. Moving on the next day, about noon on September 23, 1806, the Corps of Discovery reached St. Louis where they once again received three cheers from the citizenry. The men in the canoes responded with gunshots of joy. They would no longer need to conserve their powder and ball for defense against the unknown. For them there were no unknowns. They had traveled an 8,000-mile frontier, and they had conquered it in twenty-eight months. John Ordway, like the rest of the men, writes that he now waited for his "Settlement and then we entend to return to our native homes to See our parents once more as we have been so long from them."[160]

THE BICENTENNIAL YEARS OF THE LEWIS AND CLARK EXPEDITION: 2002–2006

Lewis and Clark together penned approximately 750,000 words in eighteen leather-bound notebooks during 863 days of the expedition. After the untimely death of Meriwether Lewis, Nicholas Biddle, a prominent Philadelphia lawyer and banker, prepared the captain's journals for publication at the request of William Clark. Instead of a verbatim account of the 3,056 pages of journals, however, Biddle recast them into a 370,000-word narrative that he called *The History of the Expedition Under the Commands of Captains Lewis and Clark* (Philadelphia, Pa., 1814). Clark earned $180 in royalties. Biddle's narrative enjoyed a renaissance in 1893 when Elliott Coues reedited the book. He also expanded the text, added chapter synopses, included maps, made an index, wrote a bibliographic essay, and greatly improved the volume with hundreds of authoritative footnotes.

It took until 1904–1905 for a verbatim edition of the journals to reach print. Reuben Gold Thwaites accomplished the task in eight volumes, including the journals of Sergeant Charles Floyd and Private Joseph Whitehouse, the only writings by the enlisted men available at the time. The series was reprinted several times until superceded by the University of Nebraska Press edition of *The Journals of the Lewis & Clark Expedition* edited by Gary E. Moulton between 1983 and 2001. Moulton's thirteen-volume, expertly footnoted edition of the captain's journals also included an atlas, a herbarium of the expedition, the journals of John Ordway, Charles Floyd, Patrick Gass, and Joseph Whitehouse, plus a general index.

The Corps of Northwest Discovery paid sizable dividends to the USA in the support it gave to America's claim to the Oregon country. Until 1989, few scholars recognized an equally important aspect of the Lewis and Clark Expedition: the scientific knowledge it gathered in the areas of natural history, geography, and ethnography. Consider that together Lewis and Clark identified a dozen fishes and fifteen reptiles and amphibians new to science. In addition, they recorded descriptions of 134 species and subspecies of birds on their journey, twenty-five of which were sufficiently unknown to rate discovery status. And they wrote the first scientific reports on forty-four new mammals, including the pronghorn antelope, mule deer, mountain lion, mountain goat, and grizzly bear. Lewis and Clark gathered 216 herbs, shrubs, seeds, roots, cuttings, and trees for transport back to the states. Many of them survive today in the Lewis & Clark Herbarium at the Philadelphia Academy of Science. Cartographic scholars have called the fifty-eight maps prepared by William Clark documents of towering significance. And the expedition observed fifty-three tribes of American Indians who lived in three distinctly different cultural divisions: the Plains, the Columbia Plateau, and the Northwest Coast.

The most insightful comment ever made about the Lewis and Clark Expedition comes from Donald Jackson, an author and editor. Each generation of Americans, he said, discover Lewis and Clark for themselves. During the nineteenth century, Americans centered what little interest they had in the expedition on the economic opportunities in the West that the journals seemed to confirm. There was no hero worship because Americans still had more wilderness than it did city streets. The low-key Lewis and Clark Centennial in the twentieth century emphasized the advancement of technology and civilization, not so much the explorers. The title of Portland, Oregon's 1905–1906 "Lewis and Clark Centennial and American Pacific Exposition and Oriental Fair" says all one needs to know about how the Corps of Discovery had to share the spotlight. The Lewis and Clark Sesquicentennial in 1955 took place without major notoriety, but it did have one great accomplishment. It called attention to the trail itself. Where was it? In 1964, Congress answered the pesky question when it authorized a Lewis and Clark Trail Commission and called upon it to identify expedition landmarks. The Commission actively placed labels along the trail, it designed a distinctive highway emblem for those parts of the trail covered by concrete, and, most importantly, it pulled several members of Congress, plus the heads of several federal agencies, however reluctantly into its twenty-eight seats. Simultaneously, eleven states formed their own Lewis and Clark Trail Committees to preserve sites and to publish educational brochures. When the federal commission expired in 1969, the Lewis and Clark Trail Heritage Foundation, a private, nonprofit organization, assumed its functions and does so to this day.

The West of Lewis and Clark underwent a transition in 2000 that was just as significant as the millennium change in the calendar. The ethnic and gender diversity of the expedition personnel caught public attention, as did ethnographic misconceptions of the captains toward American Indians. As the twenty-first century began, students of the expedition used the journals of Lewis and Clark to gauge the decline of the trans-Mississippi environment. Issues such as coal and oil exploration in the West, Indian casinos, the endangered-species list, the decline of family farms on the Great Plains, and rivers engineered more for hydroelectric power than salmon asked what would Lewis and Clark think today? Pretty soon every city with a Lewis and Clark-named landmark, school, or wildlife area got behind the idea of a bicentennial celebration. Communities without a direct connection to the Corps of Discovery enjoyed a virtual connection by joining with millions of individuals in reading Stephen Ambrose's 1996 blockbuster bestseller, *Undaunted Courage: Meriwether Lewis, Thomas Jefferson, and the Opening of the American West.*

Ambrose, a board member of the environmental group American Rivers urged the Lewis and Clark Trail Heritage Foundation to support his group's proposal for a National Council of the Lewis & Clark Bicentennial. When they did, progress really began to take place. The uniting theme would be conservation, specifically "Protecting the Rivers of Lewis and Clark." A subsequent four-hour PBS-TV documentary by Ken Burns, Lewis and Clark: The Journey of the Corps of Discovery enriched the idea of the bicentennial and expanded the agenda beyond rivers. After that tourism strategists got involved, the Advertising Council offered to provide up to $100 million in pro-bono public-service announcements, and soon politicians at both the state and national levels leaped on board. Eighty-two members of Congress formed a Lewis and Clark Caucus. The National Park Service competed for a leadership role—they had been, after all, somewhat absentee administrators of the 3,700-mile-long Lewis and Clark National Historic Trail that Congress established in 1978—and American Indian tribes recognized an opportunity to tell their side of the story to an eager audience under the umbrella of a "heritage tourism" strategy. The Sierra Club launched the largest campaign in its history, "In the Footsteps of Lewis and Clark."

The kickoff bicentennial event took place in the East Room of the White House in July 2002, and the first of fifteen Signature Events came to order in January 2003 in Charlottesville, Virginia, the home of Thomas Jefferson. The four-day attendance exceeded expectations. Attendance at the remaining Signature Events on the calendar, scheduled through 2006, should be equally high anticipated the leadership of the Bicentennial Council because, as Ken Burns said, when it comes to Lewis and Clark Americans have a sense of ownership. Trouble was, business did not feel the same. Perhaps stung by the lack of public response to the 1992 quincentennial of Christopher Columbus's voyage to the New World, no corporation answered the Bicentennial Council's plea for sponsorship. A $2.5 million grant from the Hewlett Foundation made it the bicentennial's savior at a critical time, but lots more funding was needed.

A weak financial condition limited the effectiveness of the bicentennial. Small amounts of state and federal money could not properly fund the big ideas of the Bicentennial Council. Early plans called for a $500-million federal bicentennial budget. But the events of 9/11 and the resulting economic downturn put an end to that wishful thinking. Sales of a half-million bicentennial commemorative coins, however, did provide much-needed revenue. Some state legislatures were more generous than others. Montana earmarked $200,000 for Lewis and Clark events to the Travel Montana program each year of the bicentennial. The Signature Events went as planned, and people came, but without corporate support local organizers had to rely more and more on revenue generated from the sale of Lewis and Clark T-shirts, beef jerky, pins, notecards, and the like.

The legacy of the bicentennial is not the re-editing of the expedition journals. That project of the University of Nebraska was completed before the bicentennial began. One genuine accomplishment of the bicentennial is that Euro-Americans rediscovered their history by listening to American Indians. When asked to participate, the only tribe that boycotted the bicentennial was the Lemi Shoshone, Sacagawea's tribe. The list of cooperating tribes reached 130 at one point. Tribal reluctance to revisit the past unfortunately influenced the outcome of some events. At Chamberlain, South Dakota, in 2004, for example, the Lewis and Clark re-enactors sponsored by the National Park Service were confronted by a group of Sioux who told them to halt and go back. That made the national news, and a lot of people canceled their travel plans! Attendance at the South Dakota Signature Event dropped to maybe 5,000 people, about 10 percent of the number of people who took in the Signature Events in adjacent Nebraska and North Dakota. The Chinook tribe of Indians also disapproved the Signature Event in Washington state as part of a protest that began when the Bureau of Indian Affairs in 2002 revoked their status as a recognized tribe. By contrast, in June 2006, the Nez Perce Tribe was the principal host for Idaho's Signature Event.

Another accomplishment of the bicentennial is that the Sierra Club and some sixty-odd like-minded organizations found common ground and worked together as never before on environmental and educational issues. The same can be said of federal organizations such as the National Park Service, the US Forest Service, and the Bureau of Land Management, though to a lesser degree. If, as one person familiar with the bicentennial said about attendance at the events, "it was a single, not a home run," then at least the bicentennial kept the ball in play.[161] Henceforth, the Lewis and Clark Expedition will, as Donald Jackson predicted, continue to be discovered by generation after generation of Americans.

NOTES

1. The names of the members of the permanent party were disclosed on April 1, 1804. Some adjustments would take place during the winter of 1804–1805 at Fort Clark near the Mandan villages.
2. President Jefferson asked Lewis to be his private secretary early in 1801. Lewis received the letter in Pittsburgh on March 5; he accepted on March 12; and he assumed his new duties on April 1 at the salary of $500 year.
3. Donald Jackson, *Letters of the Lewis and Clark Expedition with Related Documents,* 2nd edn (Urbana, Ill.: University of Illinois Press, 1978), vol. I, pp. 18–19.
4. References in the journals written by Lewis and Clark make it clear they brought with them about a dozen books, though the exact number is not known. They also took with them a number of manuscripts and maps. See Stephen Dow Beckham, *The Literature of the Lewis and Clark Expedition* (Portland, Oreg.: Lewis & Clark College, 2003).
5. "A Dog Named Scannon—Until Recently," in Donald Jackson, *Among the Sleeping Giants* (Urbana, Ill.: University of Illinois Press, 1987), pp. 43–53.
6. Louisiana received its name from *René-Robert Cavelier,* Sieur *de La Salle,* a French explorer. France kept possession, at least on paper, until 1762, when it ceded the claim to Spain. Forty years later, Napoleon Bonaparte regained Louisiana from Spain, an action that sparked a response from President Jefferson who feared the establishment of a new colonial empire in North America.
7. Jackson, *Letters,* vol. I, p. 61.
8. Clark's letter of agreement to participate in the expedition with the rank of captain and co-commander reached Lewis on July 29. Lewis responded on August 3 and established Clarksville as their rendezvous site.
9. In December 1783, Thomas Jefferson had asked George Rogers Clark to lead an exploring party west from the Mississippi River but Clark declined in February 1784, saying he wished to just live the rest of his life in Kentucky.
10. Lewis brought with him from Pittsburgh three good recruits. Clark added seven men who would ultimately be chosen for the permanent party. One of them, John Shields, was married, an undesirable status for an extended exploration, but an exception was made because he was a blacksmith.
11. The leaders of the expedition and most of the men who volunteered were in the army. A paperwork mistake too late to fix by the War Department commissioned Clark a second lieutenant, even though Lewis had promised him a captaincy. Clark lived with the slight until the expedition was completed and then revealed his embarrassment. The foul-up was concealed from the men, and in all respects both men were regarded as captains in rank. See William E. Foley, *Wilderness Journey: The Life of William Clark* (Columbia, Miss.: University of Missouri Press, 2004), pp. 28, 70–72.
12. Because of natural shifts in the channels of the Mississippi and Missouri rivers, the probable site of Lewis and Clark's five-month-long winter camp is now within the city limits of West Alton, Missouri.
13. See W. Raymond Wood, *Prologue to Lewis & Clark: The Mackay and Evans Expedition* (Norman, Okla.: University of Oklahoma Press, 2003).
14. James Ronda, *Lewis and Clark among the Indians* (Lincoln, Nebr.: University of Nebraska Press, 1984), pp. 2–8.
15. According to Clark's timetable estimate on January 21, 1804, it would take 190 days, or a little over six months, to reach and establish their winter camp at the source of the Missouri River in the "rock mountains." In spring 1805 they would restart the expedition and reach the Pacific Ocean in fifty-four days after 650 miles. After a delay of fifteen days at the ocean the party would return to the source of the Missouri River in fifty-four days, then to the Mandan villages in another forty-five days,

and be back in St. Louis in December 1805. *The Journals of the Lewis & Clark Expedition,* 13 vols., ed. Gary L. Moulton (Lincoln, Nebr.: University of Nebraska Press, 1986), vol. II, p. 159. Hereafter, direct quotes, except those with the journal's date of entry provided in the text, will be identified by date, the complete text of which can be found by using the search engine at University of Nebraska Press and University of Nebraska, Lincoln, Libraries Electronic Text Center, *The Journals of the Lewis and Clark Expedition* (2005), available online at http://lewisandclarkjournals.unl.edu.

16. *Journals,* May 24, 1804 (Clark), vol. II, p. 250.
17. *Journals,* May 25, 1804 (Ordway), vol. IX, p. 7.
18. Jackson, *Letters,* vol. I, p. 232.
19. Jackson, *Letters,* vol. I, p. 232. Lewis did not write in his journal between September 19 and November 11, 1803; May 14, 1804 to April 7, 1805; August 26, 1805 to January 1, 1806; and August 12 to the end of September 1806. Gary E. Moulton, "The Missing Journals of Meriwether Lewis," *Montana: The Magazine of Western History,* 35 (3) (1985): 28–39.
20. *Journals,* July 14, 1804 (Clark), vol. II, pp. 377–378.
21. *Journals,* July 19, 1804, (Clark), vol. II, p. 394.
22. *Journals,* July 19, 1804 (Clark), vol. II, p. 394.
23. Paul Cutright, *Lewis and Clark: Pioneering Naturalists* (rpt, Lincoln, Nebr.: University of Nebraska Press, 1989), lists twenty-two new plant species collected by the expedition on the journey westward across the Great Plains. See also Paul A. Johnsgard, *Lewis and Clark on the Great Plains* (Lincoln, Nebr: University of Nebraska Press, 2003), pp. 3–7.
24. The Lakota or Sioux moved into the area of the middle Missouri River in the mid-1700s where they split into four branches: Teton, Santee, Yankton, and Yanktonai. The Teton, with seven subgroups, was the largest.
25. *Journals,* August 31, 1804 (Clark), vol. III, p. 30.
26. *Journals,* September 27, 1804 (Clark), vol. III, p. 121.
27. The Mandans had lived on the Missouri River for a century before the arrival in North America of Christopher Columbus. Europeans first mentioned them in 1719, but their first visit from a Euro-American came in 1738. The tribe acquired both guns and horses in the 1750s, but neither protected them from the smallpox epidemic of 1781–1782, which reduced their population by tens of thousands. To protect themselves from the Teton Lakota they shared their territory on the Knife River with the Hidatsa Indians. Lewis and Clark did not understand that the Minitari Indians are only a division of the Hidatsa and therefore believed that there were three tribes living at the location.
28. *Journals,* April 7, 1805 (Lewis), vol. IV, p. 10.
29. *Journals,* April 7, 1805 (Lewis), vol. IV, p. 9.
30. Lewis changed his mind on April 14 and decided that the expedition had finally reached "the highest point to which any whiteman had ever ascended; except two Frenchmen . . . tho' to what place precisely I could not learn." Historians believe that French-Canadians first visited the confluence of the Yellowstone in 1802 or 1803. *Journals,* April 14, 1805, vol. IV, p. 36.
31. "In the last night we were alarmed by a Buffalow which Swam from the opposite Shore landed opposite the Perogue in which Capt Lewis & my Self were in he Crossed the perogue, and went with great force up to the fire where Several men were Sleeping and was 18 inches of their heads, when one man Sitting up alarmed him and he turned his course along the range of men as they lay, passing between 4 fires and within a few Inches of some of the mens heads." *Journals,* May 29, 1805 (Clark), vol. IV, p. 218.
32. *Journals,* May 26, 1805 (Lewis), vol. IV, p. 201.
33. *Journals,* June 29, 1805 (Lewis), vol. IV, p. 340.
34. Ronda, *Lewis and Clark among the Indians,* p. 133.
35. *Journals,* July 27, 1805 (Lewis), vol. IV, pp. 436–437.
36. *Journals,* August 12, 1805 (Lewis), vol. V, p. 74.
37. *Journals,* August 13, 1805 (Lewis), vol. V, p. 79.
38. *Journals,* August 13, 1805 (Lewis), vol. V, pp. 436–437.

39. *Journals,* August 24, 1805 (Gass), vol. X, p. 132.
40. *Journals,* September 2, 1805 (Gass), vol. X, p. 136.
41. *Journals,* September 1, 1805 (Whitehouse), vol. XI, p. 294.
42. *Journals,* September 9, 1805, (Lewis), vol. V, p. 192.
43. *Journals,* September 9, 1805, (Lewis), vol. V, p. 192.
44. *Journals,* September 14, 1805 (Clark), vol. V, p. 204.
45. *Journals,* September 14, 1805 (Gass), vol. X, p. 142.
46. *Journals,* September 14, 1805 (Gass), vol. X, p. 142.
47. *Journals,* September 14, 1805 (Ordway), vol. IX, p. 223.
48. *Journals,* September 15, 1805 (Clark), vol. V, p. 206.
49. *Journals,* September 16, 1805 (Clark), vol. V, p. 209.
50. *Journals,* September 18, 1805 (Clark), vol. V, p. 213.
51. *Journals,* September 19, 1805 (Lewis), vol. V, p. 215.
52. *Journals,* September 19, 1805 (Lewis), vol. V, p. 215.
53. *Journals,* September 19, 1805 (Gass), vol. X, p. 145.
54. *Journals,* September 20, 1805, vol. V, p. 219.
55. *Journals,* September 20, 1805 (Clark), vol. V, p. 223.
56. *Journals,* September 20, 1805 (Clark), vol. V, p. 223.
57. *Journals,* September 21, 1805 (Clark), vol. V, p. 227.
58. *Journals,* September 22, 1805 (Clark), vol. V, p. 230.
59. *Journals,* September 24, 1805 (Clark), vol. V, p. 232.
60. *Journals,* September 26, 1805 (Whitehouse), vol. XI, p. 334.
61. *Journals,* September 28, 1805 (Clark), vol. V, p. 235.
62. *Journals,* September 22, 1805 (Clark), vol. V, p. 230.
63. *Journals,* September 22, 1805 (Clark), vol. V, p. 230.
64. *Journals,* September 26, 1805 (Clark), vol. V, p. 234.
65. *Journals,* September 22 and 26, 1805 (Clark), vol. V, p. 234.
66. *Journals,* September 26, 1805 (Clark), vol. V, p. 234.
67. The branding iron was found near The Dalles, Oregon in the early 1890s and today is a possession of the Oregon Historical Society in Portland.
68. *Journals,* October 5, 1805 (Clark), vol. V, p. 246.
69. *Journals,* October 8, 1805 (Whitehouse), vol. XI, p. 342.
70. *Journals,* October 9, 1805 (Clark), vol. V, p. 253.
71. *Journals,* October 10, 1805 (Clark), vol. V, p. 256.
72. *Journals,* October 15, 1805 (Gass), vol. X, p. 155.
73. *Journals,* October 13, 1805 (Clark), vol. V, p. 266.
74. *Journals,* October 16, 1805 (Clark), vol. V, p. 278.
75. *Journals,* October 17, 1805 (Clark), vol. V, p. 288.
76. *Journals,* October 17, 1805 (Clark), vol. V, p. 286.
77. *Journals,* October 23, 1805 (Gass), vol. X, p. 161.
78. *Journals,* October 23, 1805 (Lewis), vol. V, p. 328.
79. *Journals,* October 24, 1805 (Clark) p. 333.
80. *Journals,* October 24, 1805 (Clark).
81. *Journals,* October 25, 1805 (Clark), vol. V, p. 338.
82. *Journals,* October 27, 1805 (Clark), vol. V, p. 345.
83. *Journals,* October 17, 1805 and November 1, 1805 (Clark), vol. V, p. 289, 369.
84. *Journals,* January 30, 1805 (Clark), vol. VI, p. 252. Sergeant Gass laid the groundwork for several decades of historical confusion when he wrote in his journal: "We suppose them to be a band of the Flathead nation, as all their heads are compressed into the same form" (*Journals,* October 27, 1805 [Gass], vol. X, p. 162). The Flathead tribe of Montana denies that their ancestors flattened their heads and they suggest the origin of the name relates to sign language where they are identified by pressing

both sides of the head with the flat of the hands. Gass published his journal in 1807, a year after the return of the expedition to St. Louis and seven years before the journals of Lewis and Clark reached print, so his comment was accepted for many years.

85. *Journals,* November 1, 1805 (Clark), vol. V, p. 369.
86. *Journals,* October 31, 1805 (Clark), vol. V, p. 362.
87. *Journals,* October 31, 1805 (Clark), vol. VI, p. 362.
88. *Journals,* November 1, 1805 (Clark), vol. V, p. 369.
89. *Journals,* November 3, 1805 (Whitehouse), vol. XI, pp. 382–383.
90. *Journals,* November 7, 1805 (Clark), vol. VI, p. 33.
91. *Journals,* November 8, 1805, vol. VI, p. 36.
92. *Journals,* November 9, 1805, vol. VI, p. 38.
93. *Journals,* November 9, 1805 (Clark), vol. VI, p. 38.
94. *Journals,* November 11, 1805 (Clark), vol. VI, p. 41.
95. *Journals,* November 11, 1805 (Clark), vol. VI, p. 40.
96. *Journals,* November 11, 1805 (Clark), vol. VI, p. 41.
97. *Journals,* November 11, 1805 (Clark), vol. VI, p. 40.
98. *Journals,* November 12, 1805 (Clark), vol. VI, p. 43.
99. *Journals,* November 15, 1805 (Clark), vol. VI, p. 50.
100. *Journals,* November 16, 1805 (Gass), vol. X, p. 171.
101. *Journals,* November 20, 1805 (Clark), vol. VI, p. 72.
102. *Journals,* November 20, 1805 (Clark), vol. VI, p. 73.
103. *Journals,* November 21, 1805 (Clark), vol. VI, p. 73.
104. *Journals,* November 24, 1805 (Clark), vol. VI, p. 85.
105. *Journals,* November 24, 1805 (Gass), vol. X, p. 177.
106. *Journals,* November 24, 1805 (Whitehouse), vol. XI, p. 398.
107. *Journals,* November 24, 1805 (Ordway), vol. IX, p. 256.
108. *Journals,* November 24, 1805 (Clark), vol. VI, p. 84.
109. Stephen Ambrose in *Undaunted Courage* (New York: Simon & Schuster, 1996), p. 311, states that when the expedition chose its winter campsite in November 1805 it was "the first time in American history that a black slave had voted, the first time a woman had voted." *Journals,* November 24, 1805 (Clark), vol. VI, pp. 83–84; (Gass), vol. X, p. 177; (Whitehouse), vol. XI, p. 398; (Ordway), vol. IX, p. 256.
110. More than 100 ships for which there are records visited the Northwest Coast between 1788 and 1803.
111. *Journals,* March 20, 1806 (Gass), vol. X, p. 199, (Whitehouse), vol. XI, p. 430.
112. *Journals,* March 23, 1806 (Ordway), vol. IX, p. 280.
113. *Journals,* March 23, 1806 (Lewis), vol. VII, p. 7.
114. *Journals,* March 17, 1806 (Clark), vol. VI, p. 428.
115. *Journals,* March 18, 1806 (Ordway), vol. IX, p. 278.
116. *Journals,* March 24, 1806 (Lewis), vol. VII, p. 8.
117. *Journals,* March 30, 1806 (Gass), vol. X, pp. 203–204.
118. *Journals,* April 9, 1806 (Lewis), vol. VII, p. 97.
119. *Journals,* April 8, 1806 (Lewis), vol. VII, pp. 94–95.
120. *Journals,* April 8, 1806 (Lewis), vol. VII, pp. 94–95.
121. *Journals,* April 9, 1806 (Lewis), vol. VII, p. 98.
122. *Journals,* April 11, 1806 (Lewis), vol. VII, p. 105.
123. *Journals,* April 11, 1806 (Lewis), vol. VII, p. 105.
124. *Journals,* April 11, 1806 (Lewis), vol. VII, p. 105.
125. *Journals,* April 11, 1806 (Lewis), vol. VII, p. 105.
126. *Journals,* April 11, 1806 (Clark), vol. VII, p. 109.
127. *Journals,* April 21, 1806 (Lewis), vol. VII, p. 152.

128. *Journals,* April 11, 1806 (Lewis), vol. VII, p. 106.
129. *Journals,* April 20, 1806 (Lewis), vol. VII, p. 146.
130. *Journals,* May 7, 1806 (Gass), vol. X, p. 223.
131. *Journals,* May 7, 1806 (Lewis), vol. VII, p. 222.
132. *Journals,* May 7, 1806 (Lewis), vol. VII, p. 222.
133. *Journals,* May 16, 1806 (Clark), vol. VII, p. 265.
134. *Journals,* May 21, 1806 (Lewis), vol. VII, p. 275.
135. *Journals,* June 2, 1806 (Lewis), vol. VII, p. 327.
136. *Journals,* May 17, 1806 (Lewis), vol. VII, p. 267.
137. *Journals,* June 14, 1806 (Lewis).
138. *Journals,* June 14, 1806 (Clark), vol. VIII, p. 24.
139. *Journals,* June 17, 1806 (Gass), vol. X, p. 240.
140. *Journals,* June 17, 1806 (Lewis), vol. VIII, p. 31.
141. *Journals,* June 29, 1806 (Lewis), vol. VIII, p. 64.
142. *Journals,* June 30, 1806 (Clark), vol. VIII, p. 68.
143. *Journals,* June 30, 1806 (Lewis), vol. VIII, p. 66.
144. *Journals,* July 3, 1806 (Lewis), vol. VIII, p. 83.
145. *Journals,* July 16, 1806 (Gass), vol. X, p. 254.
146. *Journals,* July 17, 1806 (Lewis), vol. VIII, p. 113.
147. *Journals,* July 25, 1806 (Lewis), vol. VIII, p. 127.
148. *Journals,* July 26, 1806 (Lewis), vol. VIII, p. 128.
149. *Journals,* July 26, 1805 (Lewis), vol. VIII, p. 131.
150. *Journals,* July 26, 1805 (Lewis), vol. VIII, p. 131.
151. Among several legendary riders for the Pony Express in 1860, Jim Moore logged 280 miles in fourteen hours and forty-six minutes. He had the advantage, however, of changing to rested mounts every 35 miles or so.
152. *Journals,* July 28, 1806 (Lewis), vol. VIII, p. 138.
153. *Journals,* July 28, 1806 (Lewis), vol. VIII, p. 137.
154. *Journals,* August 7, 1806 (Gass), vol. VIII, p. 263.
155. *Journals,* July 13, 1806 (Clark), vol. VIII, p. 180.
156. *Journals,* August 30, 1806 (Gass), vol. X, p. 274.
157. *Journals,* August 30, 1806 (Clark), vol. VIII, p. 330.
158. Elliott Coues (ed.), *History of the Lewis and Clark Expedition,* 3 vols. (rpt, New York: Dover Publications, 1965), p. 1208.
159. *Journals,* September 21, 1806 (Ordway), vol. IX, p. 365.
160. *Journals,* September 23, 1806 (Ordway), vol. IX, p. 366.
161. See David Sarasohn, *Waiting for Lewis and Clark: The Bicentennial and the Changing West* (Portland, Oreg.: Oregon Historical Society Press, 2005).

CHAPTER SIX

CONTINUED EXPLORATION OF THE AMERICAN WEST
The Fur Trade, Map Makers, Path Markers

W. R. Swagerty

Between the end of the Seven Years (French and Indian) War in 1763 and the American acquisition of Louisiana territory in 1803, Canadian, Spanish New Mexican, and St. Louis-based fur traders began integrating themselves among tribes residing on the Missouri River. Their focus: the permanent villages of the Mandans and Hidatsas in present-day North Dakota, which served as a major trade center for Indians throughout the region.[1]

Documentation of traders among the Mandans begins in 1738. A half century later, it is clear that Canadian traders were living among their hosts, elevating the importance of the Mandans and Hidatsas in the network of Plains trade, now recognized as the Middle Missouri System.[2] The number of European traders was small, but their impact large as material, biological, and cultural worlds of Europeans and Native people met face to face, bringing Missouri River and Plains tribes as far away as the Rocky Mountains into a global diaspora generally described by historians as the "Fur Trade."[3]

Driven by European and Asian depletion of their own fur-bearing animals and cultural traditions of placing high status on specific furs such as sea otter, used for fashionable collars in China; mink, sable, ermine, and fox for coats and trim of the elite in Europe; beaver, whose waterproof felt provided civilians and military alike with hats; deer and elk for gloves, shirts, and leggings; moose hides for "buff coats," the precursor to the bulletproof vest that could slow musket balls and stop a sharp blade or sword; and buffalo, whose robes warmed beds, carriages, and soldiers and whose tanned hides provided essential leather for everything from boots and bags to tension belts on wheels and gears in the earliest stages of the industrial revolution.[4]

INDIAN TRADE NETWORKS DURING "DOG DAYS"

Better thought of as an expansion of a very old "Indian Trade," the system worked well even prior to Spanish entry in the Southwest in the sixteenth century. In present-day New Mexico and Arizona, permanent villages of the Pueblo Indians served as specialized production and collection points for commodities desired by other Puebloans and by semi-nomadic Athapaskan-speaking tribes moving into the Southwest after A.D. 800. Pecos to the east; Zuñi to the west both functioned as primary centers, with Taos, Acoma, and the villages of the Hopi serving secondary roles. Pottery, storable foods—especially maize, beans, and squash—textiles made from cotton and yucca fiber, exotic shells, turquoise, feathers, and minerals (for face and body paint) from as far away as the Pacific coast made their way to these Pueblo villages through down-the-line trade, each intermediate trading culture gaining exotics in exchange for what they produced locally.[5]

This "Greater Southwest Trade System" also incorporated the Pima, Papago, and Colorado river-based tribes such as the Havasupai and Yuma, as well as tribes living below the international boundary in the modern states of Sonora, Chihuahua, Coahuila, and Nuevo Leon, Mexico. Utilizing the lower Rio Grande and Pecos rivers as conduits, Jumano pedestrian traders trudged back and forth between permanent villages of La Junta Culture, where the Conchos and Rio Grande converge, and Pecos Pueblo, the eastern gateway to the southern Plains.[6]

During "dog days," large packs of medium-sized canines dragging small travois (tree-pole sleds from the French, *travail* [or work]) and wearing backpacks strapped to either side of their small torsos, traversed alongside their human masters. In 1590, one Spanish chronicler wrote in his journal, "These people [the Apaches] had with them many loaded dogs, as is the custom in those regions, and we saw them loaded, a thing new to us, never before seen." Ten years later, Juan de Oñate, the colonizer of New Mexico, recorded an Apache trade caravan: "To carry [...] their *pinole,* or maize, the Indians use a medium-sized shaggy dog, which is their substitute for mules. They drive great trains of them. Each, girt round its breast and haunches, carrying a load of flour at least one hundred pounds, travels as fast as his master." We may quibble with Oñate's exaggeration of how many pounds an average dog can carry, but prior to the introduction and diffusion of the Euro-Asian horse into North America, scholars agree that dog transportation was an essential part of the Indian trade.[7] After introduction of the horse, but prior to its full integration, dogs remained essential in seasonal hunts and for transportation. In 1724, a Kansa hunting party of more than 1,000 people was reported by explorer Etienne Veenyard, Sieur de Bourgmont, to have traveled west on the Kansas plains with more than 300 dogs serving as beasts of burden, carrying up to 50 pounds of robes, dried meat, and camp gear between base camps and hunting grounds.[8] And as late as the 1850s, artist John Mix Stanley documented the Sioux using dog travois as well as horse travois.[9]

HORSE DAYS

The arrival of the horse marked the beginning of a transportation revolution and cultural renaissance, remembered by most tribes as a great event. With horses, the volume of baggage, including furs and robes, expanded exponentially. From the Southwest, horses spread north through trade and through theft during and after the Pueblo Revolt of 1680–1692. An expanding trade nexus was built between Apaches, Navajos, Kiowas, and Utes, who bartered with Pueblo peoples and delivered part of their newly gained wealth with Cheyennes, Wichitas, Pawnees, and Hasinai (Caddos) in the heart of the Great Plains. By 1690, the Hasanai had four to five horses per household, but the Pawnee in 1719 were reported to have only acquired one horse per man. That changed rapidly throughout the eighteenth century, and by 1800 every Pawnee village had several thousand horses.

Prior to 1735, no horses are documented north and east of the Missouri on the Plains.

By century's end, the Dakota of Minnesota had given up the canoe for the horse, and horses had become common among all Plains, elevating the stakes of raiding and warfare as villages and hunting grounds of distant tribes previously too far to reach on foot became more easily accessible. Through purchase and stealth, pedestrian tribes such as the Blackfeet and Shoshone as well as water-oriented peoples, especially those on major rivers such as the Salish (Flathead) and Kootenai, and the Nez Perce became the quintessential equestrian cultures on the West, not the Puebloans who first acquired them from the Spaniards.[10]

Ultimately, trade items from the Southwest—carried on horses and dogs—and a small quantity of horse tack and textiles, influenced by Spanish colonization and design, made their way through these Athapaskan- and Caddoan-speaking middleman traders to the upper Missouri Mandan and Arikara villages. On the James River in present-day South Dakota, the Sioux held their own seasonal trade fair, attrracting Teton, Yanktonai, Yankton, and Sisseton.

To their west, the Shoshones began hosting a seasonal summer fair in present-day southwestern Wyoming. The "Shoshone Rendezvous," as it has been since identified, provided otherwise-hostile tribes with a brief annual respite from traditional rivalries and insured distribution of horses between horse-rich and horse-poor tribes across hundreds of miles.[11]

Crows, Nez Perce, and Salish (Flathead) traders extended the network further, taking items from the Plains as well as herds of horses to other gatherings, some seasonal, others permanent villages. Most important among the latter, The Dalles on the Columbia River functioned as the largest trade emporium in the entire American West. Thousands of fishermen, Native traders from as far as British Columbia and California, and their families gathered during the annual salmon runs to harvest millions of pounds of the region's most prized fish, as well as to exchange exotic commodities from afar. Shells, human slaves acquired by aggressive tribes up and down the West Coast, whalebone (carved into clubs), euchalon (candlefish) oil, furs, and skins from terrestrial and pelagic mammals exchanged hands along with smoked fish. Pipestone (Catlinite) from a singular quarry in Minnesota (now a national monument); parfleches (rawhide folded containers carried in doubles over hind haunches by dogs and by horses); feather headdresses; buffalo bone products; pemmican, the trail-food of the day consisting of sun-dried meat mixed with fat and berries for sugar and vitamins; and bison robes arrived from the Plains across the Plateau. Baskets, canoes, minerals, and plant materials for face and body paint joined prized storable foods high in carbohydrates, especially wapatoe, camas, acorns, and bitterroot. Not surprisingly, archaeologists have found items traded at The Dalles in sites from Alaska to California and over 1,000 miles east in the Missouri River trade centers and have labeled this network the "Pacific-Plateau Trade System."[12]

No evidence exists that Native Pacific Coast and Columbia River-based traders made their way across the plains to Catlinite quarry in South Dakota, much less the earth-lodge villages on the Missouri River, but Native cosmology extended well beyond tribal and linguistic boundaries. Earliest explorers, including Peter Fidler, David Thompson, and Alexander McKenzie while working for the North West Company, as well as the Americans Lewis and Clark, made their own maps, but all benefited from Native geographic knowledge of local and extended trade territories.[13] Maps drawn on the ground, painted on hides, incised on tusks and antlers, and embossed on tree bark were common throughout the region.[14]

JEFFERSONIAN-ERA EXPLORATION

Although trade and trade systems were well established prior to the American entry into the West, Jeffersonian-era explorations (1801–1808) sped up the process of integration of tribes already engaged in extensive inter-Indian trade through contact and direct observation of traders among the Mandans and evidence of other European traders elsewhere. Along the Columbia River and on the Pacific Northwest Coast it was evident to Lewis and Clark that sailors from England, Russia, and the east coast of North America had been in contact with Chinooks, Clatsops, and others.[15]

In 1788, Americans on board the *Columbia Redivivia* had joined British and Russian vessels on the northwest coast. Four American ships with their crews, known locally among Natives as "Bostons," returned in 1789. By 1801, at least twenty American ships are documented to have contacted the coast, and in 1805, although Lewis and Clark never saw a sail on the horizon or a fellow American on Pacific shores, at least seven American vessels did business along the coast that year, trading some $257,799 in goods. Clearly Lewis and Clark were not alone on the coast although they did not find an easy ride home.[16]

Similarly, as other explorers sent out by Jefferson searched for navigable waters and extractable resources in the Southern Plains and what would become the American Southwest, tribes such as the Utes, Kiowas, and Apaches had already been contacted by Spanish and French traders operating out of Santa Fe and Louisiana. American officer and explorer Lieutenant Zebulon M. Pike's expedition of 1806–1807, and that of Thomas Freeman (1806) along the Arkansas and Red rivers respectively, stimulated further interest in tapping the Indian trade and extractable resources—especially hides and furs—from contested domain inhabited by Indians and claimed by Spain.

Figure 6.1 C. M. Russell, *Indian Hunters' Return*, 1900, oil on canvas.
Source: Courtesy of the Montana Historical Society.

Captured by a Spanish patrol in present-day southern Colorado, Pike and six of his men were conducted to Santa Fe for questioning and then sent on to Chihuahua, 550 miles to the south of New Mexico's capitol. In July 1807, the Americans were released at the Texas–Louisiana border at Natchitoches. Pike's personal papers had been confiscated by the Spaniards, but he reconstructed the expedition as best he could in an official report, released to the public in 1810. This report, along with narratives of exploration by England's Captain James Cook by sea (published 1785), and the Montreal-based North West Company's Alexander McKenzie by land (published in 1801), stimulated interest in fur-trade enterprises and encouraged entrepreneurial capital formation from St. Louis to New York and London.[17]

Thus was launched the period of greatest intensity in the Indian trade of the region beyond the Missouri following the three main conduits that became well established as pathways into the West: the Missouri River itself, which Lewis and Clark proved could not be easily navigated past the Great Falls; the overland route to the Pacific coast through South Pass, Wyoming, and on to the Columbia River (the Oregon Trail with its appendages down into California), and the Santa Fe Trail linking Missouri with New Mexico.[18]

The burst of fur-trade energy and infusion of capital began on the heels of the Jeffersonian Expeditions in 1807 and ended about 1870. Concurrently, the Indian tribes of the West suffered demographic decline, intensified intertribal warfare, and a diminished subsistence base. The Euro-American-induced trade was a prime agent in the process of biological and cultural change, although compared to later economic frontiers—especially mining, farming, and ranching—far less erosive of Indian culture owing to the accommodation of stakeholders and the creation of multicultural societies throughout the region. Labeled métis, mestizo, mixed-blood, and, often with prejudice, "half-breed," the progeny of Indian–Euro-American sexual liaisons gained numbers and power in the second and third generation of the contact frontiers with Native peoples, emerging as cultures in their own right across North America. The relationships maintained with "full blood" relatives, especially in maternal lineages, reduced tension and insured incoporation rather than rejection of these peoples "in between."[19] Still, the "fur trade" as an aspect of the Indian trade softened Native societies and reduced ability of many to hold their place in traditional trade and political systems, making it easier to rationalize participation rather than opposition to a material-based economy that embellished as well as complicated Indian life.[20]

As in most contact frontiers, the greatest culprit in Native decline was unintentional: waves of epidemic disease brought by Euro-American carriers. In the 1780s, as many as one-third of all native residents in the Northern Plains, Southwest, and Northwest died as smallpox swept the West, supplemented by cholera or bubonic plague in Texas. Streptococcal infections attacked the Sioux in 1798 and were reported among the Assiniboin and Cree by 1801. Five years later, the Mandans and Blackfeet suffered strep as well as influenza. Other "crowd diseases" swept away entire villages in the early nineteenth century. Measles and whooping cough killed thousands on the Northern Plains in 1819 and 1820, and an estimated 4,000 Comanches died from smallpox that same decade. Influenza helped depopulate much of the Columbia River Valley in the 1830s, while smallpox returned to the Plains beginning in 1831, leaving less than half of the descendants of those who had survived the pandemic of the 1780s alive by 1840.[21]

Stimulation of the trade after the epidemics that hammered tribes prior to Lewis and Clark's trek, but that had not been immediately observable to the captains, came from both private and government sources. In the wake of the Corps of Discovery's expedition from St. Louis to the Pacific and back (1804–1806), fur merchants and entrepreneurs reorganized themselves, combined capital, and launched probes into the heart of the region described favorably by Lewis and Clark as rich in fur-bearing animals. The pay-off was slow to be realized, owing to Native determination to retain middleman-trader positions that had taken decades to establish since successful incorporation of the horse between 1720 and 1800 throughout the interior West. Although many horse-rich tribes, especially the Shoshone and Nez Perce, had been friendly to the Corps, rescuing them from probable failure, others—especially the Blackfeet, Arikara, and Teton Sioux—challenged American trespass on tribally claimed hunting and trading grounds.[22]

For decades to come, conflict, as well as cooperation, permeated the record of Indian relations of fur trappers, traders, and travelers crossing between the Missouri and Yellowstone country. Many trappers and explorers lost their lives as they extracted furs and game for their own use, without government supervision or official diplomatic negotiations in the form of treaties.[23]

Unlike the relative success of the Lewis and Clark Expedition in Indian relations, nearly all of the private expeditions between 1807 and the mid-1830s, ran into conflict with one or more Indian tribes as brigades of trappers scoured rivers and ponds in search of the fur of beavers (*Castor canadiensis*), open plains and valleys in search of the hides of buffalo (*Bison bison*), and any other animal that could be eaten or whose hide or fur could be sold in China, Europe, or eastern North America. Showing little or no restraint, animals were harvested by Euro-Americans and mixed-bloods without consideration of reproductive seasons or Native taboos on species or sacred grounds. Alienation of Northern Plains tribes, especially the Blackfeet and their confederated neighbors, as well as disruption of beaver and bison ecology, led to factionalization between tribes as well as open warfare against Euro-American trespass on fur- and hide-rich grounds by those who felt most violated. Although total depletion of both major target species was ultimately averted, it is through this ecological lens that scholars are reassessing the fur trade and its impact on both wildlife and Indian societies.[24]

PACIFIC FUR COMPANY

German-born John Jacob Astor (1763–1848) arrived in New York in 1784, determined to make a fortune. His skill as a vendor of musical instruments was soon surpassed by investments in real estate and furs. Friends and acquaintances in Montreal provided vital insider information on how a private concern could compete with the two oligopolies of the British: the Montreal-based North West Company and the London-based Hudson's Bay Company (HBC)—rivals in Indian country but near hegemonous in controlling trade throughout upper and lower Canada, the Great Lakes country, and the Pacific Northwest.[25]

Formed in 1808, Astor's American Fur Company (AFC) took on the British giants but stayed south of the contested international line running through the Great Lakes. Chartered as the Pacific Fur Company, this subsidiary of the larger parent company was formed in 1810 to tap both the Indian trade along the Pacific–Plateau network, as well as the China trade between New York and Canton. Recruiting some of the best fur-trade personnel available, Astor's men sailed around South America and trod westward overland from St. Louis to the Pacific, building Astoria in 1811 near the mouth of the Columbia. Several smaller posts as far north as British Columbia (Fort She-Whaps) and as far east as Spokane (Fort Spokane) and Clarkston, Washington (the Clearwater Post) were designed to funnel furs downriver to the main depot at Astoria. Stiff competition with the North West Company and interruption of diplomacy during the War of 1812 led Astor's partners, most of whom were Canadian, to authorize sale of his operation to the North West Company rather than face seizure of all holdings as spoils of war in 1813.

Despite its short life, the Pacific Fur Company proved that parties of trappers could go overland as well as by sea to the Pacific. And, unlike Lewis and Clark, the "Overland Astorians" of 1811–1812 under Wilson Price Hunt found a gently rising saddle in the Rocky Mountains, South Pass, which would eventually serve as the main transit over the Continental Divide for west-bound fur-trade brigades and emigrants' wagon trains on the Overland Trail. Astor also proved that small investments could yield large returns. Despite his competition, the one season of 1812–1813 alone yielded an estimated $100,000 in furs. These included 17,705 pounds of beaver, sixteen grizzly bear hides, seventy-one black-bear skins, and sixty-eight sea-otter skins. In addition, he had purchased 80,000 seal skins in trade with the Russians at Sitka.[26]

Astor would reenter the Western fur trade in 1823, partnering with established firms in St. Louis. Meanwhile, in 1817 he gained control of the South West Company (also Montreal-based) and formed a "Northern Department" of the AFC with a large depot at Michilimackinac. Furs, fish, timber, and lead from galena mines poured into Astor's realm, giving him sufficient capital to tackle an even more ambitious project: the end of the US "factory" system.

Operative from 1796 to 1822, the factory system was established by Congress to better regulate trade and intercourse with Indian tribes. Official US factories (trading posts) were built at commercial crossroads on the frontier. At these posts, better thought of as general stores, American agents collected furs and hides from Indians in exchange for high-quality goods at reasonable prices. Liquor was strictly prohibited. A small number of Euro-American trappers were granted temporary licenses to enter Indian country, but no trapper—Indian or Euro-American—was extended credit at the factories. The system worked with mixed success through the War of 1812, but the floodgates opened to private enterprise, most of it illegal, once the British withdrew north of the 49th parallel. With friends in Congress and a heavy lobbying effort, Astor led private fur-company challenges to the US factory system, which ended in 1822, after which liquor flowed more openly in Indian country and fur companies operated with much less supervision from government agents. In short, standardization and regulation of trade beyond St. Louis were virtually nonexistent until military posts were built to protect Indians and Euro-Americans from each other.[27]

BRITISH ACTIVITIES IN THE INLAND NORTHWEST

Following the transfer of Astoria to the North West Company in 1813 (renamed Fort George), several Astorians were rehired. Most important was Alexander Ross, an educated Scott who had been clerk and secretary to Astor's venture in 1810. Ross was challenged to make profits for his new employer in the interior. His first strategy was to encourage Indian trappers to procure and process furs for exchange to the North West Company. Each spring, the company would collect and transport them downriver along the Columbia for transshipment to China. The problem: most tribes of the region, and especially those that were horse-rich, notably the Nez Perce, Cayuse, and Palouse, had no interest in trapping. Furthermore, they had been alienated by the Pacific fur men, who they considered stingy and arrogant.

In spite of these obstacles, in 1818 Ross built a major depot named Fort Nez Perces at a midpoint on the Columbia near present-day Walla Walla. From here, Ross implemented an alternative strategy: sending his own trappers with their families into the interior for several months to garner the furs that Indians refused to trap. For the next five years, Ross supervised field brigades, sending them as far as Idaho's Snake River country. The strategy was successful enough to warrant continuation after the North West Company and the HBC merged in 1821.[28]

Under the auspices of the HBC, Ross led the field unit himself from Flathead House in 1824, but retired at year's end, disgruntled with the new parent company. He was replaced by Peter Skene Ogden. Averaging forty-four men, women, and children, Snake River brigades consistently reaped profits for company owners while stripping streams and creating a "fur desert" to discourage Americans from competing and settling in the jointly occupied region, known as Oregon after 1818. The expeditions benefited the British in other ways as exploration of new areas brought geographic and ethnographic information from as far south as Nevada and California. In retirement at Red River, Ross wrote his memoirs, giving historians a clear view of the logistics of the systems under both the North West Company and the HBC, while venting his spleen against the latter for his forced early retirement and personnel decisions.[29]

Chartered in 1670 as a royal trading monopoly, the HBC had entered the Pacific Northwest in 1821, intent on remaining an imperial factor for the Crown and a profitable and domineering corporate giant throughout Rupert's Land, the vast region that drained into Hudson Bay, some 1.4 million square miles or around one-third of present-day Canada. To accomplish this, a governor for North America was named and sent to oversee all operations. As a result, Sir George Simpson became the most powerful man in the Pacific Northwest. Under his governorship, in 1824, John McLoughlin, physician and former Nor'wester, was brought from Fort William on the north shore of Lake Superior to build Fort Vancouver, a new main depot and decision-making center for the Columbia department. Located on a strategic bluff 6 miles above the confluence of the Williamette with the Columbia, a grander Fort Vancouver closer to the river was constructed in 1829 and served this role until 1843, when headquarters moved yet again to Fort Victoria on Vancouver Island. Called White-Headed Eagle for his long white hair and his stature at 6 feet, 6 inches, McLoughlin ruled a region larger than Great Britain for more than twenty years, eventually retiring to Oregon City, where he became an American citizen, remembered during his life and since as the "Father of Oregon."[30]

The HBC inherited the "Columbia District" from the North West Company with forts George, Nez Perces, and Spokane as sub-districts. To this area, after 1827, the HBC under Governor Simpson and Chief Factor McLoughlin redefined the "Columbia Department" as the region west of the Continental Divide from the Rockies to the "Frozen Sea" or the Arctic. Thus, modern Canadian provinces of British Columbia (called New Caledonia at the time) and a small portion of Alberta became a part of operations out of Fort Vancouver.

The lifelines of the Columbia Department were threefold. First, the Snake River brigades continued to operate until 1850, providing furs and physical presence in contested territory with Americans. Trading "over the mountains" proved lucrative, even as late as the 1840s, long after the price of beaver had declined precipitously.[31] Second, the Fraser River–Columbia Brigade connected New Caledonia with the Columbia posts and Fort Vancouver over difficult trails and water routes. By packhorse and canoe, HBC employees managed to make the system work up to 1847, when all HBC operations moved north of the Columbia to Vancouver Island.[32] Finally, there was the seaborne trade, older than all land operations, which provided the HBC with sea-otter pelts and, after 1840, a contract to provide supplies in the way of produce, cereal crops, and meats to the Russian American Fur Company operating out of Alaska. Farms on the Nisqually and Cowlitz plains in present-day Washington, as well as agricultural enterprises in Oregon's Willamette Valley served this purpose. The company also developed markets at Oahu (Hawaii), Yerba Buena (San Francisco under Mexican rule to 1848), and Sitka, Alaska, the headquarters of the Russian American Fur Company.

Overall, this diversification saved the HBC. After the Treaty of Washington in 1846 split the Oregon territory between Britain and the USA, the HBC abandoned nine posts but continued to operate select posts south of the line, supplying Indians and Euro-American settlers with blankets, goods, and services until 1871, when the last remote post, Fort Connah on the Flathead reservation in Montana, was closed. By then, all HBC properties within the USA had been abandoned, but the company retained mineral rights on town and fort sites that it had developed during the years of joint occupancy.

Much of the success of the British companies rested on their military-like organization and their highly stratified ranks. Officers ate separately from the rank and file, were paid very well for the day, and expected and normally received much respect from those below. But they also mixed well with the people of the country. Most acquired Native wives and had mixed-blood families while serving in Indian country. They also endured the same hardships in the field as the common laborers who journeyed with them. Most significantly, cooperative Natives within the region, as well as workers from outside the region, were brought into the system. Iroquois, Cree, Nipissing, and Abenaki male trappers, hunters and camp tenders joined Orkney Islanders, French-Canadian, and métis personnel under Scottish and English officers on the various brigades. Forts abounded with "country born" children and relatives of local leading Native families, whose women often provided partners to Euro-American officers and workers alike.[33]

At its peak, the workforce of the Columbia approached 600 non-Indians on payroll. The diversity of ethnic and regional backgrounds lent complexity to the trade, especially where the Iroquois worked in large numbers. Many of the latter were Christianized Caughnawaga Mohawks, who were inherited from the North West Company as "freemen" without contracts by the HBC, who insisted on signed work agreements. At the time of the merger in 1821, Iroquois made up nearly one-third of the hired hands in the Columbia District.[34] Considered among the best trappers, they were also known for their strong sense of independence, even while "engaged" on contract to the Company and their proclivity to desert when better opportunities arose. Alexander Ross considered them "sullen, fickle, cowardly and treacherous," blaming them for much of his own misfortune in Snake River country. Fellow veteran Nor'wester Ross Cox labeled them "quarrelsome, revengeful, and sometimes insubordinate." Even so, few fur trade operations in the West after 1818 lacked Iroquoian personnel, and they intermarried readily into tribes that accepted missionaries beginning in the 1830s.[35]

AMERICAN FUR TRADE ORGANIZATION AND LOGISTICS

After the transfer of Louisiana in 1803, strategies changed between 1806 and the mid-1820s, but initially the older patriarchal pattern prevailed, with top-down ownership by elite families and highly stratified tiers of workers beneath. Under American rule, Spanish and French mercantile families of Saint Louis—notably the Lisas, Chouteaus, Prattes, Gratiots, Papins, Menards, Bertholds, and Cabannes provided initial seed capital and business acumen and were joined in the 1820s by Americans such as William H. Ashley and the field representatives of New York-based John Jacob Astor. Although some of these barons of the fur trade made substantial money, losses of men, materiel, and capital was quite common as fluctuations in the business cycle, changing prices for goods and furs, changing diplomatic climates, and the ever-present problem of weather, logistics of supply and transshipment stacked the odds against success.

Manuel Lisa and the St. Louis Missouri Fur Company

Ironically, had the earliest private companies emphasized military-style garrison architecture rather than very crudely constructed outposts, their situations might have improved. The earliest American attempts to remain in Indian country after Lewis and Clark were organized by Manuel Lisa, a Spanish aristocrat born in New Orleans, who had built a successful business in Saint Louis. In 1807, Lisa seized the moment, sending a party of fifty to sixty men, including George Drouillard and John Colter, Lewis and Clark veterans, up the Missouri. Although threatened by angry Arikara middlemen as they cordelled their way upriver, the Lisa party made their way into Absaroka (Crow country) and built a small post on the Yellowstone River at the mouth of the Bighorn near present-day Hysham, Montana. From this base, called Fort Raymond (or Remon in honor of his son), traders fanned out with packhorses full of trade goods. Other employees worked streambeds and ponds, trapping and processing the furs themselves.[36]

A successful season encouraged backers in Saint Louis to further finance Lisa's trading venture, which, after 1812, was known as the Missouri Fur Company (or more popularly Manuel Lisa & Co.), with a capitalization of around $50,000. Investors included William Clark, by then Superintendent of Indian Affairs for the western region and several of the Chouteaus. The Missouri Fur Company's strategy was to funnel furs out of the middle Missouri trade network via a string of hastily built trading posts from the Mandans down to the Omaha villages near Council Bluffs. Such posts would not infringe on Indian trappers and hunters already heavily caught in the fur-trade dependency nexus and would protect company trapping parties as they headed into the upper Missouri country where Indians retained sovereignty and demanded ransom or tolls for passage through their trading zones. At its peak, the company's main party of field personnel consisted of 160 men including some Delaware Indians, who were to work in large groups for safety.

Implementation of this strategy resulted in expansion of Fort Manuel Lisa on the site of the older Fort Raymond in 1809 and construction of smaller posts to serve the Arikara–Mandan–Hidatsa (Lisa's Post, b. 1809); Teton and Yankton Sioux (Cedar Island Post, b. 1809); and Omaha, Oto, Iowa and Pawnee (Fort Lisa, b. 1812). Feeling safe in numbers, groups of Lisa's men pushed into Blackfeet country by the end of 1809, building two small posts (Fort Henry and Three Forks Post). But by 1812, Blackfeet and Gros Ventre reprisals had resulted in the loss of at least twenty men, so Lisa retreated back to the middle Missouri.[37]

The end of the War of 1812 changed Lisa's mind. In 1818, the British agreed to withdraw all trading activities south of the 49th parallel except for the jointly occupied Oregon Country. The next year, Lisa mounted another effort to tap the upper Missouri, but distrust of his partners (the Chouteaus) and illness forced him to retreat. Once back in St. Louis, he acquired new partners in Americans Joshua Pilcher, Lucien Fontenelle, Andrew Drips, Charles Bent, and William Vandenburgh, but during the summer of 1820, at age forty-eight, Lisa died, leaving Pilcher in charge.

Moving beyond the middle stretches of the Missouri with around 300 men in their employ, the new Missouri Fur Company (also Pilcher & Co.) built two major posts in 1821: Fort Vanderburgh in North Dakota near the confluence of the Knife River with the Missouri in the heart of the old Mandan–Hidatsa trade sphere, and Fort Benton in Montana. The latter endured well beyond the company, which folded after suffering an irreversible loss in 1823. In that year, seven men, including field leaders Michael Immell and Robert Jones, were killed, four others wounded, and over $15,000 in property was taken by a Blackfeet war party.[38] Losses of men and furs, instigated, so Americans claimed, by HBC traders north of the line, forced the Missouri Fur Company to abandon Fort Vanderburgh later that year, and, by 1824, the company went out of business.[39]

The Chouteaus: River Barons of St. Louis

The Chouteaus had been in St. Louis since its founding in 1764. In that year, a fourteen-year-old Auguste Chouteau ascended the Mississippi River from New Orleans with his stepfather, Pierre Laclède Liguest, to build a trading post in honor of the French king on the west bank about 20 miles below the mouth of the Missouri River. Liguest died in 1778, leaving his estate to his widow, Marie Thérèse Bourgeois Chouteau, previously married to René Chouteau, who abandoned her, leaving two sons, Renato and Auguste, in her care. Marie and Liguest had four children of their own: son Pierre and three daughters, Pelagie, Marie Louise, and Victorie, all of whom retained the surname Chouteau. The siblings married well and consolidated a family empire within St. Louis and beyond. The widow Chouteau died in 1814, the most respected and powerful woman in St. Louis. Sons Auguste (1749–1829) and his stepbrother Pierre (1758–1849) took over her business interests and expanded investments within the city and beyond to the outer boundaries of St. Louis's influence on the Mississippi, Missouri, Platte, and their tributaries. Two of Pierre's sons, Auguste Pierre (known as A. P.) and Pierre Jr., or "Cadet," both became major players in the fur trade in very different ways.[40]

A. P. Chouteau (1786–1838) graduated from West Point in 1806 and returned to Missouri in 1807 after a brief tour in the US Army. As noted, he partnered with Manuel Lisa to form the Missouri Fur Company in 1809. The company was reorganized in 1812 with A. P. Chouteau standing by Lisa, but in 1813 the company dissolved. Thereafter, A. P. struck out on his own, relying on family connections and trusts to see him through several financial failures. In 1815, A. P. and Julius DeMun joined other trappers on the Arkansas River in present-day southern Colorado, the disputed boundary between Spanish and American territory. Two years later, in the spring of 1817, after a successful trapping season, their party was arrested by Spanish soldiers, who had tolerated the Americans for two years but who officially forbade foreigners to enter the province. Arrested and imprisoned for a month, their goods and furs worth $30,381 were confiscated.

During the forty years that Spain controlled Louisiana, from 1763 to 1803, New Mexico had remained relatively isolated. Several attempts to connect New Mexico with Louisiana had failed, and no regular commercial arrangements had been possible due to Spanish resistance to opening doors to outsiders. Although trade with Utes and Comanches was common by 1800, Spaniards were xenophobic when it came to Frenchmen or Americans visiting, much less trading in their domain. Thus, like Zebulon Pike and others before them, A. P. Chouteau and Demuns were rejected but allowed to return home. Not until Mexican Independence in 1821 would the situation change, with the opening of commerce on the Santa Fe Trail between Independence, Missouri, and the capitol of Mexico's most northern state.

Back in St. Louis, another A. P. Chouteau business venture lost an additional $66,000 in the middle of the financial panic of 1819. He never rebounded entirely, and in 1822 left behind a large French Creole family in St. Louis and moved in with the Osage in present-day Oklahoma, intermarrying and directing Chouteau operations at their farthest western point for the rest of his life, keeping the Osage loyal to the family while drawing down the family fortune.

His younger brother, Cadet, entered the fur trade in 1814 after a short but success career as a general mercantilist. With partner Bartholomew Berthold, Cadet used inventory from their grocery store to finance his brother A. P., covering the latter's debts many more times in his life and financing four fur-trading posts under A. P.'s supervision known collectively as the Osage Outfit. The scale of debts claimed by his own family against A. P. on the eve of his death on Christmas Day, 1838, provide context for the power the Chouteaus exerted on the western frontier. Stepbrother Pierre filed suit the previous January for losses totaling $500,000 including $100,000 in merchandise, loans of $200,000, and $200,000 for miscellaneous, only a small fraction of which was actually collected from sales of A. P.'s estate and depredation claims against Mexico for seizing goods and furs back in 1817.[41]

Joined by Jean P. Cabanné in 1819, Berthold & Chouteau (also known as "The French Company") helped finance Manuel Lisa's 1819 adventure up the Missouri, losing over $20,000 when the party turned around, prior to reaching the Rockies. Despite this setback, Cadet persisted and ultimately succeeded in partnerships that sustained his family's wealth and prominence: Berthold & Chouteau (1813–1822); Berthold, Chouteau & Pratte (1822–1824); Bernard Pratte & Co. (1826–1834); Pratte, Chouteau & Co. (1834–1839); and finally, Pierre Chouteau & Co. (1839–1865).[42]

Chouteau's most successful move was to join forces with J. J. Astor's AFC in 1822, a move made possible by Ramsay Crooks, the general manager of the AFC. Crooks had been an overland Astorian and had endured the trek to Oregon in 1811–1812. Crooks arranged to supply the Chouteaus with high-quality goods from Europe, guaranteeing purchase of furs for New York and European markets, in exchange for a percentage of profits on Western operations. By spring 1822, a Western Department of the AFC had been formed, separating it from Northern Operations focusing on the Midwest and Great Lakes. Crooks further sealed the deal in 1825, marrying Emile Pratte, daughter of Bernard Pratte, Cadet's chief partner, further interlocking the two companies.[43]

With Crooks in charge of the Western Department, the combined interest made over $16,000 in its first year of operation. After the merger, no matter how much they tried, smaller companies ended up yielding to "the Opposition," as the AFC came to be known in the field. Experienced and seasoned traders poured into the company's employment through absorption and merger from 1823 to 1840. Most important among these, the Columbia Fur Company's managerial staff signed on in 1827. Kenneth McKenzie, William Laidlaw, James Kipp, and Daniel Lamont had run successful posts on the Missouri and were now on the Chouteau–Pratte payroll. The AFC created the Upper Missouri Outfit in 1827 with McKenzie in charge, a brilliant move that led to the building of Fort Union where the Yellowstone meets in the Missouri in 1829. Fort Clark (1831) and Fort Pierre (1832) followed. Traders were also sent to all six divisions of the Lakota Sioux, as well as to the Sacs, Iowas, Osages, Kansas, Poncas, and Otoes. Wintering houses such as the trading post of James Bordeaux among the Brulé Sioux near modern Chadron, Nebraska (reconstructed as the Museum of the Fur Trade), provided tribes at the band or village-level with utilitarian items from woolen and cotton textiles, knives, guns, tobacco, and kettles to items of finery, especially beads, mirrors, and other objects of adornment.[44]

In 1828, the AFC, through Bernard Pratte & Co., hired the surviving partners of the old Missouri Fur Company to form its mountain outfit, commonly called the "mountain business." For ten years, this unit, originally organized by Lucien Fontenelle, William Vanderburgh, and Andrew Drips, failed to make a dime, but it persisted until 1839, competing with other companies in "rendezvous country" for trappers' furs and Indian trade.

Astor retired in 1834, selling out the Western Department to Pratte, Chouteau & Co. and the Northern Department to Crooks. St. Louis firms, especially the Chouteaus, continued to rely on the New York office, paying Crooks and William B. Astor, J.J.'s son, 2.5 percent commission for selling furs in Europe through C. A. Lampson, who also marketed furs for the HBC. At this point, Cadet began to make mistakes. He allowed the Rocky Mountain Fur Company to reorganize and compete with him, and he let good post managers go without attempting to retain his brightest personnel. Most troubling to his employees, he spent most of his time in Washington, lobbying Congress and the Indian Office for payment of annuities owed his firm under US treaties with tribes within his trading arc. Pierre Chouteau Jr. & Co. lasted until 1865, weathering the panic of 1837, the high-plains smallpox epidemic of that year that halved the Indian population of the region, the collapse of beaver markets in Europe (replaced by silk hats and South American nutria), the transition to the bison robe trade, which replaced beaver as the primary target species in 1841, and the American Civil War. Throughout the 1840s, an estimated 90,000 bison robes per year made their way to St. Louis, 67,000 of them from Chouteau's firm alone in 1840. This increased to an average of 100,000 per year during the 1850s.[45]

In 1865, as Northern Plains fur country transitioned into a battleground for the Plains Indian Wars, Pierre Chouteau & Co. liquidated most operations to the Minnesota firm of Hubbell & Hawley after learning that their fur-trade license would not be renewed owing to the owners' connection with the institution of slavery. Cadet's son Charles Pierre (1819–1901) and son-in-law John Sanford (1806–1856) had been running the company for some time, and it was Sanford's "servant" Dred Scott who triggered the alarm on the old French Creole family's connection with slavery.[46]

ORGANIZATION OF FUR COMPANIES IN THE WEST

The ultimate success of the AFC rested on its parallel organization to the British companies. Initially, field partners and brigade leaders consisted of those already engaged in the Indian trade of Louisiana, many of them French-American Creoles or French-Canadian. By the 1820s, these veterans were challenged by relatively inexperienced Anglos, adventuresome Americans who emerged from Missouri, Virginia, Kentucky, and Tennessee, as well as experienced Scots and British traders drawn from large British-Canadian companies or failed smaller concerns. In search of new employment, these seasoned and tried field managers had "wintered over" and were highly valued by Americans needing business intelligence and inside information on how northern companies operated.

Men in the rank and file represented the ethnic and linguistic diversity that generations of trade had already produced in the colonial fur trade societies of New France, the "upper country" of the Great Lakes, and Louisiana. Despite their large numbers, French-Canadians and Americans of French extraction, many of whom were métis, increasingly were relegated to the lower tiers of the occupation ladder within fur-trade society. These *voyageurs* (common laborers) formed the largest numbers of workers within various concerns, but their lives are less visible in the historical record due to illiteracy among the majority and the shortness of working lives among all, an average of fifteen to twenty years at most. In 1830, their pay within the Upper Missouri Outfit averaged a mere 36¢ cents per day (compared with an average of $1 for an American urban worker), but they also received an annual "outfit" consisting of one blanket, a knife, a quantum of tobacco, and other miscellaneous articles of personal gear. They also lived off the land, supplementing their diets with wild game and had fewer opportunities to spend money than their counterparts in eastern Canada or the USA. Their "savings rate" or ability to save portions of wages may actually have been greater than their urban cousins, reversing a stereotype that they were in near-debt peonage to their employers.[47]

Above the voyageur-class, those in the craft trades—blacksmiths, coopers, gunsmiths, and carpenters—were better off than voyageurs by measure of their pay and their status. Next in rank: hunters and patroons (boat-crew leaders) at $250 and $284.50 respectively per year. Interpreters could rise even higher and were valuable for both their language skills and the diplomatic edge they could bring with specific Indian tribes, bands, or families, earning as much as $380 per annum. Traders also served as clerks, receiving around $370 per year plus other advantages including meals at officers' tables.

In the British system, "chief factors" or managers were at the top of the occupational pyramid under the governor. A successful fur-trade post had at least one honest chief factor or head clerk in residence; otherwise, material wealth disbursed without record at alarming rates to owners and share-holders. Factors were often given shares of the enterprise; on the American side, field partners joined owners as shareholders, but seldom did clerks obtain a financial piece of the larger company. Still, clerks were rewarded for loyalty and accurate accounting.

Factors in turn supervised clerks, the latter earning an average of $483 in 1830. Ship masters and mates earned $356; guides and interpreters were next at $239; sailors at $163; and tradesmen at $161; voyageurs (also contracted to hunt) at $101; common laborers at $95; and apprentices at $63. In short, both American and British systems were very hierarchical with some latitude for moving up but little reward beyond the contracted pay and perks provided in contracts.[48]

ROLE OF FUR-TRADE POSTS

Post-building as private enterprise was an old art, brought to North America by every major European colonizer but especially prized by the French and the English, whose rivalry across the entire colonial era promoted military-style garrisons in peace as well as in war. Strength of fortification was less important on the western frontier than ability to impress resident clientele of the wealth and power of the colonizing agent. Most posts had at least one blockhouse, the Western equivalent of the medieval tower or "keep," designed for tactical advantage over ground-level attack, as well as warehouse and stronghold during siege. The most impressive posts were the North West Company's Fort Nez Perces (1818), renamed Fort Walla Walla by the HBC after the merger, abandoned by the latter in 1855, and reoccupied by US troops from 1856 to 1911; the HBC's Fort Vancouver (1824), purchased eventually by the US Army in 1849 and used until 1947 as Vancouver Barracks; the AFC's Fort Union (1829), eventually abandoned in 1864 and torn down for steamboat fuel and for materials to build the US Army's nearby Fort Buford in 1866; and Fort Laramie in present-day Wyoming, originally built by the firm of Sublette & Campbell (1833), sold the following year to the Western Department of the AFC and eventually resold to the US Army in 1849. The final three are now part of the US National Park system as interpretive sites.[49]

THE ROCKY MOUNTAIN RENDEZVOUS SYSTEM

One innovator decided to avoid post-building entirely. Born in Virginia, William H. Ashley owned a gunpowder factory and was rewarded by rising to the rank of general in the militia during the War of 1812. He moved to St. Louis in 1819 and saw potential profits in the fur trade. From 1822 to 1826, Ashley, in partnership with Andrew Henry, organized and directed new strategies for extracting furs. Ashley personally led three expeditions into the mountains and instituted fundamental changes in the trade. Like Lisa before him, Ashley hired his own trappers, thus hoping to avoid paying Indian middlemen as well as native processors. He advertised in local papers for "100 adventuresome young men," and he got them. Their ranks included Jedediah Smith, James Beckwourth, and Thomas Fitzpatrick. Unlike Lisa, Ashley's men did not return to fixed fur trade posts for re-outfitting and transfer of their cache. Instead, beginning in 1825, annual rendezvous were held in parklands long used by the Shoshone and other Indian traders, after which trappers dispersed into camps at strategic parks or "holes" with sufficient game to hunt across the winter. In spring, they scouted new ground and aggressively trapped streams and ponds prior to the next rendezvous. These social and economic gatherings in the Green River country of Wyoming and elsewhere in Utah and Idaho lasted until 1840.[50]

Another innovation was the use of wagons to get goods into rendezvous country along the Platte River, the very route that would eventually serve Oregon- and California-bound emigrants on the Overland Trail. The Rocky Mountain rendezvous system lasted fifteen years and was a mixed success. The first years, 1825 and 1826, produced bonanza profits for Ashley, who brought out $50,000 in furs in 1825 and $60,000 in 1826, making him millionaire in today's equivalent money. Tired of the mountain life and interested in running for Congress, Ashley sold out to three of his own employees: Jedediah Smith, David Jackson, and William Sublette in 1826.[51]

While Ashley turned to Missouri politics, Smith, Jackson, and Sublette ranged far and wide for the rest of the decade. Their exploits, as well as those of other mountain men, individually and collectively in small companies, have led to many stereotypes of their grizzled Indianized looks as well as their lifestyle—"a reckless breed of men" as one noted historian put it, living "a life wild and perilous" in the language of another.[52] That reputation was based in part on losses. From 1823 to 1829, a total of ninety-four trappers, employees of Ashley & Henry and their successor, Smith, Jackson & Sublette, perished at the hands of Indians.[53]

Smith had been Ashley's point man, leading a brigade into far western Montana as far as the HBC's Flathead Post, land little known to Americans in 1824, prior to reaching the rendezvous prearranged at Green River in 1825. That year, Ashley made him a full partner. In 1826, Smith continued his explorations, leading an expedition from the rendezvous

Figure 6.2 C. M. Russell, *Free Trapper*, 1911, oil on canvas.
Source: Courtesy of the Montana Historical Society.

near Utah's Cache Lake to the Great Salt Lake and then southwest along the Colorado River across the Mohave desert to southern California. Eventually the party marched up California's Central Valley, crossing the Sierra Nevada Mountains and the Humboldt Sink back to the rendezvous of 1827 at Bear Lake, Utah, becoming the first Americans to traverse from west to east this difficult route.

Smith repeated his California reconnaissance in 1827, losing ten men to the Mohaves en route and reentering by way of San Gabriel, where he was placed under house arrest. Eventually allowed to sail to San Francisco (Yerba Buena), Smith rejoined his men who had avoided arrest by bivouacking beyond Mexican patrols in the San Joaquin valley. Reunited, the party of eighteen marched north to Oregon, suffering the loss of fourteen at the hands of the Umpqua Indians en route. Rescued by the HBC's John McLoughlin, Smith gained valuable intelligence on Mexican and British activities in the Far West prior to dying in yet another Indian encounter on the Santa Fe Trail in 1831. Had his maps been made available to the government and the public, Americans would have had a much clearer picture of the difficulties of travel across the arid Great Basin and might have avoided many deaths and much hardship on various cutoffs of the Overland Trail to California. As is, his accomplishment and skill as explorer and cartographer is unmatched in the annals of American fur-trade exploration, but historians had to wait until the twentieth century to reassemble his treks.[54]

Smith, Jackson, and Sublette sold out on favorable terms in 1830 to men they knew well: Thomas Fitzpatrick, Milton Sublette, Henry Fraeb, Jean-Baptiste Gervais, and Jim Bridger, who called their new concern the Rocky Mountain Fur Company (RMFC). As historian James Hanson has observed, "the new owners were all excellent trappers but poor businessmen."[55] Relying on tenuous logistics of supply and contracts for sale of furs with family member William Sublette, the RMFC lasted four years, "twisting in the wind as it slowly bled to death financially."[56] Declining demand for beaver, increased competition, and heavy debts plagued the RMFC by 1832. At the rendezvous in Idaho's Teton Basin at Pierre's Hole that year, around 500 Euro-Americans and at least 1,000 Indians attended the summer fair. The RMFC had about ninety men in its employ, but the AFC alone had 160. To this, at least 100 independents and small concerns showed up, as well as a contingent of the HBC, all of whom were challenged by the Blackfeet and some of their confederated allies. A battle broke out, leaving at least ten Blackfeet dead as well as several wounded mountain men, including William Sublette. The RMFC left the rendezvous over $5,000 in debt.[57]

Back in St. Louis, Sublette teamed up with Robert Campbell, an able businessman from Ireland, intent on giving the Western Department of the AFC stiff competition on the upper Missouri and the Plains. Toward this end, in 1834, Fort William was constructed at the Laramie fork of the North Platte, a strategic midpoint between rendezvous country and St. Louis for overland travel. The next year, Sublette and Campbell sold Fort William to the Chouteau-backed partnership of Lucien Fontenelle and Thomas Fitzpatrick, who renamed it Fort Lucien. Sublette and Campbell also sold their upper Missouri River operations to the AFC, leaving Pratte, Chouteau & Co. with a near-monopoly for several years. Sublette & Campbell continued a business relationship until 1840, selling furs gathered by other firms and buying properties in St. Louis rather than risking their own lives and capital out west.

Fort Lucien, in turn, was absorbed by Pratte, Chouteau & Co. in 1836, consolidating an entire system with Forts Union and Pierre on the other two points of the three-legged triangle. The fort was replaced in 1841 and renamed Fort John, in honor of John Sarpy, one of Chouteau's partners. Within a year, it had become an important stop for travelers on the Overland Trail to Oregon. Blacksmiths, wheelwrights, carpenters, and sutlers serviced Oregon-bound, and eventually California-bound, travelers for the next twenty years. The US Army bought the post in 1849, officially renaming it Fort Laramie, staffing it until March 1890.[58]

The Rocky Mountain System, as historical geographer David Wishart has described it, continued to operate through 1840, but it was unstable from the beginning. Relying on Euro-American and mixed-blood trappers, as well as allied Indian trappers from the East, was sure to irritate resident tribes. Losses of men and materiel reduced marginal profits to losses for most small operations. Prices at rendezvous to personnel made it difficult for trappers to save money. Companies formed and dissolved, but determined small-time investors and hearty mountain men continued to try their luck and test their skills at finding new beaver grounds and buying more time against robbery and death at the hands of hostile resident tribes.

Historians quibble over when the Rocky Mountain trade peaked, but after 1833, bison robes, not beaver, brought the bulk of returns in American warehouses. Whereas the wholesale price of beaver stood at a premium of $6 per pound in 1825, with an average of $5 at the first rendezvous, by 1850 the price had fallen to an unprofitable $2 per pound.[59] The total value of fur exports for the USA peaked in 1833 at $800,000, and never in the nineteenth century did the fur trade constitute more than 1 percent of the total gross national product (GNP) of the entire country.[60] Even so, the romantic era of the mountain men and the rendezvous continued on another seven years with best documentation in 1837.

In that year, Alfred Jacob Miller, a skilled watercolorist and painter, accompanied William Drummond Stewart, a captain in the US Army, to the rendezvous at Green River in Wyoming, not far from the old Shoshone rendezvous of yesteryear. Miller documented "about 30 wagons and charettes [two-wheeled carts] crossing the Kansas River," and he painted them several times across the next several months.[61] A few wagons had ventured to previous rendezvous, but this large contingent validated those who argued that families, as well as traders, could make the passage to the Pacific coast with relative ease by following the Platte, across South Pass, and on to the Columbia. Miller's paintings are the only firsthand account by an artist of a rendezvous. As such, they are important documents on all aspects of fur-trade society in the Rocky Mountain West.[62]

THE GOLDEN ERA OF TRADING POSTS

As bison robes replaced beaver, trading posts became more important for warehousing, heavy hides and robes, trade goods, and providing protection for fur trade personnel and their multi-ethnic families. A total of 140 forts were built in the American West as fur-trade posts, but most were small, short-lived, and forgotten within a generation of their establishment. In addition to the larger depots such as Fort Union (1829), Fort Clark (1831), Fort Pierre (1832), and Fort William (1834) also known as Lucien/Laramie, other American posts exerted regional economic and diplomatic influence beyond their smaller size. Forts McKenzie (1832–1843) among the Blackfeet, and Fort Cass (1832–1834), replaced by Fort Van Buren (1835–1843) among the Crow, are prime examples of these smaller posts that had disproportionate influence for a decade or so.

To the south, the most important on the border between Mexico and the USA was Fort Bent (Bent's Old Fort). Constructed in 1833 by the St. Louis firm of Bent & St. Vrain, the owners' object was threefold: to provide Santa Fe-bound caravans with a stopover and supply point; to serve the Southern Cheyenne and Arapaho in trade; and to serve trappers and traders in the Southwest with a reliable warehouse and a place to socialize and reprovision.

FUR TRADE OF THE SOUTHWEST

The Santa Fe Trail opened in 1821 after Mexican Independence from Spain. The lucrative traffic amounted to $190,000 in gross receipts by 1824. By 1833, trade was flourishing between the formerly isolated province and Missouri. Southern Colorado became a destination point for many trappers and traders seeking a second opportunity and for others seeking comfortable retirement among fellow mountain men and their families. Several retirement communities sprang up: Hardscrabble, Greenhorn, and Pueblo were most important of these on the Arkansas River drainage, all established in the early 1840s. During the Mexican–American War, Bent's Fort became the staging point for US troops bound for action in New Mexico and California, but it had to be abandoned and burned in 1849 after cholera infested the area.[63]

Other posts less important than Bent's Fort, but significant in local trade in the southern Rockies, include Fort Uinta on the Uinta fork of the Green River and Fort Uncompahgre on the Gunnison River, built by Santa Fe-based Antoine Robidoux in 1837, and Fort Davy Crockett, built by the three partners Phillip Thompson, William Craig, and Prewitt Sinclair in 1837 in Brown's Hole on Green River—an important wintering-over destination of mountain men during the rendezvous period. Other Colorado posts of significance include Fort Lupton, Fort Jackson, and Fort St. Vrain on the South Platte, all built in 1837 within 15 miles of each other. These posts focused on bison robes, not beaver, but had to be abandoned by 1845 for lack of animals and clientele.[64]

Santa Fe attracted many American traders well before the Mexican–American War. Those who settled and took up citizenship in Mexico after 1821 included Missourians Christopher "Kit" Carson and George C. Yount, Tennessean Ewing Young, and Kentuckian William Wolfskill, part of a larger group of 200 or so Americans prior to the Mexican–American War.[65] All were part of the "Taos Trappers," attracted to northern New Mexico and often lured on to California by opportunities in furs, as well as opportunities to own land, marry up into old Nuevo Mexicano and Californio societies, and achieve status impossible back in the USA for men often illiterate, without inheritance, skilled in outdoor survival and endurance, but destined to work as lower-class farmers or tradesmen had they remained in the Midwest or South.[66]

Lured by other trapping grounds and potential trade with Spaniards in California, several followed the Mexican example of Antonio Armijo, who blazed a trail between Santa Fe and Los Angeles in the winter of 1829–1830. The route ran from Santa Fe northwest up the Chama valley to Abiquiu and across the Continental Divide to the Green River of Utah. Crossing the Great Basin, packhorses brought Mexican and American goods, especially woolens, through Cajon Pass into the San Fernando and San Gabriel valleys, home since 1781 to Spanish town-dwellers and the sites of two important missions. Indian- and Mexican-made serapes or fresadas, very thick and almost impervious to water, were most highly desired by California rancheros and mission populations, as were sheep themselves, driven live on the hoof. Chinese goods and hides made their way east on the return annual caravans on the backs and sides of sturdy California-bred mules, more sturdy than the common mules on the Santa Fe Trail and burros of New Mexico. The "Old Spanish Trail" proved a risky but lucrative route up to 1848, after which it ceased on the cusp of the Gold Rush due to the new waterborne traffic.[67]

Carson accompanied Ewing Young on the old trail to California, trapping and trading between 1828 and 1831, but established roots in New Mexico, trapping and trading, and eventually intermarrying into the prominent Jaramillo family of Taos. In the 1840s, after serving as a hunter for Bent's Fort, Carson guided John C. Frémont on three of his five Western expeditions in the Rockies and was with Frémont during California's Bear Flag Revolt. Carson eventually was named Indian agent to the Utes, whose cause he championed in Washington. During the Civil War, Carson served in the campaigns against the Navajos and was partly responsible for their incarceration following the Long Walk of 1864 at Bosque Redondo in New Mexico. He died in 1868, a legend at the time, and the most famous of all the mountain men, through the publicity of John C. Frémont's reports of his official US expeditions in the 1840s and the latter's memoir, published in 1897 which made Kit a national hero.[68]

Young returned to California in 1832 after many exploits and trapping ventures in the Southwest, eventually settling in Oregon in 1834, where he was known by the British as an American agitator in the volatile 1840s.[69] Wolfskill originally partnered with Young in 1828 but moved on to California in 1830, seeking the lucrative sea otter. He settled down, married into a prominent Mexican family, and succeeded as a pioneer in the citrus, cattle, and wine industries in the Los Angeles and Santa Barbara areas.[70] Yount, who accompanied Wolfskill in 1830, also became a Californian, settling in the Napa valley, which became a retirement community for other mountain men on the west coast.[71]

California attracted outsiders eager to earn profits in trade from places other than New Mexico. After Mexican independence from Spain, California gradually opened diplomatic and financial doors to outsiders. Wary of trappers such as Jedediah Smith, who was openly denied the right to trap or remain on Mexican soil, others in addition to Yount and Wolfskill were welcomed or tolerated. In 1825, Finan McDonald and Thomas McKay led a HBC brigade south from newly built Fort Vancouver and probably entered California, reconnoitering the region for the prospect of beaver. The following year, Chief Trader Peter Skene Ogden led a second California brigade as far as Mount Shasta, which he described as "Mt. Sastie." Ogden returned in 1830 by way of the Colorado River, trapping the San Joaquin and Sacramento Rivers. His success prompted John McLoughlin to send an annual brigade along the Siskiyou Trail under Michel Laframboise down the Oregon coast, across Siskiyou Pass and down the Sacramento valley as far as Castoria, eventually known as French Camp, south of modern Stockton in the San Joaquin valley. Abundant beaver, far from seats of Mexican authority, justified the distance and the fact that the men traveled with their entire families reduced tensions among themselves and with resident Indians. The 1832 California Brigade consisted of eighteen Euro-American men, seventeen Indian men, eighteen Indian women, and six children. One of the Euro-American men was Alexander Carson, an engagé of the Lewis and Clark Expedition, who signed on with the HBC in 1820.

McLoughlin's strategy was to send Laframboise down the Siskiyou trail to trap the coastal region, which was virgin ground, as far as could be determined. Another HBC brigade would trap the Central Valley, and the two groups would join forces for the return trek to Fort Vancouver. To accomplish this, HBC trader and factor John Work (born John Wark in Donegal, Ireland, and a faithful HBC employee since 1814 when he signed up as a steward) left Fort Walla Walla in August 1832, with twenty-six men from the Snake River brigade and crossed into California from east of the Cascades, meeting up with some of Laframboise's men near Sacramento in December. Annoyed that Laframboise found plentiful beaver as far south down the coast as Mission Dolores near San Francisco, but that he had moved into the interior finding even more beaver near missions San Jose and Sonoma, Work spent most the winter battling the elements, healing sore Indian relations (Laframboise's men—especially his Iroquois hunters—left a trail of blood in their wake, losing two of their own), and avoiding Ewing Young and the American trappers, whose tracks had been crossed more than once.

Wintering over on the Buttes near Feather River, the Brits restocked gunpowder and shot at Fort Ross, which the Russian American Fur Company had built in 1812 north of Bodega Bay to take advantage of the sea-otter herds offshore and to attempt to grow food for the Alaska-based company. Subsequent stores were begrudgingly bartered at very high prices from the commandant of the Presidio of San Francisco.. Laframboise returned to Fort Vancouver on his own with little to show for the extended expedition while Work bivouacked at French Camp, increasingly besieged by local Indians, whose tolerance ended once disease, thought to be a virulent form of malaria, swept through California in the summer of 1833, leaving Indian villages completely depopulated as the Bay Company men struggled home. Arriving back at Fort Vancouver on October 29, Work summarized his California experience thus: "We had a great deal of trouble and some skirmishes with the natives. [...] Our hunt only amounts to 1,023 beaver and otter skins. Indeed, the country is now so exhausted that little can be done in it."[72]

Work went on to replace John McLoughlin as Chief Factor of the Columbia Department in 1846, retiring to Victoria where he died in 1861.[73]

McLoughlin would not give up, sending annual brigades south to French Camp through 1844. In 1841, he made arrangements with Mexican authorities to exempt HBC trappers from taxes on peltries and to establish a warehouse in Yerba Buena (San Francisco), run by his son-in-law, William Glen Rae. At the post, Mexican tallow and hides were stockpiled for HBC ships bound from the northwest coast to England. It was a long journey via Hawaii. Furs and hides from the Snake and Fraser brigades along with salmon, lumber, spars, and wheat were taken to Honolulu. Furs made their way to markets in Canton or were transshipped on to California along with sugar, molasses, coffee, salt, and goods from the Orient. Hawaiian sailors (known in the trade as Kanakas) along with Orkney men manned the HBC fleet, which comprised a formidable merchant-marine of ten or more vessels on the Pacific. From California, the ships sailed on to Valparaiso, Chile, where the HBC had another arrangement for warehousing and re-provisioning, and then on to the Caribbean and English ports. On the return, goods from England, France, and South America, as well as live horses, furs and hides of beaver, wapiti (elk), deer, and sea otter harvested in California filled the holds as ships such as *Cadboro, Cowlitz, Dryad, Columbia,* and *Prince Albert* made their way back to Fort Vancouver by way of Honolulu. The arrangement lasted until 1849 when the HBC liquidated its California holdings in San Francisco.[74]

The HBC entry into California irritated many older Californios such as Mariano Vallejo, a landed aristocrat with an agricultural empire headquartered at Sonoma, as well as Russian and even some newer European immigrants. One was John Augustus Sutter from Bern, who arrived by sea in 1839 after stops in New York, St. Louis, Santa Fe, Oregon, Hawaii, and Alaska. Within weeks, Governor Juan Bautista Alvarado granted him 11 square leagues of land (50,000 acres) in the Sacramento valley. Familiar with castles in his native Germany, Sutter wasted little time in building his own citadel on a strategic mound overlooking present-day Sacramento City. Sutter's Fort was completed in 1839. Its 18-feet-high walls and bastions towered above Indian villages and Mexican ranches. Sutter diversified with an agricultural and ranching operation within his "New Helvetia" (New Switzerland) and acquired Russian assets in 1841 when the Russian American Fur Company gave up attempting to grow crops and staffing its southernmost post, Fort Ross, just north of the mouth of the Russian River along the northern California coast. Also beginning in 1841, Americans who arrived by way of the California cutoff of the Overland Trail recuperated and re-provisioned at Sutter's Fort. And HBC trappers came and went, much to Sutter's irritation, paying Vallejo for the privilege of trapping on lands that Sutter considered his own land grant.

Ironically, the California Gold Rush started on Sutter's property at his sawmill on the American River near Coloma in 1848. By 1852, the trading and agricultural empire that Sutter built on charisma and credit crumbled as squatters helped themselves to his livestock for food and transportation, occupied his lands, and killed or ran off Indians who had been friendly and productive hunters and laborers for Sutter. Even his loyal "Indian army" suffered indiscriminate genocide as forty-niners and foreigners from all corners of the globe poured into California's Mother Lode region.[75]

Although California and the Southwest could be a lucrative fur frontier for a small number of trappers, its productivity paled in comparison with northern climates. Furthermore, most California and Southwestern tribes placed little value on small fur-bearing animals. Deer, elk, antelope, mountain sheep, and buffalo were easier to stalk and kill and were more prized. Even prior to the Pueblo Revolt of 1680, Indians supplied Spaniards with these coarser furs as part of the tribute expected by Spanish authorities. Never did Indians become as involved in the Southwest in the actual procurement and processing of beaver and other small furbearing animals as elsewhere in North America. And where Euro-American trappers did well, it was only in small numbers spread out over very large areas, not in large brigades such as had been successful in Canada and the Pacific Northwest.One side effect of the fur and Indian trade in the Southwest was promotion of textile arts, which Pueblo, Pima, Papago, and other horticultural-based tribes had long mastered prior to Spanish colonization.[76] Cotton and yucca-fibered blankets, clothing, *mantas* (shawls and one- and two-piece dresses), and ceremonial attire such as sashes continued to be manufactured after Spanish, Mexican, and American occupation, but woolens increasingly came to be favored for their warmth, durability, and possibilities of being walked upon as well as worn. Mexican panchos, serapes, "Saltillo" blankets based upon designs from that province in Mexico, as well as generic "Spanish-American" or "Rio Grande" blankets, now associated with the northern New Mexican community of Chimayó, and woven from coarse Churro sheep, became popular and influenced Indian weavers on the periphery of Spanish and Mexican settlements.[77]

Navajo blankets date to the mid-seventeenth century. Quite simple in earlier times and strictly utilitarian in purpose, after the Mexican War and establishment of permanent reservations in the 1860s, a cottage industry developed, encouraged by American traders and resident Indian agents. Building upon old symbols and motifs, Navajo weavers benefited from the introduction of chemical or aniline dyes after their invention in England in 1856. The color palate exploded, as did design elements, influenced by Mexican serapes, Persian and other Oriental carpets, and pictorial representations of the traditional Navajo worldview, religion, and new realities of life on reservations.[78] Traders to the Navajo—especially J. Lorenzo Hubbell, Thomas Keam, and John Wetherill—in tandem with the arrival of the Santa Fe railroad in the 1880s, created a commercial market for Navajo blankets and rugs. Hubbell Trading Post, at Ganado, now an interpretive site of the National Park Service, became a mecca for Indians and a great exporter to the world of Navajo rugs and blankets.[79] Whereas the Indian trade of the Southwest had once centered on furs and hides, after 1890 it focused on Indian arts and crafts with the Fred Harvey Company promoting Indian "curios" in shops, hotels, and railroad depots from Chicago to Los Angeles.

Ancillary Fur Trade Industries in the Pacific Northwest

The Pacific Northwest drew iconoclasts and idealists far more than any other region of the North American West in the 1830s and 1840s. One enterprise combined both: that of the American Society for Encouraging the Settlement of Oregon. Organized in 1829 by Hall Jackson Kelly, a Boston schoolteacher, the movement attracted a professional ice-cutter from Cambridge, Massachusetts: Nathaniel Jarvis Wyeth, who financed his own venture, similar to that of J. J. Astor two decades prior. Wyeth poured his own personal resources into the project, recruited subscribers and twenty employees, launching the Pacific Trading Company in 1832. By land and by sea, he determined to reach the Pacific to establish a trading emporium that would send beaver and salmon back to Boston.

Once in St. Louis, Wyeth's party joined Sublette and Campbell's brigade, bound for the rendezvous at Pierre's Hole in far-off present-day Idaho. Discouraged by summer snows and hardships on the trail, nine of Wyeth's men abandoned him, leaving him with eleven trappers by the time they reached the trade fair, just in time for a bloody encounter with the Blackfeet. Undaunted by mishaps and unfavorable odds, Wyeth persevered, trapping along the Snake en route to Fort Vancouver, where he was welcomed by Chief Factor John McLoughlin but given the news that his ship, the *Sultana,* had wrecked somewhere in the South Pacific. Fortunate to have such a tolerant host, Wyeth pressed the HBC for an agreement to divide the region geographically at the Columbia, with all furs procured by his company to be sold to the HBC at their various posts.

Ultimately, the HBC Council rejected Wyeth's brash offer. In the meantime, he had other schemes in play. One combined a group of seasoned mountain men associated with Lieutenant Benjamin Bonneville of the US Army, who would trap the region from the Columbia to California, bringing back furs to the next rendezvous near Bonneville's fort, in the heart of Green River country. One Bonneville associate, Joseph Reddeford Walker, succeeded in reaching California, describing Yosemite Valley for the first time in 1833, but fur returns were disappointing, and Wyeth had little to show for the year. In frustration, Wyeth described his competition at the rendezvous as "a great majority of Scoundrels."[80]

Wyeth's second scheme focused on fish. In November 1833, the Columbia River Fishing and Trading Company was organized in Boston with a capital of $20,000. Repeating his plan of 1832, Wyeth sent a ship, the *May Dacre,* on the long voyage around South America while he headed overland to St. Louis, where he joined the rendezvous-bound caravan with $3,000 in goods under contract for the RMFC. Competing with Sublette and Campbell, as well as the AFC, Wyeth arrived too late to sell his goods. Bitter and feeling betrayed, according to pioneer oral historian Frances Victor Fuller, he resolved to "roll a stone into [their] garden" that they would "never be able to get out."[81] The stone was a strategically located fort at the junction of the Snake and the Portneuf rivers, Fort Hall, which was completed as a crude log structure in late summer 1834.

Unable to compete successfully, the post lasted only three years as a Wyeth enterprise and was sold to the HBC in 1837. Wyeth had his own small victories, however. Dr. McLoughlin befriended him by supplying his enterprise with goods from HBC warehouses in the Northwest and tolerating his establishment of a farm on French Prairie, a fur-trade retirement community. Wyeth built a second small post near the mouth of the Columbia and named it Fort William. Still the idealist as well as the pragmatist, Wyeth reconnected with Hall Kelley, who arrived in the Willamette valley after visiting California, where he met and accompanied Ewing Young to Oregon. Unlike the ill-fated *Sultana,* the *May Dacre* arrived, damaged by lightning but afloat, in April 1835, too late for the fresh salmon runs. By now, Wyeth's sailing crew and trapping outfit were depleted through sickness, death, and desertion. A half load of barreled salmon made its way to Boston, but much of it spoiled en route, further reducing profits.

Wyeth returned to Boston in 1836, a failure in entrepreneurship out West but a success in alerting the public to the potential of the Oregon country for American business enterprises and settlement by farmers. He resumed life in the ice-cutting business, patenting fourteen labor-saving devices prior to his death in 1856, recouping much of the wealth that he had lost in his Pacific Fishing and Trading Company.[82] Not until the canning industry replaced barreling of fish in salt or brine would an enterprise like Wyeth's actually succeed.

Mapping and Scientific Exploration of the West Prior to 1870

In 1850, as California entered into the Federal Union, some areas of the West were well known and understood topographically while other portions, especially arid lands and remote mountainous regions, remained uncharted. At the beginning of the nineteenth century, the baseline for all was Philadelphian Samuel Lewis and Englishman Aaron Arrowsmith's collaborative *A New and Elegant General Atlas,* published in 1804. Four maps focused specifically on the region beyond the Mississippi: "Louisiana," "British Possessions in America," "Spanish Dominions in North America," and the generic "North America." Each of these maps incorporated new information from fur traders, explorers, and surveyors. "Louisiana" contained St. Louis surveyor Antoine Soulard's data, as well as traders who had been as far north as the Mandan Villages.[83] As Carl Wheat has noted,

> no one yet knew the location of the Mississippi's source, much less where the Missouri took its rise, the extent of its major tributaries, the true nature of the Rocky Mountains or the complexities that they encompassed, not to mention the singular charter of the Great Basin, or the vast extent of the Columbia- and Colorado river systems. West of the Mandan villages all was conjecture.[84]

The official cartographer of the USA, Nicholas King, did his best to update older maps as Jeffersonian-era data came his way. His 1805 map based on Lewis and Clark data from Fort Mandan in 1805 provided a few officials with new information on the region from Lake Michigan to the Pacific coast.[85] Zebulon Pike's map of the Southwest (1810); Aaron Arrowsmith's Mexico (1810); and Lewis & Clark's Northwest (1814) filled in large gaps.[86] Until the latter was released, according to Wheat, "the Soulard map [of 1795], in the version offered to the public by Arrowsmith and Lewis in 1804, constituted the most ambitious and—despite its many infirmities—the most informative published attempt to portray the West and Northwest of what is now the United States."[87] Arrowsmith updated his map in 1814 as "The Interior Parts of North America," patching in the new data since 1804.[88] Between 1814 and 1823, several minor maps appeared, but it was Stephen H. Long's manuscript map of 1821 that accompanied Edwin James's account of the Long Expedition, 1819–1820 (published in 1823) that moved geographic understanding to a new plateau. It was here in capital font "GREAT AMERICAN DESERT" appeared over much of the high plains from the Platte in the north to the Canadian to the south. Described by one expert as a "mother map," Long's map provided the first reliable location of rivers of the Southern Plains as well as a fairly accurate view of the front wall of the Rockies from Longs Peak to the Spanish Peaks of present-day Colorado, with all the country to the east of the Rockies correctly shown.[89]

Meanwhile, engraver Henry S. Tanner, who had worked with Arrowsmith prior, produced his own *New American Atlas* in 1822, a landmark achievement that incorporated all the best information since 1804 including Alexander Von Humboldt's "New Spain" (1811), Pike and Long's maps and Juan Pedro Walker's "Map of New California" (c. 1810, c. 1817).[90] A Louisianan by birth, Walker served in the Spanish Army as a lieutenant at various Spanish posts, including Janos. He had taken Pike's papers from him in 1807 and had a broad appreciation of the region from Louisiana west through New Mexico to the California missions and northeast to Hudson Bay, shown on his map, believed to have been first drawn in 1810 with benefit of Pike's published map, and revised in 1817, but not published.[91]

Other commercial maps followed Tanner's lead throughout the decade of the 1820s, which ended with Hall Jackson Kelley, the Boston schoolteacher and zealot's "Geographical Sketch of That Part of North America Called Oregon," repeating errors of Lewis and Clark and more of a political promotion than geographic enlightenment.[92]

Fur traders and the public at large would have benefited from Jedediah Smith's corrections and additions to the "known" West of 1831, the year of Smith's death. From firsthand knowledge, Smith revised all previous maps on the relationship of the Great Basin to the Sierra Nevada and the Central Valley of California. Smith also showed the proper distances and orientations of most of the rivers of California, Oregon, and Washington, as well as the proper course of the Willamette, which had confused all since William Clark's mythic "Multnomah" had appeared in 1814 on his final map. But Smith's data, which he planned to publish along with his journals, had to wait for rediscovery more than a century later.[93]

Two years after Smith's death on the Santa Fe Trail, David H. Burr mentioned one of Smith's maps but repeated many errors of Lewis and Clark, as did Frenchman A. H. Brué, Géographe du Roi, who did not have access to a Smith map but who had read a Jed Smith letter published in *Nouvelles Annales des voyages* in 1828 where Smith's route in California was described.[94] Arrowsmith's firm plotted on during the 1830s, putting out a new map of "British North America" in 1834.[95] The maps of fur traders David Thompson (1814, 1818) and of Alexander Ross (1821), accurate on location of posts and the region of the Columbia River basin, remained manuscripts owned by the North West Company and later the merged HBC.[96] Ross found a public audience in 1849 with *Adventures of the First Settlers on the Oregon or Columbia River* and again in 1855 with *Fur Hunters of the Far West,* which included a condensed version of his much more complex map of 1821.[97]

Figure 6.3 C. M. Russell, *Men of the Open Range*, 1923, oil on canvas.
Source: Courtesy of the Montana Historical Society.

Other Bay Company men, including clerk William Kittson and trader Peter Skene Ogden, drew maps of their own travels in the 1820s and 1830s.[98] Ogden's six Snake River expeditions, 1824–1830, as well as his one trip to California, 1829–1830, are reflected on Arrowsmith's revised map.[99] For the first time, the Columbia is extremely well drawn, as well as the rivers that feed it from Idaho, the result of HBC brigade leaders' attention to daily logs and details in mapping.[100]

Fur traders had no motivation to share their intellectual property and special knowledge of little-known places. During the 1830s, several maps, in addition to those of Jedediah Smith, could have changed the course of Western cartographic, and perhaps political, history. Warren Angus Ferris worked for the AFC from 1830 to 1835, attended a number of rendezvous, and knew the northern interior Rockies intimately. Although apparently oblivious to British exploration, Ferris drew a detailed map in 1836 that correctly identified Yellowstone Lake for the first time, showed rendezvous locations and the "parks" and "holes" where trappers wintered over, and delineated the exact range of the buffalo westward in the year 1835. His journal, *Life in the Rocky Mountains,* was published serially in the *Western Literary Messenger,* 1843–1844, but was soon forgotten.[101]

Washington Irving was no fur trader, but he enthusiastically undertook two projects that produced useful maps in the 1830s. His *Astoria* (1835) retraced Wilson Price Hunt's epoch overland trek to Astoria and Robert Stuart's return near-starvation trek, via South Pass, to St. Louis.[102] The map has since been validated as reasonably accurate, providing yet one more piece of the Western cartographic puzzle. Astor paid Captain Bonneville $1,000 to use the latter's journals and manuscripts for his *Adventures of Captain Bonneville; or, Scenes beyond the Rocky Mountains* (1837), which included two maps. The one of the Great Salt Lake region was the most accurate to date and accurately showed many interior-West rivers such as the Green, Wind, Sweetwater, Snake, and Salmon. The second map portrays the country west of the crest of the Rockies, with locations of tribes and major lakes and rivers. Both works had wide readership, making them very influential.[103]

Also influential, Protestant missionary Samuel Parker's "Map of Oregon Territory" of 1838 was made from personal observation and is considered a real advance in that it corrected locations of tribes and repositioned headwaters of many rivers where they ought to be. His journal was released with the map in 1838 and went through many editions. Parker claimed in the promotional *Oregonian and Indian's Advocate* that he had access to charts by George Vancouver, sketches by HBC trader Samuel Black, and material for the southern portion from Jedediah Smith, but he did not state where he found the Smith material, nor has Black's map surfaced.[104]

Prince Maximilian of Wied journeyed up the Missouri in 1833 with Swiss artist Karl Bodmer as his visual chronicler. Both were primarily interested in ethnography and thus left valuable descriptions and images of peoples from St. Louis to Fort McKenzie, deep in the heart of Blackfeet country. Upon his return to Germany, in 1839, Maximilian published his account with an excellent map delineating the location of tribes and posts.[105]

Another private citizen, artist George Catlin, traveled by steamboat up the Missouri in 1832, visiting posts of the AFC before heading out onto the plains, where he documented on canvas and in print tribal life, including that of the Comanches. His Letters and Notes on the Manners, Customs, and Condition of the North American Indians included his 1833 "Outline Map of Indian Localities in 1833" as well as "Map of Localities in 1840," which showed those tribes "removed" to the West under the policies of the US government in the 1830s.[106] Ethnologist John C. Ewers considered the 1833 map the most accurate of any prior to 1855 on actual location of Plains tribes.[107]

Military explorers of the American West are an underappreciated group of cartographers. The US Exploring Expedition, 1838–1842, under the command of Lieutenant Charles Wilkes, US Navy, arrived at the mouth of the Columbia in late April 1841, continuing on into Puget Sound forty-nine years after Vancouver. Surveyors fanned out by land and by water to map the jointly occupied and increasingly contested Oregon country. One group made their way overland south as far as Sutter's Fort and San Francisco in distant California, where they reunited with Wilkes, waiting for them in Whaler's Harbor (modern Sausalito). Wilkes' official report appeared in 1845 and contained two maps that elevated scientific mapping to new levels in the American West. "Map of the Oregon Territory" and "Map of Upper California" incorporated data from Jedediah Smith's maps, which Wilkes must have seen, as well as direct observation and precise measurements by the surveying crews. The objects brought back from Wilkes' circumnavigation became the basis of the natural-history collections of the newly formed Smithsonian Institution.[108]

While Wilkes navigated for the Navy at sea, Colonel John J. Abert commanded the newly formed US Army Corps of Topographical Engineers by land. Formed in 1838, Abert sent Lieutenant John C. Frémont as an assistant to Joseph Nicholas Nicollet to survey the hydrographic basin of the Mississippi River. Through Nicollet's excellent cartographic skills, one thousand miles of the well-known but poorly mapped region of the upper Mississippi, prime fur-trade grounds for centuries prior, were scientifically revealed.. A second expedition focused on the Coteau du Missouri between the Mississippi and the Missouri, which Nicollet determined to be a hydrographic basin of its own, a major geographic discovery.

Frémont would go on to command three more expeditions from 1842 to 1849, literally circling the entire region of the trans-Mississippi West and penetrating much of its interior in the process. Guided by Kit Carson, Lucien B. Maxwell, Joseph Reddeford Walker, Alex Godey, Old Bill Williams, and other mountain men with firsthand knowledge of the land, Frémont combined the science of surveying with ground-level logic and produced several important maps, used for years by the government and the public as road maps of their day. Drawn by Charles Preuss, an 1843 map was supplemented with an even better version in 1845, one that represented "an important step forward from the earlier western maps of the period as did those of Pike, Long, and Lewis and Clark in their day," according to Carl Wheat.[109] Seen from another perspective, William H. Goetzmann adds: "Though Frémont's expeditions were politically conceived and orientated toward practical objectives, as a Topographical Engineer his scientific work added a cubit to the national stature."[110]

Frémont's third (1845) and fourth (1848) expeditions coincided with the Mexican War and the contested boundary between Britain and the USA over Oregon. After Texas was admitted to the Union in 1845, tensions mounted between Mexico and the USA, prompting the Polk administration to send two expeditions to the southern Rockies that same year. Colonel Stephen Watts Kearny led the first, taking with him Lieutenant W. B. Franklin as topographer and mapmaker. Franklin corrected some of Pruess's map and suggested that the Green River flowed into the larger Colorado River system to the west, thus filling in that important piece of the geographic puzzle.[111]

Frémont marched toward California that same summer, reaching Bent's Fort on the Arkansas in early August, from which Lieutenant James W. Abert was sent south through Ratón Pass to the Llano Estacado (Staked Plains) of western Texas. Abert tracked the

Canadian River eastward, providing the first reliable map of that river system along with valuable ethnographic data on tribes of the Southern Plains, mineral and timber resources, watering holes, and native vegetation.[112]

While Abert improved knowledge of lands along the old Santa Fe Trail, guided by Carson, Maxwell, and Joseph Reddeford Walker, Frémont and sixty well-armed men headed westward along the Arkansas, across the Continental Divide to the Great Salt Lake where they made a detailed survey of the southern shore. Carson and Maxwell were sent ahead to find a route across the desert, unaware that Jed Smith had pioneered a route in 1827. This party followed the route that soon would be known as Hasting's Cutoff, a perilous shortcut to California that would cost many overlanders their lives, including members of the Donner Party in 1846. After catching up with the vanguard at Whitten's Springs, Frémont led ten men directly across the desert while Walker guided the larger party along the Humboldt River to the lake that bears his namesake. At Walker Lake, the two groups reunited only to split up again with Walker's party traveling southward to Walker Pass over the Sierras while Frémont with fifteen men crossed at soon-to-be-named Donner Pass, arriving on the American River on December 9. Eventually, both parties reunited in San Jose, alerting Mexican authorities to the potential of an American invasion, soon to become a reality.

Vastly outnumbered, Frémont escaped detention and capture, retreating toward Klamath Lake before reversing the march back into Mexican California under obscure and poorly known orders from Polk and other officials, delivered by a contingent of US Marines sent from Washington. Frémont's "scientific" expedition ended at this point, and his specimen-collectors and cartographers turned sharpshooters in the Mounted Rifles. By June 1846, Frémont's contingent was fully engaged in the Bear Flag Revolt prior to official declaration of war against Mexico that same summer.

Conflicts over authority surfaced as Frémont disobeyed orders from Kearny (now a brigadier general) and faced court martial back in Washington, where he was found guilty but was later acquitted by President Polk. Frémont's account of his third expedition reached the public in 1848, accompanied by a new map drawn by Charles Preuss, filling in areas of the Great Basin little known prior to 1846. The report proved that the desert was not as formidable as previously thought, giving false security to many who rushed to California during the Gold Rush in 1849.[113] For his part, Kearny contributed an important scientific report of the prospects of the Southwest, providing detailed topographical sketches by Lieutenant Willliam H. Emory, who had just drawn the first accurate map of Texas in 1844.[114]

THE PACIFIC RAILROAD SURVEYS

The 1850s mark a decade of transition throughout the North American West. Trade remained a matter of barter economies in remote areas of the region, but money and values based on gold and silver played more important roles in domestic and international transactions as the supply of precious metals increased due to the California Gold Rush. Just as Astor and Crooks' AFC folded with closure of their New York offices in 1849, the US government began assuming a larger role in exploring and mapping the American West. Several expeditions between 1848 and 1851 filled in cartographic and topographic holes, simultaneous with government determination to complete an accurate survey of the Mexican boundary with the USA, which was completed in stages between 1849 and 1855, with reports issued periodically and final summations of scientific data completed between 1856 and 1859.[115]

Meanwhile, in 1853, Congress authorized the Pacific Railroad Surveys. For the next six years, soldiers and civilians in government employ combed the West along strategic parallels with the goal of providing Congress the most practical route to connect the Pacific coast with the Midwest. Ironically, the route ultimately chosen along the 41st parallel was not a part of the four surveys but had been partially mapped by Stansbury and others prior to 1853. Sectional politics, as much as the need for a transcontinental railroad, entered into the government's routing of its soldiers, all of whom reported to the Secretary of War, Jefferson Davis.[116]

The Northern Survey between the 47th and 49th parallels was conducted by Isaac I. Stevens, assisted by Captain George B. McClellan. No practical route was found across the Cascades, but Stevens succeeded in securing treaties with the region's Indians from Montana to the Washington and Oregon coasts in 1855. The 38th parallel survey undertaken by Captain John W. Gunnison, with help from Edward F. Beale (both replaced by Lieutenant E. G. Beckwith after the death of Gunnison at the hands of Indians) followed one of Frémont's trails of 1848 from Fort Leavenworth along the Kansas River into Colorado and on across the Continental Divide at Cochetopa Pass (modern North Pass between Gunnison and Saguache) but struck new ground thereafter through the Wasatch Mountains into Utah. From here the route followed the Weber River and on to the coast.[117]

A second 38th-parallel expedition, privately financed and without official government sanction, left St. Louis in September 1853. Its leader was none other than John C. Frémont, determined to restore his reputation after court martial and loss of men to frostbite and starvation in the southern Rockies on his fourth expedition in 1848. On previous expeditions, beginning in 1842, Frémont had taken a camera but had limited success with capturing images of use. His last attempt was different. This time, Frémont's group included a professional photographer, the first of many to follow in capturing images of the West. From Solomon Nuñez Carvalho's daguerreotypes and journal, we know that this second ill-favored Frémont adventure turned unfortunate once the party crossed Cochetopa Pass. Rough terrain, insufficient food, death of the assistant topographer, and problems with local Ute Indians highlighted more than fifty days of struggle during an attempt in the depth of winter to get to settlements in Utah. Mormon citizens of Parowan resuscitated the party, which made its way on to California, arriving safely after crossing the desert to a point south of Bishop, from which they marched into the San Joaquin valley and down the delta to San Francisco.[118] Carvalho produced hundreds of daguerreotypes on the trip, but Frémont chose not to share them with the public or to allow Carvalho to use them for his own book. Instead, he sent them to Matthew Brady, whose studio rephotographed them onto glass-plate negatives for paper-positive prints. Thirteen of these prints were redrawn by craftsmen on engraving plates, used for illustration of Frémont's self-promoting biography, released in 1856 just as the "Pathfinder" campaigned for the presidency. These included views of Indians, a portrait of Kit Carson, a scene of Frémont hoisting the flag in the Rocky Mountains, and some landscapes including Klamath Lake and the entrance to Monterey.[119]

The third Pacific Railroad survey was conducted by Lieutenant Whipple along the 35th parallel from Cairo, Illinois, to Fort Smith, Arkansas, and then on through Indian territory along the Arkansas and Canadian rivers into New Mexico, crossing the Rio Grande near Albuquerque, on across Arizona's Gila River and up the Colorado to present-day Needles. From here, Whipple marched across the Mohave, arriving at San Bernardino.

The final route was that preferred by Southerners along the 32nd parallel, with a destination of Los Angeles. One party under Lieutenant John G. Parke explored west to east; the other under Captain John Pope moved westward. The final recommendation linked the Mississippi with California through Houston, El Paso, Tucson, and Yuma, by-passing San Diego due to the difficulty of finding a suitable pass that did not enter Mexico. Both the 1855 and the 1857 reports to Congress recommended this route over the others, but sectional politics and Jefferson Davis's stubborn resolve not to compromise at the 35th parallel led to indecision as the nation entered the Civil War in 1861.[120]

Of the many maps produced by the Pacific Railroad Surveys, the most important for the public and for government use beyond engineering of the various transcontinental railroad lines was that of Gouverneur K. Warren's "Map of the Territory of the United States," which appeared in Vol. XI. Drawn at a scale of 1:3,000,000 or 47.35 miles to the inch, the accompanying *Memoir* summarized all American exploration from Lewis and Clark up through 1857. Used as a base map for decades to come, not only all five proposed routes are shown; in addition, forts, Indian villages, places of Indian–Euro-American encounters, and well-traveled routes showed travelers and military men alike a basic reliable template for understanding Indian country and the few places Euro-Americans had colonized by the mid-1850s.[121]

THE WAGON ROADS SURVEYS

While the Topographical Corps struggled to distance its scientific efforts from sectional politics, Congress established the Pacific Wagon Road Survey in 1855. While the War Department continued to survey and build roads through the 1850s, with army engineers in charge, most of the wagon-road surveys were funded through the Department of the Interior, created in 1849. Unlike the railroad surveys, the wagon roads were territorial, not continental, projects, designed as "country roads" linking towns and forts within specific districts and future states:

> Of all the forms of federal aid, the wagon-road program was perhaps the most significant for the early territorial development. Not only did the military engineers lay out the basic lines of communication upon which ultimate economic success depended, especially in a country generally devoid of rivers affording transportation facilities, but by this means they also created the real estate values and decided the location of towns and cities and capitals.[122]

Minnesota, Nebraska, Kansas, and New Mexico received the lion's share of funding. California's rapid growth during the early days of the Gold Rush prompted mining districts and mail companies to blaze their own trails throughout the Sierra and interior valleys and prompted individuals such as Jesse Applegate to connect California with Oregon, before Congress appropriated a dime for federal wagon roads. But Oregon and Washington Territory (established in 1853) lagged behind.[123]

From the beginning of the program, a Pacific Wagon Road Office was established in San Francisco, far from the immediate sectional debate over the routing of the future transcontinental railroad. Major Hartman Bache took charge and directed surveys to improve the stage route between California and Oregon and to connect important corridors in the Pacific Northwest linking Astoria, Salem, Portland, Fort Vancouver, Seattle, Bellingham, and Walla Walla. Other federal projects in Texas and Utah provided some infrastructure in remote places, as did military roads across the plains.[124] The sharing of responsibility between the interior and war departments for these roads was harbinger of the great surveys in the post-Civil War West, where natural history, geology, photography, cartography, and military agendas merged to fill in the final geographic pieces of the West's highest, wildest, and most arid regions.[125]

The last major expeditions conducted by the Topographical Engineers in the West focused on the Colorado river region and the Dakotas. In 1857, Lieutenant Joseph Christmas Ives undertook to explore the Colorado from its mouth inland on the 50-foot steamboat, *Explorer,* sending another party overland to rendezvous at Yuma. Reaching by skiff the Black Canyon below modern Las Vegas, Ives sent the steamboat back and set off overland with twenty soldiers and four scientists, including geologist John Strong Newberry, reaching the floor of the Grand Canyon, the first Euro-American men to see the canyon from the bottom up and the first to describe its geological stratification. Climbing back onto the Colorado plateau, Ives marched across the Painted Desert to the Hopi villages and through Navajo country to Fort Defiance. Ives' legacy rests in testing the Colorado River for its navigability, in determining suitability of roads across the arid Southwest, and in reporting the canyon lands of the Colorado plateau. Furthermore, his topographer, Prussian-born Baron F. W. Egloffstein provided three-dimensional sculpted relief to his five-section map, the first to portray the Grand Canyon and plateau country and a model of the techniques to be used during the geological surveys of the 1860s and 1870s. In tandem with other expeditions of 1857–1859 under Captain John J. Macomb, sent to explore the Old Spanish Trail from Santa Fe into Utah, and that of Captain James Hervey Simpson, who set out from San Francisco to Fort Bridger, Wyoming, and back across some of Utah and Nevada's most inhospitable terrain, the Ives Expedition helped connect the Great Basin and Colorado plateau regions with the Pacific coast to the southwest and the Rockies and Plains to the northeast.[126]

The Yellowstone Expedition of 1859–1860 represents the last of the Topographical Engineers' efforts prior to the Civil War and the Army's best effort to follow up on the three expeditions commanded by General William H. Harney, but better remembered for an officer under his command: the aforementioned Lieutenant Gouverneur K. Warren. Originally a punitive expedition against the Sioux, Warren made the most of his three excursions into Indian country between 1855 and 1857, hiring excellent topographers and a scientist destined to leave a major mark on Western history: Ferdinand V. Hayden, a medical doctor and geologist who already had spent 1853 and 1854 collecting geological and paleontological samples in Dakota territory.

Of the three probes into Sioux country, the 1857 expedition was the most important; the goal: linking the Fort Snelling–Big Sioux River road with Fort Laramie and the South Pass. With an eye on logistics for wagons and future settlers and an optimistic view that the "Great American Desert" of Pike and Long could be surmounted with farms, ranches, and mines once Indian unrest had been addressed through treaties and conquest, Warren gave Congress and the Army its most practical map and guide to date with his 1857 report and map, setting the stage for a full reconnaissance of the region drained by the Yellowstone River in the next decade.

Led by Captain William F. Raynolds, with Dr. F. V. Hayden as chief scientist, the Yellowstone Expedition examined four possible wagon routes from the eastern plains through the Dakotas and Wyoming with the goal of connecting with Lieutenant John Mullan's 624-mile military road under construction from Walla Walla on the Columbia to Fort Benton, the head of navigation on the Missouri. In addition, Raynolds was instructed to survey the "numbers, habits, and dispositions" of the Indians, as well as the climate and resources of a region 250,000 square miles, or double the area of Great Britain. Although no easy road system was located between the Missouri and Bitterroot rivers, Raynolds located the advance scouts of the Mullan road, and his scientists, especially Hayden, helped to solve puzzles in geology and topography, providing a stratigraphic column in nine layers, a measuring stick in time that subsequent earth scientists could use as a gauge in their own work.[127]

TRADE IN THE WEST, 1850–1870

As government and civilian explorers surveyed, plotted, and reported Western topography, climate, and prospects for settlement, the old Indian and fur-trades continued, albeit vastly altered from the days of the beaver and early robe trade eras. Euro-American, mixed-blood, and Indian trappers, hide-hunters, and traders continued to make a living, but their options of where they worked, lived, and conducted business increasingly became restricted to smaller wildlife habitats or areas where Indians continued to exert sovereignty and relative independence.

Beaver trapping did not end with the last Rocky Mountain rendezvous after 1840. A steady decline in prices beginning in 1833 reached a nadir by 1845, but manufacturers eagerly continued to make hats of beaver even as hats made of napped silk and others made of South American nutria competed favorably. From its Canadian posts, the HBC shipped 118,000 beaver to England in 1853, peaking at 172,000 in 1867 or far more than any single year at the height of attendance at an American rendezvous in the 1830s.[128] But the days of finding uncontested streams and ponds in the American West had come to an end as Native peoples resisted conquest and forced relocation to reservations.

After 1850, the US Army expanded operations throughout the West, purchasing or taking over older fur-trade posts. The new trader often became the post sutler, whose prices were determined under government price structures requiring cash rather than trade in kind.[129] Older Euro-American companies struggled to survive in this new economic environment. Most folded. Those that continued in business competed for Indian annuity contracts, the provisioning of reservations and US military posts, and the supply of Euro-American towns and farming communities. Indian and mixed-blood producers lost purchasing power, experiencing stiff competition from Euro-American hunters over the only major resources remaining: robes and hides on the Plains; fish and furs in the Northwest; and wild and domesticated horses in the Southwest.

Throughout the 1840s, an annual average of 90,000 buffalo robes arrived in St. Louis, and by the 1850s the number rose to around 100,000. The leather trade supplying industrial belts for steam-driven power provided steady demand.[130] Still, on the Plains, in 1869, an estimated 4 million bison grazed south of the Platte River with around 500,000 remaining in the north. That year, the Union and Central Pacific railroads joined tracks, initiating an even greater slaughter of robes, hides, and tongues. In the next five years, over 3 million buffalo fell to the Euro-American hide hunters while another 500,000 were killed by Indians and settlers. By 1876, the southern herd was gone.[131] The northern herd faced a similar fate after the Battle of Little Big Horn in 1876. On the upper Missouri, I. G. Baker & Co. of Fort Benton replaced the older Chouteau operation in 1863, shipping 70,000 robes that year. By 1883, volume had declined to 5,000 robes, and in 1890, only about 1,000 animals remained.[132]

As natural resources in wildlife populations shrank or disappeared, many Euro-American and mixed-blood fur-trade personnel remained with their Indian families during these difficult years of transition, moving with wives and children to join full-blood relatives on reservations or joining other traders in residential communities where ethnicity was blurred by generations of intermarriage. These retirement communities included old voyageur towns in Missouri, especially St. Charles; Taos, New Mexico and small settlements on the Arkansas River in southern Colorado where Indians, Hispanics and Anglos intermixed; California's Napa valley; the Willamette valley in Oregon, especially the communities of French Prairie, Oregon City, and Champoeg; and freighting hubs or military posts—once fur-trading emporia such as Fort Benton, Fort Bridger, and Fort Laramie.[133]

By century's end, the Indian trade in furs and robes ceased to provide a viable living for all but a few trappers and hunters. Members of some tribes adjusted to the reservation-based economy by supplying utilitarian and ceremonial objects in the form of beaded, woven, sewn, and carved items to private collectors, tourists, and museum personnel.[134] Restoration of populations of bison and other animals depleted in the nineteenth century would require a full century of doing without. Today, that is changing as Native peoples throughout North America are taking pride in bringing back the buffalo.[135]

NOTES

1. John C. Ewers, "The Indian Trade of the Upper Missouri before Lewis and Clark: An Interpretation," *Bulletin of the Missouri Historical Society,* 10 (4) (1954): 429–446.
2. W. Raymond Wood, "Plains Trade in Prehistoric and Protohistoric Intertribal Relations," in W. Raymond Wood and Margot Liberty (ed.), *Anthropology on the Great Plains* (Lincoln, Nebr.: University of Nebraska Press, 1980), pp. 98–109; W. Raymond Wood and Thomas D. Thiessen (eds.), *Early Fur Trade on the Northern Plains: Canadian Traders among the Mandan and Hidatsa Indians, 1738–1818 — The Narratives of John Macdonell, David Thompson, François-Antoine Laroque, and Charles McKenzie* (Norman, Okla.: University of Oklahoma Press, 1985).
3. The best overviews of the fur trade are Paul C. Phillips, *The Fur Trade,* 2 vols. (Norman, Okla.: University of Oklahoma Press, 1961); and James A. Hanson, *When Skins Were Money: A History of the Fur Trade* (Chadron, Nebr.: Museum of the Fur Trade, 2005).
4. Hanson, *When Skins Were Money,* Chapter 2.
5. W. R. Swagerty, "Indian Trade of the Trans-Mississippi West to 1870," in Wilcomb Washburn (ed.), *Handbook of North American Indians,* vol. IV: *Indian–White Relations* (Washington, DC: Smithsonian Institution, 1988), pp. 351–374.
6. Carol L. Riley, *The Frontier People: The Greater Southwest in the Protohistoric Period* (Carbondale, Ill.: Southern Illinois University, 1982); Nancy Parrott Hickerson, *The Jumanos: Hunters and Traders of the Southern Plains* (Austin, Tex.: University of Texas Press, 1994); Swagerty, "Indian Trade of the Trans-Mississippi West," pp. 351–353.
7. Albert H. Schroeder and Dan S. Matson (eds.), *A Colony on the Move: Gaspar Castaño de Sosa's Journal, 1590–1591* (Santa Fe, N. Mex.: School of American Research, 1965), p. 56; Don Juan de Oñate, "Account of the Discovery of the Buffalo, 1599," in Herbert E. Bolton (ed.), *Spanish Exploration of the Southwest, 1542–1706* (New York: Scribners, 1908), p. 227; Marion Schwartz, *A History of Dogs in the Early Americas* (New Haven, Conn.: Yale University Press, 1997), pp. 52–53.
8. Bourgmont, in *Découvertes et ètablissemments des Français dans l'ouest et dans le sud de l'Amérique Septentrionale (1614–1754),* 6 vols. (Paris: Maisonneve & Cie., 1879–1888; rpt, New York: AMS, 1974), vol. VI, p. 414.
9. Stanley's art was included in Henry Rowe Schoolcraft (ed.), *Historical and Statistical Information Respecting the History, Condition and Prospect of the Indian Tribes of the United States,* 6 vols. (Philadelphia, Pa.: Lippincott, Grambo & Co., 1851–1857; rpt, New York: Arno, 1969), vol. VI, plate 30.

10. Francis Haines, The Northward Spread of Horses among the Plains Indians," *American Anthropologist,* 40 (3) (1938): 429–437; Francis Hains, *The Plains Indians* (New York: Crowell, 1976); Robert H. Lowie, *Indians of the Plains* (New York: McGraw Hill for the American Museum of Natural History, 1954), pp. 40–41; Theodore Binnema, *Common and Contested Ground: A Human and Environmental History of the Northwestern Plains* (Norman, Okla.: University of Oklahoma Press, 2001).
11. Frank Gilbert Roe, *The Indian and the Horse* (Norman, Okla.: University of Oklahoma Press, 1955); Ewers, "The Indian Trade of the Upper Missouri."
12. W. Raymond Wood, "Contrastive Features of Native North American Trade Systems," in Fred W. Voget and Robert L. Stephenson (eds.), *For the Chief: Essays in Honor of Luther S. Cressman,* University of Oregon Anthropological Papers 4, pp. 153–169.
13. Carl Wheat, *Mapping the Transmississippi West,* 6 vols. (San Francisco, Calif.: The Institute of Historical Cartography, 1958), vol. I.
14. Mark Warhus, *Another America: Native American Maps and the History of Our Land* (New York: St. Martin's, 1997); James Ronda, "'A Chart in His Way': Indian Cartography and the Lewis and Clark Expedition," *Great Plains Quarterly,* 4 (1) (1984): 43–53; *The Journals of the Lewis and Clark Expedition,* ed. Gary E. Moulton, 13 vols (Lincoln, Nebr.: University of Nebraska Press, 1983–2001), vol. I: *Atlas of the Lewis and Clark Expedition,* contains facsimiles of all known Indian-made maps collected by Lewis and Clark.
15. Lewis and Clark's official journals were published in 1814. Scholars today rely on the Moulton edition. Also see James P. Ronda, *Lewis & Clark among the Indians* (Lincoln, Nebr.: University of Nebraska Press, 1984).
16. James Gibson, *Otter Skins, Boston Ships, and China Goods: The Maritime Trade of the Northwest Coast, 1785–1841* (Seattle, Wash.: University of Washington Press, 1992), Tables 1–2, pp. 299–311.
17. Pike's narrative (1810) was published prior to the official history of the Lewis and Clark Expedition (1814) and thus takes a special place in stimulation of Western exploration and the fur trade. Most of Pike's confiscated papers were rediscovered in 1908 by historian Herbert E. Bolton while working in the archives in Mexico City. See Donald Jackson's introduction in *The Journals of Zebulon Montgomery Pike, with Letters and Related Documents,* 2 vols. (Norman, Okla.: University of Oklahoma Press, 1966); also Harvey L. Carter, *Zebulon Montgomery Pike: Pathfinder and Patriot* (Colorado Springs, Col.: Dentan Printing Co., 1956); and *"To Spare No Pains": Zebulon Montgomery Pike and his 1806–1807 Southwest Expedition,* ed. Tim Blevins et al. (Colorado Springs, Col.: Pikes Peak Library District with Pioneers Museum, 2007). Also, *Jefferson's Western Explorations: Discoveries Made in Exploring the Missouri, Red River and Washita by Captains Lewis and Clark, Doctor Sibley, and William Dunbar, and Compiled by Thomas Jefferson,* a facsimile, ed. Doug Erickson, Jeremy Skinner, and Paul Merchant (Spokane, Wash.: Arthur H. Clark Co., 2004). For excellent summaries of this era see essays in John Logan Allen (ed.), *North American Exploration,* 3 vols. (Lincoln, Nebr.: University of Nebraska Press, 1997), especially vol. III: *A Continent Comprehended.*
18. Warren A. Beck and Ynez D. Haase, *Historical Atlas of the American West* (Norman, Okla.: University of Oklahoma Press, 1989); see entries for each route in Howard R. Lamar (ed.), *The New Encyclopedia of the American West* (New Haven, Conn.: Yale University Press, 1998); John Logan Allen, "The Invention of the American West: Fur Trade Exploration, 1821–1839," in John Logan Allen (ed.), *North American Exploration,* 3 vols. (Lincoln, Nebr.: University of Nebraska Press, 1997), vol. III: *A Continent Comprehended,* pp. 132–189.
19. The foundation for these studies is Marcel Giraud, *Le Métis Canadien* (Paris: Institute d'Ethnologie, 1945). Also see Sylvia Van Kirk, *Many Tender Ties: Women in Fur Trade Society, 1670–1870* (Norman, Okla.: University of Oklahoma Press, 1983); Jacqueline Peterson and Jennifer S. H. Brown, *The New Peoples: Being and Becoming Métis in North America* (Winnipeg and Lincoln, Nebr.: University of Manitoba Press and University of Nebraska Press, 1985); John C. Jackson, *Children of the Fur Trade: Forgotten Métis of the Pacific Northwest* (Missoula, Mont.: Mountain Press, 1966); Fritz Pennekoek, "Metis Studies: The Development of a Field and New Directions," in Theodore Binnema, Gerhard J. Ens and R. C. Macleod (eds.), *From Rupert's Land to Canada* (Edmonton: University of Alberta Press, 2001), pp. 111–128.

20. John C. Ewers, "The Influence of the Fur Trade upon the Indians of the Northern Plains," in Malvina Bolus (ed.), *People and Pelts: Selected Papers of the Second North American Fur Trade Conference* (Winnipeg: Peguis Publishers, 1972), pp. 1–26.
21. Russell Thornton, *American Indian Holocaust and Survival: A Population History since 1492* (Norman, Okla.: University of Oklahoma Press, 1986); Swagerty, "Indian Trade of the Trans-Mississippi West," p. 360.
22. Ronda, *Lewis and Clark among the Indians;* Robert J. Miller, *Native America, Discovered and Conquered: Thomas Jefferson, Lewis & Clark, and Manifest Destiny* (Westport, Conn.: Praeger, 2006).
23. John C. Ewers, *The Blackfeet: Raiders on the Northwestern Plains* (Norman, Okla.: University of Oklahoma Press, 1958); Anthony McGinnis, *Counting Coup and Cutting Horses: Intertribal Warfare on the Northern Plains, 1738–1889* (Evergreen, Col.: Cordillera Press, 1990).
24. Wilbur R. Jacobs, "Frontiersmen, Fur Traders, and Other Varmints: An Ecological Appraisal of the Frontier in American History," *American Historical Association Newsletter* (November 1970): 5–11; David J. Wishart, *The Fur Trade of the American West, 1807–1840: A Geographic Synthesis* (Lincoln, Nebr.: University of Nebraska Press, 1979); Andrew C. Isenberg, *The Destruction of the Bison: An Environmental History, 1720–1920* (Cambridge: Cambridge University Press, 2000).
25. Kenneth W. Porter, *John Jacob Astor: Business Man,* 2 vols. (Cambridge, Mass.: Harvard University Press, 1931).
26. Donald Meinig, *The Great Columbia Plain: A Historical Geography, 1805–1910* (Seattle, Wash.: University of Washington Press, 1968), pp. 39–53.
27. Ora B. Peake, *A History of the United States Factory System, 1795–1822* (Denver, Col.: Sage Books, 1954); Francis P. Prucha, *American Indian Policy in the Formative Years: The Trade and Intercourse Acts, 1790–1834* (Cambridge, Mass.: Harvard University Press, 1962).
28. E. E. Rich, *A History of the Hudson's Bay Company, 1670–1870,* 2 vols. (Toronto: McClellan & Stewart, 1958–1959).
29. Alexander Ross, *Adventures of the First Settlers on the Oregon or Columbia River* (London: Smith, Elder, 1849); Ross, *The Fur Hunters of the Far West,* ed. Kenneth A. Spaulding, 2 vols. (Norman, Okla.: University of Oklahoma Press, 1956), first published 1855; Ross, *The Red River Settlement: Its Rise, Progress and Present State* (London: Smith, Elder, & Co., 1856); Gloria Griffen Cline, *Peter Skene Ogden and the Hudson's Bay Company* (Norman, Okla.: University of Oklahoma Press, 1974).
30. John S. Galbraith, *The Hudson's Bay Company as an Imperial Factor, 1821–1869* (Toronto: University of Toronto Press, 1957); Dorothy Nafus Morrison, *Outpost: John McLoughlin and the Far Northwest* (Portland, Oreg.: Oregon Historical Society Press, 1999).
31. Richard Somerset Mackie, *Trading beyond the Mountains: The British Fur Trade on the Pacific, 1793–1843* (Vancouver: University of British Columbia Press, 1997).
32. James Gibson, *The Lifeline of the Oregon Country: The Fraser-Columbia Brigade System, 1811–1847* (Vancouver: University of British Columbia Press, 1997).
33. Jennifer S. H. Brown, *Strangers in Blood: Fur Trade Company Families in Indian Country* (Vancouver: University of British Columbia Press, 1980); Jackson, *Children of the Fur Trade.*
34. Theodore Karamanski, "The Iroquois and the Fur Trade of the Far West," *The Beaver,* 312 (4) (1982): 4–13.
35. Ross, *Fur Hunters of the Far West,* p. 194; Ross Cox, *The Columbia River; or, Scenes and Adventures during a Residency of Six Years on the Western Side of the Rocky Mountains among Various Tribes of Indians Hitherto Unknown,* ed. Edgar R. Stewart and James R. Stewart, 2 vols. (Norman, Okla.: University of Oklahoma Press, 1957), first published 1831.
36. Richard E, Oglesby, *Manuel Lisa and the Opening of the Missouri Fur Trade* (Norman, Okla.: University of Oklahoma Press, 1963).
37. LeRoy R. Hafen, "A Brief History of the Fur Trade of the Far West," in LeRoy R. Hafen (ed.), *Mountain Men and the Fur Trade of the Far West,* 10 vols. (Glendale, Calif.: Arthur H. Clark Co., 1965–1972), vol. I, pp. 17–176, esp. pp. 41–46.

38. Peter W. Dunwiddie, "The Nature of the Relationship between the Blackfeet Indians and the Men of the Fur Trade," *Annals of Wyoming*, 46 (1) (1974): 123–134; Wishart, *Fur Trade of the American West*, pp. 41–46.
39. Wishart, *Fur Trade of the American West*, pp. 41–46.
40. William E. Foley and C. David Rice, *The First Chouteaus: River Barons of Early St. Louis* (Urbana, Ill.: University of Illinois Press, 1983).
41. The court awarded $66,000 from St. Louis County properties, as well as $3,000 in livestock, wagons, and furniture. In 1848, Pierre Jr. collected $81,772 ($30,380 principal, the rest interest) through the US government from Mexico on behalf of A. P. Chouteau and Jules DeMun from the confiscations of 1817 in Santa Fe. See Shirley Christian, *Before Lewis and Clark: The Story of the French Dynasty that Ruled America's Frontier* (New York: Farrar, Straus & Giroux, 2004), pp. 335–340.
42. Janet Lecompte, "The Chouteaus and the St. Louis Fur Trade," *Guide to the Microfilm Edition of Papers of the St. Louis Fur Trade: Part 1: The Chouteau Collection, 1752–1925 and Part 2: Fur Company Ledgers and Account Books, 1802–1871*, ed. William R. Swagerty (Bethesda, Md.: University Publications of America, 1992), pp. xiii–xxii.
43. David Lavender, *The Fist in the Wilderness* (Albuquerque, N. Mex.: University of New Mexico Press, 1964).
44. Hanson, *When Skins Were Money*, p. 9.
45. John E. Sunder, *The Fur Trade of the Upper Missouri, 1840–1865* (Norman, Okla.: University of Oklahoma Press, 1965), p. 17.
46. Lecompte, "The Chouteaus and the St. Louis Fur Trade."
47. William R. Swagerty, "Marriage and Settlement Patterns of Rocky Mountain Trappers and Traders," *Western Historical Quarterly*, 11 (2) (1980): 159–180. The stereotype of debt peonage was a common nineteenth-century literary image, reinforced by Hiram Martin Chittenden, a civil engineer, whose *The American Fur Trade of the Far West*, 3 vols. (New York: Francis A. Harper, 1902), remains the classic in this field.
48. William R. Swagerty and Dick A. Wilson, "Faithful Service under Different Flags: A Socioeconomic Profile of the Columbia District, Hudson's Bay Company and the Upper Missouri Outfit, American Fur Company, 1825–1835," in Jennifer S. H. Brown, W. J. Eccles, and Donald P. Heldman (eds.), *The Fur Trade Revisited: Selected Papers of the Sixth North American Fur Trade Confernece, Mackinac Island, Michigan, 1991* (East Lansing, Mich.: Michigan State University Press, 1994), pp. 243–267.
49. Robert G. Athearn, *Forts of the Upper Missouri* (Englewood Cliffs, NJ: Prentice Hall, 1967); R. G. Robinson, *Competitive Struggle: American's Western Fur Trading Posts, 1764–1865* (Boise, ID: Tamarack Books, 1999).
50. Fred R. Gowans, *Rocky Mountain Rendezvous: A History of the Fur Trade Rendezvous, 1825–1840* (Layton, UT: Peregrine Smith Books, 1985).
51. Dale Morgan, *The West of William H. Ashley* (Denver, Col.: Old West, 1964); Richard M. Clokey, *William H. Ashley: Enterprise and Politics in the Trans-Mississippi West* (Norman, Okla.: University of Oklahoma Press, 1980).
52. Robert Glass Cleland, *This Reckless Breed of Men: The Trappers and Fur Traders of the Southwest* (New York: Alfred A. Knopf, 1950); Robert M. Utley, *A Life Wild and Perilous: Mountain Men and the Paths to the Pacific* (New York: Henry Holt & Co., 1997).
53. "Names of persons killed belonging to the parties of Wm. H. Ashley, Smith, Jackson & Sublette, etc.," Sublette Papers, Missouri Historical Society, St. Louis, 1823–1829.
54. Dale L. Morgan, *Jedediah Smith and the Opening of the West* (New York: Bobbs-Merrill, 1953); Dale L. Morgan and Carl I. Wheat, *Jedediah Smith and His Maps of the American West* (San Francisco, Calif.: California Historical Society, 1954).
55. Hanson, *When Skins Were Money*, p. 158.
56. Hanson, *When Skins Were Money*, p. 160.
57. Hanson, *When Skins Were Money*, pp. 158–160; Gowans, *Rocky Mountain Rendezvous*, pp. 64–79.
58. LeRoy R. Hafen and Francis M. Young, *Fort Laramie and the Pageant of the West, 1834–1890* (Glendale, Calif.: Arthur H. Clark, 1938).

59. On prices in 1825, see Hafen, *Mountain Men and the Fur Trade,* vol. I, p. 83; on decline see Wishart, *Fur Trade of the American West,* pp. 141 ff.
60. James L. Clayton. "The Growth and Significance of the American Fur Trade, 1790–1890," in Russell W. Fridley (ed.), *Aspects of the Fur Trade: Selected Papers of the 1965 North American Fur Trade Conference* (St. Paul, Minn.: Minnesota Historical Society, 1967), pp. 62–72; p. 70.
61. Marvin C. Ross (ed.), *The West of Alfred Jacob Miller* (Norman, Okla.: University of Oklahoma Press, 1951), p. 101.
62. See illustrations 76, 132, 133, 139, 142, 147, 155, 162; Lisa Strong, *Sentimental Journey: The Art of Alfred Jacob Miller* (Fort Worth, Tex.: Amon Carter Museum, 2009).
63. The old adobe fort near La Junta, Colorado, has been replicated by the National Park Service. See David S. Lavender, *Bent's Fort* (Garden City, NY: Doubleday, 1954); Janet Lecompte, *Pueblo, Hardscrabble, Greenhorn: The Upper Arkansas, 1832–1856* (Norman, Okla.: University of Oklahoma Press, 1978).
64. Hafen, *Mountain Men and the Fur Trade,* vol. I, pp. 167–69.
65. Fray Angélico Chávez, "New Names in New Mexico, 1820–1850," *El Palacio* (Santa Fe), 64 (9–10) (1957): 291–380; pp. 300, 379.
66. David J. Weber, *The Taos Trappers: The Fur Trade in the Far Southwest, 1540–1846* (Norman, Okla.: University of Oklahoma Press, 1970).
67. LeRoy R. Hafen and Ann W. Hafen, *The Old Spanish Trail: Santa Fe to Los Angeles* (Glendale, Calif.: Arthur H. Clark Co., 1954).
68. Harvey L. Carter, *"Dear Old Kit": The Historical Christopher Carson* (Norman, Okla.: University of Oklahoma Press, 1968).
69. Harvey L. Carter, "Ewing Young," in LeRoy R. Hafen (ed.), *Mountain Men and the Fur Trade of the Far West,* 10 vols. (Glendale, Calif.: Arthur H. Clark Co., 1965–1972), vol. II, pp. 379–401.
70. Iris Higbie Wilson (Engstrand), *William Wolfskill, 1798–1866: Frontier Trapper to California Ranchero* (Glendale, Calif.: Arthur H. Clark Co., 1965).
71. Charles L. Camp (ed.), *George C. Yount and his Chronicles of the West* (Denver, Col.: Old West Publishing Co., 1966).
72. Richard Dillon, *Siskiyou Trail: The Hudson's Bay Company Route to California* (New York: McGraw-Hill, 1975); p. 213.
73. *The Journal of John Work,* ed. William S. Lewis and Paul C. Phillips (Cleveland, OH: Arthur H. Clark Co., 1923; rpt, introduction by Robert Carriker, Spokane: Arthur H. Clark Co., 2006).
74. Herman Leader, "HBC in California," *The Beaver,* Outfit 279 (March) (1949): 3–7.
75. Albert L. Hurtado, *John Sutter, A Life on the North American Frontier* (Norman, Okla.: University of Oklahoma Press, 2006).
76. Frederick J. Dockstader, *Weaving Arts of the North American Indian* (New York: T. Y. Crowell, 1978); Kate Peck Kent, *Pueblo Indian Textiles: A Living Tradition* (Santa Fe, N. Mex.: School of American Research, 1983); Marion Rodee, *Weaving of the Southwest* (Atglen, Pa.: Schiffer, 1987).
77. H. P. Mera, *Spanish-American Blanketry in Relationship to Aboriginal Weaving in the Southwest* (Santa Fe, N. Mex.: School of American Research, 1987); Nora Fisher (ed.), *Rio Grande Textiles* (Santa Fe, N. Mex.: Museum of New Mexico Press, 1994); Joe Ben Wheat, *Blanket Weaving in the Southwest,* ed. Ann Lane Hedlund (Tucson, Ariz.: University of Arizona Press, 2003).
78. George Wharton James, *Indian Blankets and Their Makers* (Chicago, Ill.: A. C. McClurg & Co., 1914); Nancy J. Blomberg, *Navajo Textiles: The William Randolph Hearst Collection* (Tucson, Ariz.: University of Arizona Press, 1988).
79. Frank McNitt, *The Indian Traders* (Norman, Okla.: University of Oklahoma Press, 1962); Kathleen L. Howard and Diana F. Pardue, *Inventing the Southwest: The Fred Harvey Company and Native American Art* (Flagstaff, Ariz.: Northland, 1996).
80. Frederick G. Young, ed., *The Correspondence and Journals of Captain Nathaniel J. Wyeth, 1831–6* (Sources of the History of Oregon, vol. I, Parts 3–6 (Eugene, 1899), 69.
81. Wyeth, as quoted by Frances Victor Fuller, *The River of the West: The Adventures of Joe Meek* (Hartford, Conn.: R. W. Bliss, 1870), p. 164.

82. William R. Sampson, "Nathaniel Jarvis Wyeth," in LeRoy R. Hafen (ed.), *Mountain Men and the Fur Trade of the Far West,* 10 vols. (Glendale, Calif.: Arthur H. Clark Co., 1965–1972), vol. V, pp. 381–401; pp. 390, 394.
83. *A New and Elegant General Atlas, Comprising All the New Discoveries to the Present Time; Containing Sixty-Three Maps, Drawn by [Aaron] Arrowsmith and [Samuel] Lewis* (Philadelphia, Pa.: John Conrad & Conrad, 1804).
84. Wheat, *Mapping the Transmississippi West,* vol. II, p. 6.
85. See discussion in Wheat, *Mapping the Transmississippi West,* vol. II, pp. 204–205. The original map has not been located. Copies were deposited with the State Department and War Department. See *Atlas of the Lewis and Clark Expedition,* Plates 32a and 32b.
86. Pike's "chart" appeared in *An Account of Expeditions to the Sources of the Mississippi, and through the Western Parts of Louisiana, to the Sources of the Arkansaw, Kans [Sic]., La Platte, and Pierre Jaun, Rivers* (Philadelphia, Pa.: C. & A. Conrad & Co., 1810). *A New Map of Mexico and Adjacent Provinces Compiled from Original Documents by a. Arrowsmith 1810* (London: engraved by E. Jones, 1810). The Lewis and Clark map was copied from Clark's manuscript map by Samuel Lewis and engraved with color separations to show outbound and inbound legs of the trip. It accompanied *History of the Expedition under the Command of Captains Lewis and Clark,* ed. Nicholas Biddle, 2 vols. (Philadelphia, Pa.: Bradford and Inskeep for Paul Allen, 1814). This is usually described as the "Allen," or first edition. See Henry R. Wagner and Charles L. Camp, *The Plains and the Rockies: A Critical Bibliography of Exploration, Adventure and Travel in the American West, 1800–1865,* 4th edn, rev. by Robert H. Becker (San Francisco, Calif.: John Howell Books, 1982), pp. 61–62. Wheat, *Mapping the Transmississippi West,* vol. II, pp. 216–217.
87. Wheat, *Mapping the Transmississippi West,* vol. II, pp. 6–7. Soulard was the official surveyor of St. Louis. See *Atlas of the Lewis and Clark Expedition,* Plate 4.
88. *A Map Exhibiting All the New Discoveries in the Interior Parts of North America Inscribed by Permission to the Honorable Governor and Company of Adventurers of England Trading into Hudsons Bay* (London: A. Arrowsmith, 1814).
89. *Account of an Expedition from Pittsburgh to the Rocky Mountains Performed in the Years 1819 and '20 ... under the Command of Major Stephen H. Long . . . Compiled by Edwin James, Botanist and Geologist for the Expedition* (Philadelphia, Pa.: H. C. Carey & I. Lea, 1823); Wheat, *Mapping the Transmississippi West,* vol. II: 80–81.
90. Henry S. Tanner, "A Map of North America," in *A New American Atlas Containing Maps of the Several States of the North American Union* (Philadelphia, Pa.: William Duffee for H. S. Tanner, 1823). Alexander de Humboldt, "A Map of New Spain," included in *Political Essays on the Kingdom of New Spain,* 2 vols., and *Atlas* (London: printed for Longman, Hurst, Rees, Orme, and Brown; Edinburgh: W. Blackwood, and Brown and Crombie, 1811). Juan Pedro Walker, Map without title, of the western part of the United States and Canada, ca. 1817. The original of 1810 has not been found but was referenced by other mapmakers in the same decade. See discussion in Wheat, *Mapping the Transmississippi West,* vol. II, pp. 83–84.
91. On Walker, see Wheat, *Mapping the Transmississippi West,* vol. II, pp. 65–66, 83–84, 219.
92. Hall Jackson Kelley, "Map of Oregon," in *A Geographical Sketch of That Part of North America, Called Oregon . . . to Which Is Attached a New Map of the Country* (Boston, Mass.: Pendleton's Lithography, 1830). See Wheat, *Mapping the Transmississippi West,* vol. II, pp. 97–98.
93. In 1953, while working at the American Geographical Society, Carl Wheat identified the data on Charles Preuss's map that accompanied John Frémont's *Report of the Exploring Expedition . . . in the Year 1842* (1845) to be that of Jedediah Smith, 1831, in the handwriting of George Gibbs, an American soldier/diplomat who negotiated treaties in Oregon in the 1850s. See Morgan and Wheat, *Jedediah Smith and His Maps;* and Wheat, *Mapping the Transmississippi West,* vol. II, pp. 119–139.
94. *A New Universal Atlas by David H. Burr* (New York: D. S. Stone, 1833). *Nouvelle Carte de l'Amérique septentrioale* (London and Paris, 1833). See Wheat, *Mapping the Transmississippi West,* vol. II, pp. 238–239.

95. *British North American, by Permission Dedicated to the Honble Hudsons Bay Company; Containing the Latest Information Which Their Documents Furnish* (London: J. Arrowsmith, 1834).
96. The original Thompson (1814) is a huge wall map now in the Archives of Ontario, reproduced for the first time in *David Thompson's Narrative of His Explorations in Western America, 1784–1812,* ed. J. B. Tyrell (Toronto: The Champlain Society, 1916). The 1818 map, titled "The Oregon Territory from Actual Survey; Corrected by Many Astronomical Observations by David Thompson, Astronomer and Surveyor," is in the British Museum. See Wheat, *Mapping the Transmississippi West,* vol. II, p. 104.
97. Alexander Ross, untitled map of the Columbia River basin, September 1821, Addition Manuscript 31, 358B, British Museum. Ross, *Adventures of the First Settlers on the Oregon or Columbia River* (London, 1849). Ross used the map in several versions for another book, *Fur Hunters of the Far West,* 2 vols. (London, 1855; rpt., ed. Kenneth Spaulding, Norman, Okla.: University of Oklahoma Press, 1956), p. 9.
98. Kittson clerked for Donald McKenzie's Snake River expedition of 1819 and drew the map in question in 1825 showing Ogden's Snake brigade route of 1824–1825 from Flathead Post to the Big Hole and Beaverhead rivers and on to Bear River, east of Salt Lake. It is reproduced in *Peter Skene Ogden's Snake Country Journals, 1824–25, and 1825–26,* ed. E. E. Rich (London: Hudson's Bay Record Society XIII, 1950). Also, Wheat, *Mapping the Transmississippi West,* vol. II, pp. 99–118.
99. *Peter Skene Ogden's Snake River Journals, 1824–1825, and 1825–1826; Peter Skene Ogden's Snake Country Journal, 1826–1827,* ed. K. G. Davies (London: Hudson's Bay Record Society XXIII, 1961); *Peter Skene Ogden's Snake Country Journals, 1827–28 and 1828–29,* ed. Glyndwr Williams (London: Hudson's Bay Record Society XXVIII, 1971).
100. "British North America, by Permission Dedicated to the Hon[ble] Hudsons Bay Company" (London: J. Arrowsmith, 1834).
101. Warren A. Ferris, *Life in the Rocky Mountains, 1830–1835,* ed. LeRoy R. Hafen (Denver, Col.: Old West, 1940; rev. edn, 1983); Wheat, *Mapping the Transmississippi West,* vol. II, pp. 155–157.
102. Washington Irving, *Astoria; or, Anecdotes of an Enterprise beyond the Rocky Mountains,* 2 vols. (Philadelphia, Pa.: Carey, Lea, & Blanchard, 1836).
103. Washington Irving, *Adventures of Captain Bonneville; or, Scenes beyond the Rocky Mountains of the Far West,* 3 vols. (London: Richard Bentley, 1837); also released as *The Rocky Mountains,* 2 vols. (Philadelphia, Pa.: Carey, Lea, & Blanchard, 1837).
104. Samuel Parker, *Journal of an Exploring Tour beyond the Rocky Mountains* (Ithaca, NY: Published for the author by Mack, Andrus, & Woodruff, 1838). Between 1840 and 1846, Parker's *Journal* appeared in four more American editions and three other editions published in Dublin, Dresden, and Edinburgh. Wagner and Camp, *The Plains and the Rockies,* pp. 172–173. Also see Wheat, *Mapping the Transmississippi West,* vol. II, pp. 165–166.
105. Maximilian Alexander Philipp, Prinz von Wied-Neuwied, *Travels in the Interior of North America by Maximilian, Prince of Wied* (London: Ackermann and Co., 1839).
106. George Catlin, *Letters and Notes on the Manners, Customs, and Condition of the North American Indians,* 2 vols. (New York: Wiley and Putnam, 1841).
107. Personal interview with John C. Ewers by the author, W. R. Swagerty; Arlington, Virginia, September, 1986.
108. Charles Wilkes, *Narrative of the United States Expedition, during the Years 1838, 1839, 1840, 1841, 1842,* 5 vols. and atlas (Philadelphia, Pa.: Lea & Blanchard, 1845). Also see William Stanton, *The Great United States Exploring Expedition of 1838–1842* (Berkeley, Calif.: University of California Press, 1975).
109. Captain J. C. Frémont, *Report of the Exploring Expedition to the Rocky Mountains in the Year 1842 and to Oregon and North California in the Years 1843–44* ... (Washington, DC: Gales and Seaton for the US Senate, 1845 [28th Cong., 2d Sess., Sen. Ex. Doc 174, Ser. Set 461]); Frémont's reports have all been reprinted in *The Expeditions of John Charles Frémont,* 3 vols., and map portfolio (Urbana, Ill.: University of Illinois Press, 1970–1984); Wheat, *Mapping the Transmississippi West,* vol. II, pp. 180–182, 195–196; p. 196.

110. William H. Goetzmann, *Army Exploration in the American West, 1803–1865* (New Haven, Conn.: Yale University Press, 1959), pp. 65–108; p. 108. Also see Vincent Ponko, Jr., "Military Explorers of the American West, 1838–1860," in John Logan Allen (ed.), *North American Exploration,* 3 vols. (Lincoln, Nebr.: University of Nebraska Press, 1997), vol. III, pp. 332–411; pp. 345–353.
111. Stephen W. Kearny, *Report of a Summer Campaign to the Rocky Mountains in 1845,* 29th Cong., 1st sess., 1846, S. Doc. 1; Goetzmann, *Army Exploration,* pp. 109–116.
112. James W. Abert, *Journal of Lieutenant James W. Abert from Bent's Fort to St. Louis.* 29th Cong., 1st sess., 1846, Exec. Doc. 438; Goetzmann, *Army Exploration,* pp. 116–127.
113. Ponko, "Military Explorers," pp. 353–365; Frémont, *Geographical Memoir upon Upper California* (Washington, DC: Wendell and Van Benthuysen, Printers, 1848); reprinted in *The Expeditions of John Charles Frémont,* vol. III, pp. 495–570; Goetzmann, *Army Exploration,* pp. 116–123. The Pruess map is analyzed in Wheat, *Mapping the Transmississippi West,* vol. III, pp. 55–62.
114. William H. Emory, *Notes of a Military Reconnaissance from Fort Leavenworth in Missouri, to San Diego in California Including Parts of the Arkansas, Del Norte (Rio Grande) and Gila Rivers,* 30th Cong., 1st sess, 1848, H. Doc. 41. Also see Goetzmann, *Army Exploration,* pp. 127–152; Ponko, "Military Explorers," pp. 361–365, and 588–589, notes 26–28. For Mexican War maps see Wheat, *Mapping the Transmississippi West,* vol. III, pp. 1–24.
115. These include Frémont's fourth expedition of 1848–1849, which was not officially sponsored by Congress but financed by Senator Thomas H. Benton of Missouri, Frémont's father-in-law; that of Captain Randolph Marcy and Lieutenant James H. Simpson, who surveyed an emigrant road from Fort Smith across Texas to Santa Fe, 1849–1850; Captain Howard Stansbury and Lieutenant John W. Gunnison's survey of a prospective emigrant road from Fort Hall in Idaho across the Great Basin to the Mormon Settlements in Utah, 1849. See Ponko, "Military Explorers," pp. 365–373. See also Goetzmann, *Army Exploration,* pp. 153 ff; also Edward S. Wallace, *The Great Reconnaissance: Soldiers, Artists, and Scientists on the Frontier, 1848–1861* (Boston, Mass.: Little, Brown, 1955). Wheat, *Mapping the Transmississippi West,* vol. III, contains extensive discussion of official government maps, personal, and commercial maps produced from the Mexican War to the Boundary Surveys, 1846–1854.
116. Goetzmann, *Army Exploration,* pp. 262–397, covers the Pacific Railroad Surveys in depth. Also see Goetzmann, *Exploration and Empire,* pp. 265–302.
117. The Pacific Railroad Surveys were reported in two stages, first as A. A. Humphreys and G. K. Warren, *An Examination by Direction of the Hon. Jefferson Davis, Secretary of War, of the Reports of Explorations for Railroad Routes from the Mississippi to the Pacific, Made under the Orders of the War Department in 1853–54,* 33rd Cong., 1st sess., 1855, H. Doc. 129. The second set of reports were printed in large quarto volumes as *Reports of Explorations and Surveys to Ascertain the Most Practicable and Economic Route for a Railroad from the Mississippi to the Pacific Ocean,* 13 vols., and atlas (Washington, DC: various publishers, 1855–1860).
118. Frémont's fifth expedition was reported in two places: John Bigelow, *Memoir of the Life and Public Services of John Charles Frémont, including an Account of his Explorations, Discoveries and Adventures on Five Successive Expeditions* (New York: Derby & Jackson, 1856); and Soloman Nuñez Carvalho, *Incidents of Travel and Adventure in the Far West, with Col. Frémont's Last Expedition across the Rocky Mountains* (New York: Derby & Jackson, 1856; London: Sampson Low, Son, & Co., 1856). Ironically, Carvalho has none of his own daguerreotypes in his book due to a prior agreement that Frémont would own all of them upon the return.
119. Charles Wentworth Upham, *Life, Explorations and Public Services of John Charles Frémont, with Illustrations* (Boston, Mass.: Ticknor & Fields, 1856). On the technology involved and the importance of the images see Martha A. Sandweiss, *Print the Legend: Photography and the American West* (New Haven, Conn.: Yale University Press, 2002), pp. 90–108. Frémont's grander project of a massive two-volume memoir never was finished. One volume, prepared by Jesse Benton-Frémont, was released in 1887, covering the first three expeditions. See Sandweiss, *Print the Legend,* pp. 106–107.
120. G. L. Albright, *Official Explorations for Pacific Railroads* (Berkeley, Calif.: University of California Press, 1921).

121. On the importance of the Pacific Railroad Survey maps see Wheat, *Mapping the Transmississippi West,* vol. IV, pp. 76–91. On the 1857 Warren map see pp. 81–84.
122. W. Turrentine Jackson, *Wagon Roads West: A Study of Federal Road Surveys and Construction in the Trans-Mississippi West, 1846–1869* (Berkeley, Calif.: University of California Press, 1952), pp. 64–65, as paraphrased by Goetzmann, *Army Exploration,* pp. 347–348. Also Wheat, *Mapping the Transmississippi West,* vol. IV, pp. 150–163.
123. Chester Lee White, "Surmounting the Sierras: The Campaign for a Wagon Road," *California Historical Society Quarterly,* 7 (1928): 3–19.
124. Goetzmann, *Army Exploration,* pp. 363–374.
125. Richard Bartlett, *Great Surveys of the American West* (Norman, Okla.: University of Oklahoma Press, 1962). Also, Goetzmann, *Exploration and Empire,* pp. 303 ff. Wheat, *Mapping the Transmississippi West,* vol. V, parts 1–2, "Civil War to the Geological Survey."
126. Goetzmann, *Army Exploration,* pp. 378–394; maps of the Ives expedition are reprinted and analyzed in Wheat, *Mapping the Transmississippi West,* vol. IV, pp. 95–102. Macomb and Simpson are discussed by Goetzmann, *Army Exploration,* pp. 394–405; and Wheat, *Mapping the Transmississippi West,* vol. IV, pp. 137–149.
127. Goetzmann, *Army Exploration,* pp. 405–426; p. 418. Wheat, *Mapping the Transmississippi West,* vol. IV, p. 187. On Hayden's subsequent role as head of the Geological Survey of Nebraska (1867) and of the Hayden Surveys of 1871–1876 see Bartlett, *Great Surveys,* Part 1 and Mike Foster, *Strange Genius: The Life of Frank Vandeveer Hayden* (Niwot, Col.: Robert Rinehart Publishers, 1994).
128. Hanson, *When Skins Were Money,* p. 161.
129. David Michael Delo, *Peddlers and Post Traders: The Army Sutler on the Frontier* (Salt Lake City, UT: University of Utah Press, 1992).
130. Sunder, *Fur Trade of the Upper Missouri,* p. 17.
131. William T. Hornaday, "The Extermination of the American Bison, with a Sketch of Its Discovery and Life History," in *Annual Report of the United States National Museum for 1887* (Washington, 1889; rpt. Seattle, Wash.: Shorey Book Co., 1971), Part 2: 367–548. Also Isenberg, *The Destruction of the Bison,* pp. 130 ff.
132. Swagerty, "Indian Trade of the Trans-Mississippi West," pp. 373–374; Hornaday, "The Extermination of the American Bison," pp. 504–525; Sunder, *Fur Trade of the Upper Missouri,* pp. 260–265.
133. Swagerty, "Indian Trade of the Trans-Mississippi West," p. 373; Rebecca McDowell Craven, *The Impact of Intimacy: Mexican-Anglo Intermarriage in New Mexico, 1821–1846* (El Paso, Tex.: Texas Western Press for the University of Texas at El Paso, 1982); Lecompte, *Pueblo, Hardscrabble, Greenhorn;* Camp, *George C. Yount;* John A. Hussey, *Champoeg: Place of Transition, A Disputed History* (Portland, Oreg.: Oregon Historical Society, 1967); Jackson, *Children of the Fur Trade;* Hafen and Young, *Fort Laramie;* Overholser, *Fort Benton.*
134. McNitt, *The Indian Traders.*
135. Ken Zontek, *Buffalo Nation: American Indian Efforts to Restore the Bison* (Lincoln, Nebr.: University of Nebraska Press, 2007).

CHAPTER SEVEN

THE "EARLIER, WILDER IMAGE"

Early Artists of the American West

Elizabeth Kuebler-Wolf

Throughout American history, the geographic West has shifted with the nation's boundaries. For those first Europeans who ventured westward across the Atlantic, the West was what is now America's easternmost region; over the centuries of European and American expansion on the continent, the westernmost reach of settlement edged progressively farther to what we now think of as the West. Earlier generations located the West first in upstate New York, then Kentucky and Indiana, then Missouri, and then points beyond, not reaching the shores of the Pacific until the 1890 census proclaimed the frontier officially closed. Although the West as geographic fact has changed over the centuries, the West as both a state of mind and a defining American characteristic has remained a fairly constant theme. This essay traces the visual imagery of the American West from its earliest manifestations in landscape art until the years just after the Civil War, when public taste for "great picture" landscape paintings waned and depictions of the West shifted away from romantic oil paintings of sweeping lands to be conquered to seemingly straightforward, unsentimental documentary photography and scientifically inspired paintings of the spectacular geography of Yellowstone and Yosemite.

What the early visual Wests had in common was not a typical geography, a single Native population, a unified set of flora and fauna, or any other feature within the physical realities of place. Rather, these Wests shared metaphoric features which have remained constant themes in American history: the West as opportunity and crisis, progress and destruction, a stage for the last chapter of world history, and, in John Sullivan's famous phrase coined in 1854, a manifest destiny. In keeping with larger historical narratives of cyclical, Hegelian history, Americans imagined their West as the stage on which the end of one civilization (Native) was a symbiotic, necessary twin of the rise of another (Euro-American) civilization.[1] The character of that stage—its rugged, majestic geographic features and fauna and flora—not only differed from the "fair scenes" of tame Europe but also determined the nature of the civilization that would settle and tame the "earlier, wilder" landscape of the American West.[2]

In the 1720s, Bishop George Berkeley gave early voice to these recurrent themes, in a poem worth reproducing in its entirety:

America, or The Muse's Refuge: A Prophecy

The Muse, disgusted at an Age and Clime,
Barren of every glorious Theme,
In distant Lands now waits a better Time,
Producing subjects worthy Fame:
In happy Climes, where from the genial Sun
And virgin Earth such Scenes ensue,
The Force of Art by Nature seems outdone,
And fancied Beauties by the true:
There shall be sung another golden Age,
The rise of Empire and of Arts,
The Good and Great inspiring epic Rage,
The wisest Heads and noblest Hearts.
Not such as *Europe* breeds in her decay;
Such as she bred when fresh and young,
When heav'nly Flame did animate her Clay,
By future Poets shall be sung.
Westward the Course of Empire takes its Way;
The four first Acts already past,
A fifth shall close the Drama with the Day;
Time's noblest Offspring is the last.[3]

In this early poem, Berkeley touched on several themes that became constant in depictions of the American West in the ensuing centuries. Empire was moving westward on a course set by destiny; Europe was on the decline as the fortunes of a new civilization arose; the West was a new, clean slate of "virgin earth"; the land itself was a more triumphant creation than any of those European monuments created by "the force of art"; that civilization would be a new, better civilization.

Berkeley wrote to encourage investors to support his scheme for a new school in Bermuda. For his English contemporaries, of course, Bermuda was a significant geographic move to the West from London. While Berkeley's school in Bermuda never was built, his imagined shift of empire proved to be an important mental relocation. It is with the changing boundaries of the mythic West in mind that we turn to Thomas Cole, one of the first visual artists to depict the American West.

Cole, born in England in 1801, immigrated westward with his family to Steubenville, Ohio, in 1819. Eventually Thomas Cole settled in the town of Catskill, New York, where he maintained a studio and painted scenes of the Catskills, the White Mountains, and other parts of the Hudson River Valley. For the early nineteenth-century American, these interior parts of New York state were the "virgin earth" and "happy climes" to which city dwellers were already turning for succor, rest, and inspiration.[4]

On the strength of the appeal of his landscapes, Cole became one of America's first commercially successful, and famous, fine artists. Together with William Cullen Bryant, the poet, and James Fenimore Cooper, the novelist, Cole was a member of the same social circles, and the three men shared an interest in depicting America's wilderness—her West—as a metaphor for the American character and destiny.[5] Indeed, Cole painted illustrations of Cooper's novels, while Bryant wrote verses on Cole's work such as that which gives this chapter its title.

For these men and for the generation that followed, the American landscape was imbued with a deep symbolic meaning, tied directly to America's future, and with important ramifications for the development of the American character. Fenimore Cooper's Leatherstocking Tales provided a template for a slew of romances of the Western landscape, with half-civilized, half-native heroes who represented a New World archetype, while Bryant's poetic imagery of American landscape remained an influential frame of reference for depictions of the American landscape well into the twentieth century.

Cole's paintings, like his writing about the American landscape, echo the lines of Governor Berkeley's poem from a century earlier. Looking into the "untamed" American landscape, Cole saw

> no gorgeous temple to speak of ostentation; but freedom's offspring—peace, security, and happiness, dwell there, the spirits of the scene [...] And in looking over the yet uncultivated scene, the mind's eye may see far into futurity. Where the wolf roams, the plough shall glisten; on the gray crag shall rise temple and tower—mighty deeds shall be done in the now pathless wilderness; and poets yet unborn shall sanctify the soil.[6]

Figure 7.1 shows such a wilderness on the brink of its transformation into a stage for the "mighty deeds" of future generations. In the foreground, a woman and child gather flowers. In the leftmost middle distance a man is engaged in breaking a wild horse; behind him a path winds through forested land toward the far distance where a small town, basking in sunlight, has found purchase. In the extreme distance tall mountains beckon. The compositional formulae of Cole's paintings—framing devices in the foreground, a winding pathway in the middle ground, and hazy distant peaks illuminated by golden sunlight—descend from the conventions of seventeenth-century landscape painters Claude Lorrain

Figure 7.1 Thomas Cole, *A View on the Catskills, Early Autumn*, 1837.
Source: The Metropolitan Museum of Art, Gift in memory of Jonathan Sturges by his children, 1895 (95.13.3). Image © Metropolitan Museum of Art.

and Salvator Rosa. In turn, Cole's use of such formulae would become a convention widely used in American landscape painting. Implicitly and often explicitly painters after Cole would continue to link visual movement through a canvas to the spread of progress and civilization's advance.

Thomas Cole continued to paint the American landscape as a stage for civilization's advance, much to the satisfaction of his urban, conservative, wealthy clientele.[7] In canvases such as *View from Mount Holyoke, Northampton, Massachusetts, after a Thunderstorm: The Oxbow* (1836) (see Figure 7.2), Cole depicted what his patrons wanted to see: a landscape on which neatly ordered fields and tidy households dotted the prosperous countryside. This patchwork of farm fields is saturated in golden light, while areas of dark storm clouds move off to the left, or west, side of the canvas. Cole's own ambivalence toward the costs of westward expansion—the overdevelopment of the natural beauty that made America so distinct from Europe in his mind—may be hinted at by the placement of the artist's easel and paints on the wilderness side of the painting; however, what made this picture popular as an engraving was the way it fit a famous American scene into the sublime and beautiful aesthetic categories of European art.

After Cole, an entire generation of American artists continued to use the same landscape formulae to depict American scenes as a stage for civilization's development. An 1853 painting by Asher B. Durand, for example, shows Euro-American settlement moving in to the trackless wilderness; this movement westward in the landscape and leftward in the painting is identified by the title simply as *Progress* (Figure 7.3). In the right foreground of the painting, cleared paths, sturdy cabins, and telegraph poles indicate the advance of civilization while on the left, amidst the wilds of uncleared land, a lone American Indian

Figure 7.2 Thomas Cole, *View from Mount Holyoke, Northampton, Massachusetts, After a Thunderstorm—The Oxbow*, **1836.**
Source: Gift of Mrs. Russell Sage, 1908 (08.228). Image © Metropolitan Museum of Art.

Figure 7.3 Asher B. Durand, *Progress*, 1853.
Source: From the Warner Collection. On display at the Westervelt Warner Museum of American Art. Property of the Westervelt Company.

observes the approach of Euro-American settlers. Painted from the point of view of Euro-Americans, who by the 1850s had begun to use the term "manifest destiny" to describe the westward press of Euro-American settlement, Durand's painting is a visual celebration of the taming of the land, turning it from ungoverned chaos to neatly ordered village.[8]

Durand's vision of expansion as progress remained a staple of midcentury American landscape scenery. Popular visual incarnations of the theme of expansion include Frances Palmer's scenes for Currier and Ives with titles such as *The Rocky Mountains: Emigrants Crossing the Plains* (1866) and *Westward the Course of Empire Takes Its Way* (1868; Figure 7.4). In these and other images, Palmer depicted movement leftward across the image from a settled, populated land on the right fading out to a plain, undeveloped vista on the left of the picture. From the east, settlers bring train tracks, schoolhouses, plowed fields, and other markers of civilization.

Another painting of the 1850s saw the same movement westward from a slightly different vantage. *The Indian's Lament* (Fig. 7.5), an anonymous folk painting, shows the same sort of landscape as Durand, with encroaching Euro-American settlement indicated by a grave marked with a cross and neatly chopped-off tree stumps indicating the beginning of a clearing. Gazing upon this scene in a melancholy posture is a lone American Indian. Although this anonymous artist lacked the polish and finesse of the academically trained Durand, he followed the same landscape conventions inherited from Claude Lorrain and Salvator Rosa. While participating in the same visual language which was for Durand unabashedly a celebration of westward movement, this artist provides a rather more sober vision of the meaning of Euro-American settlement for those who were being displaced from their ancestral homes. Where Durand's painting is titled *Progress*, *The Indian's*

Figure 7.4 Frances Palmer for Currier & Ives, *Across the Continent: Westward the Course of Empire Takes its Way*, 1868.
Source: Library of Congress Prints and Photographs Division Washington, DC, 20540.

Lament provides the viewpoint of the American Indian figure, watching glumly as westward movement transforms the landscape.

Giving voice to a similar melancholy American Indian gazing on a similar view of westward expansion in one of his poems, William Cullen Bryant observed,

> A white man, gazing on the scene,
> Would say a lovely spot was here,
> And praise the lawns, so fresh and green,
> Between the hills so sheer.
> I like it not—I would the plain
> Lay in its tall old groves again.[9]

Bryant, like the anonymous artist of *The Indian's Lament,* recognized that intertwined with the "destiny" of Euro-Americans to occupy the continent was unrelenting pressure and loss visited upon Indians. Like his contemporaries, Bryant saw this loss as inevitable, a part of the larger narrative of history that simply happened, like the weather. Inevitably as the seasons change, Bryant wrote,

> They waste us—ay—like April snow
> In the warm noon, we shrink away;
> And fast they follow, as we go
> Toward the setting day—
> Till they shall fill the land, and we
> Are driven into the Western sea.[10]

Figure 7.5 Artist unknown, *The Indian's Lament*, ca. 1850.
Source: Columbus Museum of Art, Ohio. Gift of Edgar William and Bernice Chrysler Garbisch. 1967.057.

As Bryant was voicing laments about his belief in inevitable disappearance of American Indians, some artists were taking up the task of recording those who were, in the words of artist George Catlin, "doomed to perish."[11] Catlin, Karl Bodmer, and Alfred Jacob Miller all spent time traveling through the West, usually with military or geographic expeditions, and devoted their artistic activities to painting and documenting Native peoples. In an era before photography was easily transportable, the activities of these artists was crucial for documenting the visual appearance of Native people and villages.

AN EQUAL DOOM

George Catlin (1796–1872) spent several decades collecting artifacts and painting portraits of Native leaders and representative racial "types" of various tribes. Beginning in 1831, the year after the Indian Removal Act was passed by Congress, Catlin abandoned the practice of law and commenced travels through the West to sketch and collect artifacts of civilizations he was convinced were under immediate threat of disappearing. Traveling with General William Clark, then head of the Bureau of Indian Affairs, Catlin was able to meet with representatives of several different Plains nations. Over the span of the 1830s he amassed a large collection of images and artifacts. Starting in 1836, Catlin traveled around the eastern part of the USA and to Europe, presenting his "Indian Gallery" in lecture tours, complete with staged performances of "authentic" Native dances in which his nephew Ted Catlin was the main player.

Catlin's sympathetic interest in Indians was intertwined with a sense that Euro-American progress was inexorably destroying Native culture. In *Pigeon's Egg Head Going to and Returning from Washington* (Figure 7.6), Catlin documented the acculturation of his friend and representative of the Assiniboine, Wi-jún-jon, aka The Light and Pigeon's Egg Head. In *Letters and Notes,* Catlin lamented the transformation of Wi-jún-jon into a foolish parody of a dandy when he traveled to Washington. Catlin complained that Wi-jún-jon had "exchanged his beautifully garnished and classic costume" for a suit of

> broadcloth, of the finest blue, trimmed with lace of gold; on his shoulders were mounted two immense epaulets; his neck was strangled with a shining black stock, and his feet were pinioned in a pair of water-proof boots, with high heels, which made him 'step like a yoked hog.'[12]

A desire to preserve an idealized vision of a pre-contact Native society, uncorrupted by exposure to European culture, remained a fixture of Catlin's efforts to record Native America. As he wrote in his last petition to Congress to purchase his Indian gallery, his collection represented Native culture in the 1830s and 1840s "when the Indian's modes and customs were more primitive than at the present time."[13] Contact with modern civilization had, at least in Catlin's view, sullied the pure, primitive nature of Indian culture which he had recorded years earlier in his portrait series. This desire to "preserve" Native culture as unchanging, universally, and timelessly primitive, deemed the "ethnographic present" in

Figure 7.6 George Catlin, *Pigeon's Egg Head Going to and Returning from Washington*, 1832. Source: Smithsonian Museum of American Art.

anthropological theory, remained a feature of Western exploration and study of Native cultures well into the twentieth century.[14]

For Catlin, however, more than merely contaminating Native culture, American progress was surely eradicating Native civilization. As he wrote in *Letters on the Manners and Customs of the North American Indian,*

> Many are the rudenesses and wilds in Nature's works which are destined to fall before the deadly axe and desolating hands of cultivating man. [...] Of such "rudenesses and wilds," Nature has no where [sic] presented more beautiful and lovely scenes, than those of the vast prairies of the West, and of man and beast no nobler specimens than those who inhabit them the Indian and the buffalo—joint and original tenants of the soil and fugitives together from the approach of civilized man they have fled to the great plains of the West and there under *an equal doom they have taken up their last abode where their race will expire and their bones will bleach together.*[15]

In these observations, Catlin echoed visual descriptions of progress and effects of Western settlement from his artistic forebears, which would continue to frame American understanding of the meaning of westward expansion.

George Crofutt's 1872 chromolithograph of John Gast's painting *American Progress* (Figure 7.7) gives Catlin's description of the Indian and the buffalo cowering "under an equal doom" a literal visual expression. As the light of civilization and America with the star of empire on her forehead move westward across the image, both Indians on horseback and buffalo run off to the West under a foreboding storm cloud. As Crofutt explained, Gast's image depicted the noble red man

> Fleeing from Progress and toward the blue waters of the Pacific [...] are the Indians, buffalo, wild horses, bears, and other game moving westward, ever westward—Indians with their squaws [sic], papooses [sic], and "pony lodges" turn their despairing faces toward the setting sun as they flee from the presence the wondrous vision. The "Star" is *too much for them.*[16]

Returning to the antebellum period, other artists contemporary to Catlin also documented American Indians with the same understanding that all of Native society was "doomed to perish" and thus was in need of preservation in archival, visual format. Charles Bird King (1785–1862) painted a series of portraits of important American Indian leaders who were sent to his studio through the efforts of Thomas McKenney, the head of the Department of War's Indian Affairs Office. Under McKenney's leadership, the Department of War collected the portraits created by King and a number of other artists, which became part of the illustrated text of *The History of the Indian Tribes of North America,* authored by Thomas McKenney and James Hall.[17]

This large, multi-volume folio was one of the first publications of its kind, dedicated to preserving a record of Native America at the same time that American Indians were being pushed farther to the west by the Indian Removal Act, which was enforced by the Department of War. In an essay which accompanied the publication of King's portraits, James Hall noted that

> The tribes which occupied the countries now constituting the eastern states, were annihilated, or have melted away, to make room for the Euro-Americans. The waves of population and civilization are rolling to the westward; and we now propose. [...] to

Figure 7.7 George Croffut after John Gast, *American Progress*, 1872.
Source: Library of Congress Prints and Photographs Division Washington, DC 20540.

> send them [American Indians] to a land where their existence may be prolonged, and perhaps made perpetual.[18]

Hall's tentative assertion that the existence of Native society could be "prolonged" and only *perhaps* preserved reflects a common assumption in the 1830s that Native culture was in immediate peril. Visual artists such as Charles Bird King and George Catlin understood themselves primarily as recorders of the appearance and dress of those individuals whom the artists were able to meet. As Catlin explained, "the history and customs of such a people, preserved by pictorial illustrations, are themes worthy the life-time of one man, and nothing short of the loss of my life, shall prevent me from visiting their country, and of becoming their historian."[19]

STRONGLY MARKED NATIONAL CHARACTER

In the popular imagination, the Indians of King and Catlin were relics and curiosities of a time in American history that was fast passing away, but the "disappearing" Indians of the West were only half of the story of Western expansion for mid-nineteenth-century Western artists. Frontiersmen such as Davy Crockett (1786–1836), a Tennessee politician who died at the battle of the Alamo, and Daniel Boone (1734–1820), a Revolutionary militia officer and Virginia politician who helped settle Kentucky territory, entered into the national consciousness posthumously as heroes of Western expansion.

After his death, Crockett became the star of a series of outrageous comic almanacs published beginning in the 1830s, and soon became a nearly mythological character in American culture. Daniel Boone's travails in conflict with Indians—one of his sons was killed in battle, and his daughter Jemima was captured by Shawnee in Ohio—also proved popular and compelling subjects for American artists and writers. By the middle of the nineteenth century, artists began creating scenes of generic settler and Western "types," such as flatboat river workers and fur traders, for Eastern audiences. These types embodied expansion, filled the land, and drove Natives westward toward the sea, and satisfied what Gail Husch has described as "an increasing eastern interest in specifically western subjects."[20] For American critics in Eastern cities, scenes of Western subjects also bore a "strongly marked national character" and depicted the seasoning of the American character, whom "exposure and hardships have transformed [...] into a superb looking fellow."[21]

Fur trappers, riverboat men, and other such figures from the frontiers of the American West made up the typical cast of George Caleb Bingham's (1811–1879) popular genre paintings and prints. Virginia-born Bingham moved to Missouri territory as a boy in 1820. By the 1840s, when Bingham's Western scenes became popular among art collectors, America's cultural West had shifted at least as far as Bingham's adopted home. Appetite for Bingham's scenes of lively, rowdy river boatmen, rough backwoodsmen, and brave frontiersmen confronting danger both nature and Native, was largest among the urban dwellers of the East Coast. Living vicariously through Bingham's pictures, they could experience both the thrill and the danger of Western frontier life, a life that was receding ever farther away to the West and into the past.[22]

Binghams' *Fur Traders Descending the Missouri* (1836; Figure 7.8), originally known as *French Trader and Half-Breed Son,* was one of the major successes of Bingham's career. *French Trader and Half-Breed Son* was purchased by the American Art Union (AAU), a nationwide network of artists and art-lovers who were connected via its headquarters in New York City. The AAU engraved the picture for national distribution to its membership under the title *Fur Traders Descending the Missouri,* perhaps choosing a title that would not state so boldly the "miscegenation" implied by Bingham's original description. In this picture, the two men float on a peaceful Missouri River, isolated from civilization, the very embodiment of, in Thomas Cole's words, "freedom's offspring."[23] Indeed, the father even wears a Phrygian liberty cap to underscore the uniquely unfettered position he occupies in society, compared with his Eastern counterparts. Floating on the river, termed by one writer a "highway of God's own making," these "new Men" were shaped by, as well as shapers of, America's geography.[24]

In a painting that accompanied *Fur Traders Descending the Missouri* in exhibition in 1845 at the AAU, Bingham depicted *The Concealed Enemy* (Fig. 7.9, p. 195).[25] Here the American Indian who had been included for scale in Thomas Cole's works is the central subject. Crouching for cover behind a rocky outcropping topped by gnarled trees and bush, a dark figure peers out to the right of the painting. "Offstage" right, the fur trapper and his son float down the Missouri, untroubled for the moment by the conflict that awaits them. These two pictures, when hung together, suggest the same dynamics and tensions of the West as seen in the landscape paintings of Cole and Durand: civilization and wilderness, expansion and resistance. The figures that had been markers of scale for early Western landscape painters were transformed into representative types whose activities and conflicts would shape the American West in the popular imagination.

Bingham also devoted a large-scale work to the depiction of the archetypal hero of

Figure 7.8 George Caleb Bingham, *Fur Traders Descending the Missouri*, 1836. Source: The Metropolitan Museum of Art, Morris K. Jesup Fund, 1933 (33.61). Image © Metropolitan Museum of Art.

Western expansion in his 1851–1852 painting *Daniel Boone Escorting Settlers through the Cumberland Gap* (Washington University Gallery of Art, St. Louis, Missouri). In Bingham's calm, pyramidal composition, Daniel Boone leads a party of travelers through the wilderness. Boone and his party emerge from the dark background and step onto ground bathed in the golden light of Providence, surrounded by the blasted trunks of trees that signify the "trackless wilderness" in American landscape painting. Borrowing the visual language of neoclassicism, orderly and stable, with figures posed like classical statuary, Boone's "errand into the wilderness" proceeds in a peaceful, orderly, and stately fashion. Bingham's Boone becomes an American version of Moses leading his people into the wilderness, while his wife on horseback shrouded in a blue robe calls to mind the European iconography of Mary.[26]

David Lubin has argued persuasively that Bingham borrowed the stormy landscape populated with blasted tree trunks in *Daniel Boone Escorting Settlers through the Cumberland Gap* directly from Thomas Cole. When Bingham's original, sunny version did not sell to the AAU, he repainted the landscape with a romantic background like those in Cole's religious allegories.[27] The events depicted in Bingham's painting had happened some seventy-six years earlier, but by the 1850s, Daniel Boone had taken on cultural significance as a symbol of America's conquering of the wilderness. If Bingham painted the stormy landscape as a visual symbol of the land yet untamed, his Boone is nevertheless a peaceful

Figure 7.9 George Caleb Bingham, *The Concealed Enemy*, c. 1845, oil on canvas, 29¼ × 36½ inches. Source: Stark Museum of Art, Orange, Texas.

patriarch and leader of a band of families who will not just pass through but stay and settle the land.[28]

Human mechanisms of westward expansion and settlement, generic figures as well as "branded" individuals such as Boone and Crockett, were a popular subject of mid-century genre scenes. George Caleb Bingham was one of the most prominent genre painters, whose works gained national fame through lithographed reproductions, but many other artists worked in this vein as well. William Tylee Ranney (1813–1857) depicted exclusively Western scenes of trappers, settlers, and prairie life, although he lived in New Jersey; his audience was also primarily an Eastern clientele. George Caleb Bingham's fellow Missourian Charles Deas (1818–1867) also created popular images of Western pioneers and types such as *Long Jakes, The Rocky Mountain Man* (1844, Denver Art Museum), which one reviewer described as "the literal portrait of the celebrated character of the Rocky Mountains," and *The Voyageurs* (1846, Boston Museum of Fine Arts), a scene of fur traders in a canoe done the same year as Bingham's *French Trader and Half-Breed Son*.

Although present in almanac images of Davy Crockett, the violence that often accompanied the "fading" of Native populations was not a subject that Western painters would typically embrace until the turn of the twentieth century. Painters like Frederic Remington and Charles Schreyvogel would make their livings selling pictures of violent conflicts with Indians. By the time these painters were painting, however, actual conflict with Native populations was by and large a thing of the past after the 1890 Wounded Knee massacre.

CIVILIZATION … COMING LIKE AN ARMY WITH BANNERS

As we have seen, early artists who depicted the West, like Thomas Cole and Asher Durand, moved rather easily between the poles of progress and destruction. To these artists the advance of Western settlement made inevitable, if regrettable, the ruin of Native lives and homes; both phenomena were linked in the Hegelian progress of historical development. Landscape images of the West from Cole onward utilized such Hegelian terms, and such a narrative of entwined progress and destruction is embedded in the grand manner of American landscapes painted before and during the American Civil War.

As the nation teetered on the brink of war, Congress charged Emmanuel Leutze (1816–1868), a well-respected history painter, to create a mural for a Capitol stairway.[29] Leutze, a native of Germany, moved to Philadelphia as a child with his family. As a youngster he imbibed the culture of American expansion; among his first pictures as a teenage painter was "the figure of an Indian contemplating the setting sun"—a composition that calls to mind *The Indian's Lament*.[30] As a young adult back in Germany, at school in Dusseldorf, Leutze also met William Cullen Bryant and later illustrated an edition of Bryant's poetry that appeared in 1847.[31] Leutze was well established as a painter of grand-manner scenes of American history and landscape by the time of his congressional commission.

While sectional tensions were threatening to rip apart the nation, Leutze's landscape called to mind earlier images of a nation united under divine sanction to overspread and settle the continent. Its title, *Westward the Course of Empire Takes its Way* (1861–1862; Figure 7.10), is taken from Governor Berkley's poem with which we began this chapter. Leutze's mural depicts American settlement spanning from the Atlantic, across the Grand Canyon, toward San Francisco Bay and the Pacific Ocean on the left.

Leutze understood American history to be advancing toward a Hegelian synthesis. *Westward Ho!* visually reflects that overarching historical narrative; Euro-American settlers press westward over a mountainous Western landscape toward the Pacific, while Natives are left in the past. Heroic men climb to the top of the challenging landscape and look toward the land they will soon conquer. A view of the San Francisco Bay is visible in the predella under the main scene. In Leutze's notes for the composition, he noted that his intention was "to represent as near and truthfully as the artist was able the grand peaceful conquest of the great west (paper torn off) […] without a wish to date or localize, or to represent a particular event, it is intended to give in a condensed form a picture of western emigration."[32]

Emmanuel Leutze's friend and fellow Dusseldorf student Albert Bierstadt (1830–1902) also gave vision to the grand, heroic story of America's westward expansion in the war years. Bierstadt's *Rocky Mountains, Lander's Peak* (1863, Metropolitan Museum of Art) was painted as an homage to a fallen Union soldier and sold for a record-breaking price to a British collector in 1865. *Lander's Peak* shows a majestic (though imaginary) Rocky Mountain skyline in the background, with a peaceful Shoshone village in the foreground. This splendid vista was enormously popular in its day, circulating in the American consciousness as a nationally touring single-picture exhibition, a steel engraving, a descriptive pamphlet and a companion article by Fitzhugh Ludlow which described the artist's westward travels in preparation for painting the picture.[33]

Contemporaries who saw Bierstadt's picture understood very well the implications of his majestic, wild Western landscape populated only by Indians. One Eastern reviewer of *Lander's Peak* commented that "he who lays his ear to the wild grass may perhaps hear the distant tramp, not of the buffaloes, but of civilization, coming like an army with banners."[34]

Figure 7.10 Emmanuel Leutze, *Westward the Course of Empire Takes Its Way,* 1861. Source: Smithsonian American Art Museum. Washington, DC, USA Bequest of Sara Carr Upton 1931.6.1.

Bierstadt himself expressed hope that one day "a city, populated by our descendants, may rise" on the foreground plain of his picture, to replace the Shoshone encampment.[35] In the pamphlet which advertised the engraving after Bierstadt's picture, the artist wrote that the picture was valuable to purchasers as a record of "a race which, before the advance of civilizations, fades away like the mists of morning before the rays of the rising sun."[36]

Grand landscapes in the manner of Bierstadt and Leutze were wildly popular through the end of the Civil War years but fell out of favor in the later nineteenth century. Although these artists continued to work until the turn of the century, their later careers were disappointments after the success of their pre-war, "great picture" landscapes. One later-nineteenth-century critic even complained of Bierstadt that "ambition for notoriety and money" had fuelled a certain "sensationalism" in his grand landscapes of twenty years earlier, whose popularity was inexplicable to the writer.[37] Another writer noted that Americans' taste in and need for landscape images had changed, and "The gentlemen who desire to show that we are the greatest nation that ever trod shoe leather, by the exploitation of our Western frontiers, need not find their mission gone. But mere topography will not do it."[38]

Of course the depictions of the West produced by Leutze, Church, and others were not really "mere topography." Rather, antebellum artists made visible the ideology of westward expansion, promoting the mythic vision of the West as a landscape ordained by God, unrivaled in majesty, and ripe for settlement. At the same time, the early painters of the West depicted the people who populated the story of that settlement, both those who drove expansion and those who were driven out by expansion. The Western images produced by Americans in the first eighty years of American art production do not necessarily reflect literal historical or geographical truth, but they do reflect the deeply held cultural beliefs that underwrote and governed Western expansion.

EPILOGUE

Visual images of the landscapes and people of the West, from early paintings and engravings to contemporary movies, have played a powerful role in shaping and disseminating the idealized image of the West in American culture as a place where national character was formed and as representative of America's unique, divinely sanctioned place in the world. As historians began reevaluating Western history in the 1970s and 1980s, art historians began to reassess images of the West's landscapes and people and began writing new histories that cast a critical eye over the roots of such ideas in the Western images of artists from George Catlin to Frederick Remington. In 1991, at an important exhibit at the Smithsonian Institution in Washington, these challenges to America's triumphalist Western narrative got their first major public airing in a show dedicated to visual arts of the West. *The West as America: Reinterpreting Images of the Frontier* included many major, beloved artists of the West who had always been popular with the public; however, the wall label text and the accompanying catalog included essays that incorporated the "new" Western history. This scholarship from the 1970s and 1980s reframed the settling of the West in less laudatory terms, examining the legacies of racism, colonialism, and ecological destruction that scholars argued were also part of the history of Western expansion.

Airing this dirty historical laundry to a public that was accustomed to seeing only heroic individualism in depictions of the American West proved to be a rather fraught endeavor. Protests against the show, its catalog, its curators, and the Smithsonian mounted into an avalanche of criticism from older scholars and the public alike.[39] Pressure became so great that the museum toned down the stronger language in some of the wall labels a month after the show opened.

The West as America raised issues with the Western past that many Americans found shameful, upsetting, or contrary to national self-image as a land of fairness, opportunity, and glorious expansion. As Elizabeth Broun later said, "What we didn't realize at the time was that we were not only showing western art and talking about the settling of the country, but challenging a lot of folks's sense of identity with place."[40] However, although the scholars at the Smithsonian came under fire for their critical stance, the ideas and questions that they raised were as much a part of early visual images of the West as were more celebratory odes to American expansion. The dust-up over the meaning of the work of artists such as Albert Bierstadt and George Catlin only confirms an argument central to this essay: that visual images of the American West began as and remain a crucial site for American self-understanding, national identity, and mythmaking.

NOTES

1. Jochen Weirich, "Struggling through History: Emmanuel Leutze, Hegel, and Empire," *American Art*, 15 (2) (2001): 52–71 discusses Hegel's influence on nineteenth-century American historians and artists in some detail.
2. William Cullen Bryant, "To Cole, the Painter, Departing for Europe," 1829. Reprinted in *The Poeti-*

cal Works of William Cullen Bryant, ed. Henry Cady Sturges and Richard Henry Stoddard (New York: D. Appleton & Company, 1903), pp. 127–128.

3. *The Works of George Berkeley, D.D.S., Late Bishop of Cloyne in Ireland, to Which Is Added, an Account of His Life, and Several of His Letters to Thomas Prior, Esq., Dean Gervais, and Mr. Pope, &c, &c.* vol. II (London: G. Roberson, Pater Noster Row, 1784), pp. 443–444.
4. Allan Wallach, "Thomas Cole's 'River in the Catskills' as Antipastoral," *The Art Bulletin,* 84 (2) (2002): 334–350; Allan Wallach, "Thomas Cole and the Aristocracy," in Marianne Doezema and Elizabeth Milroy (eds.), *Reading American Art* (New Haven, Conn.: Yale University Press, 1998), pp. 79–108.
5. Susan Platt, "Paradigms and Paradoxes: Nature, Morality and Art in America," *Art Journal,* 51 (2) (1992): 82–88.
6. Thomas Cole, "Essay on American Scenery," *American Monthly Magazine,* 1 (January) (1836): 1–12.
7. Alan Wallach, "Making a Picture of The View from Mt. Holyoke," in David C. Miller (ed.), *American Iconology* (New Haven, Conn.: Yale University Press, 1993), pp. 80–91.
8. The first printed appearance of the term "manifest destiny" is generally credited to John L. O'Sullivan, "The Great Nation of Futurity," *The United States Democratic Review,* 6 (23) (1839): 426–430. Roger Cushing Aikin, "Paintings of Manifest Destiny: Mapping the Nation," *American Art,* 14 (3) (2000): 78–89 discusses the leftward/westward movement in images of Westward expansion.
9. William Cullen Bryant, "An Indian at the Burial Place of his Fathers," 1824. Reprinted in *The Poetical Works of William Cullen Bryant,* ed. Henry Cady Sturges and Richard Henry Stoddard (New York: D. Appleton & Company, 1903), pp. 58–60.
10. Bryant, "An Indian at the Burial Place of his Fathers."
11. Kathryn Hight, "'Doomed to Perish': George Catlin's Depictions of the Mandan," *Art Journal,* 49 (2) (1990): 119–124.
12. George Catlin, *Letters and Notes on the Manners, Customs, and Conditions of the American Indian,* 2 vols. (Philadelphia, Pa.: Willis P. Hazard, 1857), vol. II, p. 692.
13. Petition of George Catlin to Congress, ca. 1872, George Catlin papers, 1821–1904, 1946. Archives of American Art.
14. See, for example, Shannon Egan and Edward Curtis, "'Yet in a Primitive Condition': Edward Curtis' 'North American Indian,'" *American Art,* 20 (3) (2006): 58–83.
15. George Catlin, *Letters and Notes,* vol. I, p. 895 (emphasis added).
16. George Crofutt, *Crofutt's New Overland Tourist, and Pacific Coast Guide* (Omaha, Nebr.: Overland, 1884) p. 243. Emphasis in original.
17. Shirley H. Bowers, "'Captured on Canvas': McKenney-Hall's History of the Indian Tribes of North America," *Florida Historical Quarterly,* 71 (3) (1993): 339–347.
18. James Hall, "An Essay on the History of the North American Indians, Part 2," in Thomas L. McKenney and James Hall, *History of the Indian Tribes of North America, with Biographical Sketches and Anecdotes of the Principal Chiefs. Embellished with One Hundred and Twenty Portraits, from the Indian Gallery in the Department of War, at Washington* (Philadelphia, Pa.: E. C. Biddle, 1836–1844), vol. III, pp. 79–122; p. 122.
19. Catlin, *Letters and Notes,* Letter I.
20. Gail Husch, "'Poor White Folks' and 'Western Squatters': James Henry Beard's Images of Emigration," *American Art,* 7 (3) (1993): 15–39; p. 22. See also Carrol Smith-Rosenberg, "Davy Crockett as Trickster: Pornography, Liminality, and Symbolic Inversion in Victorian America," in Carroll Smith-Rosenberg (ed.), *Disorderly Conduct: Visions of Gender in Victorian America* (Oxford, 1986), pp. 90–108; and Elizabeth Johns, *American Genre Painting: The Politics of Everyday Life* (New Haven, Conn.: Yale University Press, 1991).
21. Charles Lanman, *Adventures in the Wilds of the United States and British American Provinces* (Philadelphia, Pa.: J. W. Moore, 1856), p. 4. Lanman was reviewing the work of St. Louis-based artist Charles Deas (University of Michigan Digital Library 2005.)

22. Angela Miller, "The Mechanisms of the Market and the Invention of Western Regionalism: The Example of George Caleb Bingham," *Oxford Art Journal,* 15 (1) (1992): 3–20.
23. Thomas Cole, "Essay on American Scenery," *American Monthly Magazine,* 1 (January 1836): 1–12; pp. 11–12.
24. Anon. "The New World and the New Man," *The Atlantic Monthly,* 2 (12) (1858): 513–531; pp. 519, 523.
25. Miller, "The Mechanisms of the Market," p. 5; and Henry Adams, "A New Interpretation of Bingham's *Fur Traders Descending the Missouri,*" *Art Bulletin,* 65 (4) (1983): 675–680.
26. John Mack Faragher, *Daniel Boone: The Life and Legend of an American Pioneer* (New York: Holt, 1993), p. 338. See also David Lubin, *Picturing a Nation: Art and Social Change in Nineteenth-Century America* (New Haven, Conn.: Yale University Press, 1996), pp. 55–106.
27. David Lubin, *Picturing a Nation,* p. 62.
28. Nancy Rash, *The Painting and Politics of George Caleb Bingham* (New Haven: Yale University Press, 1991), p. 62.
29. William H. Truettner, "Storming the Teocalli Again—or, Further Thoughts on Reading History Paintings," *American Art,* 9 (3) (1995): 57–95, discusses the career of Emmanuel Leutze and the meaning of his history paintings in detail.
30. Charles Lanman, *Haphazard Personalities; Chiefly of Noted Americans* (Boston: Lee & Herald, 1886), p. 255.
31. William Cullen Bryant, *Poems by William Cullen Bryant, With Illustrations by Emmanuel Leutze, Engraved by American Artists* (Philadelphia, Pa,. Carey & Hart, 1847).
32. Emmanuel Letuze, letters and notes, undated note, Archives of American Art, Smithsonian Institution.
33. Nancy Anderson and Linda Ferber, *Albert Bierstadt: Art and Enterprise* (New York: Brooklyn Museum, 1990), discuss the history of this picture and its popular reception in great detail. See also Brucia Witthoft, "The History of James Smilie's Engraving after Albert Bierstadt's *The Rocky Mountains,*" *American Art Journal,* 19 (2) (1987): 40–51. Such blockbuster efforts were pioneered by Frederic Church, Thomas Cole's only pupil, and served as vicarious travel experiences for their many viewers.
34. *NY Leader,* April 2, 1864, quoted in Anderson and Ferber, *Albert Bierstadt,* p. 25.
35. "A. Bierstadt's Great Picture. The Rocky Mountains" pamphlet (1864), quoted in Anderson and Ferber, *Albert Bierstadt,* p. 25.
36. "Proposals for Publishing an Engraving of Mr. Bierstadt's Picture of the Rocky Mountains," Archives of American Art, quoted in Brucia Witthoft, "The History of James Smillie's Engraving after Albert Bierstadt's *The Rocky Mountains,*" *American Art Journal,* 19 (2) (1987); p. 41.
37. S. G. W. Benjamin, "Fifty Years of American Art," *Harpers New Monthly Magazine,* 59 (353) (1879): 674.
38. Raymond Westbrook, "Open Letter from New York, II," *The Atlantic Monthly,* 41 (244) (1878): 238.
39. For a sample negative review, see William Goetzmann, "The West as America: Reinterpreting Images of the Frontier, 1820–1920," *Southwestern Historical Quarterly,* 96 (1) (1992): 129–131. For the history of the controversy, see Allan Wallach, *Exhibiting Contradiction: Essays on the Art Museum in the United States* (Amherst, Mass.: University of Massachusetts, 1998), pp. 105–117; Allan Wallach, "Showdown at 'The West as America' Exhibition," *American Art,* 5 (3) (1991): 2–11; and a thoughtful book review with commentary on the public reception of revisionist history in general, Andrew Guillford, *Journal of American History,* 79 (1) (1992): 199–208. A year later, the show *Discovered Lands, Invented Pasts: Transforming Visions of the American West,* with an accompanying catalog published by Yale, covered much of the same revisionist territory without causing nearly as much controversy. See John Mack Faragher's review of the show in *Journal of American History,* 80 (3) (1993): 1007–1111.
40. Elizabeth Broun, "Conversations in Cyberspace: An Online Chat with Elizabeth Broun," *American Art,* 9 (1) (1995): 5.

CHAPTER EIGHT

THE BORDERLANDS OF THE US WEST

Sheila McManus

In Thomas King's 1993 short story "Borders," a woman and her son from a Blackfoot reserve in southern Alberta try to cross the USA–Canada border into Montana to go visit family in Salt Lake City. When the American border guard at Sweetgrass, Montana, asks the woman to declare their citizenship, she states "Blackfoot." The guard cannot accept this as an answer, and sends another guard out to question them. He asks, "Now, I know that we got Blackfeet on the American side and the Canadians got Blackfeet on their side. Just so we can keep our records straight, what side do you come from?" She replies, "Blackfoot side." An hour later an inspector comes out. When the woman still refuses to choose a side, the inspector says that they will have to go back where they came from. When they head to the Canadian border crossing at Coutts, Alberta, the woman says "Blackfoot" when the guard asks them their citizenship. The guard replies, "I know, and I'd be proud of being Blackfoot if I were Blackfoot. But you have to be American or Canadian." Since neither side will let them cross the line until they choose one or the other, the boy and his mother wind up spending the night in their car, trapped in limbo at the border crossing. They try both crossings again the next day, with the same results, and have to spend a second night in their car. By the third day the media show up, and ask the boy "about how it felt to be an Indian without a country." As a result of the media scrutiny, they are finally allowed to cross into the USA.[1]

Gloria Anzaldúa grew up at the other end of the US West, and she writes in her 1987 book *Borderlands/La Frontera,* "I am a border woman. I grew up between two cultures, the Mexican (with a heavy Indian influence) and the Anglo (as a member of colonized people in our own territory). I have been straddling that *tejas*-Mexican border, and others, all my life."[2] She describes the border as a "1, 950 mile-long open wound / dividing a Pueblo, a culture / running down the length of my body, / staking fence rods in my flesh,"[3] and concludes, "To survive the Borderlands / you must live *sin fronteras* / be a crossroads."[4]

Borders are like that. They trap and divide; they create new identities and complicate old ones. States make borders because borders define the modern nation-state, but borderlands are made by people's relationships to a border. They are the zones that stretch on either side, where the "either/or" is not as clear-cut and a person's options might be doubled or they might be cut in half.[5] They force people to make choices. Borders are not fixed; they are not ahistorical; they are never as solid as their governments might wish them to be; and they are always infinitely more complex and variable than popular representations make them out to be.

Borderlands historians analyze those zones and, as Michiel Baud and Willem Van Schendel have noted, see them as "broad scenes of intense interactions in which people from both sides work out everyday accommodations based on face-to-face relationships." This starting point "allows us to take into account the paradoxical character of borderlands. Borders create political, social, and cultural distinctions, but simultaneously imply the existence of (new) networks and systems across them."[6] Borderlands are created by people interacting with borders, and in that way the borders themselves become (in a very Thompsonian sense) actors in the creation of their borderlands.

Above all else, borderlands historians tend to stress the similarities and shared experiences that exist on both sides of a border. And yet, when it comes to the northern and southern borderlands of the US West, many borderlands historians have gotten caught in the same trap that our work normally challenges: we see three different modern nation-states, with two different kinds of modern border relationships and problems, and then read those present-day realities back in time. As Andrew Graybill and Benjamin Johnson note in their introduction to *Bridging National Borders in North America,* "historians of North America have been far more hesitant to explore the interconnectedness of Canada, the United States, and Mexico," and "until quite recently the term 'borderlands' served as a sort of shorthand to refer to the present-day US Southwest and Mexican North, with little thought to the border dividing Canada from the United States."[7] Sterling Evans states in his introduction to *The Borderlands of the American and Canadian Wests* that from 1986 to 1998 the *Journal of Borderlands Studies* published only two articles on the US–Canadian West.[8]

This historiographical state of affairs is the result of modern social, economic, and political differences that the media and political observers tend to portray as obvious, commonsense reality. Most historians have written about the northern and southern borderlands of the US West as if they are fundamentally different, perhaps because a war was fought in the Southwest and not in the Northwest, or because western Canadians are assumed to be Euro-American and English-speaking and Protestant, "just like" western Americans, while Mexicans are assumed to be brown and Spanish-speaking and Catholic, "not like" Americans. Filibusters played active roles in the history of the southwest border while the few half-hearted annexationists in the north had no real impact at all.[9] The 49th parallel makes up only 1,260 miles of the 5,525-mile border between Canada and the USA, and it has played a very minor role in the history of binational relations between Great Britain and the USA and later Canada and the USA. The national political meanings of the northern border are determined farther east. By contrast, the 1,950-mile border of the US West with Mexico is the whole border between the two countries, so the disputes aren't just regional but are truly international. The existence and placement of the Canada-US border across the West has not been an issue between the two countries, while "border issues have overshadowed all other binational concerns" between Mexico and the USA. The northern border is assumed to be peaceful; the southern one is not. As Oscar Martínez states, "by nature, border zones, especially those that are far removed from the core, spawn independence, rebellion, cultural deviation, disorder, and even lawlessness." He describes this as a "universal scheme," yet popular perceptions of the USA–Canada border rarely use words like rebellion or deviation.[10]

There are very real differences between the two borders of the US West and they occupy a prominent place in contemporary political discourse, yet the histories behind these differences are poorly understood. It is easier to see simplistic differences than to recognize that even a brief overview of the histories of the northern and southern borderlands reveals that for all their very real differences, they share a surprising range of powerful similarities. Both boundaries were cut through and inscribed on top of aboriginal territories and communities. Both are the product of North America's unique colonial history of imperial contests, nationalistic ambitions, and on-the-ground compromises. The lines are where they are not because of geography or a recognition of older aboriginal territorial boundaries, or because there is any "natural" reason for their location, but because of the specific relationships between Spain, France, England, Mexico, and the USA. The boundaries run through ecosystems and apply the words "mine" and "yours" to things like water, fish, and animals.[11] Both lines have been crossed and continue to be crossed by many different people for many different reasons.

Another thing that the two borderlands have in common is that familial metaphors are easily applied. For example, Sam Truett has described the people in the Arizona–Sonora borderlands "as siblings in an interlocking family history," while Peter Morris has written that "[r]ather than two parallel worlds divided more or less completely by a sharp, distinct boundary, the United States and Canada are more like conjoined twins each possessive of its own identity but also very much tied together as parts of the same continental entity."[12] This chapter is a brief, impressionistic comparison of these two borderlands and their shared family history. It deliberately highlights their historical similarities in an attempt to counterbalance the overwhelming scholarly and popular focus on their differences. The place to begin is where King and Anzaldúa direct us: the boundaries are, and were, the lines of outsiders drawn upon an aboriginal world.

ABORIGINAL HOMELANDS

The homelands of the aboriginal communities who got caught up in the borderlands had their own clear geographic logic and integrity that would undermine later boundary-making. Unlike the borderlines that followed, aboriginal territories generally coincided with ecosystems because aboriginal cultures were grounded in and shaped by the specific natural resources available to them.

For example, the Southwestern border runs from the Pacific Ocean to the Rio Grande and passes through nearly 2,000 miles of arid deserts, canyons, mountains, and mesas. At what would become the California–Mexico border there is little rainfall, but plenty of acorns, fish, deer, and small game sustained the Native peoples of the region. Tribes such as the Diegueño, Cupeño, and Kamia lived in houses ranging from cone-shaped pole-and-grass homes to domed earth-covered pithouses and lean-tos of bark slabs.[13]

Moving further east, the region of present-day southern Arizona, New Mexico, northwestern Texas and northern Mexico, was home to such tribes as the Yuma Cocopah, Tohono O'odham, Akimel O'odham, southern Apaches such as the Chiricahua and Lipan; and the Coahuiltec who lived along the lower Rio Grande. The region is extremely dry, averaging only 4–20 inches of rain a year, and most of that falling in a single six-week period in the summer. The Yuma Cocopah, Akimel O'odham, and Tohono O'odham were agrarian, the Apaches were nomads and raiders, and the Coahuiltec were nomadic hunters. Most of the agrarian peoples lived in distinctive apartment-like pueblos built of adobe brick or stone. Some also lived in the desert lowlands or along the rivers, where they lived in small pole-framed huts. The nomadic hunters and gatherers, who did not farm, supplemented their diet by raiding the agrarian villages for their crops. The nomadic tribes usually lived in brush-covered wickiups or earth-covered hogans.[14] As historian F. Todd Smith notes, the fact that some tribes were agriculturalists while others were nomadic hunter-gatherers means that the different tribes "arranged themselves advantageously; some moved about in small groups and others remained sedentary in populous towns." There tended to be a high level of cooperation and trade between the groups, although warfare was not unheard of.[15] The Coahuiltecans survived in "one of the harshest regions in North America [. . .] by collecting nuts and fruits, mainly prickly pear cactus and pecans. The men also hunted deer, and tribal members captured small reptiles."[16] Smith also notes that

> native groups in Sonora had been moving toward smaller, less centralized communities, for whom mobility was a crucial practice. For the Tohono O'odham, who shifted seasonally between the mouths of arroyos and upland hunting and gathering camps, mobility was a way of life. Pimas were also relatively mobile, whereas Opatas invested heavily in fixed settlements based on irrigation.[17]

The northern border of the US West is several hundred miles shorter but runs through much more diverse terrain, from the lush Pacific Coast rainforests to the high, dry, Great Plains east of the Rockies. The region that became the coastal borderlands of British Columbia and Washington has a mild climate and can get as much as 100 inches of rain or more each year. The peoples who lived in the area, groups such as the Lumni, Nooksack, and Skagit, did not need agriculture because the oceans, rivers, and forests provided more than enough sea mammals, fish, and game. The massive trees provided the cedar for the large ocean-going canoes and plank houses. This material abundance meant that the northwest-coast Native Americans developed wealthy, complex, highly stratified societies. The key social institution was the potlatch, gatherings at which the host demonstrated his or her wealth and gained prestige by giving away material possessions.[18]

Further inland, in the high plateau region between the coastal Cascade Mountains and the Rocky Mountains, tribes such as the Methow, Wenatchee, Okanagan, and Colville also did not have to farm because of the abundance provided by the rivers and streams. The people hunted some larger game at the edge of the forests, but salmon was the major food item as well as berries from the river valleys and other roots and bulbs dug up on the grasslands. Most of the people spent the winters in the river valleys in semi-underground earth-covered pithouses, and in the summers they lived in temporary homes covered with bulrush mats.[19]

Between the Rocky Mountains and the Great Lakes were the peoples of the Great Plains, from the Blackfeet just east of the Rockies to the Plains Ojibway just west of the Great Lakes. This long stretch of border runs through treeless grasslands. The short-grass plains in the west get about 10 to 20 inches of rain a year, and the long-grass prairies in the east get about twice that amount. These grasslands were the perfect habitat for the American bison, and most of the Great Plains peoples who wound up in the borderlands were hunter-gatherers who followed a seasonal round based on the bison herds' movements. Most of the people lived in portable tipis, with pole frameworks covered in buffalo hides. The group that would be affected the most by this stretch of the border were the three tribes of the Blackfeet (the Pikani/Pikuni or Piegan, the Kainah or Blood, and Siksika or Blackfoot) because their traditional territory straddled northern Montana and southern Alberta. They had vigorously defended their corner of the northwest plains against all aboriginal and European encroachments for centuries, and adapted quickly to the horse and gun in

with Mexico was about to start, the USA and Great Britain finally agreed to use the 49th parallel as the border west of the Rockies, and the 1848 Treaty of Guadalupe Hidalgo and 1853 Gadsden Purchase set the final boundary between Mexico and the USA. These borders gave each government something concrete to focus on: the US West finally had clear borders, creating borderlands that had to be watched and regulated.

Louisiana Purchase

The rapid pace of this nineteenth-century western expansion and border-making could not have been anticipated even a few decades earlier because the American appetite for land west of the Mississippi was not particularly strong at first. However, its territorial ambitions grew as rapidly as its population. In 1790, there were fewer than 100,000 Americans west of the Appalachians. In 1840, there were more than 7 million people west of the Appalachians, representing more than 40 percent of the nation's population, and three new states had already been carved out of the 1803 Louisiana Purchase: Louisiana in 1812, Missouri in 1821, and Arkansas in 1836.

At the turn of the century, the US government was concerned about the Spanish possession of New Orleans, because whoever controlled New Orleans controlled the Mississippi River, the key transportation link into the center of the continent. When Spain ceded Louisiana territory to France in 1800, the territory's exact limits were unclear, but France now undeniably had the USA completely hemmed in along its western border. Napoleon was overstretched financially and militarily, however, so he made a sudden surprising offer: for $15 million the USA could have Louisiana territory. The USA eagerly agreed, and the purchase was completed in 1803. Many Americans thought it foolish to spend so much money on what many believed was useless wilderness. It was not even exactly clear what they were buying: the only one of the territory's borders everyone agreed on was the Mississippi River in the east.

Two expeditions were promptly sent out to explore the northern and southern borderlands of Louisiana. The more famous (and arguably more successful) Lewis and Clark Expedition headed north from St. Louis in 1804. With a party of more than three dozen men and considerable assistance from local natives, they made it up the Mississippi River to the Missouri River and then west over the Rockies to the Pacific coast in 1806 to establish the USA's first tentative claim to the coast. That claim was reinforced by the 1811 Astor Expedition which founded Fort Astoria at the mouth of the Columbia River.

Before Lewis and Clark even got back to St. Louis, the less successful southern expedition under Lieutenant Zebulon Pike was sent out in 1806. He was supposed "to explore parts of the Arkansas and Red rivers," but he kept heading west and eventually entered Spanish territory. He made it all the way to the upper Rio Grande region before being arrested by Spanish troops. He was later released and returned to the USA.[28]

The official and unofficial versions of both expeditions' journals and reports brought the northern and southern edges of the US West to the popular attention of Americans for the first time. The Lewis and Clark Expedition fired up the American imagination but did not spark an immediate and overwhelming desire to conquer and populate the northern edges of the West.[29] By contrast, Pike's report made it clear that although he thought most of the northern reaches of New Spain were a wasteland, the Spanish province of Tejas was an exception. He described its fertile soil, luxurious grassland, and the excellent manners of the Tejanos or native-born Mexicans. He also said that the Tejanos were very eager to open trade relations with the USA. The Americans were now very interested in what lay beyond the southern edges of Louisiana territory.

Establishing the Northern Border

However, American attention was temporarily diverted from its new western territory by the outbreak of war with Britain in 1812. The War was the result of a host of tensions and irritations that had been building between the USA and Great Britain since the War of Independence. The USA did not really want to seize a lot of Canadian territory; more than one Congressman "dismissed Canadian lands as of little use, the climate as inhospitable, and its people as unenlightened."[30] The USA just wanted Britain to stay out of its business. The 1814 Treaty of Ghent returned North America to its pre-war status quo by forcing each side to give back the territories it had won during the war. The war's more significant (and unintended) outcome was that the two governments were forced to hammer out their western borders in the Convention of 1818.

The Convention stated that the 49th parallel would be the border from the Great Lakes to the Rocky Mountains. Historian Don Thomson notes that in its "length and durability the 49th parallel across the western interior of North America" is "unique as an international boundary" because it is based on an astronomical concept instead of "natural features" such as rivers.[31] Britain and the USA also agreed to a ten-year joint occupation of Oregon territory west of the Rockies. During the negotiations, US Secretary of State John Quincy Adams had pushed to get the Pacific boundary between the USA and the British North America also set at the 49th parallel, because he knew how lucrative the Pacific coast region could be. But the Americans' claims looked very weak compared to Britain's and Russia's substantial trading posts in the region, and Britain wanted the border set at the Columbia River to the south. The compromise agreement allowed British and American traders to operate freely north of the 42nd parallel, which was the northern boundary of Spanish California, and south of 54 degrees, 40 minutes, which was the southern boundary of Russian Alaska. Adams was not too worried that he had not been able to get the 49th parallel; he was a firm believer in manifest destiny and wrote in 1819 that "Our proper dominion" is "the continent of North America [...] The United States and North America are identical."[32] The two sides each agreed to the joint occupation in the hope that time would prove their case.

In the 1820s, Britain's claims to the Oregon territory got stronger because the two biggest fur-trading companies operating in Canada, the North West Company and the HBC, merged in 1821. The new HBC, freed from the high costs of trying to keep up with its Canadian-based rival, began to pour a lot of resources and personnel into the northwest coast to try and keep the region out of American hands. The Company founded several forts, from Fort Vancouver in the south (now in Washington state) all the way up to Fort Simpson on the boundary of Russia's territory, to strengthen their (and Britain's) claim to the Pacific Northwest. Britain and the USA officially recognized Russia's claims to the land north of 54 degrees, 40 minutes in 1825, so when the joint-occupancy deal was renewed in 1827, only the USA and Britain were left as real claimants for the region. It was clear that both were just biding their time.

America's claims to Oregon territory remained very weak until the 1830s, when Eastern promoters and missionary societies brought the region to the attention of the American public and created the first Anglo-American settlements in the region. This led to one of the biggest internal migrations in the USA's history: the Overland Trails. Between 1840 and 1860, roughly 300,000 Americans headed to the west coast, and this huge influx was all the USA needed to finally push Britain for a deal. In 1844, Britain offered to simply extend "the forty-ninth parallel to the coast to divide the region," but Democratic presidential candidate James K. Polk had campaigned on an expansionist platform combining Texas, California, and Oregon, vowing to take all of the northwest coast up to the southern boundary of Russian Alaska at 54 degrees, 40 minutes. As historian Thomson notes wryly, "This may be the only election cry ever created to make use of a latitude description."[33]

What started off as campaign bluster, however, turned into an awkward administrative commitment after Polk's victory. Polk's Secretary of State James Buchanan was willing to accept the 49th parallel proposal, but this time Britain refused. Polk briefly "escalated the dispute [...] by reiterating the Monroe Doctrine's assertion that the United States would not tolerate 'any European interference on the North American continent,'" but a looming war with Mexico and the looming presence of the British fleet convinced him to back down. In 1846, the US Senate quietly ratified the Oregon Treaty setting the boundary at the 49th parallel to the Strait of Juan de Fuca, exactly where Britain had offered to set it two years earlier.[34]

Establishing the Southern Border

The Southwest had long been the more desirable goal for the USA anyway, but it would take a war to establish that boundary. Tensions in the Southwest rose quickly in the 1810s as a result of Mexico's rebellion against Spain and the USA's clear interest in Mexico's northern provinces. This put more pressure on Spain to settle the question of where the American-owned Louisiana territory ended and Spanish territory began. Over the years, the USA offered many different alternatives, from the Sabine River to the Rio Grande itself. Adding to the tensions was the fact that Native Americans across the Southwest, who were the targets of expansionist pressure and violence from Americans and Mexicans, were also stepping up their raids into Mexico.[35]

The border issue was supposed to be settled in 1819, when John Quincy Adams and Spanish minister Luis de Onís signed the Adams–Onís Treaty. It established the border between the USA and Spanish territory "along an irregular line beginning at the Sabine, proceeding north to the forty-second parallel, and from there to the Pacific." Despite US pressure, Spain refused to hand over the province of Tejas.[36] Spain knew that it had to strengthen Tejas somehow or they would lose it to American aggression.

First Spain and then Mexico believed that allowing Americans to settle in Tejas would prevent American annexation, because the new settlers would become loyal citizens of their new country. They acted on this belief by giving Missouri entrepreneur Moses Austin, and later his son Stephen, hundreds of thousands of acres of land in Tejas on which to settle American families. Many other Americans simply drifted westward and squatted in Tejas. By 1823, the 1,500 settlers in the Austin colony had been joined by at least 3,000 in the eastern part of Tejas bordering Louisiana. The Americans called the province "Texas" and themselves Texians, and were rapidly outnumbering the Tejanos who were concentrated in the southern half of the province.

In the mid-1820s, Mexico took other steps to encourage trade and settlement in its northern provinces, still believing that this would strengthen its hold on the region. In 1823, a regular trade route opened between Missouri and Santa Fe, New Mexico, and two years later Mexico passed a Colonization Act to try and attract more foreign settlers. The Act promised land, security, and no taxes for four years. However, Mexico soon realized that these policies were actually weakening its hold in the north, and in 1830 it passed a new Colonization Law that tried to stop further American immigration. The legislation cancelled all pending land grants, banned further immigration from the USA, and authorized the Mexican Army to occupy the province. Paradoxically, this fuelled even more interest in the area, and Anglo-Americans poured in. Between 1830 and 1835, the Texian population rose from 7,000 to nearly 30,000, including 3,000 African-American slaves, and Tejanos were now outnumbered seven to one.

In 1836, a faction within the Texian community which wanted Texas to secede from Mexico and join the USA managed to start an armed rebellion against Mexico. After a year and a half of war, Mexican President Antonio López de Santa Anna was forced to sign a treaty granting Texas independence. The USA recognized the independent Republic of Texas in March 1837 but Mexico refused to follow suit. When Congress voted to annex Texas in 1845, Mexico saw it "as a serious and hostile act by the United States, for it still considered Texas part of its national domain."[37] Mexico broke diplomatic ties with the USA after the 1845 resolution. Polk aggravated the situation by insisting that the Rio Grande would be the border between Texas and Mexico and not the Nueces River. Both sides had previously accepted the Nueces as the border, but it was located an inconvenient 150 miles further north, and Polk wanted the land between it and the Rio Grande. He aggravated the situation even more by sending US troops to the Rio Grande in the spring of 1846. A skirmish between the American and Mexican troops was all the excuse Polk needed to declare war on Mexico and to blame them for starting it.

The USA used the war as a pretext to seize all of northern Mexico, not just to set the boundaries of Texas. When the two sides signed the Treaty of Guadalupe Hidalgo in February 1848 Mexico lost half of its territory, including the present-day states of Texas, New Mexico, Arizona, California, Nevada, Utah, and parts of Wyoming, Colorado, and Oklahoma. The Rio Grande was set "as the eastern half of the border" and "a line across the desert from Paso del Norte (present-day Ciudad Juárez) to the Pacific" became the western half.[38] In exchange, the USA paid Mexico $15 million.

The Treaty of Guadalupe Hidalgo was not quite the last step in finalizing the southern border, however, because pockets of land and several different options for the western border remained undecided. The negotiations over the disputed border area entered a new phase in 1853 when US representative James Gadsden took up his post in Mexico City. His instructions were to negotiate a final border settlement and get the USA out of parts of the Treaty of Guadalupe Hidalgo it had since had time to regret. He was authorized "to propose to Mexico six alternative boundaries, along with differing sums of money for purchase of land." At a bare minimum the USA wanted enough land to create a good railroad route to the Pacific and a port on the Gulf of California.[39] The Santa Anna government was in desperate need of additional funds, so it was willing to make a deal. The Americans did not get everything they wanted from the Gadsden Purchase, which was signed in December 1853 and ratified in 1854, but they got most of it.

The Border Surveys

As a result of the very different ways in which the northern and southern borders were established, the border surveys moved quickly in the south and slowly in the north. In

1850, just two years after the Treaty of Guadalupe Hidalgo, separate boundary commissions were appointed by the USA and Mexico to begin surveying the southern border. They were able to lay out the western portion of the boundary quickly, but the line between New Mexico and Chihuahua was more controversial. The treaty said that the boundary of southern New Mexico was a line that

> ran "north of the town called Paso," without specifying distance. With no precise information concerning the exact location of Paso del Norte, a crucial landmark, and unable to find an accurate map, the treaty negotiators had used an erroneous one, which placed that town 34 miles north and 130 miles east of its true location.[40]

The heads of the survey commissions worked out a compromise, but it was rejected by Washington. The Boundary Commission itself was dissolved in 1852, and a new one created in 1853 just to survey the Rio Grande from Paso del Norte to the Gulf.

The sense of urgency with which the southern border was surveyed was not matched in the north, where it took years for the governments to get around to surveying the 49th parallel. The first piece of the northern border to be surveyed was the line between British Columbia and Washington, but the work did not even start until twelve years after the 1846 agreement. Once the teams were in place, however, the work moved quickly. Instead of cutting the entire boundary through the dense trees from the ocean to the crest of the Rocky Mountains, which would have been extremely expensive and time-consuming, the teams agreed to "ascertain points on the line by astronomical observations at convenient intervals and to mark such astronomical stations by cutting a track not less than 20 feet in width on each side of the line for half-mile or so." The work was completed in 1862.[41]

The long, 850-mile stretch of the 49th parallel between the Rockies and the Great Lakes was not surveyed until the 1870s, more than fifty years after it had been established by the Convention of 1818. In December 1870, President Ulysses S. Grant proposed "a joint United States-British project to mark the boundary between the Lake of the Woods and the summit of the Rocky Mountains." Congress agreed to spend $50,000 to get the American survey underway, while the British and Canadian governments agreed to split the costs of surveying the Canadian side.[42] There were again two commissions in the field, in operation from 1872 to 1874. They cut a 30-foot gap through the few sections of the border that had enough trees, but most of the line did not have enough tree cover so they built large mounds of rocks or earth instead.

David Laird, Canada's Minister of the Interior, was pleased to announce in his annual report for 1874 that the Boundary Commission's work was done. Uncertainty about the exact location of the international boundary line had, in the past, led to "difficulties and disputes, which might have led to grave international complications," so "in the interests of peace as well as on other grounds, the authoritative determination of this missing link in our international boundary line cannot but be a source of satisfaction to the Imperial and Dominion Governments."[43]

From Pigs to *Bancos,* Geography Strikes Back

However, there was a triangle of "missing links" that challenged this "authoritative determination" of both borders. In the north, the so-called "Pig War" on tiny San Juan island off the northwest coast kept the island's status as Canadian or American territory undecided for more than a decade; the location of the northwest angle of the Lake of the Woods (the easternmost point of the western portion of the Canada-US border) provided a surprising amount of controversy for the Boundary Survey; and in the southwest, the Rio Grande's meandering created hundreds of patches of land called *bancos* which the two sides would trade and disagree over for decades.

Most of the border between British Columbia and Washington was established rather easily; the one exception was little San Juan Island in Georgia Strait. When Britain and the USA signed the treaty setting the border in 1846, it specified that the boundary should follow the "middle" of the Strait, but, as with the Rio Grande, "several channels existed" which could be seen as the "middle" one. Depending on which "middle" was chosen, San Juan could be either in Canada or the USA. The HBC had maintained a small sheep farm on the island for years to reinforce England's claim, but in the 1850s some American miners who hadn't struck it rich in British Columbia's gold fields had begun farming on the island too. The so-called Pig War began in 1859 when a pig owned by the HBC raided an American's garden one time too many and was shot. The HBC wanted the American arrested and charged, but the man "insisted on being tried in a United States court." Emotions were soon running so high that "the military commander of the US Oregon Department ordered troops to occupy the island" and Britain sent warships to reinforce Britain's claim. Throughout the 1860s, San Juan Island was jointly occupied by two separate garrisons of 100 men each. In 1871, the two governments finally signed the Treaty of Washington which sent the disagreement to arbitration. Emperor William I of Germany was chosen to decide the question, and in October 1872 he granted the island to the USA. Canada and the USA signed a protocol in the spring of 1873 in Washington, settling the last piece of the Canada-US Pacific boundary line.[44]

While the two countries were sorting out this westernmost part of the USA-Canada border, a critical mid-point was still in dispute. The men who signed the Convention of 1818 knew that the 49th parallel intersected the Lake of the Woods at some point but didn't really know where. They agreed that when the time came a line would be drawn from the "north westernmost point" of the lake north or south until it met the 49th parallel, and then the border would run westward. Nearly sixty years later, locating the northwest corner proved to be difficult, at least for the man heading the British side of the 1872–1874 Boundary Survey. For two years, the head of the British-Canadian survey team, Captain Cameron, annoyed the USA by refusing to accept the opinions of his own chief astronomer, the Americans' chief astronomer, and local native groups about the location of a monument which had been left at the spot in the 1840s. Cameron was determined to get the best possible location because he believed that the future of British interests in the West was at stake.[45] The British government ordered him to concede the point in the summer of 1874.

In the southwest, the "natural" boundary of the Rio Grande was less fixed and obvious than the original treaty-makers thought, because its banks and course shifted frequently. The Americans had repeatedly claimed that the Rio Grande was an obvious "natural" boundary, but it "was certain to sow confusion and discord because of the river's unpredictability." The Treaty of Guadalupe Hidalgo had said that the border would follow the middle of the river, and the deepest channel if there was more than one. As Martínez notes, it is hard enough "to identify 'the middle' of any river," but the Rio Grande is particularly difficult because of its

> tendency to overrun its banks and open up new meandering channels, frequently leaving detached tracts of land to the north or south of the original border. At times these shifts displaced only a few acres of rather inconsequential land, but on other occasions

> the affected terrain had considerable value because of its fertility or location at or near populated areas.[46]

In an attempt to deal with some of the problems, an ad-hoc US–Mexico International Border Commission was created in 1889 and became a permanent body in 1900. During its first ten years, the Commission "spent considerable time identifying" the *bancos* "which the meandering Rio Grande had detached from each country in the Texas-Tamaulipas border." In 1905, both sides agreed to deal with the problem by exchanging fifty-eight *bancos* according to whose side of the river they were now on. If a *banco* was bigger than 150 hectares or had more than 200 people living on it, then the original boundary was maintained and local residents could choose to keep "their original nationality or become citizens of the nation exercising jurisdiction over their locality." Another thirty-one *bancos* were traded in 1910, and the total number rose into hundreds of *bancos,* representing thousands of acres, by 1970.[47]

One of the most controversial and long-lasting disputes was over the Chamizal, a piece of land in the Rio Grande where it ran through El Paso-Ciudad Juárez. The Chamizal, named after the desert grass chamizo, was created in the mid-nineteenth century when flooding shifted the Rio Grande southward, and this piece of Mexican land was now on the north bank. The USA would not trade it back, and the issue was sent to arbitration in 1910. The three-man jury had one representative each from the USA, Mexico, and Canada. The Mexican and Canadian agreed in 1911 that the border's movement had been sudden and caused by the floods, not a slow gradual shifting, so the border should revert to where it had been in 1864. The USA refused to abide by the 1911 decision, and there was little Mexico could do but continue to protest. The issue was finally settled in 1963, when President John F. Kennedy signed a treaty that recognized the Arbitration Commission's 1911 ruling.[48]

These disputes highlight the troublesome role that water would play in the histories of both borders and the need for dedicated bilateral institutions. Nearly twenty years after the USA and Mexico set up its International Border Commission, in 1908 Canada and the USA created an International Boundary Commission. Its mandate was to clarify the boundary between western Canada and the USA. One year later, the two countries signed the Boundary Waters Treaty which created the International Joint Commission. Its sole focus was resolving water-use problems and administering the many rivers that flow between the two countries, including the St. Mary's and Milk rivers, which cross the Alberta–Montana border, and the Red River which flows out of North Dakota into Manitoba.[49]

BORDER-CROSSERS

These political narratives of the histories of the two borders provide the most common elements for the belief that the northern and southern borderlands are so different that they cannot possibly be discussed together. As Gloria Anzaldúa writes,

> borders are set up to define the places that are safe and unsafe, to distinguish US from them. A border is a dividing line, a narrow strip along a steep edge. A borderland is a vague and undetermined place created by the emotional residue of an unnatural boundary. It is in a constant state of transition. The prohibited and forbidden are its inhabitants.[50]

That "us-versus-them" objective never quite worked at either border, however, because in the north and the south the people of the borderlands never completely went along with it. As Baud and Van Schendel argue, "No matter how clearly borders are drawn on official maps, how many customs officials are appointed, or how many watchtowers are built, people will ignore borders whenever it suits them ... [and] take advantage of borders in ways that are not intended or anticipated by their creators."[51] Most of the time borderland people could find a way "to cross the border, legally or illegally. The interesting questions are when they did so and for what motives."[52] By looking at when, why, and how different groups of people used the borders, we can see that, with few exceptions, life in the northern and southern borderlands has not been so different after all.

Aboriginal Challenges/Resistance

Even while the USA was engaged in the process of fixing its western borders, those boundaries were being challenged, and agreeing on the "missing links" did not solve the problems. One characteristic that both borders shared in the nineteenth century was that only the governments had agreed to them; a consensus on the ground was nonexistent and would take many decades to forge. Even when all three nation-states believed that the borders had been established, aboriginal peoples in the borderlands did not see them the same way. Stopping Native Americans from crossing the borders when and where they wanted to became one of the key markers of a border's maturity and effectiveness, but that marker proved very difficult to achieve.

Aboriginal groups challenged the two borders in many different ways during the critical nineteenth-century border-making years. Those challenges added urgency to the process but were also the first forceful statement that these spaces, which the new national borders had just turned into contested borderlands, were still aboriginal homelands and could still be used for their purposes. The first challenge these groups posed to the borders was that many Native Americans refused to be stopped by the border and continued to move back and forth in borderlands regions pursuing their own agendas. For example, raids in the Southwest began before and continued long after the 1819 and 1848 treaties. Along the northwest coast, Euro-American settlement actually increased the amount of aboriginal movement back and forth across the border as Natives from British Columbia migrated to Washington for paid work. The second challenge was equally deliberate, as groups such as the Kickapoos and Yaquis in the Southwest and the Sioux in the North used the borders as barriers to protect themselves from something on the other side.

The cross-border raids in the Southwest which escalated in the mid-nineteenth century were the latest iteration of the raids Natives and newcomers had been engaging in for centuries. As historian Brian DeLay notes, many aboriginal groups in the Southwest increased their attacks on northern Mexico in the early 1830s because of Mexico's "declining military and diplomatic capabilities, as well as burgeoning markets for stolen livestock and captives." When they could, the Mexicans responded "with comparable cruelty and avarice."[53] The violence escalated to the point where thousands of Mexicans and natives were killed in the decade before the USA–Mexico War. The USA even used it as part of their rationale for starting the war with Mexico, arguing that since Mexico did not seem to be able to stop aboriginal raids and violence in their northern territories, it might be better off just giving all that northern territory to the USA. As part of the negotiations over the Treaty of Guadalupe Hidalgo, the USA agreed "to restrain Indians residing north of the new border from raiding into Mexico, and to rescue Mexican captives held by Indians."[54] These promises were enshrined in Article XI of the Treaty. The cross-border raids did not stop, however, and in fact they got worse in the early 1850s. As Oscar Martínez notes, the Native tribes "resented any restrictions on their movement and thought it unfair to be required to give up profitable enterprises that had been a way of life for generations. It also seemed incomprehensible to the indigenous peoples for Americans to demand that they stop attacking Mexicans, a common enemy."[55]

For example, the homeland of the Chiricahua Apaches "straddled Arizona and New Mexico at the Mexican border." They had raided into Mexico for years, and raided the Americans when they showed up. To try and stop the raids the US government established a reservation for them in 1872, but the Apaches continued their raids into Mexico. The government realized that having the reservation so close to the border was not a good idea, so they relocated more than 300 Chiricahuas further north. Two important Chiricahua leaders, Gerónimo and Victorio, refused to accept the relocation. Gerónimo and his followers fled to Mexico, while Victorio and his people chose to continue their "nomadic and predatory existence along the border." Both groups continued to attack Americans and Mexicans alike, on both sides of the border. Victorio was killed by Mexican troops in 1880. Gerónimo returned to the Fort Apache region in 1884, but went to Mexico again in 1885 "after US troops harassed Indians at Fort Apache over the manufacture of homemade liquor." The line could no longer protect him, however, because the USA and Mexico had signed an agreement that allowed US troops to enter Mexico if they were pursuing Indians. The Americans caught Gerónimo in Mexico in 1886, but as late as the 1920s the Apaches "were still using the rugged border mountains as a base from which to carry on their raids."[56]

A far less violent but equally border-threatening movement took place along the northwest coast, as aboriginal peoples moved back and forth from British Columbia to Washington. The Native peoples of the northwest coast had always followed long migratory routes up and down the coast, and the opportunities for paid work south of the border were enough incentive to add some time south of the line to their seasonal rounds.[57] The Euro-American settler population grew much more rapidly in Washington than in British Columbia, and they were not able to hire enough local Euro-American labor to work on their farms and sawmills so they were willing to hire Native workers on a temporary, seasonal basis. The so-called "Northern Indians" were mostly Kwakwaka'wakw, Tsimshian, Haida, and Tlingit, groups whose traditional territories were far to the north of the border. They started moving further south in the early 1850s, and "[a]t its peak around 1885 some six thousand aboriginal people from British Columbia, more than 20 percent of the province's aboriginal population, made a seasonal migration to Washington."[58] Native men worked in Puget Sound sawmills, while women worked as domestics, farm laborers, and prostitutes. The "Northern" women were considered more attractive than the local Native women, so Euro-American men preferred them as sexual partners and wives.[59]

Historian John Lutz argues that the Northern Indians incorporated these new jobs into their traditional practices in three ways. First, many saved their summer wages and hosted a potlatch when they got home, giving away their just-earned money and goods to increase their prestige with their home communities. Second, although the Euro-American community in Washington may have looked down on the Native women as prostitutes, the women who traded sex for money did not have a low status in their own communities and would also sometimes host potlatches when they got home. Third, this new seasonal movement created new opportunities for other important traditional activities: raiding, robbing, and enslaving other aboriginals during the trip home. Lutz notes that "[r]aiding was a long-standing part of interrelationships among aboriginal groups on the Northwest Coast."[60] Through it, Natives acquired economically useful slaves and gained prestige in their home communities. It was the very presence of the Euro-Americans and their need for aboriginal labor that "accelerated the traditional raiding and slave-taking economy" in this new cross-border movement.[61]

By the late 1860s, however, the seasonal migrations to Puget Sound began to decline. There were several reasons for this. First, slaves were becoming a less secure form of property, because the long journeys gave them more chances to escape and border officials were starting to insist that slaves had to be freed. Second, the long trips through hostile waters were increasingly seen as too dangerous, as was working in the territory of the Puget Sound Natives who'd been the victims of the raids. Third, the Euro-American settlers of Puget Sound were not comfortable with the presence of armed, highly mobile aboriginal strangers, and in 1857 Washington's territorial legislature passed an Act to try and stop the Northern Indians from crossing the border by making it illegal to hire or live with them. The legislation was rarely enforced, however, because the Northern Indians were too popular as workers and wives, and within a decade the Euro-American population was large enough that it no longer felt so threatened. Fourth, the USA had more military resources available to control the northern border after the Civil War. Finally, the USA took over Alaska in 1867 and was then in a better position to control the Tlingits' movements. The annual migrations would continue on a smaller scale until the middle of the twentieth century when, ironically, competition from Mexican and Filipino workers "finally succeeded in keeping the Northern Indians out of Puget Sound."[62]

This cross-border movement was being watched more closely, even as it was declining, because by the 1870s both governments had more officials stationed at the border and because Natives were increasingly associated with and blamed for outbreaks of diseases such as smallpox, measles, and influenza. Euro-American settlers and government officials in British Columbia and Washington each blamed the other side's Indians: the Americans said that the Indians brought the diseases with them from Canada, while the Canadians said that the Indians caught them while they were in the Puget Sound area. Officials on both sides tried to "separate and sanitize Indian and non-Indian spaces and control Indian mobility in the name of health and order." These efforts "mostly failed" because they were "improvisational and local, a matter of scolding frequent visitors." Neither side took effective action in the closing decades of the century, however; repeated calls for stricter border regulations and medical inspections were rarely implemented or enforced.[63]

The other category of aboriginal border challenges were the groups who crossed the borders to try and escape violence and find a safe place to maintain their way of life. In the Southwest, the Kickapoos moved back and forth across the border many times in the nineteenth century. They were originally from the Great Lakes region and had migrated southwards to get away from Euro-American encroachments. Some bands had migrated as far as Mexico's northern territory by the 1820s, and even more fled to Oklahoma and Mexico in the late 1830s as they became targets of Euro-American violence after Texan Independence. During the Civil War they were attacked again by Texas-based Confederate forces, and they retaliated by raiding settlements in Texas. The conflict got so bad that in 1872 the US Congress launched an investigation, which "concluded that the Kickapoos were operating under the protection of Mexican officials, and the only solution was to induce them to leave

Mexico and to migrate to Indian territory in Oklahoma." The Kickapoos did not want to leave Mexico, however, and Mexico also wanted them to stay because they were protection against tribes such as the Comanches and Mescaleros. They had often fought alongside Mexican troops in battles against the Comanches and Mescaleros; some had served in the Mexican military; and Mexico had granted them land and sanctuary in return. When the USA's initial peaceful efforts to convince the Kickapoos to move to Oklahoma failed, the US Army entered Mexico without permission from the Mexican government, kidnapped forty women and children, and took them to Oklahoma as hostages to blackmail the Kickapoos into moving north. Within a few years, most of the Mexican Kickapoos moved to Oklahoma, although some went back to Mexico in the 1890s to get away from government policies instituted on their Oklahoma reservation.[64]

On the northern border, the most famous example of a Native American tribe using the line for protection is that of the Lakota Sioux under Chief Sitting Bull, who fled into Canada after the 1876 Battle at Little Big Horn. The Sioux were not a borderlands people, and their traditional territory was concentrated well south of the border in the Dakotas. Sitting Bull was one of the leading Sioux chiefs in the mid-nineteenth century, and he'd been defying the US government's efforts to control his people since long before Little Big Horn. The 1868 Treaty of Fort Laramie had promised to set aside the Great Sioux reservation for their permanent and undisturbed use, but Sitting Bull and another Sioux leader, Crazy Horse, had refused to move their people onto the reservation. Sitting Bull was a holy man as well as a war chief, and he knew that any further compromises with the Euro-Americans were not going to be in the Sioux's best interests. When rumors began circulating in 1874 that gold had been found on the Sioux reservation in the Black Hills, miners swarmed the area, and the Army did nothing to protect the boundaries of the reservation. It was in this tense atmosphere that a huge gathering of Sioux, Cheyennes, and Arapahoe met in southern Montana in 1876 for the summer buffalo hunt and to talk about the growing threat from the Americans. Thousands of men, women, and children were camped along the Little Bighorn Creek, or Greasy Grass Creek as it was known by the Native Americans, when Lieutenant Colonel George Armstrong Custer made his fatal decision to attack. He only had 600 men with him, and they were all dead by the end of the day.

Knowing that the USA would now be seeking revenge, Sitting Bull led nearly 900 of his people across the border into southern Saskatchewan in 1877. They were met by Northwest Mounted Police officer James Walsh, who would become the key intermediary between Sitting Bull and the Canadian government and between the Sioux and the US officials who wanted them back. Canada was willing to let them stay as long as they did not cause any problems, but the Sioux soon found themselves suffering from the same hunger as Canada's First Nations were facing by the late 1870s. A combination of hunger and homesickness would drive Sitting Bull and his people back into the USA in July 1881.[65]

The Yaquis of northern Mexico were one of the few Native American groups who crossed into the US West in search of safety. Much like the Northern Indians along the northwest coast, the Yaquis had developed a pattern of moving north for paid labor in Arizona's borderlands mines and then heading south again to their home villages. Their relationship with Mexico became increasingly strained in the nineteenth century, however, because they resisted Mexico's efforts "efforts to open up their homeland to development."[66] In the 1880s and 1890s they participated in rebellions against the Mexican government, and many began to head north to escape retribution. By the early twentieth century, "battles for the Yaqui homeland raged as *broncos,* or 'hostile' Yaquis, launched raids from the mountains to defend their lands while *pacíficos,* 'peaceful' Yaquis, worked in mines and haciendas." This created problems for officials in Sonora, because while they saw Yaquis as "enemies of the state," it was hard to defeat them while "a modern Mexico valued their labor more than ever." It was even harder to find them as they moved back and forth through terrain that was intimately familiar to them. Yaquis began to develop new communities in cities such as Tucson and Phoenix, and, like "many of their Mexican and Anglo-American neighbors, they were relatively free to cross the border well into the twentieth century, and many would continue to do so to their advantage."[67]

Whether they chose to ignore the borders to maintain their traditional routes and pursue new economic opportunities or used them to try and find safety in a violent world, borderlands aboriginal groups posed some of the most daunting early challenges to the borders of the US West. During critical early decades of border-consolidation, any aboriginal resistance to the new invisible lines dismayed all three governments.

"Canadians" versus "Mexicans"

In both historical and contemporary terms, it is fair to say that Anglo-Canadians have been the most successful and least obtrusive border-crossers, while Mexicans used to be second only to Native Americans and are now the preeminent troublesome border-crossers. In the case of the former, historian Randy Widdis described it aptly when he said that Canadians made "scarcely a ripple" when they entered the USA, even though Canadians constituted "the third-largest ethnic group in the United States" in 1900, behind only the Irish and the Germans.[68] Contemporary officials made few efforts to even count them until the twentieth century, never mind to regulate or limit their numbers. As Thompson and Randall note, "Extensive transborder migration has been an ever-present fact of the US–Canada relationship."[69] Canadians headed south for the same reason that Mexicans headed north: better economic opportunities. Yet the assumed sameness in language and skin color has meant that Anglo-Canadians have never been labeled as problematic border-crossers.

By contrast, the peoples living in northern Mexico have had a much more complicated relationship with the border and its effects.[70] As historian Miguel González has observed, the northern border "is generally associated with a placid and benign coexistence," while "the US–México boundary, even today, evokes images of violence associated with poverty, illegal immigration, and drug traffic." That contrast did not develop immediately, however, nor was it ever complete. As González argues, "the relations between the border people of the two countries" have also been characterized by "multiple instances of cooperation, accommodation and negotiation. […] Conflict and cooperation, are two sides of one same reality that has typified life in the US–México border region and has involved Anglos, Mexicans (of both countries) and the two governments."[71] He stresses that for most of the nineteenth century, even in the midst of the war that would divide the region, most "of the population was willing to live in peaceful coexistence, or even active cooperation, with the Texans." As it was along the northern border, the key relationships were economic.[72] Even though the Mexican government tried to stop its northern citizens from trading across the line, "cross-border commerce grew stronger" and "new networks of interdependence" developed after 1848 because borderlanders wanted and needed them to. Thousands of Mexicans, some of whom had been displaced by the rising levels of violence in the borderlands and some of whom were *sirvientes,* or peons, escaping their debts, migrated to Texas to work in the new state's rapidly expanding economy. The ranches and farms of Texas needed cowboys, sheepherders, and shearers, and farm laborers. González observes that this created "an interesting juxtaposition. While black slaves were escaping south to gain freedom from slavery, *sirvientes* were heading north to find freedom from a similar system in México."[73]

This need for cheap labor increased rapidly in the first two decades of the twentieth century. The Mexican-born population in the USA jumped from 110,393 in 1900 to 700,541 in 1920, and

> [h]undreds of thousands more entered on a temporary basis, crossing and re-crossing the border as laborers, merchants, or casual visitors. This category of migrants, referred to by the Bureau of Immigration as non-immigrants or non-statistical entrants, outnumbered immigrants (or those entering for permanent residence) by a factor of three to one.

Southwest businesses urged the US government not to do anything that would restrict the flow of laborers, and, as a result, Mexican immigrants were exempted "from the head taxes stipulated under the Immigration Acts of 1903 and 1907" and were allowed to cross the border without inspection.[74] Not until World War I would Mexican migration start to be perceived as a border problem and the first tentative steps taken to limit and regulate that migration.

The border's pre-war openness and flexibility was suddenly called into question for two reasons. First, US officials thought that Germany might launch an invasion into the USA from Mexico and, second, because the war created a new pressure for all immigration officials "to take their jobs more seriously." The Immigration Act of 1917 and the Passport Act of 1918 were not aimed at Mexicans but did create "a new tapestry of regulations along the boundary with Mexico."[75] Officials wound up having to ignore a lot of the new rules because, as historian George Sánchez has observed, "immigration laws were rarely conceived with the realities of the border in mind."[76]

The Immigration Act was designed to restrict immigration from southern and eastern Europe and Asia, and the Passport Act tried to close loopholes in the 1917 legislation by requiring all "aliens and US citizens to present passports for inspection at the nation's ports of entry for the duration of the War." Mexicans were obviously not the target of either Act, but officials along the Mexican border did start enforcing the head tax and literacy provisions. This reduced the number of border-crossers significantly, at least at first, because the head tax and literacy tests were "real obstacles for the poor, illiterate Mexicans who made up the majority of the agricultural workers." Not surprisingly, it also increased the number of Mexicans who felt they had no choice but to try and cross the border illegally.[77]

While these Acts may have been popular with the many Americans who wanted to restrict immigration, "Immigration officials in the Southwest recognized the logistical difficulties of closing a 2,000-mile stretch of rugged terrain to unwanted immigration and the incongruity of foisting a restrictionist immigration regime upon a community long accustomed to a virtually open border." A 1918 report indicated that there were anywhere from 100,000 to 200,000 crossings per day along the border and that there were few "natural or man-made barriers" to stop "migrants from simply walking, or even swimming, across the international boundary."[78]

The restrictions weren't popular with US residents and businesses who depended on Mexican workers or customers, or Americans in the borderlands who also wanted to be able to cross the line easily for work or pleasure. As a result, the Bureau of Immigration began to create a patchwork of exemptions for border-crossers. For example, the Bureau created a temporary exemption to the passport requirement called a Section 13 certificate which allowed eligible border-crossers to avoid the passport requirement, the head tax, and the literacy test. Aliens "who lived in close proximity to either side of the border and who frequently crossed the border for 'legitimate pursuits,'" and American citizens who crossed the line regularly, could get a border-crossing card to cross the line quickly. The cards included the bearer's photograph and signature and became the easiest form of border-crossing identification. By 1924, this "*ad hoc* solution" had became "standard policy," and

> For the next forty years, Congress, courts, and the INS would demonstrate a tacit unwillingness to disturb this border crossing policy. In so doing, federal policymakers also implicitly acknowledged that the free movement of populations across the line was essential to the society and economy of the border region, if not the nation as a whole.[79]

The wartime legislation did reduce the number of Mexicans who intended to immigrate permanently to the USA but did little to reduce the number of temporary border-crossers. In 1928, the Commissioner General of Immigration "observed that the nation's borders had surpassed Ellis Island as the major ports of entry. On the Mexican and Canadian borders, he continued, 'a great change has been taking place [...] steadily are they approaching a place of first importance in the scheme of things from an immigration standpoint.'" By his estimate there had been roughly 53 million crossings of the two borders in the previous fiscal year, most of whom were commuters and visitors.[80]

The dominant motif for the northern borderlands has been cooperation, while in the Southwest it has been conflict, and Canadians have benefited nicely from the distinction. But these simplistic modern stereotypes are not ahistorical truths; examining when, where, and why each group did or did not become troublesome border-crossers reveals much about both the history and the current realities of the two borderlands.

Mormons

Anglo-Americans usually caused as few problems as Canadians when they crossed the borders of the US West. US border policy was more concerned with keeping people out than with who was leaving, and Canada and Mexico usually encouraged and welcomed American immigration.[81] One group of Anglo-Americans who used the borders for a similar reason as the Kickapoos and the Sioux, that is, to preserve their way of life, were the members of the Church of Jesus Christ of Latter-Day Saints. They had begun to settle in what is now the state of Utah in the late 1840s, trying to escape the persecution they had experienced back East. Their long-running battles with the USA came to a head in 1882 when the government passed the Edmunds Act which outlawed polygamy. Two years later, "federal marshals began making arrests in Utah, Idaho, and Arizona, where Mormons practices plural marriage as one of the cardinal Mormon doctrines." Over the next few years, members of the Church headed south to Mexico and north to Canada to establish new communities beyond the reach of the US government. In 1885, a group of Mormons who'd been living in New Mexico crossed into Chihuahua and "bought land on the Casas Grandes and Piedras Verdes rivers and in the nearby highlands, where they created six new colonies." Several years later, two more colonies were founded in Sonora on 200 square miles of land church officials purchased on the Bavispe River.[82]

Only one year later, Charles Ora Card of Logan, Utah, entered Canada along with two other members of the Church of Jesus Christ of Latter-Day Saints looking for a good site for another new Mormon community. Card had asked church officials first if he could go to Mexico but was told to head north instead to explore Canadian options. In September 1886, when the three men passed "the stone monument marking the boundary between Washington Territory and British Columbia, Card reportedly took off his hat, 'swung it around and shouted *in Collumbia We are free.*'"[83] Like Sitting Bull, Card intended to use "the border as a place of personal refuge" and hoped that he and his people would be safe north of the line. The following year, the first wave of 2,000 Mormons left Utah for what was then Canada's huge Northwest Territory. The Canadian government "welcomed these experienced dryland farmers with more enthusiasm than it had the Sioux, overlooking their conjugal idiosyncrasies for a time in favor of economic development and increased population."[84] The Canadian government certainly did not welcome plural marriage, however, and to avoid problems very few of the Mormon men who emigrated to Alberta brought more than one wife with them, and if they had others, the women were left behind.[85]

Card had hoped that the "Canadian settlement would be temporary and that he would return to the States once the polygamy issue had been settled. But church leaders denied his request to return to his home in Utah" even after the 1890 Woodruff Manifesto rejected polygamy as a central practice, and in the same year Card "was assigned to head up the new Canadian branch of the church." Not only were the Mormon settlements in southern Alberta proving to be successful and stable farming communities in the challenging semiarid environment but "the church's cooperative work with the Canadian government toward developing irrigated agriculture in southern Alberta gave it a significant foreign ally whose support Mormon leaders found helpful in dealing with the US government, with whom their relationship continued to be tenuous."[86]

Ironically, although the Mormons were initially attracted to both northern Mexico and southern Alberta because of their apparent isolation, in both places the Mormons were immediately linked economically to their new communities. In Mexico, local copper mines bought their produce, hired them to work in the mines, and paid for their wagons to haul supplies.[87] In southern Alberta in 1894, the Mormons signed an agreement "with Alexander and Elliott Galt and their Lethbridge-based Northwest Coal and Navigation Company." The Mormons provided the labor; the Galts provided the land and capital; and together they built "an irrigation canal diverting water from the St. Mary's River to Lethbridge." Not only did the canal facilitate "the settlement of the company's 20, 000-acre land grant," it is still today a critical irrigation source for southern Alberta's farmers.[88]

And, like other border-crossers, the Mormons created and have maintained a complex identity. Most of the early settlers became naturalized Canadians, and they hosted Dominion Day celebrations on July 1 for the non-Mormon Anglo-Canadians in their area. But Card and many others also returned to Utah regularly, to spend time with wives and children left behind, conduct business, and attend church gatherings. Card himself moved back permanently in 1902. The communities continued to celebrate July 4 in addition to the new Canadian holiday, and young Mormon-Canadians in southern Alberta are still more likely to attend Brigham Young University in Salt Lake City than any of the much-closer Canadian universities.[89]

Chinese and Japanese

The world beyond the continent intruded on the borderlands again with two new groups of border-crossers: Chinese and Japanese. Government efforts to regulate the movements of many other groups of North Americans across the West's borders were haphazard and paled in comparison to the efforts the USA made to regulate the movement of Chinese and Japanese immigrants. The Chinese were the original "illegal aliens," the first group of immigrants to be excluded from the USA on the basis of race, and the policies aimed at trying to keep them out of the country wound up forming the basis for the USA's modern immigration policies. The Japanese were no more welcome than the Chinese, but Japan had a very different status in the world than China which made it impossible for the USA to bar them. In both cases, the USA was frustrated in its efforts to regulate its borders by what it saw as the lax immigration policies of Canada and Mexico, and Asian migrants themselves made good strategic use of the borderlands to try and achieve their goals.

At the northern border the USA had a sympathetic ally when it came to trying to bar the Chinese completely. Canada and the USA had both relied heavily on Chinese laborers to build transcontinental rail lines in the 1870s and 1880s, and when those lines were completed both countries wanted the Chinese to leave. The USA moved first, passing the Chinese Exclusion Act of 1882 (just weeks after they passed the Edmunds Act aimed at polygamy) which barred Chinese laborers from immigrating for ten years and prevented all Chinese migrants from applying for citizenship. This was the first time in US history that a group of people was barred on the basis of race, and, as historian Erika Lee notes, it was also "the first time that illegal immigration was defined as a criminal offense in US law."[90] Three years later, Canada also passed legislation aimed at restricting Chinese immigration, which was also the first time Canada had excluded a group of people solely on the basis of their race. The Act was not as strict as the American law, however, in that it imposed a $50 head tax on Chinese migrants—a staggering fee for most impoverished Chinese laborers but not steep enough to stop all Chinese migration altogether.

This difference in the two legal systems created an opportunity for Chinese migrants who wanted to migrate to the USA but could not do so directly: pay the head tax in Canada and then take their chances crossing a land border into the USA. Even after Canada raised the head tax to $100 (in response to pressure from both the US government and vocal racists in British Columbia), a tiny trickle of Chinese were still able to afford to enter Canada, and many wound up in the USA. Until Canada passed its own Exclusion Act in 1923, illegal border crossings by Chinese were a fact of life along the northern border and often took advantage of existing smuggling networks through the Vancouver–Puget Sound area.[91]

The more cordial relationship between Canada and the USA, and the fact that Canada shared the USA's racism toward the Chinese, meant that the USA could work directly with Canadian officials to try and get them to close the northern border to Chinese migrants, legal or otherwise. The Canadian government profited enormously from the head tax (Lee estimates that the government made about $3,000 a month from 1887 to 1891), but it was always under pressure to do more to keep the Chinese out of Canada and the USA Canada did finally pass an Exclusion Act in 1923, which abolished the head tax and simply banned all Chinese except consular officials, children born in Canada, merchants, and students. Only fifteen Chinese were admitted into Canada between 1923 when the Bill was passed and 1947 when it was repealed.[92]

The USA was not able to achieve the same results along its southern border, because the Mexican government did not share its attitude toward the Chinese, pursued a very open immigration policy in the late nineteenth century to attract more workers to the country, and was not willing to work with US officials to help enforce US entry rules. Mexico desperately needed workers but was not able to attract enough Europeans, while the Chinese were willing and eager to migrate. As historian Grace Peña Delgado notes, Mexico's need for investors and workers led the government to not only welcome the Chinese but also to grant them civil rights equal to Mexican-born citizens. As a result, the "privileges enjoyed by Chinese merchants and laborers in Mexico prompted border officials to recognize, rather than deny, the ability of Chinese to assert Mexican claims of citizenship and nationality."[93] In 1899, China and Mexico signed a Treaty of Amity and Commerce which increased Chinese migration.[94]

For Chinese in the southern borderlands, Mexico's initial cordial welcome created a range of opportunities. There were well-established Chinese commercial networks on both sides of the line by the late nineteenth century, and Sonora became "the most popular Mexican destination for Chinese" because of "its relative proximity to Asian labor streams, its mining and railroad development, and, for those looking farther north, its border with the USA. From the 1890s until the early 1930s—when Sonora passed its own anti-Chinese laws—the Chinese were the first or second largest foreign group in the state."[95] They developed an extensive commercial network of small farms, "kitchens, restaurants, and groceries," that was paralleled by their "counterparts across the line, some thirty miles to the north, on the San Pedro River near Bisbee." For all their geographic proximity, however, "these spaces were also worlds apart. The Sonora farms were tied to one of the most powerful Chinese urban networks in the borderlands [...] whereas those in Arizona were pushed to the margins of urban life."[96]

The number of Chinese migrants in northern Mexico and the lack of cooperation from Mexican officials meant that the USA had to step up its own enforcement of the southern border. In 1895, it created a "Chinese Bureau" within the Bureau of Immigration, and it increased the number of inspectors along the southern border.[97] From 1907 to 1909 alone, US officials arrested nearly 2,500 Chinese for illegal entry along the Mexican border, and Mexican statistics suggest that perhaps as many as 1,000–2,000 Chinese managed to migrate illegally into the USA in the late nineteenth and early twentieth centuries.[98] The Bureau also conducted raids on Chinese communities in the borderlands looking for illegal immigrants. By the early twentieth century, not even legitimate Mexican nationality and citizenship could get Chinese across the border into the USA.[99]

An even trickier transnational border problem was created by the small number of Japanese migrants who came to North America in the late nineteenth and early twentieth century. They were able to take advantage of the so-called "transit privilege" which all "civilized" nations were supposed to grant to each other; that is, the citizens of one nation which was considered "civilized" (as Japan was and China was not) had the right in international law to enter and cross the territory of another "civilized" nation if they were headed to a third "civilized" nation.[100] What this meant was that although the USA wanted to bar all Japanese from entering the USA, and it began taking steps to do so in 1906, it could not stop Japanese from entering the country if they said they were merely passing through the USA on their way to Canada or Mexico. As historian Andrea Geiger states,

> Although the international boundaries that transect North America often functioned as barriers to the migrants' realization of their personal goals, at some stages of the Japanese migration the borders also became tools that migrants were able to use to achieve their objectives. Migrants understood that by crossing one border to access another they had access to a wider range of options in relation to the second border, because the act of crossing the first border placed them in a different position *vis-à-vis* the second.

By using both of the borders together, they became "permeable in ways that they were not when each was approached simply on its own terms."[101]

Another difference this time was that the Japanese government kept a very close eye on how its citizens were treated and would not tolerate too much blatant discrimination against its citizens abroad. To maintain a cordial relationship with Canada and the USA, in 1900 Japan agreed to limit the number of passports it issued for those two countries. Canada was also part of the British Empire, and Britain maintained a close relationship with Japan. In 1907, the two governments signed the Treaty of Commerce and Navigation, the so-called "Gentlemen's Agreement," which "provided that the subjects of both 'shall have full liberty to enter, travel or reside in any part of the dominions and possessions of the other.'"[102] Canada thus had no legal right to restrict Japanese immigration, and migrants who arrived at a US port or border crossing could claim the transit privilege and say they were heading to Canada.

US officials believed that the Japanese who landed on Canada's west coast were all going to try and enter the USA, and that most of the illegal crossings took place either between Blaine, Washington, and the foothills of the Cascade Mountains, where the "vast forest with numerous roads" and several railway bridges "offered more than one way to cross the border undetected," or through the Puget Sound and Gulf of Georgia. "No other place in the United States," one immigration inspector declared, "furnishes as many advantages for smuggling as the Puget Sound with its multitude of small islands."[103]

American immigration authorities got some help in the spring of 1907 when President Roosevelt issued an executive order "expressly barring Japanese and Korean migrants holding passports for Mexico, Hawaii or Canada from entering the United States." On top of existing laws against contract labor, this redirected even more Japanese migrants "north to British Columbia. In the fall of 1907, Canadian authorities estimated that just a third of the 4811 Japanese who arrived in Canada during the first eight months of that year had come directly from Japan; the other two thirds had come via Hawaii."[104] The USA could do what it liked when it came to directly stopping the Japanese from entering the USA, but could not deny the transit privilege to anyone who said that they were only passing through the USA to get to Canada.

A similar problem developed on the southern border. Even though Mexico and Japan did not have the close, British Empire relationship that Canada had, Mexico was reluctant to limit immigration or to cooperate too closely with the USA's exclusionary policies. The 1907 executive order had redirected Japanese migrants south to Mexico as well as north to Canada, because "the Mexico-US border offered the same kinds of opportunities as the Canada-US border and more." The Japanese government's 1900 agreement to restrict passports for Canada and the USA was not extended to Mexico, so "emigration to Mexico increasingly offered many Japanese emigrants the only chance they had to obtain a passport to any North American destination. To obtain a passport to Mexico, however, emigrants had to show that they were guaranteed work when they arrived," which generally meant signing a labor contract. In addition, the USA–Mexico border offered a similar opportunity as the northern border: being turned away by a US border official meant only that a migrant had to return to Mexico and explore other options. As Geiger concludes,

> For all these reasons, Mexico became the destination that offered the greatest number of alternatives to immigrants bound for all parts of North America during the first years of the twentieth century. In fact, US immigration officials would ultimately con-

clude that more Japanese entered the USA through Mexico during the first eight years of the 20th century than through Canada.[105]

The 1907 executive order did force most Japanese migrants to cross the border illegally, however, and each border seemed equally problematic. At the Canadian border the heavy tree cover was the problem because illegal immigrants could use it to hide, while at the Mexican border the utter lack of tree cover in the desert meant that illegal immigrants could spot the immigration inspectors' dark blue uniforms "from miles away."[106]

Japanese migrants who headed for Canada hoping to invoke the transit privilege had more problems after 1906, however, because the government issued an order-in-council requiring a "continuous journey" from a migrant's country of birth or citizenship. The Order "was incorporated into Canada's Immigration Act early in 1908 partly to cut off further immigration to Canada through Hawaii." As a result, Japanese migrants could no longer enter Canada after having traveled through the USA because that did not count as a continuous voyage directly from their home country.[107]

CONCLUSION

The 9/11 attacks brought an unprecedented level of attention to the borders of the USA, as new fears about security prompted the USA to try and regulate its borders more closely than it ever had in the past. The attacks even briefly reversed the common perception that the northern border was less dangerous.[108] The problem with this ongoing debate about border policies is that it has taken place without any historical awareness at all. For example, the widely held popular belief that the North American Free Trade Agreement, signed between Canada, the USA, and Mexico in 1992, and which came into effect in 1994, had made the borders "too open" and had compromised security in favor of trade, rests on false historical assumptions. Both borders have always been highly permeable to goods and people, and on-the-ground economic and social realities in the borderlands have always found ways to get around the policies and preferences of the distant central governments. Both borders are, quite simply, too long to ever really "close," too long to ever really be able to stop every single person that wants to cross them or every item that someone on the other side wants to buy. Borders are supposed to be a consensus—an agreement to abide by the rules of modern nation-states as to what it means to cross the edge of their territory—but with borders as long and complicated as the US West's borders with Canada and Mexico, that consensus is impossible to achieve.

This chapter has demonstrated that the familial metaphors used by Truett and Morris can be taken one step further: the northern and southern borderlands of the US West are siblings whose obvious individual differences have obscured for too long the sheer volume of shared genetic material. The northern and southern borderlands are not identical—we have only to contrast the experiences of Anglo-Canadians with Mexicans to see that—but they share many powerful similarities that could form the basis of a conversation. The borders' differences have become more evident and more politically charged in recent years, as post-9/11 fears have all but eliminated discussions of the borders that don't involve security. But, from the longer-term and more-complex perspective of a borderlands historian, those post-9/11 representations of the borders starts to look like less of an upheaval or rupture than contemporary observers, political scientists, and policy officials allow. For a little while, regulation simply re-emerged as the dominant theme, as it had many times in the past. And, as it had each time before, those regulations were protested and loosened in response to local conditions and continental economic realities.

"Illegal immigration" will, no doubt, be an American euphemism for "Mexicans" for many years to come, and perhaps not even some long-overdue study of the northern border's long history of cross-border alcohol and sex trades will make it easier to associate Canada with any illegality at all.[109] But these popular perceptions should not, and indeed must not, delay borderlands historians any longer from having a long-overdue family chat. The two borderlands aren't identical twins, and no amount of scholarship can or should try to erase their very real differences. But, perhaps, once we see that these borders' histories share as many similarities as they have differences, then we can see that both borderlands have played important roles in shaping the US West. Understanding these similarities will throw more light on why they are different, as well as cause us to ask why they aren't written about together more often.

NOTES

1. Thomas King, "Borders," in Gary Geddes (ed.), *The Art of Short Fiction* (Don Mills, Ontario: Addison-Wesley Longman, 1999), pp. 186–194.
2. Anzaldúa, *Borderlands/La Frontera: The New Mestiza,* 2nd edn (San Francisco, Calif. Aunt Lute Books, 1999), preface.
3. Anzaldúa, *Borderlands/La Frontera,* p. 24.
4. Anzaldúa, *Borderlands/La Frontera,* p. 217.
5. Peter Sahlins, *Boundaries: The Making of France and Spain in the Pyrenees* (Berkeley, Calif.: University of California Press, 1989), p. 4.
6. Michiel Baud and Willem Van Schendel, "Toward a Comparative History of Borderlands," *Journal of World History,* 8 (2) (1997): 216.
7. Andrew Graybill and Ben Johnson (eds.), *Bridging National Borders in North America: Transnational and Comparative Histories* (Durham, NC: Duke University Press, forthcoming 2010), introduction. For example, a National Endowment for the Humanities Summer Institute held in the summer of 2009 at the University of Arizona was titled "Nature and History at the Nation's Edge: A Field Institute in Environmental and Borderlands History." The "the" in the title is not accidental, as the entire institute focused exclusively on the Southwest, thus conveying that the USA only has one edge, or at least only one that matters, and that edge is in the Southwest.
8. Sterling Evans (ed.), *The Borderlands of the American and Canadian Wests: Essays on Regional History of the Forty-Ninth Parallel* (Lincoln, Nebr.: University of Nebraska Press, 2006), p. xvii.
9. See Chapter 2 of Oscar Martínez, *Troublesome Border* (Tucson, Ariz.: University of Arizona Press, 1988) and Chapter 1 of Juan Mora-Torres, *The Making of the Mexican Border: The State, Capitalism, and Society in Nuevo León, 1848–1910* (Austin, Tex.: University of Texas Press, 2001) for more on the filibusters, and Alvin C. Gluek, Jr.'s classic work *Minnesota and the Manifest Destiny of the Canadian Northwest: A Study in Canadian-American Relations* (Toronto: University of Toronto Press, 1965) for more on the Minnesota annexationist movement.
10. Martínez, *Troublesome Border,* p. 2.
11. See, for example, Michelle Morris, "Settling the 'Knotty Question' of Water Apportionment: The IJC's Role in Reinforcing American and Canadian National Sovereignty," unpublished 2008 paper, in author's possession; Stephen P. Mumme, "State Influence in Foreign Policymaking: Water Related Environmental Disputes along the United States–Mexico Border," *Western Political Quarterly,* 38 (4) (1985): 620–640; Rachel St. John, "Divided Ranges: Transborder Ranches and the Creation of National Space along the Western Mexican–US Border," in Andrew Graybill and Ben Johnson (eds.), *Bridging National Borders in North America: Transnational and Comparative Histories* (Durham, NC: Duke University Press, forthcoming 2010); Joseph E. Taylor III, "The Historical Roots

of the Canadian-American Salmon Wars," in John M. Findlay and Ken S. Coates (eds.), *Parallel Destinies: Canadian-American Relations West of the Rockies* (Seattle, Wash. and London: Center for the Study of the Pacific Northwest in association with University of Washington Press; Montreal: McGill-Queen's University Press, 2002), pp. 155–180; and Lissa Wadewitz, "The Scales of Salmon: Diplomacy and Conservation in the Western Canada–US Borderlands," in Andrew Graybill and Ben Johnson (eds.), *Bridging National Borders in North America: Transnational and Comparative Histories* (Durham, NC: Duke University Press, forthcoming 2010).

12. Samuel Truett, *Fugitive Landscapes: The Forgotten History of the US–Mexico Borderlands* (New Haven, Conn.: Yale University Press, 2006), p. 6; Peter Morris, "Charles Ora Card and Mormon Settlement on the Northwestern Plains Borderlands," in Sterling Evans (ed.) *The Borderlands of the American and Canadian Wests: Essays on Regional History of the Forty-Ninth Parallel* (Lincoln, Nebr.: University of Nebraska Press, 2006); p. 173.
13. Carl Waldman, *Atlas of the North American Indian,* rev. edn (New York: Facts on File, 2000), p. 43.
14. Waldman, *Atlas of the North American Indian,* pp. 37–38.
15. F. Todd Smith, *From Dominance to Disappearance: The Indians of Texas and the Near Southwest, 1786–1859* (Lincoln, Nebr.: University of Nebraska Press, 2005): 1.
16. Smith, *From Dominance to Disappearance,* p. 4.
17. Truett, *Fugitive Landscapes,* p. 21.
18. Waldman, *Atlas of the North American Indian,* pp. 41–42.
19. Waldman, *Atlas of the North American Indian,* pp. 39–40.
20. Waldman, *Atlas of the North American Indian,* pp. 44–45. See Theodore Binnema, *Common and Contested Ground: A Human and Environmental History of the Northwestern Plains* (Norman, Okla.: University of Oklahoma Press, 2001) for the best discussion of the lives of the Northwestern Plains peoples from pre-contact to the early nineteenth century. See also Sheila McManus, *The Line which Separates: Race, Gender, and the Making of the Alberta-Montana Borderlands* (Lincoln, Nebr.: University of Nebraska Press, 2005), Chapters 3 and 4 for a discussion of border-making in Blackfoot country; and the work of Michel Hogue on the Crees and Métis in the Canada–US borderlands: Michel Hogue, "Disputing the Medicine Line: The Plains Crees and the Canadian–American Border, 1876–1885," *Montana,* 52 (4) (2002): 2–17; Michel Hogue, "Between Race and Nation: The Creation of a Métis Borderland on the Northern Plains," in Andrew Graybill and Ben Johnson (eds.), *Bridging National Borders in North America: Transnational and Comparative Histories* (Durham, NC: Duke University Press, forthcoming 2010).
21. See Jeremy Adelman and Stephen Aron for a good overview of the three empires' strategies in North America and the effect they had on native/newcomer and imperial borderlands across the continent. Jeremy Adelman and Stephen Aron, "From Borderlands to Borders: Empires, Nation-States, and the Peoples in Between in North American History," *American Historical Review,* 104 (3) (1999): 814–842. See also W. J. Eccles, *The Canadian Frontier, 1534–1760* (New York: Holt, Rinehart & Winston, 1969) for more on the French Crescent; Chapters 1 and 2 of Pekka Hämäläinen, *The Comanche Empire* (New Haven, Conn.: Yale University Press, 2008); Smith, *From Dominance to Disappearance,* for more on imperial–Native American relations across the Southwest; John L. Kessel, *Pueblos, Spaniards, and the Kingdom of New Mexico* (Norman, Okla.: University of Oklahoma Press, 2008); Julien Vernet, "Petitions from Peripheries of Empire: Louisiana and Québec," *American Review of Canadian Studies,* 38 (4) (2008): 495–512.
22. Smith, *From Dominance to Disappearance,* pp. 10–11. See also Vernet, "Petitions from Peripheries of Empire."
23. Martínez, *Troublesome Border,* p. 9. See also Adelman and Aron, "From Borderlands to Borders" and Hämäläinen, *The Comanche Empire.*
24. Martínez, *Troublesome Border,* p. 55. See also Hämäläinen.
25. Smith, *From Dominance to Disappearance,* p. 10.
26. Smith, *From Dominance to Disappearance,* p. 18.

27. O'Sullivan, *New York Morning News,* December 27, 1845. Quoted in Julius Pratt, "The Origin of Manifest Destiny," *American Historical Review,* 32 (4) (1927).
28. Martínez, *Troublesome Border,* p. 34.
29. James P. Hendrix, "A New Vision of America: Lewis and Clark and the Emergence of the American Imagination," *Great Plains Quarterly,* 21 (3) (2001): 211–232.
30. John Herd Thompson and Stephen J. Randall, *Canada and the United States: Ambivalent Allies,* 2nd edn, Athens, Ga.: University of Georgia Press, 1997), p. 21.
31. Don W. Thomson, *Men and Meridians: The History of Surveying and Mapping in Canada,* Ottawa, Canada: Department of Mines and Surveys, 1966. Vol. 2, 176.
32. Quoted in Daniel G. Lang and Greg Russell, "The Ethics of Power in American Diplomacy: The Statecraft of John Quincy Adams," *The Review of Politics,* 52 (1) (1990): 3–31.
33. Thomson, *Men and Meridians,* vol. I, p. 261.
34. Thompson and Randall, *Canada and the United States,* pp. 31–32.
35. The two best monographs on this topic are Brian DeLay, *War of a Thousand Deserts: Indian Raids and the US–Mexican War* (New Haven, Conn.: Yale University Press, 2008) and Hämäläinen, *The Comanche Empire.*
36. Martínez, *Troublesome Border,* pp. 10–11.
37. Martínez, *Troublesome Border,* p. 12.
38. Martínez, *Troublesome Border,* p. 15. See also Mora-Torres, *The Making of the Mexican Border.*
39. Martínez, *Troublesome Border,* p. 19.
40. Martínez, *Troublesome Border,* p. 17. For more on the survey, see Robert V. Hine, *Bartlett's West: Drawing the Mexican Boundary* (New Haven, Conn.: Yale University Press, 1968).
41. Thomson, *Men and Meridians,* vol. I, pp. 261–262.
42. Thomson, *Men and Meridians,* vol. I, p. 164. For more on the survey, see Sheila McManus, *The Line which Separates: Race, Gender and the Making of the Alberta–Montana Borderlands* (Lincoln, Nebr.: University of Nebraska Press, 2005), and Tony Rees, *Arc of the Medicine Line: Mapping the Line's Longest Undefended Border across the Western Plains* (Lincoln, Nebr.: University of Nebraska Press and Vancouver/Toronto: Douglas & McIntyre, 2007).
43. David Laird, Minister of the Department of the Interior, in Canada, Department of the Interior, "Annual Report for Year Ending 30 June 1874," *Sessional Papers,* 7 (8) (1875), p. 5.
44. Thomson, *Men and Meridians,* vol. I, p. 262. See also Thompson and Randall, *Canada and the United States,* p. 40, and Scott Kaufman and John A. Soares, Jr., "'Sagacious Beyond Praise'? Winfield Scott and Anglo-American-Canadian Border Diplomacy, 1837–1860," *Diplomatic History,* 30 (1) (2006): 57–125.
45. Cameron to Secretary of State, May 19, 1874, Department of Interior, Dominion Lands Branch, Headquarters Correspondence, 1871–1946, RG 15, D-II-1, vol. 231, file 1116, NAC, Ottawa.
46. Martínez, *Troublesome Border,* p. 25.
47. Martínez, *Troublesome Border,* p. 26.
48. Martínez, *Troublesome Border,* pp. 25–29.
49. Thompson and Randall, *Canada and the United States,* pp. 72–75. See also Martínez, *Troublesome Border,* pp. 29–30, on the twentieth-century struggles over the distribution and water quality of the Rio Grande, Colorado River, and Tijuana River.
50. Anzaldúa, *Borderlands/La Frontera,* p. 25.
51. Baud and Van Schendel, "Toward a Comparative History of Borderlands," p. 211.
52. Baud and Van Schendel, "Toward a Comparative History of Borderlands," p. 220.
53. DeLay, "Independent Indians and the US–Mexican War," *American History Review,* 112 (1): 35. See also his *War of a Thousand Deserts* and Hämäläinen, *The Comanche Empire.*
54. DeLay, "Independent Indians and the US–Mexican War," p. 66.
55. DeLay, "Independent Indians and the US–Mexican War," p. 58.
56. DeLay, "Independent Indians and the US–Mexican War," pp. 68–69.
57. John Lutz, "Work, Sex, and Death on the Great Thoroughfare: Annual Migrations of 'Canadian Indians' to the American Pacific Northwest," in John M. Findlay and Ken S. Coates (eds.), *Parallel*

Destinies: Canadian-American Relations West of the Rockies (Seattle, Wash. and London: Center for the Study of the Pacific Northwest in association with University of Washington Press; Montreal: McGill-Queen's University Press, 2002), p. 80.

58. Lutz, "Work, Sex, and Death," p. 81.
59. Lutz, "Work, Sex, and Death," p. 86.
60. Lutz, "Work, Sex, and Death," pp. 87–89.
61. Lutz, "Work, Sex, and Death," p. 90.
62. Lutz, "Work, Sex, and Death," pp. 90–94.
63. Jennifer Seltz, "Epidemics, Indians, and Border-Making in the Nineteenth-Century Pacific Northwest," in Andrew Graybill and Ben Johnson (eds.), *Bridging National Borders in North America: Transnational and Comparative Histories* (Durham, NC: Duke University Press, forthcoming 2010).
64. Martínez, *Troublesome Border,* pp. 71–72.
65. See David McCrady's *Living with Strangers: The Nineteenth Century Sioux and the Canadian-American Borderlands* (Lincoln, Nebr.: University of Nebraska Press, 2006) for the best discussion of the Sioux in the nineteenth century, and Beth LaDow's *The Medicine Line* for the best discussion of the Sioux's flight to and time in Canada. See also Gary Pennanen's "Sitting Bull: Indian without a Country," *Canadian Historical Review,* 51 (2) (1970): 123–140 for an excellent analysis of how neither Canada nor the USA really wanted to deal with Sitting Bull and what he represented.
66. Truett, *Fugitive Landscapes,* p. 118.
67. Truett, *Fugitive Landscapes,* pp. 118–120. See also George Pierre Castile, "Yaquis, Edward H. Spicer, and Federal Indian Policy: From Immigrants to Native Americans," *Journal of the Southwest,* 44 (4): 383–435. for a discussion of the twentieth-century redefinition of the Yaquis.
68. Randy William Widdis, *With Scarcely a Ripple: Anglo-Canadian Migration into the United States and Western Canada, 1880–1920* (Montreal and Kingston: McGill-Queen's University Press, 1998), p. xx.
69. Thompson and Randall, *Canada and the United States,* p. 51.
70. There is a rich historiography of Mexican peoples and the border. See, for example, Antonia Castañeda et al., *Gender on the Borderlands* (Lincoln, Nebr.: University of Nebraska Press, 2007); Claire F. Fox, *The Fence and the River: Culture and Politics at the US–Mexico Border* (Minneapolis, Minn.: University of Minnesota Press, 1999); Leon C. Metz, *Border: The US–Mexico Line* (Fort Worth, Tex.: Texas Christian University Press, 1989); Oscar Martínez, *Border People: Life and Society in the US–Mexico Borderlands* (Tucson, Ariz.: University of Arizona Press, 1994); Oscar Martínez, *Mexican-Origin People in the United States: A Topical History* (Tucson, Ariz.: University of Arizona Press, 2001); Douglas Monroy, *The Borders Within: Encounters between Mexico and the US* (Tucson, Ariz.: University of Arizona Press, 2008); Samuel Truett and Elliott Young (eds.), *Continental Crossroads: Remapping US–Mexico Borderlands History* (Durham, Md.: Duke University Press, 2004); Pablo Vila, *Crossing Borders, Reinforcing Borders: Social Categories, Metaphors, and Narrative Identities in the US–Mexico Frontier* (Austin, Tex.: University of Texas Press, 2001).
71. Miguel Angel González Quiroga, "Conflict and Cooperation in the Making of Texas–Mexico Border Society, 1848–1880," in Andrew Graybill and Ben Johnson (eds.), *Bridging National Borders in North America: Transnational and Comparative Histories* (Durham, NC: Duke University Press, forthcoming 2010).
72. See, for example, Henry C. Klassen, "International Enterprise: The House of T. C. Power & Bro. in the Cypress Hills Trade, 1875–1893," *Saskatchewan History,* 43 (2) (1991): 57–71, for a discussion of one firm that operated freely and profited nicely in the Alberta, Saskatchewan, and Montana borderlands in the late nineteenth century.
73. González, "Conflict and Cooperation."
74. Kang, "Crossing the Line," in Andrew Graybill and Ben Johnson (eds.), *Bridging National Borders in North America: Transnational and Comparative Histories* (Durham, NC: Duke University Press, forthcoming 2010).

75. Kang, "Crossing the Line."
76. George J. Sánchez, *Becoming Mexican American: Ethnicity, Culture and Identity in Chicano Los Angeles, 1900–1945* (Oxford and New York: Oxford University Press, 1993) p. 58.
77. Kang, "Crossing the Line."
78. Kang, "Crossing the Line."
79. Kang, "Crossing the Line."
80. Quoted in Kang, "Crossing the Line."
81. For a discussion of some of the different kinds of American border-crossers, see Nora Faires' "'Talented and Charming Strangers from across the Line': Gendered Nationalism, Class Privilege, and the American Woman's Club of Calgary," in Elizabeth Jameson and Sheila McManus (eds.), *One Step Over the Line* (Edmonton: University of Alberta Press and Athabasca University Press, 2008); Renée G. Kasinsky, "Fugitives from Injustice: Vietnam War Draft Dodgers and Deserters in British Columbia," in Sterling Evans (ed.), *The Borderlands of the American and Canadian Wests,* pp. 270–289; Stephen T. Moore, "Refugees from Volstead: Cross-Boundary Tourism in the Northwest during Prohibition," in Sterling Evans (ed.), *The Borderlands of the American and Canadian Wests: Essays on Regional History of the Forty-Ninth Parallel* (Lincoln, Nebr.: University of Nebraska Press, 2006), pp. 247–261; and Evelyne Stitt Pickett, "Hoboes across the Border: Itinerant Cross-Border Laborers between Montana and Western Canada," in Sterling Evans (ed.), *The Borderlands of the American and Canadian Wests: Essays on Regional History of the Forty-Ninth Parallel* (Lincoln, Nebr.: University of Nebraska Press, 2006), pp. 203–221.
82. Truett, *Fugitive Landscapes,* pp. 125–126.
83. Quoted in Peter Morris, "Charles Ora Card," p. 172.
84. Thompson and Randall, *Canada and the United States,* p. 52.
85. Morris, "Charles Ora Card," p. 173.
86. Morris, "Charles Ora Card," p. 177.
87. Truett, *Fugitive Landscapes,* pp. 126–127.
88. Morris, "Charles Ora Card," p. 177.
89. Morris, "Charles Ora Card," p. 179.
90. Erika Lee, "Enforcing the Borders: Chinese Exclusion along the US Borders with Canada and Mexico, 1882–1924," *Journal of American History,* 89 (1) (2002): 54–86. See also Erika Lee, *At America's Gates: Chinese Immigration during the Exclusion Era, 1882–1943* (Chapel Hill, NC: University of North Carolina Press, 2003).
91. Lee, "Enforcing the Borders."
92. Lee, "Enforcing the Borders."
93. Grace Peña Delgado, "At Exclusion's Southern Gate: Changing Categories of Race and Class among Chinese *Fronterizos,* 1882–1904," in Samuel Truett and Elliott Young (eds.), *Continental Crossroads: Remapping US–Mexico Borderlands History* (Durham, Md.: Duke University Press, 2004), p. 184.
94. Lee, "Enforcing the Borders."
95. Truett, *Fugitive Landscapes,* p. 121.
96. Truett, *Fugitive Landscapes,* p. 123.
97. Delgado, "At Exclusion's Southern Gate," p. 191.
98. Lee, "Enforcing the Borders."
99. Delgado, "At Exclusion's Southern Gate," p. 202.
100. Andrea Geiger, "Caught in the Gap: The Transit Privilege and North America's Ambiguous Borders," in Andrew Graybill and Ben Johnson (eds.), *Bridging National Borders in North America: Transnational and Comparative Histories* (Durham, NC: Duke University Press, forthcoming 2010).
101. Geiger, "Caught in the Gap."
102. Geiger, "Caught in the Gap."
103. Geiger, "Caught in the Gap."
104. Geiger, "Caught in the Gap."
105. Geiger, "Caught in the Gap."
106. Geiger, "Caught in the Gap."

107. Geiger, "Caught in the Gap."
108. For example, after the 9/11 attacks some American commentators seized on the false rumor that the terrorists had entered through Canada. On October 3, 2001, North Dakota Senator Byron Dorgan waved an orange traffic cone around to make the point that in his state alone, fifteen of the eighteen border checkpoints were not staffed overnight and usually just a pylon or two was left in the road to block entry. Robert Russo, "For a Change, US Criticizes Its Own Security along Northern Border," available online at www.recorder.ca/cp/World/011003/w1003135.html. Not surprisingly, 9/11 drew a lot of attention to border questions and a significant amount of scholarship has been produced on the topic. For example, the articles in Peter Andreas and Thomas J. Biersteker (eds.), *The Rebordering of North America: Integration and Exclusion in a New Security Context* (New York and London: Routledge, 2003) are not historically minded but do include both borders in their analyses. See also Deborah Waller Meyers, "Does 'Smarter' Lead to Safer? An Assessment of the US Border Accords with Canada and Mexico," *International Migration,* 41 (4) (2003): 5–44; and Susan L. Bradbury and Daniel E. Turbeville III, "Are Enhanced Trade and Enhanced Security Mutually Exclusive? The Western Canada–US Borderland in a Post-9/11 World," *American Review of Canadian Studies,* 38 (3) (2008): 317–340.
109. Moore, "Refugees from Volstead"; Sharp, *Whoop-Up Country: The Canadian-American West, 1865–1885* (Minneapolis, Minn.: University of Minnesota Press, 1955); Char Smith: "Crossing The Line: American Prostitutes in Western Canada, 1895–1925," in Elizabeth Jameson and Sheila McManus (eds.), *One Step Over the Line* (Edmonton: University of Alberta Press and Athabasca University Press, 2008).

CHAPTER NINE

CALIFORNIA TRANSFORMED
Organizing a New Society, 1848–1869

Peter J. Blodgett

A SOCIETY ON THE BRINK: MEXICAN CALIFORNIA BEFORE THE GOLD RUSH

Far beyond the settled regions of a USA whose population centers lay east of the Mississippi river, on the northernmost reaches of the Republic of Mexico, the territory known in 1848 as Alta California was itself the legacy of a process begun nearly eighty years earlier in 1769. Driven by worries of possible Russian or British encroachments upon its North American possessions, the Spanish Empire had established its first permanent presence along the coast of California near present-day San Diego in order to prevent any incursions into its colonies by its European rivals. The agricultural possibilities of the region and the extensive population of native peoples (totaling perhaps 300,000 at the time of European contact) seemed to offer great opportunities for colonization on behalf of Spain and missionary work on behalf of the Catholic Church. Located at a great distance, however, from the heart of New Spain, California proved difficult to resupply and generally unattractive to potential settlers. Thus hindered, the colony grew haltingly through the remainder of the eighteenth century and the early decades of the nineteenth, first under the flag of the Spanish monarchy and then, after the overthrow of Spanish colonial rule in 1821, under that of Mexico. By the end of the Spanish era, the number of colonists probably did not reach 3,500; by 1848, the province's non-Indian population totaled perhaps 14,000.

Despite the evident obstacles in the region's path, however, the transformation in Alta California's governance that ensued from Mexico's independence sparked even greater changes in the economic and social life of the province. Abandoning the prohibitions imposed by Spain upon commerce with foreign countries, for example, Mexico opened California's ports to trade with merchants from Great Britain and the USA. California cowhides and beef tallow soon began to flow across the oceans to factory towns on both sides of the Atlantic in exchange for all kinds of manufactured goods, including shoes made from hides sent in earlier shipments. As a result, communities such as Monterey, Santa Barbara, and San Diego throughout the 1830s and 1840s played host to Britons and Americans of various backgrounds and classes, from ordinary seamen and retired ship's captains to shopkeepers and resident agents for English or American businesses.

For most of those sojourners, very little about Alta California resembled the landscape or the society back home, wherever that might be. Exotic and intriguing to some, life in

Mexican California evoked disgust and dismay among others. Richard Henry Dana, writing in his celebrated *Two Years before the Mast* (1840) about his service as an able-bodied seaman in the hide trade, condemned it as a place "at the ends of the earth; on a coast almost solitary; in a country where there is neither law nor gospel."[1] Other foreign visitors, however, found California and its opportunities sufficiently attractive that they began to settle there. Although comparatively few in numbers, perhaps 300 or so in the aggregate, they did begin to play a notable part in the province's economic and social life. Many of them, such as Americans Abel Stearns and Alfred Robinson, English-born Henry Dalton and Hugo Reid from Scotland, by converting to Catholicism and accepting Mexican citizenship, became eligible to buy property of their own. Their subsequent marriages, often to daughters of prominent local families, only strengthened the significant role they played in provincial affairs.

The commercial activities pursued by this first wave of Anglo-American emigrants drew California deeper and deeper into an international marketplace during the 1830s and 1840s, despite the province's continuing physical isolation from Mexico. At the same time, the continued presence of a small but growing population from such far-flung locations as New England and the British Isles created a cosmopolitan core of emigrant merchants and landowners within a relatively insular and provincial society. Many of them watched international affairs quite closely, especially in the 1840s, as they followed the economic expansionism of Great Britain, America's principal rival, and the frequently quarrelsome course of Mexican–American relations.

Most of the Anglo-American expatriates had maintained some ties to the mercantile world of the hide and tallow trade, since so many of them had come to California as part of it. As they settled into their new lives on the Pacific coast, however, many of them also became part of the larger Californio society around them, especially as they began to join the landowning class. Their large grants, known as *ranchos,* gave them a stake in the cattle trade that remained at the heart of the province's economic and social life. The holders of these grants, referred to as *rancheros,* benefited from the dissolution of the great mission tracts decreed by the Mexican government in 1833. Successive provincial governors in Alta California, seeking to court the region's leading citizens, had awarded large grants to friends and potential political allies, bolstering the fortunes of a well-to-do and politically influential landowning class. With vast amounts of acreage now released to private ownership, the hide and tallow trade with British and American merchants mushroomed dramatically through the 1830s and early 1840s. Much of the income generated for the rancheros from this trade went back immediately to those same merchants to pay for the often lavish way of life maintained by the rancheros as the outward sign of their position at the top of Alta California's social hierarchy.

Accompanying the rise of the rancheros in the late 1830s and into the 1840s, relations between the Mexican citizenry and California's Indian peoples grew more complicated. Exposure to unfamiliar European diseases and destruction of many familiar habitats in the course of colonization had already had a drastic impact upon the peoples of California's coastal regions. By the time that Spanish sovereignty over California ended in 1821, the total Indian population had fallen by perhaps one-third to 200,000. Now, dissolution of the missions beginning in 1833 had distributed substantial quantities of land to the mission Indians whose lack of familiarity with the buying and selling of property left them ill equipped to deal successfully with land-hungry Mexican and American rancheros. As a result, large numbers soon were convinced to sell off their lands, often at prices far below

Figure 9.1 California Vaqueros, *Returned from the Chase*, letter sheet published by Anthony & Baker. California ranchos depended upon work crews adept with horse and lariat to manage the cattle herds. Many of the tools and practices devised by the vaqueros would be inherited by the American cowboy in later years.
Source: Huntington Library: RB 48052 #188.

market value. Bereft of other resources, many of the former mission Indians, trading upon their skills as ranch hands (*vaqueros*, such as those portrayed in the pictorial letter sheet in Figure 9.1) or artisans acquired while in service at the missions, found employment on the ranchos or in the pueblos.

Given the province's minuscule non-Indian population, the new rancheros had come to depend as thoroughly upon an Indian labor force as had the mission fathers, especially since the proliferation of ranchos ensured that Alta California would remain essentially agricultural in nature. One aspiring land baron who had made particularly effective use of that resource, as he had of many other opportunities, had been the German-born and Swiss-educated John Augustus Sutter, whose peripatetic travels across North America in search of success (following his abandonment of family ties and a string of commercial failures in Switzerland) had led him eventually to Mexico's northern frontier in 1839. On his expansive land grant along the Sacramento river where he had erected the large adobe compound that would become known universally as "Sutter's Fort," Indian labor even became a commodity, used like wheat or cattle or lumber to discharge debts and settle accounts. With his peers such as San Francisco merchant William A. Leidesdorff, Sutter frequently traded in laborers as well as other goods, noting his regrets, for example, in one letter of

May 11, 1846, that "it layed not in my power to send you the 10 Indians this time, as I had only a few new hands from the mountains here" but that "in harvest time You can select them while they are all coming here to work."[2]

As the trade in the abundant natural resources of its ranchos pulled California into the global market, it also helped to realign many of the province's economic ties away from its mother country: the Mexican republic. Unfortunately, distracted by chronic financial and political instability, Mexico proved unable to respond to this challenge, further eroding the Californios' loyalties, which were already strained by the republic's routine neglect of the military and political needs of its northern frontier. As prominent Californios grew more aggressive in demanding a role in guiding the province's destiny, struggles for political power grew more frequent and the political lifespan of provincial governors grew shorter through the 1840s.

Against the backdrop of such bitter internal squabbling, the province's fate also became enmeshed in the imperial ambitions of various foreign powers during that same decade. Although Russian expansion south from their Alaska colony subsided following John Sutter's 1839 purchase of their outpost at Fort Ross north of San Francisco Bay, both England and France viewed California's abundant natural resources and its strategic position on the Pacific Coast with considerable interest. In the end, however, it would be the undisguised desire of many Americans for Mexico's northern provinces, including California, which would finally swamp Alta California's political ties to Mexico. Buoyed by a supremely confident belief in the God-given entitlement of their nation to "overspread" the entire North American continent (a sentiment often encapsulated in the phrase "manifest destiny"), a growing number of Americans during the 1830s and 1840s considered that no other people had a more deserving claim to all that California might become.

Throughout the 1840s, this increasing fascination with California propelled a small but steady parade of American emigrants who came by way of an overland crossing rather than by sea aboard the merchant ships from the east coast of the USA. Trickling illegally into Mexican California across the Sierra Nevada, perhaps 600 or 700 established themselves in the province by 1845. Many settled in the inland valleys, often in the vicinity of Sutter's Fort, where some of them, blessed like wheelwright and carpenter James Marshall with considerable experience in the mechanical arts, found work helping Sutter build his dream of a personal empire he dubbed "New Helvetia" in honor of his former homeland of Switzerland. Unwilling to abandon their American citizenship or their Protestant faith to become citizens of Catholic Mexico, they nonetheless hungered after land of their own. Dissatisfied with the unfamiliar mechanisms of Mexican law and government, they chafed under what they regarded as "foreign" rule. Some even threw themselves into the province's political upheavals, seeking to obtain political advantage by backing a winning side in one of the frequent Californio revolutions. More than a few, however, fixed their gaze upon American annexation as their permanent salvation. Thus, as war between Mexico and the USA seemed ever more likely, these restless newcomers watched for an opening.

By the middle of June 1846, some concluded that their moment had arrived. Although confirmation of a formal declaration of war could not yet be obtained, rumors of hostile encounters between Mexican and American troops in Texas had already been circulating. As the US Navy's Pacific Squadron under the cautious command of Commodore John Sloat edged closer to opening a war in California, a band of American insurrectionists launched their own rebellion against Mexican rule on June 14. These self-described revolutionaries, although rather unclear as to their final goals, did proclaim a republic in the northern California community of Sonoma under a banner emblazoned with a California grizzly, forever identifying this event as the "Bear Flag" revolt.

With Commodore Sloat's bloodless seizure of Alta California's principal ports soon thereafter, the short-lived Bear Flag Republic evaporated, making way for American occupation of the province. A steady stream of proclamations issued not only by the Bear Flaggers but also by Sloat, his successor Commodore Robert F. Stockton, and Californio leaders José Castro and Pio Pico made the initial phase of California's conquest more a war of words than of bullets and bayonets. Renewed Californio resistance brought about a bloodier series of confrontations that finally culminated in the battle of San Pasqual in late December 1846 and the American recapture of Los Angeles in early January 1847. Although another year and more would pass before news of the Mexican–American War's formal conclusion by the Treaty of Guadalupe Hidalgo in February 1848 would reach California, American occupation of this Mexican province proceeded uneventfully. Unbeknownst to all parties was the ferment that would be set off in the winter and spring of 1848 as word of the gold strike on the American river circulated through the territory.

GOLD FEVER

Having sent his jack-of-all-trades James Marshall up the American river about 40 miles to build a sawmill with a crew of discharged veterans from the Mormon Battalion who had served in the Mexican–American War, Sutter had not the slightest idea of what Marshall would discover on the morning of January 24, 1848. Stunned by the glittering particles that Marshall had scooped out of the millrace while inspecting the site, both Marshall and Sutter, though exhilarated by the discovery, proposed to keep this news confidential for a few more weeks, hoping to see the sawmill and a proposed gristmill to completion. At the same time, in an effort to secure their rights to exploit this find, Marshall and Sutter also negotiated a lease for various uses of the land surrounding Coloma with leaders of the Yalisumni Nisenan, a tribal community with which Sutter had had previous dealings. Unfortunately for Marshall's and Sutter's plans, too many people already had seen gold from the American river for the discovery to be concealed for very long. On March 15, the first notice appeared in the San Francisco-based newspaper the *Californian,* reporting that "gold has been found in considerable quantities" in the tailrace of Sutter's mill and observing that "California, no doubt, is rich in mineral wealth," which would offer "great chances here for scientific capitalists."[3] Subsequent commentary bore an ever-more passionate tone as news of the amounts of gold discovered and the relative ease with which it could be dislodged traveled further and further.

By the summer of 1848, various observers concluded that much of California had succumbed to a violent case of gold fever. Across the Pacific, as ships from California brought word of these events along with their cargoes to Hawaii, the *Sandwich Islands News,* in its June 22, 1848 issue, commented with bemusement that "it would appear that the California people are running wild after minerals."[4] Other trading vessels carried reports of Marshall's find to the British colony of Hong Kong from which the details spread throughout southern China in the spring and summer of 1848. And in Monterey, Colonel Richard B. Mason, commander of American occupation forces and military governor of California, set out on an expedition to the gold district. In the detailed report he composed and forwarded to the War Department in Washington in August, he noted estimates that as many as 4,000 men (half of them Indians) were working the gold fields, where they extracted between $30,000 and $50,000 of gold per day. Such events, he wrote, had "entirely changed the character of Upper California. Its people, before engaged in cultivating their small patches of ground and guarding their herds of cattle and horses, have all gone to the mines or are on their way thither."[5]

Dispatched to Washington in the charge of one of Colonel Mason's officers, his report (and an accompanying tea chest filled with gold samples) wended its laborious way by land and sea toward the nation's capital. Traveling in less formal channels, other versions of the news flowed north as far as the Oregon country and south to Mexico through the summer and fall. Stories first published in the San Francisco newspapers, often carried by merchant ships bound out of that port, appeared in Peruvian and Chilean newspapers by early November, confirming rumors that had been floating in Latin American cities for weeks. Other newspaper accounts, carried by ubiquitous American merchantmen, traveled as far as Australia's shores just prior to Christmas 1848, where they aroused an interest in mining the wealth of the New World through commerce as well as through prospecting. Captivated by this news, American settlers from Oregon, Mexicans from the northern province of Sonora, and Peruvian and Chilean adventurers joined the growing tide of migration. Before year's end, perhaps 6,000–8,000 newcomers had reached California, raising the former Mexican province's population to the vicinity of 20,000 people. William Reynolds, writing from San Francisco to his brother John back home in Boston on December 27, 1848, nicely summarized the character of the stampede to California that had unfolded in the eleven months since Marshall's discovery:

> Nearly all the Foreigners have left the Sandwich Islands [for California], and thousands are pouring in from all parts of South America, Guatimala [sic] and Mexico and when the mail steamers commence running which will be by Feb. next there will no doubt be thousands from the Atlantic States, if they credit it, for it is certainly almost incredible, were it not for the gold to speak for itself.[6]

East of the Mississippi, in the settled reaches of the USA, word of the stupendous discovery in the nation's newest possession penetrated the public consciousness slowly, being dismissed on occasion as created to promote emigration to California. Subsequent appearances, however, of letters from other Californians such as Thomas O. Larkin, respected merchant and the American consul in Monterey, began to arouse considerable interest. News from California thus had begun to capture the nation's fancy as 1848 dwindled away. As further details of Colonel Mason's report and his treasure chest began to leak out in late November and early December, however, it became an astonishing national obsession. With the inclusion of Mason's report in President Polk's annual message to Congress on December 5, 1848, the news about California received official validation.

Sweeping away the last doubters under a flood tide of enthusiasm, leading newspapers took up the cry only days after publication of the President's message. "The gold fever continues to rage in all sections of our country," reported the *Pennsylvania Inquirer and National Gazette* on December 15 while in the *Charleston Courier* of South Carolina on January 27, 1849, an editorialist wrote that it must be admitted that "[California's] rich mineral products have produced a state of excitement the like of which the memory of man does not compass. [...] The Almighty Dollar fever is at its height in every section of the country."[7] To any reasonable observer, it was clear that America would enter the new year of 1849 prostrated by the gold malady.

Among the thousands of Americans who fell victim to this powerful affliction, most could envision only one remedy: a pilgrimage for their health to California where they would break the grip of this fever by amassing a fortune. For all but the most ardent gold seekers, however, the complexities of actually leaving home, friends, and family raised troubling questions. Both publicly and privately, many voices spoke out in opposition to the lure of El Dorado, cautioning the prospective emigrant and warning of the moral and practical hazards that lay ahead for the Argonauts.

In many communities, opposition coalesced in the columns of local newspapers and in the pulpits of the churches. The New York City newspaper *Sunday Times and Noah's Weekly Messenger,* as it considered the spectacle of gold fever on December 17, 1848, summarized with approval the cautionary words of the *Philadelphia Public Ledger* in a column entitled "A Sober Glance at California." Applauding the *Ledger*'s application of "the test of experience and the standard of common sense to the visions of exhaustless wealth," the *Messenger*'s columnist warned that "we must know more about this golden calf, set up for universal admiration, before we fall down and worship it."[8] The Reverend Elisha Cleaveland of New Haven, Connecticut, to drive home his point forcefully, preached a sermon in January 1849 upon the folly of "Hasting to Be Rich," reminding his audiences of the warning in Proverbs, 28:20, that "a faithful man abounds with blessings; but he that maketh haste to be rich, shall not be innocent." By falling victim to "the excessive love of money," the gold seeker would eventually fall into "unprincipled and vicious conduct" and a "covetousness" that would propel him into "idolatry." Those who pursued instant wealth would give in to "the love of money [...] a passion of amazing strength [that] sweeps away the barriers of truth and morality."[9]

During the winter of 1848–1849, in thousands of homes all over the USA, these same arguments and others were taken up and deployed by anxious wives, siblings, parents, and acquaintances to convince prospective emigrants to abandon their ambitions. Warnings about the dangers to life and limb as well as the temptations of wickedness that awaited the gold miners so far from home rang in their ears. Families debated who would care for aged or infirm mothers and fathers, who would take responsibility for the management of farms and small businesses left unattended, and how the costs of the journey would be defrayed. Before coming to a resolution, angry words might be exchanged and tears shed. In the end, though, overcoming all obstacles and wearing down all resistance, they made the decision by the thousands to set off for the land of gold. Whatever their specific motive, many adopted for themselves a name harkening back to the mythology of Ancient Greece, as one song of the period put it:

> Like Argos of the ancient times,
> I'll leave this modern Greece:
> I'm bound to California mines,
> To find the Golden Fleece.

Emulating the hero Jason and his crew, the modern "Argonauts" prepared for their great quest.

PLANNING THE TREK TO EL DORADO

For most Argonauts, of course, no matter how fiercely the gold mania burned within them, setting out at once for El Dorado was simply impossible. All across the American continent, from the passes of the Sierra Nevada to the country roads of the Midwest and Northeast, winter weather in January 1849 had closed down most of the feasible overland routes to California. Along the Atlantic coast, ice-blocked harbors hampered the organization of sea-borne ventures. While a few parties from the southernmost states began their trek by land or sea just before or after the New Year, nearly all emigrants faced an unavoidable delay which they began planning their expeditions.

Confronted by a journey into the unknown of staggering proportions, many would-be gold seekers immediately cast about for sources of reliable information outlining the means of travel available and the actions that would be required to reap their fortunes once they reached California. Recognizing an attractive market when they saw one, many editors and publishers stepped forward into the breach. Rushing their words into print as quickly as possible, they unleashed a torrent of newspaper articles, travel accounts, and guidebooks that the Argonauts devoured in huge quantities. Following the appearance of a handful late in 1848, many more publications poured forth throughout 1849 from presses in the USA, England, France, Germany, and beyond.

This copious literature, as it flourished in the months after President Polk's confirmation of gold discovery, derived from many different sources, though nearly all of it shared certain common purposes. Especially in the final months of 1848 and the beginning of 1849, before many first-hand accounts had surfaced from the gold country, literary entrepreneurs set to work with scissors, paste, and stacks of previously published volumes. routinely reproduced lengthy excerpts from the official letters and reports of Colonel Mason, Thomas O. Larkin, and John C. Frémont, woven together with passages from the observations of travelers such as Edwin Bryant in his book *What I Saw in California* (1848). The resulting productions were also offered up to meet the clamor of prospective Argonauts for facts, figures, and directions.

Overseas, similar demands, rising in intensity as 1849 wore on, were met with similar measures. Publishers eager to tap into the ready audience for gold-rush treatises commissioned translations of current titles and compiled extracts from US government documents while waiting for the first English, French, or German voyagers to produce their observations of travel to and life in California's mining camps. Even as Bryant's *What I Saw in California* went through seven American editions by 1850, enterprising publishers in England, France, Holland, Belgium, Sweden, and Australia brought out edited versions as well during the same two years. The competition for commercially attractive texts grew sufficiently intense that some titles might appear in various foreign versions having been translated, edited, and sometimes even pirated with the original authors left unacknowledged.

Only months after Marshall's discovery, would-be emigrants throughout much of the world thus had access to a considerable body of literature about every aspect of life in California, including its golden riches. So compelling a topic did California remain that a steady stream of titles added to this literature year after year, continuing well into the 1850s. For those Argonauts too impatient or anxious to wait for the publishing world, however, most could find great doses of California dispatches in the newspapers of the day. Especially in America, from small-town weeklies to the great metropolitan dailies, newspapers in the winter of 1848/1849 blanketed their readers with stories from El Dorado. Struggling to feed the undiminished appetite of their readers for news about the gold country, they picked up articles from other newspapers at home or abroad, printed the anecdotes of sea captains returned from voyages to the Pacific coast, and editorialized about the likely impact of the gold mania upon the economic, political, moral, and spiritual health of the gold seeker and the American republic as a whole.

Prospective emigrants who turned to their newspapers for facts and advice in preparing for their great adventure found much to absorb as the winter months of 1849 ebbed away. Many of the papers filled their columns with stories from emigrants on their way to the gold fields and from those who were already hard at work digging for their golden reward. Gold fever even spilled into other corners of many papers as every merchant and manufacturer under the sun tried to find some way of associating his products with the California obsession. Side by side with notices advertising passage to the golden shores of California by steamer or sailing vessel were others trumpeting everything from lessons in Spanish to "camp hampers" filled with pots, pans, and plates, from six-barreled Allen pistols and rubber tents to water-tight "California" bags and portable houses made out of galvanized iron. Californians-to-be were urged to buy maps and charts, canteens and Bowie knives, shirts and socks, boots and holsters while being reminded to ensure that friends and families would not forget them by having a daguerreotype photograph taken before their departure.

Whether haunted by the tantalizing dream of uncovering a personal fortune or dazzled by the image of an American empire on the Pacific coast, the authors of these newspaper stories, guidebooks, and travelers' sagas reflected the excitement and amazement already felt by many Americans and stimulated similar feelings in thousands more around the globe. For those individuals inclined to chase their dreams, this wealth of written accounts also served as the fount of wisdom from which they drew up plans for their own expeditions to El Dorado. First and foremost, though, many discussed the merits or deficiencies of one or more of the principal routes to the realms of gold. Among the Argonauts, once they had made the decision to head for California, no other choice would be more troublesome than that of choosing a particular path to riches.

THE GREAT MIGRATION

Of course, for some travelers, practical considerations intervened to limit the range of their choices in such matters. Frenchmen, Chileans, Australians, and Chinese among others had no option other than to entrust themselves to the maritime passenger trade that sprang up during 1849 and especially 1850 between their ports and San Francisco. Americans suffering from gold fever, by contrast, could select from a great variety of routes, each with its specific advantages and drawbacks. Some of these journeys might pass entirely by sea, some might be made solely over land, and some might combine aspects of both. All required a high tolerance for hardship, tedium, and a certain amount of danger.

To many Americans, especially those from the nation's traditional maritime heartland such as New England and the Middle Atlantic states, a sea voyage seemed like the obvious choice. That voyage, whether begun in a major port such as New York or Boston or in a fishing village such as Nantucket or Newburyport, spanned a distance of more than 13,000 miles. Bearing south into the heart of the Atlantic Ocean, vessels carrying the Argonauts traveled along the eastern coast of Latin America until they finally reached the southernmost tip of the continent at Cape Horn. After safely navigating the treacherous channels and ferocious storms that often marked the passage around the Horn, they emerged into the Pacific Ocean to start the even longer leg of the voyage north to California. At the voyage's end, following a trip that might last as long as six months or more, those ships would finally pass through the Golden Gate at the mouth of San Francisco Bay to the port of the same name, where their eager passengers would disembark to continue the trek to the gold fields.

Although a few dozen ships had set out for San Francisco in the autumn and winter of 1848 from ports along the eastern seaboard of the USA, President Polk's December 1848 message clearly inspired ship owners as well as Argonauts. During the next two years, at the height of the rush to California, a veritable armada of perhaps 1,400 vessels departed from scores of harbors in the USA and Canada alone, with both their advantages and those of their destinations often aggressively promoted by handbills, broadsides, and posters such as the image shown in Figure 9.2. Individual Argonauts and organized contingents of gold seekers, responding to announcements of imminent sailings, often flocked to owners or to booking agents, pressing to purchase passage of any kind to California. Some companies of emigrants, having agreed to pool their resources, their abilities, and their labor for their mutual benefit in traveling to and then exploiting California's riches, even chartered an entire craft for the use of their members only. Whatever arrangements they made for the voyage, sea-going emigrants found it an expensive proposition; the price of individual passages might run between $500 and $1,000 as demand escalated, while the cost of memberships in emigrant companies might reach $300, $400 or $500 per head. Nonetheless, by one reliable estimate, as many as 16,000 emigrants reached California by the Cape Horn route in 1849 and nearly 12,000 more followed in 1850.

Even as the first shiploads of American emigrants destined for San Francisco stood out to sea in the winter of 1848/1849, other fortune hunters had already set other plans in motion. While many travelers appreciated the certainties of a sea voyage to California (one fixed price for their passage, guarantees of food and accommodations and a predetermined destination), thousands of others desired nothing so much as the speediest journey possible, opting instead for a trip that would combine travel by sea and by land. After a voyage south from the Atlantic into the Caribbean Sea, they would cross the North American continent by land at its narrowest point, the Isthmus of Panama, and then depart by ship again, bound north to San Francisco.

The Isthmian route and others like it (crossing Panama's northern neighbor Nicaragua or crossing Mexico at several different points) offered by far the shortest voyages to the gold country, especially in the Panamanian or Nicaraguan routes. Under the best of conditions, if emigrants encountered no delays in crossing Panama or in obtaining passage from its Pacific coast to San Francisco, they might make the entire journey in six to eight weeks, rather than the four, five, or six months required to navigate around the balance of North and South America. Charges as high as $350 or $400 for through tickets from New York to San Francisco in first-class berths tended to match or fall below costs for the trip around Cape Horn. Moreover, for those who could secure space on the steamers of the US Mail Steamship Company or the Pacific Mail Steamship Company, food and accommodations for passengers were likely to be significantly better than for most travelers on the longer ocean route. The hundreds who left American ports for Panama in the waning days of 1848 and early 1849 soon were followed by tens of thousands, with anywhere from 35,000 to 50,000 crossing each year through most of the 1850s by the Panamanian or Nicaraguan paths.

Unfortunately for the earliest Argonauts, the best of conditions rarely obtained in the first few years of the rush. Founded in 1848 with contracts from the federal government to carry the mail from the eastern states to the west coast, both steamship lines were unpre-

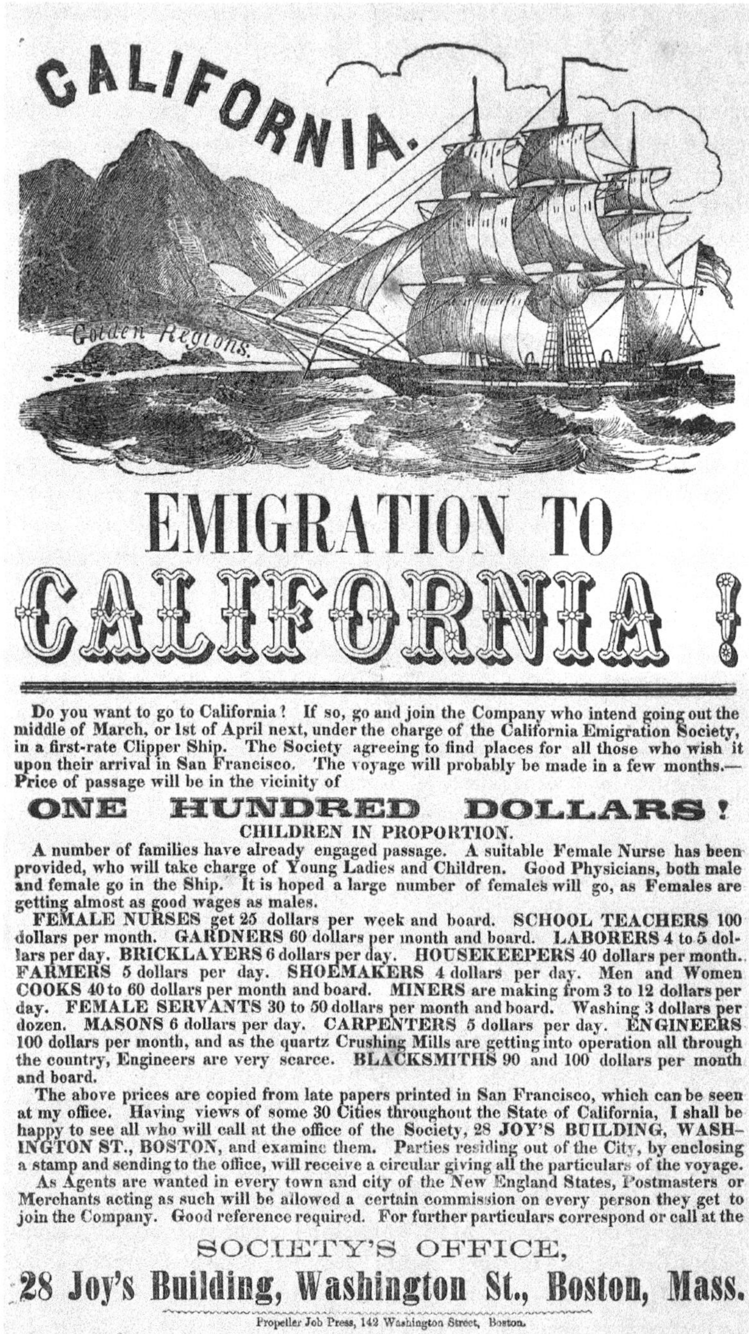

CALIFORNIA.

Golden Regions.

EMIGRATION TO

CALIFORNIA!

Do you want to go to California? If so, go and join the Company who intend going out the middle of March, or 1st of April next, under the charge of the California Emigration Society, in a first-rate Clipper Ship. The Society agreeing to find places for all those who wish it upon their arrival in San Francisco. The voyage will probably be made in a few months.—Price of passage will be in the vicinity of

ONE HUNDRED DOLLARS!

CHILDREN IN PROPORTION.

A number of families have already engaged passage. A suitable Female Nurse has been provided, who will take charge of Young Ladies and Children. Good Physicians, both male and female go in the Ship. It is hoped a large number of females will go, as Females are getting almost as good wages as males.

FEMALE NURSES get 25 dollars per week and board. SCHOOL TEACHERS 100 dollars per month. GARDNERS 60 dollars per month and board. LABORERS 4 to 5 dollars per day. BRICKLAYERS 6 dollars per day. HOUSEKEEPERS 40 dollars per month. FARMERS 5 dollars per day. SHOEMAKERS 4 dollars per day. Men and Women COOKS 40 to 60 dollars per month and board. MINERS are making from 3 to 12 dollars per day. FEMALE SERVANTS 30 to 50 dollars per month and board. Washing 3 dollars per dozen. MASONS 6 dollars per day. CARPENTERS 5 dollars per day. ENGINEERS 100 dollars per month, and as the quartz Crushing Mills are getting into operation all through the country, Engineers are very scarce. BLACKSMITHS 90 and 100 dollars per month and board.

The above prices are copied from late papers printed in San Francisco, which can be seen at my office. Having views of some 30 Cities throughout the State of California, I shall be happy to see all who will call at the office of the Society, 28 JOY'S BUILDING, WASHINGTON ST., BOSTON, and examine them. Parties residing out of the City, by enclosing a stamp and sending to the office, will receive a circular giving all the particulars of the voyage.

As Agents are wanted in every town and city of the New England States, Postmasters or Merchants acting as such will be allowed a certain commission on every person they get to join the Company. Good reference required. For further particulars correspond or call at the

SOCIETY'S OFFICE,

28 Joy's Building, Washington St., Boston, Mass.

Propeller Job Press, 142 Washington Street, Boston.

Figure 9.2 California Emigration Society, broadside entitled *Emigration to California!* Boston, 1849. As gold fever intensified through 1849 and beyond, shipowners, and brokers in every major American port produced a blizzard of newspaper advertisements and broadsides. Seeing the opportunities presented by the enormous demand for passenger berths they promoted their own golden visions of El Dorado. Source: Huntington Library: RB 82892.

pared to cope with the enormous demand for berths from Panama to California after gold discovery. Consequently, as the numbers of Argonauts arriving in Panama rose drastically through the winter and the spring of 1849, they completely outstripped the capacity of both the US Mail and the Pacific Mail's steamers to meet the need on either side of the isthmus. Soon hundreds and then thousands of Americans piled up in and around the port city of Panama, waiting with mounting desperation for the sight of *any* ship that might carry them north to California.

Taken as a whole, therefore, this maritime migration grew to vast proportions as gold fever intensified. Ships sailing for San Francisco from the east coast of the USA or from the Isthmus of Panama were joined year after year by hundreds of others that had departed from the west coast of Mexico, from China and Australia, from England, and the European continent. In the months between April 1849 and January 1850 alone, as many as 40,000 people may have reached San Francisco by sea. While much of the world rushed to the California gold fields by sea, many Americans chose to make their way to El Dorado strictly on dry land.

During the winter of 1848/1849, thousands of New Yorkers, Pennsylvanians, Ohioans, Michiganders, Kentuckians, and Missourians sat by their firesides, often in the company of friends and families, puzzling out their plans to hunt for gold in California. To reach the gold fields, however, they turned west across the continent rather than east and south across the Atlantic Ocean toward Cape Horn. For nearly a decade, a small but steadily rising number of emigrants already had been marching west overland to the Oregon country and Mexican California, pioneering the routes and techniques of transcontinental migration. Letters from relatives and former neighbors now living on the Pacific Coast, published accounts of Western travelers or explorers, and volumes of formal advice to wayfarers all contributed details (albeit on occasion contradictory ones) that were applied to organizing for the great trek.

Like so many of the sea-borne Argonauts, large numbers of overlanders to California gathered together in joint-stock companies, agreeing to pool their resources and thus to share both the risks and the rewards. In small towns and rural counties, such companies often grew out of informal associations of friends and family members; in large cities such as New York or Philadelphia, they could be assembled in response to announcements posted in leading newspapers. Even before setting out on their march to El Dorado, however, the overlanders had already grappled with the question of which path to follow across the continent. What route would lead them to California through the fewest obstacles in the shortest number of days? Balancing such variables as distance, climate, and topography based upon the incomplete or inaccurate information at hand, the overlanding forty-niners flooded into such settlements as Council Bluffs in the Iowa territory, Saint Joseph and Independence in Missouri, Fort Smith, Arkansas, and Brownsville, Texas, that became the jumping-off points for the trek.

After arriving in these frontier entrepôts, the emigrants had one last opportunity to perfect their arrangements for the great excursion. Patronizing the local merchants who did a thriving business as outfitters, the forty-niners could refurbish their equipment, replace lost or damaged supplies, or even prepare themselves completely for the journey. Oxen or mules, pack saddles or axle trees, guides or gunpowder could be found for a price at most of these commercial outposts. City-dwelling gold seekers tried to master the mysteries of handling ox teams or greasing wagon wheels while other emigrants traded gossip about the dangers of marauding Indians or cholera epidemics. Those emigrants crossing on the

Oregon–California Trail waited fretfully on the east bank of the Missouri for the spring grasses to grow up thick and lush over the prairies, conscious that too early a start would starve their livestock while one too late might strand them in California's Sierra Nevada Mountains once winter's first blizzards had blocked the passes.

Finally, with preparations complete and the last vestiges of winter dispersed, parties of emigrants took to the westward trails in the spring and early summer of 1849. Beginning as early as March of that year in the southern reaches of Texas, the bulk of the migration had set out on the northern route along the Platte river by the end of May. Especially on the Oregon–California Trail, long trains of wagons grouped together in thoroughly organized bodies ground across the prairies, covering 10 or 15 miles on a good day. Driving west across the Great Plains or the deserts of the Southwest in the summer heat, once-inexperienced emigrants (often known by the derisive term "greenhorns") grew accustomed to the daily routines of caring for themselves, their equipment, and their livestock. Though often in contact with other emigrant companies ahead of or behind them on the trails, they were now completely isolated from friends or family back in the States. Only at a handful of such military encampments or trading posts as Fort Laramie, Fort Bridger, or Fort Hall on the Oregon–California Trail or during infrequent stopovers at such established settlements as Santa Fe in New Mexico or the Mormon enclave of Salt Lake City could they expect to meet other human beings not bound for California.

Week after week, as the emigrants trudged westward during a journey that regularly lasted four months or more, other disasters both great and small might claim the lives of fellow travelers or at least inspire some to abandon their quest for a golden reward. For most of the Argonauts, however, the journey proved far more tedious than dangerous, often testing their patience most of all, given their burning desire to reach California's gold fields. Among the more than 20,000 emigrants who followed just the Oregon–California Trail in 1849, for example, the best informed estimates suggest that 1,000 or fewer died during the trek, with the largest number probably succumbing to the cholera that plagued the 1849 migration in its early phases. By the fall of that year, perhaps 35,000 emigrants had crossed the continent on one of the various overland routes, followed by possibly well over 100,000 in the first three years of the 1850s. Like their comrades tramping through the feverish jungles of Panama or enduring eight months at sea sailing around Cape Horn, overlanders year after year pressed on, caught up by the lure of adventure or by their hopes for golden prospects in California.

LIFE IN THE GOLD COUNTRY

> I have been up to this Placer and it Surpasses all expectation Men are washing their twenty dollars per day per man with ease and Some have washed even one hundred fifty dollars in one day. [...] It is found in great abundance about the Sacramento river and all its tributaries and as these rivers fall the Gold is found more plenty. [...] I [...] enjoyed good health until very lately having suffered severely from an attack of the 'Yellow or gold fever' which as of yet has not abated.[10]

By the time Henry Richardson wrote this description of the gold country for southern California rancher Abel Stearns on July 9, 1848, many of the first Argonauts had already begun to realize their own golden prospects. Months before the great migrations from the eastern USA or Europe even started to take shape, hundreds and then thousands of gold seekers were scouring hillsides, canyons, and streambeds up and down the Sierra Nevada.

Before year's end, as many as 5,000 miners had clustered at likely sites along a 150-mile stretch running north to south in the Sierra foothills. Rancheros and townsmen from all parts of Mexican California, Indians from the Sierran tribes, veterans from the US Army's garrisons in California, Mexicans, Chileans, and Hawaiian Islanders (known as "Kanakas") prospected and dug with a vengeance. Halted only by winter's onset in the mountains, these miners resumed work in the following year with the first signs of spring, their numbers greatly augmented by waves of migrants arriving from Latin America, Australia, and the USA. By the summer of 1849, for instance, the San Francisco newspaper *Alta California* estimated that as many as 5,000 Mexicans had already made the overland trek while more than 1,500 Chileans and Peruvians had come by sea from their homelands. Observing the unfolding frenzy during his own excursion to the gold regions, southern California rancher Hugo Reid was appalled by what he saw. Writing to his fellow rancher and dear friend Abel Stearns from Monterey on July 14, 1849, Reid noted scornfully that "the whole of the diggings from the numbers and continual movement is well represented by an ants nest.—only the latter possess order & morality which is sadly lacking among the former."[11]

Upon arrival in the gold fields, the Argonauts found themselves instead reduced to a dismayingly primitive level of technology. In the early weeks and months, quite a few discoveries could be dug out of crevices or clefts in and around the creeks and streams. Once those had been exhausted, however, gold hunters then faced a much more demanding prospect. To separate finer specimens of gold from dirt, gravel, and other debris, they had to find ways in which to run earth and water together, relying upon gold's unusual weight to pull the ore to the bottom of any receptacle where it would await discovery. Thus, individual miners quickly adapted the shallow metal pan in which dirt and gravel would be swirled together carefully with stream water, discarding all the contents except a sandy gold-bearing residue—if the miner was fortunate. After the results of each washing were carefully picked over, the miner would fill his pan from a pile of dirt previously dug up with pickax and shovel, bend back down to the stream beside him, and begin the process again for as many repetitions as his body could stand in a day, for as many days in a row as he could tolerate.

Gold pans, being simple to handle and eminently portable, allowed miners to depart at a moment's notice for new diggings and to begin anew somewhere else just as quickly. Restricted to the use of one man at a time, however, they could not take advantage of the gains in efficiency realized in the division of tasks among several men. Eager to find other means that would allow the speedier processing of greater volume to improve the return on their investment of time and labor, miners soon began to improvise and to collaborate. Devices known as "cradles," "rockers," "long toms," and "sluice boxes" appeared in the mining country, with each one improving upon its predecessors in the struggle to separate gold from the sand, gravel, and dirt that usually concealed it. Simple contraptions usually cobbled together from wood and iron, they mimicked the action of natural streambeds, carrying a slurry of earth and water through boxes or troughs lined with small barriers that would catch the heavier particles of gold as they flowed by. Operated by teams of miners who could dig up piles of promising soil, shovel it into the machines, turn the water into them, and sort through the dregs left behind for gold, these machines allowed the Argonauts to sift through far greater quantities of earth in a day than an individual miner might have washed in a week.

Miners all over the gold country adopted such advances as quickly as they could, for in these same years they found themselves in a desperate competition. By the fall of 1849, with 25,000 gold seekers flooding over the Sierras via the transcontinental routes and tens of thousands more streaming through the port of San Francisco, one estimate asserts that the number of miners had increased a remarkable eightfold in just a year to 40,000. Only a year later, by the close of 1850, that figure rose two-and-a-half times to the staggering total of 100,000. In the face of this vast army of rivals, most miners working established claims grasped any advantage they could find to squeeze out the very last grain possible.

Prospectors, ranging over hundreds of square miles in search of the next great strike, could ill afford to waste time, for gold mining had its seasons. Like the emigrants who had crossed the continent by wagon to reach California, the miners soon learned to watch for the changes in weather that marked climatic turning points in the year. They watched for the first signs of spring when the warmth of the sun would start to melt the great mountain snowpacks, restoring the streams and rivers that supplied the water for cradles, rockers, and long toms. As summer's hot, dry months approached, they watched the last spring freshets dry up, reducing the quantities of water available. Above all, as the Argonauts grew accustomed to California's seasons, they learned that little could be accomplished during the wet, chill months of winter in the Sierras, when torrential rains might wash out an entire canyon full of diggings and miners with little warning.

To anxious miners, climatic obstacles posed yet one more challenge to realizing the golden reward that had lured most of them to California in the first place. Many, especially among those arriving after 1849, discovered that the search for riches demanded hard, unremitting toil and privation with no assurance of success. With steady depletion of the most accessible placer deposits as the number of miners grew and grew, various gold seekers had to pursue bigger, costlier, and more complicated projects. Some sought to provide a more certain supply for long toms and sluice boxes no matter the season through the diversion of natural water courses into manmade ones, requiring the construction of canals, ditches, and flumes. Others diverted rivers and streams entirely, uncovering potentially gold-laden beds for mining operations. And still others actually excavated tunnels and shafts into the earth in search of the veins of gold locked within the quartz rock.

Such enterprises demanded large crews of laborers, heavy tools and machinery such as winches, pulleys, and drills, experienced engineers to direct everyone's efforts and significant amounts of outside capital to pay all the bills before reaping any profits. Other capital-intensive ventures appeared: the great iron presses known as "stamp mills," used to crush gold-bearing quartz into tiny pieces, and the mercury mines at New Almaden, New Idria, and elsewhere, whose product proved essential in extracting gold from the quartz rubble. Two inventive miners, working independently of one another, developed the technique known as "hydraulic mining" to channel water from stream beds through narrow canvas hoses capped with iron nozzles. Directing the resulting high-pressure flow onto gold-bearing hillsides, such set-ups would blast away thousands of cubic yards of earth and carry it through sluices to capture the golden particles. Although many miners continued to devote themselves to the pickax, the pan, or the cradle alone or with a partner or two, the intensification of the mining enterprise represented by hydraulic and quartz mining became predominant in the early 1850s. Moreover, the frenzied search for gold across a broad swath of California's landscape left large portions of the gold fields thoroughly despoiled by such intrusive techniques.

As the technology of mining evolved, other changes came to California's gold country as well, driven in large part by the enormous expansion in the population of would-be miners during the early 1850s. The small clusters of tent camps or log cabins first erected by miners near their claims now proved, in many mining districts, the seedlings from which grew scores of ragtag mining towns. Those towns, in turn, merged into an elaborate network stretching all the way to San Francisco, already claiming for itself the grand title of "Queen City of the Pacific." Across San Francisco Bay and then up the Sacramento, the San Joaquin, or the Feather rivers flowed a steady procession of sailing ships and steamboats, destined for the smaller port cities of Sacramento, Stockton, or Marysville. Laden with pork, pickaxes, window sashes, flannel shirts, letters from home, and newly arrived gold seekers, these vessels disembarked their cargoes which then were distributed to Sonora, Columbia, Volcano, Placerville, Coloma, Nevada City, and dozens of other mining towns by stagecoaches, freight wagons, or pack mules.

In the gold camps themselves, the future was always uncertain. Camps located in booming districts might mushroom in size, attracting hundreds or thousands of new residents, including merchants selling goods and services of every description to feed, clothe, house, and entertain the miner while he was in town and to keep him fully supplied when he was in the diggings. Played-out mining claims, conversely, often resulted in the abandonment of whole towns as the miners and the businesses that depended on them moved on to more attractive prospects. Winter or spring floods might inundate an entire town, drowning every home and business in silt and mud, or a sudden blaze might reduce its flimsy wooden shacks to glowing cinders overnight.

Confronted with such obstacles, entrepreneurs in mining camps and towns from Sacramento down to the littlest hamlet made every effort to ensure that they provided everything the miners might require in order to attract their patronage. Consequently, restaurants, dry-goods stores, and hardware shops shared the scene in many of these towns with saloons, gambling halls, and brothels, all intending to meet the miner's every need. Having so many attractions handy, many of these towns became magnets for miners from every surrounding claim, throwing into close proximity a remarkably cosmopolitan array of men and women. A half-dozen nationalities might be represented on any street and might be briskly engaged in buying and selling or drinking and gambling. The cosmopolitan character of the crowd, however, never guaranteed tolerance or mutual good will among all parties.

For thousands of Argonauts, mining camps and towns such as Sacramento, Sonora, Placerville or Volcano had become a near-permanent abode. Very few, however, completely abandoned those hopes of eventually departing from El Dorado with a hefty share of its wealth in their pockets, for the parting from loved ones had involved too many hardships and tribulations. Most Argonauts remained deeply bound to friends and family left behind, sensing keenly the barriers erected by time and distance. Tangible expressions of this commitment from the more successful Argonauts usually consisted of shipments of gold dust or bills of exchange drawn on one of the different express companies that had sprung up for just such purposes. From others, frequent letters filled with encouragement, consolation, and admonition sought to preserve the strongest possible links between separated family members.

While struggling to attend to their affairs in California, many Argonauts thus struggled to retain some part in the conduct of family affairs back home. Often, however, like Jonathan Heywood writing from Jackson Creek on March 8, 1851, they could only sympathize with the difficulties spawned by their departures. Heywood, commiserating with his wife Jane, wrote "You must have had a hard task to manage the farm and take care of the children and built so much as you have."[12] As months and years slipped by, with news of more and more changes communicated in every letter from home, some even grew concerned that they might have no place left for them, as Heywood suggested when he wrote that "I am almost afraid you are making more improvements and more money for the farm than I could if I were there. I don't expect," he continued wistfully, "to know the place when I return."[13]

ACCOMMODATING SOCIAL CHANGE

Even as they reflected upon the transformations taking place at home in their absence, many of the Argonauts were struggling to cope with the precipitous pace of change in their adopted homeland as well. California's explosive population growth in the first few years after gold discovery (from approximately 15,000 in the summer of 1848 to more than 100,000 by the end of 1849, reaching nearly 225,000 by 1852) set loose upon the land a restless and often combative multitude, impatient of any restraint upon their efforts to grasp the golden reward that had drawn them to California. Thrown together in the search for El Dorado from many different nations, they often found the peaceful reconciliation of conflicting cultures and beliefs difficult to achieve under the best of circumstances.

Exacerbated by the ferocious pressure, many felt to succeed in their gold digging at all costs, confrontations between gold seekers from the USA and those from Latin America or Europe, between the Argonauts and the Indian peoples who occupied the gold country, and between newly arrived emigrants and Californios who claimed large ranchos across northern California frequently escalated into violent collisions. On other occasions, around the mining claims, in the gold camps, and in the flourishing cities, repeated instances of robbery, assault, and murder incited such widespread feelings of fear and anger that groups of outraged citizens sought their own remedies outside the scope of traditional courts and law enforcement. The resulting episodes of lynch mobs enacting vigilante justice convulsed entire communities in the name of forestalling further occurrences of crime and violence. For the vast majority of Argonauts, however, daily life in Gold Rush California revolved around the struggle to fulfill their dreams of success in fortune hunting and to escape the moral and financial failures they saw ensnaring many of their peers.

With the great upsurge in the numbers of miners in 1849 and 1850 as the first large waves of emigration from the eastern USA began to arrive, fears arose among many that they might be crowded out by new mobs of competitors. For many American Argonauts, this anxiety was compounded by a sense of their republic's proprietary rights to newly conquered California, acquired in their victorious conflict with their Mexican neighbor. Resentment of foreign intruders in the gold fields also drew strength from the rampant hostility toward peoples of different races and religions rooted in many quarters of American society at a time of rising immigration from various European nations to the USA. Various confrontations during 1849 and 1850 with Anglo-American gold seekers in mining districts or camps drove some Chileans, Peruvians, Mexicans, and Frenchmen into the southern reaches of the gold country, around the community of Sonora, and hastened the departure from California of many more.

If the influx of French- and Spanish-speaking gold seekers proved unsettling to many Anglo-American emigrants, raising in their minds the specter of a California occupied by racially inferior and religiously suspect peoples, the arrival of more and more Chinese laborers during the early 1850s stirred even greater concern. Although probably totaling

Figure 9.3 *Chinese Camp in the Mine*, lithograph in J. D. Borthwick, *Three Years in California*, Edinburgh, 1857. Although few Chinese arrived in California during 1849 or 1850, they were numbered in the thousands by 1852, as the enthusiasm for "Gold Mountain" took root. Some of the Caucasian Argonauts appreciated Chinese energy and determination, but most found the Chinese ways of life disturbingly alien.
Source: Huntington Library: RB 257.

less than 100 in the first eighteen months of the rush, the number of Chinese emigrants grew into the thousands by 1852 and the tens of thousands soon thereafter. Completely outside the experience or understanding of nearly all Americans in the 1850s in terms of everything from food and clothing to language and spiritual beliefs, the Chinese wayfarers to what they termed "Gold Mountain" puzzled or disturbed most Euro-Americans who came in contact with them (as suggested in Figure 9.3 by this illustration by the Scottish artist J. D. Borthwick for his book about Gold Rush California). Unlike the Californios or the Argonauts from Europe or Latin America, they had no piece of the common heritage of Western culture or Christianity; unlike African Americans, they did not share a common language, a common history, and a common belief in the possibilities of the American republic; unlike the Indians, they were not even equipped to defend themselves by force if necessary. The fact that many came as laborers indentured to the Chinese merchants who paid for their transportation further alienated many Americans, who regarded the use of forced labor of any kind in the gold fields as grossly unfair competition with hardworking free Euro-American men. To many Americans, they represented a people fundamentally incompatible with all features of American society at that time.

To others in California during the Gold Rush era, however, the presence of these sojourners appeared in a less bleak light. Among the Protestant missionary societies of the era, for example, the Chinese in America represented an opportunity to make progress in their efforts to bring the Christian gospel to China's millions. Bodies such as the Presbyterian Board of Foreign Missions assigned clergymen like the Reverend William Speer to the task. Speer himself became an eloquent and ardent advocate for what he believed were the rights of the Chinese. Other observers of the Chinese, less spiritually oriented, had noticed their exceptional capacity for hard work under adverse conditions with little or no visible protest. Speculating about how desirable such a labor force might be for many different tasks, various mining companies and agricultural enterprises began recruiting Chinese workers. Successful contractors during the 1850s continued to bring in contingents of laborers, expanding the already sizable Chinese population to the dismay of many Anglo-Californians.

Within a few years after gold discovery, California thus had become the arena in which the hopes of Argonauts from many nations were thrown into contrast with the powerful nationalism of mid-century America as expressed in the spirit of "manifest destiny." In the same way that American merchants such as Thomas O. Larkin had worried before the Mexican–American War about British and French intrigues to acquire Alta California, many Americans in California after 1848 seemed to worry about the presence of foreigners who might interfere with the absorption of the Golden State into the Union. Securing the place of California in that scheme of manifest destiny, however, would be further complicated by the presence of other peoples who had a firmer, if still contested, claim to a part of that American future. Mid-nineteenth-century America had begun anew its struggles as a nation over the role that African Americans and American Indians would play in the life of the nation.

Like so many of their peers, some free African Americans found in the news of the gold discovery in California sufficient inspiration to abandon their previous lives and set out in search of opportunities to improve their circumstances, start a fresh life, and escape the legacy of slavery. Though the numbers who came to California were not large (possibly 600 or 700 in the first eighteen months and 1,500 in the first three or four years), they provoked comment and concern. Their very presence incited much hostility, especially from Argonauts who hailed from the slave states and aroused discussion and debate in the framing of the California state constitution in 1849. The promise of California nonetheless proved as irresistible to some African Americans as it did to many others of every region, religion, or ethnicity.

In the meantime, the steady upsurge of American emigration west to California during the 1850s brought more and more pressure to bear upon the native peoples of the Sierra Nevada foothills as gold seekers roamed further and further afield in search of rich diggings. Americans had been engaged for generations in the dispossession of Indian communities from one part of the landscape to another, culminating with the designation of large chunks of land west of the Missouri river as permanent reserves for various tribes relocated from the states east of the Mississippi. Now, launched on the frenzied treasure hunt that followed gold discovery, they encountered indigenous peoples numbering in the tens of thousands who resisted efforts to push them aside so that more claim sites could be opened. Parties of prospectors blundered into bands of Indians as the former searched for gold and the latter for food.

Reversing the reliance upon Indian labor, recruited from such Sierra Nevada tribes as the Miwoks that had characterized the first months of the Gold Rush, the Argonauts of 1849 and thereafter drove them out of the gold fields. Indian resistance only provoked more vigorous responses from the miners, initiating an increasingly violent cycle of skirmishes, ambushes, and massacres. Ad-hoc militia companies, similar to the bodies of miners who gathered to discipline recalcitrant foreigners, marched off into the mountains, tracking down Indians suspected of crimes ranging from stock theft to murder. In their haste to inflict punishment, many of these part-time soldiers swept away the innocent with the guilty, burning out rancherias and shooting down their inhabitants indiscriminately. Throughout most of the 1850s, private and state-sponsored expeditions battled the Native peoples of California's interior again and again in murderous encounters.

The proprietary regard with which American Argonauts often viewed California's gold extended in the minds of many of them to California's land as well. Reconciling the rights of current landholders with the aspirations of land hungry settlers involved sorting out bitter conflicts similar to those between American miners and their foreign counterparts over access to the gold fields, frequently to no one's ultimate satisfaction. For the American settlers who had crossed the continent to California in the early 1840s before the Mexican–American War, the remarkable potential of certain California landscapes such as the Sacramento valley around John Sutter's New Helvetia had been the great attraction. Their ambitions to acquire land of their own, unencumbered by the need to become naturalized Mexican citizens, had played a part in the unrest leading up to the Bear Flag revolt in the early phases of the American conquest of California.

Following gold discovery in 1848, the hordes of gold seekers who came in 1848, 1849, or 1850 at first seemed far less concerned with ownership of land than with access to its mineral wealth. Some of the gold seekers, however, soon began to see a brighter future in other pursuits than the endless search for El Dorado, while other pioneers came to California with every intention of finding some fertile plot of land to claim for themselves. In either case, the rules and regulations governing the acquisition of land in California grew to have a pressing importance for them. Unfortunately, like so much else in the wake of the Gold Rush, the control of land became a subject for continuing controversy.

Under the terms of the Treaty of Guadalupe Hidalgo at the conclusion of the Mexican–American War, Mexican nationals remaining in the conquered territories, whether they retained their citizenship or adopted that of the USA, were promised that "property of every kind [...] shall be inviolably respected" and that they would be "maintained in the free enjoyment of their liberty and property."[14] In the ensuing effort to incorporate California into the Union, however, the encounter between Mexican and American systems of outlining and occupying property led to lengthy legal and political wrangling. Unlike the mathematically precise surveys and formal deeds common under American land law, many of the Mexican land grants in Alta California rested upon indistinct boundaries and casual registration of the final documents. Grants of varying size issued under varying circumstances heightened the confusion about who held what property and who had what rights to dispose of it.

Determined to acquire part of California's landed bounty for themselves, many emigrants refused to wait for the settlement of these questions through the legal or political process. Instead, bands of "squatters" seized parcels of property for their own and forcibly resisted the efforts of landowners and civil authorities to remove them. Dismissing their defiance of the law, squatters and their supporters appealed to a higher moral authority that justified such behavior. The continuing pro- and anti-squatter agitation spilled over into the courts, the legislature, and even the street, generating law suits and riots throughout the 1850s.

In an effort to resolve this continuing turmoil, the federal government in 1849 commissioned reports to examine the legal procedures under which the Spanish and Mexican governments had granted land to private parties and therefore lay a groundwork for determining the legitimacy of specific land grants. Unfortunately, the two principal surveys came to very different conclusions about the legality of many existing grants, leaving it up to the Congress to propose a course of action. After much debate, that body in turn adopted the plan submitted to it by William M. Gwin, one of California's first two senators following its admission to the Union in September 1850. Gwin's Bill established in 1851 a three-member commission empowered to examine all private land claims issued under the authority of the Spanish or Mexican governments and to pass upon their legitimacy, subject to appeal to the federal court system.

Though Gwin and his partisans assured all who would listen that such a commission would put an end to the rampant uncertainty currently troubling California land titles, the end results proved far more troublesome. By requiring all holders of Spanish or Mexican land grants to prove a particular grant's legitimacy, the US Land Commission's enabling legislation threw an aura of suspicion around such grants and their claimants from the beginning. Despite the stipulations of the Treaty of Guadalupe Hidalgo about the inviolability of the property of those Mexican citizens now under the authority of the USA, the Commission's activities opened many titles to challenge. Moreover, the conduct of its proceedings under the rules of American jurisprudence forced many of the Californios to defend their claims in a legal system and a language completely foreign to them. Although the vast majority of the claims eventually were confirmed either through the commission's hearings or subsequent court cases, many landowners were financially ruined by the costs of lawyers, translators, and repeated litigation. Even after the Commission's disbanding, frequent court challenges to its decisions led to interminable litigation, prompting one group of Californio landholders to speak out in a February 1859 petition forwarded to the US Congress. Invoking the promises made to them during the Mexican–American War by the occupying forces and reiterated in the Treaty of Guadalupe Hidalgo that their property "*shall be inviolably protected and insured,*" they argued that, on the contrary, their lands and their way of life had never been less secure.[15] The procedures of the land commission, the hordes of law-defying land seekers brought by the Gold Rush, and the crushing burden of taxation imposed by the new state all acted to threaten many with financial ruin or dispossession. For the Californios, the tumultuous Gold Rush era had brought disappointment and, in many cases, disaster, transforming dreams into nightmares.

As the decade following gold discovery unfolded, many people in California thus demonstrated their willingness to resort to violence, or at least the threat of violence, in order to advance a particular purpose. With the state hampered by weak and impoverished government that spent its first few years in search of a permanent seat of residence and local administrations little better off, resolute individuals often did not hesitate to pursue their own ends through other means than the formal channels of law and order. Many Argonauts, on the other hand, having no interest in setting down roots in California, ignored government and civic life as much as possible, devoting themselves only to fulfilling their dreams of finding El Dorado. In such an environment, the unscrupulous and the wicked, attracted to California by their own particular vision of a golden reward, found significant scope for their activities. Throughout the 1850s, the letters and diaries of many Argonauts offered heartfelt indictments of the deteriorating state of order and morality, repeating stories of lawlessness occurring on all sides, both in the mining camps and in the cities.

With courts, prisons, and peace officers frequently lacking in the mining country, the perception of rampant crime and violence drove many Argonauts toward vigilante justice as the only remedy left to them. In some instances adopting the forms of civil authority by organizing vigilance committees and constituting their own courts, vigilantes were often a specific group of men coalescing in response to a particular outrage in a particular community. Dispensing with the structures of judge and jury, they might settle guilt or innocence and then impose punishment by popular consensus, brushing aside any one, whether public official or private citizen, who dared to oppose them.

In the Gold Rush cities generally and in San Francisco particularly, vigilantes also would be active as agents for suppressing crime and violence. First in 1851 and then again in 1856, large and well-organized vigilance committees would actually supplant the municipal government of San Francisco for weeks at a time, resolved to usurp the police and the courts in apprehending, convicting, and punishing criminals. Hundreds of men, especially from the merchant classes of the city's population, enrolled each time in paramilitary bodies. Committee detachments drilled under arms and, in 1856, gathered at a fortified headquarters known variously as "Fort Vigilant" or "Fort Gunnybags" (the latter title because of the number of sandbags surrounding the building). Also in 1856, with many merchants and other members of the middle classes concerned about political and social corruption, the murder of newspaper editor James King of William by local politician James Casey not only brought the second vigilance committee into being but pushed it into pursuing electoral fraud and deceit. Although its activities eventually concluded peacefully and the city returned after much excitement to the rule of the elected government, the 1856 vigilance committee's role in civic affairs represented a climax of vigilante activity not equaled in the Gold Rush period.

Shocking as such events could be, many of the emigrants throughout the Gold Rush decade focused first and foremost on achieving the financial success that they had promised their families would justify their sacrifices. For nearly all the Argonauts, painfully aware of the financial, emotional, and psychological costs inflicted upon their loved ones by these expeditions to El Dorado, the burden of realizing their dreams under the looming shadow of failure wore them down with its weight.

Among the women who had joined the rush to El Dorado, most became involved in the same struggle between success and failure in pursuit of a golden reward as had their male counterparts. While some braved it only vicariously, through the careers of their husbands, many of them were pulled into the fray directly. Lucy Stoddard Wakefield, writing to "Lucius and Rebecca" from Placerville on September 18, 1851, proclaimed that "I love California and probably now shall make my home and final resting place here," but continued, "I have been toiling hard for the last two and a half years and am still doing an almost incredible amount of work." That would be the lot of any of her sex who came, she added, for "there is no way for a woman to make money except by hard work of some sort," requiring "years of toil, hardship, and in some respects severe privation." Echoing many other women and men after lengthy exposure to the hard labor of making money, she concluded that she expected to stop working in a few months since "I am tired of work and though I have not a very big pile, yet I am not ambitious of wealth. A competency is all I look for, for myself."[16] Strictly limited by the moral standards of the Victorian era as to the occupations that might be considered "proper" for their sex, women like Wakefield found their options few. In Gold Rush era California, with its emphasis upon manual labor in the gold fields and commerce in the cities and towns, respectable women were relegated to a few roles that mirrored their culturally prescribed responsibilities within the family as homemaker.

The unremitting toil and drudgery so many men and women found in El Dorado and wrote of so frequently in their diaries and journals, whether as prospector or miner, cook or charwoman, convinced some that California was nothing more than a snare and a delusion. For many of the Argonauts, such complaints would have struck a sympathetic chord, for they found themselves struggling against the same challenges, sometimes with the same lack of success. The majority, putting on the bravest front possible, wrote home of their undiminished resolution to pursue their golden dreams, even as the months and years flew by. Others, struggling with various frustrations, made their peace with the circumstances of separation and disappointment. Writing to his sister and brother-in-law in Farmington Falls, Maine, on December 17, 1857, William Farnsworth outlined his travels in the previous few months before coming to rest on a claim 200 yards from the mining town of Columbia, "where I have remained ever since as busy as a bee trying to catch some of the shining specks." His health, he reported, had remained excellent over the past three years but, he pensively admitted, "it would be more agreable [sic] for me to live whare [sic] I would see my friends a little oftener without being to so much expense of traveling." Nonetheless, he asserted, if he was going to be a Westerner, California was the place for him: "you will not catch me to go to Kansas or any of those new countries to beat the bush and let somebody [else] catch the bird." He would move out of the mountains, he thought, to one of the coastal valleys: "It is some of the finest climates in the world and cant be beat by your Nebraska nor no other wild land."[17] In so saying, Farnsworth reflected yet another transformation he and like-minded Argonauts had worked upon California during its golden age: they had begun to turn it into home.

DEFINING A NEW CALIFORNIA: THE CONSTITUTION OF 1849 AND POLITICAL LIFE

Other than Marshall's discovery of gold itself, perhaps the most important event in the transformation of California during the Gold Rush era occurred on February 2, 1848, when the Mexican and American commissioners initialed the Treaty of Guadalupe Hidalgo, bringing the Mexican–American War to a conclusion. At that point, as hostilities between the two North American republics ended, California's metamorphosis from Mexican province to American state began. The process unfolded awkwardly at first, since the federal government had made no provision for the administration of its new possession after the end of hostilities. Lacking any detailed guidance, the successive US military commanders in California improvised. They sustained such Mexican institutions as the alcalde (a figure whose closest American counterpart might be the mayor of a town or city) to govern on the local level while retaining for the senior American army officer the effective position of military governor.

The perpetuation of military government in California, even when amalgamated with some elements of civil authority, soon excited suspicion and hostility in some quarters. Although many of the gold seekers had no immediate interest in the finer intricacies of California's governance, some of those Americans who had come before the Gold Rush grew restless quickly under what they regarded as the "arbitrary" rule of the army. These "old settlers" (as they sometimes referred to themselves) had wished for American rule to be extended over California as early as the mid-1840s. Now, with the fulfillment of that goal near at hand, they found their fate absorbed in the great national convulsion over the question of allowing slavery in the nation's new Western territories.

Unwilling to wait upon the resolution of such questions, Californians in various places, including the city of San Francisco, launched local governments of their own design in opposition to military rule. More and more Americans, disenchanted with military control, supported such movements. A perception that anarchy was rampant in the land increasingly took hold. Soon after his arrival in April 1849, General Bennett Riley, the new military governor, decided that, before the situation escaped his control completely, he would issue a call for a constitutional convention in return for the dissolution of the various ad-hoc local governments. Although Riley's authority to take such action seemed uncertain at best, his proclamation of June 3, 1849 met with a generally favorable reaction and led, in turn, to elections of delegates from various districts around California.

The forty-eight men chosen to devise a written charter for the state-to-be arrived in Monterey to begin their work on September 3, 1849. Californios, Anglo-American "old settlers," veterans from the American army of occupation who had settled in California after their release from the service, and Argonauts from various parts of the USA, they possessed a formidable array of backgrounds, occupations, and interests. Included among them were men who had stood on opposite sides of the Mexican–American War, men whose ties to California reached back decades, and men who had only just arrived from the USA. Drawn heavily from the ranks of merchants, lawyers, farmers, and ranchers, the membership of the convention represented segments of the population who all desired the establishment of stable, orderly government, although the degree and scope of its power could be debated.

In trying to give the new government shape and form, the delegates relied heavily upon preexisting models such as the constitutions of Iowa and New York state as well as consulting those of several other states. The constitution's provisions, therefore, represented the sort of institutions familiar to most of the participants and to thousands of the Argonauts who had come to El Dorado from those states. The document touched upon various issues of surpassing significance to many Americans in the middle of the nineteenth century, such as the suspicion with which the general public regarded the power exercised by banks and corporations.

No matter how routine the process or familiar the end results might seem to the Anglo-American delegates, for the Californios the convention's efforts constituted a fundamental redrawing of their political landscape according to the vision of others. Nonetheless, they did manage to secure certain measures to protect matters of importance to their constituents, such as ensuring that the new government's public documents would have to be published in Spanish as well as English. Though not a solidly unified voting bloc, they and their Anglo-American allies such as Abel Stearns and Stephen Foster represented a quarter of the convention's strength.

Reshaping the political world of California, however, did pull Anglo-Americans and Californios into fractious debate on more than one occasion during the convention. When the convention began to lay out the standards for who could exercise the right to vote, for example, it followed the standard American practice of the time and limited suffrage to Euro-American male citizens over the age of twenty-one. Some members of the convention pointed out, however, that under the stipulations of the Treaty of Guadalupe Hidalgo citizens of the Mexican Republic who accepted American citizenship had to be accorded the full rights and privileges of any other citizen. Moreover, while the notion of granting the right to vote to African Americans and Indians received no strong support, other delegates noted that to limit the franchise only to *Euro-American* males might well act to bar some of the Californios of mixed ancestry and dark complexion. As a compromise, the final version of the suffrage article included a provision that the legislature, at a later date, could choose to grant Indians the right to vote, though such a measure would need the support of two-thirds of the legislature. Furthermore, it specifically recognized the right of citizens of Mexico to vote after they had become citizens of the USA as long as they were Euro-American.

The same sense of racial prejudice concerning people of color drove forward a proposal that the first legislature be instructed to prohibit free African Americans from moving to the new state. Although the convention had accepted a provision that outlawed slavery in California, such an action hardly represented an acceptance of racial equality or, on the part of some delegates, a belief in racial tolerance. Even those who believed that slavery was an unmitigated evil often accepted the notion of African-American inferiority and felt that putting the races in proximity would drag down the superior Euro-American race. In the end, only a fear of the impact that such a measure might have upon Congress and the opposition that it might arouse there to California's admission as a state forestalled such a restriction. In both instances, the convention's debates became embroiled in the question of defining who might participate and how in certain aspects of California's public life and, by extension, who might even be defined as a Californian and who might be excluded. Thus, as the constitutional convention struggled to lay out the blueprint of a successful government, it also influenced whose dreams might be part of California's future and whose might be excluded.

Completion of the convention's work on October 12, 1849, only five weeks after it had begun, provided that blueprint. Building a government from that plan began almost immediately, with military governor Riley calling an election for November 13 of that year, to ratify the constitution and to elect the first civilian officials to such posts as governor and the two houses of the legislature. With its approval by a margin of more than twelve to one at that election, the Constitution came into force, and the first legislature to meet under its authority began on December 15, 1849, to grapple with all the mundane details of organizing local governments, courts, and executive departments to conduct the people's business. The legislature also took up the cause of statehood for California, with the selection of William M. Gwin and John C. Frémont as US senators-designate to present a petition to the US Congress urging California's admission. Once again, however, the hopes of statehood's advocates became enmeshed in the ongoing national debate over the extension of slavery. Another eight months would pass after the arrival of California's free-state constitution before the Congress would finally pass an Act in September 1850 to admit the Golden State as part of the elaborate program of legislative measures known as the Compromise of 1850.

In Gold Rush California, as in most other parts of the Union, party politics held a great and continuing fascination for many Americans. Although far away from their familiar surroundings, they still carried with them their political allegiances and sentiments, just as they carried with them memories of home and family. Some, of course, hoped for a share of the spoils to be distributed by successful candidates for elective office; others shared a commitment to the principles espoused by one party or another. Various ambitious individuals saw politics in California as their quickest path to advancement on the state and even national level. Similarly, many Californians, ambitious for the development of the state, saw the political parties as the mechanisms to achieve such results as the creation of a transcontinental railroad. Moreover, as the debates in the constitutional convention over the exclusion of slavery and the proposed exclusion of free African Americans demonstrated, many Argonauts were drawn into politics because of their devotion to one side or the other in the great slavery controversy.

The continued agitation of the slavery question in national politics, embroiling most issues of public life by the end of the 1850s, was mirrored in California by other divisive controversies revolving around matters of race and ethnicity. Although these questions arose from peculiarly Californian circumstances, they exemplified many of the same issues embedded within the debate over slavery. In trying to resolve them during the decade, various Californians quarreled about who would be included and who would be excluded from full participation in the economic, political, or social life of the Golden State.

For many of California's Indian peoples, the 1850s only intensified the struggle between them and the enormous influx of gold-seeking outsiders that had begun after James Marshall's discovery. In many parts of the Sierra Nevada Mountains and especially in the far northern forests and mountains that comprised Shasta, Trinity, Klamath, and Siskiyou counties, hostile encounters between Euro-American prospectors and Yuroks, Klamaths, Karoks, and other tribal communities escalated into bloody warfare with little quarter shown. With the meager allotment of federal troops for California stretched thin across the state, militia expeditions authorized by various governors and independent bands of Indian fighters took up the cause of suppressing Indian resistance. That warfare, combined with the effects of disease and starvation as many of California's Indians were exposed to new illnesses and crowded off their traditional habitats, led to a catastrophic population collapse among California Indians that may have cut down their numbers from 150,000 in 1848 to 30,000 by 1860.

The unrelenting hostility commonly felt toward California's Indians reflected the understanding shared among most Californians that the Indians could have no place in the evolving society of the 1850s. Of course, as we have already seen, some of the Argonauts themselves also felt such hostility directed toward them from Anglo-American emigrants, though usually not to such a vitriolic degree. Peoples of color, however, rarely escaped discrimination of one form or another; Chileans, Peruvians, and Mexicans, for instance, all suffered under bitter waves of prejudice and even mob violence.

African-American Californians, among Californians of color, had been the target of a particularly virulent animosity, for their very presence in the state had been an object of controversy during the constitutional convention. Furthermore, even after the exclusion of free African Americans had been defeated, few public schools were open to them, and the legislature in 1850 passed an Act specifying that "no black or mulatto person or Indian shall be permitted to give evidence in favor of or against any white person."[18] With such de-facto and de-jure disabilities imposed upon them, African-American Californians were thus consciously assigned a status of permanent inequality despite the increasing success of various African-American entrepreneurs. Refusing to accept such a status, however, leaders of the African-American community mobilized in subsequent years, especially against the prohibition on giving testimony in cases where Euro-American people were involved. Having absorbed many of the precepts of American democracy in the struggle for the liberation of their people, they hammered away eloquently at the fundamental injustice of the law. Petitions from African-American and Euro-American Californians alike arrived in the legislature year after year, despite that body's steadfast refusal to consider the issue until during the Civil War, nearly fifteen years after the first enactment of the prohibition. Notwithstanding the repeated failures of their efforts, African-American Californians continued the struggle in pursuit of the rights they regarded as part of their natural inheritance.

Among the many peoples who had come to California in search of fortune, however, perhaps none received a more hostile reception than the Chinese, nor were any others so totally isolated from the Anglo-Americans who had come to dominate the gold fields. As we have already seen, their visibility in the mines and mining camps led many Euro-Americans to regard them as dangerous competitors, especially with the exhaustion of the easily accessible placer deposits. For a substantial number of Californians during the 1850s, neither appeals to California's economic well-being nor to the principle of the Golden Rule could overcome the essential feature of their encounter with the immigrant Chinese, that of the chasm separating one race from the other. As State Senator Wilson Flint argued in 1856 during debate over legislation to increase the Foreign Miner's Tax, far more was at stake than merely increasing the commercial prospects or material wealth of Californians: "It cannot have come to this, that the philanthropist will make no remonstrance when he sees the God endowed Euro-American man forced into degrading equality with races who can never share with him the duties and burdens of self-government, the responsibility and glory of a free citizen!"[19] To Senator Flint and those who shared his conception of California's unfolding destiny in the 1850s, the Chinese represented an intrusion of barbarism from the shores of Asia that could only retard California's march to greatness.

BUILDING THE COMMONWEALTH: GOLD DISCOVERY AND CALIFORNIA'S DEVELOPMENT

However bitterly Californians might debate who would be included in the brilliant future of the Golden State, many seemed to accept without question that greatness did lay ahead in the 1850s. To thousands of Argonauts, of course, the boom had long since collapsed and would never be resurrected, either due to the obstacles of race or nationality or due to persistent bad luck in the hunt for treasure. Other observers simply rejected the notion that any good could arise out of the endemic iniquity and sinfulness that characterized life in the Gold Rush era. Even among the ranks of those Argonauts who failed to pick up a pocketful of nuggets, however, there were many who found California's possibilities intriguing.

Before the great rush reached its second anniversary, it was clear that the wealth of El Dorado would galvanize economic enterprises of all kinds. As gold production vaulted upward from $250,000 in 1848 to more than $10 million in 1849 to beyond $40 million in 1850 and the number of miners increased tenfold in the same period, many astute men and women saw endless opportunities for profit from the combination of vast amounts of ready money and enormous demand for nearly every commodity and service imaginable. Many entrepreneurs gravitated to the mining camps and towns to meet the demands of the miners themselves for shirts and blankets, flour and beans, shovels, and long toms, not to mention the perennial threesome of wine, women, and song. Frank Marryat, commenting upon the mining business in his account, *Mountains and Molehills,* noted that "it is an extraordinary fact that, let the diggings fall off as they will, the miners will still require *bread* and *breeches,* and will find the money to pay for them."[20] Outside of the cities, the possibilities of exploiting California's fertile landscapes to feed the state's booming population attracted the attention of newcomers and longtime residents alike. With precisely such a calculation in mind, many of the Californios in the 1850s found that the cattle they had once raised for hides and tallow now held far greater value as groceries on the hoof to nourish hungry miners. As early as the summer of 1849, rancheros such as Hugo Reid were contemplating the market for their stock in the northern counties. Within another year, Abel Stearns, Cave Couts, and many others were running cattle drives north to the gold country, fifteen years before the more celebrated trail drives from Texas to the Kansas rail towns. The cattle trade became the major industry of the southern half of the state, providing a welcome financial infusion for the rancheros in their battles before the federal government's land commission. Unfortunately, like so many other enterprises driven by the speculative frenzy of this era, the cattle boom had already begun to wear out even as more investors scrambled to get a piece of the action. With huge herds of stock attracted from other Western territories, the supply soon so overwhelmed demand that prices had collapsed before the end of the decade, not to recover again.

The ruin of the cattle trade underscored the economic misfortune that could result from the constant speculation, crushingly high interest rates and an erratic boom-and-bust business cycle that characterized much of the 1850s. California's magnetic attraction to fortune hunters, however, remained too powerful for many to resist. For many, the golden dream of California was hardly a dream at all but rather a matter of applying good sense and cold hard cash to the opportunities that abounded. Moreover, with the unquenchable demand from the mining counties year after year for so many products, an immense import trade sprang up. The port of San Francisco mushroomed into a bustling commercial hub only months after the gold discovery with hundreds of ships arriving and departing each year well into the 1850s.

Unlike their laxity in supplying California with a full-fledged territorial government, members of Congress speedily extended the authority of US customs regulations over their new possession in the spring of 1849. The resulting duties on imported goods paid to the collector of customs at the US Customs House in San Francisco provided a significant source of federal income (and also provided an irritating example of taxation without representation as Congress continued to drag its heels in establishing a California territorial government during 1849 and 1850). With a burgeoning population in both the gold-mining regions and such cities as San Francisco, Sacramento, and Stockton, the federal employees at the Customs House remained very busy. In the autumn of 1854, for example, the customs ledger kept to record activity in the port noted the arrival of many vessels, large and small, hailing from Great Britain, Denmark, Holland, Germany, France, Sweden, and Chile as well as the USA. Aboard those well-traveled merchantmen were kegs of paint and casks of porter, pigs of iron and pigs of flesh and blood, bags of rice and cases of cigars, barrels of vinegar and barrels of molasses, baskets of champagne, boxes of dried oysters, coal by the ton, oranges by the tens of thousands, nails, rope, bricks, shawls, squashes, bananas, and tea. Even as California's links to international trade and commerce expanded, its importance to the development of the trans-Mississippi West grew deeper and stronger as well. With the exhaustion of the placer deposits during the early 1850s and the subsequent focus upon more expensive and complex mining techniques, many experienced miners were left at loose ends. As bonanzas or even rumors of bonanzas circulated, these "old Californians," as some described them, became an important reservoir of talent and expertise. Many first rushed north in 1858 to the Fraser river in modern-day British Columbia, a boom that quickly turned mostly to bust. The discovery a year later of incredibly rich deposits of silver ore to the east in Nevada territory in what became known as the "Comstock Lode" ignited the first great post-California boom. To help extract that wealth from the earth, California's resources of capital, industrial plants, and entrepreneurial energy proved absolutely indispensable.

Men and money from the Golden State thus helped launch the great era of mining bonanzas over the quarter-century after 1860 that proved so critical to the development and settlement of the mountain West. Other California entrepreneurs threw their initiative behind other endeavors. Perhaps the grandest of them all was to be the Central Pacific Railroad, organized by Sacramento merchant Collis P. Huntington, his business partner Mark Hopkins, and their associates Charles Crocker and Leland Stanford. Inaugurated in 1862 as the western half of the first transcontinental railroad, the Central Pacific capitalized upon the long-standing agitation in California for a railroad stretching from the eastern USA to the Pacific.

Even years before the Comstock Lode or the Central Pacific occupied the energies of many Californians, however, the dream of an imperial California seemed nearly at hand. Contemplating the astounding pace of California's evolution just in the few years since 1848, it became a commonplace for observers to speak in tones of reverence and wonder. William Van Voorhies, delivering the Admission Day anniversary oration to the Society of California Pioneers in September 1853 contrasted the California of his day with that of just a few years ago: "But now, how changed! The busy hum of active industry, like the tones of 'applauding thunder,' rings from the mountain top to the bosom of the valleys; cities and villages, as if by magic, spring up; mountain fastnesses are penetrated and subdued," while the world itself had become the backyard of the Golden State, where "harbors whiten with the sails of a world-wide commerce; the remote West grasps the hand of the remote East while Hail Columbia is sung under the walls of Nankin, and Yankee Doodle whistled at the gates of Japan."[21]

The great jolt delivered by gold discovery and the subsequent worldwide frenzy thus had energized California's domestic and international markets, reshaping the entire nature of California's economy in the process. Such a shock also struck many of California's cities and towns with a similarly galvanizing effect very shortly after the rush to El Dorado began. As we have seen, small and not-so-small towns sprouted throughout the mining districts, adding a distinctively urban aspect to the essentially rural character of the mining experience. Cities of considerable size and scope such as Sacramento and Stockton also shot up, seemingly overnight, while San Francisco erupted into a major metropolis. Far to the south of the gold regions, even the Pueblo of Los Angeles and its surrounding districts felt the impact on several different fronts.

Pre-war capital of Alta California and center of a vast cattle-raising region, Los Angeles was still a village of fewer than 2,000 people by the beginning of the Gold Rush era. The boom in the cattle trade that followed would enrich many of the rancheros in the south (although its subsequent collapse would cripple many of them just as quickly) and would make stock raising the centerpiece of the Southland economy for a few years. Meanwhile, large numbers of overlanders from the eastern USA who set out for El Dorado along one of the southern routes passed through or near Los Angeles, exposing them to the delights of its climate and its agricultural possibilities.

If even so distant a locale as Los Angeles could feel some of the ripples of the Gold Rush's impact, then certainly San Francisco lay at the very center of what travel writer Bayard Taylor referred to as the "dizzy vortex."[22] In many ways, San Francisco exemplified the process of nearly instant urbanization, especially in the startling pace and stunning consequences of such growth. One forty-niner, Orval Comstock, writing home to his parents on June 5, 1849, described the boom town in progress:

> On ascending a hill on the road from the landing we caught a view of the town beautifully situated in the valley below. Many of the houses are about as substantial as

> the usual board & cloth tents at a *Camp Meeting*. Stores, Groceries, Eatinghouses, & gambling shops are generally nothing more than temporary tents. [...] I said that San Francisco was a miniature world & it may be called so for every nation that has any knowledge of the *Gold Mines* is represented here.[23]

Already by then the gateway to El Dorado for the vast majority of the gold seekers and the goods that sustained them, San Francisco was in the midst of an astounding expansion that saw its population increase by thirtyfold or more in only two years, from 1,000 or fewer in 1848 to approximately 30,000 in 1850. The ever-observant Mary Jane Megquier, writing home from San Francisco to friends and family in the spring of 1853, after she and her husband had gone home to Maine and then returned to California, could write on April 8 with great enthusiasm that

> I have seen so much of things a little more exciting I fear I shall never feel perfectly satisfied with their quiet ways again [at home in Maine]. Here you can step out of your house and see the whole world spread out before you in every shape and form. Your ears are filled with the most delightful music, your eyes are dazzeled [sic] with every thing that is beautiful.[24]

Mary Jane Megquier's rapturous description of San Francisco's irresistible allure aptly reflected the seductive appeal to so many Californians of a belief in the Golden State's great possibilities. Although the vast majority of those who attempted to earn their golden reward from the mines had failed to do more than pay their expenses, many of the Argonauts still found aspects of life in California too tempting to abandon easily. Some, like Maine nurseryman Jesse Smart, who saw golden visions of California's future, remained. Absolutely overcome by California's possibilities, Smart could write to his son from Jacksonville in Tuolumne County on January 19, 1853, with unimpaired enthusiasm: "I advise you to rent the farm and make your arrangements to come to California by Oct. next if practicable. [...] This is a country where an industrious man can earn money." "The miners," he noted wryly, "will not farm and they must eat."[25] As in Smart's case, such visions often grew out of endeavors far removed from the life of the gold miner.

Intoxicated by the wonders of its climate, the beauties of its landscapes or the richness of its natural resources, various Argonauts found new employment for their talents and their energies. John S. Hittell, a careful student of the Golden State's progress, proclaimed in an 1869 address to the Society of California Pioneers on the anniversary of Admission Day, that California

> gave to all a cherished home, a sunny and genial sky, a fertile soil, a delightful landscape, a clime suited to the development of every energy, the companionship of the most intelligent and enterprising people, and a site suited for the great city and for the concentration of the commerce of a wealthy coast. [...] the favorite place of residence for many thousands from abroad. They will fill the land with wealth, luxury, and art.[26]

Writing with equal fervor in the preface to the 1874 edition of his detailed analysis, *The Resources of California,* he admitted freely that "I am so much attached to California, that I could not live contentedly elsewhere; and I imagine that neither the earth, the sky, nor the people of any other country, equal that of this State."[27]

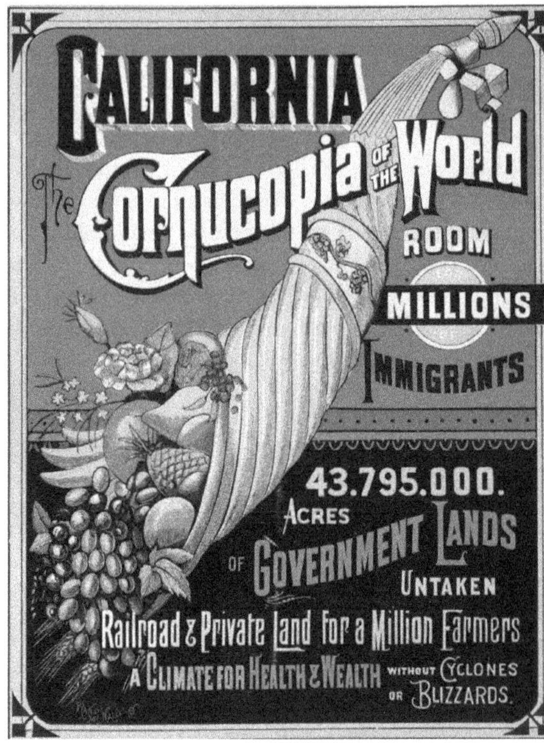

Figure 9.4 California Immigration Commission, *California, the Cornucopia of the World*, Chicago, 1883. Despite gold mining's continued importance to California's economy through the 1860s and 1870s, California's agricultural possibilities took on greater significance. During the 1870s and 1880s, broadsides and brochures describing the marvelous opportunities available for farmers proclaimed that a new golden age was at hand for the Golden State. Source: Huntington Library: RB 248676.

By the 1880s, forty years after gold discovery, California's advantages were the stuff of legend, promoted around the world to attract tourists, investors, and especially immigrants. The railroads that served the Western USA collaborated with local authorities and the state government to produce pamphlets and brochures to champion the proposition that California could be "Cornucopia of the World," a veritable paradise whose staggering natural abundance was matched only by the equally staggering amounts of land available for the taking (as illustrated by the accompanying reproduction of its cover illustration in Figure 9.4). California possessed, in the words of one 1883 pamphlet, "Room for Millions of Immigrants," including more than 43 million acres of unoccupied government land, "railroad & private land for a million farmers," and, not least important for many of the Eastern and Midwestern farmers who would read such brochures, "a climate for health & wealth, without Cyclones or Blizzards." Here, the boosters of California argued, was opportunity on an unparalleled scale "for immigrants desiring to secure lands at cheap rates, on which to make happy and prosperous homes."[28] For the immigrants of the 1880s, like the Argonauts of 1849, here was El Dorado in all its glory, where they might reach for dreams once thought permanently out of their grasp.

Even among those individuals who honored and celebrated the transformation of the Golden State since the "days of '49," however, some had begun with more than a tinge of regret to reckon the costs of what most Californians regarded as unalloyed progress. Josiah Royce, California-born Professor of Philosophy at Harvard University, offered perhaps the most striking assessment of the age as a whole in his celebrated 1886 study, *California, from the Conquest in 1846 to the Second Vigilance Committee in San Francisco.* Intrigued by what the events in California meant for his country as a whole, Royce had chosen to describe his work with the subtitle "A Study in American Character." In pondering the period at length, Royce argued that the wholesale abandonment by nearly all who rushed to California of their responsibilities to help maintain civil society aggravated the tensions that had begun to fracture what little social order existed in these frenzied years and thus empowered the worst of them to seize the moment. Americans in California, he continued, "exhibited a novel degree of carelessness and overhastiness, an extravagant trust in luck, a previously unknown blindness to our social duties, and an indifference to the rights of foreigners, whereof we cannot be proud."[29] At the same time, however, Royce found that, in time, most Americans redeemed many of these failings through the persistent reappearance of

> our best national traits. [...] As a body, our pioneer community in California was persistently cheerful, energetic, courageous, and teachable. In a few years, [...] it was ready to begin with fresh devotion the work whose true importance it hand now at length learned—the work of building a well-organized, permanent, and progressive state on the Pacific coast.[30]

As we reflect upon the extraordinary events of the Gold Rush years and upon the ordinary people who lived through them, Royce's measured appraisal of the era still has much to commend it to our attention more that a century and a half later.

NOTES

1 Richard Henry Dana, *Two Years Before the Mast* (New York: 1840), p. 112.

2 This quotation is taken from an original letter by John A. Sutter to William Leidesdorff (LE 129, May 11, 1846) housed at the Huntington Library.

3 This quotation is taken from an article entitled "Gold Mine Found" in the March 15, 1848, issue of the *Californian*, San Francisco.

4 This quotation is taken from a photostatic copy of the June 22, 1848 issue of the *Sandwich Islands News* (Honolulu), p. 2, located in the Ralph Bieber Collection at the Huntington Library (hereafter, Bieber Collection).

5 The quotation is from the copy of Colonel Mason's report reprinted in Owen C. Coy, *Gold Days* (San Francisco/Los Angeles, 1929), pp. 329–330.

6 This quotation is taken from an original letter by William P. Reynolds to John Reynolds (HM 4157, December 27, 1848) housed at the Huntington Library.

7 The first quotation is from an article entitled "The Gold Fever" in the December 15, 1848 issue of the *Pennsylvania Inquirer and National Gazette* (Philadelphia), p. 1 while the second is from an article entitled "The Gold Excitement" in the January 27, 1849 issue of *Charleston Courier* (Charleston, SC), p. 2 (both photostatic copies in the Bieber Collection).

8 This quotation is taken from the aforementioned article in the December 17, 1848, issue of *Sunday Times and Noah's Weekly Messenger* (New York City), p. 2 (photostatic copy in the Bieber Collection).

9 The quotations are from Elisha L. Cleaveland, *Hasting to be Rich. A Sermon, occasioned by the present excitement respecting the Gold of California, preached in the cities of New Haven and Bridgeport, Jan. and Feb. 1849* (New Haven, CT: 1849), pp. 4, 7.

10 This quotation is taken from an original letter from Henry Richardson to Abel Stearns (SG Box 51, July 9, 1848) housed at the Huntington Library.

11 This quotation is taken from an original letter by Hugo Reid to Abel Stearns (SG Box 53, July 14, 1849) housed at the Huntington Library.

12 This quotation is taken from an original letter written by Jonathan Heywood to his wife Jane (HM 4191, March 8, 1851) housed at the Huntington Library.

13 This quotation is taken from the letter mentioned in note #12.

14 These quotations are taken from Articles VIII and IX, respectively, of the Treaty of Guadalupe Hidalgo as printed in Appendix 2 of Richard Griswold del Castillo, *The Treaty of Guadalupe Hidalgo: A Legacy of Conflict* (Norman, OK:, 1990), pp. 189–190.

15 This quotation is taken from an original document entitled "Petition of Citizens of California to the U. S. Congress" (HM 514, February 21, 1859) housed at the Huntington Library.

16 All these quotations have been taken from the same original letter by Lucy Stoddard Wakefield to "Lucius and Rebecca" (HM 16386, September 18–25, 1851) housed at the Huntington Library.

17 All of these quotations are taken from the same original letter by William Farnsworth to Increase Blake and Sara Farnsworth Blake (HM 43127, December 17, 1857) housed at the Huntington Library.

18 This quotation from *Statutes of California 1850* is reproduced in Rudolph Lapp, *Blacks in Gold Rush California* (New Haven, CT: 1977), p. 192.

19 The quotation is taken from Wilson Flint, *Speech ... in the Senate of California, March 21, 1856, on the Bill to Reduce the Chinese Mining License Tax* (Sacramento: 1856).

20 The quotation is taken from Frank Marryat, *Mountains and Molehills, or, Recollections of a Burnt Journal* (London: 1855), p. 376.

21 The quotation is taken from William Van Voorhies, *Oration Before the Society of California Pioneers...* (San Francisco: 1853), p. 5.

22 The quotation is from Bayard Taylor, *Eldorado; or, Adventures in the path of empire: comprising a voyage to California, via Panama, life in San Francisco and Monterey, pictures of the gold region, and experiences of Mexican travel / with illus. by the author* (1850; repr., New York: 1949), p. 88.

23 The quotation is taken from an original letter by Orval Comstock to his parents (HM 24503, June 5, 1849) housed at the Huntington Library.

24 The quotation is taken from an original letter by Mary Jane Megquier to "My Dear Children" (MQ 39, April 8, 1853) housed at the Huntington Library.

25 These quotations are taken from an original letter by Jesse Smart to his son (HM 63845, January 19, 1853) housed at the Huntington Library.

26 The quotation is from Society of California Pioneers, *Nineteenth anniversary of the corporate Society of California pioneers. Oration: by Hon. John S. Hittell* (San Francisco: 1869), pp. 16, 20–21.

27 The quotation is taken from John S. Hittell, *The resources of California: comprising the society, climate, salubrity, scenery, commerce and industry of the state* 6th ed. (San Francisco: 1874), p. v.

28 These quotations are taken from the cover illustration of California Immigration Commission, *California, the Cornucopia of the World* (Chicago: 1883).

29 The quotation is from Josiah Royce, *California, from the Conquest in 1846 to the Second Vigilance Committee in San Francisco: A Study in American Character* (Boston: 1886), p. 363.

30 The quotation is from Royce, *California*, p. 2.

CHAPTER TEN

THE URBANIZATION OF THE AMERICAN WEST

The Processes and People in the Rise of Instant Cities and Their Evolution into the Twenty-First Century

Brent M. Rogers

The history of the American West is largely an urban story. In his classic work *The Pacific Slope,* historian Earl Pomeroy demonstrates that Western American society was, from its beginnings, remarkably urbanized.[1] Eminent historian William Cronon has declared that the central theme of Western US history is an "expanding metropolitan economy creating ever more elaborate and intimate linkages between city and country."[2] For the West, the metropoles remain the most important centers of economic organization, political activism, and cultural diffusion. With vast stretches of barren and rural land comprising much of the region, urban settings connect distant places, peoples, and ideas. Inter-urban connections, urban and rural connections, and connections to a national and global framework occur via city-centered networks in the West.[3] People, technology, communities, politics, economics, and cultures merge in the city to create extraordinary spaces and places where the currents of Western history intersect. Fueled by economic pursuits, demographic changes, political ideologies, cultural interactions, and pressures on the environment, the process of urbanization has transformed the West in profound material, political, and cultural ways. Since the arrival of Anglo-Americans in the nineteenth century, natural mineral resource extraction, technological change, and spending by the federal government put into play the processes that shaped the West. Anglo-American capitalism introduced swift spurts of economic and spatial growth that brought diverse peoples, structures, and interactions that generated great social and cultural change. The social landscape in urban areas was the result of human choices and politics. In the West, which now comprises the most urbanized region in the USA, the metropolitan milieu provided women and minority groups the ability to voice dissent against a Euro-American male power structure. Such processes wrought by urbanization make the urban West a crucial component to the American experience.

Urbanization refers to a continual process of making a place urban through population growth at a given locale. It also refers to a cultural shift, which implies the dominance of urban attributes over rural ones. Looked at another way, urbanization is the process of change resulting from human activity, politics, and the intersection of class, race, and gender. The urban population threshold according to the US Census Bureau has varied over time. The 1880 census indicated that an urban place consisted of 4,000 or more people in a geographically defined space denoted by a city or town. By 1910, the Census Bureau dropped that figure to 2,500. In 1950, the urban definition changed again to include three categories: first, incorporated cities, towns, boroughs, and villages with 2,500 or more inhabitants; second, unincorporated territory on the "urban fringe" of 50,000 or more inhabitants; and third, unincorporated places of 2,500 or more inhabitants.[4] Therefore, the urban West consists of the city, suburb, and nearby hinterlands. The urban West organizes, controls, and influences substantial amounts of geographic space.

Cities have played a crucial role in national development and provide valuable historical context for regional growth. In the West, there was space to grow, and physical change of the natural lands into a built setting transformed the landscape quickly into an urban environment. The ready availability of vast quantities of land also provided the impetus for the in-migration of peoples from all over the globe. Hinterlands—the regions that lay beyond the major metropolitan or cultural centers—supported the urban growth structure and provided valuable natural resources and places for agricultural production. In the mid-nineteenth century, city boosters and capitalists often took advantage of rich hinterlands, exploiting and extracting them at will. With exploitable support structures in place, the metropolitan West expanded and increased its regional influence almost immediately. Metropoles provided a vibrant and lively place, dichotomous from the notion of a rural, empty West often emphasized in nineteenth-century Western American history.

Popular notions still conceive of the West and its growth as a languid process of Anglo-American conquest over American Indian and Hispanic peoples accompanied by agrarian settlement. However, historians of the US West have strenuously challenged that notion during the past four decades. Historians emphasize the region's rapid urbanization and the critical role of its cities. Such conflicts in perceptions arise from a most interesting paradox. While the American West has remained one of the most barren regions of the USA in terms of land size and spatial usage, it has existed as the most urbanized. Nevertheless, rapid urban growth is a hallmark of Western history. Historian Bradford Luckingham noted, "Even the Rocky Mountain mining camps and the Kansas cattle towns served as agents of an urban civilization, despite the myth of the 'Wild West' popularized in the media."[5] Settlers who arrived in the trans-Mississippi West tended to congregate in areas that had geographic advantages, typically near bodies of water or locations with visibly fertile soil. The thrust for growth in specific localities came from the available space, geographic features, and ecology of the region, all of which proved crucial in the West's urbanization process. In 1899, Adna Ferrin Weber, a labor statistician and fellow in economics and social science at Columbia University, published a salient statistical essay entitled, "The Growth of Cities in the Nineteenth Century." In that report, he argued that the US West owed its penchant for urbanism partly to geography. Weber discerned that an austere, arid terrain and harsh climate—particularly in the Southwest—forced people to live in cities in closer proximity to the most advantageous lands, thereby leaving much of the West's acreage uninhabited. In essence, the geography of the region encouraged the development of urban centers for sustainability in the West.

The historian Earl Pomeroy also emphasized the dominance of cities in the region. Pomeroy proposed that isolation created by the expansive Great Plains and the towering Rocky Mountains inhibited the West from easily drawing on Eastern urban areas as earlier Midwestern frontiers had done. Instead of expanding westward from Eastern urban centers, Western mineral rushes initially lured people from other regions. In order to send goods, services, people, and capital to the isolated region, transit routes to the west coast became essential and raised the need for seaports and later railroad centers, which, according to Pomeroy, meant that towns and cities on the west coast preceded farms and other forms of agrarian settlement.[6] Hinterland agriculture buttressed urban and regional development. However, the urban centers arose as the engines of economic and demographic growth as well as political power capable of controlling and reorganizing space for hundreds of miles beyond their municipal limits.[7] During the second half of the nineteenth century, the direction of American expansion shifted from the countryside to the city, from agricultural to industrial output. People from all over the globe migrated to American cities. Whereas immigrants found small and crowded cities in the East, spacious urban areas that accommodated the necessities of economic development and population growth awaited them in the West.[8] Cities were able to supply their growing populations with foodstuffs and other essential resources from their rural tributaries.

From social and political reform movements to modern technologies, the urban West, like the rest of urban America, underwent much change through the early decades of the twentieth century. Urban people made revolutionary efforts to transform their social spaces. By the middle of the twentieth century, the threat of a world war prompted an unprecedented increase in federal involvement in the region. The World War II era stimulated extraordinary economic and demographic growth in the West as the federal government pumped capital into the cities in order to maximize their industrial potential. Since that period, Western cities have become pacesetters in terms of national and international economics, culture, and urban planning.[9] Historian Arthur R. Gómez has postulated that remote hinterland sub-regions have significantly contributed to continued "metropolitanization" of the dominant region in the post-1940 period. He states, "Endowed with abundant natural resources—minerals, petroleum, natural gas, and timber—hinterland communities became the principal source of energy commodities required for the protracted urbanization and modernization of the West."[10] These cities controlled the intervening spaces by converting miles of open rural lands into a sprawling capitalist landscape.[11] During the second half of the twentieth century, the West continued its urbanization process and by the dawn of the twenty-first century, approximately 90 percent of Westerners resided in an area defined as urban. Western cities therefore dominated in an influential network of capitalism, information, transportation, and culture. An evolution from rural beginnings to urban endings cannot characterize the American West. Rather, instant cities and the profound processes of urbanization triggered Western US growth and development and still have a determining influence on the region, the nation, and the world.

MINING: THE TRIGGER OF INSTANT CITIES IN THE NINETEENTH CENTURY

City building in the nineteenth-century West depended on a ferocious extractive economy. Triggered by gold, silver, or copper discoveries, people moved west to profit from the seemingly abundant mineral wealth. As a result, "instant cities" materialized. Instant cities, the main title historian Gunther Barth gave to his 1975 archetypal study of urbanization, indicates a place growing spontaneously from auspicious beginnings to city size within months. San Francisco, Denver, and Butte, all built on mining, fit into this category.[12] Mining was an urban industry. Where people discovered minerals, towns and cities sprang up to process the ore and to provide services for the thousands of workers who came to labor in the hinterland minefields. San Francisco, for instance, built its initial fortune on the gold from the Sierra Nevada foothills and soon pulled hinterlands into its economic orbit from Alaska, Panama, and across the Pacific into China. Prior to the 1848 California Gold Rush, San Francisco had no more than 1,000 residents. According to California historian J. S. Holliday, by August 1848, seven months after the discovery of gold the city's population lingered around 6,000 and by November had jumped to 15,000.[13] By 1852, the California State Census revealed that San Francisco's population had grown to a tremendous 34,776, due primarily to the onslaught of gold seekers and the emerging service industry.[14]

Within ten years of the initial mining onslaught, San Francisco was no longer a gold-rush city of transient single men but a booming metropolis with a diversified economy based solidly on commerce, cattle, and light and heavy manufacturing. The sudden conversion of San Francisco from tiny pueblo to international city brought frantic urbanization to the entire bay area and its inland tributaries. The result was the rapid development of an urban network to all points of value on the periphery.[15] The availability of land provided the means for capitalized companies to expand across the Far West. In San Francisco's hinterlands, large firms, such as Henry Miller and Charles Lux's cattle-ranching empire, monopolized grazing lands west of the San Joaquin River. The large cattle industry remained urban-based through an integrated meatpacking enterprise. San Francisco also provided a large market and distribution outlet for the product. Such large industries employed a diverse base of wageworkers including Chinese cooks, Mexican vaqueros, and Italian laborers and effectively linked San Francisco's processing centers to hinterland properties.[16] Portland, with approximately 18,000 people, and Los Angeles with fewer than 12,000 according to the 1880 US Census, existed as minor commercial centers that remained economically dependent on San Francisco for most of the second half of the nineteenth century. Mining provided the economic base that allowed San Francisco to maintain a strong control over the entire region's commerce and manufacturing due to its geographic position on the water and terminus of rail-transportation routes in the West until the first decades of the twentieth century.

Mineral discovery attracted a diverse population to each locale. By 1880, San Francisco boasted a population near 233,000 and had the highest proportion of foreign-born residents of any city in the nation at 45 percent. Such astounding ethnic diversity included Chinese, Mexicans, Italians, Irish, and others from around the world. The development of the urban landscape added to the difficulty in adjusting to the cities growing diversity, which generated great racial and social tension. San Francisco's compact layout and the closeness of Chinatown to the downtown Euro-American business district heightened social divisions. San Francisco's customs and the city's Euro-American elite power structure did not permit Chinese to live outside Chinatown. The Chinese found themselves segregated residentially and humiliated occupationally. Ordinances and social norms forced Chinese males into positions most Euro-Americans thought of as "women's work." These positions consisted of work in laundries, the garment and textile industries, and domestic service. Working-class Euro-American voters, led by Dennis Kearney and the Workingman's Party, considered the Chinese a primary source of their economic troubles and advocated that the Euro-American-controlled governments pass anti-Chinese measures. The legislation passed prevented Asians from becoming naturalized citizens and enforced Chinese immigration restrictions as put forth in the federal 1882 Chinese Exclusion Act.[17]

Denver did not experience the same meteoric growth as San Francisco, but a gold discovery in 1858 by a party of Cherokee and Euro-American settlers near the confluence of Cherry Creek and the South Platte River similarly sparked the area's growth. Historians Lyle W. Dorsett and Michael McCarthy conclude that the discovery of valuable natural resources made the development of a regional metropolis inevitable.[18] Shortly after the gold discovery, newcomers, such as William N. Byers—who brought a printing press with him and founded the *Rocky Mountain News*—aggressively spearheaded efforts to boost property values and the growing urban greatness of Denver. Byers and others sought to attract people to the area by noting Denver's central location as one that could connect the growing West as an important distribution and transportation point.[19] The importance of urban centers such as San Francisco and Denver in the development of a regional economy facilitated investment from the East and Europe, which helped organize access to regional markets.[20] With a significant population and economic base at the crossroads of the Great Plains and Rocky Mountains, Denver would soon become a crucial entry point to the West.

Another important mineral propelled the exponential growth of Butte, Montana, in the late nineteenth and early twentieth century. As technologies and electricity became widespread in the late nineteenth century, copper became a required material resource. Rich copper deposits in Butte created an instant city, the population grew from 3,000 in 1880 to 90,000 in 1916, complete with a monolithic enterprise, the Anaconda Mining Company, and could boast its own extraordinarily diverse workforce. The Irish formed the base of an immigrant working class in Butte and made up more than a quarter of the city's population, a higher percentage than any other city in the USA in 1900.[21] Butte also boasted a large Finnish population, and the city's east side served as residence for Lebanese, Serbs, Croatians, Slovenes, Montenegrins, Slavs, Swedes, Norwegians, Germans, and French, all living in relatively close quarters.[22] While San Francisco and Denver diversified their economic base, Butte remained strapped to the copper-mining industry, which ensured the city's dramatic rise and its eventual fall.

THE RAILROAD AND LATE NINETEENTH-CENTURY URBANIZATION

Parallel steel bands traversing the North American continent effectively tied together the Pacific and the Atlantic and, in so doing, transformed America. With the railroad in place, the West could realize its commercial, political, and cultural potential. The iron horse accelerated the pace of urban development in the West by creating a network for growth, which, after 1869, had a pervasive influence on the Western economic, political, and social order for the next fifty years. Railroads lessened isolation, brought settlers, provided access to markets, extended national culture and ideas, and simultaneously provided an urban network to coordinate the supply and demand of economic and social services. While the city of San Francisco continued to grow after the mining boom it also benefited from its position as the terminus of the western portion of the transcontinental railroad; this helped the city diversify economically and maintain its power on the coast. The significance of the railroad can be seen in Los Angeles, Dallas, and Phoenix, which were diminutive agricultural settlements until transcontinental railroad connections arrived and provided outlets to markets. Railroads encouraged urbanization through economic and demographic stimulation in the forms of land sales, enhanced mobility, and relation with the region and nation. However, the railroad also brought negative components to urban life, namely excessive prices and political corruption.

Urban boosters initiated a process for luring investment capital, industry, and people based on the railroad. This process enabled cities to become centers of population, economic activity, and political activism and to claim spheres of influence. In coastal cities, such as San Francisco, Portland, and Seattle, boosters were influential in attracting capital to filling in tidelands and wetlands, which encouraged larger-scale growth. In addition, in the burgeoning urban areas of Los Angeles, Phoenix, and Dallas, boosters promoted horizontal real-estate development across vast geographic space.[23] Upon the railroad's arrival in Western settlements, Anglo-American city boosters vigorously promoted the area through advertisements and proclaimed the availability of cheap land. These promoters realized that the all-pervading technology of the railroad would encourage urban growth.

Portland emerged as a city built around river and ocean commerce, but not until two transcontinental railroads arrived with real-estate speculators and boosters did the city reach regional supremacy.[24] The region's early urban specialty of resource processing and export, based on the wet and mild climate for lumbering and farming in its hinterlands, made it an ideal place for shipping and transporting goods and people. Portland was the natural gateway between the growing agriculture of the Columbia Basin and the vast markets of the Orient and Latin America. However, the introduction of rail transit connected it and made it a dominant economic player in the region. After a generation of dominance in the Pacific Northwest, Portland businesses struggled in the aftermath of the 1893 financial panic, which allowed Seattle to surge ahead in the region. Seattle's emergence made it the perfect staging point for the Klondike gold rush.[25]

After the initial growth brought by mining, an expanding network of railroads helped Denver structurally realize its potential as the banking, minting, supply, and processing center for Colorado and neighboring states. Railroad links emerged quickly; by the 1880s, the Atchison, Topeka & Santa Fe; the Chicago, Rock Island & Pacific; the Chicago, Burlington & Quincy; the Missouri Pacific; and the Union Pacific railroads all operated in or passed through the city. These several railroads linked central Colorado, the Southwest, and the Great Plains to the rest of the country.[26] Gradually, a host of small-scale manufacturing and service industries entered the scene: lumberyards, pipe works, breweries, commercial laundries, and small factories that produced soap, cigars, boxes, trunks, clothing, and food products. Heavier industries, such as smelting and meatpacking, developed near railroad yards for efficient movement of processed goods. Railroads and the capital investment that followed brought a mix of urban occupations, particularly in construction and with railroad companies, which afforded opportunities for work for minorities and immigrants. Denver became the distribution and transportation hub for the interior West as national and international products and capital passed through the Mile High City.

Thanks in most part to the railroad establishing a more dynamic and complex industrial-urban complex, the city experienced a population boom from 1870 to 1890 from 4,759 to 106,713, making it only second in population to San Francisco among Western cities. In 1870, the city's annual industrial output lingered around $250,000. Thirty years later, Denver's 134,000 residents were producing commodities valued at more than $50 million, an immense increase in volume and productivity, which allowed the city to establish a direct relationship with Eastern financiers and assume a commanding role in the mining ventures of the Rockefeller and Guggenheim families. Mineral wealth helped start the city, but railroads made Denver, the "Queen City of the Plains" and the heart of the Rocky Mountain empire.[27]

In the 1870s, Dallas acquired the Houston & Texas Central and the Texas & Pacific railroads. These railroad mainstays enabled the city to develop a mixed, diverse economy. The economic base consisted of wheat and corn processing, brick and tile factories, publishing, and the saddle and harness industries.[28] The railroads primed the city for growth by forging connections to larger markets. Such elements precipitated a jump in the city's population from 3,000 in 1870 to 10,358 in 1879.[29] By the 1880s, Dallas had established itself as a regional leader in the production of saddles and harnesses, becoming a primary distribution center for farm equipment with total industrial output second only to Houston. In a discussion of the impact of the railroad on the development of Dallas, historian Patricia Evridge Hill states that a "decade after the arrival of the railroads, Dallas possessed six flour mills, two grain elevators, and two cotton compresses as well as six banks, two foundries, and over seven hundred other commercial structures."[30] Property values in the city appreciated swiftly as the railroads, industry, and capital permitted boosters to identify Dallas as a booming metropolis.

Kansas City, like Denver and Dallas, existed on the border of the Great Plains. In this city, the railroad encouraged urban growth and urbanization processes. Kansas City owes its place among Western urban centers primarily to its geography, located precisely where the Missouri River begins flowing east. The city's ability to exploit this feature became essential to Kansas City's infrastructure. The construction of the Hannibal Bridge across the Missouri River in 1869—the first railroad bridge to span the river—allowed Kansas City to blossom into a meat-processing and grain-distribution center. The city became a crucial place for supply lines to the East, connecting the Great Plains and places west to the national network via the railroad.[31] Kansas City became an important exchange point, a regional gateway city built on and wedded to the railroads. Such a connection boosted the area's economy and enabled Kansas City to achieve metropolitan status early. Private enterprise dominated finance in the city and helped the Hannibal and St. Joseph Railroad construct an extension of its main line through Kansas City into the Southwest. By 1880, Kansas City was one of the most important transportation hubs in the nation west of Chicago and had established itself as the "Gateway to the Southwest" with important rail junctions that lured investors to its burgeoning industries. As regular distribution and transportation exchange points, Kansas City and Denver became routine outlets for social, technological, and industrial change in the West.[32]

Historian Bradford Luckingham likewise characterizes the coming of the railroad to Phoenix and the Southwest as crucial to the area's development. By 1880, Anglo-American promoters arrived in Phoenix and cultivated Euro-American migration to the area; this process dwarfed the Hispanic and Indian population and consolidated the balance of power into Euro-American hands as it did in many Western cities. Railroad construction proved the key to Phoenix's emergence as the principal city of the Southwest.[33] By 1895, the city enjoyed the use of two transcontinental outlets—the Southern Pacific and the Atchison, Topeka, and Santa Fe—which prompted the movement of more people and capital into the area. The railroads, as in other Western cities, linked Phoenix to the national economy and allowed it to expand as a trade and distribution center for a productive hinterland that included farming, ranching, and mining.[34] City boosters helped Phoenix become the southwestern sanctuary for health seekers and tourism through vigorous promotion of the city's climate and sun culture. New industries thus developed and provided more opportunities for growth.

The "Valley of the Sun" had a major drawback: a scarce natural water supply could not support massive growth. Federal legislation played an important role in Phoenix, as in other Western cities, in developing agricultural and water practices that could help sustain development. In 1902, the Newlands or National Reclamation Act funded irrigation and reclamation projects for the arid lands of the American West, which Phoenix used to construct Roosevelt Dam and to organize the Salt River Valley Water Users' Association on February 7, 1903, to manage the water and power supply. Water management proved crucial to

Phoenix's sustained, long-term growth, as it did in many other Western cities. During the late nineteenth century and early twentieth century, as the American West continued to develop, it became clear that the region's economic future depended increasingly upon politics and governmental policies for storing water for cities and agriculture.[35] The storage and availability of water in an arid land proved critical for residents and particularly acute for city leaders who realized that progress remained outside their grasp without water control.[36]

Los Angeles grew in much the same way as the other southwestern cities. By the late 1870s, investors had secured the route of the Southern Pacific and Santa Fe transcontinental railroads through the region, which promised an effective link for the small town of Los Angeles to the national railroad network by the 1880s. The land boom followed when the two rail lines were completed and the city began growing; its population multiplied fivefold to 50,395 that decade.[37] Land values and growth soared once the interurban electric streetcar system, named the Los Angeles and Pasadena Railway, established by Henry Edwards Huntington—nephew of Southern Pacific President Collis P. Huntington—arrived in the 1890s and connected surrounding areas to downtown through mass, rapid transit. Favorable publicity by *Los Angeles Times* editor and publisher Harrison Gray Otis promoted his own and the city's real-estate speculations and encouraged greater urbanization.[38] Tourism, health treatment, and land speculation in southern California combined as climate, scenery, and location became increasingly crucial for attracting more people and raising land value.

Similar to Phoenix, by the early 1900s, Otis, Huntington, and other city leaders observed that their southern California land of sunshine lacked sufficient water to support the swelling masses of a great city. The humble Los Angeles River, periodic droughts, and the area's underground aquifers could hardly sustain the population at the turn of the century, let alone the envisioned growth. In finding a solution to the water deficit, former Los Angeles mayor Fred Eaton, city engineer William Mulholland, and Reclamation Service engineer Joseph Lippincott masterminded the effort to make the Owens River and accompanying valley the source of Los Angeles' water supply. In 1905 and 1907, the Los Angeles Department of Water and Power's Board of Water Commissioners proposed a bond for more than $25 million to finance the construction of an aqueduct that would divert the flow of water from the Owens River and its surrounding valley over 200 miles away to the city. Voters approved the massive reclamation project to provide the city with a sustainable water supply. The Owens Valley and Los Angeles water project also served the interests of city leaders and elites who stood to become incredibly wealthy from the growth promised by the available and lasting water supply.[39] The building of the Los Angeles Aqueduct brought almost immediate controversy as the growing metropolis extended an intense impact on over 225 miles of hinterlands. Problems grew initially as the federal Bureau of Reclamation failed to assist the farmers and ranchers of Owens valley improve their irrigation system. Rather, it worked to remove the valley's water to the city. Not only did the aqueduct remove the natural water from the hinterlands and prevent growth therein, it also became a symbol for urban power. Controversy over the aqueduct's effects on the hinterland population brought protest and violence, including occupations and bombings of the aqueduct. Disagreement over the Los Angeles water project lingers in the twenty-first century. Los Angeles and Phoenix remain among the most dependent on fixed sources of water and constantly struggle for the scarce and precious resource in the arid West.[40]

Railroads prompted rapid growth and accelerated urbanization processes, which meant that city and corporate leaders soon faced growing pains in the form of water and political control. Within the issue of water control also rests an important point to make regarding spatial segregation and racism in urban planning. In Los Angeles, city-building Euro-Americans upheld a view of race and ethnicity that sought to distance the Euro-American communities from other groups, particularly from those of Mexican ancestry. City leaders attempted to remove the city's Hispanic inhabitants from areas they had long occupied within the city by relocating and segregating them in less desirable locations on the outskirts of the city. Where they could not relocate Hispanics, the historian William Deverell posits that Euro-Americans used the specter of floods along the Los Angeles River to control the Hispanic population. These potential floods required the creation of a concrete river channel through the city, which simultaneously formed a dividing line between races, classes, and neighborhoods and established rigid pockets of Mexican housing and neighborhoods effectively isolated from Euro-American society.[41] Such exercises in urban planning pronounced Euro-American cultural authority and ethnic stratification within Western cities.

The railroad also brought cohorts of people with deep political and economic interests in seeing that the iron horse remained king. While urban inhabitants appreciated the railroad technology and its ability to cut travel time, they disapproved of the railroad magnates for charging exorbitant rates for moving people and goods. Hostility toward the railroad corporations and the people who ran them increased. Westerners looked to government at many levels to combat the unfair prices and hegemonic influence of the railroads. Unfortunately, in most urban centers, the people received little assistance in reform from local and state governments that were often led by people under the influence of the railroad interests. A nationwide depression during the 1890s sparked an increased and multifaceted opposition to the railroads. By the turn of the twentieth century, urban political corruption had become rampant in the West because of the pervasive railroad. Urbanites targeted the railroad for social ills and began movements demanding political reform, summarily indicting state politicians for looking after railroad interests rather than the interest of the people. Newspaper editors and writers such as Frank Norris and his novel *The Octopus* (1901) depicted the extraordinary impact of the railroad and the fissures it created amongst the people who had aligned themselves with different forms of opposition.[42] Such voices advocated change within urban politics and communities. These early movements against the railroad set the stage for the Progressive movement and subsequent community political endeavors, which fundamentally altered the urban West's commercial, political, and social landscape. The railroad generated many positives in the form of economic and population growth, but that came with a high price. It brought more Anglo-Americans into formerly ethnically distinct areas, which increased diversity and changed the demographics and power structure of the region. These shifting dynamics introduced escalating racial, spatial, economic, and political tension, which permeate Western American history.

WOMEN'S IMPACT ON URBANIZATION

During the second half of the nineteenth century, the direction of American expansion shifted from the countryside to the city. As mining and railroads encouraged urban growth in the first half of the nineteenth century, new industries and opportunities for employment emerged in the urban centers. Simultaneously, the development of new farm machinery pushed agriculturalists into long-term debt as interest rates rose and incomes from wheat, corn, and livestock plummeted during the farm crises of the 1870s and the 1890s. Tens of thousands of families abandoned their farms and ranches and headed for urban areas in the West, introducing significant numbers of women into the previously male-dominated cities. Such movement into urban areas amounted to a stunning reversal of the traditional pattern of Western settlement, which previously featured the presence of many young, unmarried men and substantially fewer single women and entire families.[43] While patriarchic attitudes existed in the urban realm, women discovered more opportunities to exercise their voice in public roles and venues for their activities. Women—primarily Euro-American middle-class women—found degrees of economic, social, and political freedom in the West as they helped establish the social and cultural infrastructures, public institutions, and agencies crucial in the modern urban setting. Women used their power in many ways to shape the urban landscape.

Rapid growth during the late nineteenth century stimulated market demand for traditional female skills in the labor force. Throughout the nineteenth century—until the 1920s when clerical and sales work overtook it—domestic service reigned as the single most important form of wage labor for women, particularly young single women. Only a handful of women practiced medicine or law in the early period, and a proportion of women filled occupations in food service, laundry work, and prostitution.[44] Despite opportunities for a greater variety of employment for women, a segmented, and often segregated, workforce prevailed. Most Euro-American women often found employment but remained trapped in blue-collar, unskilled positions and traditional areas of domestic service.[45] Similarly, African-American women found employment in urban areas in domestic service and dressmaking, but in smaller numbers.

The early West's women were often young, single, and of foreign birth or parentage and faced the effects of class and gender prejudice, which determined the kinds of jobs women in the West could and would hold. For instance, working-class women typically comprised the female factory workers. Wage labor rested on the deep-seated idea that work only takes place in a public setting—where men dominate and women largely do not exist.[46] Still, as US historian Elaine Tyler May suggests, women's labors inside the home and out remained essential for men's and their families' economic lives.[47] Men maintained a definition of womanhood based on domesticity and ensured that the pay of any work done by women earned less than a living wage in order to ensure their dependence on men. Keeping women economically dependent served to bolster the significance of men in society; particularly as industrialization and modernity tested notions of manhood. Further, the West became crucial to constructions of masculinity and gendered forms of power. National symbols for manliness—such as Theodore Roosevelt—either went to or came from the West. Urban Western women challenged traditional gender norms finding ways to use the variegated forms of power they possessed.

San Francisco's commercial superiority over its hinterlands made industrial investment and capital accumulation easier and created greater demand for women in a variety of positions. It became a venue, along with Denver, for intercultural, cross-class activity; women's efforts established firm cultural foundations in these cities. In her book *Relations of Rescue,* historian Peggy Pascoe examines the growth of women's clubs and charitable organizations in the two cities. In Denver, local charities combined into a Charity Organization Society in 1889, about the same time a parallel organization, the Associate Charities of San Francisco, was established in the city by the bay. Pascoe writes,

> Both Denver and San Francisco charity organization programs reflected the techniques of 'scientific philanthropy': advocates favored moral exhortation over financial aid, hoped to end duplication of services by sharing information about aid recipients,

> and tried to avoid promoting the 'pauperization' they felt resulted from letting recipients depend on charitable aid.[48]

For the short period of time in which class structures were relatively flexible, some Western cities provided unusual opportunities for new female residents to make their mark as community leaders, often through social and charitable work.

As women became increasingly visible in the male-dominated urban center, their drive for moral reform seemed to thrive on the fluidity of emerging class formations. In the 1870s and 1880s, Pascoe suggests that cities "offered a minimum of established charities and a maximum of opportunities for middle-class women to gain community standing through benevolent work."[49] Women's religious groups and literary and social clubs sought to expand social services and to upgrade the cultural and physical environment as a form of "city building" comparable to the activities of male entrepreneurs who focused on commerce. Women's civic activities altered and shaped the physical and social conditions of living in an urban place.[50] The institutions of domesticity, homes, education, fine arts, and religion combined to develop ideas about social community in urban settings, all led by women's efforts.

In several more ways, women were deeply involved in early city growth. Urban women showed an excited interest in the process of city growth and development. Historian Lee M. A. Simpson has argued that Western women established themselves as economic actors before they became political actors. In one of her articles on women's roles in city growth in California, Simpson posits that middle-class Euro-American women owned property in metropolitan areas, which essentially made them personal and financial stockholders in urban ambitions for growth. Essentially, Simpson articulates that women were crucial participants in urban competitions for regional dominance, where their economic independence endowed them with experience in the overall process of city building beginning in the late nineteenth century.[51] Developing a variety of social and moral services such as churches, clubs, and parks, female city boosters demonstrated the growth potential of the city in order to attract residents who would invest in property and enhance the wealth of all city property owners. With personal wealth based in real property, female property owners found their personal fortunes dependent on successful urban expansion. Most Western urban histories that address the issue of rapid urbanization depict a male-centered society where men receive the credit for vision and benefits of economic growth.[52] Through paid employment, work at home, the creation of social institutions, and involvement in real-estate development, women of every class, race, and ethnic group participated in the development of the city's economy, society, and politics. Women organized for a variety of causes, and, in so doing, they brought about innovative social improvements and expanded their roles in public spaces to improve the quality of urban life.[53] In these and other ways women exerted power and control over local politics before they had the vote.

As male and female residents struggled to define their urban communities, issues of class, gender, and race intersected in movements for political reform. Despite their involvement and use of power in social and economic interests, women sought a direct political influence so that they could use their power in civil society. In his study of Progressive-era Seattle, historian John C. Putman states, "As key figures in arguably the two greatest social movements of their age—feminism and labor unionism—working-class women in Seattle facilitated, if only temporarily, the formation of cross-class and cross-gender political alliances."[54] Although a seemingly masculine region of loggers, dockworkers, and gold hunters, women exerted a significant social and political influence in shaping the contours of Seattle. Middle-class and working-class women came together to build the foundation of feminist politics in Seattle. Both classes of women increasingly viewed direct political activity as a way to both defend its status and to secure a voice in the city's future. The class-consciousness exacerbated by America's entry into World War I deeply polarized Seattle residents and tested the progressive impulses that had connected middle-class and working-class women for more than a decade. The war years did help improve the image of women as public citizens and political equals. While Seattle women recoiled somewhat from political activity after World War I, one woman reached the pinnacle of power when voters elected Bertha Knight Landes as the city's mayor in 1926, the first female mayor of a major American city. By 1930, the Seattle Federation of Women's Clubs had changed priorities from city politics to social and personal improvement. Nevertheless, the role of women in city politics still challenged the era's traditional gender norms.[55]

In Butte, and other Western cities, women assertively made their presence known in public settings and conducted themselves in a manner that proclaimed their right to share the street, dance halls, restaurants, and nightclubs with men. Western women pushed the envelope on established gender roles by ascertaining new rights, privileges, and extraordinary roles in the previously male-dominated society. New social spaces emerged in the cities to allow the inhabitants, regardless of gender, to redefine their roles and identities.[56] During the 1920s and 1930s, the increasing power of a consumer culture and the empire of commodities heightened the strain on the hard-pressed working-class families who had middle-class aspirations. In Butte, for instance, as the price of copper dropped during the Great Depression forcing mining men out of work, women stepped into new positions as wage laborers and mobilized social networks to keep their families and community going during hard economic times.[57] Women positioned themselves not only as guardians of the moral and social order of the community but often as guardians of its economic stability as well. In short, Butte historian Janet L. Finn states, "as masculinity was variably constructed in relation to production, the convergence of paternal, patriarchal, and capitalist forces structured women's maternal place and practice as guardians of social reproduction and stewards of community consumption."[58] With men idled, women worked overtime, sometimes earning the respect, sometimes the resentment of their partners. Women negotiated daily with grocers, bankers, property owners, and utility companies to make minimal payments and to hold off debts. Women worked hard behind the scenes in an effort to maintain routines while fulfilling their obligations as caretakers by extending the meaning and power of "women's work" both publicly and privately.

Women made continued efforts to contribute to local political movements, and they advocated for social-welfare legislation during the New Deal era. Challenging gender distinctions further in World War II, women poured into "men's" work in US cities and in unprecedented numbers in the West. This era provided liberating potential for women in the city and those coming to the city. Women assumed labor and economic roles traditionally ascribed to men. They left the home in massive numbers heeding the call of Rosie the Riveter. In San Francisco, the female labor force grew from 138,000 to 275,000 between 1940 and 1945. A time for new experiences and social change enabled female laborers the opportunity to demand higher wages. Women were permitted to join unions with fewer barriers and benefited from the improved pay and working conditions provided by those unions. Many women employed in traditionally female fields—retail trade and domestic service—also benefited from the new competitive labor market as they experienced improvements in wage levels and working conditions.[59] For the decades following World War II, women's roles in America were bifurcated between their experiences of public work in wartime and their forced transition back into traditional roles. Historians Richard Etulain and Michael Malone state: "Caught between such earlier ideals of womanhood and the pressures of the workplace, women suffered mounting pressures, evidenced in soaring divorce rates and addictions to tranquilizers and alcohol."[60] Civil Rights and student movements brought about a rebirth of feminist activism in the 1960s. In those movements, women often encountered sexism and a patriarchal social order. In addition, literature such as Betty Friedan's *The Feminine Mystique* (1963) sparked feelings of discontent in the nation's women. Such displeasure prompted many women to band together to form associations such as the National Organization for Women (NOW) in 1966. Western women organized local chapters of NOW in several cities, including San Francisco and Los Angeles, to foster newfound community amongst the region's women in fighting for female equality. Western women, like their counterparts nationwide, refuted the idea of domesticity that they considered an obstacle to equality. Still, they mobilized as women of earlier generations had via neighborhood gatherings, religious organization, and other grassroots campaigns to raise feminist consciousness in the cities. These mid-1960s and 1970s female movements for reform brought tremendous challenges to and sought to transform male structures in the cities to make them more democratic and less patriarchal. Resulting from the political mobilization of women in the urban West, in the last decades of the twentieth century women became much more prevalent in high municipal political offices. Some of the region's largest cities elected female mayors, including Dianne Feinstein of San Francisco from 1978 to 1988 and Annette Strauss of Dallas from 1987 to 1991. Furthermore, in the last half of the twentieth century, more women entered the full-time workforce. By the late 1980s, more than 70 percent of women aged twenty-five to fifty-four were in the labor force. Three-quarters of those women worked full-time.[61] However, they still did not receive the same pay for the same job as a man and had few opportunities for upward mobility. While labor and market activity of women continue to increase, and while the problems of inequality in the workplace have been somewhat ameliorated, urban women continue to face the specter of lower pay and limited opportunities for advancement in many fields.

URBAN COMMUNITIES, 1870S–1930S

In a time of great significance for the growth of modern America, the nation's metropolises hosted large-scale immigration and movements for political and labor reform. Of all the dynamic changes of the late nineteenth century, the rise of the huge, impersonal metropolis was among the most striking to the people of the USA. In writing about the impact urban living had on social alienation, David Paul Nord writes, "Urbanization perforce created the paradox of people who were increasingly strangers but who were also increasingly dependent upon one another and increasingly affected by one another's private behavior."[62] Governments made gestures within the city to create communities amongst the inhabitants, particularly through the urban planning and the segregation of public space. Race, class, and gender hierarchies informed the definition and organization of spaces within the urban realm. While urbanization transformed lifestyles and emphasized class differences in an urban political arena, it also created an atmosphere and a place that allowed people to adopt new ways of thinking and acting politically. The practice and the ideology of working-class and urban community activism emerged as a crucial feature of the early and mid-twentieth century. The roots of this urban class activism derived from Dennis Kearney's San Francisco based Workingman's Party of the late 1870s. The Workingman's Party was a labor organization that produced rallies against the Chinese immigrant labor force and its primary employer, the Central Pacific railroad. These rallies found significant support amongst other Euro-American urbanites who found themselves without work due to immigrant labor. The Workingman's Party movements eventually led to the Chinese Exclusion Act.

Urban social problems and politics were central to defining and contesting national public spaces from the 1880s to the 1940s. The Progressive era solidified the intellectual urban climate that led to public activist movements. In urban areas, people who lived in close proximity to others but remained isolated at the same time could turn to newspapers to help break down that isolation and connect urbanites in thinking and acting; the press served as a medium for urban communities to connect with their constituents.[63] African Americans, for example, established newspapers as a central part of their community building in many Western cities. Historian Robert V. Hine noted, "A prime element of community was the newspaper, subscribed to by practically everybody and probably read more avidly, if not more frequently, than the Bible."[64] The press joined in the chorus of city boosters by extolling urban virtues. In addition to exhorting the virtues of urban living and promoting real estate, newspapers also called attention to political and social inequalities through muckraking and editorials that exposed government corruption and the ills of society.

Where urban living created problems, urbanites also found the city to be the place for solutions. Newspapers and other paper media helped spark movements for reform in urban centers by repudiating vice, social ills, and inequality, and by espousing methods for change. In addition to the newspapers, urban literary figures created an awareness of the urban West and its problems. Frank Norris's aforementioned novel, *The Octopus,* described the callous and egregious impact of the railroad and the political problems associated with it, which brought an awareness of the problems to the public, who then generated movements for reform.[65] Sexuality, a most private and intimate concern for most people, became blatantly public when prostitution became an issue for public policy and regulation. Butte joined 200 other American cities that shut down their red-light districts by 1920. In Butte, as in many other Western cities, closing of red-light districts failed to eradicate problems. Rather, these crusades only changed the structure of vice; it moved underground and dispersed through the city into speakeasies and clubs. Historian Mary Murphy writes, "The closing of the red-light district ended a particular form of commercialized vice, affecting both the public and private lives of prostitutes, but it certainly did not extinguish Butte's demimonde."[66] Stamping out prostitution and vice was not just an urban Western phenomenon. Nationwide urban reformers sought to clean up their cities' brothels and saloons.[67]

Another issue central to the Western urban public was race. Urban areas influenced regional social structures, and World War I had a decided impact on the region in this regard. The outbreak of the war increased the diversification of the Western population, especially in hastening the influx of minority groups such as Mexican and African Americans.[68] African Americans viewed World War I as an opportunity to gain the important goal of social justice and freedom at home, particularly as increased urbanization offered jobs and possibilities, however few. The African-American community fervently backed the nation with patriotic zeal even though the federal government and mainstream America continued viewing them as second-class citizens. While the West, with some notable exceptions, largely escaped the 1919 race riots rampant in the East, racial and ethnic tension mounted in its cities. Most Western cities already had a strong hierarchy of social, political, and economic power with Euro-Americans on top. These power structures compelled Mexicans, Asians, and African Americans into a lower class for menial labor and allowed Euro-Americans to acquire more wealth and influence from a position of strength. Urban Euro-Americans dictated and controlled the ethnic arrangement and were able to configure a form of spatial segregation that kept non-Euro-Americans, and even lower-class Euro-Americans, on the "wrong side" of town.[69] Euro-Americans employed the racial hierarchy to consolidate markets in land, labor, capital, and resources and to deny access to minorities. Minorities embraced the ideal of racial egalitarianism and the promises of unfettered opportunity that seemed to characterize the West. However, historians William Deverell and Douglas Flamming note that the ideal of promised opportunity was based on the racial hierarchy and the interests of Euro-American control; these competing visions of opportunity brought continual tension to the region.[70] Euro-American elites were able to structure urban space into distinct districts by exploiting minority labor, offering few possibilities for advancement, which meant that minority workers could only afford second-rate housing on marginal lands in and around the cities.

In the 1920s, the city of Los Angeles experienced truly astounding demographic, geographic, infrastructural, and cultural growth. During that decade, Los Angeles' population grew from 577,000 to 1.24 million. The city's boundaries sprawled dramatically and forced its inhabitants to rely on the automobile to traverse the area. Such massive growth ensured that Los Angeles would emerge as the business, financial, and cultural center of the region. In essence, Los Angeles assumed its modern form during the decade.[71] During the interwar period, Los Angeles remained a Euro-American haven, where non-Euro-Americans faced the disconcerting realities of false opportunity and the certainty of discrimination.[72] The rigid social culture of Los Angeles resulted in a place of dynamic movement overflowing with racial volatility and an ethos of Euro-American power. Despite the racial hostility, the city was becoming tremendously diverse with a coherent African-American community and culture. While Los Angeles African Americans had no voice in city politics, they tried to provide services within their community. Los Angeles allowed greater progress for African Americans than other large cities as evidenced by the thriving cultural life, success of African-American business, and early civil-rights initiatives introduced by legislators. Nevertheless, the same manifestations of racism found in the rest of the country including the Klan, restrictive real-estate covenants, and police brutality that segregated swimming pools and public places, also characterized Los Angeles and other Western cities.[73] African Americans gained substantial victories such as the desegregation of Los Angeles beaches, but they also suffered crushing defeats such as losing the right to swim in public pools. However, African-American residents did not remain passive. In his detailed study of African-American communities in Los Angeles in the early 1900s, Douglas Flamming asserts that leading the charge in denouncing racism and demanding equal rights was the striving African-American middle class.[74]

The African-American community's early optimism for freedom, equality, and economic advancement contrasted with patterns of everyday racism. African Americans nationwide soon found themselves in a constant fight to maintain and extend their basic human rights.[75] Charlotta Bass and other African-American community leaders pushed for the development of racial equality through boosterism and civil-rights campaigns in Los Angeles. One element of life that African-American Los Angelinos strove for was home ownership. Historian Greg Hise suggests that many people who came west viewed as a positive and magnetic force the possibility of home-ownership in a self-contained community.[76] However, such communities also exacerbated the existing social and racial divisions amongst a city's population. Urban historian Becky Nicolaides writes, "In interwar

Los Angeles, the politics of neighborhood took on immense importance in the lives of White working-class families, rivaling if not overshadowing workplace struggles."[77] Working-class Euro-American homeowners protected the modest economic security that home-ownership provided by restricting African-American residents in their neighborhoods. Euro-American residents, urban planners, and city governments reinforced spatial segregation by forcing minority and ethnic groups, including African Americans, in Los Angeles into separate communities.[78] Similarly, in Denver, the growth of the suburbs of Montclair, Cherry Creek, Country Club, and University Heights deepened social divisions for new housing due to covenants and racial restrictions. Desirable housing in the city became increasingly inaccessible to the growing numbers of European immigrants and African Americans. While Euro-Americans acknowledged African-American settlers in the Mile High City for their economic and labor contributions and their good citizenship, they remained walled off within the city.[79]

Caught in the same tensions, African-American residents of Kansas City faced a troubling environment in asserting their own identity prior to the outbreak of World War II.[80] African-American labor helped fuel the growth of Kansas City from a way station into a thriving metropolis. The African-American community of Kansas City also made major national cultural contributions, particularly in the development of its own style of jazz in the 1920s.[81] Still, racial tensions and conflicts abounded. Social historian Sherry Lamb Schirmer investigated the racial landscape of Kansas City in this period. Her examination reveals that Euro-American middle-class reformers sought to uproot vice in Kansas City but settled for a quarantine line around the neighborhoods in which saloons, brothels, and gambling places thrived, primarily in African-American neighborhoods. Schirmer notes that this quarantine had a twofold impact. First, Euro-American reformers' efforts helped reinforce spatial segregation in city housing prior to World War II as they connected vice and African-American residents in popular perception. These connections extended racial discrimination and provided more ammunition for Euro-American segregationists to distance their homes from those of African Americans. Second, vice became an unpleasant fact for every resident of the African-American community, regardless of their personal values, as Euro-American Kansas Citians found all African Americans guilty by association.[82] In short, African-American Kansas Citians as a whole could not avoid the injurious effects of the criminal actions of relatively few African-American men and women.[83] Unfortunately, in the West, this form of city-building and neighborhood formation was not exceptional.

An endemic problem nationwide, Euro-Americans programmed African Americans into a set of collective images so negative and widely held that they made an effective and politically acceptable rationale for racial exclusion and isolation within cities. African-American newspaper weeklies, such as the *Kansas City Sun, Kansas City Call,* and *Kansas City American* articulated concerns, grievances, and practical remedies for urban African Americans, hoping to move the African-American community into action.[84] *Kansas City Call* Editor Chester Arthur Franklin carefully chronicled the city's racial injustices but could not mount a movement to rectify them against the city's Euro-American power structure. The dominant Pendergast political machine in Kansas City, which controlled the city's politics and planning and which included future President Harry S. Truman who would desegregate the US military, provided no help to African-American Kansas Citians and even exacerbated problems of spatial segregation.[85] Politics offered no relief for African Americans as lower-class Euro-Americans, executives, union officials, and influential Euro-Americans continued the creation and enforcement of brutal racial policies in accordance with the racial hierarchy.[86]

In the interwar years, African Americans began exercising their political voice, criticizing urban inequalities and joined a rising civil-rights community that sought glimpses of the Western urban ideal of opportunity and equality. However, the Western ideals were founded on the myth of the West as Euro-American country where only Euro Americans could enjoy opportunity and equality. While Euro-Americans made efforts to Americanize immigrant and minority peoples, they sought to dominate and show supremacy in the multicultural environment they encountered.[87] The resulting tension continued for decades to come. All ethnic and racial groups felt the consequences of the urban Euro-American hierarchy and made little effective reform against it prior to World War II.

WORLD WAR II: THE CATALYST FOR ACCELERATED URBANIZATION IN THE TWENTIETH CENTURY

The post-1940 period is distinctive for the urban West because it witnessed the transmutation of Western cities from regional metropolises into major national and global centers. While the West was already attractive to migrants and the federal government had set a precedent for spending in the region, World War II introduced federal spending, industrialization, and population growth on a new scale. Western cities moved beyond their extractive economies. The power, wealth, size, and reputation of Western cities grew exponentially following the USA involvement in World War II. The war profoundly stimulated urban growth in terms of an unprecedented increase in population, diversity, economic expansion, demands for housing and services, social dislocation, racial tension, and environmental problems.[88] The war years ushered into the urban West more problems directly related to increased heterogeneity. In this period, urbanization processes already underway accelerated onto a dramatic new scale. Ramping up these processes transformed human relations within urban areas that brought enduring negative consequences.

In *The Metropolitan Frontier,* urban historian Carl Abbott links these changes to long-wave cycles of economic growth in American capitalism, with federal spending encouraging large-scale industrial production in Western urban centers. The onset of war brought defense industries to coastal cities and military bases to the interior West, and the federal government utilized the space to construct military facilities and to develop an aerospace industry on the coast. The development of nuclear weapons—the mining of uranium, building of missiles, and weapon tests—likewise spoke to the federal presence in the region. As a result, historian Gerald Nash argues the federal government became the dominant force in the Western economy.[89] Funding for harbors, airports, and, later, a national highway system transformed transportation and concepts of space and time in the West. The growth of defense industries accompanied by housing subsidies accelerated urban growth, and urban growth promoted recreation and environmental amenities that influenced the region's economy. Federal spending further spurred the growth of electronics, communications, and aerospace all centered in the West.

Economic expansion brought demographic growth, more influential political movements, and quickly transformed the urban West into a national and international economic, cultural, and political pacesetter. Heavily affected by war and defense, cities in the West completely replaced the countryside as the preferred place of residence during the post-1940 era.[90] While rural areas certainly contributed to the war effort, desperately needed industrial output came from urban factories. In the West, the level of urbanization was such that the federal government could quickly, efficiently, and cheaply assemble and amass resources for the war effort. As historian Roger W. Lotchin posits, "The freeways, city infrastructures, power plants, aqueducts, excess housing, and recreational spaces allowed the government to evade many of the costs of war."[91] Lotchin makes clear that each of the major cities were underutilizing their resources, which enabled the federal government to redirect them toward fighting the war.[92] However, Lotchin also maintains that, at least in California, World War II did not have a decisive transformative impact.

Nevertheless, as historians Gerald Nash, Marilynn Johnson, and others have articulated, World War II indeed marked a critical turning point in the urban American West as the federal government began to subsidize large-scale building of defense industries, military bases, and eventually housing. While the federal government had subsidized the railroad industry and growth in the nineteenth-century West, the 1940s-era federal subsidies sparked Western economic growth in previously unthinkable ways. Defense spending proved essential to growth following 1940. In Denver, military facilities and defense arsenals employed approximately 20,000 people as the city grew by 20 percent during the war, attracting 100,000 new residents. The proportion of Coloradans living in cities rose in the next thirty years from 53 to 79 percent. At the nexus of the Great Plains and Rocky Mountains, Denver now boasts more federal employees than any other city in the USA, save Washington, DC.[93] Federal aerospace and defense contracts in the postwar period often went to California firms and provided annual financial support to the state's economy of roughly $20 billion during the immediate postwar decades. These defense industries drove California's suburban development and served as the state's economic engine during the early Cold War period.[94]

Long associated with the lumber business, Seattle became a center for aircraft manufacturing with the prodigious rise of the Boeing Company, which dominated the economy and accounted for 40 percent job growth in the 1940s.[95] Boeing built unfathomably large buildings and inside them assembled row upon row of planes. These included larger planes such as the B-29, which was a longer-range plane than any that the USA had produced to that time. Literary critic and author Roger Sale notes: "By 1944, the peak war year, Boeing employment was up to 50,000 in Seattle, and its total sales were over 600 million dollars."[96] Portland also achieved substantial growth, doubling in size by attracting 160,000 new citizens during the war years. Portland and Seattle in the twentieth century, however, pursued significantly different economic goals and roles that have strongly affected their character and their position as a regional city and a global city respectfully. Urban historian Carl Abbott concluded that following the growth of the World War II era, downtown Seattle looked out toward the Pacific to Alaska, national competitors, and world markets, while Portland looked east toward Mount Hood and inward to continental resources and markets.[97]

By 1939, Phoenix had officially received the title the Valley of the Sun and attracted more than 35,000 health seekers and winter visitors a year. However, World War II catapulted Phoenix into the economic and demographic stratosphere with the establishment of three major air bases and a premium location for defense industries. With warm weather nearly year around, business could meet production schedules without fear of interruption by adverse weather. The cotton fields that bloomed in the irrigated environs of Phoenix now competed with airfields, army bases, and defense plants for land and workers. Manufacturing became the city's principal source of income by 1955, emerging as the metropolitan center of commerce and industry in the southwest. Electronics and aerospace dominated the industrial landscape, marking the area as a high-technology center by 1980.[98] The arrival of Motorola in 1949 and other electronic firms, as well as military bases and a booming tourist industry propelled the expansion in white-collar jobs and high-technology industries. Meanwhile, the city's population grew exponentially from 65,414 in 1940 to 106,818 in 1950 and catapulting to 439,170 by 1960.[99]

Since 1940, a demographic shift and rising economic power have pointed toward the rising urban centers of the Inter-Mountain, Southwest, and California.[100] Science, the military, and industry all played a significant role in bringing wartime immigrants to the West. The tremendous demographic shift represented the greatest numeric inflow of humanity in the region's history. Historian John M. Findlay states,

> It had taken the American Far West one hundred years, from 1840 to 1940, to attain a population exceeding 13 million; in the next twenty years, 1940–1960, it added its second 13 million, and then in the following decade, 1960–1970, it added still another 7 million. This burst of growth dwarfed that of a century earlier.[101]

The immense population growth brought with it further economic expansion with the introduction of vast amounts of private capital. Simultaneously, the combination of new cultures generated racial conflicts and continued attempts for spatial order and segregation in the cities.

An infusion of racial minorities arrived, which exacerbated the existing racial and social tensions in the city that set the stage for a highly political and tension-filled postwar urban experience. Seattle, for instance, saw its African-American population expand from under 4,000 to nearly 30,000 from 1940 to 1945.[102] National racial mores demonstrated intolerance during the war years. Few people protested when Japanese Americans from all over the West, urban and rural, received adverse treatment in the form of mass arrests and deportations to concentration camps. The question of how members of many different peoples managed to coalesce, interact, and function as civic entities in metropolitan communities and their role in twentieth-century urbanization came into stark relief. During and especially after the war, urban Westerners battled over race and its meaning.

POST-WORLD WAR II URBAN MINORITY EXPERIENCE

In the decade and a half after World War II, the rapid regional growth not only presented enormous opportunities for Westerners but also confronted them with a variety of continuing problems. During the 1960s, two central issues that absorbed Westerners' attention were the acculturation of minority groups in their midst and the alarming pollution of the natural environment. The West harbored the majority of Mexican Americans in the USA, most of its American Indians, and an increasing proportion of its African Americans.[103] The war experience brought new opportunities to many urban minority groups for accessing power and movements toward gaining equality in the social structure. Roger Lotchin notes, "The *San Francisco Chronicle* did a magnificent job of publicizing the war activities of the Italians, the Chinese, the French, the Germans, the Irish, the Jews, and, in fact, every neighborhood. They praised each ethnic group and insisted on its legitimacy within the community."[104] However, the ways in which cities and their people put aside racial and ethnic differences for the war effort were temporary. Entrance into the workforce and the acknowledgment of their efforts, followed by their quick dismissal upon the war's completion, ramped up the activism that had started in minority communities during the interwar period. Historian Bradford Luckingham succinctly noted an important point for the post-1940 Western urban political arena: "For African Americans, as well as Mexican Americans and Chinese Americans in Phoenix, World War II worked as a catalyst [. . .] Having fought a war to defeat racism, they felt it was time for them to be treated equitably as first-class American citizens."[105] Unfortunately, Euro-American government officials and city leaders typically did not share this sentiment. Historian Kevin Leonard argues in *The Battle for Los Angeles* that few Anglo-Americans were willing to give up their perceived privileges based on the urban social hierarchy. That meant, in particular, that Euro-Americans defended their right to live in segregated neighborhoods and, in the case of Los Angeles, voted against racial equality laws such as fair employment practices.[106] White supremacist proponents argued that social conditions such as poverty, criminal behavior, and subversion existed as inherent racial characteristics. Advocates of this traditional racial notion fought with supporters of a modernist racial ideology, which articulated that physical race had no bearing on a person's beliefs, behaviors, abilities, or place in the social order.[107] Racial violence increased and further exposed the anti-democratic nature of politics and the social order in Western urban centers. Minority communities struggled for political and economic power in the dynamic postwar West, particularly regarding decisions of who governed them.

As more people from many different racial groups moved to cities, the struggles for power and equality intensified. American Indians, who were moving en masse to Western urban centers following World War II, entered the swelling animosities and debate. In 1952, the federal government initiated the Indian relocation program as part of the larger federal Indian termination policy advocated by such Westerners as Senator Arthur V. Watkins of Utah. Relocation encouraged, nearly forced, the movement of reservation-based Indians into urban environments. The goal was to improve American Indian standards of living. The impetus for relocation came in part from the war. About 40,000 American Indians left their homes for the cities to find jobs in the booming war industries. Over the next twenty years, approximately 80,000 Indians moved to Western cities in search of new opportunities. According to Commissioner of Indian Affairs Dillon S. Myer, the goal was to accelerate Native assimilation into the dominant society. Four of the relocation centers, Denver, Los Angeles, San Francisco, and Dallas, brought American Indians to Far Western cities, while others moved into cities such as Phoenix and Albuquerque. The USA had become a predominately urban nation, and American Indian poverty would never be resolved, according to relocation logic, as long as the majority of the American population continued viewing American Indians as an agricultural people. Native urbanization, however, was not a new phenomenon in post-World War II America, but rather it had a long history. In fact, historian Larry W. Burt states, "the rate of farm-to-city movement among American Indians before the 1930s was roughly the same as for the whole United States population at comparable levels of industrialization."[108]

Upon arrival in the cities, American Indians tried to deal with complex social and political realities presented by the urban setting. The federal government's relocation program, coupled with its termination policy, attempted to usurp indigenous lands and indigenous cultures. Terminating "Indianness" sent a clear message that the federal government and Euro-American America at large intended for American Indians to cease to exist as cultural and self-determined peoples by sending them to alienating urban areas. Seeking to absorb indigenous peoples into mainstream society, the relocation program sent them from reservations to urban areas where the federal government would provide vocational training, a place to live, and a job. The supposedly favorable opportunities in the cities came with heavy opportunity costs. Relocation to urban areas removed American Indians from their traditional support services. They had no medical care, little education, and lived amongst a much larger non-Indian population. Essentially, Indians that moved to cities in the early years of the relocation program faced a cultural loss of tribal and familial support, which made urban living inherently difficult.[109]

Indians supported the relocation program initially and saw it as a stride toward freedom from Indian Bureau paternalism, economic insecurity, racial injustice, segregation, and second-class citizenship. Regrettably, the majority of American Indians facing urban situations for the first time found them disruptive and distressing. Forced to make many personal adjustments to their new surroundings, American Indians faced cultural alienation, racism, identity problems, ghetto housing, and employment problems. Working and living among non-Indians was a foreign experience, and Indians were considered the minority amongst many other races, ethnicities, and cultures. The sociocultural transition from communalism to individualism inhibited many American Indians and made them feel lost, insecure, and inferior to the majority population of Euro-Americans in large.[110] A tremendous culture shock, most Indians who previously lived on reservations had never owned an automobile or knew traffic rules, found modern technologies and city noise terrifying, had not mastered the English language, and suffered from homesickness. In most relocation cities, often when American Indians lost their jobs, they could not find other employment. While employment opportunities were much higher than on the reservation, the urban Indian unemployment rate in 1970 "was more than twice the national average," which rested at approximately 6 percent.[111] They were often unable find adequate housing as promised. Desirable rental property often cost too much for low-paid, unskilled Indian workers who faced frequent layoffs or seasonal work. Consequently, Indians usually had to move into public housing projects or shabby unfurnished ghetto apartments in lower-class neighborhoods where slum property owners charged excessive rents.[112] Such was the case in Los Angeles, where American Indians often lived in shacks and frequented skid-row bars, which led to many arrests of Indians for public drunkenness. The result was the growing perception that Indians were poor, inner-city slum dwellers with a proclivity for alcoholism and violence.

The high rates of poverty, underemployment, and inadequate education that characterize the urban Indian population are associated with conditions such as homelessness and substance abuse, which increased the risk for injury and arrests.[113] Historian Donald Fixico noted that nationwide in 1971, reports on urban arrests indicated that 75 percent of all arrests among Indians were alcohol-related, compared to 33 percent of all arrests for non-Indians. In some Western cities, the percentage increased. For example, in Denver in 1971 an estimated 95 percent of Indian arrests were for alcohol-related issues. Further Fixico states, "Feelings of inadequacy have been a compelling motive for many urban Indians to drink. From their relocation experience, they quickly realized how their status had changed."[114] Urban Indians felt vulnerable due to their lack of knowledge of urban society upon their arrival, and alcohol became an escape from the physical and psychological feelings of vulnerability and depression.[115]

While facing alienation and homesickness, Indians chose to associate with other Indians and soon developed pan-Indian activities. In urban community centers, created as a part of the relocation program, Indians of many different tribes came together, discussed their experiences, realized their shared histories, and consciously sought to better their situation. Federal officials hoped that relocation would weaken tribalism and promote assimilation. However, the cross-tribal affiliations growing from Indian community centers fostered an urban pan-Indianism that provided a mechanism for maintaining a separate cultural heritage.

The advancement of a sense of Indian nationalism gave Indians the impetus to protest the injustices they endured throughout the USA and effectively transformed urban space into an intertribal space. Political mobilization and activism followed and articulated Indian concerns. The significance of this pan-Indianism in general terms is that it provides a social and cultural framework within which acculturating Indian groups can maintain their sense of identity and integrity as Indians as long as the dominant larger society assigns them to subordinate status. Pan-Indian activity in urban centers generated larger associations such as the American Indian Movement. This activity enabled Indians to exercise a loud political voice and to attempt to gain access to different forms of power. Indian activism spurred organized efforts to call attention to the American Indians plight—as was the case in the nineteen-month-long Alcatraz takeover in 1969. The Seattle Indian movement, for instance, helped not only to build a pan-Indian community but also to establish activist relationships with city African Americans. The appearance of Indian enclaves in America's inner cities made them and their problems more visible than on isolated and remote reservations, earning attention in Lyndon B. Johnson's War on Poverty programs. The period from 1960 through 1980 witnessed an increasing tendency to include Indians in social-welfare legislation, and their presence in cities helped them to enter the mainstream of public social concern.[116] Since the 1980s, a more secure urban Indian population has resided in large cities as third- and fourth-generation urban Indians began coming of age in the cities.[117] The 1990 US census indicated that 53 percent of Indian people lived in cities, with Los Angeles, Denver, Seattle, and San Francisco serving as major metropolitan destinations.[118]

Other urban minority groups also forged dynamic communities as they mobilized and became active in exercising their political voice through protest against the racial subordination embedded in urban housing, education, and labor markets. Economic and political marginalization in the urban milieu meant that residential segregation continued. The postwar environment engendered a fight for racial liberalism. World War II brought to the American West an unprecedented increase in the region's African-American population. In just ten years, the West's African-American population soared from roughly 170,000 in 1940 to over 570,000. This watershed era accelerated the drive for equality. Euro-American hierarchical structures pervaded daily life even in urban areas where African-American residents did not experience ghetto conditions or encounter overt acts of racist harassment. In San Francisco, the African-American population grew from under 20,000 in 1940 to over 130,000 in 1950.[119] African-American San Franciscans made persistent attempts to achieve racial equality by building social institutions and developing community resources to overcome the deleterious effects of the American racial caste system. Historian Albert Broussard suggests that San Francisco had a liberal and progressive façade that overshadowed the city's racist reality. The color line remained highly restrictive, principally in the economic, political, and housing spheres. Following the war, there was an overall lack of industrial niche for African Americans, and union discrimination discouraged minorities from seeking certain jobs.[120] Similarly, in Seattle, the African-American population had grown enormously since 1945. The city experienced a 71 percent increase in its African-American population between 1940 and 1960.[121] African Americans in Seattle were drawn into intense competition with the city's Chinese, Japanese, and Filipino populations for low-wage employment and minimal housing opportunities.[122]

As minority-group populations soared in the post-1940 urban West, adequate and sufficient housing became a serious concern. Los Angeles needed millions of new housing units in the postwar era, which made livable space highly contested. African Americans and other groups faced acute housing shortages. Even while the African-American and Mexican-American populations expanded exponentially in Western cities such as Los Angeles following the war, restrictive housing covenants barred any "non-Caucasians" from nearly 95 percent of the new housing built in the immediate postwar period. Such patterns forced African Americans and Mexican Americans into already crowded urban neighborhoods, known as ghetto and barrio housing. Los Angeles became one of the most segregated urban areas in the nation. As more employment for Euro-Americans moved to industrial parks on the urban fringe and into suburbia, unemployment amongst African Americans and Mexican Americans rose. Over 40 percent of African-American families in Los Angeles lived at or below the poverty level in 1950. Downtown political leaders refused efforts to move African Americans and Hispanics into areas occupied or desired by Euro-Americans, but they understood the great need to improve the racial dialogue. The uninspired efforts to ameliorate racial tensions in most Western cities demonstrated the maintenance of an ethos of spatial segregation, which gave many minority groups a significant reason to voice their dissatisfaction with urban problems. However, public displays of disaffection often exacerbated racial tensions, which made the urban West a proverbial tinderbox in the postwar decades.

Systematically discriminated against and denied equal access to jobs, housing, services, and educational facilities, Mexican Americans, Chinese Americans, and African Americans in America's Sun Belt found a place that largely replicated the Deep South's racism. Continued residential segregation into the Cold War era foiled national racial-equality victories, such as *Brown* v. *Board of Education of Topeka* (1954). Issues of urban land use and attitudes concerning city space were central to continued segregation.[123] Where urbanites condoned segregated housing they also found educational and employment segregation because ethnic neighborhoods did not have the tax base to support educational facilities or to have the transportation to get to distant employment centers. By 1960, de-facto spatial segregation emerged as a primary obstacle to African-American progress in the aftermath of *Brown* v. *Board of Education.* In Phoenix in 1960, the African-American population reached 20,919, up from 4,263 in 1940. According to historian Matthew Whittaker, in 1960, the Phoenix Urban League concluded that approximately 95 percent of African-American Phoenicians lived south of Van Buren Street in the "worst housing areas in the city."[124] The founders and early boosters of Phoenix were former Southerners who embraced many of the anti-African-American attitudes that dominated race relations in the South. To them, African Americans represented the antithesis of Whiteness, while they continually deemed Mexican Americans as culturally inferior. Residential segregation continued to keep minority groups at a disadvantage economically and politically.[125] The effect of substandard schools, unskilled low-paying jobs, and discriminatory real-estate patterns led to continued minority isolation and poverty.[126] De-facto segregation in schools and housing remained through the 1980s in some of the West's urban centers.

In the 1950s and 1960s, despite ongoing racial prejudice and discrimination, improvements began to appear following the growing number of militant and peaceful protests. Successful legal actions to desegregate the schools and end discrimination in employment, housing, public accommodations, and voting were celebrated. The Western civil-rights movement and political activism found support not only in the African-American community but also amongst other minority groups. Latinos just as much as African Americans were crowded into substandard housing without running water, heat, or electricity; forced to attend inferior schools; chastised in the major newspapers; and subjected to Americanization campaigns. However, public policy and public opinion geared against any progress for Mexican Americans halted positive community building.[127] In Phoenix, due to the city's small African-American population, African-American leaders were compelled to form alliances with progressive Euro-Americans and Mexican Americans. As a result, the multiracial coalitions that were formed as part of twentieth-century urbanization processes "pushed civil rights beyond black and white."[128] In many of the West's cities, including Phoenix, minority communities remained decentralized post-1940, and despite the traditional tendency of minority groups to dislike each other and to see each other as rivals, cross-race coalitions in response to housing segregation began changing the social makeup of the West's cities.

Already intolerant from the long history of poor treatment they endured, racial groups in the Vietnam War era began building movements toward justice, many of which took place in metropolitan spaces and against metropolitan structures.[129] The civil-rights movement of the 1960s gave the examination of urban African-American America new urgency. By 1965, the urban African-American experience became center stage of national attention with the riots in Watts—one of Los Angeles's most impoverished African-American neighborhoods. The postwar African-American social-justice and freedom movement took several unique forms and trajectories against the race-embedded structure. While the movement was not cohesive, the reality was that many African Americans hoped that the riots would draw more attention to the racial problems of urban life in America.[130] Quintard Taylor notes that more than any previous racial conflagration, "Watts forced the nation to confront the disturbing reality of the Black urban experience. Northern ghettos, bypassed by the civil rights legislation the nation so proudly embraced, remained angry, ignored, and alienated, and expressed their dissatisfaction in unfocused urban violence."[131] Similarly, Mexican Americans, or Chicanos, the vast majority of whom lived in Western cities, confronted urban dilemmas and their place in American society. Arising from the civil-rights rhetoric and other Vietnam-era protests, Mexican American discontent coalesced around such leaders as César Chávez, a principal figure in the United Farm Workers. The Chicano movement of the 1960s and 1970s became a vital element in the evolution of Mexican-American social and political thought. The Chicano Movement stood behind the slogan, "Our front line is not in Viet Nam but in the struggle for social justice in the United States."[132] The movement culminated with the National Chicano Moratorium March on August 29, 1970, when between 20,000 and 30,000 protestors gathered in Los Angeles to voice their concerns. While such protests did not eradicate the institutionalized racial and social order or gain the level of social justice desired, they made important incursions and generated identity and community in the Hispanic urban population.[133] Non-Euro-Americans no longer stood idly by and accepted their second-tier status on the hierarchy; rather, they generated public protests where their voices would be heard the loudest, in the cities and on their streets.

Changing ethnic demographics have continued in Los Angeles and the rest of the urban West during the 1990s and early 2000s. Engineers and technicians from eastern and southern Asia, especially India, came for high-tech jobs and added to and altered the urban social and cultural landscape.[134] Western urban metropolises have often remained ambivalent about commitment to racial equality. However, by the last decade of the twentieth century, the election of racial minorities to some of the highest and most prestigious positions in city governments signify change in a positive direction. As historian Quintard Taylor highlights in the introduction to his study of the African-American community in

Seattle, Norman Rice's 1989 election as the city's first African-American mayor marked a "dawning of a new age in African-American politics where black office seekers could successfully appeal to nonracial issues and themes to garner support across the political spectrum despite their race."[135] African Americans in Seattle confronted the uncertainty generated by the public endorsement of equality and the private practice of discrimination. Likewise, in Denver, the 1991 election of African American mayor Wellington Webb, as well as his Mexican-American predecessor Frederico Peña, speaks to the urban West's slow but growing commitment to racial equality. Urban Westerners have had to learn to live and coexist in a multiracial society, an ongoing adjustment process wrought by urbanization; though in many Western urban cities imbalance continues to pervade society based on the foundations established by the Euro-American elite. In Los Angeles, for instance, Euro-Americans constituted only 37 percent of the population by the close of the twentieth century, but they made up 70 percent of the registered voters, 80 percent of the jury pool, and controlled 90 percent of the fixed wealth.[136]

SUBURBAN GROWTH AND URBAN RENEWAL

As minority groups fought for equality in housing, politics, and economic opportunities in cities following World War II, a significant portion of urban Euro-Americans began moving out of the cities proper and into "new" instant cities called suburbs. Urban cities began losing ground to the suburbs in the late 1950s and early 1960s as jobs, tax revenues, and people left the city in large numbers. Most Euro-American Westerners began considering cities as congested, impoverished, filthy, immoral, transient, uncertain, and heterogeneous. Conversely, they viewed the growing trend of suburbs as spacious, affluent, clean, and homogenous. Nevertheless, the sprawl associated with suburban growth came with major problems including polluted air, congested freeways that divided ethnic neighborhoods, and blighted urban centers.[137] Suburbia rearranged downtown land values, housing, tax bases, job opportunities, schools, industry, retailing, and population concentrations.

After 1945, young people rushed into marriage, childbearing, and searching for a place to live. Young couples in the immediate postwar period had little choice about where to live because of stagnation in the domestic construction industry during the Depression and the war. The solution came in the rapid development of entirely new tracts or subdivisions of houses on the outskirts of cities, far enough away from the city centers that the price of land was cheap but close enough for those who worked in the city to commute. Suburban developers adapted the assembly line to home building. The 1944 Servicemen's Readjustment Act, known as the GI Bill of Rights, in addition to providing educational grants, offered returning veterans low-interest mortgages and business loans, which subsidized the growth of the suburbs.

Suburban booms began in Seattle, San Francisco, Los Angeles, and San Diego long before the significant urban racial problems that characterized postwar urban growth existed from which to flee.[138] However, as more non-Euro-Americans moved into urban centers, Euro-Americans moved out in increasing numbers. Expanding blight hastened the Euro-American middle-class exodus from the urban core and became a real concern to downtown business.[139] Retail followed the outward flowing population and resulted in 26 percent fewer stores in the central business districts in 1958 from ten years earlier. Phoenix, like Dallas, undertook an aggressive annexation campaign and kept the decentralizing population within city limits.

New industries furthered the dispersal of metropolitan areas. In Denver, Euro-Americans in the post-1940 era typically moved to nearby suburbs or cities such as Boulder, Longmont, Loveland, or Colorado Springs. During the 1950s, the median income in the suburban counties surpassed that in the central city for the first time. The gap widened in the 1960s as Denver increasingly became an island of elderly people, poor people, and minority-group members surrounded by a sea of middle-class Euro-American families who found that suburban living allowed the greatest enjoyment of Colorado's space and climate.[140] These communities were not only homogenous in appearance but also in their population. Over 95 percent of the adults were married Euro-Americans, between twenty and thirty-five years of age with young children; all made roughly the same income in white-collar professions.

New Deal-era tax reforms and labor unions protected American workers financially, especially the middle class. Such protection provided a larger proportion of their income for discretionary spending. In fact, in 1950, discretionary spending totaled approximately $100 billion compared to $40 billion in 1940, and that $100 billion increased steadily during the 1950s as a mass consumer and leisure culture exponentially expanded. Newly affluent Americans itched to spend their wealth on goods and services that made life more comfortable and convenient. One such commodity was the automobile. The growth of suburbia caused a sharp increase in automobile demand. More automobiles required more roads and highways. In 1956, Washington responded with the Interstate Highway Act, through which the government spent a total of $1 billion per year in road construction, and by 1960, this expenditure rose to $2.9 billion a year. Within a ten-year period following the Interstate Highway Act, over 41,000 miles of new roads spanned the country coast to coast. Within that figure, more than 5,000 miles of freeway were urban, which connected suburbs to big cities, a road network that encouraged more urban sprawl as well as more urban spatial segregation. Highways carved through cities, destroying or relocating minority-group neighborhoods, and created residential islands of peoples who could not afford to move out of the poor conditions generated by traffic, noise, and pollution. Without a strong middle-class tax base, some Western cities deteriorated physically and suffered in terms of poor schools, hospitals, and crime prevention.

Throughout the 1950s and 1960s, highway-construction and urban-renewal programs reshaped and dislocated neighborhoods and communities. Highway construction did not require the replacement of destroyed housing or a guarantee that displaced peoples had a place to live. In Denver, urban renewal and rebuilding of the downtown area in the 1950s and 1960s dislocated Chicano and African-American residents, but still two-thirds of voters approved measures to destroy and rebuild parts of the city center.[141] City elites, bent on cleaning up what they viewed as troublesome neighborhoods, found several excuses for urban renewal. Communities made efforts to fight back, but government or business held the upper hand. The importance of social power in the urban setting demonstrates that one community that succeeded in uniting for a political voice was the political conservatives. Conservativism in the West, reinforced by the presence of military and defense installations, grew strong and became a primary reason why de-facto segregation in housing and schools remained in the West through the 1980s.

Since 1950, cities obliterated large tracts of land and many minority or ethnic neighborhoods for freeway construction, urban renewal, and industrial and commercial use. Much of the literature on federal urban renewal either has focused on its failure to resurrect the city or has documented its horrendous impact on the city's neediest people. Cities experienced not only slum housing but also the blighting of downtown caused by rapid decentralization and suburban growth.[142] Historian Michael F. Logan adds to this the deep roots of resistance that emerged during urban renewal projects and increasing sprawl. While the booster-government pro-growth attitudes remain a large emphasis in historical literature, Logan depicts the dissenters and reveals that great contention rather than consensus existed in urban planning issues, particularly those concerning annexation and zoning. Moreover, much ethnic and racial resistance to redevelopment and renewal had some success in modifying urban renewal plans.[143] Nevertheless, urban renewal brought continued ethnic group and neighborhood displacement, as managerial, professional, and other white-collar personnel moved back into the newly redeveloped inner cities.[144]

Cities have recently tried to remake their images or identity through rebuilding and transforming blighted areas into places of amenity and desirable land. More recent urban renewal efforts have sought to change the poor areas into profitable and beautiful areas, to attract affluent residents and tourists. Some cities have tried renewal in the forms of energy preservation and environmentalism, such as emphasizing public transportation or walking and biking in cities. Promoters and boosters remain active into the twenty-first century, as they had been in the nineteenth, in watching over their images.[145] Kansas City, for instance, now designates itself the "City of Fountains," boasting more than 250 fountains in the city. The former gateway rail city is undergoing urban revitalization in many downtown neighborhoods and in the late 1990s, its famous Union Station was remodeled into a science museum.[146]

During much of the American West's urbanization, city government, business, and social leadership have attempted to create and maintain a spatial order by altering and transforming urban geography through railroad development, positioning of ethnic neighborhoods, and elaborately planned homogenous communities. Such attempts at controlling space, and thereby controlling group behavior, were not novel. However, spatial coherence in the post-1940 era allowed white-collar Westerners to come to terms with unwieldy city growth and afforded them a new sense of community and stability in making the metropolis seem more manageable. Historian John Findlay describes the postwar reshaping of urban forms as the creation of "magic lands." He provides examples such as Anaheim's Disneyland, a Phoenix retirement community labeled Sun City, and the Seattle World's Fair of 1962, all of which produced innovative Western cityscapes. These were highly controlled, exhaustively planned enclaves whose design pioneered the new polycentric city, which fragmented the metropolis with numerous self-sufficient nodes of social activity, rather than the singular downtown model often associated with a city. In the process, they managed to restore a semblance of order and a sense of community without sacrificing the suburban character of metropolitan life. The Sun City development became typical of urbanization in greater Phoenix and contributed to the region's reputation for amenities and leisure. As with most suburban developments and urban renewal projects, "The process that created new communities and alternative downtowns often worked directly against the interests of blue-collar workers and the poor, who were least able to afford to live near the urban villages" and new shopping centers.[147] In catering to an affluent clientele, which emphasized amenities and extolled middle-class values, these "magic lands" presented a stark alternative to the polluted, congested, and impoverished downtowns that characterized Eastern industrial cities. Western cities have clung to an ideal of city life, despite the real and severe problems of environmental degradation, racial tension, poverty, and a weakening sense of community. Nevertheless, the Western city model took hold, based on twentieth-century culture and a distinctly different natural setting. Ultimately, the Western city became the example for urban development across the globe.

URBANIZATION AND ENVIRONMENTAL ISSUES

Urban society has maintained a close relationship to the natural environment, one that is often destructive as development degrades the environment. Transmutations of nature often have an urban underpinning. Whether from extractive industries, the over-taxation of natural resources by an ever-expanding population, poor efforts at water control, or the introduction of many polluting machines, cities have played a major role in creating and continuing environmental problems. Mining in general had a highly destructive impact on the environment and left a visible legacy. Looking at current aerial photographs of or simply driving around Folsom, California, a San Francisco mining hinterland in the late nineteenth and early twentieth centuries, the rows of massive undulations on the landscape left by area hydraulic mining remain visible. Hydraulic mining introduced vast toxic waste into waterways while simultaneously altering watercourses and ecology.

Cities, likewise, engineered some of the earliest attempts to reorganize, and thus potentially threaten, Western geography. The Los Angeles Owens Valley Project, Hoover Dam, and other damming of Western water sources for urban use significantly altered regional ecosystems, but for urban boosters these were necessary. While Western cities grew originally from land with the most natural value, the insufficient resources could not sustain substantial populations. In the arid West, without major water sources, many Western cities were destined to no-growth status. Inadequate natural resources prompted city leaders to enhance their surroundings through an artificial environment that consequently harmed the natural environment. Flood threats resulted from water-development policies that favored big farmers and urban developers rather than flood control. Instead of eliminating problems, the projects seemed to compound them and created the demand for yet more water projects and other pieces of infrastructure. Since westward expansion commenced, empire builders of the Southwest pushed for more large-scale energy development, bigger water projects, and greater cities all in naturally unsustainable places.[148]

Recently, scholars have undertaken studies to explore and examine the often-ignored costs of city building on the environment, thereby adding a rich and extremely complex layer to the urban scene in Western history. One such study is Gray Brechin's *Imperial San Francisco: Urban Power, Earthly*. In San Francisco, the city's wealthiest citizens quietly conspired to draw resources from increasingly distant hinterlands into their empire with little regard for the land or for mining or timber laborers. Brechin's study depicts the foreboding consequences that surface when people fail to appreciate the destructive reach of the metropolis. In describing the urban maelstrom, Brechin states,

> Far more than whales and bears were pulled toward extinction as the city's transformative power grew. Forests were leveled on all Pacific shores, rivers and lakes vanished, and the bay from which the city took its name was filled, poisoned, and plundered, while wildlife and natives within the vortex were speedily exterminated. Secondary cities were established to feed San Francisco.[149]

In addition, San Francisco city leaders made efforts to dam the Tuolumne River in the Hetch Hetchy valley in the 1910s to create a water reservoir to serve its growing population.

Eminent environmentalist John Muir worked tirelessly for seven years to fight the building of the O'Shaughnessy Dam, but to no avail.

Since the advent of the automobile, Western cities have remained far too reliant on it as the preferred mode of transportation. Such reliance grew exponentially in the post-1940 era when suburbs and new highway systems made the automobile king. Denver, Phoenix, Dallas, and Los Angeles remain highly dependent on cars, each of which has among the highest per-capita vehicle ownership rates in the country, which has caused more highway congestion, freeway construction, and pollution. Growing and continuing urbanization in each Western city brought increasing air pollution. Denver had long publicized its clean, crisp air, but beginning in the 1970s, horrendous urban smog embarrassed the city. One account notes the thick brown skies hiding the scenic panorama and the dirtying of the once-pure mountain air, which caused, by 1978, the Environmental Protection Agency to label Denver's air as the second-worst in the country, trailing only Los Angeles.[150] Denver's problem stemmed from heavy automobile emissions trapped by front-range temperature inversions. However, vehicles and transportation are not the only reasons for smog and air pollution. Currently, power plants as far away as Utah and Phoenix, which export electricity to Los Angeles and other Southwestern cities, also generate substantial amounts of smog that drifts into the Grand Canyon and other places, disrupting the natural environment. Urbanization processes have thus made Western cities primary gross polluters.

While cities have produced much of the environmental concerns that face the USA, urban areas have also fashioned most of the notable twentieth- and twenty-first-century environmentalists and green movements. Butte, Montana, once known as the richest hill on earth with its copper reserves, served the Anaconda Copper Mining Company's interest between 1920 and 1985. Historian Janet L. Finn notes, "The city has declined from its glory days and copper mining has lost its central place in the local economy, though remnants of the industry still dominate the terrain."[151] Those remnants include polluted waters, namely the headwaters of the Columbia River, polluted lands in a most biologically diverse area, and exposed toxic mine waste at Berkeley Pit—the Anaconda Copper Mining Company's open copper mine. Historian Michael P. Malone indicates that the Anaconda mining hegemony remains *the* key fact of Montana's history. An environment exploited by their operations helps explain why a state that is politically conservative is also environmentalist and anti-corporate in its sympathies.[152] Likewise, San Francisco and Denver's rapid growth had a tremendously cruel impact on the hinterland, as people exploited the land without regard. Perhaps because of such a legacy of destructive forces, these cities have become central places for organizing the environmentally conscious during the latter part of the twentieth century and into the twenty-first.

Continual growth and few solutions have exacerbated environmental problems since 1940. The limits of Western urban developments remain unknown but seem close to a breaking point, particularly in terms of resources and the problems originally associated with instant city expansion. Many Western cities have more recently created beautiful open, green spaces interrupting high-rise renewal districts to demonstrate their environmental consciousness. However, these efforts provide an implicit illustration of the continued tension between environmentalism and urbanism.[153] Urbanization has produced immense environmental issues, but those same processes have generated human efforts toward environmental policies that aim to coexist with and improve the ravaged natural world. Movements for restoration, conservation, or preservation of species and entire ecosystems often come from urbanites who now seek to once again manage, rather than brutally exploit, Western natural resources. The West has always boasted a wonderful diversity of flora and fauna, and for such natural diversity to continue, it is up to urban Westerners to hold in check the calamitous reach of the metropolis.[154]

CONCLUSION

Central to the urbanization of the American West are questions of and control over space, politics, economy, housing, and the natural environment. Urban areas have given shape to and confined politics and culture in material ways that reveal extensive consequences for understanding the West and the nation. As John Findlay suggests, "the ability of the West to influence the nation grew with its population and its urbanization."[155] Early in the twenty-first century, urban populations seem to still grow, but less gradually, resting more on a plateau than on a sharp upward slope. In California, over 90 percent of the population resides in urban areas. With more than 85 percent of the population west of the Mississippi living in cities, the West is the most urbanized region in America.[156] Following 1940 and the development of military industry in the West, the cities therein achieved national stature, measured by population growth and expansion. These metropolitan areas had the physical space to accommodate growth, and many of their civic and economic leaders built in anticipation of growth. The availability of land remains a determining factor for Western urbanization as metropolitan regions now stretch up to or exceeding 150 miles across and new divisions arise from religious divisions, suburb-to-suburb competition, or industrial affiliation. Los Angeles, Dallas, and Phoenix are among the top metropolitan areas covering the largest land space in square miles. By 1990, Los Angeles, Dallas, and Phoenix all ranked among the nation's ten largest cities. Further, in 1990, greater Los Angeles, San Francisco–Oakland–San Jose, and Dallas–Fort Worth ranked second, fourth, and ninth in the nation in population respectively. As of 2000, amongst the nation's fifty largest cities, the West accounted for twenty-three, with San Francisco proper, Denver, Portland, and Seattle among the top thirty, and Kansas City, Missouri ranked number 39.[157] One thing the West still has is land to accommodate urban growth, but environmental miscues can make sustainability impossible if precious natural resources are not consumed acutely.

By 2000, the Denver metropolitan area reached a population of 2.1 million, three-quarters of whom lived in neighboring suburban counties.[158] Denver proper expanded during the second half of the twentieth century to 316 square miles with a population of 435,000 in 2001. The city still maintains its position as a transportation hub, cemented when the city opened a $5-billion airport in 1995. Denver International Airport is the largest in terms of area and capacity for growth, giving credence to the boosters' moniker of the "World's Largest" airport. Through its evolving urbanization processes, Denver continues its dominance of the Interior West with its skiing and tourism, blossoming high-tech industry, aviation, transportations of processed foods overseas, communications, energy, and finance that make it one of the nation's fastest growing sub-areas.[159] Modernization in the West occurred through urbanization, and the emergence and continued presence of cities such as Denver make the region more central to national historical narratives.

Cities have remained at the core, the essence of the West, making connections with hinterlands, other cities, suburbs, a global economic framework, and people. People comprise the core of the city; they are the crucial component of what constitutes a city; they remain the major resource of and the actors in the economic, political, cultural, and social choices for metropoles. The people form the urban social and cultural landscape, which in turn generates the urban political, economic, and architectural landscape. Western cities are the place where cultures and people meet each other to create new and extraordinary spaces. Urbanization processes in nineteenth- and twentieth-century Western cities contributed to the shape of the American metropolis, and in the twenty-first century, they not only shape but also have become the model for global metropolitan growth. The plethora of problems and changes arising out of Western urbanization should prove a cautionary tale for current-era urban politicians, planners, and inhabitants. Critical urban processes often underlie the issues of race, class, and gender.

Cities grew on the "Old West" economy of resource extraction, production, and direct resource processing. The "New West" economy adds to the old and overlays the new information economy, high finance, intellectual products and services, electronics, and professional and recreational services, energy, and tourism and related services. In Western urban centers, as part of the urbanization process, minority groups struggled heroically to survive and surmount inequities imposed by a Euro-American-dominated society. Luckingham writes, "In the nineteenth and twentieth centuries the struggle and the difficulties proved to be continuous; progress was slow, and equity and parity were elusive."[160] The forces of urbanization have fractured communities, altered the physical, cultural, political, and economic landscape, and generated racial and cultural tension in defining ways that inform the way we live today. Cities are, as Gray Brechin writes, "humanity's most complex artifacts."[161] In the West, the people and processes that established those complex urban areas comprise the entire spectrum of Western American history. The mythic nineteenth-century West should no longer mesmerize popular notions of the West. Rather, thought must concentrate on the West's people within a regional context of urban growth and development. Such a context reveals one of the most complex and urbanized places in the country that remains a model in terms of culture, economics, and growth. The ongoing processes of urbanization continue to shape the American West and the overall American experience.

NOTES

1. Earl Pomeroy, *The Pacific Slope: A History of California, Oregon, Washington, Idaho, Utah, and Nevada* (Lincoln, Nebr.: University of Nebraska Press, 1991), p. 120. First published 1965.
2. William Cronon, *Nature's Metropolis: Chicago and the Great West* (New York: W. W. Norton & Co., 1992), p. xiii.
3. John R. Borchert, "America's Changing Metropolitan Regions," *Annals of the Association of American Geographers*, 62 (2) (1972): 352–373; p. 352.
4. Campbell Gibson, "Population of the 100 Largest Cities and Other Urban Places in the United States: 1790 to 1990," Population Division Working Paper No. 27 (Washington, DC: Population Division, US Bureau of the Census, 1998), p. 2.
5. Bradford Luckingham, *The Urban Southwest: A Profile History of Albuquerque–El Paso–Phoenix–Tucson* (El Paso, Tex.: Texas Western Press, 1982), p. xi.
6. Pomeroy, *The Pacific Slope*, pp. 123–165.
7. Eugene P. Moehring, "The Comstock Urban Network," *The Pacific Historical Quarterly*, 66 (3) (1997): 337–362; p. 337. Moehring has lamented that historians tend to de-emphasize the importance of hinterland towns as crucial support structures for the city. The relationship between hinterland and city was important in nineteenth-century growth but seems to have fallen to the more important inter-city relationships that emerged during the middle to late twentieth century.
8. Gerald D. Nash, *Creating the West: Historical Interpretations, 1890–1990*, The Calvin P. Horn Lectures in Western History and Culture, University of New Mexico, September 30—October 3, 1990 (Albuquerque, N. Mex.: University of New Mexico Press, 1991), p. 159.

9. Carl Abbott, *The Metropolitan Frontier: Cities in the Modern American West* (Tucson, Ariz.: University of Arizona Press, 1993).
10. Arthur R. Gómez, *Quest for the Golden Circle: The Four Corners and the Metropolitan West, 1945–1970* (Lawrence, Kans.: University Press of Kansas, 2000), p. xvi. Gómez concentrates on the Southwestern Four Corners municipalities of Durango, Colorado; Moab, Utah; Flagstaff, Arizona; and Farmington, New Mexico. Their significant functions made enduring contributions to the "postwar economic expansion of not only the West but the entire nation" (p. xvii).
11. Eugene P. Moehring, *Urbanism and Empire in the Far West, 1840–1890* (Reno, Nev.: University of Nevada Press, 2004), p. 177.
12. Gunther Barth, *Instant Cities: Urbanization and the Rise of San Francisco and Denver* (Oxford: Oxford University Press, 1975); Gordon Morris Bakken and Brenda Farrington (eds.), *The Urban West* (New York: Garland Publishing, Inc., 2000), p. xv.
13. J. S. Holliday, *The World Rushed In: The California Gold Rush Experiment: An Eyewitness Account of a Nation Heading West* (New York: Simon & Schuster, 1981), pp. 51, 301.
14. California State Census, 1852, California State Archives, Sacramento, Calif.
15. Moehring, *Urbanism and Empire,* p. 41.
16. For more on the Miller & Lux corporation see David Igler, *Industrial Cowboys: Miller & Lux and the Transformation of the Far West, 1850–1920* (Berkeley, Calif.: University of California Press, 2001).
17. For more details see Yong Chen, *Chinese San Francisco, 1850–1943: A Trans-Pacific Community* (Palo Alto, Calif.: Stanford University Press, 2000); Kevin Starr, *California: A History* (New York: The Modern Library, 2005), p. 121; and Robert W. Cherny, Richard Griswold del Castillo, and Gretchen Lemke-Santangelo, *Competing Visions: A History of California* (Boston, Mass.: Houghton Mifflin, 2005), pp. 180–181.
18. Lyle W. Dorsett and Michael McCarthy, *The Queen City: A History of Denver,* 2nd edn (Boulder, Col.: Pruett, 1986), p. 4.
19. Carl Abbott, Stephen J. Leonard, and David McComb, *Colorado: A History of the Centennial State,* 3rd edn (Boulder, Col.: University Press of Colorado, 1994), p. 54; Lawrence H. Larsen, *The Urban West at the End of the Frontier* (Lawrence, Kans.: The Regents Press of Kansas, 1978), p. 4.
20. Carl Abbott, "The Urban West and the Twenty-First Century," *Montana: The Magazine of Western History,* 43 (2) (1993): 62–68; p. 68.
21. David M. Emmons, *The Butte Irish: Class and Ethnicity in an American Mining Town* (Urbana, Ill.: University of Illinois Press, 1989), p. 13.
22. For more on Butte, Montana's early history as a demographically diverse instant city and the repercussion of copper in its political history see Michael P. Malone, *The Battle for Butte: Mining and Politics on the Northern Frontier, 1864–1906* (Seattle, Wash.: University of Washington Press, 1981). See also Bakken and Farrington, *The Urban West,* p. xxv.
23. Robert O. Self, "City Lights: Urban History in the West," in William Deverell (ed.), *A Companion to the American West* (Malden, Mass.: Blackwell Publishing, 2004), pp. 412–441; p. 415.
24. Carl Abbott, *Portland: Planning, Politics, and Growth in a Twentieth-Century City* (Lincoln, Nebr.: University of Nebraska Press, 1983), p. 11; Earl Pomeroy, *The Pacific Slope,* p. 146; Carol O'Connor, "A Region of Cities," in Clyde A. Milner II, Carol A. O'Connor, and Martha A. Sandweiss (eds.), *The Oxford History of the American West* (Oxford: Oxford University Press, 1994), pp. 535–564; p. 537.
25. Abbott, *Portland,* p. 33.
26. Lawrence H. Larsen, *The Urban West at the End of the Frontier* (Lawrence, Kans.: The Regents Press of Kansas, 1978), p. 17.
27. Abbott et al., *Colorado,* pp. 240–250.
28. Patricia Evridge Hill, *Dallas: The Making of a Modern City* (Austin, Tex.: University of Texas Press, 1996), p. xxiii.
29. Texas Writers Project, *Dallas Guide and History* (Austin, Tex.: Briscoe Center for American History, University of Texas at Austin, 1940), Chronology, pp. 700–701.
30. Hill, *Dallas,* p. xxiii.

31. David J. Wishart, "Cities and Towns," in David J. Wishart (ed.), *Encyclopedia of the Great Plains* (Lincoln, Nebr.: University of Nebraska Press, 2004), pp. 151–157; p. 151.
32. Mark H. Rose, *Cities of Light and Heat: Domesticating Gas and Electricity in Urban America* (University Park, Pa.: The Pennsylvania State University Press, 1995), p. 15.
33. Bradford Luckingham, "The American Southwest: An Urban View," *The Western Historical Quarterly,* 15 (3) (1984): 261–280; p. 265; see also Bradford Luckingham, *Phoenix: The History of a Southwestern Metropolis* (Tucson, Ariz.: University of Arizona Press, 1989), pp. 2–4.
34. Luckingham, *Phoenix,* p. 3.
35. Donald J. Pisani, "Water Planning in the Progressive Era: The Inland Waterways Commission Reconsidered," *Journal of Policy History,* 18 (4) (2006): 389–418.
36. Luckingham, *The Urban Southwest,* pp. 40–41.
37. Kevin Starr, *Inventing the Dream: California through the Progressive Era* (Oxford: Oxford University Press, 1985), p. 64.
38. Starr, *Inventing the Dream,* pp. 70–72.
39. "Titanic Project to Give City a River," *Los Angeles Daily Times,* June 29, 1905; John Walton, *Western Times and Water Wars: State, Culture, and Rebellion in California* (Berkeley, Calif.: University of California Press, 1992), pp. 143–155; Robert M. Fogelson, *The Fragmented Metropolis: Los Angeles, 1850–1930* (Cambridge, Mass.: Harvard University Press, 1967), p. 85; Marc Resiner, *Cadillac Desert: The American West and Its Disappearing Water,* rev. edn (New York: Penguin, 1993), p. 77.
40. See the several essays on community history available in Wesley G. Johnson, Jr. *Phoenix in the Twentieth Century: Essays in Community History* (Norman, Okla.: University of Oklahoma Press, 1993).
41. William Deverell, *Whitewashed Adobe: The Rise of Los Angeles and the Remaking of Its Mexican Past* (Berkeley, Calif.: University of California Press, 2004), pp. 93–116.
42. William Deverell, *Railroad Crossing: Californians and the Railroad, 1850–1910* (Berkeley, Calif.: University of California Press, 1994), pp. 139–146.
43. For more on the changing demographics and movements of people to the West, see Walter Nugent, *Into the West: The Story of its People* (New York: Vintage Books, 2001).
44. Alan Lessoff, "America's City: Dallas and United States Urban History," *Journal of Urban History,* 26 (4) (2000): 530–544; p. 539.
45. Mary Lou Locke, "Out of the Shadows and into the Western Sun: Working Women of the Late Nineteenth-Century Urban Far West," *Journal of Urban History,* 16 (2) (1990): 175–204; p. 176.
46. Elaine Tyler May, "Expanding the Past: Recent Scholarship on Women in Politics and Work," *Reviews in American History,* 10 (4) (1982): 216–233; p. 222.
47. May, "Expanding the Past," pp. 222–223.
48. Peggy Pascoe, *Relations of Rescue: The Search for Female Moral Authority in the American West, 1874–1939* (Oxford: Oxford University Press, 1990), p. 188.
49. Pascoe, *Relations of Rescue,* p. 11.
50. Elizabeth York Enstam, *Women and the Creation of Urban Life: Dallas, Texas, 1843–1920* (College Station, Tex.: Texas A&M University Press, 1998), p. 177.
51. Lee M. A. Simpson, "Women, Real Estate, and Urban Growth: A Case Study of Two Generations of Women Property Owners in Redlands, California, 1880–1940," *California History,* 76 (1) (1997): 24–43; p. 27.
52. Simpson, "Women, Real Estate, and Urban Growth," pp. 29–30.
53. Enstam, *Women and the Creation of Urban Life,* p. 180.
54. John C. Putman, *Class and Gender Politics in Progressive-Era Seattle* (Reno, Tex.: University of Nevada Press, 2008), pp. 2–5.
55. Putman, *Class and Gender Politics in Progressive-Era Seattle,* p. 215.
56. Mary Murphy, *Mining Cultures: Men, Women, and Leisure in Butte, 1914–1941* (Urbana, Ill.: University of Illinois Press, 1997), p. 74.
57. Janet L. Finn, "A Penny for Your Thoughts: Stories of Women, Copper, and Community," *Frontiers: A Journal of Women Studies,* 19 (2) (1998): 231–249; p. 232. See also William R. Leach, *Land of Desire: Merchants, Power, and the Rise of a New American Culture* (New York: Vintage, 1994).

58. Finn, "A Penny for Your Thoughts," p. 232.
59. Karen Anderson, *Wartime Women: Sex Roles, Family Relations, and the Status of Women during World War II* (Westport, Conn.: Greenwood Press, 1981), p. 23.
60. Richard W. Etulain and Michael P. Malone, *The American West: A Modern History, 1900 to the Present,* 2nd edn (Lincoln, Nebr.: University of Nebraska Press, 2007), p. 162.
61. Susan E. Shank, "Women and the Labor Market: The Link Grows Stronger," *Monthly Labor Review,* 62 (3) (1988): 3–8.
62. David Paul Nord, *Communities of Journalism: A History of American Newspapers and Their Readers* (Urbana, Ill.: University of Illinois Press, 2001), p. 127.
63. Thorin Tritter, "The Press and its Place: How Newspapers Fit into Urban History," *Journal of Urban History,* 31 (4) (2005): 571–578.
64. Robert V. Hine, *The American West: An Interpretive History,* 2nd edn (Boston, Mass.: Little, Brown & Company, 1984), p. 270.
65. Deverell, *Railroad Crossing,* p. 139.
66. Mary Murphy, "The Private Lives of Public Women: Prostitution in Butte, Montana, 1878–1917," *Frontiers: A Journal of Women's Studies,* 7 (3) (1984): 30–35; p. 35.
67. Paul S. Boyer, *Urban Masses and Moral Order in America, 1820–1920* (Cambridge, Mass.: Harvard University Press, 1978), p. 199.
68. Gerald D. Nash, *The American West in the Twentieth Century: A Short History of an Urban Oasis* (Albuquerque, N. Mex.: University of New Mexico Press, 1977), p. 71.
69. Nash, *The American West in the Twentieth Century,* p. 269.
70. William Deverell and Douglas Flamming, "Race, Rhetoric, and Regional Identity: Boosting Los Angeles, 1880–1930," in Richard White and John M. Findlay (eds.), *Power and Place in the North American West* (Seattle, Wash.: University of Washington Press, 1999), pp. 117–143; p. 133.
71. Tom Sitton and William Deverell (eds.), *Metropolis in the Making: Los Angeles in the 1920s* (Berkeley, Calif.: University of California Press, 2001), p. 2.
72. Deverell and Flamming, "Race, Rhetoric, and Regional Identity," p. 132.
73. Douglas Flamming, "The Star of Ethiopia and the NAACP: Pageantry, Politics, and the Los Angeles African American Community," in Tom Sitton and William Deverell (eds.), *Metropolis in the Making: Los Angeles in the 1920s* (Berkeley, Calif.: University of California Press, 2001), pp. 145–160; p. 158; David Wrobel, "Paradise Pondered: Urban California, 1850–1920," *Journal of Urban History,* 34 (6) (September 2008): 1029–1043; p. 1037.
74. Douglas Flamming, *Bound for Freedom: Black Los Angeles in Jim Crow America* (Berkeley, Calif.: University of California Press, and the George Gund Foundation Imprint in African American Studies, 2005), p. 49.
75. Flamming, *Bound for Freedom,* p. 2.
76. For more on the home-ownership aspect to urban growth in Los Angeles, see Greg Hise, *Magnetic Los Angeles: Planning the Twentieth Century Metropolis* (Baltimore, Md.: Johns Hopkins University Press, 1997).
77. Becky M. Nicolaides, "The Neighborhood Politics of Class in a Working-Class Suburb of Los Angeles," *Journal of Urban History,* 30 (2) (2004): 428–451; pp. 428, 446.
78. Flamming, *Bound for Freedom,* p. 300.
79. Abbott et al., *Colorado,* p. 251.
80. Charles E. Coulter, *"Take Up the Black Man's Burden": Kansas City's African American Communities, 1865–1939* (Columbia, Miss.: University of Missouri Press, 2006), p. 11.
81. Karen J. De Bres, "Kansas City, Kansas and Missouri," in David J. Wishart (ed.), *Encyclopedia of the Great Plains* (Lincoln, Nebr.: University of Nebraska Press, 2004), p. 171.
82. Sherry Lamb Schirmer, *A City Divided: The Racial Landscape of Kansas City, 1900–1960* (Columbia, Miss.: University of Missouri Press, 2002), p. 126.
83. Coulter, *Take Up the Black Man's Burden,* p. 16; Schirmer, *A City Divided,* p. 126.
84. Coulter, *Take Up the Black Man's Burden,* p. 9.
85. See Lyle W. Dorsett, *The Pendergast Machine* (Oxford: Oxford University Press, 1968).

86. Schirmer, *A City Divided,* pp. 63–147.
87. See more detailed arguments and articulations on White assimilation programs and the White movements to maintain dominance and supremacy in the West in Frank Van Nuys, *Americanizing the West: Race, Immigrants, and Citizenship, 1890* (Lawrence, Kans.: University Press of Kansas, 2002).
88. The standard account for the immense transformations brought to the American West by World War II is Gerald D. Nash, *The American West Transformed: The Impact of the Second World War* (Lincoln, Nebr.: University of Nebraska Press, 1985). For more on the alterations in human relations and social dislocations see Marilynn S. Johnson, *The Second Gold Rush: Oakland and the East Bay in World War II* (Berkeley, Calif.: University of California Press, 1993).
89. Gerald Nash, *The Federal Landscape: An Economic History of the Twentieth-Century American West* (Tucson, Ariz.: University of Arizona Press, 1999).
90. Roger W. Lotchin, "California Cities and the Hurricane of Change: World War II in the San Francisco, Los Angeles, and San Diego Metropolitan Areas," *The Pacific Historical Review,* 63 (3) (1994): 393–420; p. 395.
91. Roger W. Lotchin, *The Bad City in the Good War: San Francisco, Los Angeles, Oakland, and San Diego* (Bloomington, Ind.: Indiana University Press, 2003), p. 70.
92. Lotchin, *The Bad City in the Good War,* p. 70.
93. Abbott et al., *Colorado,* p. 309.
94. Kevin Starr, *California: A History* (New York: The Modern Library, 2005), p. 242; see also Roger W. Lotchin, *Fortress California, 1910–1961: From Warfare to Welfare* (Oxford: Oxford University Press, 1992); and James L. Clayton, *The Economic Impact of the Cold War: Sources and Readings* (New York: Harcourt, Brace & World, 1970).
95. Carol O'Connor, "A Region of Cities," p. 556. Roger Sale, *Seattle, Past to Present* (Seattle, Wash.: University of Washington Press, 1976), pp. 181–216.
96. Sale, *Seattle,* p. 181.
97. Carl Abbott, *Greater Portland: Urban Life and Landscape in the Pacific Northwest* (Philadelphia, Pa.: University of Pennsylvania Press, 2001), p. 40.
98. Luckingham, "The American Southwest," p. 272.
99. Census of Population 2000, United States Bureau of Census.
100. Luckingham, "The American Southwest," p. 279.
101. John M. Findlay, "Far Western Cityscapes and American Culture since 1940," *Western Historical Quarterly,* 22 (1) (1991): 19–43; p. 19.
102. Etulain and Malone, *The American West,* p. 115.
103. Nash, *The American West in the Twentieth Century,* p. 264.
104. Lotchin, "California Cities and the Hurricane of Change," p. 397.
105. Bradford Luckingham, *Minorities in Phoenix: A Profile of Mexican American, Chinese American, and African American Communities, 1860–1992* (Tucson, Ariz.: University of Arizona Press, 1994), p. 205.
106. Kevin Allen Leonard, *The Battle for Los Angeles: Racial Ideology and World War II* (Albuquerque, N. Mex.: University of New Mexico Press, 2006), p. 18.
107. Leonard, *The Battle for Los Angeles,* p. 297.
108. Larry W. Burt, "Roots of the Native American Urban Experience: Relocation Policy in the 1950s," *American Indian Quarterly,* 10 (2) (1986): 85–99; p. 85.
109. J. Matthew Shumway and Richard H. Jackson, "Native American Population Patterns," *Geographical Review,* 85 (2) (1995): 185–201.
110. Donald L. Fixico, *The Urban Indian Experience in America* (Albuquerque, N. Mex.: University of New Mexico Press, 2000), p. 3.
111. Fixico, *The Urban Indian Experience in America,* p. 76.
112. Kenneth R. Philip, "Stride toward Freedom: The Relocation of Indians to Cities, 1952–1960," *Western Historical Quarterly,* 16 (2) (1985): 175–190; pp. 185–186.

113. Jonathan R. Sugarman and David C. Grossman, "Trauma among American Indians in an Urban County," *Public Health Reports,* 111 (July 1996–August 1997): 321–327.
114. Fixico, *The Urban Indian Experience in America,* p. 104.
115. Fixico, *The Urban Indian Experience in America,* p. 104.
116. Vine Deloria, Jr. "Legislation and Litigation Concerning American Indians," *Annals of the American Academy of Political and Social Science,* 436 (1978): 89–96; 86.
117. Fixico, *The Urban Indian Experience in America,* p. x.
118. Fixico, *The Urban Indian Experience in America,* pp. 4 and 14.
119. Campbell Gibson and Kay Jung, "Historical Census Statistics on Population Totals by Race, 1790 to 1990, and by Hispanic Origin, 1970 to 1990, for the United States, Regions, Divisions, and States," *Population Division, US Census Bureau,* Washington, DC, September 2002, Working Paper series no. 56, Tables 5 and 19.
120. Albert S. Broussard, *Black San Francisco: The Struggle for Racial Equality in the West, 1900–1954* (Lawrence, Kans.: University Press of Kansas, 1993).
121. Sale, *Seattle,* p. 217.
122. Quintard Taylor notes that a need exists for studies on Western urban communities to explore the impact of competition and cooperation among various peoples of color in urban settings. See Quintard Taylor, *The Forging of a Black Community: Seattle's Central District from 1870 through the Civil Rights Era* (Seattle, Wash.: University of Washington Press, 1994), p. 6.
123. Schirmer, *A City Divided,* p. 2.
124. Matthew C. Whitaker, "'Creative Conflict': Lincoln and Eleanor Ragsdale, Collaboration, and Community Activism in Phoenix, 1953–1965," *Western Historical Quarterly,* 34 (2) (2003): 165–190; p. 173. Whitaker has since expanded the arguments put forth in this article, particularly regarding the quest for racial equality in Phoenix, into a monograph entitled *Race Work: The Rise of Civil Rights in the Urban West* (Lincoln, Nebr.: University of Nebraska Press, 2005).
125. *Los Angeles Times,* January 10, 1995; See also Whitaker, "Creative Conflict," pp. 165–190; Quintard Taylor, *In Search of the Racial Frontier: African Americans in the American West, 1528–1990* (New York: W. W. Norton & Co., 1998), p. 280; Luckingham, *Minorities in Phoenix,* p. 57; and Gerald Horne, *Fire This Time: The Watts Uprising and the 1960s* (Charlottesville, Va.: University of Virginia Press, 1995), p. 249.
126. Whitaker, "Creative Conflict," pp. 165–190.
127. Luckingham, *Minorities in Phoenix.*
128. Matthew C. Whitaker, "Creative Conflict," p. 166; see also Taylor, *In Search of the Racial Frontier,* pp. 179–280;
129. See Robert O. Self, *American Babylon: Race and the Struggle for Postwar Oakland* (Princeton, NJ: Princeton University Press, 2003). Using Oakland, California, as his examination site, Self weaves together two stories, one of Cold War-era political culture and the other of metropolitan space in highlighting the African-American struggle for racial equality.
130. Horne, *Fire This Time,* p. 31.
131. Taylor, *Forging of a Black Community,* p. 4.
132. Lorena Oropeza, *¡Raza Sí! ¡Guerra No!: Chicano Protest and Patriotism during the Viet Nam War Era* (Berkeley, Calif.: University of California Press, 2005), p. 147.
133. See also Stephen J. Pitti, *The Devil in Silicon Valley: Northern, California, Race, and Mexican Americans* (Princeton, NJ: Princeton University Press, 2003), pp. 178–186.
134. Cherny et al., *Competing Visions,* p. 417.
135. Taylor, *The Forging of a Black Community,* p. 3.
136. David Reiff, *Los Angeles: Capital of the Third World* (New York: Simon & Schuster, 1991), pp. 190–195; "Los Angeles," in Linda Schmittroth (ed.), *Cities of the United States: A Compilation of Current Information on Economic, Cultural, Geographic, and Social Conditions,* vol. II*: The West* (Detroit, Mich.: Gale Group, 2001), pp. 117–131.
137. Patricia Nelson Limerick, *The Legacy of Conquest: The Unbroken Past of the American West* (New York: W. W. Norton, 1987), p. 165.

138. Roger W. Lotchin, "The Impending Urban Past: An Essay on the Twentieth-Century West," in Gerald D. Nash and Richard W. Etulain (eds.), *Researching Western History: Topics in the Twentieth Century* (Albuquerque, N. Mex.: University of New Mexico Press, 1997), pp. 53–81; p. 70.
139. Robert B. Fairbanks, "The Failure of Urban Renewal in the Southwest: From City Needs to Individual Rights," *Western Historical Quarterly,* 37 (3) (2006): 303–326.
140. Abbott et al., *Colorado,* p. 301.
141. Abbott et al., *Colorado,* p. 313.
142. Fairbanks, "The Failure of Urban Renewal in the Southwest," pp. 303–304.
143. Michael F. Logan, *Fighting Sprawl and City Hall: Resistance to Urban Growth in the Southwest* (Tucson, Ariz.: University of Arizona Press, 1995), pp. 1–10, 168.
144. Peter Wiley and Robert Gottlieb, *Empires in the Sun: The Rise of the New American West* (New York: G. P. Putnam's Sons, 1982), p. 134.
145. Limerick, *The Legacy of Conquest,* p. 167.
146. De Bres, "Kansas City, Kansas and Missouri," p. 171.
147. John M. Findlay, *Magic Lands: Western Cityscapes and American Culture after 1940* (Berkeley, Calif.: University of California Press, 1992), p. 292.
148. Wiley and Gottlieb, *Empires in the Sun,* p. xiv.
149. Gray Brechin, *Imperial San Francisco: Urban Power, Earthly Ruin* (Berkeley, Calif.: University of California Press, 1999), p. xxiv.
150. Wiley and Gottlieb, *Empires in the Sun,* p. 133.
151. Janet L. Finn, "Intimate Strangers: The Interlocking Histories of Butte, Montana, and Chuquicamata, Chile," *Montana: The Magazine of Western History,* 48 (3) (1998): 22–39; p. 23.
152. Michael P. Malone, "The Close of the Copper Century," *Montana: The Magazine of Western History,* 35 (2) (1985): 69–72; p. 70.
153. Abbott, *Greater Portland,* pp. 1–2.
154. Historian Dan Flores does not refer specifically to this point on the role of urbanites and the ruinous urban reach in his study entitled *The Natural West: Environmental History in the Great Plains and Rocky Mountains* (Norman, Okla.: Red River Books of the University of Oklahoma Press, 2001), pp. 185–195, but he does discuss preservation, conservation, and restoration of the Western environment. For more on direct connections to the destructive reach of the metropolis, see Brechin, *Imperial San Francisco.*
155. Findlay, *Magic Lands,* p. 2.
156. Carl Abbott, "The Urban West and the Twenty-First Century," *Montana: The Magazine of Western History,* 43 (2) (1993): 62–68; p. 63; Census 2000, US Census Bureau; Linda Schmittroth (ed.), *Cities of the United States: A Compilation of Current Information on Economic, Cultural, Geographic, and Social Conditions,* vol. II: *The West* (Detroit, Mich.: Gale Group 2001).
157. Census, 2000.
158. Thomas J. Noel, "Denver, Colorado," in David J. Wishart (ed.), *Encyclopedia of the Great Plains* (Lincoln, Nebr.: University of Nebraska Press, 2004), p. 165.
159. Noel, "Denver, Colorado," p. 165.
160. Luckingham, *Minorities in Phoenix,* p. 191.
161. Brechin, *Imperial San Francisco,* p. xxv.

FURTHER READING

Abbott, Carl, *Greater Portland: Urban Life and Landscape in the Pacific Northwest* (Philadelphia, Pa.: University of Pennsylvania Press, 2001).

—— *The Metropolitan Frontier: Cities in the Modern American West* (Tucson, Ariz.: University of Arizona Press, 1993).

—— *Portland: Planning, Politics, and Growth in a Twentieth-Century City* (Lincoln, Nebr.: University of Nebraska Press, 1983).

—— (1993) "The Urban West and the Twenty-First Century," *Montana: The Magazine of Western History,* 43 (2): 62–68.
Anderson, Karen, *Wartime Women: Sex Roles, Family Relations, and the Status of Women during World War II* (Westport, Conn.: Greenwood Press, 1981).
Bakken, Gordon Morris and Farrington, Brenda (eds.) *The Urban West* (London and New York: Garland, 2000).
Barth, Gunther, *Instant Cities: Urbanization and the Rise of San Francisco and Denver* (Oxford: Oxford University Press, 1975).
Borchert, John R. "America's Changing Metropolitan Regions," *Annals of the Association of American Geographers,* 62 (2) (1972): 352–373.
Boyer, Paul S., *Urban Masses and Moral Order in America, 1820–1920* (Cambridge, Mass.: Harvard University Press, 1978).
Brechin, Gray, *Imperial San Francisco: Urban Power, Earthly Ruin* (Berkeley, Calif.: University of California Press, 1999).
Broussard, Albert S., *Black San Francisco: The Struggle for Racial Equality in the West, 1900–1954* (Lawrence, Kans.: University Press of Kansas, 1993).
Burt, Larry W., "Roots of the Native American Urban Experience: Relocation Policy in the 1950s," *American Indian Quarterly,* 10 (2) (1986): 85–99.
California State Census, 1852, California State Archives, Sacramento, Calif.
Cherny, Robert W., del Castillo, Richard Griswold, and Lemke-Santangelo, Gretchen, *Competing Visions: A History of California* (Boston, Mass.: Houghton Mifflin Company, 2005).
Coulter, Charles E., *"Take Up the Black Man's Burden": Kansas City's African American Communities, 1865–1939* (Columbia, Mo.: University of Missouri Press, 2006).
Cronon, William, *Nature's Metropolis: Chicago and the Great West* (New York: W. W. Norton & Co., 1992).
De Bres, Karen J., "Kansas City, Kansas and Missouri," in David J. Wishart (ed.), *Encyclopedia of the Great Plains* (Lincoln, Nebr.: University of Nebraska Press, 2004), p. 171.
Deloria Jr., Vine, "Legislation and Litigation Concerning American Indians," *Annals of the American Academy of Political and Social Science,* 436 (1978): 86–96.
Deverell, William and Flamming, Douglas, "Race, Rhetoric, and Regional Identity: Boosting Los Angeles, 1880–1930," in Richard White and John M. Findlay (eds.), *Power and Place in the North American West* (Seattle, Wash.: Center for the Study of the Pacific Northwest in association with University of Washington Press, 1999).
—— *Railroad Crossing: Californians and the Railroad, 1850–1910* (Berkeley, Calif.: University of California Press, 1994).
—— *Whitewashed Adobe: The Rise of Los Angeles and the Remaking of Its Mexican Past* (Berkeley, Calif.: University of California Press, 2004).
Dorsett, Lyle W. and McCarthy, Michael, *The Queen City: A History of Denver,* 2nd edn (Boulder, Col.: Pruett, 1986).
Emmons, David M., *The Butte Irish: Class and Ethnicity in an American Mining Town* (Urbana, Ill.: University of Illinois Press, 1989).
Enstam, Elizabeth York, *Women and the Creation of Urban Life: Dallas, Texas, 1843–1920* (College Station, Tex.: Texas A&M University Press, 1998).
Fairbanks, Robert B., "The Failure of Urban Renewal in the Southwest: From City Needs to Individual Rights," *Western Historical Quarterly,* 37 (3) (2006): 303–326.
Findlay, John M., "Far Western Cityscapes and American Culture since 1940," *Western Historical Quarterly,* 22 (1) (1991): 19–43.
—— *Magic Lands: Western Cityscapes and American Culture after 1940* (Berkeley, Calif.: University of California Press, 1992).
Finn, Janet L., "A Penny for Your Thoughts: Stories of Women, Copper, and Community," *Frontiers: A Journal of Women Studies,* 19 (2) (1998): 231–249.
Fixico, Donald L., *The Urban Indian Experience in America* (Albuquerque, N. Mex.: University of New Mexico Press, 2000).

Flamming, Douglas "*The Star of Ethiopia* and the NAACP: Pageantry, Politics, and the Los Angeles African American Community," in Tom Sitton and William Deverell (eds.), *Metropolis in the Making: Los Angeles in the 1920s* (Berkeley, Calif.: University of California Press, 2001).
—— *Bound for Freedom: Black Los Angeles in Jim Crow America* (Berkeley, Calif.: University of California Press, and the George Gund Foundation Imprint in African American Studies, 2005).
Fogelson, Robert M., *The Fragmented Metropolis: Los Angeles, 1850–1930* (Cambridge, Mass.: Harvard University Press, 1967).
Flores, Dan L., *The Natural West: Environmental History in the Great Plains and Rocky Mountains* (Norman, Okla.: Red River Books of the University of Oklahoma Press, 2001).
Gibson, Campbell, "Population of the 100 Largest Cities and Other Urban Places in the United States: 1790 to 1990," Population Division Working Paper No. 27. Washington, DC: Population Division, US Bureau of the Census, 1998.
Gómez, Arthur R., *Quest for the Golden Circle: The Four Corners and the Metropolitan West, 1945–1970* (Lawrence, Kans.: University Press of Kansas, 2000).
Hill, Patricia Evridge, *Dallas: The Making of a Modern City* (Austin, Tex.: University of Texas Press, 1996).
Hine, Robert V. *The American West: An Interpretive History,* 2nd edn (Boston, Mass.: Little, Brown & Co., 1984).
Hise, Greg, *Magnetic Los Angeles: Planning the Twentieth Century Metropolis* (Baltimore, Md.: Johns Hopkins University Press, 1997).
Holliday, J. S., *The World Rushed In: The California Gold Rush Experiment: An Eyewitness Account of a Nation Heading West* (New York: A Touchstone Book published by Simon & Schuster, 1981).
Horne, Gerald, *Fire This Time: The Watts Uprising and the 1960s* (Charlottesville, Va.: University of Virginia Press, 1995).
Igler, David, *Industrial Cowboys: Miller and Lux and the Transformation of the Far West, 1850–1920* (Berkeley, Calif.: University of California Press, 2001).
Jackson, Kenneth T., *Crabgrass Frontier: The Suburbanization of the United States* (Oxford: Oxford University Press, 1985).
Johnson, Marilynn S., *The Second Gold Rush: Oakland and the East Bay in World War II* (Berkeley, Calif.: University of California Press, 1993).
Johnson Jr., G. Wesley (ed.), *Phoenix in the Twentieth Century: Essays in Community History* (Norman, Okla: University of Oklahoma Press, 1993).
Larsen, Lawrence H., *The Urban West at the End of the Frontier* (Lawrence, Kans.: Regents Press, 1978).
Leonard, Kevin Allen, *The Battle for Los Angeles: Racial Ideology and World War II* (Albuquerque, N. Mex.: University of New Mexico Press, 2006).
Leonard, Stephen J. and McComb, David, *Colorado: A History of the Centennial State,* 3rd edn (Boulder, Col.: University Press of Colorado, 1994).
Lessoff, Alan, "America's City: Dallas and United States Urban History," *Journal of Urban History,* 26 (4) (2000): 530–544.
Limerick, Patricia Nelson, *The Legacy of Conquest: The Unbroken Pact of the American West* (New York: W. W. Norton, 1987).
Locke, Mary Lou, "Out of the Shadows and Into the Western Sun: Working Women of the Late Nineteenth-Century Urban Far West," *Journal of Urban History,* 16 (2) (1990): 175–204.
Logan, Michael F., *Fighting Sprawl and City Hall: Resistance to Urban Growth in the Southwest* (Tucson, Ariz.: University of Arizona Press, 1995).
—— *Desert Cities: The Environmental History of Phoenix and Tucson* (Pittsburgh, Pa.: University of Pittsburgh Press, 2006).
Lotchin, Roger W., *The Bad City in the Good War: San Francisco, Los Angeles, Oakland, and San Diego* (Bloomington, Ind.: Indiana University Press, 2003).
—— *Fortress California, 1910–1961: From Warfare to Welfare* (Oxford: Oxford University Press, 1992).
—— "California Cities and the Hurricane of Change: World War II in the San Francisco, Los Angeles, and San Diego Metropolitan Areas," *The Pacific Historical Review,* 63 (3) (1994): 393–420.

—— "The Impending Urban Past: An Essay on the Twentieth-Century West," in Gerald D. Nash and Richard W. Etulain (eds.), *Researching Western History: Topics in the Twentieth Century* (Albuquerque, N. Mex.: University of New Mexico Press, 1997).
Luckingham, Bradford, *Phoenix: The History of a Southwestern Metropolis* (Tucson, Ariz.: University of Arizona Press, 1989).
—— *The Urban Southwest: A Profile History of Albuquerque–El Paso–Phoenix–Tucson* (El Paso, Tex.: Texas Western Press, 1982).
—— *Minorities in Phoenix: A Profile of Mexican American, Chinese American, and African American Communities, 1860–1992* (Tucson, Ariz.: University of Arizona Press, 1994).
—— "The American Southwest: An Urban View," *The Western Historical Quarterly,* 15 (1984): 261–280.
May, Elaine Tyler, "Expanding the Past: Recent Scholarship on Women in Politics and Work," *Reviews in American History,* 10 (4) (1982): 216–233.
Malone, Michael P., "The Close of the Copper Century," *Montana: The Magazine of Western History,* 35 (2) (1985): 70.
Moehring, Eugene P., "The Comstock Urban Network," *The Pacific Historical Quarterly,* 66 (3) (1997): 337–362.
—— *Urbanism and Empire in the Far West, 1840–1890* (Reno, Nev.: University of Nevada Press, 2004).
Murphy, Mary, *Mining Cultures: Men, Women, and Leisure in Butte, 1914–1941* (Urbana, Ill.: University of Illinois Press, 1997).
—— "The Private Lives of Public Women: Prostitution in Butte, Montana, 1878–1917," *Frontiers: A Journal of Women's Studies,* 7 (3) (1984): 30–35.
Nash, Gerald D., *Creating the West: Historical Interpretations, 1890–1990* (Albuquerque, N. Mex.: University of New Mexico Press, 1991).
—— *The American West Transformed: The Impact of the Second World War* (Lincoln, Nebr.: University of Nebraska Press, 1985).
—— *The American West in the Twentieth Century: A Short History of an Urban Oasis* (Albuquerque, N. Mex.: University of New Mexico Press, 1977).
—— *The Federal Landscape: An Economic History of the Twentieth Century American West* (Tucson, Ariz.: University of Arizona Press, 1999).
Nicolaides, Becky M., "The Neighborhood Politics of Class in a Working-Class Suburb of Los Angeles," *Journal of Urban History,* 30 (2) (2004): 428–451.
Noel, Thomas J., "Denver, Colorado," in David J. Wishart (ed.), *Encyclopedia of the Great Plains* (Lincoln, Nebr.: University of Nebraska Press, 2004), p. 165.
Nord, David Paul, *Communities of Journalism: A History of American Newspapers and Their Readers* (Urbana, Ill.: University of Illinois Press, 2001).
Nugent, Walter, *Into the West: The Story of its People* (New York: Vintage Books, 2001).
O'Connor, Carol A., "A Region of Cities," in Clyde A. Milner, II, Carol A. O'Connor, and Martha A. Sandweiss (eds.), *The Oxford History of the American West* (Oxford: Oxford University Press, 1994), pp. 535–64.
Oropeza, Lorena, *¡Raza Sí! ¡Guerra No! Chicano Protest and Patriotism during the Viet Nam War Era* (Berkeley, Calif.: University of California Press, 2005).
Pascoe, Peggy, *Relations of Rescue: The Search for Female Moral Authority in the American West, 1874–1939* (Oxford: Oxford University Press, 1990).
Pisani, Donald J., "Water Planning in the Progressive Era: The Inland Waterways Commission Reconsidered," *Journal of Policy History,* 18 (4) (2006): 389–418.
Pitti, Stephen J., *The Devil in Silicon Valley: Northern, California, Race, and Mexican Americans* (Princeton, NJ: Princeton University Press, 2003).
Philip, Kenneth R., "Stride toward Freedom: The Relocation of Indians to Cities, 1952–1960," *Western Historical Quarterly,* 16 (2) (1985): 175–190.
Pomeroy, Earl, *The Pacific Slope: A History of California, Oregon, Washington, Idaho, Utah, and Nevada* (Lincoln, Nebr.: University of Nebraska Press, 1991).

Putman, John C., *Class and Gender Politics in Progressive-Era Seattle* (Reno, Nev.: University of Nevada Press, 2008).
Reiff, David, *Los Angeles: Capital of the Third World* (New York: Simon & Schuster, 1991).
Resiner, Marc, *Cadillac Desert: The American West and Its Disappearing Water* (New York: Penguin, 1993).
Rose, Mark H., *Cities of Light and Heat: Domesticating Gas and Electricity in Urban America* (University Park, Pa.: The Pennsylvania State University Press, 1995).
Sale, Roger, *Seattle, Past to Present* (Seattle, Wash.: University of Washington Press, 1976).
Sánchez, George J., *Becoming Mexican American: Ethnicity, Culture and Identity in Chicano Los Angeles, 1900–1945* (Oxford: Oxford University Press, 1993).
Schirmer, Sherry Lamb, *A City Divided: The Racial Landscape of Kansas City, 1900–1960* (Columbia, Mo.: University of Missouri Press, 2002).
Schmittroth, Linda (ed.), *Cities of the United States: A Compilation of Current Information on Economic, Cultural, Geographic, and Social Conditions, Vol. 2: The West* (Detroit, Mich.: Gale Group, 2001).
Self, Robert O., *American Babylon: Race and the Struggle for Postwar Oakland* (Princeton, NJ: Princeton University Press, 2003).
—— "City Lights: Urban History in the West," in William Deverell (ed.), *A Companion to the American West* (Malden, Mass.: Blackwell Publishing, 2004).
Shank, Susan E., "Women and the Labor Market: The Link Grows Stronger," *Monthly Labor Review*, (1988): 3–8.
Simpson, Lee M. A., "Women, Real Estate, and Urban Growth: A Case Study of Two Generations of Women Property Owners in Redlands, California, 1880–1940," *California History*, 27 (1997): 24–43.
Starr, Kevin, *Inventing the Dream: California Through the Progressive Era* (Oxford: Oxford University Press, 1985).
—— *California: A History* (New York: The Modern Library, 2005).
Takezawa, Yasuko I., *Breaking the Silence: Redress and Japanese American Ethnicity* (Ithaca, NY: Cornell University Press, 1995).
Taylor, Quintard, *The Forging of a Black Community: Seattle's Central District from 1870 through the Civil Rights Era* (Seattle, Wash.: University of Washington Pres, 1994).
—— *In Search of the Racial Frontier: African Americans in the American West, 1528–1990* (New York: W. W. Norton & Co., 1998).
Thomas, Tracy, and Bodine, Walt, *Right Here in River City: A Portrait of Kansas City* (Garden City, NY: Doubleday & Company, Inc., 1976).
Tritter, Thorin, "The Press and its Place: How Newspapers Fit into Urban History," *Journal of Urban History*, 31 (4) (2005): 571–578.
Van Nuys, Frank, *Americanizing the West: Race, Immigrants, and Citizenship, 1890–1930* (Lawrence, Kans.: University Press of Kansas, 2002).
Walton, John, *Western Times and Water Wars: State, Culture, and Rebellion in California* (Berkeley, Calif.: University of California Press, 1992).
Weatherby, James B., and Witt, Stephanie L., *The Urban West: Managing Growth and Decline* (Westport, Conn.: Praeger Publishers, 1994).
Whitaker, Matthew C., *Race Work: The Rise of Civil Rights in the Urban West* (Lincoln, Nebr.: University of Nebraska Press, 2005).
—— "'Creative Conflict': Lincoln and Eleanor Ragsdale, Collaboration, and Community Activism in Phoenix, 1953–1965," *Western Historical Quarterly*, 34 (2) (2003): 165–190.
Wiley, Peter, and Gottlieb, Robert, *Empires in the Sun: The Rise of the New American West* (New York: G. P. Putnam's Sons, 1982).
Wishart, David J., "Cities and Towns," in David J. Wishart (ed.) *Encyclopedia of the Great Plains* (Lincoln, Nebr.: University of Nebraska Press, 2004), p. 151.
Worster, Donald (1985) *Rivers of Empire: Water, Aridity, and the Growth of the American West* (Oxford: Oxford University Press, 1985).
Wrobel, David (2008) "Paradise Pondered: Urban California, 1850–1920," *Journal of Urban History*, 34 (6): 1029–1043.

CHAPTER ELEVEN

THE HYDRAULIC WEST

The History of Irrigation

Charles R. Porter

> Water helped ancient man learn those first difficult lessons about the rights of others and responsibility to a larger society. Even the most rudimentary irrigation system required organization, discipline, cooperation, and a measure of social cohesion. Mutual need begets mutual aid. Notions of sharing, of equity, of compromise, and of the common good first floated precariously on this liquid foundation to be later cemented in philosophical thought and codified law.[1]

Irrigation is critical to the success of agriculture, which is especially true in the Western USA. Known collectively as "the West," the area is comprised of the states of Arizona, California, Colorado, Idaho, Kansas, Montana, Nebraska, Nevada, New Mexico, North Dakota, Oklahoma, South Dakota, Texas, Utah, Washington, and Wyoming. Most of the West is arid, short of reliable rainfall, and prone to drought at times. Settlers from the generally moist East were confronted with a very different climate upon crossing the 98th meridian. Not all areas of the West suffer from a lack of rainfall; it varies from state to state, and in some areas of some states it can be plentiful. However, for most of the West, agriculture, industry, and communities exist on water drawn from the diversion of streams and rivers through irrigation canals, from water wells, or from rainfall and flood waters stored in reservoirs.

Agriculture is successful not only when rainfall is plentiful but also when the soil moisture is adequate to support the particular crop being grown. The total amount of rainfall is not always a reliable indicator of the success of any growing season. Ask most professional farmers about the weather in their area and typically the reference will be to levels of "moisture" instead of inches of "rain." "Moisture" factors in the needs of the specific crops and grasses they depend on in their area. For example, all farmers know that too much rain at the wrong time can be as destructive to agriculture as too little rain. When their response to questions about the weather is "we have had good moisture this year" they really mean they have had adequate rainfall to make their crops and grasses grow to an economically successful harvest. In the West, many years go by when moisture is simply not available or reliable for successful farming. Irrigation is essential to the survival of agri-business in the West.

The history of irrigation in the Western USA can be divided into three distinct periods: the era prior to Spanish contact or generally prior to 1540; the era under the claim of Spain and Mexico from 1540 to 1848; and the period after Texas independence and the Treaty of

Guadalupe Hidalgo of 1848. The Spanish system of water rights, laws, and irrigation dominated the West for three centuries, twice as long as the current "Anglo" system. The Spanish system continues to influence the region's laws, especially those pertaining to water.

IRRIGATION IN THE WEST PRIOR TO SPANISH CONTACT

For thousands of years farmers have tried to "dry farm" or grow crops dependent entirely upon rainfall as their source of moisture. According to William E. Doolittle, "Dry farming was certainly a common practice in the American Southwest at the time of European contact. It was, however, not a simple and straightforward practice and the evidence is at best checkered."[2] Dry farming is usually done on more or less flat land since steeply sloped land makes successful growth of crops difficult due to the destructive runoff of rain when it comes.

In the rolling plains, hills, and mountains of the West, terracing was first practiced by pre-contact American Indians in order to slow and get better use from runoff water. Runoff is not always destructive; concentrated water flow can deliver new silt which refreshes and enhances existing top soil. Terracing controls and diverts water flow to its highest use. According to Doolittle, "five functions can be ascribed to terracing: (i) create nearly level planting surfaces; (ii) increase soil depth; (iii) control soil erosion; (iv) manage the flow of water; and (v) to prevent frosts."[3] Archeological evidence of pre-contact terracing has been discovered across the West, mostly in the Southwestern states of California, Arizona, New Mexico, and Utah. Some pre-contact American Indians "harvested" water by digging short canals and building rock diversion weirs. The water-harvesters used seasonal streams instead of constantly flowing rivers or springs and the creeks flowing from them.[4]

What is meant by the term "irrigation"? As Doolittle says, "in a very basic sense, irrigation is defined as the artificial application of water to otherwise dry lands in order to facilitate cultivation."[5] The most basic method of watering is filling water in a jug and pouring it directly on a plant. However, for the purposes of this chapter, irrigation is considered to be a total system of diversion dams, gates, and canals which are built to water large fields of crops. Typical irrigation systems distribute diverted or captured water to crops by gravity-flow canals, underground pipes, or overhead via center-pivot sprinklers. The earliest form of irrigation in the West was done by open gravity flow canals.

The most well-known early farmers of the West were the Hohokam Indians whose irrigation canals, discovered southwest of Phoenix, Arizona, date to 300 B.C. The source of these canals was the Gila River. The canals were dug and maintained with wooden and stone tools and with back-breaking hand labor.[6] Their work in irrigation was extensive. For example, Hohokam farmers eventually dug 135 miles of irrigation canals which watered 100,000 acres in the Phoenix area.[7] Their culture thrived until A.D. 1200 when it then declined, disappearing completely over the following 200 years. The first Spanish explorers observed the extent of the Hohokam canals and the success these canals brought to the land. In 1697, a missionary priest named Eusebio Francisco Kino observed the ruins of Casa Grande in the Gila Valley:

> that it had a large aqueduct [canal] with a very great embankment, which must have been three varas [approximately 10 feet] high and six or seven wide—wider than the causeway of Guadalupe of Mexico. This very great aqueduct, as is still seen, not only

conducted the water from the river to the Casa Grande, but at the same time, making a great turn, it watered and enclosed a plain many leagues in length and breadth, and of very level and very rich land.[8]

Although the Hohokam were perhaps the most significant irrigators in the pre-historic West, recent discoveries prove there were other ancient unnamed American Indians practicing canal irrigation in southern Utah, Arizona, northern Mexico (Sonora), and New Mexico. Recently, the oldest canal in the West unearthed so far was found in the Santa Cruz River Valley near Tucson known as the Las Capas site. It was discovered by Jonathan B. Mabry of the Tucson Historic Preservation Office. Radiocarbon techniques date the canals to between 1000 and 1100 B.C.[9] According to Mabry, the Santa Cruz River Valley supported a farming culture "that flourished between 2100 B.C. and 50 A.D. (long before the Hohokam!)"[10] Other canals in the immediate area may date as early as 1500 B.C.

Father Kino found Sobaipuri Indian villages with irrigated fields along the Santa Cruz River; he estimated their population to be 1,680 residents, the highest population density in southern Arizona at the time.[11] Irrigation agriculture made this level of population density possible.

The irrigation systems of ancient ancestors of American Indians, in Doolittle's opinion, "were on par with some of the great waterworks from other parts of both the ancient and modern worlds."[12] Some of the ancient canals were renovated by historic irrigators which disguise their true age. Irrigation practices in the West likely date much earlier and were used much more extensively than previously considered by historians and archeologists. Father Kino noted the Subaipuris and Pimas grew a large variety of irrigated crop such as maize (corn), beans, squash, cantaloupes, watermelons, and cotton used for their clothing.[13]

Earlier Spanish explorers such as Francisco de Ibarra and his chronicler Baltasar de Obregón noticed irrigation in this area of modern-day Arizona as early as 1565 and said of the people in the valleys that "their plantations are well provided with canals used for irrigating them."[14] On another Spanish expedition into northern New Mexico along the Rio Grande River in 1582, Antonio de Espejo observed that the people had: "fields planted with corn, beans, calabashes, and tobacco (piciete) in abundance. These crops are seasonal, dependent on rainfall, or they are irrigated by means of good ditches."[15]

The techniques of canal irrigation used by pre-contact native peoples varied from place to place. Some canal systems were large; others were much smaller. The Hohokams irrigated thousands of acres over hundreds of years. The archeology to date indicates that all ancient peoples, including the Hohokam, tended to abandon canals which were damaged by flooding or for other reasons and simply constructed new canals at other locations.[16]

Another group of American Indian irrigators were the Anasazi of the Four Corners region (the intersection of the corners of New Mexico, Arizona, Colorado, and Utah). They used canal irrigation with dams for diversion of water to build a large irrigation system that supported 10,000 people until climate change and soil salination caused their culture to disappear by the end of the fifteenth century.

Charles C. Mann, author of *1491: New Revelations of the Americas before Columbus,* contends that the true extent of American Indian agriculture is just now being discovered and that these new discoveries will prove that farming was practiced in their ancient world much more extensively and over wider areas.[17] It is important to note that irrigation was as much the key element of success in agricultural endeavors then as it is now.

The Spanish were the first Europeans to begin settlement of the West. They relied upon their own ancient irrigation expertise to successfully establish communities. The Spanish throughout history always sought irrigable land first for their communities. They classified land in three categories: land irrigable for farming, land for farming without irrigation, and pasture land. "Respect for water was bred in the bone of the Spaniard."[18]

The Spanish developed the mission system to settle their lands in the West. The mission system consisted of three elements: mission grounds and buildings to Christianize the Indians which were operated by either Franciscan or Jesuit missionary priests; a presidio to protect the mission inhabitants and keep the mission Indians in the mission pueblo; and a local town with settlers from the interior of Mexico or, at times, directly from Spain. Immigrant settlers also came from Spain's overseas lands such as the Canary Islands, Cuba, or other locations in their Caribbean holdings.

Four social groups, missionaries, Indians, settlers, and soldiers, lived together in frontier communities, in tension with one another but nonetheless as Spanish subjects in a Spanish culture. The Spanish irrigation system was a communal system with water as the focal point of the initial relationships between the groups. Wherever water was plentiful, reliable, and easily distributable by a working irrigation system, the Spanish communities flourished. Indian labor was under the exclusive control of the missionaries which was an issue of continuous dispute with the soldiers and settlers; everyone in the community wanted the Indians' labor for their own use. Exclusive Indian labor allowed the missions to successfully farm large tracts of land. Some missions such as those in San Antonio set up elaborate and complex agri-businesses. Irrigation systems allowed the Spanish some control over nature in the arid lands of the West.

In good years, the friars and settlers could depend upon harvesting two or more crops because of their irrigation system. Often the mission farms generated surpluses which they sold to the presidio under exclusive contract, much to the distress of the other non-mission farmers, the settlers, or other less successful missions in the area. The stable food supply for humans also allowed the raising of livestock to develop into another important area of agribusiness activity for the missions. Other trades sprang up in the missions such as leatherworking, blacksmithing, and flour milling. Until the secularization of the missions, or distribution of the mission lands, ostensibly to the Indian converts, some of the missions in the West were periodically successful. Other missions failed because of the limited quantity and poor quality of the available water. Another reason for mission failure was the attitude of the area Indians; some native groups simply would not congregate in the missions for long periods of time. Sadly, all over the Spanish North American system Indian neophytes in the missions were consumed by European diseases as were many other indigenous peoples in the New World. A properly built and well-maintained irrigation system gave the missions and communities a chance to succeed and to dig a precarious foothold into the hostile and challenging West. What were the main elements of a typical Spanish irrigation system in the West, who built it using what methods, how was it maintained, and how was it managed?

EARLY SPANISH IRRIGATION SYSTEMS

Donald Worster's description of the history of the American West defines the Spanish experience: "one of people encountering difficult environments, of driving to overcome them through technological means, of creating the social organization to do so, of leading on and on to indigenous bureaucracy and corporatism."[19]

For roughly three centuries, the Spanish system of irrigation dominated the social and legal culture of water in the West under direct Spanish rule and also under the Republic of Mexico. According to borderlands historian David J. Weber, the Spanish judicial system still remained intact with few modifications after Spain lost Mexico in the revolution of 1821.[20] The Spanish influence on irrigation in the West continued from the early 1500s to the Mexican War of 1846–1848. Irrigation and water-distribution systems were vital to community formation in all the Spanish settlements. San Antonio, founded in 1718, while not the earliest Spanish community in the West, was one of the most successful and is indicative of a typical Spanish frontier irrigation community.

The initial act of social cooperation in irrigation, water management, and distribution in San Antonio was the construction of the first *acequia* or irrigation canal dug by the settlers, soldiers, missionaries, and Indians. In January 1719, the first ditches in San Antonio were begun. The construction of these irrigation ditches was essential to the survival of everyone in the new community. Typical of all new Spanish communities in the West, before any permanent structures were constructed an initial irrigation canal was completed.[21] Confirming that this practice was the standard across New Spain, Michael C. Meyer said:

> With amazing regularity, the construction of an irrigation system for the new communities of the north began even before the houses, public buildings, and churches were finished. Prior to the completion of the ditches, water had to be conveyed by *aquadores,* who carried heavy buckets hanging from yokes across their shoulders.[22]

The historical record confirms this practice. For example, Juan de Oñate, the original colonizer of New Mexico, arrived at the pueblo of San Juan in 1598 and immediately constructed an irrigation ditch. Governor Pedro de Peralta established the headquarters of Spain in New Mexico in Santa Fe in 1610 next to a stream for irrigation. In 1706, Governor Valdez reported to Philip V of Spain about the founding of Albuquerque that "Irrigation ditches are in place and the water running."[23] The establishment of water-distribution systems for irrigation of crops and use by citizens was the initial duty of all Spanish colonizers from Texas to California.

The Spanish learned how to use scarce water resources to irrigate from the ancient conquerors of their own Iberian peninsula, and from that education, technology, laws, and practices were developed to share and distribute in common any available water resources. Areas of Spain have been irrigated for almost over 2,000 years. The Romans and the Visigoths established new irrigation practices, and these practices were modified and improved by the Moors who entered Spain in the eighth century.[24]

S. P. Scott, author of *History of the Moorish Empire in Europe,* said of the transformational power of irrigation: "A considerable portion of the country (Spain) which had never been subjected to tillage because of its aridity became suddenly metamorphosed, as if by the wand of an enchanter. [...] In districts where, according to ancient tradition, no water had ever been seen, now flowed noisy rivulets and broad canals."[25]

The innovative Moors used every known device to water the land: reservoirs, wells, sluices, tunnels, siphons, and aqueducts. They improved the irrigation methods of the Egyptians; from the Persians they learned the use of a wheel with rows of jars to pump water to lands higher than the irrigation ditches.[26] These were called *norias* in Spanish.[27]

When the Spanish arrived in the West they had been implementing and adapting their irrigation systems for over 200 years to meet the harsh arid conditions in northern Mexico. An integral part of their irrigation technology was the social and political organizations they developed to manage and conserve the precious water resources. Another positive outcome of Spanish expertise and experience in irrigation systems was the opportunity for settlers to live together in larger groups to allow better defense in the event of enemy attack.

In 1962, Texas Supreme Court Chief Justice Robert W. Calvert described concisely the typical Spanish *acequia* system's basic elements:

> Water was diverted from the river by a dam, built in the bed of the river, which raised the water level and permitted water to enter the ditch by gravity flow through a headgate. Water from the river was thus made available to the mission and for irrigation of lands adjacent to the ditch.[28]

Eventually the Spanish in San Antonio built a lasting gravity-flow irrigation system of over 50 miles, parts of which is still in use today. The water-distribution system was built by the hand labor of mission Indians, citizen-soldiers, and other settlers, but the bulk of the system was supervised by Franciscan missionaries, typical of systems across the West.

As the missions proved often in eighteenth-century San Antonio and elsewhere in the Spanish West, when water was used with diligence and hard work (albeit the forced labor of the Indians), crop surpluses could be generated to fill the granaries and storehouses, providing stability during the inevitable droughts which from time to time encompassed the arid lands. Yet, even with an irrigation system that reliably delivers fresh water, other problems happen when farming that make it a very risky enterprise. Crops can fail due to rodent attacks, insect swarms, late freezes, hail storms, too much rain at the wrong times, and a myriad of other nightmares which keep all farmers awake at night.[29] Irrigation, at least, eliminates the risk of too little water or moisture, and water is the only weapon against aridity.[30]

A working irrigation system was then and is now very labor intensive. Gates must be opened and closed, lateral ditches to the fields must be prepared each and every time water is sent to them, and the topography of the field must be maintained for gravity flow of the water to reach all the plants in the field. Not only did it take strength and time to prepare the field, it took skill to coordinate gate openings along with continual modification of the water's path once in the field for the irrigation system to be effective as shown in Figure 11.1 below. It is an art to irrigate a field of any size so that all plants can enjoy moisture without some being over-watered while others are under-watered; consistent levels of moisture must be delivered to all the plants.

DESCRIPTION OF A TYPICAL SPANISH SYSTEM

The typical Spanish irrigation system was from 5 to 7 feet wide and 2 to 11 feet deep.[31] Building the system itself was hard labor performed by hand, digging up soil, rocks, and cutting roots with tools such as wooden shovels, iron pry bars, and axes. Frederick Chabot said of the Alamo Madre *acequia* in San Antonio, "They had worked four years in bringing the water from the river to the fields. All the work had been done with bars, and the missionaries had not lacked a single day of work."[32] This first section of the Alamo Madre *acequia* was less than 2 miles in length, indicating the difficulty and time-consuming nature of digging an *acequia* in the San Antonio rocky soil. The soil conditions across the West made the digging of ditches the most difficult of work. The ditches in the Spanish times were rarely straight but meandered to avoid trees, large rocks, and hills, following the natural topography to seek the easiest way possible to flow with gravity.[33]

Figure 11.1 San Elizario in El Paso, Texas irrigation activity. This photo illustrates the benefit of irrigation in the arid climate predominate across the West.
Source: Photo by author.

The grades of ditches had to be carefully planned, and the plan carefully followed so as not to either erode the banks by being too steep or to create stagnant "dead" water by being too shallow.[34] Depth at the beginning of an *acequia* has to be "cognizant" of depth at the end, as the water had to eventually return to the river. This was the Spanish understanding based upon the concept gained from the Romans, "Aqua currit et debet curere ut currere solebat," or "water runs and ought to run, as it has used to run."[35]

Acueductos (Aqueducts) and *Presas* (Dams)

Throughout the Spanish colonization period dams and weirs were constructed by moving and placing rocks, logs, and debris in the stream to lift or divert water into the head gate of the *acequia*. At many places along the typical system, *acueductos* were constructed to carry a ditch over other water courses.[36] Some of these *acueductos* were no more than a dug-out tree trunk laid from one bank to the other; some were marvelous rock-and-mortar structures rivaling those in Europe. There is no finer example of Spanish colonial engineering and construction that exists in its original state than the aqueduct which still carries the water of the Espada Ditch over Six-Mile or Piedras Creek in San Antonio, Texas (see Figures 11.2 and 11.3).[37] This aqueduct was made of stones held together by ground limestone mortar which traditional stories claim was made with "the whites of thousands of eggs and much goat milk."[38]

Dams or *presas* were placed along the river to lift the water into the head gate of the *acequia*. These dams were weir or diversion dams; they did not stop the flow of the entire river, only siphoned off a portion of it to direct and lift the water into the ditch. After each flood they had to be rebuilt. One of the only Spanish colonial-period dams which exists today is Espada Dam in San Antonio; it was one of the few dams built in the Spanish West using stone and mortar. It continues to lift water into the head gate of the Espada Ditch today.

WERE THE SPANISH FRIARS AND JESUITS TRAINED IN IRRIGATION?

The early irrigation systems in the West were built and maintained under the supervision of the Spanish missionaries. In 1741, in a written response to a claim made by the Canary Islanders to the Viceroy for the use of mission Indian laborers, President of the Querétaran missions, Friar Benito Fernández said:

> the Fathers used effective means to teach the Indians how to use the axe and clear the lands, and also how to plow and plant and irrigate the fields, how to build houses, become carpenters and tailors; and this the Fathers also achieved and with greater effort; they have taught the Indians the mysteries of our holy Faith.[39]

His statement is consistent with the writings of the scholars backed up in the primary sources that the friars supervised the digging of irrigation ditches for their missions.[40]

The missionaries in the West were trained, after 1683, at the colleges of Querétaro and Zacatécas in the heart of Mexico, close to Mexico City. A close examination of the curriculum of these colleges does not find any mention of temporal or practical training such as how to build irrigation systems, only spiritual and humility training.[41] However, much of what the Indians knew about agriculture, which included irrigation techniques and

Figure 11.2 Aqueduct over Six-Mile Creek in 1896.
Source: From author's private collection.

Figure 11.3 Aqueduct over San Piedras Creek.
Source: Photo by author.

practices, planting, and tending crops, were taught to them by the friars and the friars alone.[42] The friars also taught them about herding cattle and other domestic animals, weaving cloth and spinning thread, producing leatherwork, operating water-driven machinery to grind corn in the mill, manufacturing ironwork, and blacksmithing.[43] The friars were the teachers of the Indians, but if the colleges did not teach the friars the techniques, then who did and how? So far, no manuals or reference to the friars' training have been found.

The closest document found to date is a guideline for the missionary at Mission Concepcion in San Antonio, Texas which minutely detailed the way in which the friar should manage both temporal and spiritual life in the mission.[44] Some of the instructions referred to duties of the friars in irrigation such as rules 60, 61, and 62 which said,

> When the time for planting nears, the irrigation ditches are cleaned, stopping the water where the river enters. The missionary also will see to it that the bridges and the dam are repaired, when needed.
>
> When the irrigation ditches are cleaned, the natives will mend the gaps in the fence around the field and repair what is in poor condition [...] While some do repair work, others are busy preparing the soil for the planting of cotton, fruit, chile, the seedlings having already been prepared [...] He [the missionary] must also provide for irrigation in due time, and for the hoeing and weeding.[45]

However, this manual still does not mention how to determine where to dig a new ditch, how to properly set the grade, or how to place and build head gates.

Friars from the earliest times in Mexico practiced and supervised the construction of irrigation systems. According to Robert Ricard in his classic *The Spiritual Conquest of Mexico:*

> The development of agriculture is necessarily linked to irrigation. Indeed, it would have been astonishing if the Spaniards had not kept their veneration for water which is still today so striking on the Peninsula. The perpetual fruits and greenery of Uruapan were the work of Fray Juan de San Miguel, who captured and canalized the water of the nearby streams; and in the Mixteca, a dry country of scanty rainfall, the cultivation of the prickly pear for cochineal would have been impossible without the irrigation system directed by Fray Francisco Marín.
>
> Augustinian Fray Antonio de Aguilar [...] transformed the village of Epazoyuca [...] into a healthful and pleasant town by means of an abundance of irrigation water.[46]

Without doubt, in later years, the Royal Corps of Engineers did assist in irrigation planning and implementation in some areas of Mexico and the northern frontier, but they were not founded until 1711 in Spain, at least 140 years after the latest date Ricard's book covered.[47]

The irrigation projects described by Ricard were accomplished during 1523 and 1572 in Mexico and clearly show the success of missionary-built irrigation systems. Their efforts to turn barren, arid land into beautiful food-producing gardens and fields parallels their prime directive perfectly, which was "to spread the Faith, first, last, and always."[48] The "gift" of living water to barren ground in order to give it a chance to become a garden runs hand in hand with the "gift" of the living spirit of Christ which the Friars brought to the native population so as to, in the Friar's minds, give otherwise lost souls a chance to be with Christ in Heaven.

There is little specific documentation yet found that describes exactly the technique used to accomplish the proper gradient. However, there are some indications of the manner in which it could have been accomplished in various articles and books by scholars. Wells A. Hutchins wrote in 1928 in the Southwestern Historical Quarterly:

> Methods of Construction—Judged by present day methods, the early settlers undoubtedly expended great labor in building their *acequias*. In many places, owing to the absence of steel implements, digging was done with wooden spades, shaped with knives, the dirt being removed on rawhides drawn by a yoke of oxen or suspended from poles carried by two men. Surveying instruments were generally unknown, the procedure being to estimate the fall and to turn water into the ditch at frequent intervals during construction to test the grade. Consequently the ditches were frequently winding and complicated, for cuts and fills were avoided and a ditch would be brought close to a tree and carried around it rather than cut down the tree or alter the course of the ditch to keep it straight. While some of the work was being done guns were kept close at hand owing to the menace of hostile Indians. These settlers were usually very poor, and in those cases where livestock was their chief means of support and required much attention they could not work on the ditches for many days at a time, with the result that many of these ditches were years in the building.[49]

According to Arneson, since Fray Mariano had lived at the San Antonio missions as president during the twenty years that most of the *acequias* were dug there, his instructions Bolton found to build an irrigation system at the new San Xavier mission could mirror the way a typical *acequia* was dug across the Spanish West. Fray Mariano determined the location and tools needed for the new ditch at San Xavier in 1750, as, Bolton says:

> he proceeded to order the ministers to be prepared to assist in the work on the fifteenth, each mission providing as many yokes of oxen as it might have, seven bars, fifteen picks, four axes, and one cauldron. [...] he appointed Father Ganzabal superintendent of the work, with the duty of assigning the tasks and taking care of the tools and supplies. [...] With this preface Fray Mariano asked that one soldier be charged, during the work on the ditch, with caring for the tools, and another with looking after the oxen sent to work on the ditch and to haul stone for the dam. [...] Such were the preparations for opening the irrigating ditch and building the dam.[50]

According to Bolton, Fray Mariano, "appointed Father Ganzabal superintendent of the work," which gives proof, at least in this case, that the Father was to be personally involved at the site of the *acequia* system's construction in the day-to-day work, in fact, directing it.[51] Bolton's source was a letter from Fray Mariano to the missionaries dated October 17, 1750, in *Diligencias practicadas parel R. P. Presidte.* The tools required by Fray Mariano concur with the description of the methods mentioned by Hutchins, also.[52]

It is certainly possible that irrigation expertise was common knowledge to the friars from Spain, because many of them came from areas there that were totally dependent upon irrigation to generate food production. The earliest Franciscan to appear in Mexico was Friar Martin of Valencia in 1523.[53] Friar Antonio Margil, founder of San Jose mission in San Antonio and President of the Zacatécas College after spending twenty-three years at the Querétaro College, was born in Valencia in 1657.[54] Valencia, a key city and area in Spain, had depended upon irrigation for hundreds of years before the births of Friar Martin and Friar Margil, so it is very likely both could have observed the water distribution methods so critical in that province and that knowledge was "in their minds" by experience if not by training.[55] According to ancient irrigation scholar Thomas F. Glick, "Irrigation practices, that is, migrated from Spain to America in the *minds of men* and not in law treatises, as many students of water law appeared to believe."[56] None of the known descriptions of the daily curriculum at the missionary colleges mention any formal irrigation training at the colleges.

The achievement of the correct depth of a ditch covering several miles of varying ground required a skillful supervisor. The work on the frontier of Spanish West at this time was still done without the benefit of any professional surveying tools such as a transit.

The *acequia* systems of San Antonio and other Spanish communities in the West were more than just irrigation systems for agricultural use. The systems distributed water for all uses by all the settlers, including personal consumption and other household use, were managed and regulated by the local community and were the first municipal water systems in today's continental USA. The American Public Works Association defines public works as:

> works developed by and for the benefit of people. Designed to protect and enhance the human environment, they represent investments in the future for the people who create them and for succeeding generations. [...] Government institutions, land, capital, labor, and public support are prerequisites for the development of virtually all public works facilities.[57]

The *acequias* of the Spanish West meet this definition well and were truly "public works." For example, Villa San Fernando de Bejár was formed by authority of the Viceroy in 1731 as a municipality and of its main duties, the construction, operation, and management of their *acequia* system was foremost in the duty the municipal government owed the settlers.

The earliest public water system in the USA was in Bethlehem, Pennsylvania, and it consisted of a pump forcing spring water through bored hemlock logs into a wooden reservoir that provided the town with water: however, this did not come into existence until 1754.[58] This was more than thirty years after the settlers of the original Villa de Bejár completed their *acequia* from San Pedro Springs.[59] The city of Boston saw no need for a publicly owned municipal water system until fire fighting in the city was clearly hampered by private water-well companies and people fighting fires with bucket lines. This makeshift system had resulted in the recent loss of fifty houses and stores, so the city passed a resolution to establish a public water system in 1825.[60] Other cities suspected that epidemic diseases came from contaminated public water wells or reservoirs and were unsure of public water systems in general until the late nineteenth century.[61]

WATER LAW CONCEPTS IN THE SPANISH WEST

The rules, regulations, laws, and management systems in the Spanish West, while varying locally from community to community, were essentially the same across the land and were especially similar in all the Indian pueblos. Since the fundamental laws and regulations continued to be valid for three centuries and portions are still honored today, an exhaustive examination is important to understand the history of irrigation in the West.

The combination of water in an irrigation system with the wise Spanish concept of *derecho,* right or justice, formed the foundations of the new communities in the Spanish West.[62] *Derecho* has no exact equivalent word in English; "justice" comes closest.[63] The legal system of Spain which was applicable to its colonies was designed to be flexible to fit local needs; therefore, when adapted and modified it is referred to as *derecho vulgar* and is a "legitimate expression of local self-government."[64] Justice was considered in court first, not law. According to Charles Cutter, "One rarely argued that the *ley,* law, was on his or her side; instead a litigant approached the court seeking one's justice or '*su derecho.*'"[65] Justice was found by the written law, local custom, and *equidad* or a locally defined sense of fairness.[66] Cutter writes "Local usage and long-standing practice also carried the weight of authority under the Hispanic system."[67] As Cutter goes on to say, "this respect for local particularism, even when *contra legem,* contrary to the law, has been perhaps the most overlooked dimension of the Spanish colonial legal system."[68] The Spanish concept of *derecho vulgar* allowed local regulations to be developed to meet local customs and to match the unique needs of a local population.[69] From the arrival of the earliest settlers, rulings about surface water-sharing in the new settlement were based on the Spanish concept that water was to be used in common by everyone in the community.[70] This idea is an example of the overall fairness in the Spanish legal and administrative system for which they are rarely, if ever, given their due credit. According to Meyer:

> In a system of exaggerated local control, abuse in water allocation was constant, and when it became intolerable the judicial mechanisms of the state were called into place.
>
> But there are enough examples of Indians, mestizos, and poor Spaniards coming out of the courts with more water than when they entered to conclude that the voluminous legislation designed to protect the interests of the disadvantaged, both before and after the Mexican independence, was not completely in vain. Compromise and concern for

> the common good were not merely lofty goals rejected cavalierly in the courts of the Hispanic Southwest. [...] They were fundamental principles brought to bear even in the most complex water adjudications and even when the status of one of the litigants would have suggested that his opponent stood no chance in the impending case.[71]

Mutual dependence upon shared water forged a sense of community in the people; the methods they practiced in sharing this water helped form one of the most basic principles in Western community water law as we know it today: Water is to be protected for community use and benefit and is to be shared fairly by all. From a legal perspective, Spanish ideas that water was a common resource to be used *by* everyone and administered with justice *for* everyone comprised an ethical cornerstone for the development of their communities. For example, these just ideas, established almost 300 years ago, live today in much of the water law in the state of Texas and elsewhere in the West. As Dr. Andrés Tijerina said, "the public nature of water ownership is probably the most pervasive of the Hispanic traditions in Texas water law."[72]

The irrigation systems built in the Spanish West to distribute the water became the focal point, practice field, and classroom for diverse people with competing interests to learn how to live with one another in community.

ANCIENT SPANISH WATER LAW

> Spanish water law, if not representing the triumph of reason over will, was an ingenious system that provided the moral mechanism for bridging the gap between the self-interest of the individual and the larger concerns of society. It combined the reasonableness of private property with the justice of serving the common good.[73]

The laws of Spain stem from codes, regional laws, decrees of the king, and decisions of the Supreme Court of Spain. A royal order of the king is supreme law that supersedes and repeals all preceding laws inconsistent with it. The decrees of the king were codified, and, as a new code repealed only a portion of an earlier code, it was important to search former codes as well as recent ones. Las Siete Partidas was first published in 1263 and became the recognized law of Iberian Spain after 1348.[74]

There was a discussion of water in Titles 28, 29, 31, 32, Part 3. These stated that the waters of navigable streams for stated purposes may be used by all persons in common. Among those uses were navigation, mooring of boats, making repairs on ships or sails, landing merchandise, and fishing and drying nets. As to irrigation, Title 31 stated that there were restrictions upon mills and canals, "because it would not be wise that the benefit of all men be hindered by the interest of some individuals." Title 28 stated that while common use of waters from navigable streams for stated purposes was the rule, springs and wells on a man's premises were not for common use. Notice in all these rules, only "use" of the water was consistently discussed, never ownership. The ownership of the water remained in the crown.[75]

Land transfer started by an initial grant from the king. Even if the land granted was riparian, or adjacent to flowing water, the grantee did not have an automatic right to take water from the stream or river unless the king specifically granted that right.

Because of this requirement that use of water in a stream had to be granted by the king in each individual case, another important base concept arises. Rights to water in times of scarcity were based upon the earlier the grant: first in time granted, first in access. Later grants to water in the same stream were subservient to earlier grants. This is known in modern terms as prior appropriation. If water was very scarce due to drought, upstream users had to first be sure downstream users with prior grants were satisfied *before* taking their water. The idea of a turn to water based upon age of the grant still rules in most Western water law.

In 1520, a scant twenty-eight years after discovery of the Indies, the King created the Council of the Indies to try to cope with the myriad of challenges in his colonies. The Council of the Indies' decrees governed New Spain, and the King transferred "supreme jurisdiction over the Indies" to it; his will for the colonies was expressed through the Council exclusively. Laws "poured" out from the Council in "profusion," and, finally, in 1680, all the former decrees were codified into the Recopilación de las Leyes de las Indias. The decree that promulgated the edition is important to note as it confirms the overall concept of justice in the Spanish world. It stated, "That the laws contained in this book are given for good government and administration of *justice* [emphasis added] by our Council of the Indies." As subsequent orders from the Recopilación were made from the Council of the Indies they were again controlling expressions of the absolute sovereign's will.[76]

The intentions of the King that were expressed through the Council of the Indies via the Recopilación were further expressed in individual grants of water rights and land. The Recopilación made it clear that the King intended that the waters from rivers and streams were to be for the common use of all men unless and until there was a right to use given to an individual by a royal grant. Don Felipe II in 1563 and Don Felipe IV in 1631 established a system of water judges for the colonies, similar to the ancient water judges of Spain. Water was to be apportioned "according to need," and colonial officials were authorized to make grants in the King's name not merely of lands but also of lands and waters.[77]

The wording of the establishment of the judges in New Spain further shows the justice in the Spanish system and the concept of flexibility and faith in local administration to fit the unique need of any locality in the New World. It is a rare trait of kings to give away their absolute power to local administrations, and in this concept, Spanish kings showed their wisdom along with their pragmatism.

> We decree that the assembled audiencias name judges, unless a contrary custom shall exist, as nominated by the Viceroy or President City or Corporate Town who shall apportion waters to the Indians for irrigation of their farms, orchards, and seed beds, and to water their cattle. They are to be such as shall offend no one and shall apportion waters according to need [...] We order the Viceroys to inform themselves as to the lands which there may be that are irrigable; and that they order the cattle to be taken therefrom, and that said lands be sown in wheat; unless the owners have titles to maintain this type of farm.[78]

Land classification was required to meet the duty of "according to need" and lands were first distinguished as "irrigable" land, non-irrigable land for farming or "cultivated land" and "pasture" lands. An example of the importance of these classifications was the special order above that was sent to the viceroys to remove cattle from irrigable lands.[79] In normal progression, the quantity of the land granted and its price depended upon the classification. Irrigable land cost more and was granted in lesser quantities than non-irrigable lands or pasture lands.

Law 9, Title 5, Book 4 of the Recopilación turns to water: "Each one shall be given the water he should have, alternating from one to another."[80] The idea of a turn or *dula* to use water gets its legality from this provision. The concept of a turn in itself is just and coupled with the basic concept that "water runs and ought to run as it used to run" helped ensure everyone had a chance to share the water especially when rainfall was scarce and the flow of the springs and river was weak.[81] During weak flow periods, the *mayordomo* was responsible for reporting to the *cabildo* that it was time to place conservation limits on users.

In 1761, the Viceroy of New Spain authorized the publication of Reglamento General de la Medidas de Aguas, a commentary on water written by Lasso de la Vega. Key phrases of his commentary clarify further the rights the people have to personal and domestic use of water and that using public water for irrigation requires a specific grant from the king:

> [I]t needs to be said, for the right interpretation of this subject that all the waters of public rivers being, as they are, of public and common use, they are not so with reference to personal and domestic use, which is in accordance with the general freedom with which one can take the water he wants to succor his domestic needs [...] it is a legitimate conclusion that no one without permission of the Prince can conduct public waters to his lands for irrigation, especially in New Spain, where, let it be known, His Majesty has conceded very ample powers to the most excellent and high Viceroys and Presidents of the Audiencia Real of this New Spain, so that in the aforementioned terms they may issue grants of lands and waters, as things appertaining to His Royal Crown, and for which today there is a special Court.[82]

The primary title to land or water in the Spanish West was always the grant by the king or his authorized representatives. If the grant did not mention waters, a right to the water was never considered to exist solely by reason of proximity; hence location of riparians in the Spanish West, those simply adjacent to the stream or river, did not give such lands any right to use the water.

The riparian system of water rights was honored in the English colonies on the east coast of North America and later in the thirteen states comprising the early USA. The riparian landowner had no ownership in the water that flowed by his property but was entitled to reasonable use of it, subject to the rights of other riparian owners on the stream.[83] When individual rights are subject to the rights of other individuals, these rights are said to be correlative. The Spanish water rights system of the West, however, was not a riparian system.[84]

Keep in mind that if one was the recipient of water rights by Spanish royal grant, the grant was only a usufruct right or right to use the water, never ownership of the water. The Spanish system of grants of water usage made their system of water rights an appropriative system. The preference in the use of water of certain landholders over others originated not in the location, low or high, near or far, of the tracts of land but in the antiquity of the grant known as prior appropriation. The older grantees had first access to their water in times of scarcity, no matter the location on the river or stream. Rights among individual grantees in a prior appropriative system such as that of the Spanish system were not correlative. Waters that were not specifically granted by the King remained in the ownership of the King. The states in the West generally follow the prior appropriative system of water law.

EXAMPLES OF EARLY WATER GRANTS AND MANAGEMENT

Indicative of land and water grants in the Spanish west, immigrants and the Franciscan or Jesuit missions received their irrigation rights by specific or express grants. For example, irrigation grants were made to persons who were granted land adjacent to the San Antonio River or San Pedro Creek or both, and other grants for irrigation were made to persons not adjacent to any water course. Land grants to those riparian to both streams were many times made *sin agua* (without water). These water rights were sold and exchanged without regard to riparian status: water grants were made apart from lands; waters were mortgaged apart from land; and in case of dispute, the quarrel was determined by the title one could exhibit.

Spanish colonial laws were also recognized by the Republic of Mexico in the 1824 Constitution, by the Constitution of Coahuila and Texas in 1827, by the Constitution of the Republic of Texas, by the State of Texas, and by the USA in the Treaty of Guadalupe Hidalgo. Courts across the West since have modified and the state legislatures have made changes that impacted the early Spanish laws, but in general the courts to this day honor Spanish land grants, and portions of the Spanish water laws are incorporated into modern law.

From its beginning, the Spanish West's human–water interface determined relationships between groups, the work people did, spatial boundaries of groups of people, and the shape of individual land grants. An integral part of the *acequia* system was the development of social and political organizations and laws to manage and conserve the precious water resource.

Since most Western state water laws and adjudication proceedings usually allow Spanish land and water grants to prove water rights and their priority, samples of these grants are germane to this chapter. Use of the water in a Spanish irrigation system was usually divided into discrete segments of time that an individual's head gate could be opened to let water from the *acequia* into his field or lateral ditch. Division of access to the ditch required intense voluntary cooperation among the users; upstream irrigators must be conscious of downstream irrigators, and all had to participate in overall maintenance of the canals.[85] This consciousness is normal under reasonable times, but in times of scarcity, the older grants under Spanish law, no matter of location on the river or stream, had first access.

Formal distribution of land and water rights was carried out after formal establishment of a community in the Spanish West. The Governor or his assign, at this time, publicly declared some percentage, usually 20 percent, of all lands in this distribution belonged to the community as their *propios* to rent or sell to support the villa.[86] The communities through their *cabildos* were granted their rights or *aquas de propios* to a typical 20 percent of the water in a stream.[87] The *cabildo* within a short time rented its rights to turns to the water in its *acequia* to individuals, thereby generating revenue for the villa, a universal practice in Spanish villages in the West.

The division of access to water in the ditches required supervision by some authority, and this duty fell to the *mayordomo de los bienes y propios de la republica*.[88] For example, the first *mayordomo* of San Antonio was an eighteen-year-old Canary Islander named Antonio Rodriquez Mederos whose responsibilities included management of public funds and the general supervision of public works within the town, works which for many years were comprised mostly of the *acequias*.[89] He was authorized to police the use of the town *acequia* and perform inspections of the entire irrigation system.[90] His duties were indicative of the duties *mayordomos* performed across the West and were similar to today's water master's.

The entire community was required to help maintain the system annually, and individuals were subject to fine or incarceration if they did not comply. The ditches had to be cleared of silt, limbs, leaves, and releveled to keep the water flowing. The work created a further sense of community for the citizens.

The water-distribution system became a social organization, and its managers projected the power of the controlling group of the *cabildo* in the community. Here lies the main difference between American Indian and European views of water. As Meyer aptly puts it,

"While the native American populations viewed the precious liquid as the medium of life, the conquerors, missionaries, and settlers viewed it as an instrument of control, a source of power, and most importantly as the fount of accumulated wealth."[91]

Since the local authority marketed its own water rights, there was no way to keep other citizens from trading or renting their water rights, which led quickly to the severance of those water rights from the land to which they had been originally adscripted.[92] The time allotments or dulas to the water were sold or rented elsewhere also throughout the Mexican north.[93] Michael C. Meyer said "Just as with any other commodity, the price of water was dictated by what the market would bear."[94]

A typical example of a sale of land and detailed water rights occurred on September 15, 1747, in San Antonio and is indicative of land and water transactions across the Spanish West. Antonio Rodriguez Mederos, the original *mayordomo* of San Fernando, sold two *suertes* of land to Miguel de Castro and one day of water every twenty days.[95] An example of a sale of water rights without the land was Antonia Armas to Francisco Amangual and Jose Antonio Bustillos. Armas was owner of three hours of water and corresponding land and sold the three hours to Amangual and the land to Bustillos.[96]

Another early example of a subsequent sale typical of the manner in which property transactions were completed under Spanish law was from Joseph Padron and his wife to Geronimo Flores on March 22, 1738, of "two suertes or portions of irrigable land [...] and the water from San Pedro Creek to which we are entitled."[97] The description of the property is described as "the two suertes or portions of *irrigable* [emphasis added] land" clearly establishing a classification of the land as *irrigable*.[98] The continued description of the land and water in the deed located the *suertes* and detailed the source of the water right they were granted. Next the deed further located the four neighboring sides. One was San Pedro Creek; one was the San Antonio River; one was the land of Joseph Cavrera; and the final side was owned by Don Juan Leal Goras. "These lands and *water* [emphasis added] are free from ground rent, entail, right of primogeniture, debenture, mortgage." It is undeniable when reading this deed that the sellers owned the land and water right and were selling both to Geronimo Flores. The deed is witnessed by four settlers and notarized by one of them prior to being filed of record. The express grant of the land and water rights were sold. The land could have been sold separately and the water rights could have been sold separately, but Joseph Padron and his wife chose to sell both.

Another example of a similar sale, Angel Navarro on June 8, 1789, acting for himself, his brother, and sister, sold to Jose Salvador Diaz two *suertes* with four and a half hours water rights for 10 pesos cash, one horse, stone for a house, and a pair of handmade spurs.[99] It is important to note that by 1789 in San Antonio, water was being apportioned to the hour and even half hour, proving the extent of its value to the landowners. This began to occur elsewhere in the Spanish West.

At times, transactions were trades of the water rights themselves, without any land attached such as in the case of the trade on March 23, 1782, between Juan Joseph Montes de Oca and Juachin Flores de Sandeja where Montes traded straight up his twenty-three-hour water right to Flores for a sixteen-hour water right.[100] As time went on, the transactions became more complicated, such as the July 11, 1802, sale from "Bueno" to "Valle" of half a *suerte* and 4 hours of water plus three-quarters of another *suerte* for 100 pesos.[101] Water rights were described in the land deeds and could be traded or sold separately. The purchase and sale of water rights, without title to the land above or adjacent to them, is still done regularly today across the West but is subject to great controversy and complications.

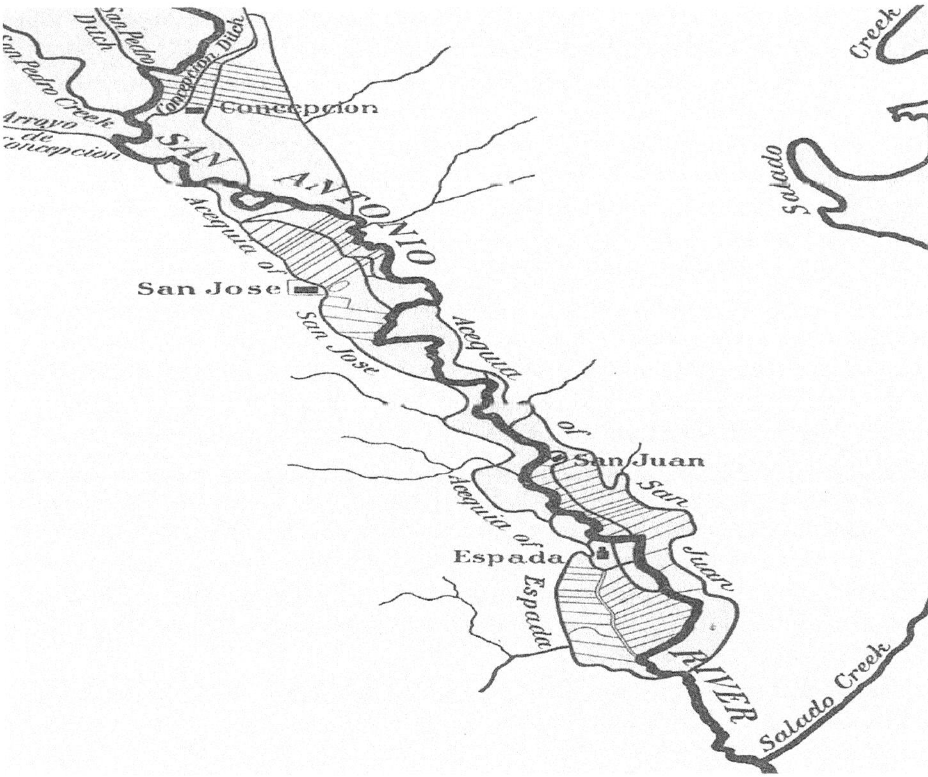

Figure 11.4 Eighteenth-century lot configuration, San Antonio.
Source: From author's private collection. Department of the Interior. Water Supply and Irrigation Papers of the United States Geological Survey No. 13 (Washington, DC: US Government Printing Office, 1896).

The Spanish grants of land were surveyed and staked out in long rectangles so more people could access the water (see Figure 11.4). This, of course, aided colonization: the more access to water, the more settlers, hence the better collective defense against hostiles that led eventually to more secure national claims.

Spain and Mexico ruled the West until 1836 in Texas and ruled the remainder of the West until 1848 after the loss of the Mexican–American War with the USA. As the USA tried to settle the West to fulfill its "manifest destiny," each area, territory, and eventually state developed its own water laws to govern irrigation through much consternation and seemingly never-ending dispute.

WATER RIGHTS BECOME CONFUSED IN THE POST-SPANISH WEST AFTER 1850

In the words of David J. Weber, "There is a saying in the West that 'water does not run downhill. It runs toward money.'"[102] The story of irrigation in the West is one fraught with confusing and conflicting laws and court rulings, a general lack of overall direction toward any common goal for the entire country's benefit, short-sighted legal experimentation which resulted in dual systems of law in many states and territories, and a disjointed at best approach to water conservation. The myriad of laws, regulations, and court rulings which vary from state to state make the understanding of the history of irrigation in the West since the Mexican–American War a challenge for even the most learned jurist or legislator.

The basic causation of the confusion in the West was the application and misunderstanding of riparian rights under the blended Spanish and English legal concepts. Indicative of this problem across the West, "In its wisdom," wrote water-rights expert Garland F. Smith, "the Legislature [in Texas] adopted the Common Law of England, and the courts (with the aid of the Bar) for over a century tired to adapt the non-consumptive riparian doctrine to the consumptive use of irrigation."[103]

As mentioned, the word "riparian" simply means adjacent to a flowing river or stream. However, riparian rights under English common law are made up of a bundle of rights, which include the right to flow (for industries such as mills), to fish, to navigate, and to use the water for irrigation and for domestic purposes.[104] Irrigation is but one right in this package of rights but is a highly consumptive use of water and certainly the most controversial in arid climates.

In England and the eastern coast of the USA, home to naturally well-watered lands with ample rainfall, riparian rights are rarely controversial; there is generally plenty of water for everyone. This is not true in the arid West: riparian rights of irrigation create serious water shortages during the recurring and, at times, lengthy periods of drought. When water is ample, few disputes arise, but when it becomes scarce due to drought of excessive use by one party over another, conflict explodes.

Hans W. Baade wrote in 1986 in his *St. Mary's Law Journal* article on the historical background of Texas water law:

> "Therefore, while Spanish and Mexican water *law* is entirely unrelated to the substantive contents of the Texas Water Code, water *rights* based on the law of the pre-Independence sovereigns of Texas [Spain and the Republic of Mexico] not only continued to be valid, but indeed had to be ascertained, whenever and wherever there were Spanish or Mexican land grants within a stream or stream segment to be adjudicated."[105]

Baade continued: "water rights claimed to have arisen under Spanish and Mexican era grants had to be ascertained, even at this late date [1967], in terms of the water law of Spain and Mexico at the time and place of the grant of the surface estate pertinent to them."[106]

Since title to almost one-sixth of all the land in Texas was traced to valid pre-Independence grants, the question of the water rights associated with these claims was not just one of historical interest but was and still is critical to the livelihood of the landowners and also to the market value of the property for both the owners and the various government taxing authorities such as school districts. The adjudication of these rights in Texas was easier since there is virtually no land owned by the federal government in Texas. But elsewhere in the West, the land holdings of the US government complicated the already difficult issues of water rights.

How did the USA acquire the vast lands of the West? The first large acquisition was the annexation of Texas in 1845. Along with it came portions of New Mexico, Oklahoma, Kansas, Colorado, and Wyoming, which had been claimed from Mexico by Texas in 1836. The Oregon Compromise of 1846, coinciding with the beginning of the Mexican War, annexed the areas now known as Washington, Oregon, Idaho, and Montana into the USA from a settlement with Great Britain. The land comprising California, Nevada, Wyoming, Colorado, Arizona, and New Mexico came into the USA after the conclusion of the Mexican War in 1848. In 1853, through the Gasden Purchase, the USA acquired an additional 19 million acres from Mexico. The land that would eventually comprise the Western seventeen states was fully acquired by the USA by the end of 1853.

In 1848, the USA agreed to the Treaty of Guadalupe Hidalgo that ended the Mexican–American War. One of its terms was an agreement that the USA would honor all then-existing claims to property in the newly gained lands of the West owned by citizens of Mexico. "Mexicans now established in the territories previously belonging to Mexico [...] shall be free to [...] remove at anytime to the Mexican Republic, retaining the property they possess in the said territories [...] In the said territories, property of every kind, now belonging to Mexicans not established there, shall be inviolably respected."[107]

The US Senate failed to ratify Article X, which specifically discussed land grants causing enough concern in Mexico that President Polk sent envoys to Mexico to negotiate an international protocol that resulted in an explanation detailed in the Protocol of Querétaro Article II:

> The American government by suppressing the 10th article of the Treaty of Guadalupe Hidalgo did not in any way intend to annul the grants of land made by Mexico in the ceded territories. These grants, notwithstanding the suppression of this article of the treaty, preserve the legal value that they may possess and the grantees may cause their legitimate titles to be acknowledged before American tribunals. [...] [L]egitimate titles to every description of property, personal and real, existing in the ceded territories, are those whose legitimate titles under the Mexican law of California and New Mexico up to the thirteenth of May, 1846, and in Texas up to the 2nd of March, 1836.[108]

The amended treaty won ratification by Mexico, and the documents confirmed valid Mexican land grants in the West.

WATER RIGHTS AND WATER LAW IN GENERAL

Water by its very nature presents a unique ownership problem as well described by Sir William Blackstone in 1807,

> For water is a moveable wandering thing, and must of necessity continue common by the law of nature; so that I can only have a temporary, transient, usufructuary [sic], property therein: wherefore, if a body of water runs out of my pond into another man's, I have no right to reclaim it.[109]

Property ownership includes a bundle of rights, one of which is a right to water. Water rights under English common law were different than under the civil law of Spain and the Republic of Mexico. From 1850 to date, some Western state legislatures and courts have struggled with the conflict between the two legal systems over the rights to water of riparian lands, or lands adjacent to flowing water in rivers and streams. The Western states adopted English common law in which riparians had the right to take water from flowing streams for irrigation based solely upon their adjacency to the water. Under Spanish and Republic of Mexico law, water could not be taken from a river or stream for irrigation without an *explicit* grant from the sovereign; adjacency to the stream or being *riparian* in the Spanish West did not guarantee a water right for irrigation. The Spanish and Mexican system was an appropriative system, and many of the courts in the West have relied upon the customary practices and application of Spanish laws to support their rulings and set legal precedents on water issues since 1850, and in Texas since 1840.

From the beginning, the laws relating to water rights in the West sent many attorneys and more than a few judges scurrying for help. Federal Judge James V. Allred, former Attorney General and Governor of Texas said in 1955:

> For years it has been a matter of common knowledge that the Texas water laws and decisions are in hopeless confusion; that even if they are clear as some attorneys profess to believe them, their application and administration would be difficult [...]; that the Board [Texas State Board of Water Engineers] has granted permits on many streams in the state, very few of which have been cancelled, in such numbers and for such quantities, that if riparian rights are given the full effect, practically every drop of water, normal flow or flood, is "bespoken."[110]

Logic-defying and seemingly unfair judicial rulings on water issues in the West have been rendered in the past and continue today. The same holds true for guidance from the legislatures; the lack of legislation on various important water issues stands out most noticeably in Texas and Kansas. Two of the most controversial water issues are rights to irrigate from a flowing stream of surface water and the use of large volumes of groundwater in such quantity that neighboring water wells, springs, and creeks go dry. A primer in water uses and the types of legal "geological containers" in which water is held under the law helps to clarify the issues that are in dispute.

There are three basic ways water is used under the law: for households and livestock, for irrigation, and for industry. There are three types of geological containers under most Western state law: surface water in streams and rivers, groundwater or water underground in aquifers or streams, and "diffused" surface water or water run-off from rains that does not flow in a stream with a defined bed. Spanish, Republic of Mexico, and Western state laws have always generally concurred on protection of household and livestock use of any type of water; of this there has been some but little dispute.

The most heated and challenging disputes have always been over irrigation use, whether from surface water or groundwater. Why?

Irrigation of land requires huge amounts of water and is considered as an artificial use. Domestic use for households and watering livestock is considered a natural use. Until the late nineteenth century, most irrigation water came from rivers and streams. By the end of the first decade of the twentieth century, more water than flowed in most of the rivers of the West had been claimed. With late-nineteenth-century technology in which deep drilling and large pumping equipment became available across the West, irrigation by groundwater became increasingly popular. For example, irrigation today accounts for almost two-thirds of all groundwater use in Texas; other states similarly use high quantities of groundwater for irrigation purposes. Groundwater irrigation puts pressure on aquifers and can dry up springs and their resulting streams all around the West during times of drought.

Since some of the states in the West officially recognize their mixed Spanish/Anglo heritage, the confusion and frustration over irrigation water and riparian rights to use it could be said to be a central issue in water-rights disputes. In Texas, in fact, it was not until the fine clarifying work of Justice Jack Pope in 1962's *State* v. *Valmont Plantations* that the Spanish water-rights system was undeniably proved to be an appropriative system. Pope proved their system was based only upon specific grants to the use of the surface water. His ruling backed up by exhaustive historical research ended the riparian right to irrigation for good in Texas law. Subsequently the Texas Water Rights Adjudication Act of 1967 was passed in order to settle for once and for all the issue of surface-water irrigation rights in Texas. After 120 years of disputes in the courts, legislature, and government agencies statewide, the Act required that all claims to surface water for irrigation purposes had to be presented to the Texas Water Rights Commission for adjudication between July 1967 and July 1969. The goal was to verify the true claims with a certificate of adjudication to settle the years of costly conflict between parties claiming surface water rights. Another goal was to fairly and responsibly allocate the finite amount of Texas surface water. Some 10,000 claims were presented during this time, and most claims were adjudicated as confirmed or denied, but even today a few remain in dispute.[111] Texas is one of the few Western states which today have fully adjudicated surface water rights.

IRRIGATION IN THE WEST AFTER 1850

> The problem facing the settlers of the West was [...] to apply a limited quantity of available water to an unlimited quantity of available lands in the hands of a constantly increasing number of people to get the maximum of benefit and good to the maximum of people, the community as a whole.[112]

Gold was discovered in California in 1848, drawing thousands of new immigrants virtually overnight. Gold and silver were also discovered in the mountains of Colorado which led to the new city of Denver exploding into existence in the same manner. Silver was discovered in Arizona. The large influx of the miners mostly came from the east coast of the USA where they understood water rights to be of a riparian nature.[113] Water was desperately needed for these early mining ventures, and riparian concepts were inappropriate for their work. According to Hildebrand:

> Water was necessary to mining and was taken from streams on public lands, often at great expense and over long distances. It became the custom to recognize that he who first diverted the water from a stream was first in right as to so much water as he put to beneficial use. The federal government was the only riparian owner, and its silence was regarded by the settlers as acquiescence to the custom.[114]

The activity of miners was considered to be for beneficial use, and early on their rights were given priority across the West. The early California miners, called "forty-niners," are given credit for introducing the concept of prior appropriation to the West. Miners in the states of Colorado and Wyoming may also have a claim to being the first to introduce the doctrine of prior appropriation in the West. The early Mormons in Utah also were practitioners of an appropriative doctrine.[115]

Congress officially passed into law the doctrine of prior appropriation of water on federal land in the Mining Act of 1866 which read in part, "by priority of possession, rights to use of water for mining, agricultural, manufacturing, or other purposes, have vested and accrued."[116] On July 9, 1870, Congress strengthened prior appropriation on federal lands with: "All patents granted, or pre-emption or homesteads allowed, shall be subject to any vested and accrued water-rights, or rights to ditches and reservoirs used in connection with such water-rights, as may have been acquired under or recognized by the preceding section."[117]

Water-law scholar Donald J. Pisani opined about prior appropriation,

> The greatest legal innovation in the history of the arid west was the doctrine of prior appropriation, which made water as much of a commodity as land, minerals, trees, crops and livestock. [...] The California mining industry, and specifically its enormous need for capital, transformed water into a species of private property.[118]

Miners needed water, and lots of it. By the late 1880s the use of surface water had significantly declined, but the concept of prior appropriation stayed across the West as the dominant legal idea in water rights.

Mining brought innovative water laws, but mining activities had unintended consequences. According to Western legal scholar Gordon Morris Bakken:

> Water and gold mining were joined when the first pan filled with pay dirt and water left a trace of color in the pan. Water was flumed to gravel and processed more dirt until 1853 when the invention of the penstock concentrated water at high pressure. From there American invention abounded until rivers of water were run through hoses, nozzled at even higher pressure into mountains of gravel, and sluiced away downstream leaving more gold [...] Downstream riparians saw something else. Polluted water, dead cattle, destroyed pastures, and buried private property caught their eye. Lawsuits resulted, and mining engineers went to work gathering evidence.[119]

Large portions of hills and mountains were simply washed away, leaving death and destruction behind. Edward E. Mattison is given credit for inventing "modern" hydraulic mining in 1853. Descriptive of the effectiveness of the technique, Pisani comments, "The toughest clay dissolves like wax, this disintegrating much fine gold, a greater part of which has hitherto been lost. [...] The advantages [...] The reduction of manual labor at least two-thirds."[120]

The typical cycle of business efficiencies in America were in place in early hydraulic mining ventures in the West. Capital investment demanded profits that resulted in increasing returns on investment. Profit and returns only increase when more revenue flows into a business at less cost. "Modern" hydraulic mining, while the ultimate in destructiveness to the environment nonetheless was the inevitable step in early stage mining in the West. As capital investments increased, the providers of the capital required protection for their most important raw material used to pull the gold from the dirt: water. The doctrine of prior appropriation was the logical way for the investment community to best guarantee their efforts could continue at the cheapest cost and give their existing operations an edge over the competition. Competitors were of little consequence if they were unable to obtain access to water for their hoses.

The prior-appropriation doctrine had overall advantages according to Pisani:

> First, no value judgments had to be made about how or where the water was used. As long as the water was claimed for a "beneficial use," no one could overturn the original right—even if a subsequent claimant needed the water more or promised greater wealth with it. Prior appropriation offered equal access to wealth, if not equal opportunity; it was consistent with American ideas about political liberty. [...] It did not require an expensive bureaucracy to administer or maintain, a bureaucracy whose cost would have been prohibitive in frontier California [...] Finally, it was consistent with the cherished American ideal that individuals, not society, should control their

> destiny. The welfare of the community became the product of a multitude of discrete decisions made in the marketplace, with no sense of planning and little of collective responsibility.[121]

Prior appropriation has been followed on public and private lands in Texas, Arizona, Colorado, Utah, Idaho, Nevada, New Mexico, and Wyoming. According to Hildebrand in its most basic form, "It [prior appropriation] rests upon the general proposition that he who is first in time is first in right."[122] Pisani summed up the importance of prior appropriation in the West: "Without prior appropriation, the capital needed to transform the West into a region of dams and canals could not have been raised, and it is easy to forget that private capital reclaimed most of the West's irrigated land."[123]

However, over the years, some legal scholars have opined that the freedom of the appropriator has decreased and the control of state authorities who issue licenses for diversion of water from the stream or river has increased to the point that the doctrine of prior appropriation has been all but effectively abandoned. According to White and Wilson, "The tendency toward public supervision is manifest in the other arid States, and it seems only a question of time when the doctrine of appropriation will give way to complete public supervision."[124] Their 1955 comments have proven to be correct.

GROUNDWATER LAW

Irrigation for the past century has depended heavily upon groundwater sources and is its major use over most of the West, putting the resource under intense pressure in many states. Most Western states have strong groundwater laws similar to their surface-water laws; however, Texas is the only remaining state that honors the rule of capture. The "rule of capture" in Texas as it pertains to groundwater states that a landowner who drills water wells on his or her land can pump all the water from underground that can be "captured" without any liability for damage to any neighboring property owner. If the landowner does not waste the water from the well or dig the water well to intentionally harm a neighbor, even if the neighbor's water well dries up completely, the neighbor has no claim for compensation or damages.

The "rule of capture" does not recognize correlative rights or rights with mutual or reciprocal relation in such sense that the existence of one necessarily implies the existence of the other.[125] As between landowners pumping groundwater in Texas courts there have consistently upheld the rule for over 100 years. A dissenting Texas justice wrote in 1955: "In the field of water law there is no consolation in the rule of capture [...] If the law of capture has any true application to underground water, it is an extremely limited one. No one can live in a vacuum. Therefore, all property rights are to a certain extent correlative."[126]

The question of the rule of capture has been generally resolved across the West except in Texas. Yet, modern-day supporters of absolute rights to groundwater could find solace in Spanish water law. In the words of Spanish water-law expert Michael C. Meyer:

> Water that originated on a piece of land, that ran solely within its confines, or that lay *under* [emphasis added] it was automatically alienated from state ownership with the sale or grant of the land. It was appurtenant to landownership. No special water right or additional permission was required to use it, and no limits were set on the amounts that might be used.[127]

The *Siete Partidas,* in fact, specified it was an obligation of all inhabitants to make their land productive, and it further indicated that "man has the power to do as he sees fit with those things that belong to him according to the laws of God and man."[128] Hispanic groundwater law was designed to protect individual rights, to encourage private initiative and entrepreneurship, to stimulate economic development, and even to accumulate personal wealth.[129]

Meyer went on to sum up that, "It [Spanish water law] combined the reasonableness of private property with the justice of serving the common good."[130] For whatever reason, the courts ignored the much older, more appropriate, and actually, at that time, more recognized law in Texas, that of Spanish water law.

In upholding the original trial courts decision of 1904 to allow the rule of capture to become the law in Texas, the Texas Supreme Court relied primarily upon two cases: the English case, *Acton* v. *Blundell,* and a case decided by the Supreme Court of Ohio, *Frazier* v. *Brown,* which held that:

> In the absence of express contract and a positive authorized legislation, as between proprietors of adjoining land, the law recognizes no correlative rights in respect to underground waters percolating, oozing, or filtrating through the earth; and this mainly from considerations of public policy: (1) Because the existence, origin, movement, and course of such waters, and the causes which govern and direct their movements, are so secret, occult and concealed that an attempt to administer any set of legal rules in respect to them would be involved in hopeless uncertainty, and would be practically be impossible. (2) Because any such recognition of correlative rights would interfere, to the material detriment of the commonwealth, with the drainage of agriculture, mining, the construction of highways and railroads, with sanitary regulations, building, and the general progress of improvement in works of embellishment and utility.[131]

By upholding the district-court decision and reversing the appellate-court decision, the justices of the Texas Supreme Court effectively ruled out the "American rule of reasonable use" and chose the "rule of capture" as the law in Texas beginning a journey through decades of law suits and rulings consistently in favor of the landowner with the "larger pump." The result of reluctant Texas Supreme Court rulings confirming the "rule of capture" has been the establishment of ninety-four groundwater conservation districts across Texas which have the right, as yet in most cases unexercised, to control takings of groundwater from their districts.[132] Short of any major edicts from the Texas legislature weakening or eliminating the rule of capture, the groundwater-conservation districts will be Texas's main hope for the regulation of scarcer groundwater resources.

Other states in the West regulate groundwater in the same manner in which they regulate surface water and closely manage it. For example, in 1931, New Mexico enacted its groundwater code. The state engineer was given jurisdiction over groundwater in the groundwater basins, which were declared at varying dates throughout the state. The code requires a permit for any appropriation of groundwater in the declared basins.[133]

One of the weaknesses that must be addressed in the West is development of a coordinated groundwater policy not only between the states but also with Mexico. The Eighth Report of the Good Neighbor Environmental Board reported "Knowledge about groundwater resources in the border region lags far behind what is known about surface waters. And yet, in the meantime, there are sizable areas throughout the arid southwest where groundwater is being withdrawn at rates which some consider alarming."[134] The most pressing future issue of irrigation will revolve around groundwater, especially from aquifers that travel under several states and nations.

EXAMPLES OF HOW THE STATES AND TERRITORIES ADDRESSED WATER ISSUES

As the Western states came into the Union, they convened constitutional conventions. Water rights, rules about allocation, and resulting management practices were in the forefront of their fundamental policy statements. The key water policy chosen by most of the state constitutional conventions in the West was to reward those who put water to "beneficial use."[135]

For example, New Mexico's constitution was drafted in 1910. Article XVI, paragraph 1 reads in part, "all existing rights to the use of any waters in the state for any useful or beneficial purpose are hereby recognized and confirmed." Note the use of the words "any waters," which include all waters—surface water and groundwater.[136] According to Dumars,

> The New Mexico Constitution makes no mention of groundwater. While at first this appears to be an oversight, in fact the drafters may have had more wisdom than we understood. Water is water [...] Groundwater is a reserve source of surface water encased in soils and protected from air and wind-born contaminants.[137]

Article XVI, Paragraph 2 of the New Mexico Constitution states that water rights available to the public are "subject to [prior] appropriation for beneficial use, in accordance with the laws of the state. Priority of appropriation shall give the better right. [...] [B]eneficial use shall be the basis, the measure, and the limit of the right to use the water."

In 1932, the New Mexico Supreme Court made clear the public ownership of water in their state with this statement: "We already had a policy, also time-honored, as to waters. We had nationalized them. Not as a source of public revenue, as minerals are retained for royalties; but as an elemental necessity, like air, which must not be allowed to fall under private control."[138]

Nevada is another state that has similar water laws to New Mexico. Nevada is the driest state in the nation with an average precipitation of only nine inches; some parts of the state receive four inches or even less. According to Sylvia Harrison, Ph.D., "The scarcity of Nevada's water is the defining circumstance of its water laws."[139] Nevada's water laws are based upon the principle of prior appropriation. The second decision of the Nevada Supreme Court found in part that "where the right to the use of running water is based upon appropriation and not upon ownership of the soil, the first appropriator has the superior right as against subsequent appropriators."[140] In 1913, Nevada passed the 1913 Water Law, which made clear that all water "within the boundaries of the state, whether above or beneath the surface of the ground, belongs to the public."[141] Section 2 provided, "subject to existing rights, all water may be appropriated for beneficial use as provided in this chapter and not otherwise."[142] The laws of Nevada and New Mexico are quite similar, as are the laws of the other Western states except the usual outlier, Texas.

Territorial and state courts gradually assumed the brunt of resolving water-right disputes, not their legislatures. As states entered the union, the rules and laws were applied based upon the English common law of the USA, yet in states such as Texas, which adopted common law but honored Spanish and Mexican water rights, dual systems of riparian rights and appropriative rights were supported by many of the courts resulting in confusion and frustration. Determining jurisdiction between courts in the individual states started the expensive nightmare for many litigants.

No matter where the disputes occurred, water by its nature coupled with the prior appropriative doctrine created hard-to-solve procedural problems in the courts. For example, does a junior appropriator who has a claim of injury against a senior appropriator need to bring in other senior appropriators as defendants? Who is the actual damaging party to the junior appropriator? In what county does the venue lie? Does a single county judge have jurisdiction to adjudicate water rights of parties in the many counties through which the water runs? Many times, the private litigant could find no justice simply because he or she could not afford the process and fees.[143] Those with money could better control the disputes from the first filing at times due to their influence in the procedural processes. These procedural problems have not been fully resolved even in today's state court systems. Assuring these basic problems will continue, the state legislatures still prefer that the courts address water issues. Partly because of their ignorance of the complicated laws pertaining to water and partly because they simply do not have the political will and courage to address long-term solutions, the legislatures are as much to blame for water-rights disputes as are the litigants.

Early Irrigation Communities, Colonies, and Unincorporated Mutual Associations

Before the miners of 1849, the Mormons, religious refugees led by Brigham Young, arrived in what then was Mexico but eventually became Utah. The Mormons were farmers and had scouted the lands for water observing the Spanish/Mexican irrigation systems that influenced their methods of construction and management. In the summer of 1847, the first group of settlers arrived in the Salt Lake valley and found the "grass had nearly dried for want of moisture."[144] Another settler, Erastus Snow, wrote, "We found the land so dry [...] that to plough it was impossible, and in attempting to do so some of the ploughs were broken. We therefore had to distribute the water over the land before it could be worked."[145] They built diversion dams across the creeks and canals as their first irrigation system. They divided the land into wards and appointed water masters to manage the system as it related to the landowners. The Mormons believed their land would be cultivated individually but the fields were fenced and irrigated cooperatively.[146] They did not accept the riparian water-rights doctrine; their streams were owned and managed by the public for the public good. They established exclusive grants and licenses to waters of streams and rivers so they practiced a modified appropriative system of water management. The water masters apportioned the water by a system of rotation, delivering it for a certain amount of time dependent upon the user's needs. In that manner, the public water was shared. These early organizations were unincorporated mutual associations that depended upon trust and common interests. Eventually these communities became incorporated mutual associations.

In Mormon communities, the rule was, "In water disputes, let the Priesthood rule." Mormon communities spread to California, Arizona, Idaho, and Wyoming. The Utah territorial legislature codified many of the water practices put into place by the Mormons in 1850. But in 1880, the Utah legislature, all Mormon, rescinded the earlier statutes. Gustave O. Larson called their actions, "the Americanization of Utah." The legislature agreed on a law of capture proven by the user constructing a canal with "diligence" and diversion of the stream or river "for any useful purpose." Rights were "proved up" non-judicially and ostensibly were given a certificate of recording of the "proved-up" right. According to Dunbar, "Unfortunately, this statute was enforced in only a few counties and Utah became like the other Western states and territories, a jungle of uncontrolled appropriations and undetermined water rights."[147]

One of the negative consequences of this uncoordinated approach as opposed to the original Mormon method was to make it impossible to reasonably determine the total appropriations of the limited water in the streams. Other colonies and communities were tried in the West with irrigation systems as the main feature to attract new settlers.

IRRIGATION COLONIES AND COMMUNITIES

Irrigation-based communities and colonies were established in Anaheim, California, in 1857, at the confluence of the Platte and Cache la Poudre rivers in Colorado in 1870 named the Union Colony and its community was named Greeley after Horace Greeley of the New York Tribune, who was an early sponsor. The Chicago–Colorado Colony was established near Longmont, Colorado.[148] These were cooperative communities roughly similar to the Mormon communities. These communities and others like them were based upon utopian ideas of communal living. Some were simply new business schemes.

The first of these irrigation-based colonies was organized by a group of Germans from San Francisco, California, by the Los Angeles Vineyard Society in 1857 to produce grapes and wine. The settlers named their colony Anaheim. The leader, George Hansen, laid out the settlement along the Mormon model and built the irrigation ditches and laterals for individual farms prior to the arrival of the first fifty colonists. The colonists were issued one share of stock each of which represented the water needed to irrigate their 20-acre farms and half-acre town lots.[149] Other irrigation-based colonies followed across the West, but few were successful in the long run due to the massive amount of capital needed to build a large irrigation system and the difficulty of maintenance with its relentless expense. Irrigation as a basis for community formation continued a predominant feature of settlement in the West with other new forms of business entities and organizations. Private business took the lead for the next stage of irrigation development in the West.

PRIVATE IRRIGATION COMPANIES

Private companies and corporate entities began to view irrigation as a business opportunity to develop and sell land at a profit, not without unanticipated problems, however. A problem in any large irrigation system from the beginning was the right of way for the canal system. The importance of irrigation to the states of the West can be seen as most of the new states' constitutions gave persons and corporations the right to condemn public and private land for irrigation ditches which were of beneficial use. The most obvious accepted use was for agricultural pursuits and condemnation rights from the states seemingly solved the problem on the surface. Condemnation though required payment of compensation to those whose land was being condemned which added to the up-front capital outlay. The condemnation authority varied from state to state, but the message was clear from the governments. The states viewed agricultural irrigation as a critical factor in their future growth.[150]

As the West grew in the nineteenth century, a myriad of private companies such as the Highland Ditch Company of Colorado, The Boise City Land & Irrigation Company in Idaho, and the Pecos Irrigation & Investment Company of New Mexico sprang up. The private companies rapidly found the profitability of operating a costly irrigation system not to mention the start-up costs of construction was all but impossible.[151] Small, undercapitalized farmers and the fickle nature of agriculture even with irrigation simply could not afford the high prices these companies needed to charge to be profitable and good investments.

Incorporated mutual ditch companies were tried in many states with the help of state and territorial legislatures. Companies such as the Hardy Irrigation Canal Company in Arizona and the Big Ditch Company in the Yellowstone Valley near Billings, Montana, issued stock to farmers and assigned to them proportionate shares of irrigation water. One of the private enterprise efforts that was tried in a few locations was that of the corporate developer of land for sale which had access to irrigation ditches. The developer mostly looked to the land sales for profit, not the operation of the irrigation system. The combination of mutual ditch companies that built irrigation ditches and land developers selling land and mutual stock to individual farmers worked well in some places.[152]

The mutual irrigation companies became so numerous in Colorado that they were able to influence the legislature to pass a law that exempted them from all ad-valorem taxes from any taxing entity which was ultimately added to the Colorado Constitution.[153] The states of Utah, Arizona, New Mexico, Idaho, and Wyoming added similar language in their constitutions.[154] To add to the positive aspects of mutual irrigation companies, The Capper–Volstead Act of 1922 made them exempt from federal income taxes.[155]

Still, the capital outlay for construction and maintenance of a major irrigation system was huge. Katherine Coman wrote in 1911, "Nothing goes to wreck more quickly than irrigation works where repairs are not maintained; the ditches fill with sand and silt, the flumes warp in the sun, and cement dams disintegrate under the alternate action of frost and heat."[156] Irrigation systems were and still are capital-intensive enterprises. Farming success varies due to a highly volatile business cycle and rarely can support itself without patient, long-term capital availability. Even with adequate capital, the "boom and bust cycles in the markets for agricultural products, poor business judgments, and flawed business models, especially those formed in the boom years prior to the Panic of 1893, led to the failure of many commercial irrigation companies."[157] Another form of entity was formed to advance irrigation across the West, the irrigation district.

IRRIGATION DISTRICTS

The most successful efforts to expand irrigation systems came from the creation of irrigation districts with the power to issue bonds. In 1887, California's Wright Act led the way for sixteen other states to follow this practice.[158] The bonds were secured by the taxes imposed on all agricultural land in the county of the district, a use of the credit of the entire county to develop irrigation systems for all farmers.[159] When state supervision became the rule after 1897, the irrigation districts thrived. Millions of acres are under irrigation in all seventeen Western states today due to irrigation districts.[160] According to Bertsen, "Established as public institutions, irrigation districts were formed to benefit specific groups of private landowners. This fact have irrigation districts a chameleon-like quality through their ability to 'affect public colors when advantageous, but resort to private camouflage when needed.'"[161] Using public credit to finance private enterprise is most noticeable today in local construction of our professional sports arenas.

A major advantage given to irrigation districts by Congress in 1922 and 1926 ensured their continued success. The 1922 Act authorized the Bureau of Reclamation to contract with irrigation districts for federal funds to repay project costs, and the 1926 Act made the irrigation districts the sole and exclusive irrigation organization that could contract for federal funds to repay costs.[162]

Irrigation districts' ability to issue bonds and tax via assessments made them viable. They were political subdivisions of the states and exempt from federal income taxes. Most states certified the bonds while not directly guaranteeing them. The strongest advantage of certification was that the states declared "the bonds eligible as investment vehicles for private trust funds and the funds of insurance companies, banks, and public institutions."[163] The district owned the appropriative rights to water. Through the combination of water rights coupled with the authority to issue bonds and to levy taxes, irrigation districts became the most viable institutions to build irrigation systems.

The Wright Act has had long-reaching impact across the West. The Turlock Irrigation District, the first district established in California under the Wright Act, along with the Modesto Irrigation District, provides irrigation for over 200,000 acres today.[164] These two districts alone equal 2.5 percent of the total irrigated acreage in California.

FEDERAL EFFORTS TO ENHANCE IRRIGATION

The California Gold Rush was short, as were other gold and silver rushes in the West, and while they heightened the immigration to the West after 1848, the Civil War effectively delayed public land distribution. The Homestead Act was passed in 1862 and started into motion an eventual transfer of over 285 million acres to individuals through a variety of federal legislation; an additional 131 million acres of public land was granted to the railroads. The Homestead Act initially offered land in tracts of 160 acres to adult individuals for settlement and farming. The Timber Act of 1873, which survived until 1891, also granted 160 acres to individuals for the specific purpose of planting trees on 40 acres with the obligation to keep them alive for at least ten years. In 1877, the Desert Land Sales Act offered 640 acres of land to individuals who agreed to irrigate otherwise unproductive land within three years.

The federal government to date has given away over 1 billion acres, or 70 percent of the public domain in the West. John Wesley Powell was Director of the US Geological Survey, and he recommended that a way to better divide the Western lands would be based upon watersheds and not by section, a rectangle of 640 acres. His recommendations went unheeded. The Carey Act of 1894 gave incentives to Western states to reclaim land through irrigation. Due to the aridity of the lands distributed, water conflicts were frequent. Displaced Indian tribes were given land as they were pushed further and further into the West. Yet, the federal government ignored their rights as a whole and to water in particular until Winters v. United States in 1908. The Winters ruling gave the Indians on the Fort Belknap Reservation in Montana water in the Milk River and gave the federal government the power to "reserve" water in any treaties with Indians. The federal government preferred the individual states establish their own water laws and water rights even at times when federal lands were involved.

As the states became more involved, financial efficiencies became a byproduct, especially when it came to the development of reservoirs to store water in the off seasons. These reservoir projects, expensive due to right-of-way and condemnation issues, not to speak of construction costs, saved the water flowing by in the non-growing seasons. Few if any private ditch companies had the power or financial support to accomplish the construction of large reservoirs and other efficient storage systems. The states needed the financial assistance of the federal government, and, in 1894, the Carey Act gave eleven of the Western states up to 1 million acres of the federal desert lands in any of the states to be divided into 160-acre tracts and to sell them to settlers for reclamation. Settlers purchased the land and water rights and would not receive their patent to the land until they had irrigated at least one-eighth of their tracts. While not as successful as hoped, the Carey Act was used extensively in Idaho and Wyoming. California, Washington, and Arizona did not participate at all. The federal government continued to cede federal lands to states "for the establishment of permanent reservoirs for irrigation purposes" in 1896 and 1898.[165] The states built dams and reservoirs throughout the late nineteenth and into the twentieth centuries for irrigation use.[166]

The federal government also encouraged settlement on its lands in the West, beginning in earnest with the Desert Land Act of 1877. Proof of irrigation was required usually on one-eighth of the section (640 acres) within three years. At first the Act only included eleven Western states plus the Dakotas, but Colorado was added in 1891. It was very difficult for small individual farmers to construct working irrigation systems within three years. Enough interest was developed by the Carey Act to point to the need for large reservoirs to store water. Senators Henry M. Teller of Colorado and William M. Stewart of Nevada began the process in Congress of studies to determine the potential locations of the reservoirs.[167]

Congressman Francis G. Newlands of Nevada introduced the Reclamation Act of 1902 to construct major irrigation projects across the sixteen Western states and territories; Texas was not initially included but was added as the seventeenth state in 1905. According to Pisani, "federal reclamation promised to 'subdue worthless land,' turn the desert wilderness into a garden, and convert the West into a commonwealth of small farms."[168] The "commonwealth of small farms" did not happen. Pisani continued, "Virtually every large transfer of land in the nation's history resulted in huge monopolies and rampant speculation. The Reclamation Act was no exception."[169] The traditional duty of American business is to become more efficient for more profit, and the small farms do not attract capital investment simply because they cannot produce an adequate return on investment.

Once Newlands agreed to amend the Bill to add protections for vested rights in water, the Bill passed. President Theodore Roosevelt was a supporter more for the Act's impact on the preservation and enhancement of the habitat of the animal kingdom of the West; more reservoirs in desert lands meant more crops and more green land which could feed and water wild creatures. Roosevelt designated the first twenty-four irrigation projects.[170] According to Clay Jenkinson, "Roosevelt was not afraid to employ the federal government as a land and resources steward. He did not believe that local populations were capable of the best stewardship without the supervisory hand of the national government."[171]

The projects under the Reclamation Act of 1902, commonly known as the Newlands Act, included complete irrigation systems including diversion dams, reservoirs, and canals. The Truckee–Carson Irrigation District was formed to work with the US Reclamation Service in 1918. Churchill and Lyon counties in Nevada claim to be the first sites of federal reclamation authorized under the new act. Figure 11.5 is an advertisement for free land in Churchill County, in farms of 40 and 80 acres, with a requirement of residence on the farm for three years to secure title and cultivation of one-quarter of the irrigable acreage in three years and half of the irrigable acreage in five years.

This ad is typical of others used by the Reclamation Service in the West. Note the comment on the ad, "The Opportunity of a lifetime for the homeseeker exists right now in Churchill County."[172]

The Truckee–Carson project was typical of many of the projects across the West, and its problems were challenging to say the least. The limited farmers who tried to settle had little or no experience in irrigated farming. The delivery of water was unreliable in the early years. As in Spanish days, farming is subject to disasters other than moisture shortages. Flooding, wind storms, and the fickleness of nature caused the ruin of many homesteads. Reservoirs were slow to build and of absolute necessity to any chance for farming to succeed. Maintenance was always a problem as elements of the irrigation system deteriorated over the years. Droughts came and went, and the system faced more deterioration which the best efforts of the federal agency simply could not remedy. A July 6, 1954, earthquake damaged much of the system, and, just as repairs were being completed, another quake struck on August 23, destroying much of the work and water stopped flowing in the system yet again.

The Truckee–Carson Project is a good example of the difficulties any large irrigation system faces, and of the Herculean effort which must be undertaken to overcome the unending challenges. The Bureau of Reclamation assesses the success of the Reclamations Act of 1902, and the Truckee–Carson Project in particular, in these words on its website:

Figure 11.5 Advertisement for "Free Land" in Churchill County, Nevada, under the Reclamation Act of 1902. Source: Courtesy of Truckee-Carson Irrigation District.

> as the Bureau of Reclamation enters its second century, it seems unlikely that the controversies and conflicts surrounding the Newlands Project and the Truckee River will be any less complicated than in the previous hundred years. [...] The problems are complicated, and the solutions promise to challenge Reclamation's expertise and innovativeness.[173]

The Bureau of Reclamation's second commissioner was Elwood Mead. He expanded the work of the Bureau to include a screening process for farmers and provided them with the means of credit and education to build the infrastructure to make their farms work. The work also expanded into flood control, hydroelectric plants, and all kind of land-improvement projects. The Bureau has constructed hundreds of reservoirs; the dams of many also generate hydro-electric power.[174]

MODERN IRRIGATION, WATER MASTERS, AND LICENSES

Major John Wesley Powell, a one-armed Civil War veteran, explored the West for over fifty years, beginning in 1869. He encouraged regional oversight of all resources, especially water. He was a proponent of self-governing water districts set up based upon watersheds not sections of land with water masters overseeing individual water rights. The states of California, Colorado, Montana, New Mexico, and Oregon have water-master programs that can trace their concept to Powell's theories. Texas has a water-master program, also, but it is based on the unique Texas laws that stemmed from the deluge of litigation over surface water rights.[175]

The states of the West chose prior appropriation when the federal government chose to allow them to make up their own rules about water allocation in their state. In 1978, the Supreme Court of the USA in two separate cases, California v. United States of America and United States of America v. New Mexico, the court confirmed its historical attitude of deference to state law on water issues in comments written by Associate Justice William Rehnquist about the history of federal and state relations over irrigation:

> [in California v. United States of America] is both long and involved, but through it runs the consistent thread of purposeful and continued deference to state water law by Congress. [In United States of America v. Mexico,] [w]here congress has expressly addressed the question of whether federal entities must abide by state water law, it has almost invariably deferred to state law.

Justice Rehnquist fully supported his comments with documentation.[176]

From state to state, water laws and management practices vary but little from a general theme of prior appropriation overseen by local water masters or state agencies or, in most cases, a combination of both. In 1890, its first year of statehood, Wyoming's Board of Control adjudicated water rights at the recommendation of its territorial engineer, young Elwood Mead.[177] As mentioned before, Texas "fully" adjudicated its surface-water rights in the mid-1960s, which eliminated one of its major sources of litigation. States which do not have fully adjudicated surface-water rights are either in the process of doing so or will do so soon.

Water masters are the field managers and regulators of irrigation systems in the West and have similar jobs across the West. For example, in Texas, water-master duties include measurement of reservoirs and Doppler measurements on diverters' water pipes, with the option of setting stream-flow markets to help water-right-holders comply with guidelines. They also keep databases on the amounts of water authorized and used and update and maintain water-right ownerships and assessments on each water-right account.

The Texas Commission on Environmental Quality describes the program:

> The South Texas Watermaster Program is responsible for an area that encompasses 50 counties in south central Texas. Five field deputies patrol these counties from offices located near their jurisdictional areas. Water-right management is based on "run of the river rights." That means the deputies oversee a constantly changing, dynamic surface-water system of rivers and tributaries that allow diversions as water is available and as it passes individual diversion points.[178]

In New Mexico, the job responsibilities of the Middle Rio Grande water master include:

- evaluation of water rights;
- determining those who lose water rights due to abandonment, the relinquishment of the right by the owner with the intention to forsake and desert it, or forfeiture, the involuntary or forced loss of a water right, caused by the failure of the appropriator or owner to put the water to beneficial use;
- distribution of surface waters within the middle Rio Grande River;
- maintenance of a water-rights inventory; and,
- management of a geodatabase that maps all water declarations, all transfers of pre-1907 surface water rights to groundwater, and all irrigation wells. The geodatabase is an important tool for the water master's administration of water rights in the basin.[179]

Water masters have similar but varying duties across the water basins and irrigation systems of the West. They have far-reaching power and influence on the irrigation systems of the West, and their authority will continue to be expanded in the upcoming years.

Most Western states now have or are in the process of developing complex licensing systems to allow diversion of waters from their rivers or streams for irrigation. The surface waters of most states are likely over-appropriated in dry years. In Texas, it is all but impossible today to obtain a permit to divert water for irrigation purposes from any river or stream from the ruling agency, the Texas Commission on Environmental Quality, except in all but the rarest of temporary circumstances. Today, across the West, there are many highly vocal supporters of government ownership and total control of all water-related resources whether surface or groundwater. The proponents of total government control strongly criticize the prior appropriation doctrine, especially when it interferes with environmental issues such as endangered species protection.

FEDERAL INVOLVEMENT IN IRRIGATION IN RECENT TIMES

Even though the federal government and courts continue to defer to the states on most water issues, it has been and is more involved in water issues than it may appear. In 1899, through the courts, the federal government blocked a dam across the upper Rio Grande. In 1901, the Supreme Court asserted its control of allocation of interstate waters. In 1902, the Supreme Court held in favor of Indian treaties and implied reservations of water in

Winters v. United States of America. In *Federal Power Commission v. Oregon* in 1955, the "Winters Doctrine" was used by the Supreme Court to indicate the USA had the power to reserve water any time it designated lands for a particular federal purpose. In *Arizona v. California*, the Supreme Court confirmed that reserved water rights are not limited to Indian lands but may be established by the federal government when it designates the land for other specific purposes.

The Endangered Species Act (ESA) of 1973 is another example of increased federal activity in water issues and has gradually had increasing impact. As more fish species are listed as endangered, the government is requiring more water to remain in stream. According to Jennie L. Bricker and David E. Filippi:

> While many view the ESA as a tool for trumping long-established state water rights and returning water to streams in order to protect threatened fish species, others view the ESA as effectively undermining states' rights and usurping local communities. [...] To require proportionate reductions in water use arguably places a disproportionate burden on senior water users.[180]

This area of federal interest will surely strengthen the government's involvement in water issues around the nation. The Act allows petition by any interested person to list a species as endangered or threatened. "Any direct or incidental death or injury to a listed fish is actionable under the ESA. [...] For example, NMFS [National Marine Fisheries Service] has identified water withdrawals, unscreened diversions, and grazing in riparian areas as 'activities that may constitute a take' [a take is capturing or destroying an endangered species]."[181] There have been and will continue to be conflicts between state water rights and this federal statute. The overall impact on water rights across the West is yet to be seen, but certainly it will be significant, with the states' rights more than likely weakening in the federal onslaught. Yet, the Clean Water Act of 1977 does include, "nothing in this chapter shall be construed to supersede or abrogate the rights to quantities of water which have been established by any State."[182]

The Fifth Amendment to the US Constitution keeps the federal government from taking private property without just compensation. A question arises: Does enforcement of the ESA against irrigators, especially when protecting the flow of any stream to protect the endangered fish or aquatic plants, which results in inadequate flow for appropriative irrigation water use to occur trigger a duty by the federal government to compensate the irrigator? This issue will doubtlessly become more important as water resources grow more and more scarce and the number of listed endangered fish species inevitably increase.

Federal involvement in water rights will continue in conflicts between states and in international water disputes. The sheer magnitude of federal agencies with jurisdiction over water rights in the West is overwhelming. A partial list of key agencies include the US Army Corps of Engineers, the US Bureau of Reclamation, the US Department of Agriculture–National Resource Conservation Service, the US Department of Agriculture–Forest Service, the US Department of the Interior, and the US Environmental Protection Agency. The federal government is also an active participant in the US Section of the International Boundary and Water Commission and oversees the US rights and duties in the Rio Grande and Colorado River Compacts, the New Mexico–Texas Water Commission, the Pecos River Compact, and the Upper Colorado River Basin Compact. The federal government is involved in the protection of the water rights of the Indian tribes in the West; the Indian tribes are considered sovereign nations.

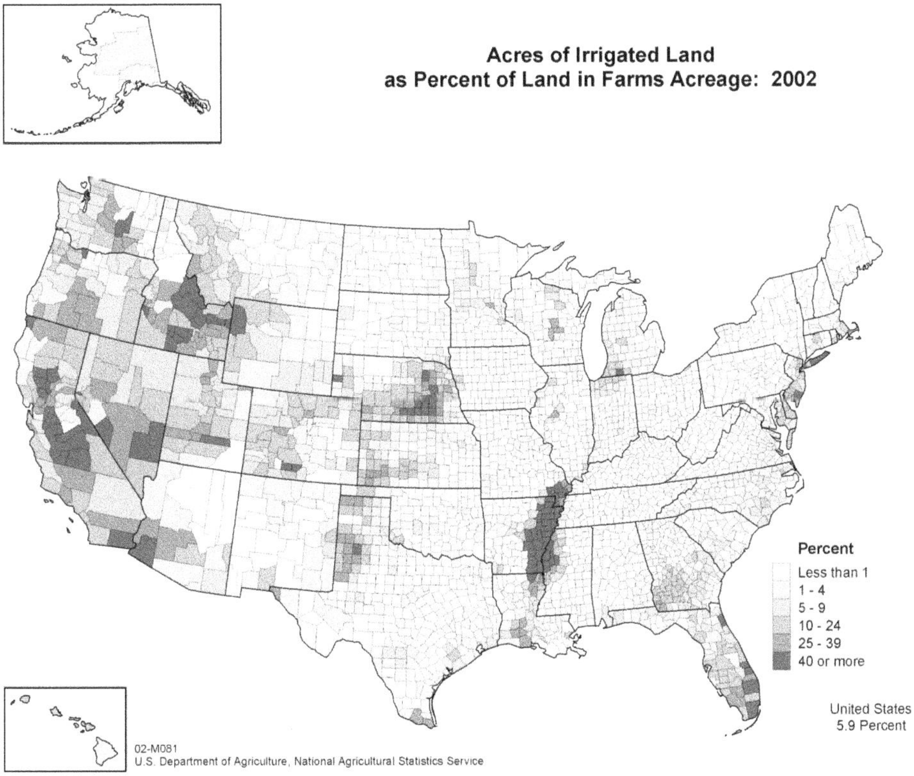

Figure 11.6 Irrigated land in the USA.
Source: Public Domain.

MODERN IRRIGATION SYSTEMS

Modern irrigation techniques are a vast improvement over early systems but continue to have the basic function of delivering adequate moisture to crops evenly at the proper time. Notice the vast amount of irrigated farm land in the West as shown in Figure 11.6. While surface water is used across the West for irrigation, groundwater is an important source also. Groundwater use began in the late nineteenth century when water-drilling technology improved to allow deep-water wells to be put in service. As the drilling technology improved over time, coupled with better and stronger electric and diesel pumps, groundwater became in many areas of the West the preferred source for irrigation and has had the most major impact on the aquifers and groundwater resources of the West. It is preferred by farmers because they have autonomy over their own system; generally, they can water when they want anyway they desire. Even in the states where groundwater use is tightly restricted, a farmer has much less involvement with his neighbors than he would have participating in a surface-water canal system.

FUTURE CHALLENGES FOR IRRIGATION AND WATER IN THE WEST

There seems to be no end to the growth of the West. With this growth there will be increasing pressure on its scarce water resources. As water resources become more and more lim-

Figure 11.7 Field readied for irrigation.
Source: Photo by author.

ited per individual user, especially in times of drought, the importance of water laws and water rights may manifest into the most critical natural resource issue of our time. The state legislatures, the US Congress, and state and federal agencies will find themselves in a highly sensitive and challenging national debate.

Surface-water laws, rights, and management practices have been addressed and are in the process of final adjudication across many areas of the West. Groundwater is yet another matter. The ways in which groundwater interfaces with surface water, the flows of groundwater in aquifers that cross under several state lines and the interrelationship of rights to this groundwater from state to state, and the needs of the individual rural groundwater user versus the needs of the ever-expanding urban sprawl cannot be ignored. A uniform and fair state and national policy must be addressed. The outcome of the myriad of groundwater issues cannot be predicted and will be determined by decades of political debate and thousands of lawsuits in the local, state, and federal courts.

Kansas draws 97.25 percent of its irrigation water from groundwater sources, Nebraska 89.05 percent, and Texas 87.38 percent (see Table 11.1). Texas has the rule of capture which means irrigation from groundwater is all but unregulated in the state. These three states share the Ogallala aquifer which is under tremendous pressure these days during dry years, and with the growth of population, especially in Texas, the pressure will surely increase. Groundwater management remains a key issue for these states.

The cities of the USA have sovereignty and are in the process of exercising their power to obtain water for their needs, creating a crisis across the West for those who claim individual water rights. The cities have the financial resources to challenge both corporate and

Figure 11.8 San Elizario sluice gate.
Source: Photo by author.

individual water rights and are doing so aggressively. The city of Del Rio, Texas, purchased a small tract from a landowner on the outskirts of the city with the water rights retained by the seller. The city immediately proceeded to drill three water wells for municipal use nonetheless. Even though the city of Del Rio stipulated in the subsequent lawsuit with the seller that the seller explicitly did not transfer any water rights to them in the purchase contract and deed, they claimed the right to the water anyway since the landowner had not "captured" it as yet. This suit is an example of the determination of cities to increase their water resources and their willingness to challenge even the most obvious water-rights concepts.[183] The entire West will likely see hundreds of battles between cities and individuals, especially individuals and companies whose agricultural activities depend upon irrigation from groundwater, over water in the next century. The outcome of these conflicts will shape the future of irrigation.

In 1996, The National Research Council Committee on the Future of Irrigation in the Face of Competing Demands summed up the three major "forces of change" facing irrigators and the agricultural community. These are: "competition over water supplies, changing economic conditions, and environmental concerns [...] The underlying premise of this study is that given the increasing competition for water supplies, changes in how water is managed, allocated, and valued are inevitable."[184] Perhaps these issues have faced irrigation from its beginning in the West. Irrigation transformed the landscape, shaped the culture, and built the society and economy of the West. It will surely continue to do so.

Figure 11.9 Step lateral irrigation technique flooded.
Source: Photo by author.

Table 11.1 Irrigation percentage from groundwater in the West

State	*Total irrigated acres*	*Acres irrigated from groundwater*	*Percentage from groundwater*
Arizona	873,589	243,313	27.85%
California	8,139,834	3,071,740	37.74%
Colorado	2,942,230	1,331,615	45.26%
Idaho	3,188,406	1,226,924	38.48%
Kansas	2,650,486	2,577,611	97.25%
Montana	1,740,873	46,606	2.68%
Nebraska	5,692,215	5,069,036	89.05%
Nevada	694,930	223,747	32.20%
New Mexico	720,319	425,530	59.08%
North Dakota	164,741	103,714	62.96%
Oklahoma	451,788	355,414	78.67%
Oregon	1,534,961	304,579	19.84%
South Dakota	297,205	137,788	46.36%
Texas	5,237,584	4,576,456	87.38%
Utah	1,076,346	139,693	12.98%
Washington	1,554,813	445,927	28.68%
Wyoming	1,533,468	90,730	5.92%
Total: Western States	38,493,788	20,370,423	52.92%

Source: Farm and Ranch Irrigation Survey (1998), National Agricultural Statistics Service, U.S. Department of Agriculture (1999). Data were summarized by the Economic Research Service, USDA. Table prepared by author.

Figure 11.10 Pecan trees irrigated.
Source: Photo by author.

NOTES

1. Michael C. Meyer, *Water in the Hispanic Southwest* (Tucson, Ariz.: University of Arizona Press, 1984), p. ix.
2. William E. Doolittle, *Cultivated Landscapes of North America* (Oxford: Oxford University Press, 2000), p. 252.
3. Doolittle, *Cultivated Landscapes of North America,* p. 257.
4. Doolittle, *Cultivated Landscapes of North America,* pp. 300–301.
5. Doolittle, *Cultivated Landscapes of North America,* p. 347.
6. Robert G. Dunbar, *Forging New Rights in Western Waters* (Lincoln, Nebr.: University of Nebraska Press, 1983), p. 3.
7. Dunbar, *Forging New Rights in Western Waters,* p. 3.
8. Doolittle, *Cultivated Landscapes of North America,* p. 369.
9. Doolittle, *Cultivated Landscapes of North America,* pp. 371–372.
10. See Jonathan Mabry, "The Ancient Oasis: 4,000 Years of Agriculture and Irrigation in Tucson," Tucson Historic Preservation Office, available online at www.tucsonpimawaterstudy.com/meetings/meeting_Archives/meetings/8.27.08/presentations/08-27-2008%20Mabry_The%20Ancient%20Oasis.pdf (accessed April 27, 2010).
11. Mabry, "The Ancient Oasis."
12. Doolittle, *Cultivated Landscapes of North America,* p. 369.
13. Dunbar, *Forging New Rights,* p. 3.
14. *Obregón's History of 16th Century Exploration in Western America Entitled Chronicle, Commentary, or Relation of the Ancient and Modern Discoveries in New Spain and New Mexico, Mexico, 1584,* translated, edited and annotated by George P. Hammond and Agapito Rey (Los Angeles, Calif.: Wetzel, 1928), p. 160.

15. William E. Doolittle, "Agriculture in North America on the Eve of Contact: A Reassessment," *Annals of the Association of American Geographers,* 82 (3); "The Americas before and after 1492," *Current Geographical Research* (1992); p. 389.
16. Doolittle, "Agriculture in North America," p. 389.
17. Personal interview with Charles C. Mann in August 2008.
18. Edwin P. Arneson, "Early Irrigation in Texas," *Southwestern Historical Quarterly Online,* 25 (2). Available at www.tsha.utexas.edu/publications/journals/shq/online/v025/n2/contrib_DIVL1538_print.html (accessed June 6, 2007).
19. Donald Worster, *Rivers of Empire: Water, Aridity, and the Growth of the American West* (New York: Pantheon Books, 1985), p. 11.
20. David J. Weber, *The Mexican Frontier, 1821–1846: The American West under Mexico* (Albuquerque, N. Mex.: University of New Mexico Press, 1982), p. 38.
21. Jesus de la Teja, *San Antonio de Bexar: A Community on New Spain's Northern Frontier* (Albuquerque, N. Mex.: University of New Mexico Press, 1995), p. 96.
22. Meyer, *Water in the Hispanic Southwest,* p. 37.
23. Dunbar, *Forging New Rights,* p. 4.
24. Betty Eakle Dobkins, *The Spanish Element in Texas Water Law* (Austin, Tex.: University of Texas Press, 1959), p. 64.
25. Dobkins, *The Spanish Element,* p. 64.
26. Dobkins, *The Spanish Element,* p. 64.
27. A *noria* is a waterwheel.
28. *San Antonio River Authority, Petitioner* v. *G. Garrett Lewis, et al., Respondents,* No. 1–8304, The Supreme Court of Texas, 363 S.W., 2nd 444, 1962.
29. Robert H. Jackson and Edward Castillo, *Indians, Franciscans, and Spanish Colonization* (Albuquerque, N. Mex.: University of New Mexico Press, 1995), p. 13.
30. Meyer, *Water in the Hispanic Southwest,* p. 8.
31. The Water Resources Center at Texas Tech University *The Acequias of San Antonio, Historic American Engineering Record TX-1* (Washington, DC: The National Parks Service 1973).
32. Frederick C. Chabot, *With the Makers of San Antonio* (San Antonio, Tex.: Privately published and printed by Artes Graficas, 1937), p. 148.
33. Meyer, *Water in the Hispanic Southwest,* p. 41.
34. I. Waynne Cox, *The Spanish Acequias of San Antonio* (San Antonio, Tex.: Maverick Publishing Company, 2005), p. 3.
35. *San Antonio River Authority, Petitioner* v. *G. Garrett Lewis,* p. 3.
36. Cox. *The Spanish Acequias of San Antonio,* p. 78.
37. Cox. *The Spanish Acequias of San Antonio,* pp. 32–33.
38. Andrés Tijerina, *Tejanos and Texas: Under the Mexican Flag, 1821–1836* (College Station, Tex.: Texas A. & M. Press, 2003), p. 53.
39. Friar Benito Fernández, "Memorial of Father Benito Fernández Concerning the Canary Islanders, 1741," trans. Friar Benedict Leutenegger, OFM, *Southwestern Historical Quarterly,* 82 (3) (1979): 264–296; p. 290, emphasis added. Friar Fernández swears to the friars teaching the Indians how to be farmers, irrigators, carpenters, and tailors.
40. Herbert E. Bolton, *Texas in the Middle Eighteenth Century* (Austin, Tex.: University of Texas Press, 1970), pp. 234–237; Arneson, "Early Irrigation in Texas," p. 1.
41. Charles R. Porter, "Querétaro in Focus: The Franciscan Missionary Colleges and the Texas Missions," *Catholic Southwest, A Journal of History and Culture,* 19 (2008): 9–51; p. 27; Fr. Zephyrin Engelhardt, OFM, *The Missions and Missionaries of California* (Santa Barbara, Calif.: Mission Santa Barbara, 1929), p. 729.
42. Herbert E. Bolton, "The Mission as a Frontier Institution in the Spanish-American Colonies," *The American Historical Review,* 23 (1) (1917): 57–58; David J. Weber, *The Spanish Frontier in North America* (New Haven, Conn.: Yale University Press, 1992), p. 106; Carlos Castañeda, *Our Catholic*

Heritage, 1519–1936, vol. III: *The Missions at Work* (Austin, Tex.: Von Boeckmann-Jones Co., 1938), pp. 27–28.

43. Fernández, "Memorial," p. 290.
44. Unknown, *Documents Relating to the Old Spanish Missions of Texas,* vol. I, *Guidelines for a Texas Mission: Instructions for the Missionary of Mission Concepción in San Antonio,* trans. Benedict Leutenegger, OFM (San Antonio, Tex.: Our Lady of the Lake University, 1994). Originally written in 1787. This manual has eighty-three specific rules with ten additional supplementary rules for operating Mission Concepción. The detail is all but unbelievable. Rules 60, 61, and 62 discuss irrigation-ditch-maintenance techniques.
45. Unknown, *Documents Relating to the Old Spanish Missions,* p. 33.
46. Robert Ricard, *The Spiritual Conquest of Mexico: An Essay on the Apostolate and the Evangelizing Methods of the Mendicant Orders in New Spain, 1523–1572,* trans. Lesly Byrd Simpson (Berkeley, Calif.: University of California Press, 1966), p. 145. Originally published as *Conquête spirituelle du Mexique* (Paris: Université de Paris, 1933).
47. Janet R. Fireman, *The Spanish Royal Corps of Engineers in the Western Borderlands, 1764–1815* (Glendale, Calif.: Arthur H. Clark, 1977), p. 54.
48. Bolton, "The Mission as a Frontier Institution," p. 47.
49. Wells A. Hutchins, "The Community *Acequia:* Its Origin and Development," *Southwestern Historical Quarterly,* 31 (3): 261–284.
50. Bolton, *Texas in the Middle Eighteenth Century,* pp. 234–237.
51. Bolton, *Texas in the Middle Eighteenth Century,* p. 235.
52. Hutchins, "The Community *Acequia,*" p. 282.
53. Thomas P. O'Rourke, CSB, A.M., *The Franciscan Missions in Texas (1690–1793)* (Washington, DC: Washington Monotype Composition Co, 1927), p. 3.
54. O'Rourke, *The Franciscan Missions in Texas,* p. 44.
55. Thomas F. Glick, *Islamic and Christian Spain in the Early Middle Ages* (Princeton, NJ: Princeton University Press, 1979), pp. 71–73.
56. Thomas F. Glick, *The Old World Background of the Irrigation System of San Antonio, Texas* (El Paso, Tex.: Texas Western Press, 1972), p. 3.
57. Ellis L. Armstrong (ed.), *History of Public Works in the United States, 1776–1976* (Chicago, Ill.: American Public Works Association, 1976), p. 1.
58. Armstrong, *History of Public Works,* p. 217.
59. De la Teja, *San Antonio de Bexar,* p. 32.
60. De la Teja, *San Antonio de Bexar,* p. 219.
61. De la Teja, *San Antonio de Bexar,* p. 218.
62. Charles R. Cutter, "Community and the Law in Northern New Spain," *The Americas,* 50 (1994): 467–480; p. 467.
63. Cutter, "Community and the Law," p. 468.
64. Cutter, "Community and the Law," p. 467.
65. Cutter, "Community and the Law," p. 467. This is also the position of Dr. Thomas Glick of Boston University. See his paper "Irrigation Rights and the Limits of Civil Law in the Spanish Legal Tradition," Proceedings of the III World Conference on Water Law and Administration, Valencia, Spain, December 11–14, 1989.
66. Cutter, "Community and the Law," p. 467.
67. Cutter, "Community and the Law," p. 467.
68. Cutter, "Community and the Law," p. 467.
69. Cutter, "Community and the Law," p. 467.
70. El Marques de Casafuerte to Fray Gabriel de Vergara, December 25, 1731. Bexar Archives Translations, the Center for American History, the University of Texas at Austin, Series 1, General Manuscripts 1717–1789, Box 2C14, Vol 2. December 25, 1731—June 25, 1733.
71. Meyer, *Water in the Hispanic Southwest,* p. 166.

72. Andrés Tijerina, *Tejanos and Texas under the Mexican Flag, 1821–1836* (College Station, Tex.: Texas A. & M. Press, 1994), p. 51.
73. Meyer, *Water in the Hispanic Southwest,* p. 180.
74. *State of Texas et al., Appellants* v. *Valmont Plantations et al., Appellees,* 346 S.W. 2nd 853.
75. *State of Texas et al.* v. *Valmont Plantations et al.*
76. *State of Texas et al.* v. *Valmont Plantations et al.*
77. *State of Texas et al.* v. *Valmont Plantations et al.*
78. *State of Texas et al.* v. *Valmont Plantations et al.*
79. *State of Texas et al.* v. *Valmont Plantations et al.*
80. Recopilacion de Leyes de Indias, lib. IV, tit. V, ley 9.
81. *State of Texas et al.* v. *Valmont Plantations et al.*
82. *State of Texas et al.* v. *Valmont Plantations et al.*
83. Ira P. Hildebrand, "The Rights of Riparian Owners at Common Law in Texas," *Texas Law Review,* 6 (1927–1928): 19.
84. Riparian rights were based upon reasonable use which only applies to artificial uses. Irrigation is considered to be an artificial use. Use of water for household use by humans and for consumption by livestock are considered natural uses, and water flow may generally be taken without regard to the effect on others. In some states in the West, water for household and livestock use can be diverted from streams and rivers into reservoirs without even obtaining a permit, as in Texas, where a user may divert up to 200 acre-feet (an acre foot of water is 325,851 gallons, or the amount of water need to flood one acre of land one foot deep).
85. Elinor Ostrom and Roy Gardner, "Coping with Asymmetries in the Commons: Self-Governing Irrigation Systems Can Work," *The Journal of Economic Perspectives,* 7 (4) (1993): 96–97. Olstrom and Gardner provide an interesting discussion of the obvious opinion that "headenders" and "tailenders" are in very different positions, but if selfish "headenders" ignore the scarcity they generate for "tailenders" it is understandable that those at the tail end have less reason to contribute to the continual maintenance of the common system. Allocation of the water (appropriation) and maintenance of the irrigation system (provision) are the two problems any common-pool resource faces. "In addition (to these problems) irrigation common-pool resources also have an asymmetry between 'headenders' and 'tailenders,' which increases the difficulty of providing irrigation systems over time" ("Coping with Asymmetries in the Commons").
86. Carlos Castañeda, *Our Catholic Heritage, 1519–1936,* vol. II: *The Winning of Texas* (Austin, Tex.: Von Boeckmann-Jones Co., 1938), p. 306.
87. Glick, *The Old World Background of the Irrigation System of San Antonio,* p. 37.
88. "Steward of assets owned by the 'republic' or town." Mary Alice Austin, "The Municipal Government of San Fernando de Bexar, 1730—1800," *Southwest Historical Quarter Online,* 8 (4). Available online at www.tsha.utexas.edu/publications/journals/shq/online/v008/n4/article_1_print.html. Castañeda, vol. II: *The Winning of Texas,* p. 308. *Mayordomo* is defined as "administrator of public lands."
89. Glick, *The Old World Background of the Irrigation System of San Antonio,* pp. 39–40.
90. C. Allan Jones, *Texas Roots: Agriculture and Rural Life before the Civil War* (College Station, Tex.: Texas A. & M. Press, 2005), p. 34.
91. Meyer, *Water in the Hispanic Southwest,* p. 167.
92. "Adscripted" is defined as a right that is permanently attached to the land.
93. De la Teja, *San Antonio de Bexar,* p. 87.
94. Meyer, *Water in the Hispanic Southwest,* p. 91.
95. Bexar Archives Translations, The Center for American History, The University of Texas at Austin, Series 1, General Manuscripts 1717–1789, Box 2C15, Vol. 2, December 25, 1731–June 25, 1733, p. 17. Glick, *The Old World Background of the Irrigation System of San Antonio,* p. 39. Bexar Archives Translations, 2C17, pp. 129–132.
96. De la Teja, *San Antonio de Bexar,* p. 87.

97. Bexar Archives Translations. March 22, 1738, Box 2C15, vol. II, p. 45.
98. Bexar Archives Translations, March 22, 1738, Box 2C15, vol. II, p. 45.
99. Carlos Castañeda, *A Report on the Spanish Archives in San Antonio, Texas* (San Antonio, Tex.: Yanaguana Society, 1937), p. 48.
100. Castañeda, *A Report,* p. 48.
101. Castañeda, *A Report,* p. 52.
102. David J. Weber, "Mexico's Far Northern Frontier, 1821–1854: Historiography Askew," *The Western Historical Quarterly,* 8 (3) (1976): 279–293; p. 282.
103. Garland F. Smith, "The Valley Water Suit and Its Impact on Texas Water Policy: Some Practical Advice for the Future," *Texas Tech Law Review,* 8 (1977): 578–636; p. 580.
104. A. A. White and Will Wilson, "The Flow and Underflow of Motl v. Boyd: The Problem," *Southwestern Law Journal,* 9 (1) (1955): 377–433; p. 379.
105. Hans W. Baade, "The Historical Background of Texas Water Law: A Tribute to Jack Pope," *St. Mary's Law Journal,* 18 (1) (1986): 1–98; p. 11.
106. Baade, "The Historical Background of Texas Water Law," p. 21.
107. US Senate, *The Treaty between the United States and Mexico,* 30th Congress, 1st Session, Executive Document 52 (Washington, DC, 1848), p. 47.
108. US Senate, *The Treaty between the United States and Mexico,* p. 47.
109. Sir William Blackstone, *Commentaries on the Laws of England* (Portland, Oreg.: Thomas B. Wait & Co., 1807), p. 18.
110. White and Wilson, "The Flow and Underflow of Motl v. Boyd," p. 2.
111. Author's interview of Russell S. Johnson, Attorney, one of the foremost water-rights experts in Texas on June 10, 2008 in Austin, Texas.
112. Unknown writer, cited in White and Wilson, "The Flow and Underflow of Motl v. Boyd."
113. White and Wilson, "The Flow and Underflow of Motl v. Boyd," p. 3.
114. Hildebrand, "The Rights of Riparian Owners," pp. 19–20. "In 1866, by act of the Congress, the new system was expressly recognized in the following language: Whenever, by priority of possession, rights to the use of water for mining, agricultural, manufacturing, or other purposes, have vested and accrued, and the same are recognized and acknowledged by the local custom, laws, and the decisions of the courts, the possessors and owners of such vested rights shall be maintained in the same; and the right of way for construction of ditches and canals for the purposes herein specified is acknowledged and confirmed; but whenever any person, in the construction of any ditch or canal, injures or damages the possession of any settler on the public domain, the party committing such injury or damage shall be liable to the party injured for such injury or damage."
115. White and Wilson, "The Flow and Underflow of Motl v. Boyd," pp. 20–21.
116. Act of Congress of July 26, 1866. 14 Stat. at Large, 253.
117. Act of Congress of July 9, 1870. Stat. at Large, 218.
118. Donald J. Pisani, *To Reclaim a Divided West: Water, Law, and Public Policy 1848–1902* (Albuquerque, N. Mex.: University of New Mexico Press, 1992), pp. 11–12.
119. Gordon Morris Bakken, *The Mining Law of 1872: Past, Politics, and Prospects* (Albuquerque, N. Mex.: The University of New Mexico Press, 2008), pp. 94–96.
120. Pisani, *To Reclaim a Divided West,* p. 17.
121. Pisani, *To Reclaim a Divided West,* p. 31.
122. Hildebrand, "The Rights of Riparian Owners," p. 20.
123. Donald J. Pisani, *Water, Land, and Law in the West* (Lawrence, Kans.: University Press of Kansas, 1996), p. 1.
124. White and Wilson, "The Flow and Underflow of Motl v. Boyd," n. 17.
125. *Black's Law Dictionary* (St. Paul, Minn.: West, 1990), p. 344.
126. Russell S. Johnson, "Texas Water Law: The Next Century," speech delivered September 1, 2001 at the Fourteenth Annual Legal Seminar on Ad Valorem Taxation, San Antonio, Texas, p. 9.
127. Meyer, *Water in the Hispanic Southwest,* p. 178.

128. Meyer, *Water in the Hispanic Southwest,* p. 178. *Siete Partidas,* Partida 3, Titulo 28, Ley 1.
129. Meyer, *Water in the Hispanic Southwest,* p. 179.
130. Meyer, *Water in the Hispanic Southwest,* p. 180.
131. *Houston & Texas Central Railroad Company* v. *W. A. East,* 98 Texas Reports, 146, June 13, 1904.
132. Another more recent case which yet again confirmed the rule of capture was 1999's *Sipranio* v. *Great Spring Waters of America, Inc.* A similar ruling based upon the *East* case was made by the judges in this case. The bottling company purchased land neighboring Sipranio, constructed wells, and began producing groundwater for bottling purposes which in turn depleted Sipriano's water wells. The Texas Supreme Court found in favor of the bottling company and against Siprianio even though his wells were ruined. In *Sipranio,* as in other cases such as *City of Corpus Christi* v. *City of Pleasanton* in 1955, the Texas Supreme Court expressed strong comments in the record about the legislature's reluctance to address the rule of capture such as "but it [the legislature] has not seen fit to do so" and "I concur in the view that, *for now—but I think only for now—East* should not be overruled."
133. Gary Stansifer, "Role of the Middle Rio Grande Water Master," PowerPoint presentation by the Office of the State Engineer to the Interstate Stream Commission on October 7, 2008.
134. Good Neighbor Environmental Board, *Water Resources Management on the US–Mexico Border: Eighth Report to the President and Congress of the United States* (Washington, DC: US Environmental Protection Agency, February, 2005), p. 24.
135. Charles T. Dumars, "Changing Interpretations of New Mexico's Constitutional Provisions Allocating Water Resources: Integrating Private Property Rights and Public Values," *New Mexico Law Review,* 26 (summer) (1996): 367–397; p. 2.
136. In neighboring Texas, groundwater is drawn according to the "rule of capture" and is, to this day, with very limited exceptions, available to the "landowner who has the biggest pump" without liability to other landowners.
137. Dumars, "Changing Interpretations," p. 12.
138. Dumars, "Changing Interpretations," p. 2.
139. Sylvia Harrison, "The Historical Development of Nevada Water Law," *University of Denver Water Law Review,* 5 (fall) (2001); p. 4.
140. Harrison, "The Historical Development of Nevada Water Law," p. 4.
141. Harrison, "The Historical Development of Nevada Water Law," p. 8.
142. Harrison, "The Historical Development of Nevada Water Law," p. 9.
143. Smith, "The Valley Water Suit and Its Impact on Texas Water Policy," p. 583.
144. Dunbar, *Forging New Rights,* p. 9.
145. John E. Thornton, Ramsey Laursoo Kropf, Dar Crammond, and Andrea K. Gerlak, "Dividing Western Waters: A Century of Adjudicating Rivers and Streams," *University of Denver Water Law Review,* 8 (spring) (2005): 148–189; p. 401.
146. Thornton et al., "Dividing Western Waters," p. 402.
147. Dunbar, *Forging New Rights,* p. 17.
148. Thornton et al., "Dividing Western Waters," p. 378; Dunbar, *Forging New Rights,* pp. 21–23.
149. Dunbar, *Forging New Rights,* pp. 20–21.
150. Stephen N. Bretsen and Peter J. Hill, "Irrigation Institutions in the American West," *UCLA Journal of Environmental Law,* 25; p. 285.
151. Thornton et al., "Dividing Western Waters," p. 379; Dunbar, *Forging New Rights,* p. 26.
152. Thornton et al., "Dividing Western Waters," p. 381; Dunbar, *Forging New Rights,* pp. 29–30.
153. Dunbar, *Forging New Rights,* p. 29.
154. Dunbar, *Forging New Rights,* p. 29.
155. Dunbar, *Forging New Rights,* p. 29.
156. Katherine Coman, "Some Unsettled Problems of Irrigation," *American Economic Review,* 1 (5) (1911): 1–19.
157. Bertsen, p. 294.
158. Thornton et al., "Dividing Western Waters," p. 381.

159. Dunbar, *Forging New Rights*, p. 33.
160. Dunbar, *Forging New Rights*, p. 34.
161. Bertsen, p. 315.
162. Bertsen, p. 316.
163. Bertsen, p. 322.
164. "Liquid Gold: California's Water—An Exhibit by the Water Resources Center Archives," see www.lib.berkeley.edu/WRCA/exhibit.html (accessed April 27, 2010).
165. Dunbar, *Forging New Rights*, p. 42.
166. Dunbar, *Forging New Rights*, pp. 36–45.
167. Dunbar, *Forging New Rights*, pp. 36–45.
168. Donald J. Pisani, *Water and American Government: The Reclamation Bureau, National Water Policy, and the West, 1902–1935* (Berkeley, Calif.: University of California Press, 2002), p. 272.
169. Pisani, *Water and American Government*, p. 275.
170. Clay S. Jenkinson, "Theodore Roosevelt's Footprint on North Dakota," available online at www.theodorerooseveltcenter.com/TR_Footprint.asp (accessed April 27, 2010).
171. Jenkinson, "Theodore Roosevelt's Footprint."
172. www.tcid.org/history2.html.
173. www.usbr.gov/dataweb/html/newlands1.html.
174. Dunbar, *Forging New Rights*, pp. 46–58.
175. Augustus L. Campbell, "Texas Watermasters: A Legal History and Analysis of Surface Rights Enforcement," *Texas Tech Administrative Law Journal*, 7 (spring) (2006); p. 2.
176. Reed D. Benson, "Deflating the Deference Myth: National Interests vs. State Authority under Federal Laws Affecting Water Use," *Utah Law Review*, (2006): 241.
177. Elwood Mead was Wyoming Territorial and State Engineer, 1888–1889, and head of the USDA Office of Irrigation Investigations, 1898–1907. He was a major influence on Wyoming's Water Code and on federal irrigation legislation in the 1890s.
178. www.tceq.state.tx.us/compliance/field/field_ops/wmaster/stwr/southtexas.html.
179. Stansifer, "Role of the Middle Rio Grande Water Master."
180. Jennie L. Bricker and David E. Filippi, "Endangered Species Act Enforcement and Western Law," *Environmental Law*, 30 (fall) (2000); p. 736.
181. Bricker and Filippi, "Endangered Species Act Enforcement and Western Law," p. 747.
182. Bricker and Filippi, "Endangered Species Act Enforcement and Western Law," p. 751.
183. *City of Del Rio, Appellant* v. *Clayton Sam Colt Hamilton Trust, Appellee*, Court of Appeals of Texas, Fourth District, San Antonio, Texas, February, 2008.
184. Committee on the Future of Irrigation in the Face of Competing Demands, Water Science and Technology Board, Commission on Geosciences, Environment, and Resources, National Research Council, *A New Era for Irrigation* (Washington, DC: National Academy Press, 1996), pp. 18, 120.

CHAPTER TWELVE

LEISURE IN THE WEST

Randal Fulkerson

Westerners used anything at their disposal to relieve the boredom of everyday life. Hunting and fishing began as economic activities designed to supplement the food supply. By the twentieth century, hunting and fishing became major industries involving advertising, science, and government involvement.

Dietary considerations and religion motivated American Indian hunting. Early Americans hunted what they needed for survival. Boys received instruction in hunting, which led to manhood rites that tested their hunting ability. Elders taught them to consider animals brothers. They endowed animals with moral characteristics. Elders reminded their charges not to kill a morally superior animal. This education prepared children for the male-only hunting fraternity. Conservation rested on pragmatic grounds. Plentiful game ensured a balance in nature. The primarily nomadic tribes had no motive to store meat.[1]

Virgin-soil epidemics precipitated religious apostasy in Native America. Generally, American Indians believed animals caused the diseases. This relieved them of their duty to respect animals thus making hunting rituals meaningless. Some believed animals attacked them for trading with Euro-Americans. On the Plains, American Indians killed animals without mercy. Many tribes began hunting for money.[2]

Buffalo provided protein and an easy target for hunters. Early American Indians used fires to force stampedes destroying numerous beasts for a few kills. Frequently they left meat wasting on the plains. During the buffalo-robe fad, American Indian hunters killed for hides. A good hunter could kill 300 per day. Hunters brought hides to Bent's Fork, Colorado, where they sold them relatively cheaply. The discovery of better tanning processes and factories made the business highly profitable, but hunters received little for their work. In the horse era, hunters swarmed around a single buffalo shooting into it. They relied on their horse to flee if the buffalo attacked. After the slaughter of the southern herd, American Indians returned to the belief in Father Buffalo as gamekeeper.[3]

In early America, people despised hunters. Eastern Americans considered hunting crude and inhumane. Few Americans hunted in the 1700s. Many Americans held to a four-stage theory of social development that taught hunting was of the lowest order. Americans believed hunting led to excessive laziness and individualism, which, in turn, led to disrespect for community concerns.[4]

The glorification of hunting began in the nineteenth century. By the 1830s, Americans portrayed hunters as symbols of freedom. Writers glorified hunters in novels and poems. This glorification coincided with the expanding political rights and the Jackson era's go-ahead mentality. Part of this development was in reaction to Eastern intellectualism and perceived feminizing tendencies. Americans now considered hunting for food chivalric. Easterners, such as Washington Irving, made pilgrimages to the Western prairies to enhance their manhood and become American.

The Lewis and Clark Expedition provided hunters with a cleaner, nobler image. With new scientific justification hunters could now be perceived as virtuous and pure. The crew hunted for meat and scientific specimens. Lewis allowed men to hunt to relieve boredom. When the expedition made its return trip they regularly met hunters coming west.[5]

By 1846, hunting became embedded as a ritual to eliminate the feminizing effects of urban society. A boy's first kill became a rite of passage in which he escaped the effect of being raised by a woman. When hunters replaced farmers as democratic symbols, Americans became more aggressive. The hunting ethic replaced selflessness and cooperation with selfishness and competition. This change was an evolutionistic misapplication of the muscular Christianity movement popular during the same era.

People viewed hunting as patriotic during the development of the free-market system. Americans needed a leisurely outlet, and hunting provided one where gambling and alcohol were not required. After the Civil War, more Americans hunted because the factory system made guns available for middle-class Americans. Hunting, especially buffalo hunting, became a means for men to prove their Americanness. To Americans, hunting buffalo was similar to dragon-slaying myths in ancient England. This development encouraged self-reliance as a manly virtue. Hunting appealed to many of the man-is-god philosophical bent such as Transcendentalists. Hunting became a secular religion where one communed with nature. Americans viewed guns as a part of the natural realm. Many American hunters engaged in taxidermy to give themselves a scientific image.[6]

Early pioneers considered animals a detriment to progress. Old ideas of killing only in need seemed irrational. Ranchers killed antelope to make room for cattle and sheep. Most early settlers lacked guns and had no time for sport hunting. In 1835, John C. Fremont warned of buffalo disappearing. By 1840, the animal population had been decimated. When Euro-Americans came West they had no fear of game bosses. The American backwoodsman was not constrained by Christian stewardship or animism. He often killed without discrimination.[7]

Because of its numbers, the buffalo provided an easy target, but settlers had to convert the old Kentucky rifle into a percussion weapon. The older weapons lacked the power to bring down a buffalo. Initially, Westerners killed buffalo for meat. After buffalo robes became popular, Euro-Americans slaughtered buffalo for profit. Railroads provided the impetus for wholesale slaughter. Professionals hunted to provide meat for the railroad crews. William "Buffalo Bill" Cody allegedly killed 4,200 bison to supply food for workers. Trains often slowed down and opened windows so people could shoot buffalo. Market hunters approached hunting from an economical perspective. They despised sportsmen who had nothing to lose if they failed. For economic reasons, professionals shot only what the market could sell. Killing more would only raise their costs. Once they had their kill they needed to sell their meat.

Still-hunting and horse-running were the two methods of hunting buffalo. The latter method involved charging a particular bison. Hunters shot the animal from close range. Professionals preferred the more economical still method. They would operate with a three or four man crew and a cook. Keeping low and quiet, they shot animals from about 300 feet. They limited kills to the amount skinners could handle. When they reached their limit, the chief hunter called off the hunt. Professionals originally hunted the southern herd because the Santa Fe Railroad had means of shipment. The southern herd, separated from the northern when the Union Pacific reached Cheyenne, was decimated first.[8]

Subsistence hunters, looked down upon by sportsmen, hunted for their own food. American legal rulings gave hunters rights not enjoyed under the English flag. The law did not consider subsistence hunters poachers while in pursuit of game. Additionally, the hunter who killed an animal received preference over one in pursuit. Subsistence hunters feared the coming of English-type laws advocated by sporting clubs. By 1860, many anti-hunters considered subsistence hunting the only legitimate form.[9]

Sportsmen hunted for trophies and prestige. This resembled the American Indian rites of manhood. The difficulty of the hunt raised the prestige factor for sportsmen. They considered hunting superior to blood sports such as boxing. Grizzly bears ranked highest on kill lists due to their power and aggressiveness. Hunters portrayed grizzlies as the knights of nature. To be maimed by a grizzly was considered a badge of honor. Killing grizzlies held greater prestige because they could kill men.

Wealthy Europeans and Easterners hunted in the American West. These hunters sought high kill numbers to promote a type of cultural imperialism and disdain for the West. Accompanied by over fifty dogs and gun-loading servants, Sir George Gore killed about 2,000 bison for fun. Government Indian agents stopped his slaughter on behalf of Wyoming tribes. The agents contended that Gore had too many helpers to be a sportsman. Russian Archduke Alexis toured Dakota with numerous escorts seeking his share of kills. English sportsmen demanded special trains and accommodations for their dogs when hunting. American sportsmen ridiculed Europeans who required servants and flashy outfits.[10]

Railroads provided hunting tours, which allowed the leisure class to display their wealth. The completion of the Great Northern and Northern Pacific opened up big-game country. The Northern Pacific ran through the Flathead–Kootenai region, which hitherto had been virtually inaccessible. Outfitters rented guides, horses, and equipment to wealthy hunters. Circular tours led sportsmen into the region to hunt moose, elk, and grizzly, but the new prize was the bighorn mountain sheep. Due to its quickness, agility, and alertness, the bighorn proved a most difficult kill. The popularity of big game led to slaughter and an anti-hunting movement.[11]

Hunting clubs began during the antebellum era and grew powerful during the late 1800s. They developed rituals that brought hunters into local fraternities which had national appeal. Clubs claimed to lift hunting to an artistically higher plane. Organizers taught members to kill mercifully (meaning quickly), to lift themselves above the common pot-hunter. Clubs glorified individualism and woodsmanship. Members considered clubs part of the American reform movement. They developed rituals and codes designed to mold members into model Americans. Highly nationalistic, clubs sought to end hunting's indolent image while creating an anti-urban wilderness mentality.[12]

As subsistence hunters feared, clubs became a haunt for the wealthy. They became increasingly exclusive, adopting an English style in their corporate culture. They excluded the poor from membership and, by association, from manhood. These clubs coincided with and fostered a divide between the educated upper middle class and the rural poor. Wealthy urbanites purchased prime hunting lands for their clubs. Additionally, they agitated for laws that excluded the poor from hunting. In Denver and Los Angeles, clubs controlled nearly all duck-hunting grounds. San Francisco clubs hired English-style private game-keepers. Clubs mirrored the elite belief in social Darwinism.[13]

Hunting contributed to the Gilded Age manliness cult. Theodore Roosevelt personifies the movement, which advocated aggressive, strenuous behavior. He argued that hunting made America strong while farming made America weak. He advocated game laws to channel mechanical tendencies in men living in an era of machines and corporations. Roosevelt hunted whenever he wished to celebrate, even when he learned of President McKinley's death. He argued that two types of men existed: one was the courageous hunter and the other the cowardly anti-imperialist. In the manliness era, men founded boys' clubs such as the Sons of Daniel Boone to encourage hunting.[14]

Western hunting was not an exclusively male activity. Women hunted for food during pioneer times. Like men, women originally hunted for subsistence and later for recreation. They drew distinctions between market and sports hunting. They founded girls' groups, such as the Camp Fire Girls, to promote woodsmanship. In 1936, Bess Kennedy became the first woman to receive a bronze badge as an official predator hunter. Today about 10 percent of Western hunters are women.[15]

Western political and economic leaders learned they could promote their region as a sportsmen's paradise. Using romantic legends and grand scenery, they advertised the region to those living in the urban East and the boring Midwest as a spectacular escape and haven to pursue manliness. The sounds and excitement of grizzly hunts and the opportunity to walk in the footsteps of pioneers such as Kit Carson and Jed Smith provide easy marketing for Western entrepreneurs wishing to cash in on the region's image. Hotels and private reserves catered to wealthy, urban hunters. They promoted the Western hunt as a type of sacred American rite.

Rail and air travel brought tourists to areas that had been unreachable. Automobiles and better roads gave the average hunter better access to hunting sites. State magazines, quite amateurish in the early days, trumpeted local hunting and fishing spots. Western magazines usually operated on a more professional level than their Eastern counterparts. In the 1950s, many states thought of using television. Some magazines, such as *NebraskaLand,* appealed solely to tourists. Wyoming ran an annual big-game issue. What these magazines all have in common is an intention to draw wealthy sportsmen into their state.[16]

The years 1870 to 1900 was the golden age for Western big-game hunting. The American West became a hunter's paradise at this time. Regions that could not provide dramatic grizzlies and moose promoted other types of hunting. In the plains, the sharp-tail grouse and prairie chicken replaced the buffalo, resulting in the massive slaughter of grouse. Pacific Coast promoters advertised quail as a special target for hunters. Even songbirds became targets for hunting parties. In Texas, over 10,000 robins died courtesy of a single hunting party.[17]

Predator, or bounty hunting, is a controversial part of hunting history. As America expanded westward, animals needed to hunt for food. The loss of natural food sources meant animals had to attack livestock and farm animals to survive. Hunters worked the West, receiving bounties for each kill. Ranchers and farmers considered wolves and pumas a detriment. State and federal bounties made these predators marked beasts. In 1915, the national government paid hunters to kill predators. Hunters have killed over 3 million since then. Today, most predators are nearly extinct. As late as 1958, the government killed 300 pumas. Today, only the coyote remains, but in small numbers. An ironic twist to predator hunting is that the animals that did the most damage tended to escape hunters. Those killed for bounty were the ones that could not get away.[18]

Hunters initiated the American conservation movement. Observing the demise of numerous species, hunters realized their passion depended on saving wildlife. Americans have spent a great deal of money in species restoration. Early groups such as the American Wild Fowlers Society, now Ducks Unlimited, have agitated for laws to protect animals. Prominent hunters such as George Caitlan, George Crook, and Theodore Roosevelt pushed for protective legislation. In 1890, Wyoming banned big-game sports hunting. By 1900, numerous groups existed to lobby for federal protection. As President, Roosevelt used the Dawes Severalty Act to seize land for preserves and scientific management. This set a precedent for federal controls. In the past, states could not enforce conservation laws. One problem was that animals do not recognize state boundaries.[19]

The federal government began rebuilding the buffalo population. It shipped buffalo to the Wichita National Wildlife Refuge in Oklahoma to breed a new herd. In the twentieth century, the government passed laws to protect and restock wildlife. In 1913, a law was passed to protect migratory birds. In 1933, the University of Wisconsin opened a program to train conservationists. In 1937, the Pittman–Robertson Act instituted an excise tax on guns and ammunition which provided funds to purchase preserves. A federal bird-hunting stamp and state fees are used to protect and replenish species. By far the most extensive legislation is the Richard Nixon-initiated Endangered Species Act of 1973. This controversial measure allows bureaucrats to decide which species are endangered. However, the law has no safeguards, and, according to a Supreme Court ruling in 1995, the US Fish and Wildlife agency can take land by fiat.[20]

American hunters have never been without critics. Anti-hunting is almost as much a part of American tradition as hunting itself. In the nineteenth century, opponents attacked sports and market hunting. Today, anti-hunters allege that hunters just hate wildlife. They portray hunters as inhumane and selfish. Many anti-hunters have been raised on Disney's *Bambi* and cutesy-pooh anthropomorphic cartoons. It is now fashionable for intellectuals to mindlessly consider all hunters guilty of bloodlust. Some activists exhibit outright hatred of hunters as exhibited by an early twenty-first-century Simon & Schuster game, Deer Hunter Avenger, in which children vicariously killed hunters. The tendency of anti-hunters to exaggerate effects and to refuse to acknowledge hunters' humanity has retarded conservation and compromise.[21]

FISHING

Fishing in the West used to be a province of the rich. In the early days of Western settlement, fishing provided a means of supplementing the dinner table and relieving boredom. However, the time involved prevented the average person from fishing. Western fishermen used live bait and preferred bass. Fly-fishing became the major attraction for fishers traveling the West. Pole-fishers mocked fly-fishers who demanded beautiful views and fish that jumped into their nets.[22]

Fly-fishing involved fishermen trying to trick trout into believing the artificial fly is an edible insect. Deceivers and lures are the two basic forms of fly. The deceiver tricks the trout into thinking the fly is food. Making flies has become an art form. The successful fly-maker needs a rudimentary knowledge of entomology. The lure simply makes movement in the water to attract the fish. The lure need not be anatomically accurate because it relies on movement. Since the 1930s, fishers have used spinning tackle and casting reels for bass.

The main reason fly-fishing is more prevalent is that sports writers promote the form. They make fly-fishing appear more difficult and artistic. Their magazines advertise fly-fishing gear which is more expensive than the conventional pole-fishing equipment. In the early twentieth century, Western hotel owners promoted fly fishing to attract Eastern tourist. In the 1920s, marketing focused on California and the Pacific Northwest. Tourists and promoters ignored the Rockies until the late 1930s.[23]

When the Rockies became the Eden of fly-fishing, hotels and fly factories urged Easterners to fly-fish in the unspoiled wilderness and escape the polluted East. Fly-fishing was the ultimate Western experience according to the promoters. They advertised the Western fishing trip as an American rite of passage and promoted famous fishermen such as Bud Lilly to lure tourists. Lilly set up lodges and provided guidance, turning his vocation into a full-time business. Factories distributed catalogues and opened schools to preach fly-fishing, hoping to convert wealthy novices into fanatics. Advertisements in Eastern papers appealed to wealthy urbanites. Frequently they contrasted the beauty of fly-fishing with the disgusting use of live bait. State magazines promoted their own fishing hotbeds to pull in tourist dollars. Films and advertisements featuring famous fishermen such as Bing Crosby and Ted Williams were a part of the campaign.[24]

The ability to actually get to a fishing spot was a major factor in developing the tourist-fishing industry. Increased air travel and road construction made travel more efficient. Air-travel efficiency made fishing lodges an economically feasible investment. This was especially true in Alaska where air travel is often the only means of reaching a fishing spot.[25]

As in hunting, fishing has relied on conservation to maintain its viability. In 1875, *Forest and Stream* called for restocking fish populations and condemned the mounting of fish. Conservationists needed to reeducate fishers. In the past, fishers kept as many fish as they could catch. In the latter nineteenth century, the only-take-what-you-can-eat mindset developed. The fish-and-release mentality, initiated in 1936 by the National Parks Service, did not predominate until the late twentieth century. The federal government now operates hatcheries where fish culture is developed and then placed elsewhere. The government developed new restrictions to ensure fish populations continue. Access to fishing areas is now limited in time and place. Additionally, there are limits on the number of boats in a given area at given times. Most states now require licenses. Funds from licenses are used to restock fish.[26,27]

GAMBLING

Traditionally, Americans have had a love–hate relationship with gambling. They loathe endorsing activities that cater to greed and laziness yet love the excitement and possibility of financing government programs without taxes. Gambling crosses social and cultural lines in its appeal.[28]

Long before European contact, American Indians became addicted to gambling. Like Euro-Americans, American Indians were known to risk anything, including their wives, and lose. Frequently games involved high-stakes purses of guns and horses. The Yakima often played for slaves. Often bets continued on a double-or-nothing system. When tribes played team games, a separate area was set aside for betting.

Native gaming was rarely honest. Many games required sleight-of-hand skills or the ability to mislead an opponent. Guessing games, such as an Arapahoe button game, involved singing silly songs or meaningless conversation to confuse opponents. Tribes gambled against other tribes. The Paiutes would cheat the honest Shoshones in various games whenever the Shoshones had surplus goods. Horsemen were not averse to cheating to win their races. Like Euro-American gamblers, American Indians did not consider it cheating if deception required skill.[29]

Different tribes favored different gambling activities. The Zuni and Arapahoe played hiding games where opponents tried to guess the position of a ball or button. In the West, running and throwing contests predominated. In one game, organizers rolled a hoop as contestants threw spears through it. The further the hoop was from the thrower, the greater the score. Apaches used smaller hoops for similar games. The Mandan practiced a rapid-fire archery game in which arrows were prizes. The Pawnee stressed accuracy in their archery contests. The Yakima allowed Euro-Americans to participate in their shooting matches. Horseracing was prominent among plains tribes. The Apache played various card games with hand-painted cards. American Indians embraced poker. Numerous leaders became quite adept at poker.

Most tribes incorporated religion into their gambling. The Wichita only gambled during religious ceremonies. Many tribes believed the twins of creation gambled over the settlement of the planet. Often races, especially foot-races, were considered religious acts. Tribes employed shamans for divine aid to defeat an opponent.[30]

Social problems often accompanied American Indian gambling. American Indians have been known to die during a battle rather than interrupt their gambling. In some tribes, winners returned a percentage so losers could avoid destitution. If a loser became destitute that person would be part of the non-producing element of society. In eastern Washington, losers depended on Euro-American largesse for survival. Different tribes treated losers in various manners. In some the losers remained destitute while in others one could work back up through success in hunting, fishing, or battle.[31]

In Anglo-America, Robert Fulton's steamboats provided professional gamblers with both a venue and clientele. From the 1840s to the 1860s the Mississippi river served as a floating gambling den. Most gamblers, frequently disguised as farmers or preachers, emanated from New Orleans, working upstream. Natchez-Under-the-Hill served as a haven for dishonest gamblers. Steamboats carried a large number of bored, gullible passengers. People got on and off steamers at various stops. This provided fresh victims for the professional. Additionally, it kept the gambler from traveling a long time with the victim. River travel put gamblers outside legal jurisdictions provided the captain was friendly. Captains often knew gamblers since they were regular passengers. Often captains would be in on the cheating.[32]

When trains replaced steamboats, gamblers sought less mobile accommodations because fast-moving trains provided little profit. Many gamblers left the profession and wrote, or lectured, about their exploits. Much of our information comes from reformed gamblers. Retired professional Cole Martin acknowledged gambling was a cutthroat profession. Gamblers tended to be violent men who generally carried concealed weapons such as derringers. Yet the successful professional was usually conversant in numerous topics. This attribute allowed him to misdirect suckers with his conversation. Ironically, many gamblers, while quite skilled in their chosen games, were suckers in other games. Something in their personality rendered them incapable of quitting when losing outside their field. Riverboat gamblers faced some hazards. Most boats carried warnings about their presence.

Some captains tried to stop the cheating. They would eject cheaters at nearby stops, depositing them with the local constabulary. Pickpockets and thieves targeted gamblers who generally had more money than other passengers.[33]

Gambling entered mining towns almost simultaneously with the original strikes. Men who mined led lonely, rigorous lives. They were generally young and restless. Mining for gold or silver tended to be a gamble itself. Gambling was an extension of their normal activity. Although miners needed little temptation, gambling halls used nefarious means to entice the miners. Early gambling halls were little more than tents with overturned barrels as tables. Gaudy roulette wheels were designed to attract men. Saloons used alcohol and women to entice miners.[34]

Miners extracted over $250 million in gold during the California Gold Rush. Professional gamblers flocked to boomtowns to fleece the miners. Professionals insisted on small stakes, hoping to grow rich off volume gambling. Customers despised the professionals, but saloon owners needed them to run the games. Roulette and faro required three men to operate.[35]

After cattle drives, cowboys had plenty of money and a desire to spend it. Saloons offered excitement after the long, boring cattle drive. Cowboys rushed in to gamble and drink. Female companions helped cowboys pursue excitement. Saloon keepers, welcomed by townsfolk, arrived in town with the cattle drives. Gambling houses paralleled the boom–bust cycle of cow towns. Saloon keepers relied on small stakes and volume gambling for success. When towns became respectable, professionals either went underground or left town. Towns such as Abilene and Dodge City, Kansas, had a notorious reputation in their hey-day. When towns grew, cattle terminals were pushed to new cities further north. Cow towns in Kansas and Nebraska catered to cowboys with money.[36]

Dodge City had a raucous image before the cattle era. Originally called Buffalo City, it catered to hunters and soldiers. Gambling halls lined Front Street. Saloons such as the Comique even provided entertainers such as Eddie Foy to lure customers. Dodge City served as a prototype for Las Vegas. People saved money to come and gamble in Dodge City. Saloons promoted a good-times concept to tourists.[37]

Professional gamblers worked in teams, often in association with confidence men. Generally, teams consisted of two to six men, or women. Playing a cutthroat style, gamblers cooperated to ensure a victim's loss. They divided winnings after games closed down. Usually the active cheater was an innocent-looking player dressed as a farmer or preacher. The fancy-dressed gambler usually lost to create an aura of honesty. The stereotypical gambler wore black suits, fancy shirts, and a variety of shiny accessories. The numerous, often fake, diamond and gold accessories served to distract victims. Because professionals needed soft hands to manipulate cards, they avoided work that might callous their hands. A gambler's affable personality distracted victims and provided a friendly atmosphere.[38]

Gamblers could not earn a living honestly. Statistical probability prevented success in honest gambling. Claims of professionals playing honestly cannot be supported by evidence. When professionals spoke of being good they were referring to their ability to deceive people. They did not consider hiding cards or cutting decks cheating because skill was required. Successful cheating required years of practice and teamwork. Professionals palmed cards, dealt from different sides, and hid cards in their specially designed clothing. Saloon owners purchased special faro boxes and roulette wheels to cheat gamblers. Fraudulent lotteries could be lucrative. James Pattie, the King of Lotteries, ran numerous fraudulent lotteries in 1870s Nebraska. He moved to Laramie when Nebraska authorities suspected his enterprise. Cheating was so predominant that businesses advertised cheating devices in major newspapers.[39]

Victims usually shot cheaters they caught. Cheating and gunplay occurred frequently in Dodge City and Tombstone. The mere suspicion of cheating destroyed the friendly atmosphere of card games. This does not mean townsfolk despised gamblers. Indeed, locals usually held resident professionals in high esteem. Young people emulated their talents. Their presence provided evidence of local prosperity. Saloon owners even hired them for a percentage. Some professionals, called pluggers, played for the house.[40]

Usually gamblers led a dismal life. Many committed suicide or were addicted to alcohol or morphine. Some, like John Quinn and Cole Martin, reformed and wrote or lectured about gambling's pitfalls. Not all professionals were men. Former Confederate spy Belle Siddons and Poker Alice roamed the West as professional gamblers. The most famous gambler, Wild Bill Hickok, was shot through the head while playing poker in Deadwood. His hand, a pair of black aces and a pair of black eights, became known as the dead man's hand.[41]

Gamblers played a wide variety of games. The three favorite games were faro, poker, and roulette. Faro, by legend invented in Egypt, was a fast moving, no-skill game. Players bet on which card would spring up from the faro box. The first card, the soda, was not used. Dealers limited bets as the game progressed to ensure the house won. "Lookers-out" watched to ensure that winners put a chip in a slot for the house. Dealers used fake boxes to know what cards were coming.[42]

Draw poker came to be a national pastime. In an honest game, the dealer has no advantage. Players maintain straight faces and engage in small talk, called chaff, to distract opponents. In early days, players drew their own cards. Only the dealer had to say how many he drew. Card tables had slots where players put in a check, now called chip, with each win. To stop extensive bluffing, a Jacks-or-better rule was adopted in the late nineteenth century. The straight flush became different from a flush at the same time. Earlier, four aces or four kings with an ace were unbeatable hands.[43]

Roulette held a mystical attraction to gamblers, particularly the inebriated. The roulette wheel consists of red and black numbers. Players bet in a number of fashions such as numbers, colors, or blocks. In an honest game, a single number paid 35 to 1 yet the odds were 39 to 1. Roulette was nearly always crooked. Bankers used hidden levers to manipulate the ball or slots. The wheel rotated in reverse direction of the ball which meant the gamblers, usually drunk, could not see the cheating.[44]

Horseracing was and remains immensely popular. It has the added prestige of being socially acceptable. In the Old West, races generally involved town favorites in impromptu matches. During the Mexican rule of California, Governor Pío Píco raced against landowner Don Andréas Sepulveda. Sepulveda won the enormous bet which included over 1,000 cattle. Arguments over races often led to fistfights. When this occurred, spectators bet on the fistfight. Judges in Los Angeles and the Dakota territory declared racing debts legally binding.[45]

The story of Elias "Lucky" Baldwin tells of the development of the West's premier racetrack. Baldwin owned worthless stock in Nevada land. When he left for an Asian holiday, he instructed his attorney to sell the stock if it reached the original price. While away, his stock shot up in value, but the attorney could not sell because Baldwin forgot to leave the key to his safe. The impulsive Baldwin sold the stock at a higher price and purchased Rancho Santa Anita. He hoped to build the finest track in America. Santa Anita opened in 1907 before a substantial crowd including numerous bookies. These bookies brought order to the track by replacing the memory system. Tracks such as Santa Anita would later incorporate bookmaking as part of the business.[46]

As a Western gambling center, no town equaled San Francisco in its prime. After Thomas Chambers opened the El Dorado, the city's first casino, numerous others opened. Due to intense competition, casinos offered music, showgirls, and complimentary meals to attract customers. When California banned gambling in 1872, casinos went underground as private clubs such as the Baldwin Club. Underground operators used a variety of ruses to avoid the law.[47]

California reformers attacked gambling before statehood. In 1846 and 1848, anti-gambling drives took place in Monterrey and San Francisco. They linked gambling with drunkenness, bribery, and gangs. In 1872, California officially banned gambling hoping to drive professionals away. In 1878, all games of chance were banned. Saloon owners renamed games to avoid prosecution. Following numerous 1890 scandals, reformers raised the stakes by making owners and players equally culpable. In 1909, San Francisco banned slot machines hoping to close one last legal loophole. This drove the industry underground. Nevada, embarrassed by its image, banned all gambling the same year.[48]

California gambling survived underground for a number of years. In the 1940s, after legal gambling, or gaming, resurfaced in Nevada, San Bernardino was known as the California kasbah. The town, east of Los Angeles, competed with Las Vegas before reformers shut it down. Casino operators then moved to the Morongo Valley. Many Nevada casino owners learned their trade in the California underground.[49]

Slot machines, "one-armed bandits" in popular parlance, were invented to circumvent reform. The original one-reel games failed because patrons could see they were being cheated. Bavarian immigrant Charles Fey designed three-reel machines, equally crooked, that caught on. Chicagoan Herbert Mills pirated Fey's slots and forced him out of the business. Law officers ignored slots because they allegedly paid in cigars or fruit gum. Customers responded enthusiastically to the slots that used patriotic or whimsical images and music to entice participation.[50]

On March 19, 1931, Nevada legalized gaming in licensed casinos. This followed a high-profile propaganda campaign led by Las Vegas developer Tom Carroll. But it was Reno that initially succeeded. Reno combined quickie divorces and filthy casinos on Center Street to become the gaming leader.[51]

Two Californians seeking asylum from reformers invigorated Reno's gaming industry. In 1935, Harold Smith converted a Virginia Street bingo parlor into Harold's Club casino. The innovator brought in women dealers, air conditioning, and a friendly atmosphere. He targeted smaller gamblers and curious tourists to launch his casino. He inundated highways with his now-famous "Harold's or Bust" signs to lure customers. His success pulled the casino district to Virginia Street.[52]

Venice, California operator William Harrah arrived in Reno in 1937. The introverted Harrah wanted the best possible image for his industry and pushed customer comfort and friendly service. He refused to hire anyone trained in Las Vegas where dealers had a reputation for belligerence. He advertised in California and provided buses for seniors. Harrah replaced wood chairs with padded stools. His philosophy was to keep a pleasant atmosphere and to let statistical probability provide profits.[53]

The depression and the Hoover Dam contributed to Las Vegas's growth as a gaming Mecca. Las Vegas was an obscure railroad stopover before the Hoover Dam. Its initial casino area was originally zoned as a red-light district. Fremont Street already had illegal clubs when the state legalized gaming. Within two weeks of legalization, Las Vegas approved four licenses without background checks. The coming war and the dam brought people and money to town. Many dealers and bankers moved to Las Vegas to work legally.[54]

Postwar prosperity initiated a change in Las Vegas gaming. *Hollywood Reporter* publisher Billy Wilkerson, a gambling addict, began building a casino in Las Vegas. Mobster Bugsy Siegel forced Wilkerson to make him a partner. Eastern investors forced Wilkerson out of town when the project struggled. Mobsters Moe Sedway and Gus Greenbaum took over the Flamingo after mismanagement forced its closure. Unknown assailants murdered Bugsy Siegel. The murder led to the myth of Siegel as a visionary, yet it was Wilkerson who had the idea of a resort casino.[55]

Four other casinos were known to be owned by organized crime. Mob activity in Las Vegas was common knowledge in the 1950s. The Kefauver Committee and Kennedy crusades forced many gangland figures to Las Vegas. Felons such as Tony Stralla and Guy McAfee, a former Los Angeles vice-cop, used dummy owners to purchase casinos. The underworld moved out of Las Vegas in the 1970s.[56]

In the 1950s, the Rat Pack, entertainers Frank Sinatra, Dean Martin, and Sammy Davis, drew wealthy visitors to Las Vegas by promoting a party atmosphere. The Copa Room at the Sands Hotel served as the focal point for their carousing. These entertainers initiated a new casino concept: use big-name entertainers to draw new customers.[57]

When Paul Laxalt served as governor, Nevada changed its laws allowing corporations to invest in casinos. The new laws allowed subsidiary companies to manage casinos. The state could investigate the subsidiary managers when granting licenses. Previously, laws required the impossible task of investigating every stockholder. The investments of wealthy aviators Kirk Kerkorian and Howard Hughes opened the door for corporate-owned casinos. After other states and Indian gaming became competitors, Las Vegas turned to a more family-style tourist venue. Gaming remains the key element, but the family attractions give Las Vegas a more diversified tourist economy.[58]

In the 1970s, many states legalized gambling to avoid raising taxes. These states have had mixed experiences with gaming. Some, like Colorado and South Dakota, have used it in conjunction with Old West tourism. States with gaming received an initial boom of tax money followed by an inevitable bust when neighboring states joined the gaming industry. State lotteries have been a mixed blessing. Most of the profits go to gaming corporations. Winning the lottery has meant sheer misery for most winners. Unprepared for wealth, they fall prey to numerous confidence operators and telemarketers. By 2003, only Utah and Hawaii were without gaming.

Indian gaming has been controversial since its inception in the 1980s. The Reagan administration encouraged gaming as a step toward self-sufficiency. In 1987, a Supreme Court ruling allowing Cabazon Tribe casinos in California opened the door for widespread gaming. By 1995, twenty California tribes operated casinos. Some tribes own casino land in trust away from the actual reservations. Often managed by Nevada corporations, these casinos do little for Native unemployment. Legal casinos complain that unregulated American Indian casinos have an unfair advantage over heavily regulated casinos.[59] [60]

ENTERTAINMENT

In the early American West, any diversion sufficed to entertain the bored, often lonely inhabitants. When towns grew, the newer more family-oriented population sought social stability. Entertainment provided a focus for communal gatherings. Civic leaders, usually women, tried to improve culture through literary societies and church socials. Incredibly curious, Westerners welcomed informational events such as lectures, readings, or stereo-optical lectures. Western entertainment generally followed Eastern trends. Audiences, once wildly participatory, grew passive after World War I. After television became popular, what was once communal became a solitary pursuit.

Cowboys and miners had money to spend and little to do during their down times. Saloons served as local gathering spots. Saloon keepers recognized they could compete economically if they provided entertainment. They even allowed fistfights, which served the double purpose of settling scores and drawing crowds. Female dancers came to boomtowns to shill drinks. These girls drank with the lonely men and received a percentage of the sales. Itinerant musicians played these hurdy-gurdy houses for drinks and tips. Dancing girls performed in halls that extended from the bar. Owners built these extensions that required patrons to go through the bar to see the show. After World War I, dance halls had a brief revival on the Pacific coast as creep-joints. In these places "instructerers" charged patrons 10¢ a lesson (dance). The system worked so dancers built up a thirst, drank soft drinks which fueled energy and returned to dance, thus building a cycle.[61]

In the 1870s, rigidly structured minstrel shows toured the West. In the opening act, minstrels sat in a semi-circle. The interlocutor, a pompous, well-dressed man, bantered with two endmen called Mr. Tambo and Mr. Bones. The black-faced endmen would always outwit the interlocutor, to the audience's delight. The banter satirized educated lecturers and introduced a song or dance number. The second part, the olio, featured specialty performances and dumb acts such as tumbling or magic. A short play, called the after-piece, completed the show. Minstrel humor required special knowledge of the local area.[62]

Traveling lecturers, called professors, garnered enthusiastic support from the incurably curious Westerners. Lecturers spoke on diverse subjects. Saloon owners actually booked temperance lecturers. On the rare occasion a patron signed the pledge, the excess crowds made up for any lost revenue. Numerous debates, sponsored by women's groups, drew large crowds. Spectators openly heckled speakers and debaters in the West. Performers accepted audience participation as a legitimate part of their craft.[63]

Lecturers used props to supplement their presentations. Professor James McDaniel used a stereopticon to show women at various stages of undress in his educational presentations. Fined for obscenity in Wyoming, McDaniel continued a successful career with his traveling program. Enthusiastic cowboys supported McDaniel in their pursuit of anatomical education. Civic leaders deemed traveling-wagon shows as morally superior to stage shows where prostitutes and gamblers provided peripheral entertainment. Audiences responded enthusiastically to panoramas, "pan-stereoramas," and magic-lantern shows. Wagon showmen lectured while people viewed images. Panoramas were large paintings of world sites. The pan-stereorama, made of wood, cork, or pasteboard, provided three-dimensional views of famous cities such as Jerusalem or told stories such as the Life of Christ. The Mormon Church used pan-stereoramas in their visitor's centers to proselytize. Similarly, mechanical dioramas demonstrated famous battles moving the armies with unseen devices. Artwork for these programs was generally quite exquisite.[64]

Westerners loved the magic-lantern shows. The Bonheur Brothers of Kansas managed a successful magic-lantern tour on the Plains. The three brothers began as children in Ohio when they purchased a magic-lantern outfit. The machine consisted of a sperm-oil lamp, small tin box, and glass slide. When operators maneuvered pictures on the slide, the lantern projected a moving image against a stationary scene. After initial success, the brothers purchased a camera which provided greater variety of pictures and longer gigs. Customers preferred shows about notorious people such as the James Gang. Copy-cat operators purchased scripted lectures which were sold with the stereopticon machine. Trade papers advertised the equipment to potential lecturers into the early twentieth century.[65]

Lecturers risked ridicule and disdain on Western tours. Yet audiences supported those who could take the heckling. Oscar Wilde tended to disdain his audience when he spoke on aesthetics. His effete manners and dress made him an easy target. Yet Leadville miners respected Wilde after he proved his ability to out-drink the miners. Westerners also regularly attended the lectures of teetotaling William Jennings Bryan who once spoke to over 100,000 people without a microphone in San Francisco. Conversely, popular author Charles Dickens offended people with his arrogance.[66]

The medicine show symbolizes a bygone era. These theatrical confidence men entertained and fleeced audiences throughout rural America. They brought their shows and alcohol-laced serums to any sized town. They used a variety of motifs in their pitches. These "doctors" used supposed medical lectures, religious preaching, or Native-American lore to legitimize their trade. Most shows had side entertainment such as tumblers and sword-swallowers to drum up business. Some enterprises maintained a large payroll. The Big Sensation Medicine Company roamed Nebraska with a twelve-piece band in the late nineteenth century. The more conservative Hamlin's Wizard Oil Company provided free religious concerts by its resident quartet, the Lyceum Four. In 1898, about 150 medicine shows worked the rural landscape. Medicine shows declined with the advent of vaudeville and movies.[67]

In the late nineteenth century, the circus ruled the Western entertainment world. Like other forms, the circus received economic stimulus from the transcontinental railroad. Circuses dazzled patrons with exotic animals, freak shows, and dumb acts. These huge, bigger-than-life troupes overpowered the senses of rural people. A circus entered town like an invading army. Fights often broke out between locals and dishonest gamblers who worked outside the tents. Circus owners allowed crooked games, such as three-card monte, for a percentage. Refreshments served were often spiked with alcohol to help the gamblers. Performers and employees formed a type of trained militia anytime the warning call of "Hey Rube" was shouted. Outnumbered by the armed and trained circus troupe, the locals usually had no choice but to back off. A brawl, known as a clem, occurred outside David City, Nebraska, in 1887. Townsfolk returned one night after a local had been cheated in gambling. The resulting clem ended with four people being killed and the main tent incinerated.[68]

Carnivals traveled the West, generally hitting the smaller towns. Carnivals were basically a circus without the big top. Gambling and freak shows were usually the only attraction. Some carnivals had lecturers and pitch men. In the twentieth century, some carnivals featured a unique "Crime does not pay" circuit featuring parents of famous gangsters. Carnival promoters often bribed local officials or set up just outside a jurisdiction. They often worked with towns at street fairs and town days serving as entertainment venues.[69]

Wealthy Westerners wanted socially respectable entertainment to enhance their town's reputation. Cities brought in companies featuring serious drama such as *Richard III* and *Our American Cousin*. These troupes defrayed some of their costs by hiring locals for minor roles. Actors performed on small stages before mostly male audiences. Spectators thought little about cheering or jeering a performer. It was not uncommon for an inebriated spectator to actually go on stage during a play. Performers would usually be called upon to give a speech on closing nights. Spectators would break up plays to announce hangings or a possible fistfight at nearby saloons. After such announcements, audience, cast, and crew would leave to watch the spectacle then return and pick up where they left off. Westerners treated favorite actors like royalty. Locals greeted actors on their arrivals and escorted them to the depot on departures. Audiences threw money to particularly popular actors. They even bought drinks and provided food and lodging.[70]

Numerous actors such as the Booth Brothers and Faye Templeton toured the West. Frontier audiences supported melodrama and Shakespeare. However, legitimate theater did not fare as well as vaudeville. Arrogant legitimate actors rarely concealed their contempt for Westerners. Had it not been for the inflated Western money, many actors would not have come to the West. Small town theaters struggled against the competition of legitimate theater. Broadway's monopoly forced many small concerns out of business and suffocated artistic creativity. Few businessmen would invest in troupes not associated with Broadway. Most Western towns did have local repertoire groups that played courthouses and private homes.[71]

Cities desired the status that accompanied legitimate theater. Salt Lake City led the way in establishing a theater program. The Church of Jesus Christ of Latter-Day Saints established a theater featuring clean entertainment and audiences. The Deseret-sponsored theater struggled initially, but following a strange fire that destroyed the privately owned Walker House, the Mormon state-owned theater had no competition. The Mormon theater was the first religious-backed house in American theater. Mormons established the Deseret Dramatic Association when they first arrived in the Salt Lake valley. By 1864, Eastern troupes were coming to Salt Lake City. Salt Lake's location made it a rewarding stop for troupes wishing to avoid the costly "long jump" from Omaha to San Francisco. Serious drama usually headlined shows in Salt Lake City, but they often contained a comic after-piece. Church officials prescreened all performances. Leaders, such as Brigham Young, sat in private boxes armed with opera glasses. They used the glasses to search the audience. The Mormon Church discouraged children at the theater. Ironically, it was here that nine-month-old Maude Adams starred in *The Lost Child,* setting a youth record for a lead performer.[72]

San Francisco went theater crazy in the 1880s. The economic elites spent freely, luring troupes to town. The desire for respectability can be seen in the term "direct from New York," an advertising ploy that guaranteed success. San Francisco became the arbiter of theater taste in the West. Patrons unable to gain a seat in the better houses supported a wide array of lesser playhouses. Denver sought to line up regular tours in its desire to be accepted as a cultural center. Successfully replacing Salt Lake City as a base, Denverites responded enthusiastically to Shakespeare and Gilbert & Sullivan. Denver theaters featured special boxes for the wealthy and served alcohol.[73]

Vaudeville replaced legitimate theater as the main attraction, but it was the movies that devastated the theater economically. Westerners preferred movies because they were cheaper and did not require enduring the obnoxious attitude of stage actors. Western legitimate theater virtually collapsed after talkies. In 1910, 349 legitimate theaters operated in the West but in 1925 only 150, mostly restricted, theaters remained. Legitimate theater enjoyed a brief resurgence during the New Deal. Financed by the Federal Theater Project and the Federal Writers Project, acting troupes performed optimistic shows such as *Uncle Sam Time* and *The Skin of Our Teeth* before captive audiences in federal institutions. Congress closed down the program contending that it was a haven for radicals.[74]

Theater experienced a revival after World War II. Many performances occur within other venues such as theme parks like Disneyland. Touring, made economical with air travel, regularly brings Broadway shows to Western cities. Westerners have been more innovative in theater, creating concepts such as dinner theater and theater in the round. These venues, with virtually no backstage area, bring audiences closer to the performers again.[75]

The Western tradition of accepting actors aided vaudeville. From 1894 to 1904, vaudeville exploded in popularity, especially in mining towns. Early troupes traveled as somewhat unorganized tent shows. Once successful, vaudeville attracted wealthy investors who established new theaters. Established in 1895, Mott's Hall in Los Angeles was the first true vaudeville house in the West. In 1896, the Grand Opera House, later called the Orpheum, opened in that same city. Westerners more readily identified with the down-to-earth vaudevillians than with the arrogant stage actors. Performances included frequent interchanges between audience and actor. Westerners showered performers with gifts, meals, and letters of support. Vaudeville possessed an aura of good fellowship where performers ridiculed etiquette and the educated elite before an appreciative assembly. Together, performers and audience laughed at life's frustrations. Troupes tried to play cities with large immigrant populations who could better understand ethnic humor. After World War I, ethnic humor was forced out of the show.[76]

The West's wide-open spaces presented special problems for theater owners. It took a troupe eight days to travel from New York to San Francisco. Troupes needed to avoid costly long jumps where they traveled far without a performance. By 1901, there were no strictly vaudeville houses between Omaha and San Francisco. Early troupes usually centered on a veteran star. The star would book lesser performers with him or her, making booking easier for minor performers. Troupes tried to have a base city such as Denver or San Francisco.[77]

Entrepreneurs devised means to organize the booking system. These booking trusts, called circuits, developed simultaneously with the star system. Theater manager Michael Leavitt organized the Western theatre circuit. His company received a percentage of both salaries and gates for the service. Small-town theaters formed syndicates to expedite their booking arrangements. In Missouri and Nebraska, the syndicate possessed enough power to force troupes to play small towns. With big business and syndicates involved, small companies collapsed when they experienced economic difficulties.

One of the colorful characters of Western show business emerged in the development of circuits. Greek immigrant Alexander Pantages, an eccentric illiterate, made his money in the Alaskan gold rush. Working as a dishwasher in a Klondike saloon, Pantages noticed miners often spilled gold dust on the floor, especially when drunk. Pantages swept up the gold dust and sifted dirt beneath the wooden floors to amass enough money to buy a 10¢ house in Seattle. He built a chain by purchasing struggling theaters. He feuded with competitor John Considine, but made peace when his daughter married Considine's son. Pantages owned forty houses before selling out to RKO.[78]

Performers who played local audiences needed to know the audience. Vaudevillians that joined in local customs and mannerisms succeeded. When Charles Vivian noticed his *Oliver Twist* bored the audience, he switched to singing Irish drinking songs. The Taylor Family Troupe, actually a one-man show, sang local loyalty songs. Mart Taylor researched every town he played. He used a standard script like a template and filled in local names for songs mocking locals. His performance worked like a universal roast of town leaders.

Western theater owners were innovative. In Oakland, the Liberty Theater installed a revolving stage so five scenes could be performed simultaneously. California had airdromes, forerunners of drive-ins, where the show went on inside during bad weather and outside in fair weather. The Los Angeles Orpheum printed programs for customers to identify performers. In the West, the 10¢ houses proved quite popular. At these venues the audience paid less, but the program played twice, even three times, a day. Family theater became a lucrative Western venture.[79]

Eddie Foy is the most recognized vaudevillian to tour the West. His ability to take a hazing in Dodge City added to his appeal. Once when he ridiculed town leaders such as Bat Masterson, local cowboys decided to return the favor. The fancy-dressed Foy proved an inviting target, but after he accepted a dunking in a horse trough, he became a local favorite. Foy made several trips to Leadville and San Francisco during his career. He played billiards and boxed at the San Francisco Boxing Club. His willingness to be one of the boys endeared him to patrons. Foy's gambling made him a favorite of saloon owners. He often worked to pay off gambling debts, an activity that remains part of entertainment life today in Nevada. Foy's contributions have not all been positive. In order to keep his children in his act, Foy teamed with Samuel Gompers to pursue a legal ruling that child actors do not work for a living and cannot be protected by child-labor laws.[80]

San Francisco was the Broadway of the West. When the legitimate theaters closed, women went home and men went to vaudeville houses. The Orpheum would be sold out late night to early morning. Smaller concerns profited by serving those locked out. Western audiences, influenced by San Francisco, were considered more savvy than Midwestern, or even rural Eastern audiences.[81]

Many successful athletes have embarked on acting careers. It is something of a tradition and a mark of achievement for athletes. Even taciturn Charley Gehringer engaged in a vaudeville career. San Francisco boxer Gentleman Jim Corbett was one such athlete. A Wells Fargo teller, Corbett trained as an instructor at the San Francisco Boxing Club. His Irish manager, William Brady, used his connections with boxer-turned-actor Maurice Barrymore to ease Corbett into vaudeville. Corbett became a popular attraction after defeating John L. Sullivan for the title. He played in a number of skits and farces. The opportunity to work in the entertainment field remains a status symbol for athletes and their sport today.[82]

Vaudeville began collapsing in the first decade of the new century. A 1906 fire destroyed the San Francisco theater district. It took a year for the first theater, the Colonial, to reopen in San Francisco. A 1903 Chicago theater fire, where Eddie Foy was performing *Bluebeard,* created a public outcry for stricter regulations. As fire codes moved west, numerous theaters closed. The popularity of movies and tabs, truncated musical comedies geared for less-educated patrons, further undercut vaudeville. The 1912 Weberfields Tour that played Omaha and Kansas City was the last all-star tour. By 1928, the Los Angeles Orpheum was the only strictly vaudeville theater remaining. In spite of its imminent collapse, vaudeville overtook legitimate theater as the leading entertainment venue. Despite losing its hold on the populace, vaudeville succeeded financially into the early 1920s. Economically, 1924 proved to be its best year, but a general two-per-day performance schedule inflated those figures. The star system took an economic toll on vaudeville. In 1924, dancer Johan Eltinge earned $40,000 at the Los Angeles Grauman's Metropolitan. In 1927, Al Jolson pulled in $17,000 a week at the same house.[83]

Changes in audience behavior signaled vaudeville's demise. Previously, audiences responded vocally, loudly, and often to stage activity. In many ways, the audience was a part of the show. In the 1920s, audiences passively applauded at specific times. Theaters banned smoking within the structure and actually had signs informing patrons how to behave. Americans became sophisticated in the jazz age. People frowned on boisterous audience behavior at a time when private behavior degenerated. The fact that vaudeville was now performed in a darkened theater possibly curtailed responsive activity.[84]

In 1925, Ocean Park, California, amusement-park operator Charles Lick created an applauseograph machine with the advertised purpose of discouraging inappropriate noise such as whistling. This disapproval of audience responsiveness created a wedge between actor and fan. Today, fans no longer buy drinks or share meals with actors. The emotional need to share with favorite actors is carried out in the form of cards, fan letters, and anonymous gifts. However, many modern actors are unaware that this response is a vital part of their craft.[85]

The stock-market crash and emergence of radio dealt the final blow to vaudeville. Vaudevillians could not compete with a venue that broadcast to large audiences with fresh material daily. Vaudeville's demise marks a major turn in Western life and the end of audience–performer camaraderie. The vaudevillian considered any town he or she played as important as Broadway. Newer forms of entertainment not only separated audiences from performers but also separated audiences themselves.

In the nineteenth century, moving pictures seemed an unlikely competitor to vaudeville and the stage. The early one-reel peep-shows served as peripheral entertainment for carnivals. Itinerant Englishman Eadweard Muybridge filmed the first known moving picture to settle a bet for California Governor Leland Stanford in 1872. Muybridge filmed a running horse to see if all four hooves touched the ground simultaneously. He stringed together twelve shutters which opened when the horse tripped the wires. Stanford won his $25,000 bet but spent $100,000 to do so. Muybridge tried to develop his process without success.[86]

Movies did not become an entertainment magnet until some producers moved to California. The free artistic and economic climate of southern California plus the proximity of Mexico made California a haven for producers escaping the Edison-led trust. California offered varied topography, consistent sunshine, and the unique light available in Mediterranean and desert climates. Southern California possessed large belts of useless real estate that would be used by studios which required land for backlots, sound stages, and storage.[87]

Scholars have debated which film initiated the California movie business. Colonel William Selig filmed parts of *The Count of Monte Cristo* and *The Heart of a Rare Tout* in Los Angeles in 1907 and 1909 respectively. He then established California's first studio north of Los Angeles. George Anderson produced weekly *Broncho Billy* westerns in Niles, California, from 1908 to 1915. As the industry prospered, numerous entrepreneurs joined the bandwagon. Many immigrants entered the booming industry which challenged oil and agriculture as California's dominant industry. By 1915, California produced half of all films.[88]

Once established, California producers implemented the studio system. This system made efficient use of facilities through rotating schedules and executive control. Studios developed specialized departments for various aspects of the production process. With stages and backlots continually in use, producers spent less money for the same product. George Westmore developed the first make-up department standardizing that craft for the industry. Westmore created the famous Mary Pickford sausage curls at a cost of $50 per curl.

To the producers' surprise, the Hollywood star system coincided with the rising studio system. Audiences demanded more of their favorites such as Mary Pickford and Francis X. Bushman. By 1915, Americans developed an intense fascination with movie stars. Sensing this, the stars played producers against each other to escalate salaries. In 1917, Charlie Chaplin earned over $1 million. Producers reacted by signing legitimate actors to replace stars. Legitimate and vaudeville actors leaped at the opportunity. Animosity developed between film and legitimate actors, especially when producers pushed stars into second-banana roles. Producers regretted their tactics when most legitimate actors flopped at the box office. A combination of arrogance and an inability to perform in the more subtle media provided the downfall for most legitimate stars.[89]

David Wark Griffith deserves the most credit for turning movies into an art form. From 1908 to 1913 Griffith directed over 400 films in which he experimented with angles and lighting. His three-shot system remains the standard today. Although not the first to use different angles, he was the first to let the scene dictate the shot. Like Renaissance painters, Griffith used light and shade to bring out human emotions. Understanding the reality of screen presence, he developed typecasting and underacting. He believed actors could not perform when a director screamed at them. His monumental *Birth of a Nation* opened in Los Angeles at Clune's Auditorium in 1915. Condemned as a racist film, it was used by the Ku Klux Klan to initiate a revival. Griffith's *Intolerance* served as a call for freedom of the screen.[90]

Hollywood experienced major changes in the 1920s. Like the nation, Hollywood embraced moral relativism and spiritual degeneracy. Hollywood was, and is, no worse than any other American city, but people watched Hollywood more. Scandals rocked the business, forcing producers to seek a guiding hand. America's sweetheart Mary Pickford married Douglas Fairbanks after a quickie Nevada divorce. Numerous deaths, many drug-related, embarrassed producers and shocked fans. The exploits of Roscoe "Fatty" Arbuckle inspired a customer backlash. In July 1921, news stories appeared about a wild 1917 Massachusetts orgy arranged by Arbuckle. In the stories, information surfaced about a $100,000 gift Arbuckle gave to a Massachusetts district attorney. In September 1921, police discovered actress Virginia Rappe dead in her San Francisco hotel. She had been at an Arbuckle party. Charged with her murder, Arbuckle was acquitted after previous juries were hung. When Will Hays allowed him to return to work the public outcry forced Hays to rescind his decision.[91]

Former Postmaster Will Hays served as the producers' guiding force in the 1920s. Falsely portrayed as a czar, Hays recommended changes to producers for business reasons. The Producer's Code served as a voluntary guideline for filmmakers. Hays worked to clean up the industry and established central casting and regularized business practices to end sexual exploitation. He instituted a publicity campaign to clean up the industry's image. The establishment of the Academy Awards was a part of this campaign. Producers voluntarily sent films for pre-release review to make them suitable for the average patron. The Breen Code, developed in reaction to the 1920s debauchery, served as a moral compass. Developed by Catholic layman Joseph Breen, the code advocated avoiding brutality and sexual promiscuity while sanctifying marriage. Producers abided by their code for business reasons. Producers could, and did, avoid pre-release reviews, but the lack of the code seal in a picture doomed it with audiences. The code sought to end gangster glorification but producers simply made the gangsters journalists and continued.[92]

Hollywood continued to sell stars in the 1920s. The idea was to have an extravaganza or show. Producers filmed costly spectacles with expensive costumes. Theater owner Sid Grauman built elaborately designed movie palaces such as the Egyptian and Chinese theaters. The theaters became a part of the show. Theater owners worked to control audience behavior. They used Charles Lick's applausograph to eliminate rude cheering. Rouben Mamoulian developed an applause machine so theaters could cue audiences when to applaud. Houses installed cooling systems, making them more comfortable.[93]

The coming of sound, or talkies, provided a major economic boost for Hollywood. In 1927, Warner Brothers' *The Jazz Singer* used sound as part of the story. Later, that same company made *The Lights of New York* a full talkie. Talkies came to stay, making Warner Brothers a production power and forever changing the industry. Most silent stars failed to make the switch. Charlie Chaplin refused to try. In a sense, Mickey Mouse replaced Chaplin. Similar in basic concepts, Mickey did the best he could while Chaplin wanted to rule. Mickey saw success as good while Chaplin saw it as evil. Mickey, the optimist, would carry through the Depression fitting the new American mood.[94]

The movie industry collapsed in the early Depression, sinking to an all-time low in 1932. Industry leaders made drastic cutbacks, including a 60 percent pay cut for Hays. The Depression plus overpaid actors crippled the business. Producers reacted by buying out contracts, loaning out stars, and operating with daily contracts. Theater owners sold their houses back to producers, hoping to stabilize the system. A British film invasion forced producers to reorganize. Actors called on Washington to end foreign infiltration of the market. The Big Five and Little Three studios emerged from the Depression to solidify the studio system. Actors fought back with unionization efforts in the American Federation of Radio Artists (AFRA), forerunner of the American Federation of Television and Radio Artists (AFTRA), and the Screen Actors Guild (SAG) behind the efforts of Boris Karloff, Ronald Reagan, and Groucho Marx. The National Recovery Act eventually established a minimum wage for talent.[95]

The Depression sobered Westerners, making for liberal politics and conservative mores. Family values and belief in God rose in the 1930s. This change made people more optimistic. Audiences craved wholesome products. Hollywood responded with optimistic movies. Uplifting songs, singing cowboys, and Shirley Temple became immensely popular. Temple was the most popular star in the late 1930s. Yet more changes were needed to bring back a relatively impoverished audience. The 1930s theater manager acted like a modern game-show host. In Denver, a theater offered an automobile lottery between features. The National Recovery Act tried to stop the practice, but audiences demanded it. Theaters avoided gambling charges by having spelling bees, screeno, and keno, which required skill. Americans supported studio films in record numbers during the 1930s. In 1937, studios produced over 500 feature films. The previous year studios paid over $100 million in taxes. American films of the late 1930s were not unlike British drama from 1576 to 1642. Studios operated a repertoire system designed to service a socially and culturally diverse public. Amidst the optimism and musicals Hollywood produced anti-Fascist movies in the late 1930s.[96]

Hollywood embraced World War II patriotism with the rest of America. After early concentration on war messages, producers followed Roosevelt's advice to lighten up the content. Later wartime films were either patriotic or musicals. *Going My Way* and *Yankee Doodle Dandy* best exemplifies wartime films. Movies provided escape from the drudgery of rations and war news. Canteens, places to entertain soldiers, emerged throughout the West. The Pacific coast served as a major troop-staging area. At the Hollywood Canteen, stars washed dishes, served meals, and entertained. In San Francisco, an open-stage canteen provided the same service. Bay-area theaters regularly ran free ads for shipbuilders needing workers. The Hollywood Victory Committee sent stars to the front to entertain troops. When feasible, Hollywood films premiered at the front. Nearly 500 leading actors served in the wartime military. Producers paid servicemen-employees during the war. The studios all lost money due to patriotic sacrifices. All was not patriotic however. Civilian stars had to be pressured into performing under the federal salary cap. Self-centered zoot-suiters regularly vandalized theaters and harassed customers.[97]

The studio system died after the war. Producers made so-called adult pictures featuring suffering and atrocity. The Supreme Court forced producers to divest themselves of all theaters in *USA* v. *Paramount*. It is noteworthy that the federal government went after the weakest of the Big Five studios. Drive-ins dotted the postwar landscape, providing a different outlet for movies. Offering bottle-warmers, laundromats, and snack bars within sight of screens, the drive-in allowed families to view the show from within their car. This further separated audience members from one another. Drive-ins increased profits by using saturation bookings in which films ran up to two weeks. Beginning in the late 1950s, drive-ins ran dated, high-budget films.[98]

Television cut drastically into movie revenues. It totally separated audience members since patrons remained in their own homes. Television allowed people to view sports, movies, or vaudeville without mingling with others. Producers fought television with extravagant films such as *Ben Hur* and *Giant*. They introduced larger screens and gimmicks like 3D to lure audiences. At the opposite end, companies such as American-International distributed cheap movies such as *The Blob* and *I Was a Teenage Werewolf*. Movies fought television for supremacy but eventually made peace.

Several producers went to a strictly television venue. Warner Brothers and Disney led the way in embracing television. Live television and gambling shows featuring comedic hosts such as Groucho Marx garnered huge audiences. In 1951, *I Love Lucy* changed television for ever. Desi Arnaz's three-camera system and permanent sets made situation comedies an economic bonanza. Television blended commercials from sponsors such as Phillip Morris and Carnation into the storyline. Americans developed lasting friendships with fictional television characters. The mass demand for "coonskin" caps after *Davy Crockett* aired proved television's influence.[99]

Postwar films tended to be cynical. Many glorified rebellion and delinquency. Supposedly this was an era of adult westerns; however, films such as *Shane* and *High Noon* were not the norm. The most popular westerns were the "B" pictures patronized by children coming to see straight-talking heroes such as Roy Rogers and Randy Scott. Rogers became so popular that he embarked on a famous marketing campaign to finance the cost of answering fan mail. The postage cost actually exceeded his salary. Rogers became a national icon who provided standards of decency at a time when schools and churches abandoned their roles of moral instruction.[100]

Beginning in the 1960s, audiences and directors developed new values that clashed with American life. Directors, leaving the heartland to television, concentrated on coastal fans. They sought out a minority within the market. Films, especially after the Protestant and Catholic "censors" voluntarily closed, aimed at younger, emotionally immature audiences. Producers and directors no longer respected America or the people. Directors used gimmicks such as quick cuts and slow motion to replace plots and innovation. Style became more important than substance. Films such as *Dr. Strangelove* featured sex–violence combinations and ironic juxtaposition of sight and sound. Films by new directors such as

Arthur Penn and Robert Altman hated America, order, and the establishment. Filmmakers glorified New York City and hated real people and suburbs. In films such as *Bonnie and Clyde* and *Little Big Man,* bad was good and good was bad. Films in this era glorified the moment of death.[101]

Ironically, clean movies such as *Oliver!* and *Funny Girl* fared quite well in this era. Studios rediscovered plots and reality in the 1970s. Generally clean movies such as *Rocky* and *Star Wars* drew large crowds that lined up hours in advance of the show. Audiences became more involved, literally giving standing ovations for *Star Wars.* Some filmmakers grew up to realize that clean movies do better. The newer films looked for goodness and combated hatred.

Major changes came with cable and satellite television. Movies now sought out specialized audiences. The direct-to-video concept allowed filmmakers to make profits without enormous expenditures. On-demand movies further segregated the audience. Once communal, patrons often barricaded themselves at home when watching movies.[102]

In 1955, innovator Walt Disney drastically affected American culture when he opened Disneyland in Anaheim, California. Disney, a successful producer of animated and nature films, brooded over the fact that visitors to Hollywood had no place for amusement. Disney originally planned to put an amusement park south of Riverside Drive in Burbank. This site, currently the location of the Disney animation, proved too small for his growing vision. His ideas proved too large for the location. Disney brilliantly used cross-marketing in his vision. His television show ran three episodes pitching the park. The park advertised his movies and television show. All Disneyland amusements required a story. He designed different lands that coincided with the lands on his show. Over 1 million visitors came the first seven weeks of operation.

Disney concentrated on image and happiness. He studied other parks, especially Knott's Berry Farm, to see what worked best. He noticed people became frustrated in lines, so he created double-back lines to give the appearance of progress. At his theme park, the audience participates in the attraction. Patrons ride through the exhibits to avoid the fatigue associated with museums. He banned the word "educational." Disney used what he called "wienies," visible sites to draw people toward the attractions. Kids going to the park sought to be the first to see the Matterhorn. Sleeping Beauty's Castle served as weenie for Fantasyland. Disney quickly closed a circus that operated within the park because circus performers tend to be surly. He also learned that audiences preferred clean attractions when given a choice. Disney influenced a number of copycat amusement concerns that operate across the West.[103,104]

PROFESSIONAL SPORTS

The world of sports interconnects with entertainment in a number of manners. Many Americans preferred watching sports over playing them. Professional athletic teams existed earlier but they became institutionalized in the 1920s.

Baseball

Baseball has its roots in various British games such as rounders. Puritans brought some of their games with them to America. Americans carried these sports with them throughout the westward movement. Nearly all Western towns had a baseball team. There were numerous attempts to establish profitable minor leagues in the West. In 1903, the Pacific Coast

League (PCL) formed, replacing the outlaw California League. The original six teams operated in Los Angeles, San Francisco, San Diego, Oakland, Seattle, and Portland. Five of these cities now have a major league club. Originally, the Oakland Oaks played their games in Emeryville. The PCL made numerous franchise shifts but the original cities remained. In 1926, Salt Lake City moved to Los Angeles as the Hollywood Stars. After a move to San Diego, the Stars returned to play at Gilmore Field in Hollywood. The Stars regularly traveled by air in 1928. A California League team once used air travel in 1921, but the Stars were the first to do so regularly.[105]

The PCL generally led the minors in attendance. Major-league teams complained that the PCL paid better wages. In 1939, PCL president W. C. Tuttle agitated for major-league status. After World War II, the league regularly petitioned to become a major league. The postwar population boom gave credence to the PCL's demands. In 1952, the Major Leagues granted open status to the PCL. This made them a quasi-major league, allowing them to pay higher salaries. The league had threatened to form an independent league if nothing was done to protect their rights. When the Dodgers and Giants moved west, the PCL slipped to AAA status. The entrance of the majors eliminated the PCL's four most lucrative franchises.[106]

From 1944 to 1956, both the New York Giants and Brooklyn Dodgers experienced declining attendance. Playing in dilapidated parks and crime-ridden neighborhoods, both teams wanted out of New York. In 1950, Walter O'Malley gained full control of the Dodgers. Seeing the success of the PCL, O'Malley worked to move to Los Angeles after Brooklyn balked on a new stadium. Giant's owner Horace Stoneham had been trying to relocate before 1956. Both men believed improved air travel and the expected arrival of cable television made California a golden land. Both men wanted parks in a crime-free neighborhood.[107]

O'Malley purchased the Los Angeles PCL franchise and Wrigley Field from the Cubs. Additionally, he paid $900,000 to the PCL when he moved the Dodgers. Stoneham visited San Francisco before moving. He believed the move would keep the teams' rivalry going while benefiting from the natural Los Angeles–San Francisco rivalry. The National League, which refused the Dodgers' initial request, allowed both teams to relocate for the 1958 season.[108]

Both clubs played in inadequate stadiums in 1958. The Dodgers played in the spacious Coliseum which was unsuitable for baseball. The 251-feet left field and deep right field made a star out of Wally Moon but effectively ended the career of Californian Duke Snider. The Giant's Seal's Stadium, with its 22,000 seating capacity, was too small for the majors.[109]

The Dodgers needed a city referendum and court ruling to build Dodger Stadium. Voters, encouraged by a telethon featuring George Burns, Ronald Reagan, and Groucho Marx, supported the stadium with a 52 percent vote. The city moved a landfill and Hispanic neighborhood out of Chavez Ravine to build the facility. Homeowners received $150,000 compensation in the action.

Dodger Stadium, a beautiful, fan-shaped creation, initiated the symmetrical era in stadiums. Boasting eight levels, open bleachers, and a stadium club, it inspired architects of future stadiums. On the field, the Dodgers played a unique style of baseball emphasizing speed and defense. The Dodgers paid for the park, the first privately funded stadium since 1923, but Los Angeles provided access roads. The Giants did not fare so well with their new park. Candlestick Park would be plagued by cold nights and winds that made fielding adventurous.[110]

The Dodgers usually led the majors in attendance. Their successful teams won three World Series and another pennant in their first seven years. They maintained an aura of stability not generally known in sports. Incoming manager Walter Alston remained at his post until 1976. O'Malley remained the sole owner until his death in 1979 when his son Peter replaced him. Even broadcaster Vin Scully was a Los Angeles fixture. Peter O'Malley eventually sold the Dodgers to Fox Broadcasting. The Stonehams sold the Giants to Robert Laurie in 1976. Laurie sold the club to grocery magnate Peter Magown in 1992. In 1962, the Giants defeated the Dodgers in a playoff to win their first San Francisco pennant.[111]

The American League expanded to ten teams in 1961. Actor Gene Autry and partner Bob Reynolds purchased the Los Angeles Angels franchise mainly to broadcast games on their radio station KMPC. Ford Frick banned the Angels from using the Coliseum. They played their first year in city-owned Wrigley Field. This small park served as the home for the popular Home Run Derby show.

In their second year, the Angels were forced into a one-sided deal to play in the new Dodger Stadium. The lease gave the Dodgers all parking and concession sales. Autry insisted the stadium be called Chavez Ravine when the Angels played. Their animosity became public in an extra-inning game when the snack bars ran out of food. Autry ordered the message board to direct complaints to the Dodgers.[112]

On the field, the Angels were relatively successful, perhaps too successful for an expansion team. They remained in the pennant race until September of their second season, eventually finishing in third place. They played over 500 three of their first seven years. This gave them a relatively poor drafting position when baseball instituted the draft. Additionally, Autry's generosity tended to hurt the team.[113]

In 1966, the Angels moved to Anaheim Stadium where they became the California Angels. Locally called the Big A, Anaheim Stadium was a cheaper version of Dodger Stadium with coloring influenced by Detroit's Tiger Stadium. In 1980, the large A-shaped scoreboard was moved outside the stadium when the NFL Rams moved in. The stadium, now enclosed for Ram's attendance needs, lost much of its atmosphere and the breeze that once cooled the stadium. The scoreboard itself would crash into the bleachers in a 1994 earthquake.

The Angels seemed snake-bitten in the 1970s when only Nolan Ryan's pitching gave fans something to cheer for. They achieved some success in the late 1970s after free agency began. Disney purchased a minority portion of the team when Jackie Autry ran the club. Jackie Autry insisted the club remain in her ailing husband's possession while he lived. In 1995, after the aged and beloved owner died, Disney purchased the club outright. The Angels became competitive again, eventually winning the World Series in 2002.[114]

The colorful and bombastic Charles Finley purchased the Kansas City As in the early 1960s, vowing to make it a legitimate club. They had been a quasi-Yankee farm club prior to Finley's purchase. Finley feuded with local officials, the league, and especially the Yankees. He introduced a mule mascot, colored uniforms, and a controversial pennant porch in hopes of drawing attention. After years of fighting, Finley finally received permission to move to Oakland in 1968. Rumors surfaced that Gene Autry persuaded Yankees CBS President William Paley to allow the move.

Finley continued his feuding ways in Oakland, but his uncanny ability to judge young talent supplied Oakland with baseball's best team of the 1970s. While his team won three straight titles, Finley fought with Oakland, his players, and baseball commissioner Bowie Kuhn. He lost star pitcher Catfish Hunter when he breached an annuity proviso in the contract. Seeing free agency coming, Finley tried to sell off his stars and regroup. This is not uncommon in sports, but a vengeful Kuhn blocked the sale. Additionally, Kuhn interfered with Finley's contract negotiations. Frustrated, Finley sold out to Levi-Strauss heir Walter Haas in 1980. Haas sold the team to local businessmen in 1995. Finley made a major mark on baseball. He brought in night World Series games, colorful uniforms, and a showmanship long missing in the stodgy world of baseball.[115]

Major League baseball continued to expand after the 1961 round. Western franchises included Kansas City, Denver, Arizona, San Diego, and Seattle. Kansas City owner Ewing Kaufmann experimented with a two-year baseball academy that trained possible future players. The San Diego Padres, after initially suing to stop the gimmick, provided the San Diego Chicken, prototype for modern sports mascots. The Seattle Mariners sold out to Nintendo in 1989, becoming the first foreign-owned franchise.[116,117]

Football

Like baseball, football had its beginnings in English games brought over by the Puritans. Western colleges adopted football to promote camaraderie and draw students. Several Western powers emerged in the early days of college football, mostly notably Stanford, Southern California, and Santa Clara. The Plains states turned football into an unofficial public religion. Graduates of these colleges often went on to play professional football.

In its beginning the National Football League (NFL) operated in the East. In 1926, the NFL awarded a Los Angeles franchise to a group of film executives but banned games on the West Coast. That same year, Los Angeles fielded an American Football League (AFL) team, the Wildcats, which played in the East as well. In 1937, Los Angeles tried again with the new AFL but only lasted one year. In order to gain professional teams, the West had to deal with travel expenses and Eastern arrogance. A successful Pacific Coast League operated until the late 1940s. In 1946, the Cleveland Rams moved to Los Angeles where they competed with two other professional teams for the Coliseum.[118]

After World War II, a group of businessmen met in Chicago to form a new league. This meeting, instigated by *Chicago Tribune* writer Arch Ward, formed the All-American Football Conference (AAFC). The AAFC included the San Francisco 49ers and Los Angeles Dons. The Dons, owned by Ben Lindheimer, and 49ers, owned by lumber magnate Tony Marabito, succeeded, but six of the seven Eastern teams failed. The Dons led all leagues in attendance and set a single-game record in 1948 by drawing 105,000. Lindheimer subsidized the failing Chicago Rockets with a $300,000 gift in 1949. It was Lindheimer who met with Chicago Bears owner George Halas to initiate a merger.[119]

The 49ers finished second to the Cleveland Browns every year. The 49ers had a West Coast slant to their roster. Marabito hired legendary Santa Clara coach Buck Shaw to lead the team. Stanford graduate Franky Albert, a diminutive left-hander, quarterbacked the team. The 49ers used the revolutionary T-formation, invented for Albert by Stanford coach Clark Shaughnessy. The 49ers made a profit in 1949 the last year of the AAFC.

When the two leagues merged, the 49ers and Cleveland Browns joined the NFL. The Dons merged with the Rams who had been decimated when they moved to Los Angeles. When owner Dan Reeves moved the club, a Cleveland judge ruled all Ram contracts invalid because they changed the team name. The Cleveland Browns signed a number of Rams who did not wish to relocate.[120]

Initially, the Rams fared well on the field, winning the NFL championship in 1951. When the Rams and Dons merged, they possessed an abundance of talent. Dan Reeves' passing game, prototype of the West-Coast offense, produced record-setting numbers. The presence of two talented quarterbacks, Bob Waterfield and Norm Van Brocklin, led to the first Ram's quarterback controversy which became a Los Angeles tradition. The

Rams consistently led the league in attendance. At one game against the 49ers they drew an NFL record 102,368 with over 10,000 being turned away. The Rams and 49ers regularly led the League in attendance and salaries. The Rams began slipping on the field when Pete Rozelle was general manager. Rozelle continually hurt his team by trading good, high-salaried players. When he became commissioner, the Rams began to regroup and eventually recover.[121]

The 49ers struggled in the NFL. Their teams were skilled and competitive but too small for the new league. Like all Western teams, the 49ers struggled with the fatigue associated with greater travel distances. Both the Rams and 49ers received heavy pressure from Eastern teams to raise ticket prices. The 49ers paid players well, made profits, charged the lowest prices in the game and offered seats on a first-come-first-served basis. The 49ers used numerous innovations on the field. They signed sprinter Joe Perry out of high school, invented the shotgun formation and alley-oop pass, and reinstituted the exciting West-Coast offense in the 1980s. During the 1980s, San Francisco, under new ownership, was the league's most successful team.[122]

In August 1959, organizers announced the formation of a new AFL. Initially, two Western teams, the Los Angeles Chargers and Denver Broncos, were part of the league. A third team, the Oakland Raiders, haphazardly formed when the Minnesota franchise reneged and joined the NFL. The Chargers, owned by Barron Hilton, succeeded on the field but moved to San Diego after the season. They won the 1964 AFL championship. The Broncos and Raiders struggled mightily. The Raiders, who lost nineteen straight at one point, were so bad the League held a special draft from existing teams to make them and the New York Titans competitive. Evidently this helped since the Raiders went ten and four in 1963.

The AFL struggled for recognition. In 1966, Al Davis resigned from Oakland to become commissioner. Davis instituted a raid on NFL talent and a bidding war. This led to peace talks and an eventual merger. The peace talks led to the first Super Bowl hosted by the Los Angeles Coliseum after the 1966 season. In August 1967, the Broncos became the first AFL team to beat an NFL team when they defeated Detroit in an exhibition game.

The NFL revamped its divisions in 1968. The Rams tied the Baltimore Colts in the supposed weak division. Both teams lost only one game. The Rams won the division by outscoring the Colts in direct competition. On the day after Christmas, owner Dan Reeves fired popular Coach George Allen. In January, protests from players and fans led Reeves to rescind his decision. After Reeves' death, his estate and Warner Brothers, the minority owners, sold the Rams to Robert Irsay. In a strange, yet anticipated move, Irsay traded teams with Baltimore Colts owner Carroll Rosenbloom. The Rams moved to Anaheim in 1980 and then to St. Louis in 1995.[123]

The Broncos and Raiders both succeeded in the new NFL. The Raiders moved to Los Angeles in 1984 and returned in 1995. In 1998, Phoenix entered the NFL when the St. Louis Cardinals relocated there. Los Angeles has somehow survived without an NFL team. Efforts to move a team to Los Angeles have failed, perhaps because many Angelinos realize the obvious, namely that NFL football on television is entertaining because of the extra trappings. NFL football at the stadium is pitifully boring.

Basketball

Western colleges competed quite successfully against Easterners in basketball. Under legendary Coach John Wooden, UCLA won a string of national titles from 1964 to 1973. The only team to interrupt this run was another Western college, Texas-Western. Prior to the Wooden era, squads from San Francisco and California won national titles.

Denver fielded a weak team in the National Basketball League before the National Basketball Association (NBA) formed. The club, which included future Congressman Morris Udall, failed to survive the merger into the NBA. The first and most successful Western NBA team was the Los Angeles Lakers. When Minneapolis team owner Bob Short saw the Dodgers' success, he made preparations to move to Los Angeles. He believed the glamour and population of Los Angeles provided a winning combination for his struggling franchise. Los Angeles had already built a 14,000 seat arena before Short announced his move.[124]

The Lakers came west with a young, talented team. Elgin Baylor revolutionized basketball with his powerful, yet graceful, moves. Rookie Jerry West supplied a flashy outside shooter to compliment Baylor. Additionally, the Lakers featured opinionated announcer Chick Hearn who lent excitement to every contest. The early LA Lakers did well competitively and economically but could never unseat the dominant Boston Celtics from the NBA title. They came close in 1962 when guard Frank Selvey's last-second shot went in-and-out of the basket. In 1962, Baylor played only on weekends while serving in the army.[125]

Sports entrepreneur Jack Kent Cooke purchased the team in 1962. Cooke opened his privately financed arena, the Forum, in 1967. Cooke spared no expense seeking a championship for Los Angeles. He brought in Wilt Chamberlain, the game's most dominant center, hoping to capture a title. The Lakers still kept coming up short. In the 1971–1972 season, under new coach Bill Sharman, the Lakers finally captured the elusive title. They set a league record during the season with a thirty-three-game winning streak. Sadly, the streak began the game after popular Elgin Baylor retired.

Former Southern California chemistry professor Jerry Buss purchased the Lakers in 1979. Buss, who owned a tennis team also, initiated a season-ticket system that provided seats for all events. His variable price system, featuring some remarkably low prices, allowed less affluent fans to attend Laker games. He actively promoted the star system with general manager Jerry West. The Lakers became a dominant NBA franchise after 1979.[126]

After the Lakers' financial success, the league looked favorably on a second West Coast team. In 1962, the Philadelphia Warriors relocated to San Francisco. League officials believed San Francisco would provide a natural rival for the Lakers and allow visiting teams to save money on Western trips. Executives from the Diner's Club invested $850,000 for the Warriors. They recruited Gene Autry as a California investor, but Autry balked at a non-LA franchise. Autry suggested bay-area broadcaster Franklin Mieuli. Mieuli became sole owner during the 1970s. In spite of having Wilt Chamberlain, the Warriors struggled at first. In 1970, they changed the name to Golden State Warriors. The Warriors did win an NBA title in 1975.[127]

In 1967, a group of entrepreneurs founded the American Basketball Association (ABA). The ABA started with three Western teams: the Anaheim Amigos, Oakland Oaks, and Denver Rockets. They tried to create more interest with a 3-point shooting range and red, white, and blue ball. The Amigos, who moved to Los Angeles in 1968, never managed to sign any star players. They moved to Utah in 1970. The Oaks won the first championship. They signed Warriors star Rick Barry their first year, but the courts made Barry sit out a year. Barry's father-in-law coached the team. Former rock-and-roll singer Pat Boone was part owner. The Oaks moved to Washington, DC after the first year.[128]

Denver created controversy when they signed sophomore Spencer Haywood. Citing a hardship clause for justification, the Rockets regretted their decision. In 1971, Haywood walked out of his contract in midseason over a dispute. He signed with the NBA Seattle franchise.[129]

The ABA expanded into San Diego in 1972. The San Diego Conquistadors signed aging Wilt Chamberlain as player-coach in 1973. Chamberlain sat out the first year and did not play for San Diego. San Diego renamed themselves the Sails in 1975 and folded that same year. The ABA, unable to succeed in large cities, merged with the NBA.

The NBA expanded or transferred teams into numerous Western cities. In the late 1960s they opened franchises in Seattle, Phoenix, and San Diego. The San Diego club moved to Houston in 1971. Portland opened a franchise in 1970. In the late 1970s, teams from Buffalo and New Orleans moved to San Diego and Utah respectively. The Kansas City Kings moved to Sacramento in 1985 two years after the San Diego Clippers moved to Los Angeles. Each of these clubs experienced periods of success, but none have matched the Lakers.[130]

Ice Hockey

In 1967, the National Hockey League (NHL) expanded westward. Westerners have not embraced hockey like other sports. Jack Kent Cooke's Los Angeles Kings enjoyed limited success, yet even with two of the game's top stars, Marcel Dionne and Rogie Vachon, the Kings struggled financially. The chief cause of financial problems came from the NHL's goon mentality of the 1970s. The public viewed hockey as professional wrestling on skates. The Kings are the only Western team from initial expansion still in its original city.

The World Hockey Association (WHA) began operations in the 1970s. WHA clubs signed numerous young players. This enabled one team, the Edmonton Oilers, to develop without relying on goon tactics. The NHL absorbed Edmonton when the WHA folded. The Oiler's success brought the goon era to a close.[131]

Oiler's superstar Wayne Gretzky moved to the Kings in a controversial trade. His star power created fan interest throughout the West. In 1991, the San Jose Sharks started to play in the Daly City Cow Palace. They moved to the Shark Tank the following year. In 1993, Disney CEO Michael Eisner used cross-marketing techniques to develop the successful Anaheim Mighty Ducks. In 2007, the Ducks became the first Pacific coast team to win the Stanley Cup since the NHL took over the trophy. Disney's franchise did not come without controversy. The Minnesota franchise was negotiating to move to southern California until Eisner intervened. The North Stars, who helped Eisner film the movie he used to cross-market his team, moved to Dallas instead.[132]

Two Canadian teams have relocated to the West since expansion. Quebec moved to Denver where they operate a successful franchise, winning the cup their first year. The Winnipeg team relocated to Phoenix where they have had occasional success. The Kings have suffered from financial mismanagement and are now a struggling franchise. Hockey's future, in spite of successes, is precarious in the West.

Western Regional Sports

Rodeo

The true story of Western sports lies not in the major sports. The West differs little from the East in this area. True Western sport lies in games developed to coincide with Western environment. One such sport is rodeo. Rodeo's beginnings stem from the Spanish colonizers who let horses and cattle roam the open range. The vaqueros developed riding and roping skills to round up stock. Vaqueros and later American cowboys devised games to relieve boredom. The Wild West shows featured these games before Eastern audiences.

Rodeos now feature six, sometimes seven, events. Bareback riding requires cowboys to remain on a horse while continually spurring. Saddle bronc-riding, a residue from taming work, requires cowboys to ride a saddled bronc. Since 1956, the Professional Rodeo Cowboy's Association has honored a horse of the year in bronc riding. Bull riding, the most dangerous event, is a fascinating mixture of competition, camaraderie, and teamwork. Unlike broncs, bulls attack riders after they throw them. Steer wrestling requires a cowboy to ride down a steer and wrestle it to the ground. Cowboys rope calves or steers in other events. Team-roping is a continuation of the branding process from the open range. In 1936, cowboys unionized into what is now the Professional Rodeo Cowboy's Association. This body works for the safety and financial benefit of cowboys.

Pecos, Texas, hosted the earliest known rodeo in 1883. Prescott, Arizona, and Cheyenne, Wyoming, have hosted rodeos since 1887 and 1897 respectively. Towns such as Burwell, Nebraska, and Red Bluff, California, sponsor popular midsize rodeos. Rodeos begin with a colorful entrance where riders come in bearing their state flags. Rodeos often contract other acts to entertain between events. Rodeos require meticulous planning and cooperation from their respective communities. The mere presence of a rodeo is a testament to the cooperative spirit of the community.[133]

Surfing

Hawaiians have practiced he'h nalu for centuries. Today we call it surfing. Hawaiians used wooden boards carved from fallen trees. The sport was brought to California in the early 1900s. Malibu and Rincon were the dominant spots for early California surfing. In 1925, Thomas Blake developed a lighter, hollowed board. In the 1940s, California Institute of Technology student Robert Simmons invented the Simmons Board, or Malibu Chip. This lighter board is made from polyurethane foam and fiberglass covering and is glued with resins. Surfing became more popular after the development of this better balanced board. In northern California, Berkeley student Hugh Bradner invented the wetsuit, thus making surfing a year-round sport. By 1976, there were over 1 million surfers in California.

Professional surfing developed in part to combat the image and reality of surfer hedonism and gang activities. Professional surfing helped the sport become a constructive activity. In 1966, US San Diego faced US International in the first intercollegiate meet. Six California community colleges formed a surfing league after the meet. Surfing is a major sport at many high schools. At some schools, surfing draws more participants than football or basketball.[134]

Skateboarding

Not everyone in California lives close enough to the beach to surf. In the postwar era, California youths developed skateboarding. The freer social atmosphere coupled with the remarkably smooth sidewalks in southern California provided an environment for skateboarding to develop. Unfettered by adult coaches and normative standards, skaters freely developed their sport and skills at their own chosen speed. As with surfing, the development of polyurethane boards and paraphernalia has boosted the sport. Yet boards today, which come in a variety of widths and lengths, are remarkably similar to the boards of the 1970s.[135]

In the 1970s, the West blossomed forth, with skateparks providing skaters with different style runs to develop skills and improve techniques. The skateboard industry sponsored events and skaters to promote both the sport and product. The new prestige of skating was best demonstrated when legendary Tony Hawk told a "boy's" magazine his "proudest... accomplishment" was being featured in *The Simpsons*.[136] This appearance validated Hawk as a legitimate sports icon and skateboarding as a legitimate sport and cultural activity. Today skateboarding is a regular part of the Xgames, a series of "extreme" sports that began touring the country in 1994.

Iditarod

In January 1925, Alaskan doctor Curtis Welch issued a call for diphtheria serum because he feared for Eskimos who had never been exposed to the disease. Pilots could not fly into Nome due to weather conditions. Men brought the serum to Nome on dogsleds. In 1973, Alaska began the Iditarod Race to commemorate this event.

Dog-racing is as old as dogsled transportation. The Iditarod Trail is approximately 1,200 miles long. The race is held in March when the weather is still quite frigid. Racers use one of three breeds: the Siberian husky, the Alaskan malamute, and a mongrel breed called the sled dog. All sleds must contain survival supplies to compete. There are twenty-two stops along the route from Anchorage to Nome. Each stop is required to provide veterinarians and adequate food supplies. A race can be frozen, temporarily stopped, if an upcoming stop lacks sufficient supplies. Humane treatment of dogs and fair play are the basic rules of Iditarod. In 1987, Libby Riddle became the first woman to win the Iditarod.[137]

Western leisure differs little from other areas of the country. It frequently involved adaptation to the local environment. Westerners generally are concerned with relieving boredom in their activities. The future of Western recreation lies in the people's inventiveness and ability to maintain the environmental integrity of the land.

NOTES

1. James B. Whisker, *Hunting in the American Tradition: The Western Hunting Tradition* (Lewiston, ME: Edwin Mellon Press, 1999), 99.
2. Whisker, *Hunting Tradition*, 115.
3. Whisker, *Hunting Tradition,* 113.
4. Daniel Herman, *Hunting and the American Imagination* (Washington D.C.: Smithsonian Institute Press, 1999), 44.
5. Herman, *American Imagination,* 96,100,120.
6. Herman, *American Imagination,* 136, 146.
7. Whisker, *American Tradition,* 125; Herman, *American Imagination*, 53.
8. Herman, *American Imagination,* 86, 205; Robert Richmond and Robert Murdock eds., *A Nation Moving West* (Lincoln, NE: University of Nebraska Press, 1966), 275; Editors of Outdoor Life, *The Story of American Hunting and Firearms* (New York: McGraw-Hill, 1959), 99.
9. Herman, *American Imagination*, 65, 123.
10. Herman, *American Imagination,* 201–204, 217.
11. Outdoor Life, *American Hunting*, 111.
12. Herman, *American Imagination*, 157.

13. Herman, *American Imagination*, 250–253.
14. Herman, *American Imagination,* 218.
15. Mary Zeiss Stange, "Women and Hunting in the West," *Montana: The Magazine of Western History* (Autumn 2005), 55 no. 3, 16–20.
16. James Bashline and James Saults, eds., *America's Great Outdoors* (Chicago: J. B. Ferguson, 1976), 309, 334.
17. Herman, *American Imagination*, 205; Outdoor Life, *American Hunting*, 124, 144.
18. Whisker, *American Tradition*, 126.
19. Herman, *American Imagination*, 222; Whisker, *American Tradition*, 129.
20. Whisker, *American Tradition,* 130, 137.
21. Whisker, *American Tradition*, 131; Herman, *American Imagination*, 273.
22. Bashline, *Great Outdoors*, 116.
23. Ken Owens, " Fishing the Hatch," *Montana: The Magazine of Western History* 52, no. 2 (Summer 2002), 10–19.
24. John Byorth, "Trout Shangri-la," *Montana: The Magazine of Western History,* 52, no. 2 (Summer 2002), 38–47; Bashline, *Great Outdoors,* 310.
25. Bashline, *Great Outdoors,* 309.
26. Byorth, "Trout Shangri-la," 38–47.
27. James B. Whisker, *Hunting in the American Tradition,* vol. I, *The Western Hunting Tradition* (Lewiston, ME: Edwin Mellon Press, 1999), 96–129; Daniel J. Herman, *Hunting and the American Imagination* (Washington, DC: Smithsonian Institute Press, 2001), 57–256, passim; James Bashline and Dan Saults (eds.), *America's Great Outdoors* (Chicago, Ill.: J. B. Ferguson, 1976), 65–331, passim; Mary Zeiss Stange, "Women & Hunting in the West," *Montana: the Magazine of Western History,* 55 (3) (2005): 16–20; Robert Richmond and Robert Murdock (eds.), *A Nation Moving West* (Lincoln, Nebr.: University of Nebraska Press, 1966), 275; John Byorth, "Trout Shangri-la," *Montana: The Magazine of Western History,* 52 (2) (2002): 38–47.
28. Edward A. Riconda, The Reference Shelf 67, no. 4, *Gambling* (New York: H. W. Wilson, 1995), 7.
29. Jim Hicks, ed., *The Old West: The Gamblers* (Alexandria, VA: Time-Life, 1995), 7, 21, 34.
30. Hicks, *Old West*, 20, 23, 41.
31. Hicks, *Old West,* 22.
32. Riconda, *Gambling,* 20; Hicks, *Old West*, 49.
33. Hicks, *Old West*, 54, 61.
34. David G. Schwartz, *Roll the Bones: the History of Gambling* (New York: Gotham Books, 2006), 26; Hicks, *Old West*, 149.
35. Hicks, *Old West*, 83; John Philip Quinn, Criminology, Law Enforcement and Social Problems V. 48 *Gambling and Gambling Devices* (Montclair, NJ: Patterson-Smith Reprint Series, 1969), 52, 58.
36. Hicks, *Old West*, 152.
37. Hicks, *Old West*, 159.
38. Riconda, *Gambling,* 20; Schwartz, *Roll the Bones*, 254.
39. Quinn, *Gambling Devices,* 10; Schwartz, *Roll the Bones,* 254; Hicks, *Old West*, 141, 132.
40. Hicks, *Old West,* 112.
41. Riconda, *Gambling,* 21; Hicks, *Old West,* 61; Schwartz, *Roll the Bones,* 66.
42. Quinn, *Gambling Devices*, 52.
43. Quinn, *Gambling Devices*, 29; Hicks, *Old West*, 168.
44. Quinn, *Gambling Devices*, 58–59.
45. Hicks, *Old West*, 197, 200.
46. Hicks, *Old West,* 212.
47. Schwartz, *Roll the Bones*, 260; Hicks, *Old West*, 103.
48. Schwartz, *Roll the Bones*, 260.
49. Schwartz, *Roll the Bones,* 381.
50. Schwartz, *Roll the Bones,* 331; Hicks, *Old West,* 188.

51. Riconda, *Gambling,* 22.
52. Schwartz, *Roll the Bones,* 358.
53. Schwartz, *Roll the Bones,* 290,359.
54. Schwartz, *Roll the Bones,* 355, 362.
55. Riconda, *Gambling,* 22; Schwartz, *Roll the Bones,* 393.
56. Riconda, *Gambling,* 22.
57. Schwartz, *Roll the Bones,* 290.
58. Schwartz, *Roll the Bones,*408, 419; Riconda, *Gambling,* 30.
59. Riconda, *Gambling,* 23–27.
60. Riconda, *Gambling,* 7–22; Jim Hicks (ed.), *The Old West: The Gamblers* (Alexandria, Va.: Time-Life, 1978), 61–210, passim; Quinn, *Gambling and Gambling Devices,* 9, 52, 58; Schwartz, *Roll the Bones: The History of Gambling,* 330–411, passim.
61. Keith Wheeler, *The Old West: The Townsmen* (Alexandria, VA: Time-Life Books, 1975), 168; Armand Fields, *Eddie Foy: A Biography* (Jefferson, NC: McFarland and Co. Publishing, 1999), 28, 38.
62. Armand Fields and L. Marc Fields, *From Bowery to Broadway: Lew Fields and the Roots of American Popular Theater* (New York: Oxford University Press, 1993), 27, 60.
63. Wheeler, *Townsmen,* 163.
64. Wheeler, *Townsmen,* 176; Joseph Csida and June Bundy Csida, *American Entertainment: A Unique History of Popular Show Business* (New York: Watson-Guptill, 1978), 23, 40.
65. Csida, *American Entertainment,* 50, 175.
66. Wheeler, *Townsmen,* 184; Csida, *American Entertainment,* 65; Anthony Slide, *The Encyclopedia of Vaudeville* (Westport, CT: Greenwood Press, 1984), 76.
67. Wheeler, *Townsmen,*163; Csida, *American Entertainment,* 71, 137.
68. Fields, *Bowery,* 75.
69. Csida, *American Entertainment,* 71, 315.
70. Wheeler, *Townsmen,* 176–180, 188.
71. Wheeler, *Townsmen,* 178, 181; Fields, *Bowery,* 344.
72. Fields, *Eddie Foy,* 62; Wheeler, *Townsmen,* 176; Csida, *American Entertainment,*65; Fields, *Bowery,* 88.
73. Fields, *Bowery,* 88, 90; Abel Green and Joe Laurie Jr., *From Vaude to Video* (New York: Henry Holt & Co., 1951), 109; Fields, *Eddie Foy,* 41.
74. Csida, *American Entertainment,* 216,269; Green, *Vaude to Video,* 109.
75. Wheeler, *Townsmen,* 173.
76. Csida, *American Entertainment,* 69; Fields, *Bowery,* 15, 37; Slide, *Vaudeville,* 325.
77. Fields, *Bowery,* 94; Fields, *Eddie Foy,* 23.
78. Slide, *Vaudeville,* 387.
79. Fields, *Eddie Foy,* 40; Wheeler, *Townsmen,* 173; Green, *Vaude to Video,* 39.
80. Fields, *Eddie Foy,* 29, 61; Wheeler, *Townsmen,* 173.
81. Fields, *Bowery,* 89; Fields, *Eddie Foy,* 115.
82. Fields, *Eddie Foy,* 61, 95.
83. Csida, *American Entertainment,* 140; Green, *Vaude to Video,* 72, 277; Slide, *Vaudeville,* 325.
84. Fields, *Eddie Foy,* 223.
85. Green, *Vaude to Video,* 468.
86. Joe Laurie Jr., *Vaudeville: From the Honky-tonks to the Palace* (New York: Henry Holt & Co., 1953), 4; Csida, *American Entertainment,* 231.
87. Gerald Mast, *A Short History of the Movies 2nd ed.* (Indianapolis: Bobbs-Merrill, 1976), 117.
88. Marc Wannamaker and Robert Nudelman, *Images of America: Early Hollywood* (Charleston, SC: Arcadia Publishing, 2007), 31; Csida, *American Entertainment,* 219.
89. Csida, *American Entertainment,* 213.
90. Mast, *Movies 2nd ed.,* 53, 82.
91. Mast, *Movies 2nd ed.,* 123.

92. Mast, *Movies 2nd ed.*, 125,266.
93. Green, *Vaude to Video*, 247, 271.
94. Green, *Vaude to Video*, 238, 266.
95. Green, *Vaude to Video*, 368, 395.
96. Green, *Vaude to Video*, 389, 400; Mast, *Movies 2nd ed.*, 265.
97. Green, *Vaude to Video*, 484, 497.
98. Mast, *Movies 2nd ed.*, 315.
99. Mast, *Movies 2nd ed.*, 327.
100. Robert Phillips, *Roy Rogers: a Biography* (Jefferson, NC: McFarland and Co., 1995), 19; William O'Neill, *American High: The Years of Confidence, 1945–1960* (New York: The Free Press, 1994), 26.
101. Mast, *Movies 2nd ed.*, 475, 484.
102. Gerald Mast and Bruce F. Kawin, *A Short History of the Movies, 6th ed.* (Boston: Allyn and Bacon, 1996), 547.
103. Bob Thomas, *Walt Disney: An American Original* (New York: Simon and Schuster, 1976), 218, 243, 274.
104. Green, *From Vaude to Video*, 230–531, passim; Csida, *American Entertainment*, 23–71, 213–269, passim; Fields, *Bowery*, 15, 27, 60, 89, 344; Slide, *Vaudeville*, 76, 325, 387; O'Neill, *American High*, 12; Mast, *Movies*, 2nd edn, 53–82, 475–490, passim; Thomas, *Walt Disney*, 260–289, passim; Wannamaker, *Images of America*, 31.
105. Robert Finch, ed., *The Story of Minor League Baseball* (Columbus, OH: The Stoneman Press, 1953), 6; Jonathan Fraser Light, *The Cultural Encyclopedia of Baseball* (Jefferson, NC: McFarland and Co., 1985), 7, 51.
106. Light, *Cultural Encyclopedia*, 533.
107. Leonard Koppett, Kopett's *Concise History of Major League Baseball* (Philadelphia: Temple University Press, 1998), 254.
108. Light, *Cultural Encyclopedia*, 121.
109. Kopett, *Concise History*, 281.
110. Kopett, *Concise History*, 301; Light, *Cultural Encyclopedia*, 132.
111. Kopett, *Concise History*, 423; Light, *Cultural Encyclopedia*, 648.
112. Light, *Cultural Encyclopedia*, 130; Kopett, *Concise History*, 294.
113. Kopett, *Concise History*, 294.
114. Kopett, *Concise History*, 308, 347; Light, *Cultural Encyclopedia*, 132.
115. Kopett, *Concise History*, 296, 347.
116. Light, *Cultural Encyclopedia*, 644.
117. Koppett, *Concise History*, 255–294, passim; Light, *Cultural Encyclopedia*, 121–179 passim; Finch, *Minor League Baseball*, 41–53.
118. Richard M. Cohen, Jordan A. Deutsch, Roland T. Johnson, and David S. Neft (eds.), *Scrapbook History of Pro Football* (Indianapolis: Bobbs-Merrill, 1976), 35, 94.
119. Dan McGuire, *San Francisco 49ers* (New York: Coward-McCormack, 1959), 23, 51.
120. McGuire, *49ers*,55; Cohen, *Scrapbook History*, 95.
121. Cohen, *Scrapbook History*, 95, 143.
122. McGuire, *49ers*, 59, 100.
123. Cohen, *Scrapbook History*, 218, 230, 246.
124. David S. Neft, Richard M. Cohen, Roland T. Johnson, and Jordan A. Deutsch (eds.), *The Sports Encyclopedia: Pro Basketball* (New York: Gross and Dunlap, 1975), 76, 158.
125. Neft, *Basketball*, 163.
126. Glenn Dickey, *The History of Pro Basketball since 1896* (New York: Stein and Day, 1982), 205.
127. Dickey, *Basketball since 1896*, 115
128. Dickey, *Basketball since 1896*, 133.
129. Neft, *Basketball*, 246.
130. Neft, *Basketball*, 205, 216.

131. Stan Fischler and Shirley Fischler, *The Hockey Encyclopedia: the Complete Record of Professional Hockey* (New York: MacMillan, 1983), 12, 125.
132. Leigh Montville, "Spleen for Green," *Sports Illustrated* (19 April 1993) 78, no. 15, 36–43.
133. M. S. Robertson, *Rodeo: Standard Guide to the Cowboy Sport* (Berkeley, CA: Howell-North, 1961), 12, 35, 65.
134. Gary R. Fairmont Filosa III, *The Surfer's Almanac: An International Surfing Guide* (New York: E. F. Dutton, 1977), 8, 11, 73.
135. Skateboarding section is based on personal experience and observations.
136. Dylan T. Sprouse and Cole M. Sprouse, "What it's Like Being . . . Tony Hawk," *Sprouse Bros. Code* (16 October 2006) 1, no. 1, 20–21.
137. Libby Riddles and Tim Jones, *Race Across Alaska: The First Woman to Win the Iditarod Tells her Story* (Harrisburg, PA: Stockdale Books, 1988), 49, 130.

CHAPTER THIRTEEN

LEARNING IN THE WEST

Western Women and the Culture of Education

Andrea G. Radke-Moss

In 1889, Grace Fairchild departed her Wisconsin home to marry an older widower, and together they settled on a homestead in South Dakota. Grace willingly accompanied her husband, but the shock of living in a rural and isolated Plains state took its toll on her, and she remembered her new surroundings as a "desolate unknown."[1] Almost immediately, she went to work improving her homestead with furnishings, gardening, and making social connections with her neighbors. Eventually Grace also led the way in establishing a school for the local children, by first contacting the county superintendent, then helping to purchase a cabin from a vacated homestead, and eventually hiring a seventeen-year-old girl as the teacher. Although the young woman had little training and experience, Grace admitted, "The children seemed to be learning. Just having a school available made life look a lot better to me."[2] Her reaction echoed the sentiments of many Western women who diligently sought educational opportunities for themselves and their children, especially as they attempted to bring their version of civilization to a difficult environment.

As Euro-American settlers moved west, women such as Grace Fairchild often took the initiative in getting schools started in their rural communities. Education was an integral part of the civilizing process, which, for nineteenth-century men and women, meant morality, Christianity, hard work, cultural refinement, and patriotism. Amy Kaplan has argued that women implemented their ideals of domesticity, both at home and in the public sphere through educational institutions and cultural associations, in order to partner with Western men in their efforts to tame the American West.[3] Since schools were considered appropriate extensions of women's feminine sphere, then "[d]omestication in this sense is related to the imperial project of civilizing, and the conditions of domesticity often become markers that distinguish civilization from savagery."[4] In spite of women's profound efforts, the establishment of educational institutions in the rural American West still came with many challenges. These included the great distances between settlements and schools, the lack of training of early teachers, and the inadequate or nonexistent facilities for schools. In spite of these impediments, Western women understood the importance of learning as a connecting force to the larger American civilization from which they had come. While male administrators and officials were often behind the official founding of many Western schools, Western women played their own organizational, fundraising, and support roles for these institutions. In the post-Civil War West, Western women participated in many types of educational opportunities, including primary and secondary schools, boarding and day schools for American Indian youth, and various institutions of higher education, like normal schools, junior normal schools, land-grant colleges, state universities, and sectarian colleges. Whatever the purpose or availability of these schools, women used these familiar institutions of learning and adapted them to their unique Western circumstances and civilizing goals. Western women took advantage of these educational opportunities to increase their own hopes for learning, personal self-advancement, economic independence, and political enfranchisement.

In 1958, Dee Brown's *Gentle Tamers* set up a few straw women that have taken feminist historians of the American West over forty years to knock down.[5] Among them was the "Madonna of the Prairie"—an angelic, idealistic, sunbonneted Euro-American woman, who went west with her family and aided in the Turnerian process of subduing the wilderness. Her name is significant: "Madonna" brought the civilizing tools of religion, morals, and culture to tame the West. The frontier schoolteacher also participated in Brown's taming process. By 1958, enough time had elapsed since the era of the one-room schoolhouse frontier that Brown easily drew upon Western nostalgia to create the romantic myth of the "schoolmarm." Like the "Madonna," the "schoolmarm" was also Anglo-European, idealistic, moral, and mostly unmarried. The young teachers in the West, like all other women settlers, were "bent primarily upon taming their wild environment."[6]

Prior to the publication of *Gentle Tamers,* Western historiography had previously given very little attention to the importance of female schoolteachers. A few published reminiscences, biographies, and local histories had narrated the lives of teachers and literary works such as Owen Wister's *The Virginian* (1902) and Laura Ingalls Wilder's *Little House on the Prairie* books had further romanticized the image of the frontier teacher. In *The Virginian,* Miss Molly arrives in Wyoming as the idealistic New England teacher, bent on using her civilizing influence to tame both the Virginian and the unlawful West. While she ultimately failed to prevent the Virginian from killing his foe, Trampas, in a shoot-out, still her civilizing efforts saw minor successes in other ways: her moralizing speeches about justice and legal order, discussions of literature, Shakespeare and music, and, of course, her attempts to correct the schoolchildren's fascination with extralegal violence.[7] And in Wilder's *Little House* series, just as "Ma" Ingalls easily fit the role of the sunbonneted "Madonna," so her daughters grew up to become the moralizing "schoolmarms" of Brown's *Gentle Tamers.* His teachers were predominantly single, Euro-American, moral, and a bit spunky. Still, the presence of schoolteachers on the Western frontier should not be dismissed as a mass stereotype of moral and virginal civilizers. These anecdotal and literary narratives, while interesting and appealing to nostalgia, failed to adequately demonstrate the social, economic, and political realities of frontier teaching and learning.

Brown's inclusion of schoolteachers in a book about Western women was in itself a necessary recognition. The image of Western schoolteachers as maternal civilizers certainly fit within the late-nineteenth- and early twentieth-century notions of female teaching as an extension of the domestic sphere. Indeed, while social historians such as Suzanne Schrems, Glenda Riley, and Sandra Myres have recognized this same maternal self-definition among female educators on the frontier, other historians have tied women's civilizing impulse in education to a larger imperialist motive in the West.[8] However, limited characterizations of teachers have masked the complexities and diversity of Western female teachers as social and economic groups that have demanded greater examination, especially on questions of teachers' varied degrees of marital and maternal status; the diverse ethnic, religious, race, and age groups involved in teaching; the often bleak social and physical conditions of the environment; unequal pay between male and female educators; the move to professionalize education; the political and social opportunities available to female teachers and accessibility to administrative positions; and, perhaps most importantly, the complex motivations of women—young and old, married and unmarried—to enter the profession. These questions remained to be answered by a later generation of social historians of the American West.

Most historical research of women teachers in the American West has focused on the post-Civil War period from 1870 to the 1920s. However, the earliest women teachers to the frontier are worthy of examination, and some historians have devoted greater scholarly attention to this pre-Civil War period. Susanne Schrems has examined the earliest schoolteachers to the Pacific Northwest and their dual roles as missionary-teachers among the American Indians in the late 1830s and the two decades following. These included both single women as well as missionary wives who accompanied their husbands to their missionary assignments in places such as the Whitman Mission, or Waiilaptu, near Walla Walla, Washington. These women saw missionary work not only as an extension of their civilizing and moralizing role but also as a step above the traditional domestic sphere. Missionary work afforded them a greater involvement in civic and religious decision making among the American Indian tribes of the Pacific Northwest.[9] For instance, at Waiilaptu in 1838, missionary wife Mary Walker had hoped to be able to teach the American Indians at the mission, but when her first child was born she had to set aside her missionary ambitions for motherhood and domesticity.[10] Although never able to return to her goals of missionary teaching, Walker's other female colleagues, such as Narcissa Whitman, managed to successfully teach basic reading and writing to the Cayuse Indian children and to any Euro-American immigrants that arrived at the mission, while also instilling Christian virtues through bible instruction. Some wives of missionaries acted as educators alongside their husbands, but a few of these early women teacher-missionaries eschewed marriage and family in favor of what they considered to be a higher cause of advancing Christian civilization. Ironically, these unmarried female missionaries instructed native women in the proper domestic expectations for nineteenth-century women, showing how, as Amy Kaplan has argued, that Euro-American women educators played an integral role in national efforts to imperialize and assimilate American Indians in the West.[11]

Polly Kaufman has also sought to bring increased attention to the pre-Civil War period of education in the West. In the 1840s and 1850s, the National Board of Popular Education recruited and assigned young female teachers to work in Illinois, Iowa, Missouri, and Oregon, as these areas constituted the main pockets of Western Anglo-American settlements in the antebellum era. Like the idealistic missionary-teachers of Schrems' study, Kaufman's subjects were also mostly Euro-American, Protestant, middle class, and had received at least some limited training from seminaries and colleges in New England and the Northeast, such as the Troy Female Seminary in New York.[12]

In spite of missionaries' early efforts to educate their own children and neighboring American Indians, most Western communities prior to the Civil War had rudimentary common school offerings at best. According to Julie Jeffrey, "even bustling Kansas City [...] had only one public school in the late 1850s [...] [and] 90 percent of school-age children were not in school." Educational institutions were often low on the priority list of pre-Civil War Westerners, especially in the midst of other community hardships during the early settlement period. Jeffrey argued, "[o]nly a minority of the pioneers (often of New England backgrounds) worked to set up public, quasi-public, or private schools, thus ensuring that some children at least would be exposed to the moral, religious, and educational values central to the nineteenth-century curriculum."[13] And when schools were started, women usually stood behind those efforts, either by providing a location for the school in their homes or by serving as the first teachers. Since state-financed and tax-supported education was not yet the norm in the West, frontier communities had to compensate by obtaining revenue from other sources. Many women contributed monetarily to schools, as in the case of wealthy Oregon widow Tabitha Brown, who "donated land, a house, $550 of her earnings, and a bell before her death in 1857."[14] In one Oklahoma town, "the Ladies Aid Society raised the money to pay the teacher." Elsewhere, "women put on box suppers, theatricals, and other entertainments to raise funds for schools," and even helped to physically build the schoolhouse itself.[15]

If school attendance was poor for Euro-American children in the West, educational opportunities proved even bleaker for Hispanic children. In the Southwest, Catholic parochial schools still provided a majority of the education in future Southwestern states such as New Mexico, Arizona, and California. Wealthier Hispanic families sometimes sent their children to private institutions in Mexico City or St. Louis, leaving the local nuns, priests, and friars to teach the poorer families of Hispanic, American Indian, and mixed-blood children. Young Hispanic girls were particularly neglected in this system, or as one California girl remembered, "Owing to [...] the great difficulty of obtaining teachers, most of the girls of the time had scanty education."[16] So Euro-American Protestant children remained the primary beneficiaries of any early attempts at public education, and Euro-American women teachers increasingly led the way in these efforts toward schooling.

These teachers also represented the growing transition toward the feminization of education that was already occurring back East, but was felt more acutely in the West, where teacher shortages drove a greater demand for young women. Even before the Civil War, young men began turning their attention away from low-paying teaching positions in favor of other more lucrative careers in the growing market economy, but the war itself expedited this trend. Thomas Morain has shown that prior to 1880, young men had used teaching as a jumping-off point to other careers in business and industry, and women flocked to fill in those gaps.[17] Willing to accept the subsistence salaries offered by rural districts, women gradually took over the teaching profession between 1870 and 1920. Some numbers from the Great Plains and mountain states are indicative of this trend. Whereas men still made up 52 percent of teachers in Nebraska in 1871 and 47.2 percent in Kansas the same year, those numbers had decreased to 12 percent for Nebraska and 18.7 percent for Kansas by 1914.[18] Or as Glenda Riley has demonstrated, by 1910 in Montana, female teachers outnumbered males by 2,217 to 449; while in Kansas, the ratio was 12,273 women to 2,763 men. In general, men represented about no more than about 15 percent to 18 percent of teachers in the West by the end of the century.[19]

Suzanne Schrems has further examined the feminization of education in the West, in that "[d]espite the low wages, women sought employment as schoolteachers in the West over higher-paying domestic employment because it offered more respectability. It also enabled educated women, who were usually barred from professional opportunities, to use their skills."[20] Thus, the opportunities that teaching afforded women greatly outweighed the economic limitations, discrimination, and emotional hardships they experienced. For most Western women, the benefits of independence, intellectual stimulation, travel, professional autonomy, and an expanded political voice all combined to drive women further toward education as an appropriate profession.[21] Further, as Kathleen Underwood has argued, "teaching did not revolutionize women's lives; rather, the decisions they made reflected individual social and familial milieu."[22] So, women might spend a few years teaching while working toward others goals such as marriage, family, another profession, or more schooling.

The Civil War had a profound impact on the gendered nature and demographics of education, both in Western states and territories as well as the nation as a whole. Since men outnumbered women during the early periods of Western settlement, men were still greatly in demand as schoolteachers, especially during the winter months when the increased presence of male adolescent students required more rigid discipline standards. But as Western schools increasingly filled with women in the years after the War, the reality of the female teacher stood behind the stereotype of the "schoolmarm." Ronald E. Butchart has argued that "schoolmarm" became a derogatory appellation suggesting the inexperience and lack of professionalism associated with young female teachers.[23]

The feminization of education had numerous positive and negative consequences for Western women in their roles as teachers. First, since the feminization of education held that women were to be the spiritual and moral role models for children, women discovered great empowerment in teaching as an extension of their maternal and domestic duties. According to Underwood, "[w]estern women's involvement in education, as students, teachers, and electors, afforded little challenge to nineteenth-century domestic ideology; after all, classrooms were extensions of the home and, in the popular mind, no one was better qualified to staff them than women."[24] But with that status, however, came the accusations of the loss of intellectualism and professional standards, as well as the criticism that women could not discipline as well as men. For all of these reasons, but mainly for how the market devalued women's work, the worst consequence of the feminization of education was that female teachers received far less pay than their male counterparts.[25]

By the mid-nineteenth century, women already made up a majority of sectarian and public primary teachers in many of the earliest Euro-American settlements in the American West, including the Willamette Valley of Oregon and the Salt Lake Valley of Utah. Not without the hardships expected of any early settlement to the West, still these communities claimed a few useful resources, as well as immigrant populations from which they could draw some experienced educators. As early as 1849 in the new state of Oregon, Governor Joseph Lane supported a law that provided tax money for public schools. Partly due to this climate of educational support, Oregon also enjoyed some of the first institutions of secondary learning in the American West.[26] Willamette College in Salem was founded in 1853 as the educational institution for Jason Lee's Methodist Episcopal Church and was followed by the establishment of other colleges along the Willamette River Valley during the 1850s. Usually sectarian and always coeducational, by 1860 Oregon claimed over a dozen of these "colleges"—most of them in name only. Rather than the modern designation as baccalaureate-degree-granting universities, these institutions mostly served as early preparatory schools for primary-age children and secondary-age adolescents, with course offerings in reading, writing, arithmetic, spelling, or more advanced studies in English grammar, trigonometry, algebra, Latin, and geometry, for those who could afford the higher tuition.[27]

Mormon-settled Utah territory also experienced some promising educational beginnings prior to the Civil War, and many Mormon women and men brought with them teaching experience from their previous homes in Illinois, Missouri, and Ohio, as well as England and Wales. For instance, only three weeks after her arrival in the Salt Lake valley in 1847, Mary Jane Dillworth began conducting school in a military tent in the old fort; and in Ogden in 1849, Clarilla Browning taught in a log house.[28] Utah's education struggled without adequate public funding in the early years, so many teachers turned to more lucrative private schools or sectarian "ward" schools that were organized according to the boundaries of local Mormon congregations.[29] Utah also enjoyed a handful of unique "family schools," established by affluent polygamist Mormon leaders such as Brigham Young and Heber C. Kimball to educate their many children.[30]

Jill Mulvay has shown how Mormon women teachers often worked under limited physical conditions, sometimes teaching in mud dugouts and wagon boxes.[31] Like other Western women teachers, Mormon women adapted their educating role to the demands of their environment and their unique cultural-religious expectations. These women certainly accepted the prescribed role of teaching as a continuation of their domestic sphere and even called themselves "mothers in Israel," with the added responsibility of educating Mormon children in their preparation for exaltation in the hereafter.[32] Still, in spite of an emphasis on education that is central to the Mormon faith, even Mormon women teachers felt the criticism that often came with the feminization of education in the nineteenth century. Lu Dalton remembered an incident with her male teacher, who reminded her of the inferiority of women's intellects, especially in regards to their hopes for becoming teachers.

> I remember my disgust when I asked a gentleman teacher if, in his opinion, I was sufficiently advanced in mathematics to study algebra. [...] [H]e replied that it would be wasted time for me to ever study it, because I already had more learning than was necessary for a good housekeeper, wife and mother which was a woman's only proper place on earth.[33]

So, in the minds of some male contemporaries, a woman could gain just enough education for motherhood or to become a teacher, but any "quest to be more knowledgeable in advanced subjects" was misspent time and effort.[34]

These attitudes would not last long in the West, especially because the demand for increased numbers of teachers would push most states and territories toward improving teacher preparation and licensing through normal schools, land-grant colleges, and state universities. In spite of unequal treatment and low wages that usually accompanied the feminization of education, women still sought teaching as an acceptable profession. These women also sought innovative ways to overcome the challenges they faced in frontier schools.

While the conditions for rural schoolteachers in the West were certainly not monolithic, still certain typical circumstances characterized a majority of teachers' experiences. For most rural Western teachers, facilities were rudimentary at best. In the early years of teaching, schools were held in wagons, tents, dugouts, sod houses, log cabins, brush arbors, and even railroad boxcars.[35] Only as districts acquired more money through taxation or fundraising could they afford to build a solid frame, brick, or stone schoolhouse. In some communities, the church building served as both the location for worship services on Sunday and as the schoolhouse during the week. This shared arrangement not only cut expenses but also maintained the sectarian influence on early primary-school curricula. The famed "one-room schoolhouse" endured in the West well into the twentieth century, especially in the Plains and Intermountain states. Even if a teacher enjoyed a more modern structure, she still might have lacked basic school supplies such as books, dictionaries, and pencils, to enhance the children's learning. Some examples from Montana schools demonstrate the diversity of facilities available to Western teachers.

One teacher remembered her first year in the late 1870s teaching at the Newman School in Sweetgrass County:

> I had 15 students to start with. The room was rather dark and there was a fireplace in the corner which was poorly ventilated and filled the room with smoke whenever it was lited [...] The benches were made of cottonwood logs with sticks for legs and there were no desks. We sent to Bozeman for all the books and there were no blackboards, every pupil using a slate and sponge. Each child was required to buy his own books. The subjects that I taught were reading, writing, grammar, history and arithmetic.[36]

In contrast to the conditions at the Newman Ranch School, the Terry School in Custer County was rebuilt in 1886 after a devastating fire had destroyed the original school. The new building "was a story and a half frame building (30 × 50 feet) with a small organ, hanging lamps, and 'the most modern seats then in use.'"[37]

Along with these primitive conditions came the domestic chores necessary to maintain a semblance of a civilized learning environment. Teachers and students together performed numerous required chores such as collecting firewood for heating the school, or gathering buffalo and cattle "chips," if the school was in the treeless Great Plains or desert ranch areas. Teachers also hauled water for drinking and washing, cleaned the privy, and cared for the school itself. Muirl Dorrough in Nebraska filled such requirements for school care: "Talk about work, I have a big piece to perform. The stove pipe fell down and so my school room floor is black as a stack of black cats. My! But I do dread to scrub it."[38] Beating rugs, sweeping sod floors, and killing local vermin—snakes, mice, rats, insects, and even gila monsters—were not above the duties of a Western teacher. Indeed, the domestic chores

Figure 13.1 Muirl Dorrough with her students, Whitman, Nebraska, 1912 or 1913. Source: Photo courtesy of David E. Radke, Elizabeth, Illinois.

associated with keeping up a school sometimes required that teachers live in a small home or cabin attached to the schoolhouse.[39]

For those who did not have the luxury of their own home, they had to board with homesteader families who lived close to the school. Because of the distances between home and school, teachers had to walk or ride a horse, or they might accompany the boarding family in the wagon or buggy. Wintertime cold and blizzards often made the travel even more dangerous for both students and teacher, who might simply cancel school until the weather improved. For these reasons, school schedules and attendance were inconsistent during times of extreme weather. Indeed, requiring mandatory student attendance in farm and ranch communities was next to impossible, and many schools only held classes during the winter and summer terms, leaving spring and fall for planting and harvesting.

Even if the distance was close enough for a somewhat easy commute, still the boarding in itself was often a great hardship to young teachers. If they were lucky, young women might end up in a loving and supportive home with comfortable conditions, as in the case of one Kansas teacher who boarded with a mixed-race couple. She found herself surprised by the hospitality and unexpected comforts:

> I had to take my turn staying there for fear of offending them. As we had never seen negroes before we came to Kansas, I was naturally prejudiced. But my experience there was most pleasant. I was given a spare room, waited upon and treated like a queen, and fed upon the choicest of foods, for "Black Ann," as the woman was known, was a famous cook.[40]

But in many cases, the boarding families' homes might be crowded and unclean, with meager meals and even tense family relations. Muirl Dorrough declared in 1910 that "I did not mind very much to stay at Upton's for one night but deliver me from more than one. They were all nice to me. So I had not better tell tales." Muirl, aged twenty-two, had also been expelled from one boarding family's home because she had developed a romantic attachment to the family's eighteen-year-old son—also one of her pupils.[41] Wayne Fuller acknowledges that dating or courting students was common among teachers of both genders, but that there is very little recorded evidence of physical or sexual assaults by older male students on female teachers.[42] Still, discipline could be a common problem for Western schoolteachers, especially considering the wide range of students' ages. Muirl Dorrough declared in 1911 that

> I don't like my school at all. [...] I have only eight pupils. One girl is twenty-two years old and only in the fifth grade. My next oldest pupil is (or was) a boy that is twenty years old. He said I was so mean to him that he quit school. Oh! Harry, I did just hate him. He is so sneaky & mean. He would not let my horse eat their hay: and tied it up to the fence post. I am glad he has quit. Would you not be too?[43]

Other teachers showed similar frustration regarding discipline problems, and one woman admitted, "I have only 11 scholars, but I am glad I have no more for it is all I can do to manage them. I have to use my ruler pretty often."[44] To Julie Jeffrey, it appeared that many women were torn "between an image of woman as nurturer and the reality of teacher as disciplinarian," but to critics of women teachers, discipline problems just provided more fodder for challenging the effects of the feminization of education.[45]

In the process of trying to adapt a productive learning environment to often primitive conditions, Western teachers often showed great ingenuity and resolve. Teaching and learning in the West was not always about the challenge of overcoming difficult circumstances. In fact, many teachers and students found great joys in the learning process, school activities, and interactions with each other and within the community. Most taught basic reading, writing, and arithmetic, and also expected memorization and recitation from primers such as McGuffey's Readers. And many teachers experimented with fun and cre ative activities such as school music and recitation programs, field trips, outdoor games and picnics, and special book orders to expand students' literacy.[46]

But the greatest problem for rural schoolteachers was their lack of training and professional development, and, in so many cases, this challenge also led to criticisms of the feminization of education: that female teachers were less intellectual, less rigorous, and less professional than male teachers. Indeed, it was the very problem of inexperienced and unlicensed teachers in the rural West that drove a demand for more improved teacher training. Although normal schools had first appeared in New England in the early nineteenth century and then spread throughout middle-American communities in the decades prior to the Civil War, Westerners also accepted the urgent need for more normal schools, scattered throughout states and territories, in order to meet a wide dispersal of population. Many states established their normal schools in smaller communities close to agricultural sub-regions where the farmers' children could attend without the sacrifice of boarding costs or extreme distances away from the family farm or ranch.[47]

The first teacher training appeared in the earliest settled regions of the West in the late 1850s, usually in the form of smaller sectarian colleges, such as Oregon's Corvallis College in Corvallis and Willamette College in Salem. While not normal schools, these colleges

Figure 13.2 Women students in front of Benton Hall, Oregon Agricultural College, Corvallis, 1898. Source: #905. Courtesy of Oregon State University Archives.

offered young teacher hopefuls courses that would qualify them for employment in most schools. Perhaps the most famous of these, Corvallis College, was started and restarted as a Protestant college amidst much difficulty and financial troubles during the late 1850s; it struggled to survive during the Civil War and then was later revived as the coeducational Oregon Agricultural College in 1868.[48] Women of Utah benefited from the University of Deseret, founded in 1850 by Brigham Young as a coeducational "college" that was really the equivalent of a secondary high school or academy. The university struggled for funding and support, but between 1869 and 1871, Deseret offered a two-course program for normal and collegiate students. Of the 223 students, over half were women seeking the normal course.[49] In spite of its financial and institutional struggles, the University of Deseret survived later in the nineteenth century as the University of Utah, with a continuing emphasis on teacher education.

By the 1870s, the Church of the Latter Day Saints, fearing the outside influence and persecution of non-Mormon "gentiles" in the territory, established a unique system of academies in the main pockets of settlements along the Mormon corridor. These sectarian high schools offered some limited teacher training for Mormon youth. The earliest of these, the Brigham Young Academy in Provo (1875) and the Brigham Young College in Logan (1877), were coeducational and included early normal courses. But after the devastating anti-polygamy 1887 Edmunds–Tucker Act that disenfranchised the Church, the Mormon community turned increasingly inward as a political and social defense against persecution. As an alternative to public education, "every Mormon stake was called upon to establish an academy in its area."[50] Among the twenty-two academies along the Mormon corridor, Utah enjoyed such coeducational academies as Weber in Ogden, Dixie in St. George, and Snow Academy in Ephraim, all of which later became colleges. In Idaho, the Mormons founded the Fielding Academy in Paris, the Oneida Academy in Preston, Cassia in Oakley, and the most famous Ricks Academy in Rexburg, which later became Ricks College, a two-year church-supported junior college.[51] The Mormon academy system educated youth between the 1870s and 1930s, even reaching down as far south as the Mormon colonies in Mexico.[52]

These early efforts in Oregon and Utah represented an important start for teacher training in the West. But official state-supported normal schools emerged in the post-Civil War communities of the West where population density demanded such institutions. Adding to that community support, political farmers' associations such as the Grange, Farmers' Alliance, and the Populists supported the growth of public schooling and even included demands for improved teacher certification on their political platforms. Nebraska provides an example of the evolution of normal training in that state over time, as well as the diverse origins of its various normal institutions. In 1865, the Methodist Episcopal Church founded the Mount Vernon Seminary in Peru, Nebraska, primarily as a normal training institution for teachers in eastern Nebraska and western Iowa. The sectarian origins of the Mount Vernon Seminary parallel the origins of other early normal schools in Oregon and Utah, whose churches usually led the way in improving teacher training. Then, when Nebraska became a state in 1867, the state legislature immediately purchased the institution as the first public Nebraska state normal school. From 1867 until 1903, it served as the main teacher training institution for Nebraska, demonstrating how only teachers who lived and taught closer to more heavily populated areas in the West actually benefited from the earliest normal training and state licensing opportunities, when these were available.

For rural teachers in Nebraska and other rural Western states, the distance and cost of attending a normal training institution proved so formidable that thousands went without that needed education. Homesteader Grace Fairchild in South Dakota described the difficulties of hiring teachers with any training at all, especially during the population boom of the 1880s in the Great Plains. During one school year, a young local woman started as the teacher, but, after failing her exams for recertification, a neighboring farmer had to finish the year. For the next two years, another local woman taught the school as it continued to fill with homesteaders' children, even reaching as many as twenty-nine students for one small school. Fairchild remembered, "Families were moving in to take up claims faster than the teacher supply increased. That little log house was filled to the rafters and the only teachers we could get were homesteaders' wives or daughters, and very few had ever taught school before. Sometimes the blind led the blind."[53] Teachers were often young people or married women who were chosen from among the locals just out of sheer desperation to have a warm body in the school. The difficulty of having few experienced teachers was common to many rural Western states, which led them to attempt unique adaptations to their teacher-training problems.

Beginning in the 1870s, some Western states and territories offered limited and sporadic summer institute programs in communities where teachers could receive a few weeks' worth of subject coursework or practical educational methods in between the school years. The success was limited but promising: in 1897, one county in Arizona could brag that nearly every teacher in the county attended the summer institute.[54] By 1890, the Western counties of Nebraska—where there was no state normal school—began to offer two- to six-week summer programs called "county institutes" that could offer basic teaching instruction and preparation in elementary subjects for teachers.[55] Union normals were the next step in the evolution of Nebraska's teacher training. These allowed for two, three, or four counties to unify their resources and funding for more efficient summer-training programs.[56] Still, the "county institutes" and "union normals" were quite limited in their availability and offerings and only further reinforced the need for Nebraska to institutionalize its teacher-training efforts on a state level. In 1891, a group of developers founded the first private normal college in Wayne, Nebraska. Now with two normal schools—one public and one private, Nebraska could better provide training opportunities for its teachers, but still only for the eastern portion of the state. In 1902, educational reformers noted that in eastern Nebraska counties, 46 percent of teacher had received some kind of training, whereas in the west, only 22 percent had attended a normal institution.[57]

So, in 1903, the legislature established the Nebraska State Normal School at Kearney, located more centrally to accommodate ranch and farm communities that were farther away from the eastern urban centers. In conjunction with the Kearney State Normal, the state also mandated the creation of "Junior Normal Schools" in the far panhandle towns of Alliance and McCook, as temporary and less expensive alternatives to the more established normal schools in the East. The State Superintendent of Education, William Fowler, clarified his goals for these new "junior normals": "By a junior normal I mean a normal school conducted at the expense of the state ten or twelve weeks each year during the summer months."[58] The junior normals accommodated the unique needs of Western teachers by offering classes between June and August every year and by paying for students' admission, fees, and travel expenses. This was a truly revolutionary idea—indeed, even "republican education at its very best," since the cost to the state was very low, and no teacher was excluded due to an inability to pay.[59] Between 1903 and 1914, the "junior normals" grew to include eight communities and provided training to hundreds of teachers, most of whom were women. The junior-normal idea demonstrated how Western states could meet their unique teaching challenges with a bit of ingenuity and cost-effectiveness.

Nebraska finalized its efforts to consolidate state normal schools between 1909 and 1911. First, the state finally realized the necessity of a teacher-training institution for the far western part of the state, and set aside funds in 1909 for a fourth state normal school. In January 1910, Nebraska legislators selected Chadron from among six panhandle communities as the location of the Nebraska State Normal School.[60] Also in 1910, the Nebraska legislature followed an earlier pattern it had set at the State Normal School at Peru, Nebraska, by buying out the private Nebraska Normal College in Wayne. Renamed the Nebraska State Normal School and later Wayne State College, it joined the three other institutions for providing public teacher training for the whole state. In 1921, all four of Nebraska's state normal schools earned the status of state teachers' colleges, and during the period between the 1960s and the 1980s, these teachers' colleges expanded their curricular offerings and became state colleges. As the Nebraska example demonstrates, the move to bring normal training to Western states and territories often progressed slowly but still demonstrated how Westerners sought improved teacher education and were willing to make adaptations to their unique geographical conditions.

Other Western states followed the example of trying to establish normal training for their widespread populations and distant farm and ranch areas. For example, the Idaho state legislature created two normal schools in 1893 to meet the training demands for its rural school districts. Similar to other teachers throughout the West at the time, Idaho's rural teachers had only needed an eighth-grade education and a county examination certificate.[61] The Lewiston State Normal School was founded in 1893 but did not have the funds or facilities to admit students until 1896. As the only normal school for the panhandle region of northern Idaho, it offered a two-year teaching certificate to elementary or primary teachers. Most of these teachers were women who completed the course in preparation to "teaching in one-room country school[s]."[62] In 1943, it began granting a four-year baccalaureate degree in elementary education. Later closed for lack of legislative funding, it was reopened in 1955 as the Lewis–Clark Normal School. For southern Idaho, the Albion Normal School was established in 1893 to meet the needs of farm youth from the dairying and ranching communities of south-central Idaho around Twin Falls and Burley, without requiring students to go as far north to the Lewiston State Normal School. It also closed in 1951 due to lack of state funds but never reopened. By the mid-twentieth century, the teacher-training demands for southern Idaho would be absorbed by Idaho State University, Utah State University, and Ricks College in Rexburg.

In most cases, the local citizens themselves led the way in the move to bring institutional training to their areas. In Colorado in 1880, the citizens of Greeley circulated a petition to create a school for teacher education. Many of these citizen movements received foundational and ideological support from their local Grange organizations, which believed in public education and teacher certification as part of improving the overall republican experiment for rural farmers and for the nation as a whole. The Colorado legislature responded with the founding in Greeley of the Colorado State Normal School in 1889, which then became the Colorado State Teachers College in 1911 and later the University of Northern Colorado in 1970. Over 350 women graduated from the school between 1890 and 1900, with over 80 percent of graduates able to get hired in various types of educational settings.[63] These normal schools had profound effects on teachers in their states and territories. The Northern Arizona Normal School in Flagstaff led to improved rates of certification for the rural teachers in northern counties of Arizona and also to increased pay for both male and female teachers, although men's salaries exceeded women's salaries well into the twentieth century.[64]

During the Progressive Era, reformers demanded better standards of teacher training, not only to improve the quality of public education but also to increase monthly salaries for teachers. After the turn of the century, teachers were increasingly exposed to the theories of John Dewey, which required exposing children to experiential methods and age-appropriate curriculum. As teachers departed from the memorization and recitation-based instruction methods of the nineteenth century to the more experience-based methods of the twentieth century, evaluations of schools also reflected these changes. The demands for the professionalization of education also led to calls for school consolidation, with the expectations that combining rural school districts into more centralized schools would lead to greater efficiency and improved educational outcomes. For the West, with its great distances, consolidation efforts occurred haphazardly through the mid-twentieth century. Some rural one-room schools have persisted even to today, while a majority of school districts in rural areas began the process toward consolidation. Consolidated schools benefited from new buildings, updated technology, and access to state-determined curriculum materials, as well as class divisions based upon age and grade.

Still, Wayne Fuller has questioned the overall effectiveness of consolidated schools for the West and has argued that rural, one-room schools were much more effective than their detractors believed. A comparison of the standardized test scores in the 1920s between rural one-room schools and urban consolidated schools showed only a minimal difference. Fuller has also argued that early teacher training in the West was not as rudimentary as often thought. Certainly, county institutes in the 1880s and 1890s were more "practical rather than theoretical," with an emphasis on "what to teach rather than how to teach." But even with that, Fuller suggested, the certification exams were quite challenging and in themselves set a "standard of achievement." Teachers were also encouraged by their leaders to subscribe to educational journals, participate in readers' circles, book clubs, and teachers associations. Plus, the fact that "third-grade" certificates were only good for one year and subject to renewal showed that educational leaders expected continued progress from their teachers.[65]

The increasing professionalization of education led to expanded political opportunities for women. Many Western states voted for school suffrage for women long before they allowed full suffrage for women, and a few female educators even began to run for higher administrative positions by the mid-1870s. For example, since Utah women enjoyed the right to vote from 1870 through 1882, their representation in school administration was an appropriate correlative benefit of their rights of suffrage. Mormon teachers' gradual appearance as county superintendents and on school boards mirrored a similar trend occurring in many Western states after the Civil War. Still, only Emma J. McVicker served as the Utah Territorial Superintendent of Education for three months in 1895, showing that women were still mostly kept out of the highest political offices, in spite of their progress in suffrage and educational leadership on local levels.[66] Colorado gave women the vote for school elections in 1876, which opened the way for many women to serve on local and state boards of education. Even more notably, Colorado also elected five women as state superintendents of education between 1894 and 1909.[67] Other Western states experienced similar victories for their women administrators. Wyoming teacher Mrs. T. F. Caldwell was first elected the Fremont County Superintendent of Schools in 1884, closely followed in 1886 by

Mrs. Mary Mason. The position was no small task: In a county that was one-fourth the size of the present state of Wyoming, Mason spent twelve years as the superintendent. She covered fifteen districts, often riding on horseback, and observed schools all the way from the Sweetwater River to the Montana state line. By 1896, she administered twenty-five schools scattered across the northwestern corner of Wyoming, and performed her duties with the same limited budget she had back in 1887.[68] And in Colorado in 1906, "thirty-four of the fifty-nine counties had women superintendents."[69] For women in the West, school political representation was more than just a token offering as an extension of their domestic sphere. According to Sandra Myres,

> Since the teacher was a prominent and very public figure in small Western communities, her role as both civilizer and community leader could be an extremely important one. [...] As women began to participate in public affairs as teachers and as voters on school issues, Western residents could clearly see for themselves that public life need not cause the defeminization of women or change their basic domestic instincts, and women could enter public life in other areas as well.[70]

The demand for more teacher-training opportunities also fed into the emergence of other institutions of higher education in the West, such as state universities and land-grant colleges. Not surprisingly, most Western women's entrance into higher education grew out of the demand for normal training. Some state universities were rooted in sectarian origins, such as the University of Deseret, founded in 1850 as an Mormon coeducational institution that offered normal training. It later became the public-supported University of Utah. Mixed-gender higher education gained such acceptance in the West that "in 1872, there were ninety-seven major private and public coeducational institutions in the country, and the majority of them, sixty-seven, were located in the West."[71] After the University of Iowa admitted women in 1856, other Western states followed with coeducational state universities, and "[b]y 1870 Wisconsin, Kansas, Indiana, Minnesota, Missouri, Michigan, and California all admitted women to their state universities, although not always on equal terms with men."[72] As opportunities for higher education in the West expanded in the years following the Civil War, women could enter the state universities or land-grant colleges for other avenues of professional development. That the West led the way in coeducational offerings for women was significant in the broader national spectrum of educational reform, especially as some Eastern institutions continued to resist this important change.

When President Abraham Lincoln signed the Morrill Act in 1862, neither Lincoln nor the Bill's sponsor, Senator Justin Morrill, had any idea of the great impact the law would have on thousands of college women in the American West. Indeed, the Act itself made no mention whatsoever of the admission of women or their equal education. Thus, the decisions to implement coeducation would be left to individual states, and many of the first states to do so lay in the American West. For example, soon after the Civil War, many Western states immediately took advantage of the Morrill Act's generous land grants and chartered their first land-grant colleges, sometimes with only a few dozen entering students during the inaugural year. The first land-grant colleges appeared in states with long-established populations of settlers, such as Oregon Agricultural College in Corvallis (1868), Kansas State Agricultural College in Manhattan (1863), and the University of California (1870).[73] Others would follow by the 1880s in Montana, Utah, Washington, and Colorado, especially spurred on by the successes of earlier land grants. Many Western states today claim two public universities (one land-grant and one state university)—as in Utah State

Table 13.1 Western states: land-grant and state or public universities

State	*Territory status*	*Statehood*	*Land-grant established*	*Woman's suffrage granted*	*State or public university*
WYOMING	1868	1890	1886—Univ. of Wyoming (Laramie)	1869 (territory) 1890 (state)	1886—Univ. of Wyoming (Laramie)
UTAH	1850	1896	1888—Utah State University (Logan)	1870 to 1887 (territory, revoked); 1896 (state, restored)	1850— Univ. of Deseret or 1892 Univ. of Utah (Salt Lake City)
COLORADO	1861	1876	1879—Colorado State University (Ft. Collins)	1893	1876—Univ. of Colorado (4 campuses)
IDAHO	1863	1890	1889—Univ. of Idaho (Moscow)	1896	1901—Idaho State Univ. (Pocatello)
WASHINGTON	1853	1889	1890—Washington State University (Pullman)	1883-87 (territory) and 1910 (state)	1861 – Univ. of Washington (Seattle)
CALIFORNIA	1848	1850	1868 (co-ed in 1870)—Univ. of California (Berkeley)	1911	1868—Univ. of California System (10 campuses)
KANSAS	1854	1861	1863—Kansas State University (Manhattan)	1912	1865—Univ. of Kansas (Lawrence)
OREGON	1848	1859	1868—Oregon State University (Corvallis)	1912	1872—Univ. of Oregon (Eugene)
ARIZONA	1863	1912	1885—Univ. of Arizona (Tuscon)	1912	1885—Arizona State Univ. (Tempe)
ALASKA	1863	1959	1917—Univ. of Alaska (Fairbanks)	1913 (territory) 1959 (state)	1917—Univ. of Alaska System (3 campuses)
MONTANA	1864	1889	1893—Montana State University (Bozeman)	1914	1893—Univ. of Montana (Missoula)
NEVADA	1861	1864	1874—Univ. of Nevada (Reno)	1914	1951—Univ. of Nevada, Las Vegas
NEW MEXICO	1850	1912	1888—New Mexico State Univ. (Alamagordo)	1920	1889—Univ. of New Mexico (Albuquerque)

(*continued*)

Table 13.1 Continued

State	*Territory status*	*Statehood*	*Land-grant established*	*Woman's suffrage granted*	*State or public university*
NEBRASKA	1854	1867	1869—Univ. of Nebraska (Lincoln)	1920	1869—Univ. of Nebraska (Lincoln)
SOUTH DAKOTA	1861	1889	1881—Dakota Agricultural College (Brookings)	1920	1862—Univ. of South Dakota (Vermillion)
NORTH DAKOTA	1861	1889	1890—North Dakota Agricultural College (Fargo)	1920	1883—Univ. of North Dakota (Grand Forks)
OKLAHOMA	1890	1907	1890—Oklahoma State Univ. (Stillwater)	1920	1890—Univ. of Oklahoma (Norman)

Source: Warren A. Beck and Ynez D. Haase, *Historical Atlas of the American West* (Norman: University of Oklahoma Press, 1989), maps 41 and 42; and Myres, 219-230.

University (the land grant) and the University of Utah (the state university). In a few cases, like the University of Nebraska–Lincoln (1869) and the University of Wyoming, these universities served as both the land-grant college and the state university together.

The first and most practical reason behind Western coeducation was economic necessity. It was simply less expensive to educate men and women together, rather than to establish two separate institutions for males and females. Since the land-grant colleges and state universities received funds from either federal land grants or from state monies, taxpayers often demanded the equal availability of higher education to their daughters as well as their sons. Early land grants did not require tuition, but the expense of hiring faculty and staff for one coed institution was considerably less than for two sex-segregated schools. Second, larger numbers of students also brought added economic stimulus into communities, through housing, transportation, and town's offering of food, theater, and consumer sales of newspapers, clothing, shoes, and bicycles. Third, progressive ideology favoring women's rights certainly played a large part in the pioneering efforts of education leaders in the West. Land-grant administrators and supporters often came from Midwestern states such as Ohio and Michigan, where mixed-gender education had already begun to gain acceptance, at least in principle. In the West, legislators and educational leaders found environments favorable to trying new ideas. According to Sandra Myres, "because the trans-Mississippi states were relatively new, [...] [they] had no deeply entrenched tradition of restriction, [and] it was easier to convince Western legislators to pass women's rights legislation."[74] Finally, coeducation at land-grant colleges allowed both male and female students to join together in a mutually beneficial education, where each could learn the skills necessary for their sphere of life. Coeducation promoted an environment of educated sociality, where students hopefully would seek and prepare for "companionate" marriage full of love, respect and mutual understanding. After all, said Thomas Woody in 1929, "[i]t is the natural arrangement that boys and girls should be together in school as they are in the family."[75] Land-grant administrators proudly heralded coeducation as a desirable alternative to the gender-segregated education of the East:

> Well done, O East, but not the best!
> Here in the fresh and fearless West,
> We smile to think of monks and nuns.
> We dare to trust our noble sons; [...]
> The manly and the maiden mind
> Together grow more bright, refined.
> That place is holy ground and sweet,
> Where earth and heaven together meet.[76]

Western coeducation was at first imitative, especially as institutions borrowed ideas from a growing culture of education reform and women's rights in America. But the West was also innovative, since educators took the idea of coeducation and adapted it to fit their unique needs. In fact, the unique goals of land-grant education in the West received further impetus from the Morrill Act itself. Because the land-grant mission explicitly sought to provide scientific training to the agricultural and mechanical classes, Western educators translated this goal for women to also gain scientific training for their domestic sphere. Eventually, this developed into the domestic-science or domestic-economy course of study for women that many Western universities offered, since the Western environment particularly demanded a need for educated farm wives. According to Virginia Gunn, "life on the Plains worked best as a cooperative venture between the sexes."[77] President John Anderson of Kansas State Agricultural College in 1875 tried to convince reluctant critics of the importance of a woman's course, recognizing as he did that Eastern educators might "'purse their classic lips at the idea of including such topics in a collegiate course, but they made sense in a Kansas girl's education."[78]

Since land-grant colleges sought to fulfill the educational purposes unique to the rural and agricultural West, and especially unique to Western women, the correlative course work for women that eventually emerged was the domestic-economy or domestic-science course. While some feminist critics have derided domestic science as simply teaching women to do household chores, others heralded a course of study that taught women the scientific theories and practical training behind the administration of a more efficient, hygienic, moral, and productive home. In other words, women would learn chemistry that might be applied to household nutrition or cooking, or they might learn biology and entomology that could be adapted to agricultural production, pest control, and simple gardening. At first, land-grant administrators and educators balked at this proposed course of study for women, instead steering their female students toward a general education in the classics, mathematics, literature, history, science, and the genteel arts, music, and language courses expected of all proper and well-bred American young ladies. But the stronger reason came from economic class expectations in nineteenth-century America, because domestic labor was what middle-class women—and even farm wives—could expect from their "hired girls," who were often poor, young, and of immigrant or African-American parentage. But at the land-grant colleges women were always expected to abide by both standards of middle-class gentility and refinement, as well as educational training that would teach them the practical skills of running a proper farm house. One commenter of the Utah Agricultural College felt the same confusion regarding the expectations for

Figure 13.3 Oregon Agricultural College domestic science class, ca. 1905. Source: #848–1. Courtesy of Oregon State University Archives.

women: "Through all those early catalogs, the write ups of courses for women, seemed—I do not know just how to express it, should I say—torn between training a woman to be a Victorian Lady and a practical housewife."[79]

While Iowa Agricultural College and Kansas State Agricultural College were two of the first land grants to introduce domestic-science coursework in the late 1870s, Oregon Agricultural College did not establish its domestic-science program until the 1890s and the University of Nebraska waited until 1908 to offer women their own domestic-science course. However, as domestic science gained mainstream acceptance by the end of the century, even among Eastern women's colleges and universities, land grants turned to this course program as a de-facto expectation for a majority of middle-class women. Although some feminist scholars have often portrayed domestic science or economy as the reinforcement of traditional female gender roles, nonetheless domestic economy, especially to female students, instructors and reformers, became an important step in their own educational progress.[80] Domestic-science programs validated the work that women had been doing for centuries as something worthy of scientific training, professional education, and the earning of wages. It recognized the right of women to be able to work for their self-support, even if that work was an extension of the domestic sphere. Andrea G. Radke-Moss has suggested that domestic-science courses might have "assisted women graduates who wished to be self-supporting in professional trades such as dressmaking, millinery, and hostelry management," but most land-grant graduates in domestic science did not pursue those fields.[81] Instead,

> Women most often chose one of two professions. [...] [M]ost went on to teach domestic or home economics at other coeducational institutions. Graduates could also pursue

the growing field of home economics extension programs, designed by the state and federal governments to take practical domestic science training to rural farm wives.[82]

The success of land-grant domestic-science education was calculable, for when President Woodrow Wilson signed the Smith–Lever Act in 1914 that provided federal support for rural extension programs, "the first women leaders and extension agents came from among land-grant graduates."[83]

Although domestic science became the main course for female land-grant students by 1900, that did not preclude women from pursuing other courses such as science, literature, commerce (business or secretarial), pharmacology, law, and medicine.[84] One Utah Agricultural College student, Almeda Perry, realized in 1897 that she found no fulfillment in domestic-science courses. "At first I registered with home economics as my major, but a few weeks of making little seams, hems, and patches, and of washing and ironing clothes, including men's shirts with superstiff bosoms, as were in style at the time, I decided I was not getting my money's worth." She decided instead to major in general science, because "I had not fully decided what I wanted to do after graduation, [and] general science would probably provide the best background for whatever line of work I should want to follow."[85] With a decision like this made by a female student, one might expect a certain lack of support by faculty for this choice of study. On the contrary, when one male mathematics professor leaned of Almeda's intention to take public-speaking courses, he protested, "A girl with your mind settling for elocution!"[86] Thus, by providing women with a variety of course selections, through which they could fairly compete with male students, land-grant universities offered another validation of the coeducation experience for women.

One of the unanticipated results of some women's land-grant training was that they found an environment favorable to the pursuit of scientific study. Alice Edwards was a senior at Oregon Agricultural College in 1907 and 1908, where she had discovered her own aptitude for science. After graduation, she worked as an assistant instructor in entomology, zoology, and physiology. She eventually went on to graduate school at various Eastern institutions and finally found a position in 1926 as the Executive Secretary of the American Home Economics Association in Washington, DC. She served in that position for ten years and helped to do research in the growing home-economics movement, including high-school courses, extension programs, and other government studies on nutrition and health. In fact, Edwards "represented the ideal woman graduate of land-grant education."[87]

Coeducation at land-grant colleges and state universities allowed women to challenge nineteenth-century gender segregation in other ways, especially as male and female students worked out the actual practice of mixed-gender interactions. Women students sought participation in activities such as literary and debate societies, athletic teams, campus military regiments for women, newspaper editorial staffs, woman's suffrage activism, and other areas of reform, such as temperance and education. For instance, the University of Nebraska in the 1890s claimed a mixed-gender Tennis Association and a bicycle club, and later established sports for women such as track and field, basketball, and field hockey. Basketball was such a popular sport for both Euro-American and American Indian women in the West that turn-of-the-century teams vied for long-distance competitions with other schools, and games often drew sold-out crowds.[88] Land-grant women also enjoyed opportunities to form military companies, suffrage associations, and other reform organizations that demonstrated the connections between higher education for women and the broader claims for women's expanded rights in the nineteenth and twentieth centuries.[89]

These educational institutions provided an ideal to which young women in the West might aspire, especially as they came of age in communities that celebrated the benefits of education. Through the domestic-science coursework, young women students of land-grant colleges and state universities became agents of the process of "imperial domesticity," as outlined by Amy Kaplan, or that "[d]omestication in this sense is related to the imperial project of civilizing, and the conditions of domesticity often become markers that distinguish civilization from savagery."[90] In other words, domestic science helped to imbed young women with the values of civilization, stability, and virtue in an untamed environment; in turn, it was hoped that those ideals might firmly establish Euro-American, middle-class culture to the West. In another way, land-grant domestic-science courses eventually offered an important model to American Indian schools of how to instruct and train young women in domestic-science courses, especially for the purposes of teaching American Indian youth the proper gender and class expectations of Euro-American middle-class farm families.

As most American Indian tribes in the American West were forcibly moved to reservations in the 1860s and 1870s, the American government fully pursued its policy of assimilation through education and Christianization. The end of the 1870s Indian Wars and the beginning of President Ulysses S. Grant's 1869 Peace Policy left native populations throughout the West subject to the assimilation efforts of reservation officials, missionaries, government agents, the military, and most especially teachers. Since adult American Indians

Figure 13.4 Eduma Buffum in front of Old Main, University of Wyoming, 1899. Source: Old Main Photo File, American Heritage Center, University of Wyoming.

were often seen as too old to reform from their native ways, the government's assimilation policies were directed primarily at American Indian children. Reformers saw primary and secondary education as the preferred method of Colonel Richard Henry Pratt's goal "[to] kill the Indian [...] to save the man," especially as young children could more easily unlearn their native languages, religions and cultural habits.[91]

Education for American Indian children occurred through many different institutions. At first, Protestant religions took on the responsibility of educating American Indian youth through mission schools, or reservation day schools, where priests, ministers or female missionaries combined traditional education in English language, reading, and writing with bible instruction, Christian moral lessons, and goals for ultimate conversion. The Methodist Episcopal Church, Congregationalists, Roman Catholics, and Presbyterians had the largest sectarian presence on the reservations, with female missionary-teachers acting as the main agents of cultural and educational efforts. Many pro-assimilation reformers criticized these day schools, since children went home to their parents at night, often undoing the acculturation efforts that had taken place during the school hours. But this very fact made the reservation day schools a much less traumatic experience for children, since they could maintain emotional and physical ties to their parents and extended families while still receiving the benefits of education and training. Thus, in spite of historical criticism for their roles as agents of assimilation, female mission teachers were often sincere and well-intentioned women who approached the children with sympathy and even a level of respect for their tribal cultures.

For example, from 1875 to 1885, Mary Clementine Collins taught at the Oahe Mission near Yankton in the Dakota Territory, where she learned the Dakota language in order to better communicate with her students.[92] In 1885, she moved to the Standing Rock Agency to teach school while she provided medical care to the Sioux people, from whom she earned the name "medicine woman." Collins taught and ministered among the Dakotas until 1910 and later advocated for some reforms "that were designed to modify United States governmental policies toward Indian groups."[93] Where American Indian children were taught close to home by teachers such as Collins who knew their language and understood their culture, the emotional impact was far less negative than for students who were later removed to off-reservation boarding schools.

Elaine Goodale Eastman arrived at the Pine Ridge Agency as a single woman teacher, was also bilingual, and eventually served as the Dakota Superintendent of Indian Education from 1885 to 1891. Goodale met and married a Dakota doctor, Charles Eastman (Ohiyesa), with whom she had six children, demonstrating that some Euro-American women challenged the social and marital expectations of Euro-American society to embrace some aspects of American Indian life.[94] Goodale (Eastman), like many Euro-American women teachers, felt some ambivalence about her role as educator-civilizer to American Indian children, since she "accepted the superiority of white culture yet also came to see Indians as complex and creative individuals."[95] That uncertainty opened the door for some contradictions in her responses to the Sioux, or, as Glenda Riley has argued, "She was often torn by [...] the stated governmental policy of rejecting and destroying 'everything characteristically native without regard to intrinsic values.'"[96] For example, she met and married a Dakota doctor, Charles Eastman (Ohiyesa), with whom she had six children, demonstrating how a Euro-American woman might challenge the racial and marital expectations of Euro-American society in order to embrace some aspects of American Indian life.[97]

Regarding the influence of these reservation female missionary-teachers, particularly among the Episcopalians, Ruth Ann Alexander has argued that "[a]lthough imbued with assimilationist ideas of their times and with Christian missionary zeal, the women usually liked the people they served, advocated gradual persuasion in encouraging them to become Christian and 'civilized,' and tried to alleviate their suffering in adapting to reservation life."[98]

In spite of these female missionaries' hard work and sincere motives to bring education to the reservation, their efforts did not go far enough for some governments officials. In fact, many teachers' attempts to bridge two cultures even undermined what government officials had hoped for in reservation policy. Without the language and cultural immersion that was necessary for total assimilation into the dominant society, American Indians experienced a shift in the late 1870s and early 1880s regarding the US government's educational policy, from an emphasis on reservation day and mission schools toward the idea of off-reservation boarding schools. By achieving the complete isolation and removal of American Indian children from their families, languages, and tribal cultures, the government hoped to better enact its plans for native children to become productive and civilized American citizens.

Lieutenant Richard Henry Pratt originated the idea of the off-reservation boarding school in the mid-1870s, when he saw the benefits gained from offering some formal instruction to a few Southern Plains war captives. In 1878, he arranged for a few dozen American Indians to attend the Hampton Normal and Agricultural Institute in Virginia, which had been founded to educate former slaves. But in 1879 Pratt recruited Sioux Indian students from the Rosebud and Pine Ridge Reservations in South Dakota to attend the Carlisle Indian School in Pennsylvania as the first off-reservation boarding school exclusively for native students.[99] Often considered the flagship of the American Indian boarding-school system, Carlisle educated thousands of American Indian children until its closure in 1920, most notably the world-class athlete, Jim Thorpe. Between 1880 and 1902, the US government built twenty-five off-reservation schools throughout the West and Midwest, with 20,000 to 30,000 American Indian children attending, or "roughly 10 percent of the total Indian population in 1900." All told, estimates place the numbers of American Indian children who went through the boarding school system between 1879 and the 1950s at over 100,000.[100] Besides Carlisle, students were also sent to Minnesota, Wisconsin, or Michigan.

Not all native children had to go back east for boarding-school education. In fact, a majority of boarding schools were located in proximity to the more populous tribes, or at least situated at an agency that represented numerous tribes and reservations within a region. Besides the Carlisle Indian School, the Bureau of Indian Affairs' (BIA's) next important boarding institution was the Haskell Institute in Lawrence, Kansas, with an annual enrollment of close to 900 students who came from various reservations throughout the Great Plains.[101] California had three boarding schools, with another in neighboring Carson, Nevada, for the California and Great Basin Indians; and Colorado, New Mexico, and Arizona boasted two boarding schools each.[102] In South Dakota, four boarding schools accommodated the entire Sioux population—in Flandreau, Pierre, Chamberlain, and Rapid City.[103] The Fort Shaw Indian School near Sun River, Montana, outside of Great Falls, served American Indian tribes from numerous reservations in Idaho, Montana, and Wyoming, including Colville, Spokane, Coeur d'Alene, Flathead, Blackfeet, Fort Belknap, Fort Peck, Crow, Northern Cheyenne, Wind River, Lemhi Shoshone, and Fort Hall in Idaho.[104] Emphasis on boarding schools as the main venues for American Indian education did not signal an end of the other reservation schools. During the boarding-school era, close to 500 schools of various types met the education and assimilation demands of the American Indian education program. Whether they were local reservation mission schools, agency day schools or on- and off-reservation boarding schools, these schools represented a widespread effort by the US government to terminate American Indian resistance and tribal sovereignty through assimilation efforts and educational institutions.[105]

Off-reservation boarding schools often traumatized American Indian children through harsh procedures and punishments and thus became the longest-enduring negative symbols of US government assimilation policies. To recruit young children for boarding-school enrollment, BIA agents and missionaries visited reservation homes to convince parents and children of the efficacy of "going East" to receive the white man's education. Agents played upon children's vulnerability and curiosity by promising unlimited food, interesting sightseeing, and free rides on railways. Zitkala-Ša remembered the missionary recruiters in her home for their promises of "the great tree where grew red, red apples; and how we could reach out our hands and pick all the red apples we could eat." The ensuing conversation and Zitkala-Ša's reaction showed how successful agents were at capturing the imaginations of young American Indian children.

> "Mother, ask them if little girls may have all the red apples they want, when they go East," I whispered aloud, in my excitement.
>
> The interpreter heard me, and answered: "Yes, little girl, the nice red apples are for those who pick them; and you will have a ride on the iron horse if you go with these good people."
>
> I had never seen a train, and he knew it.
>
> "Mother, I'm going East! I like the big red apples, and I want to ride on the iron horse! Mother, say yes!" I pleaded.
>
> My mother said nothing. The missionaries waited in silence; and my eyes began to blur with tears, though I struggled to choke them back. The corners of my mouth twitched, and my mother saw me.[106]

The masked devastation that Zitkala-Ša's mother felt at the likelihood of losing her child still did not prevent the mother from allowing her daughter to go East. She consented reluctantly but recognized the necessity and importance of education for American Indian children.

> [M]y daughter, though she does not understand what it all means, is anxious to go. She will need an education when she is grown, for then there will be fewer real Dakotas, and many more palefaces. This tearing her away, so young, from her mother is necessary, if I would have her [be] an educated woman. The palefaces, who owe us a large debt for stolen lands, have begun to pay a tardy justice in offering some education to our children. But I know my daughter must suffer keenly in this experiment.[107]

In 1884, eight-year-old Zitkala-Ša left her mother and home on the Yankton Sioux Agency in South Dakota to attend White's Indiana Manual Labor Institute in Wabash, Indiana. Although her boarding school was not located in the West, Zitkala-Ša's reminiscences of her first day at the school would ring familiar to the experiences of thousands and thousands of American Indian children who were either forced or enticed to leave their families for boarding schools. Some children and families found ways to resist the recruiting efforts of government agents. Helen Sekaquaptewa remembered the Hopi parents encouraging the children to play a game of "hide and seek" to avoid police. And in one dramatic incident, the Hopis in Arizona "surrendered a group of men to the police, rather than voluntarily relinquish their children." As a result of their resistance, "[t]he Hopi men served time in federal prison at Alcatraz."[108] But not all parents resisted. One Ojibwe mother allowed her children to go, because she, like some American Indian families, "viewed government boarding schools as the only alternative for their children, one of the few opportunities for young people from rural areas to be educated and develop skills for future employment."[109]

Arrival at the boarding school was not the idyllic dream that American Indian children expected. Instead, young American Indian students often endured debilitating loneliness, strict regimentation, and harsh physical punishments for speaking their native languages, practicing tribal traditions, or for trying to run away. Euro-American teachers often replaced the children's American Indian names with anglicized names, only adding to their emotional separation from their tribes and families. Historians have demonstrated that, rather than passively accepting these attempts to extinguish American Indian children's cultural ties, many students enacted forms of resistance, such as whispering to each other in their native tongues while in their dormitory rooms at night or retaining use of their tribal names among close friends and siblings.

In the worst cases, children experienced brutal physical and sexual abuse and sickness or death from the rampant spread of diseases like "influenza, whooping cough, measles, smallpox, tuberculosis, and trachoma."[110] Trachoma, a highly contagious eye disease, was second only to tuberculosis in the difficulty to contain, because affected children would be sent to the boarding schools without a prior medical examination or preventative care. Brenda Child has demonstrated how boarding-school officials were so bent on guaranteeing the effects of total assimilation that they often refused to send children home at their parents' requests, in spite of devastating loneliness, overcrowded conditions, and the children's risks of disease contagion. For example, when one father wrote to officials at the Flandreau Boarding School in South Dakota, informing the school of his wife's imminent death to tuberculosis and requesting that his daughter be allowed to return home, he was told "it would be pretty bad for her school interests to leave now." And another father was informed that "it is not the policy of the school to permit pupils to go home on account of sickness in the family."[111] It was only after John Collier became the head of the BIA in 1933 that "boarding school students [were] actually encouraged to spend summer vacations with their families."[112]

Perhaps the most shaming experience for all American Indian children—both male and female—was to have their hair cut or shaved. Mostly done for delousing the children and for facilitating easier hygienic care, the cutting of hair was done without regard for the importance of long hair to many American Indian cultures as a symbol of cultural pride. As Zitkala-Ša remembered, "Our mothers had taught us that only unskilled warriors who were captured had their hair shingled by the enemy. Among our people, short hair was worn by mourners, and shingled hair by cowards!" Since she knew what to expect, she tried to run away and hide in a dormitory room. However, there was little that she could do to avoid the inevitable.

> I cried aloud, shaking my head all the while until I felt the cold blades of the scissors against my neck, and heard them gnaw off one of my thick braids. Then I lost my spirit.

> [...] And now my long hair was shingled like a coward's! In my anguish I moaned for my mother, but no one came to comfort me. Not a soul reasoned quietly with me, as my own mother used to do; for now I was only one of many little animals driven by a herder.[113]

Indeed, it was the final and shaming act of hair cutting that signaled the complete separation from parents and tribes and that children now belonged to the US government.

The actual educational curriculum for American Indian children consisted of basic primary instruction in reading, grammar, English, math, history, and some sciences. Students also received instruction in music, arts, and athletics. Secondary-age pupils took more applied coursework in the manual arts, as in agriculture and mechanics for young men, and domestic-science skills for women. The fact that young men learned farming and young women learned housekeeping undermined American Indian women's traditional roles as agriculturists and producers within their tribes. Instead, by training young women in domesticity, boarding schools imposed the class and gender expectations of middle-class Euro-American society on their young American Indian charges (Child, p. 51). Both genders received some teacher training, and many young American Indian women used their normal training to become teachers at boarding schools or reservation day schools, which ironically allowed them to contribute to the assimilationist policies against their own people.

That Euro-American women often participated in the various stages of removing American Indian children to the boarding schools, from the initial recruitment of young children to the actual educating and acculturating in the schools, shows how women themselves became agents of American imperialism. It is a stark contrast to Ruth Ann Alexander's "gentle evangelist" image. In fact, as Margaret Jacobs has argued, "An emphasis on white women as agents of colonial control in the American West may also be jarring in a field that has been so focused on westering white women's triumphs and tribulations."[114] But, by 1900, Euro-American women made up 82 percent of teachers in American Indian schools, and many of these missionary-teachers regarded the removal of American Indian children as a "maternalistic duty to rescue indigenous children from what [they] considered a savage background and to raise them instead in a 'civilized' environment."[115] Thus, Euro-American women felt justified as so-called superior maternal figures in removing American Indian children from their homes and families.

This familial separation left the most enduring negative impact on tens of thousands of American Indian children. By being severed from parental influence while still very young, many American Indian children were robbed of emotional connectedness with their mothers and fathers and also lost their own essential parenting skills. Genevieve Williams attended the Tulalip School, Cushman Indian Hospital, and the St. Georges Mission School near Tacoma, Washington. By the time she left boarding schools at the age of fourteen, she "didn't recognize her own mother [and] [...] the two never did bond."[116] Zitkala-Ša suffered from a similar irreparable emotional distancing from her mother. American Indians who experienced boarding-school life often became ill-equipped to form positive nurturing bonds with their own children, resulting in domestic problems such as alcoholism, child and spousal abuse, and even incest, "behavior that originated in large part in the boarding schools."[117]

Many tribes have tried to recover from their emotional damage through counseling programs and government-sponsored studies on emotional health. In 2002, American Indians from various tribes founded the Boarding School Healing Project, which is "documenting abuses so communities can seek redress from the government and churches, in the form of laws and money to improve Indian education."[118] Among the Tulalip Indians in Washington, for example, members are "weaving lessons about the boarding school era into child-rearing programs for parents and into decision-making classes for teens." For example, Martina Whelshula, a psychotherapist and member of the Colville Indian nation, recognized patterns of abuse, coldness, and lack of affection in three generations of her own family. To repair the damage, Whelshula got her family together in a group counseling session, "so every generation could hear grandma Alice Stewart's story about growing up in boarding schools. That way the family could understand its parenting style in the context of history."[119]

Still, in spite of the traumatic separation of children from parents, the grouping together of students from different tribes had unintended and ironic positive consequences for later American Indian activism. Many historians partly trace the origins of pan-Indianism of the twentieth century to the boarding schools, where students overcame some of their historic rivalries to form important inter-tribal relationships. In part, the mandatory English language requirement also fostered "the beginnings of the pan-tribalism that would one day prove invaluable to American Indians in their united push toward changes in federal policy, which no one tribe could have brought about on its own."[120] The resulting kinship and group identities formed in the boarding schools, combined with the post-World War II relocation of thousands of American Indians to urban centers, led to the formation of the American Indian Movement of the 1960s and 1970s, with some of its founding members such as activists Dennis Banks, George Mitchell, and Leonard Peltier all gaining their early reform ideas and connections from their boarding-school experiences and sympathetic teachers.[121]

Historians are looking more fully at the complex results of boarding-school education for American Indian youth. Indeed, many American Indian children claimed other positive benefits from their boarding-school experiences, such as acquiring useful education and job skills, leadership training, and college preparation. For some American Indian youth, boarding schools were considered a release from difficult family and economic situations back on the reservation, while others escaped abusive and alcoholic parents, starvation, and hopeless tribal conditions, and even insisted that "the best years of their lives were spent in boarding schools, where they got three meals a day and met lifelong friends."[122]

Zitkala-Ša's experiences with Euro-American education demonstrate the contradictory nature of boarding-school life for American Indian children. While she later rejected the negative forms of assimilation and punishment imposed upon American Indian children, Zitkala-Ša also benefited from the foundations she gained from Euro-American education. She earned renown for her achievements as an author, poet, lecturer, musician, and composer, but "her success was dearly purchased. Caught in a tug-of-war between two civilizations, [she] sacrificed her relationship with her mother, yet was never fully accepted into white society."[123] Zitkala-Ša remained a harsh critic of US assimilationist policies for her adult life. Her later reform efforts included advocating citizenship for American Indians, "exposing corruption in the BIA, [and] insisting on the dignity of Indian religions."[124]

The young women of the Fort Shaw Indian School's 1904 World Championship basketball team offer an alternate image to that of abuse and humiliation often associated with the boarding schools. Many of the team members and their fellow students were of mixed-blood or métis heritage, usually with a white European father and an American Indian mother who had married "á la façon du pays." Boarding-school life offered a positive middle-ground environment where they could blend both parents' cultures, especially by taking advantage of the benefits of Euro-American education, while also experiencing a non-threatening group association with other American Indian and mixed-blood children. These students adapted to boarding-school life more easily because they already spoke some English and had been exposed to certain aspects of English or French-Canadian culture, encouraged by their fathers and mothers who wanted them to enjoy the benefits of both European life and native traditions.

Because of their boarding-school experience at the Fort Shaw Indian School, the basketball team members enjoyed the opportunity to develop their athletic talents on the court and then won their way to the Women's World Championship Basketball Tournament at the St. Louis World's Fair of 1904. Their successes also brought them fame, adventure, some limited respect, and the opportunity to travel. Beyond that, during their time at Fort Shaw, they also gained a high-school education, learned valuable skills, and made lifelong friends. To demonstrate these successes at the 1904 Fair, the Indian Department constructed and displayed a "Model Indian School" which acted as a live exhibit of American Indian children performing their various manual skills and household labor. Ironically, while enjoying the empowerment that came from their athletic experiences, the basketball girls were also used by the Indian Department as living artifacts to display the successes of US assimilation policies.[125]

The contradictions of boarding-school life have been ever present in the historical memory of those who experienced the boarding schools, as well as in the minds of the historians who have studied them. Linda Peavy and Ursula Smith have adequately captured the inherent incongruities for the Fort Shaw Indian School Basketball Team members:

> That success, of course must be measure against its cost. The school—and the government—that fostered their achievements also sought to deprive them of their customs, cultures, and language and assimilate them into a world based on the customs, cultures, and language of the dominant white society. The losses incurred by the girls [...] must never be underestimated.[126]

Thus, for the Fort Shaw girls as well as for Indian children throughout the West,

> [t]herein lies the irony of their situation. Had they not been in an off-reservation boarding school environment, without the responsibilities and distractions of family and tribal life, they would likely never have become stellar athletes, musicians, performers. Nor would they have forged ties across tribal lines, emerging as a tight-knit group with a shared goal.[127]

As the twentieth century dawned, states and territories in the American West enjoyed an entrenched system of institutionalized education, from the most rural of primary schools to dozens of institutions of higher education. With sectarian, private, and public schools to benefit a growing Western population, youth of all races, religions, and backgrounds could enjoy some form of educational offering. While the roots of all these institutions were often rudimentary, and the early processes of growth were haphazard at best, the eventual success of schools demonstrated Westerners' absolute commitment to the importance of education. Almost from the earliest introduction of education to the West, women benefited from these offerings in numerous forms: as primary students, teachers in training, veteran "schoolmarms," and as enthusiastic college coeds. For American

Indians as for Euro-Americans alike, both were subject to the broader expectations of Euro-American civilization, and both became agents of that same process of assimilation. Indeed, while education served the purposes of expanding the goals and opportunities for Western youth, education also served to entrench the imperial purposes of imposing Euro-American civilization on a "savage" environment. Within this process, Euro-American and American Indian women both found opportunities to stand out with their individual successes and to use education for their own personal benefit. Indeed, if anyone wishes to examine the history of female education in the West, it must always be with an eye to the women themselves, and how they chose to adapt their educational goals to their unique Western environment.

NOTES

1. Walker D. Wyman (ed.), as retold from the original notes and letters of Grace Fairchild, *Frontier Woman: The Life of a Woman Homesteader on the Dakota Frontier* (River Falls, Wisc.: University of Wisconsin-River Falls Press, 1972), pp. 7–8 and 13.
2. Wyman (ed.), *Frontier Woman,* p. 31.
3. Amy Kaplan, "Manifest Domesticity," *American Literature,* 70 (3) (1998): 581–606.
4. Kaplan, "Manifest Domesticity," p. 582.
5. Dee Brown, *Gentle Tamers: Women of the Old Wild West* (New York: Putnam, 1958; and Lincoln, Neb., Bison Books, 1968).
6. Brown, *Gentle Tamers,* p. 284.
7. Owen Wister, *The Virginian* (New York: Penguin, 2002).
8. Kaplan, "Manifest Domesticity," pp. 581–606; see also Glenda Riley, *The Female Frontier: A Comparative View of Women on the Prairie and the Plains* (Lawrence, Kans.: University Press of Kansas, 1988) and Sandra L. Myres, *Westering Women and the Frontier Experience, 1800–1915* (Albuquerque, N. Mex.: University of New Mexico Press, 1999).
9. Suzanne H Schrems, "Teaching School on the Western Frontier: Acceptable Occupation for Nineteenth Century Women," *Montana: The Magazine of Western History,* 37 (1987): 54–63.
10. Schrems, "Teaching School," p. 57.
11. Kaplan, "Manifest Domesticity," and Schrems, "Teaching School," p. 58.
12. Polly Welts Kaufman, *Woman Teachers on the Frontier* (New Haven, Conn.: Yale University Press, 1984).
13. Julie Roy Jeffrey, *Frontier Women: "Civilizing" the West? 1840–1880,* rev. edn (New York: Hill & Wang, 1998), pp. 109–110.
14. Jeffrey, *Frontier Women,* p. 111.
15. Myres, *Westering Women,* p. 184.
16. Myres, *Westering Women,* p. 183.
17. Thomas Morain, "The Departure of Males from the Teaching Profession in Nineteenth-Century Iowa," *Civil War History,* 26 (1980): 161–170.
18. Wayne E. Fuller, "Country Schoolteaching on the Sod-House Frontier," *Arizona and the West,* 17 (1975): 121–140.
19. Riley, *The Female Frontier,* p. 122.
20. Schrems, "Teaching School," p. 56.
21. Kathleen Underwood, "The Pace of Their Own Lives: Teacher Training and the Life Course of Western Women," *Pacific Historical Review,* 55 (1986): 513–530.
22. Underwood, "The Pace of Their Own Lives," p. 515.
23. Ronald E. Butchart, "The Frontier Teacher: Arizona, 1875–1925," *Journal of the West,* 16 (1977): 54–67.
24. Underwood, "The Pace of Their Own Lives," p. 517. See also Myres, *Westering Women,* p. 185; and Jeffrey, *Frontier Women,* pp. 196–199.

25. Schrems, "Teaching School," pp. 56–58. See also Fuller, "Country Schoolteaching," pp. 121–140.
26. George Edmonston, Jr., and Tom Bennett, "The Birth of Corvallis College," Oregon State University Alumni Association; available online at www. osualum.com.
27. Andrea G. Radke-Moss, *Bright Epoch: Women and Coeducation in the American West* (Lincoln, Nebr.: University of Nebraska Press, 2008).
28. Bruce L. Campbell and Eugene E. Campbell, "Early Cultural and Intellectual Development," in Richard D. Poll, Thomas G. Alexander, Eugene E. Campbell, and David E. Miller (eds.), *Utah's History* (Logan, UT: Utah State University Press, 1989), p. 300.
29. Jill Mulvay, "Zion's Schoolmarms," in Claudia Bushman (ed.), *Mormon Sisters: Women in Early Utah* (Salt Lake City, UT: Olympus, 1976), pp. 67–87.
30. Campbell and Campbell, "Early Cultural and Intellectual Development," p. 300.
31. Mulvay, "Zion's Schoolmarms."
32. Mulvay, "Zion's Schoolmarms."
33. Lu Dalton to Mrs. Emmeline B. Wells, December 27, 1875, Bancroft Library, University of Califor nia, Berkeley, quoted in Schrems, "Teaching School," p. 61.
34. Schrems, "Teaching School," p. 62.
35. Ronald E. Butchart, "The Frontier Teacher: Arizona, 1875–1925," *Journal of the West,* 16 (1977): 54–67; Myres, *Westering Women,* p. 194.
36. Cheri Reynolds, "Education on the Frontier," *Hoofprints from the Yellowstone Corral of the Westerners,* 31 (1) (2001): 17.
37. Norm Clark, *Tracing Terry Trails* (1982), quoted in Reynolds, "Education on the Frontier," p. 11.
38. Muirl Dorrough, Elva, Nebraska, to Harry Shimmin, Dunwell Ranch, Tryon, Nebraska, December 17, 1912. Transcript in possession of the author.
39. Butchart, "The Frontier Teacher," pp. 54–67.
40. Joanna L. Stratton, "Classrooms and Schoolmarms," in *Pioneer Women Voices from the Kansas Frontier* (New York: Simon & Schuster, 1981), p. 161.
41. Muirl Dorrough, Nebraska, to Harry Shimmin, Dunwell Ranch, Tryon, Nebraska, 1910. Transcript in possession of the author. David E. Radke, oral correspondence with the author, 1998–1999.
42. Fuller, "Country Schoolteaching."
43. Muirl Dorrough, Alliance, Nebraska, to Harry Shimmin, Dunwell Ranch, Tryon, Nebraska, July 14, 1911. Transcript in possession of the author.
44. Jeffrey, *Frontier Women,* p. 114.
45. Jeffrey, *Frontier Women,* p. 114.
46. Jeffrey, *Frontier Women,* p. 111.
47. See Underwood, "The Pace of Their Own Lives."
48. See Radke-Moss, *Bright Epoch.*
49. Campbell and Campbell, "Early Cultural and Intellectual Development," p. 302.
50. Campbell and Campbell, "Early Cultural and Intellectual Development," p. 301.
51. Leonard Arrington, *History of Idaho,* vol. II (Boise, ID: University of Idaho Press and the Idaho State Historical Society, 1994), p. 166.
52. Campbell and Campbell, "Early Cultural and Intellectual Development," p. 303.
53. Wyman (ed.), *Frontier Woman,* p. 32.
54. Butchart, "The Frontier Teacher," pp. 54–67.
55. Andrea Radke, "'I Am Very Aspiring': Muirl Dorrough and the Alliance Junior Normal School," *Nebraska History,* 81 (2000): 2–11.
56. Radke, "I Am Very Aspiring," p. 5.
57. Radke, "I Am Very Aspiring," p. 5.
58. Superintendent William K. Fowler, quoted in Radke, "I Am Very Aspiring," p. 5.
59. Radke, "I Am Very Aspiring," p. 6.
60. Andrew Elkins, "Chadron State College and Allies Bring Humanities to the Panhandle," *Nebraska Humanities,* 9 (1) (1999): 19–21.
61. Arrington, *History of Idaho,* p. 177.

62. Arrington, *History of Idaho,* p. 177.
63. Underwood, "The Pace of Their Own Lives," p. 521.
64. Butchart, "The Frontier Teacher," pp. 54–67.
65. Fuller, "Country Schoolteaching."
66. Maureen Ursenbach Beecher and Kathryn L. MacKay, "Women in Twentieth-century Utah," in Richard D. Poll, Thomas G. Alexander, Eugene E. Campbell, and David E. Miller (eds.), *Utah's History* (Logan, UT: Utah State University Press, 1989), p. 570.
67. Underwood, "The Pace of Their Own Lives," p. 518.
68. "Pioneer Profiles: Mrs. Mary Mason," *Wind River Mountaineer,* 8 (2&3) (1992): 2, 40–46.
69. Myres, *Westering Women,* p. 185.
70. Myres, *Westering Women,* p. 185.
71. Jeffrey, *Frontier Women,* p. 233.
72. Myres, *Westering Women,* p. 185.
73. For more on the coeducational experience at the University of California, see Lynn D. Gordon, *Gender and Higher Education in the Progressive Era* (New Haven, Conn.: Yale University Press, 1990), pp. 53–84.
74. Myres, *Westering Women,* p. 233.
75. Thomas Woody, *A History of Women's Education in the United States,* vol. II (New York and Lancaster, Pa.: The Science Press, 1929), p. 226.
76. H. H. Parker, "The Ideal Farmer and His Wife," *Inaugural Address Delivered at the Opening of the Iowa State Agricultural College* (Davenport, IA: Gazette Premium Book and Job Printers, 1869), p. 47. Iowa State University Archives and Special Collections Department, Ames, Iowa.
77. Virginia Railsback Gunn, "Industrialist not Butterflies: Women's Higher Education at Kansas State Agricultural College, 1873–1882," *Kansas History,* 18 (1995): 5.
78. John Anderson, quoted in Gunn, "Industrialist not Butterflies," p. 6.
79. Allie Peterson Burgoyne, "Our University as I Know It," lecture presented to the Utah State University Faculty Women's League, November 7, 1958, Utah State University Archives, Logan, Utah.
80. For feminist critiques of domestic economy, see Jill K. Conway, "Perspectives on the History of Women's Education in the United States," *History of Education Quarterly,* 14 (1974): 1–12; and Linda Marie Fritschner, "Women's Work and Women's Education: The Case of Home Economics, 1870–1920," *Sociology of Work and Occupations,* 4 (1977): 209–304.
81. Radke-Moss, *Bright Epoch,* p. 150.
82. Radke-Moss, *Bright Epoch,* p. 172.
83. Radke-Moss, *Bright Epoch,* p. 172.
84. Radke-Moss, *Bright Epoch,* pp. 181–186.
85. Almeda Perry Brown, *Memories,* ed. Charles F. Brown (Downey, Calif.: n.p., 1993), p. 62.
86. Brown, *Memories,* p. 63.
87. Radke-Moss, *Bright Epoch,* p. 187.
88. Radke-Moss, *Bright Epoch,* pp. 197–201, 209–216.
89. For a thorough look at the practice of coeducation at land-grant colleges in the West, see Radke-Moss, *Bright Epoch.*
90. Kaplan, "Manifest Domesticity," p. 582.
91. Linda Peavy and Ursula Smith, *Full-Court Quest: The Girls from Fort Shaw Indian School Basketball Champions of the World* (Norman, Okla.: University of Oklahoma Press, 2008), p. 23.
92. South Dakota State Historical Society, "Mary C. Collins, 1846–1920," in *Dakota Profiles,* available online at www.sdhistory.org/rp/dp/dp_collins.htm; see also "Dakota Images: Mary C. Collins," *South Dakota History,* 7 (winter 1976).
93. Riley, *The Female Frontier,* pp. 126–127.
94. Theodore D. Sargeant, *The Life of Elaine Goodale Eastman* (Lincoln, Nebr.: University of Nebraska Press, 2005).
95. Riley, *The Female Frontier,* p. 126.
96. Riley, *The Female Frontier,* p. 126.

97. Sargeant, *Elaine Goodale Eastman.*
98. Ruth Ann Alexander, "Gentle Evangelists: Women in Dakota Episcopal Missions, 1867–1900," *South Dakota History,* 24 (fall/winter) (1994): 174.
99. "Indian Boarding Schools," Indian Country Diaries (Native American Public Telecommunications, 2006); available online at www.pbs.org/indiancountry/history/boarding.html (accessed June 7, 2009).
100. "Indian Boarding Schools."
101. Donal F. Lindsey, *Indians at Hampton Institute, 1877–1923* (Champaign-Urbana, Il.: University of Illinois Press, 1995).
102. "Indian Boarding Schools."
103. Scott Riney, *The Rapid City Indian School, 1898–1933* (Norman, Okla.: University of Oklahoma Press, 1999).
104. Peavy and Smith, *Full-Court Quest,* p. 24.
105. For more on the Indian boarding-school experiences, see Brenda Child, *Boarding School Seasons: American Indian Families, 1900–1940* (Lincoln, Nebr.: University of Nebraska Press, 1998); David Wallace Adams, *Education for Extinction: American Indians and the Boarding School Experience, 1875–1928* (Lawrence: University of Kansas Press, 1995); and K. Tsianina Lomawaima, *They Called it Prairie Light: The Story of Chilocco Indian School* (Lincoln, Nebr.: University of Nebraska Press, 1994).
106. Zitkala-Ša (Gertrude Simmons Bonnin), from "Impressions of an Indian Childhood," "The School Days of an Indian Girl," and "An Indian Teacher among Indians," in Linda K. Kerber and Jane Sherron De Hart (eds.), *Women's America: Refocusing the Past,* 6th edn (Oxford: Oxford University Press, 2004), p. 284.
107. Zitkala-Ša, p. 284.
108. Child, *Boarding School Seasons,* pp. 11–12.
109. Child, *Boarding School Seasons,* pp. 12–13.
110. Child, *Boarding School Seasons,* pp. 55–56.
111. Child, *Boarding School Seasons,* p. 50.
112. Child, *Boarding School Seasons,* p. 44.
113. Zitkala-Ša, p. 285.
114. Margaret D. Jacobs, "Maternal Colonialism: White Women and Indigenous Child Removal in the American West and Australia, 1880–1940," *Western Historical Quarterly,* 36 (2005): 454.
115. Jacobs, "Maternal Colonialism," pp. 455 and 466.
116. Marsha King, "Tribes Confront Painful Legacy of Boarding Schools," *Seattle Times,* February 3, 2008.
117. King, "Tribes Confront Painful Legacy."
118. King, "Tribes Confront Painful Legacy."
119. King, "Tribes Confront Painful Legacy."
120. Peavy and Smith, *Full-Court Quest,* p. 38.
121. "Indian Boarding Schools."
122. King, "Tribes Confront Painful Legacy."
123. Gertrude Simmons Bonnin (Zitkala-Ša), "Impressions of an Indian Childhood" in Susan Cummins Miller (ed.), *A Sweet, Separate Intimacy: Women Writers of the American Frontier, 1800–1992* (Lubbock: Texas Tech University Press, 2007), 385.
124. Zitkala-Ša, p. 282. For more on boarding-school education, see Robert Trennert, "Educating Indian Girls at Nonreservation Boarding Schools, 1878–1920," *Western Historical Quarterly,* 13 (July 1982): 271–290.
125. Peavy and Smith, *Full Court Quest,* pp. 276 and 314–316.
126. Peavy and Smith, *Full-Court Quest,* p. 345.
127. Peavy and Smith, *Full-Court Quest,* p. 345.

CHAPTER FOURTEEN

DEFENSE INDUSTRIES

Susan Tschabrun

INTRODUCTION

One of the ironies of the popular notion that individualism and self-reliance built the American West is the extent to which much fundamental economic change can be traced back to the actions of the federal government. Whether it was the Homestead Act, the Highway Act of 1916, or the Works Project Administration of the New Deal, it is hard to imagine the emergence of the contemporary West without seeing the government hovering in the background. One area of exceptional importance is the role that government spending on defense had on the industrial development of the region. Largely concentrated on the Pacific coast but having an impact throughout the West, twentieth-century defense-related industries evolved from small beginnings into a veritable military-industrial complex involving close ties between government and big business in the research, development, and production of military weapon systems. The influence of the military in this formative development has been neither an unalloyed good, nor an unmitigated evil, but its transformative impact cannot easily be overestimated. The story has its beginning in the opening years of the twentieth century.

THE MILITARY AND THE WEST

Military spending by the federal government has played a pivotal role in the Western economy since the Western territories began to be incorporated into the USA. A study by Robert Frazer showed that military spending was the most important factor in the economic development of the Southwest region between the Mexican War and the Civil War. Other research traces how army payrolls and contracts stimulated regional economies in the same region from 1861 to 1885. Nevertheless, the role of the federal government in the development of military-related *industry* in the American West took a qualitative leap only in the World War I period. The Spanish–American War of 1898 showed for the first time that war could become an economic stimulus rather than a drain on the economy, but the economic effect of World War I was even more comprehensive, causing the USA to become the Allied powers' arsenal. The wartime demand for industrial products in turn had a sizable impact on postwar industrial efficiency and production. While other regions of the USA shared the benefits of this wartime stimulus, the unindustrialized West particularly felt the impact, especially on the nascent aircraft industry, and to a lesser extent on shipbuilding.

AVIATION TAKES OFF

In 1908, five years after the Wright brothers' famous flight, the US Army grudgingly began to recognize the significance of aircraft to modern warfare and took the first steps toward creating an air division. Early army aviation tests took place in Virginia and Maryland, but the open expanses and the favorable flying conditions of the West soon attracted army aviators to Western locations. In 1909, a few men from the Army's Aeronautical Section took a crated up Wright aircraft to San Antonio, Texas. Several years of testing and successful support of American forces in Mexican skirmishes in 1911 appear to have persuaded Congress to allocate a modest amount ($125,000) for army aviation.

In 1915, a growing concern that the USA would be drawn into World War I caused Congress to pass legislation creating the National Advisory Committee for Aeronautics, later renamed the National Committee on Aeronautics (NACA). The NACA mainly served as an advisory and consultative body concerned with encouraging research in aeronautics, which led to some development in the design of propellers at Stanford University and the construction of a small wind tunnel there. However, federal funding for aviation research and development (R&D) did not have much of an impact before the late 1920s and 1930s. Instead, the growth of the aircraft industry relied on the individual efforts of several aviation pioneers—a colorful mix of tinkerers, aviation exhibitionists and barnstormers, engineers, and wealthy flying enthusiasts—whose fortunes were made and sometimes unmade by the procurement decisions of the US military.

THE EARLY AVIATION PIONEERS

Glenn H. Curtiss

In late 1910, Glenn H. Curtiss opened an army and navy aviation training center on North Island in San Diego. Curtiss, of the Curtiss Aeroplane and Motor Corporation in New York, had one year earlier made the first commercial sale of an aircraft and, as a result, was sued for patent infringement by the Wright Bros. Curtiss had come west to lease remote North Island as a location for winter flight training and to develop a hydroplane. Hoping to interest the military in his aircraft, he invited both the Army and Navy to send pilots for free training. As a result, San Diego became location of the US Army's first air operational unit, the 1st Aero Squadron, in 1914. The Squadron soon moved to San Antonio, where it participated in the Pershing Punitive Expedition against Pancho Villa in 1916. Although existing aircraft were not sturdy enough to make much of an impact on the war itself, the experience led to the passage of the National Defense Act of 1916, which gave the military over $13 million to spend on aircraft, training, and associated personnel. The role of San Diego in the early history of the Army Air Force was but a footnote to the much more important part the city would play for the Navy after World War I.

Glenn L. Martin

One company positioned to take advantage of the wartime demand for aircraft was the Glenn L. Martin Company, formed in Santa Ana, California in 1912, but relocated to Los

Angeles that same year. Martin had built his first plane in 1909, but the company's fortunes still relied on the pioneer aviator's exploits as a daring exhibition pilot in southern California air shows and aviation meets, his appearances in Hollywood films, and flying lessons and coyote-hunting excursions he conducted for wealthy air enthusiasts, such as timber magnate William Boeing of Seattle. In order to interest the military in his aircraft, Martin sunk considerable funds into prototypes that the Army never purchased, including an armored scout and a destroyer, but he did win an army contract to build "Model TT" trainers, designed by Donald Douglas, a young aeronautical engineer recently employed by the company. In 1916, Martin was approached by the Wright Company with a merger offer that would result in a new company headquartered in New Jersey. Martin agreed but soon regretted the decision because the new company did not emphasize aircraft construction but rather focused on engines. After delivering the last of the Model TTs in early 1917, Martin closed his Los Angeles factory and moved to Cleveland, Ohio, where he incorporated a new Glenn L. Martin Company, after which the company ceased to figure in Western history until the 1990s when its namesake, Martin Marietta, acquired Lockheed.

William Boeing

Another Western aircraft company that benefited from procurement contracts during World War I was the Pacific Aero Products Company founded by William Edward Boeing in Seattle, Washington, in 1916. Boeing's enthusiasm for aircraft dated from his attendance at the Los Angeles Aviation Meet in 1910, his experience flying a Curtiss seaplane in 1914, and as a student at Glenn Martin's flying school in 1915. Boeing had inherited his parents' wealth, which included large holdings of timber and iron. His wealth allowed him to experiment with aircraft design as well as to support aeronautical research by financing the construction of a wind tunnel at the University of Washington.

When the USA entered the war in 1917, Boeing renamed his company the Boeing Airplane Company as he turned his attention to the manufacture of military aircraft rather than the seaplanes that had been his earlier focus. In an era of wooden aircraft, Boeing's supply of lumber from his timber business was an advantage. In early 1918, the Company finally landed a navy contract to build 50 Model C trainers for shipment to the naval base at Pensacola, Florida. Since the planes were too small to fly all the way to Florida, they were dismantled, packed in crates and shipped there where Boeing employees reassembled them. Like other airframe manufacturers, the Boeing Company suffered from the canceling of contracts at Armistice and shifted production temporarily away from manufacturing wooden airplanes into furniture and sea-sled boats. After almost a decade of slow business, Boeing won a federal contract, which he underwrote with his own money, to build twenty-six Model 40A airplanes for the Chicago to San Francisco mail route by 1927. By 1929, Boeing had evolved into the largest manufacturer of aircraft of any type and established its dominance in commercial aviation.

Donald Douglas

Donald Douglas was a twenty-three-year-old Massachusetts Institute of Technology (MIT) science graduate when he joined the Glenn L. Martin firm in California in 1915. Later, after a stint as an engineer with the Army Signal Corps in Washington, DC, Douglas returned to work for Martin, who had meanwhile relocated his company to Cleveland. It was in

Cleveland that Douglas met Bill Henry of the Los Angeles Times, who introduced him to the Los Angeles business community. With funds raised by Harry Chandler, owner of the Los Angeles Times, Douglas was encouraged to establish own company in southern California in 1920, first in a partnership which he soon bought out. Using his contacts in Washington, DC, and the eastern military establishment, Douglas succeeded in running a thriving defense business in the 1920s. In his large production facility in Santa Monica, he built the Douglas World Cruiser, which brought the company worldwide fame for making the first circumnavigation of the world in 1924. It was there he also built torpedo airplanes, cargo airplanes, and observation airplanes for military use.

Allan and Malcolm Loughead (Lockheed)

Allan and Malcolm Loughead were brothers who moved their seaplane company, founded in San Francisco in 1912, to Santa Barbara (north of Los Angeles) in 1916. There, they organized the Loughead Aircraft Manufacturing Company and won a contract in 1917 to built two flying navy boats, with design assistance from a young draftsman, Jack Northrop. In the wake of World War I, the Loughead Company closed its doors and liquidated its assets, but the Loughead brothers returned in 1926, pairing up with Northrop once again, to relaunch the company, now named Lockheed Aircraft Corporation, in Hollywood. It was there that Northrop designed the Vega, a high-speed monoplane that became the choice of top aviators, such as Amelia Earhart and Wiley Post. In 1927, the company was moved to Burbank for more space. But in 1929, Allan Loughead sold the company to the Detroit Aircraft Corporation, after which it promptly went bankrupt. Jack Northrup, meanwhile, had left in 1928 to establish his own Avion Corporation in 1928.

The demand for aircraft for both US and other Allied forces during World War I infused money into the nascent aircraft industry, a trend that benefited all of the aviation pioneers to a greater or lesser degree. At the same time, the late entry of the USA into the war meant that the procurement boom was squeezed into a mere eighteen months. Moreover, the retreat to a foreign policy of isolationism at the conclusion of the war caused the government to cancel all federal contracts in late 1918, depressing the US economy for the next two years, and, as we have seen, forcing many early aircraft fabricators to cut back on production. Some pioneers, such as Boeing, found ways of getting business through federal civilian airmail contracts. Others, like Douglas, continued to supply the military, while attempting to develop commercially viable versions of their aircraft designs. Some soldiered on, only to succumb to the mergers of the late 1920s, or failures of the early 1930s.

THE NAVY, SHIPBUILDING, AND WORLD WAR I

Before the creation of a separate air force, all branches of the military had some level of interest in the development of an air capability, but the Navy's mainstay was, of course, its fleets. The opening of the Panama Canal in 1914, the growth of trade and commercial tensions with Japan, and the need to protect territories in the Pacific region caused the federal government to view with favor the idea of a two-ocean fleet, which finally came into being with the Pacific Fleet in 1919. Since 1907, the year when the Navy had stopped at Pacific ports during a worldwide cruise, port cities up and down the West Coast saw attracting new naval facilities to accommodate this fleet as an opportunity for growth and prosperity, but none worked as tenaciously toward this goal as San Diego. Over more than a decade, the city funded several dredging projects to make the harbor deep enough for Navy ships, as well as creating oil-fueling stations, an advanced marine base at Camp Pendleton, docks for mothballing old destroyers, and the air station on North Island. Meanwhile, wartime orders for ships led to a boom in shipbuilding at major harbors on the Pacific coast, where over 100,000 workers labored for the Navy or the Emergency Fleet Corporation, a federal agency supervising private shipyards. Just like the aircraft industry, the cessation of hostilities brought on a serious depression to the shipyards, especially in Seattle and San Francisco, where shipbuilding industry was most concentrated. In fact, the large Skinner and Eddy Shipyard in Seattle never recovered and was sold in 1923, only to become the site of a "Hooverville" during the 1930s Depression.

AIRCRAFT INDUSTRY GETS A TOEHOLD IN CALIFORNIA

Once the government had committed to a second fleet in the Pacific, the location of shipbuilding facilities in some of the best harbors along the West Coast during wartime needs no special explanation, but the question of why so many pioneer aircraft companies congregated in the southern California area has consumed a good deal of ink. Many writers have emphasized the region's mild climate which permits both year round flying and fabrication in the large, drafty facilities required for airframe manufacture. Others have found a random assortment of personal reasons and coincidence. Perhaps the most persuasive explanation stresses the work of city boosters, who early on saw the future potential in aviation and helped the pioneers find cheap land and buildings for their operations in the underdeveloped areas ringing Los Angeles. Said another way, Los Angeles in the 1920s and 1930s possessed "the tradition of civic boosterism allied to land speculation and development on an epic scale, in which private profit joined with public spirit to promote growth."[1] Other booster-developers had a similar impact in San Francisco and San Diego.

The last half of the 1920s was a period of fevered growth in the embryonic aircraft industry, just like the economy at large. In 1926, military aviation received a boost with the creation of the semi-independent Army Air Corps and the start of five-year aircraft-procurement programs, while civilian airframe production benefited, as we have seen, from the Airmail Act of 1925 and the Air Commerce Act of 1926; both Acts went far to develop aviation infrastructure and to subsidize airmail. Charles Lindbergh's solo flight across the Atlantic in 1927 captured the country's imagination. The last two years of the boom saw numerous takeovers and mergers and the creation of large holding companies backed by Wall Street. For instance, North American Aviation was created as a holding company that included the Curtiss Aeroplane and Motor Corporation, Boeing's United Aircraft and Transport Corporation acquired Northrop's Avion Corporation, and Stearman Aircraft Company of Wichita, Kansas, among other manufacturers, and Detroit Aircraft Corporation acquired Lockheed.

When the stock market crashed, the prices of holding companies fell further than most other types of company, and there were many failures. Those holding companies that did not fail were broken up in 1934, when antitrust legislation was passed that prohibited a single company from owning both airlines and aircraft-manufacturing operations. Out of the breakup of the United Aircraft and Transport Corporation, Boeing Airplane Company reemerged as an aircraft manufacturer, while United Air Lines continued the airmail service that had been originally pioneered by the Varney Airlines in 1926 in Boise, Idaho, and acquired by Boeing in 1927. Other commercial airlines, including American Airways and Trans World Airlines, were also taking shape in the early 1930s through the same processes of merger and break-up.

As the country focused on economic recovery, and military orders slowed, the Depression years looked, on the surface at least, inauspicious for defense-related industries in the West. Nevertheless, key foundations were being laid during the 1930s for the unprecedented defense-related growth the West experienced during World War II. One area of change concerned the respective roles of government and industry in mobilizing the economy for war. During World War I, a lack of coordination between the procurement efforts of the War Department and the newly created War Industries Board (WIB), led by business leaders, caused high prices, shortages and other economic dislocations. In response, President Wilson had taken over control of the WIB and strengthened its ability to regulate the economy through greater collaboration between industry and government. Later, the National Defense Act of 1920 authorized the government to work with industry to plan for future wars and to set the stage for a number of economic mobilization plans, which, over the course of the 1930s, gradually prepared the way for a level of military cooperation, some charged collusion, with industrial "merchants of death," that presaged the full-blown "military-industrial complex" decried by Eisenhower in 1961.

In the aircraft industry, the 1930s was also a period during which aircraft design improved dramatically. Although the NACA had declared metal a superior building material for aircraft as early as 1920, it took federal collaboration with Alcoa to develop aluminum alloys that were strong and flexible enough to make all-metal airplanes common. Other NACA-sponsored breakthroughs were in the area of drag reduction through airfoils and engine cowling. These developments, together with monocoque construction, enclosed cockpits, and retractable landing gear, were among the many improvements that made possible the development of the Martin B-10 (1934), the first modern bomber; the Boeing 247 (1933), the first modern airliner; the Boeing B-17 bomber (1935); and the Douglas DC-3 (1935), in both its civilian and military versions.

Howard Hughes's early contributions to the aircraft industry also date to the 1930s. As a young man of eighteen years, Hughes inherited his father's wealth and a controlling interest in the Hughes Tool Company, which had brought the elder Hughes a fortune through the manufacture of oil drill bits in Texas. In the 1920s, Hughes Jr. focused his energies on the motion-picture industry, where his fascination with aeronautics became manifest through his production of Hell's Angels, a movie about World War I fighter pilots released in 1930. The making of the film's air-combat scenes resulted in Hughes amassing the largest private fleet of airplanes anywhere in the world. His main interest was not making but rather flying airplanes, and his goal was to break world speed records, which he did in 1934, 1935, 1937, and 1938. However, his desire to use high-performing airplanes encouraged him to pursue research and to develop his own plane, the H-1 racer, with help from Cal-Tech and its engineers. In 1935, Hughes founded Hughes Aircraft Company as a division of the Hughes Tool Company in a hangar at the Glendale airport north of the city of Los Angeles.

In addition to Hughes's new company, other aircraft manufacturers set up or returned to the southern California region during the 1930s. In 1932, the Lockheed Corporation was rescued from receivership and by 1934 was back in operation at the airport in Burbank, California. With Douglas's help, Jack Northrop was able to establish the Northrop Corporation in El Segundo near the future LAX in 1932; in 1939, Northrop founded yet another

Northrop Corporation in Hawthorne California, a few miles south along the coast, in order to become totally independent from Douglas. North American Aviation was also attracted to the LAX area when its chief executive, James H. "Dutch" Kindelberger, a former Douglas employee, moved the company to southern California in 1935. That was the same year that Consolidated Aircraft moved its operations to San Diego from Buffalo, New York, in order to develop flying boats for the Navy and aircraft for the Army. By the late 1930s, a dense concentration of aircraft-manufacturing operations and aeronautical expertise had formed in southern California, especially in the Los Angeles area. Boeing, then unveiling its new line of Stratoliner passenger planes featuring pressurized cabins, represented a Pacific coast outlier in Seattle.

The New Deal period, with its large federal presence, laid the foundation for the transformative role that defense industries would play in the West in other ways. Historian Gerald Nash has written that Henry J. Kaiser, a New Deal industrialist par excellence "[m]ore than any other individual person [...] was responsible for the wartime manufacturing boom in the West."[2] Kaiser worked throughout the 1920s and 1930s paving roads, building tunnels and bridges, laying petroleum and gas pipelines, and constructing huge federal dams, which helped create the infrastructure and cheap energy that made industrial expansion possible. His role as the chairman of the executive committee of the Six Companies gave him close ties to key bureaucrats in Washington, ties which he continued to nurture throughout the 1930s. Kaiser's Depression-era experience in managing large projects and his vision of an industrialized West helped set the stage for what would come with the outbreak of World War II.

WORLD WAR II

As significant as New Deal spending was to the West, the federal outlay during World War II was enormous, amounting to $60 billion between 1940 and 1945. Since half of this amount was devoted to war manufactures, it represented a huge boost to certain sectors of Western industry, in particular the shipbuilding and aircraft industries. Since half of all Western federal spending during the war went to California, of which about $16 billion was spent directly on major war-supply contracts, the Golden State was the big winner, but other Western states saw economic growth as well.

Shipbuilding

German U-boat attacks and other acts of war in the North Atlantic created a large demand for new cargo ships in Britain even before the USA entered the war. Shipbuilding had atrophied on the West Coast during the interwar years, but Kaiser, who was wrapping up work at Grand Coulee Dam, saw an opportunity to redeploy resources and labor from the dam operation to his newly acquired Seattle–Tacoma Shipbuilding Corporation, while landing a navy contract to build five ships. That early shipbuilding experience helped Kaiser to secure a British contract for thirty vessels, forerunners of the Liberty Ships, in late 1940. Immediately after Pearl Harbor, President Roosevelt issued a call to the existing East Coast shipbuilders for almost 20 million tons to be built in 1942 and 1943. When the Eastern shipbuilders balked, the government looked to the Six Companies who had long impressed Washington with their ability to tackle huge projects and who had been involved in the construction of military installations since 1935.

As it turned out, the government's reliance on Western industrialists paid off. Initially, Kaiser partnered with Todd Shipbuilding Corporation in his early British contract, hoping to gain from that company's years of shipbuilding experience. In fact, the shipbuilding methods of Todd and other existing shipbuilders were not suited to the huge output required by the USA during World War II. Using management techniques honed in the big construction projects of the New Deal, Kaiser and Bechtel (another Six Companies partner) were able to apply mass-production techniques to shipbuilding, including assembly-line organization, the employment of unskilled and semiskilled labor in lieu of craft labor and prefabrication. They also dedicated entire shipyards to building a single type of ship which also increased productivity. Between 1940 and 1945, Kaiser alone was able to turn out almost 1,500 cargo ships, aircraft carriers, and landing craft for the US Maritime Commission, and he became a national and international hero in the process. Older shipyards could not compete with the newcomers.

The San Francisco Bay area dominated West Coast shipbuilding with Kaiser's Richmond shipyards, Bechtel's Marinship, the Vallejo Navy Yard, and the Bethlehem Alameda Shipyard. Bechtel's Calship and Kaiser's shipyard, both in Los Angeles and the latter's Oregonship in the Portland/Vancouver area also made significant contributions. Taken together, Pacific Coast shipyards built 52 percent of all the vessels built during the war, not a small percentage considering the industry was nearly nonexistent prior to the war. Shipbuilding stimulated population growth by providing high-paying jobs, which employed African Americans, Mexican Americans and women, as well as Euro-American men. Although shipbuilding was a coastal activity, the industry employed large numbers of subcontractors, elsewhere in the Western coastal states, but also inland. Even Denver became a prosperous center of small shipbuilding.

Aircraft Industry

As important as shipbuilding was to the West during the war, its effects were far less transformative than those of the aircraft industry. Despite the strides that it had made throughout the 1930s, the aircraft industry was still small compared to the major Eastern industries of the era. By 1939, California possessed 44 percent of the floor space and 33 percent of the employment in the industry nationwide; nevertheless, the industry was still in its infancy and mass-production methods were unknown. Over the thirty-year history of the industry since 1908, only about 50,000 aircraft had been produced, and one-third of that number was built during World War I. When World War II began, the USA still did not recognize the central role that aircraft would play in the eventual Allied victory. However, World War II soon showed that the airmen had been right about the need for a powerful air force whose fleets of bombers could inflict punishing damage behind enemy lines. The bombing of Hiroshima and Nagasaki using Boeing-constructed B-29 bombers were the logical culmination of the new strategic reliance.

The buildup in the aircraft industry took place almost overnight with President Roosevelt's 1940 call to build 50,000 planes by 1941. Western airplane manufacturers mostly clustered in southern California undertook the bulk of the expansion, growing from 33,000 airframe workers in early 1940—about half of the nation's total—to 280,000 three years later. Defense Plant Corporation loans funded much of the growth and enabled companies to expand their operations dramatically. Douglas was the largest airframe producer in the Los Angeles region, building bombers, attack planes, and military transports in his plants in Santa Monica, El Segundo, and eventually Long Beach. Northrop operated his plant in Hawthorne as a subsidiary of Douglas and produced flying boats. Lockheed grew during the course of the war into the biggest aircraft manufacturer with plants in Burbank, Long Beach, Santa Monica, and a few other sites in California. San Diego's airplane industry also expanded quickly with Ryan Aeronautics, Consolidated-Vultee, and Consolidated Aircraft. Up in Seattle, Boeing churned out both the shorter range B-17 bombers and the long-range B-29 Superfortress.

Although Roosevelt favored dispersal of the aircraft industry to protect it from attack, the urgent need for more planes led the government to initially concentrate aircraft production on the coast. As time went on, however, most companies took advantage of the government's need for corporate management expertise to run government-built plants in the interior. For instance, North American Aviation had its main plant in Inglewood but eventually operated DPC-built plants in Dallas, Texas, and Kansas City, Kansas. Consolidated-Vultee had a plant in Downey near Los Angeles but also oversaw government-built plants in Dallas–Fort Worth. Douglas ran government plants in Oklahoma City and Tulsa, in addition to the company's numerous plants in the Los Angeles area. Boeing had its Seattle operations but also operated plants in Wichita, Kansas, and Vancouver, British Columbia. Eastern airframe manufacturers, such as Martin and Curtiss, also operated plants in Western locations, such as Omaha, Nebraska and Saint Louis, Missouri.

The government rarely intervened in the affairs of even the plants it owned. Government contracting officers were usually recruited from business and limited their activities to finding locations, materials, and supplies and managing contracts. Private companies were in charge of everything else. The Boeing Wichita division had some pre-World War II history behind it and was one of the most successful World War II operations, but most of the other plants were temporary phenomena only. Dallas–Fort Worth and St. Louis were the only two aircraft centers that parlayed their World War II manufacturing experience into a significant postwar role. With the cessation of hostilities, the aircraft industry retracted back to the coast.

The main chapter of the story of the wartime defense industries was written in California, which supplied the US war effort with about one-third of all military planes by weight and one-third of all merchant ships. Over the course of the war, California received $16.4 billion in major war-supply contracts and over $2.5 billion in federal investment in military and industrial capacity, making California top in the nation in terms of military installations, aircraft contracts and shipbuilding, and third in total federal spending. Key to California's success was its ability to rapidly expand its labor force through a mixture of high levels of migration and the rise in the participation rates of new categories of labor, specifically housewives, students, retirees, and other workers who were brought back into the labor force after a decade of depression. The high migration rates—3.5 million people over the course of the war—had the effect of diversifying and urbanizing the state, especially in the southern California region. Los Angeles alone absorbed a half a million new people, and San Diego grew by about 150 percent. Two-thirds of all people involved in manufacturing on the West Coast labored in factories or shipyards in southern California.

Gerald D. Nash has famously argued in his *The American West Transformed: The Impact of the Second World War* and other works that the massive federal spending of the World War II era, in California and in the West more generally, compressed decades of development into four short war years and made the West a pacesetter in the postwar period. Paul Rhode, among others, argues that Nash overstates the underdevelopment of the prewar

California economy, pointing out that by 1940 Los Angeles was already emerging as the nation's second industrial core having become the leading producer of aircraft and second leading producer of automobiles and rubber tires. In this view, the huge infusion of federal defense dollars during World War II consolidated and expanded an existing economic base but did not create it.

Winding Down World War II

What happened after the end of the war was even more crucial to the history of the West than the war itself. The decommissioning of the shipbuilding industry after the war showed that even a massive infusion of defense dollars could not create an industry that was viable in the long term in the absence of demand. In the case of aircraft, it was in late 1943 that the certainty of eventual victory caused the Joint Chiefs of Staff to begin to scale back defense contracts for aircraft manufacturers and other war suppliers. That made 1944 the peak production year with 95,272 aircraft, followed by ever-decreasing numbers over the subsequent years: 48,912 aircraft in 1945, 36,000 in 1946, and 18,000 in 1947. As a result of this precipitous decline, employment in the aircraft industry declined drastically even as the war continued to be waged. Even healthy aircraft manufacturers went under, leaving a few giants, almost all concentrated on the West Coast, to weather the initial postwar period, in which, by 1947, the industry had "run out of money, ideas, courage, and hope."[3] However, in the middle of this demobilization, the National Security Act of 1947 created the military infrastructure that would guide future defense procurement policy through the creation of the position of Defense Secretary, the Pentagon, and the separate departments of the Army, Navy and Air Force. Even more crucial to the region's subsequent history were postwar developments that artificially maintained the demand for military hardware and equipment. National employment levels in the aircraft industry fell precipitously from its high of 2.1 million people in 1944 to 185,000 in late August, 1945, with 100,000 of the remaining employees working in plants in the West. But these laid-off employees carried their expertise into the postwar consumer industries, and many were available to be re-employed when the Korean War and the Cold War galvanized an aircraft industry that was rapidly morphing into aerospace.

THE ATOMIC WEST

Some of key the characteristics of the new Cold War-driven aerospace industry had precedents during the wartime drive to produce atomic weapons. Through the Manhattan Project, President Roosevelt assigned the Army Corps of Engineers the task of building the atomic bomb. The success of this project produced one of the most significant innovations during World War II. It also helped spur the elaboration of the private contractor system, which stood in contrast to the old public armory system in which the military pursued defense-related developments in-house. In the new system, the military establishment mobilized and organized private corporations and the country's scientific and technical infrastructure to pursue R&D in support of military-related projects, a model that would become dominant in the postwar period.

The role of the West in the development of atomic weapons was one key to the success of the Manhattan Project. The feasibility of producing controlled nuclear fission had been demonstrated at the University of Chicago in 1942 on the basis of German breakthroughs in the late 1930s. In 1942, Colonel Leslie Groves took charge of the Manhattan Project and set about creating an atomic scientific, technical, and industrial infrastructure. The West, with its relatively empty spaces, presented an ideal location for the dirtier and more dangerous atomic tasks, which included the production and testing of atomic bombs and the storage of hazardous materials.

During the first period of the project from 1942 to 1945, the Army Corps of Engineers constructed a weapons laboratory to design and assemble atomic bombs at Los Alamos, New Mexico. It also built and operated the Hanford Engineer Works for manufacturing and storing plutonium at a site on the Columbia River, a location that had already received federal attention during the New Deal. The first detonation of an atomic bomb in July 1945 was also at a Western site, namely the Trinity site near Alamogordo, New Mexico. By the time the bomb was tested, the infrastructural complex that Groves had created included a huge network of scientific laboratories, facilities to produce plutonium, uranium separation units, and test reactors. World War II famously came to a close when Little Boy, the first uranium bomb, was detonated over Hiroshima on August 6, 1945, and Fat Man, a plutonium bomb, was detonated over Nagasaki three days later.

The first three years of the Manhattan Project cost the government $3 billion, but it proved that massive federal spending could not only speed up innovation and conjure up new cities out of thin air but also bring an entirely new defense industry into being. It has been estimated that the Manhattan Project created a manufacturing complex that was almost the same size as the automobile industry of that time. Although the same slowdown that hobbled the aircraft industry brought a brief decline to the atomic weapons industry in 1946, the onset of the Cold War accelerated the development of nuclear weapons and general atomic R&D. In 1946, the authority to promote and regulate military and civilian nuclear research and production was transferred from the Manhattan Project to the US Atomic Energy Commission (AEC). In 1947, the AEC increased activity at both Hanford and Los Alamos. Later in 1949, the agency founded the National Reactor Testing Station in Idaho and the Nevada Test Site in 1951. A year later, the AEC signed an agreement with the University of California to establish a second weapons lab in the West at Livermore, California. By 1952, an extensive production networked had developed whereby plutonium mined at Hanford was milled in a Rocky Flats plant outside Denver to be shipped to a facility in Texas where nuclear weapons were assembled, many of which ended up as warheads on missiles stationed throughout the Great Plains and Mountain states.

The unprecedented government-financed technical and scientific effort to win the atomic arms race during a time of war evolved in the postwar era into a permanent, long-term mobilization with economic and environmental impacts all across the Western landscape. Moreover, the Manhattan Project's success helped show the power of the three-way collaboration between government, the science and research establishment, and private enterprise that would have equally far-ranging effects throughout the Western economy.

COLD WAR AND EISENHOWER'S MILITARY-INDUSTRIAL COMPLEX

The relationship between the federal government and business in the Cold War era moved beyond the cordial collaboration of both World War I and II into a new, more intensive phase with the transition from the "hot" war against Germany and Japan to the "cold" war with the Soviet Union. Tensions between the USA and the Soviet Union soon escalated after the end of World War II. In rapid succession, events such as the Communist coup in Czechoslovakia, the Berlin blockade by the Soviets, the ascendancy of Mao Zedong in China, and the successful testing of a Soviet atomic bomb in 1949 all underscored the real and imagined dangers of the postwar world. In the foreign-policy realm, the defense posture became one of "containment," which committed the USA to a permanent wartime footing even in a time of peace and pitted the country against a foe that could potentially manifest itself in all areas of the globe.

At the same time, the massive expenditures of the early Cold War years had a domestic impact that eventually put to rest the lingering concerns of those who worried that the cessation of wartime hostilities would inevitably usher in a return to the hardships of the Great Depression. Cold War spending was the solution to the concerns of many constituencies: military leaders who wished to maintain their budgets; manufacturers who were being squeezed by the postwar downturn; congressional representatives hoping to bring federal largesse to their districts; and communities that were suffering from unemployment due to the closing of defense plants and the generalized postwar economic slump. In the interwar period, military suppliers could be kept afloat with such devices as airmail contracts, but Cold War defense requirements necessitated much larger government outlays and the expansion of war industries during peacetime.

In addition to the economic rationale for defense spending, the Cold War benefited from the widespread feeling in the USA that the country could not risk entering another war as unprepared as it had been for the conflict that had just ended. Moreover, as the clear victors in the World War II, Americans felt a responsibility to maintain global peace and were willing to devote large sums to meet both existing and potential threats. Fifteen years after the end of World War II, President Eisenhower gave a label of "military-industrial complex" to the resulting interdependent network of government agencies, large corporations, and research institutions that grew out of perceived Cold War defense needs, while warning of "the acquisition of unwarranted influence" and potential for "misplaced power" that might result.[4] Leaving aside the political consequences of the "complex," there is no doubt that the prolonged infusion of federal money into defense-related research and industry over a forty-year period had a transformative impact on the landscape of the West.

From Rocketry to Missiles

The emergence of the Soviet Union as the main challenger to US hegemony presented American military planners with a continental-sized adversary, whose potential reach no longer favored either coast nor exempted the interior of the country from possible attack. Furthermore, the development of the atomic bomb during the war had rendered much World War II military strategy obsolete. Conventional aircraft no longer occupied center stage; the new and in many cases unproven technologies associated with rockets, satellites, and missiles now dominated defense thinking and procurement policy. Like the Manhattan Project, these new Cold War weapons systems required massive infusions of scientific and technical expertise into both research and production. On the West Coast, the origins of a scientific establishment ready and willing to pursue these new defense-related agendas had already taken shape as early as the 1920s and 1930s around aviation, aeronautics, and rocketry.

The earliest collaborations between the defense industry and the research establishment in the West involved wind tunnels at educational institutions such as the California Institute of Technology and the University of Washington. In 1929, the Guggenheim Aeronautical

Laboratory at the California Institute of Technology (GALCIT) was founded near Los Angeles and under Theodore von Karman's leadership became involved in rocket research in 1936. When General Hap Arnold visited GALCIT in 1938, he became fascinated with the work being pursued there and assisted the laboratory to get army funding for further research, even though the Eastern military establishment was skeptical of rockets. World War II intensified the interest in rocketry and resulted in the foundation of the Jet Propulsion Laboratory in 1942 with a mission to research and develop long-range jet-assisted missiles. In 1944, JPL received $3 million in federal funding while simultaneously working on a variety of projects with aircraft manufacturers. At the same time, the educational infrastructure was being enhanced in the Los Angeles region, including UCLA's College of Engineering (with an emphasis on aeronautical engineering) and Northrup's Aeronautical Institute, both founded in 1941, and University of Southern California's aeronautical engineering program a few years later. The research environment in southern California was even further enhanced in 1946 when, under General Arnold's advice, Donald Douglas set up an independent research center known as RAND (for Research and Development) in Santa Monica. In 1948, RAND became its own non-profit organization with air-force funding.

In 1944, von Karman chaired an Army Air Force committee charged with using the best research of the era to forecast twenty years into the future of defense. In the words of General Arnold,

> You know, we've been winning this war by brute force and mass production and I don't like it. Maybe next time we won't be able to produce as fast and we'll get snowed under. I want you people to look ahead 20 years in aviation and tell us where we're going and how we are going to get there.[5]

Von Karman's report, "Where We Stand," has been described as "the brief for the aerospace revolution."[6] It described an unmanned air force composed of long-range intercontinental ballistic missiles, nuclear warheads, multi-staged satellites, and supersonic flight.

THE AIRCRAFT INDUSTRY MORPHS INTO AEROSPACE

Missiles are created out of the same basic components as aircraft, namely an airframe, a source of power, and a guidance-and-control mechanism, so it is not surprising that the aircraft manufacturers were the first to move into missile production in the 1940s and 1950s. Douglas Aircraft was an early entrant, with its Roc I guided bomb in 1941 and the Roc II in 1944. In 1944, Douglas began work on the Corporal E missile, which was later taken over by the Firestone Defense Products Division. Work on missile development intensified in 1945, with the start of work on the Douglas Nike Ajax surface-to-air missile, followed by the Sparrow II in 1950. In 1946, North American Aviation began work on the Navaho missile. In the same year, the Army Air Force paid Consolidated Vultee Aircraft Corporation (Convair) $1.9 million to develop a 1,500–505,000-mile range missile to carry a 5,000-pound nuclear warhead, the MX-774. Although that project was terminated, and Convair moved down to San Diego, the work on MX-774 eventually led to the development of the Atlas intercontinental ballistic missile (ICBM) in the 1950s. In 1947, the newly constituted Air Force contracted with Hughes Aircraft to produce what became the Falcon guided air-to-air missile in the 1950s.

The military's interest in missiles during the late 1940s, while keen, was not matched by funds sufficient to bring projects to early fruition. In addition, the aircraft companies' resources were inadequate in terms of capital, organizational capability and personnel for the highly technological and systems-based projects envisioned by the new ideas now dominating air-force strategic thinking. Two events in 1950 helped power the transition to aerospace. First, the National Security Council report, NSC-68, predicted war with the Soviet Union, and, in contrast to George F. Kennan's containment policy, emphasized military action over diplomacy. Moreover, NSC-68 envisioned a costly rearmament, consuming as much as 20 percent of the GDP. The second key event was the Korean War, which broke out a few months later in June 1950 and gave the government a "hot" war and a more urgent reason to pump money into the defense industry. The postwar economic contraction of the late 1940s had centralized the bulk of the aircraft industry in the southern California region after its expansion into other mainly Western states during the war. Despite President Truman's 1951 "Memorandum and Statement of Policy on the Need for Industrial Dispersion," the preeminence of California in aviation and aerospace would continue as manufacturers of missile and space systems pleaded for, and were granted, exemptions from the dispersal policy.

In 1953, the incoming Eisenhower administration charged the Air Force with doing a comparative analysis of long-range guided missiles. The Strategic Missiles Evaluation Committee, better known as the "Teapot Committee" and composed of eleven prominent scientists and engineers, brought out its first report in early 1954. The report included a sharp critique of both the design and the management of the Convair Atlas missile project, complaining that Convair made too much use of airframe technology and techniques. Instead, the Committee proposed the creation of a new "development-management" group that would accelerate and complete the Atlas program and be made up of an "unusually competent group of scientists and engineers capable of making systems analyses, supervising the research phases, and completely controlling the experimental and hardware phases of the program."[7] It was unclear from the report was whether such a group would be drawn from government, industry, or the universities, or a combination of the three, but it was clear that an in-house program such as the Army's Redstone Arsenal in Huntsville, Alabama was never envisioned.

The Teapot Committee's report had an immediate effect in southern California through the creation of an Air Research and Development Command (ARDC) field office on the West Coast. The ARCD commander in turn ordered the creation of the Western Development Division (WDD) in Inglewood in July 1954. The commander, Brigadier General Bernard A. Schriever, immediately set to work developing a new organization that put leading scientists in a position to help guide the development of the Atlas ICBM. The Ramo–Wooldridge Corporation was brought in due to their expertise in systems engineering and integration. Ramo–Wooldridge had been founded just a year earlier and possessed the advantage of having no prior defense contracts or existing production capability which would present either a conflict of interest or impede their ability to bring in the best scientific and technical talent.

Simon Ramo and Dean Wooldridge, both Cal Tech Ph.D.s, got their start in the California aircraft industry in 1947 as the directors of research of the Hughes Aircraft Electronics Division, helping to put the company in the forefront of the emerging aerospace revolution. The company founder, Howard Hughes, had early on possessed an interest in aeronautical R&D that can be traced back to his days as an aviator, his quest to build aircraft able to break aviation speed records in the 1930s, and his independent wealth which allowed him the luxury of developing innovative aircraft that never made it to the assembly lines. During World War II, Hughes Aircraft grew from a small outfit to a major defense supplier employing 80,000 people. His most famous project was the H-4 Hercules, more popularly known as the "Spruce Goose," a plane that only flew once but that demonstrated the feasibility of large plane construction. At the end of the war, Hughes was left with only one ongoing defense contract, a $500,000 Army Air Force project to work on an experimental air-to-air missile designated as the JB-3. The use of radar in sophisticated missile guidance systems was the innovation that caused Hughes to search out Ramo, then an expert in microwave transmission and detection technology at General Electric. Around the same time, Hughes' wartime work on the H-4 Hercules and X-F 11 photo reconnaissance plane project led to him being investigated by a special Senate investigative committee in 1947. Although eventually cleared, the experience embittered Hughes and may have contributed to his declining mental health.

It was Hughes's erratic management style that eventually drove Ramo and Wooldridge to leave Hughes Aircraft to found their own company in 1953. But that was after they had built up the company into the second largest scientific organization after the Bell Laboratories in New Jersey, employing over 400 Ph.D.s. By 1952, Hughes employed 15,000 workers, of which 1,000 were scientists. Hughes continued its growth in the new defense-related electronics sector even after the loss of Ramo and Wooldridge, securing about 20 percent of all California electronics business by the end of the 1950s.

Meanwhile by 1954, Ramo–Wooldridge, together with the WDD and the Special Aircraft Project Office (SAPO) of Air Material Control, the air-force procurement arm, were intensively pursuing development of the Atlas ICBM. But this new organization with its direct air-force oversight was opposed by aircraft industry leaders who felt that they were capable of developing the ICBM themselves and therefore supported Convair against the government. To mollify these leaders, Ramo–Wooldridge was prohibited from ICBM hardware production. Despite that restriction, Ramo–Wooldridge was overseeing 220 prime contracts by 1957. In 1958, Ramo–Wooldridge merged with Thompson Products, Inc., a Cleveland aircraft engine firm, to create Thompson Ramo Wooldridge Inc. (TRW). Southern California aircraft companies benefited from the air-force decision to locate its weapons development liaison body in their midst, but not all of them were able to adapt to the new requirements for ongoing R&D, systems engineering, and aviation electronics, or avionics. In 1959, there were eight aircraft companies among the top sixteen missile contractors. The other eight firms included six electric and electronics firms, one automobile manufacturer, and one tire maker. However, the top five aircraft companies were the most successful in getting into missiles, and each had its own missile project: Convair (Atlas), Martin (Titan), Boeing (Minuteman), Lockheed (Polaris), and North American Aviation (Hounddog).

By 1961, 61 percent of the total US missile production was the work of the top five aircraft companies. Douglas and Northrop were also involved in missile production, but to a lesser extent than the "Big Five." Douglas was one of the first out the gate (1950) to establish a missile division, followed in 1954 by Lockheed's Missile Systems Division in Van Nuys and Convair's Astronautics Division in San Diego, North American Aviation's Space and Information Systems Division in Downey and its Rocketdyne Division in Canogo Park, both in 1955, and Ford Motor Company's Aeronutronic Systems Division in Glendale in 1956. By the mid-1950s, the missile production largely located in the southern California region had created a dense network of linkages that included eight prime missile contractors, eleven major subcontractors and 175 subcontractors supplying electronic guidance, tracking, and other missile-related electronic systems.

Missiles were only part of the emerging aerospace industry. The role of satellites in Cold War military strategy had been foreseen by some scientists and military leaders by at least 1945. In 1946, RAND, still operating as a department of Douglas Aircraft, completed a feasibility study on satellites, titled "Preliminary Design of an Experimental World-Circling Spaceship," which forecast their development as potent scientific tools and hypothesized that they would become a stepping stone to space travel. Unlike missiles, however, the work on satellites work proceeded slowly at first. A 1951 RAND report proposed the Military Satellite System (WS 117L) project, which was taken up by the WDD. In 1956, Lockheed was awarded the contract, but funding was not forthcoming until after October 4, 1957, the date when the Soviet Union launched Sputnik I. The Soviets followed up that accomplishment a month later with the launch of Sputnik II. The successful launching of the Sputnik satellites came as a shock to Americans who had believed that the Soviets were far less technologically advanced than now appeared to be the case. Fear that Sputnik augured a more potent Soviet ballistic missile capability lent the topic of satellites an urgent tone and helped fuel the USA–USSR Race to Space and the opening chapter of the "Space Age."

As funding materialized for space-related projects, the Army and Air Force brought differing procurement philosophies to the table. The Army space effort built on the work of the NACA and the Army Ballistic Missile Agency (ABMA) established at Redstone Arsenal in Huntsville, Alabama, in 1956. As Director of ABMA, German-born Wernher von Braun had developed the rocket that launched Explorer 1, the first US satellite. Looking to create an umbrella agency to oversee space-related projects, President Eisenhower created the National Aeronautics and Space Administration (NASA) in October 1958, incorporating the NACA operation and a number of other R&D centers, including the ABMA and Wehrner von Braun, who served as the first Director of the Marshall Space Flight Center, NASA's original home at Redstone Arsenal in Huntsville, Alabama. Although Redstone had long had an in-house development philosophy, NASA found itself joining the Air Force in contracting out a good deal of its work to commercial aerospace firms, many of which were located in southern California.

By 1959, the defense industry in southern California had completed its transition from World War II-era aircraft manufacturers to Cold War aerospace corporations. As such they were shaped by an economic reality quite different from that experienced by normal civilian corporations, or even the aircraft companies of the war years. In a conventional war, military strength could be measured, at least in part, by the output of defense-related assembly lines. By the late 1950s, most weapons, satellite, and space systems were designed, developed, and tested, but almost never mass-produced. The Cold War arms race demanded that new weapons be developed for their hypothetical ability to deter aggression, not for actual use. Furthermore, innovation was not just valued but an ongoing necessity, since weapons succeeded only if they were perceived as able to preempt the other side's weapons. The complexity of the new weapons and space systems also encouraged interdependence among defense contractors; in order to speed up production, the new field of systems engineering evolved to enable each subsystem of a complex space capsule or missile to be designed and produced, even when multiple firms, with their managers and engineers, were involved.

The need for aerospace firms to constantly innovate and collaborate helped concentrate the industry despite the fact that congressional representatives from other states eyed California's proportion of total US prime contracts with envy. In 1959, New York Senators Jacob Javits and Kenneth Keating co-sponsored a Bill, called the "Armed Services Competitive Procurement Act," in an effort to divide military contacts more equitably between geographical regions, but the California delegation fought back, and the Bill never made it out of committee. During and after the Korean War, US defense spending never dipped below about 5 percent of total gross national product throughout the Cold War era, and the West got an increasingly large share of that money. California had been third in the list of states with the most prime contracts during World War II and the Korean War, but in the period 1945–1982, California remained the number one state with around 20 percent of total US prime contracts. The other Western state in the top ten was Texas at number 3 with around 6–9 percent of all US prime contracts. For a short time in the late 1950s, the state of Washington also made it into the top ten list with about 6 percent of all prime contracts.

After the "hot" Vietnam War, defense spending contracted during the mid- to late 1970s. However, in the 1980s, the Reagan administration's Star Wars initiatives led once again to a boom in aerospace defense spending, with a focus on advanced radar, surveillance, communication systems, and information processing. This second phase built on the first and tended to reinforce the geographical patterns that had taken shape during the first phase of the Cold War. By the mid-1980s, 69 percent of all federal R&D spending was devoted to defense, and those contracts tended benefit a few large firms. Geographically speaking, R&D contracts became even more concentrated in California than prime contracts. During this era, a number of aerospace corporations spun off SETA (Systems Engineering and Technical Assistance) firms which specialized in the development, review, and analysis of aerospace and weapons systems concepts or designs, but did not produce hardware. Concentrations of scientific and technical expertise produced a different pattern of urbanization than the mass-production industries of the Midwest and Eastern seaboard.

"Tilt the world on its side and everything loose will land in Los Angeles."[8] This quotation, variously attributed to Will Rogers and Frank Lloyd Wright, may have, unbeknownst to the speaker, described the role of defense spending on patterns of urbanization in the West. The impact of the aircraft and aerospace industries on population growth can hardly be disputed. The growth of the population of the greater Los Angeles area got its first big boost during the decades of the 1940s and 1950s in association with the development of the aircraft industry, increasing from around 3.5 million people in 1940 to almost 9 million in 1960. Most of this growth took place in Los Angeles County but after 1950 population also grew quickly in neighboring Orange County. Labor statistics reinforce the picture of a region increasingly dominated by federal dollars. In the early 1950s, around 13 percent of the Los Angeles labor force was employed directly by defense-related firms. By the early 1960s, the direct, indirect, and induced effects of defense expenditures totaled almost 44 percent of total employment in the Los Angeles metropolitan area.

By the end of the 1950s, southern California had become the largest high-tech industrial region in the world, and it retained that preeminence at least into the early 1990s. By 1990, almost 500,000 workers labored for defense-related industries in over 2,000 individual establishments. The need to stay innovative tended to keep defense contractors tethered to the large pool of scientific and technical labor that over time had concentrated in southern California. However, as areas of Los Angeles that had once been on the unpopulated fringes of the city began to be built up, aerospace manufacturers tended to push outwards from the center in a search for cheaper space, or due to the pull of labor anxious to live in residential areas away from the inner city. Some firms moved to El Segundo south of LAX airport (Rockwell, Hughes, TRW), while others expanded down to Long Beach (Douglas), or westward along the San Fernando Valley (North American Aviation, Litton Industries).

However, the most common route out of Los Angeles was into Orange County, which was built up almost entirely during the aerospace and electronics expansion of the 1950s–1980s. Prior to the 1950s, Orange County had been a predominantly agricultural region, as reflected in its name. The first defense-related firm to move into the area was Northrop's Nortronics Division, which relocated from Los Angeles County to Anaheim in 1951. In 1957, Hughes Aircraft moved its ground systems division to Fullerton, a location from which it later expanded into Newport Beach and Irvine. By the early 1960s, several other aerospace firms established Orange County facilities: Douglas Aircraft's Space Systems installation in Huntington Beach, Ford Motor Company's Aeronutronics facility in Newport Beach and Irvine, and North American Rockwell's Autonetics Electronic Systems Division in Anaheim. These firms worked on a variety of projects during this early period, including advanced radar systems, guidance systems, and missiles.

Orange County defense work tended to concentrate in the area of electronics, which brought into being a flourishing electrical equipment and supplies industry in the county. The number of electronics workers rose rapidly in the 1950s to 11,000 workers in 1957, exploding by 1964 to 42,000 workers, or 46 percent of all Orange County manufacturing employees. In 1964, Orange County possessed almost 90 percent of all electronics employment in the area of defense-related advanced communications and detection equipment in the USA. The source of this rapid growth was the expansion of the US C3I budget (command, control, communications, and intelligence); a major recipient was Rockwell International, which had become the county's largest manufacturer by the mid-1970s. At the end of the 1970s, over 500 electronics firms had located in Orange County, and 35 percent of Orange County's manufacturing jobs were in the area of aerospace and defense. The Reagan military buildup injected even more money into the industry, so that by 1988, there were 96,000 people employed in defense. By that time, Orange County had diversified economically to the point that the region was no longer so dependent on defense manufacturing that it could not weather the sharp declines in defense spending of the early 1990s.

THE ROLE OF DEFENSE IN WESTERN (SUB)URBANIZATION

The centrifugal pattern of dispersal of the aircraft and aerospace industries in the southern California area helped create relatively low-density, automobile-dominated, multi-centered urban spatial configurations most famously associated with Los Angeles but common in other Western cities. Of course, suburbanization in the West was driven by numerous factors, including the invention of the automobile and air conditioning, the GI Bill of 1944, which helped veterans to purchase homes, the Interstate Highway Act of 1956, and "white flight." Federal dispersal policies as they evolved during the 1950s helped the process by providing defense manufacturers with incentives to locate outside of major urban agglomerations, such as the requirement for a firm to secure a "certificate of necessity" before building new facilities and tax write-offs that compensated firms that moved to dispersed locations. In 1956, a Commerce Department report stated that "All things being essentially equal, a firm with well-dispersed facilities has a definite advantage in securing military electronic business over competitors whose facilities are in critical target areas."[9] Given that during the 1950s, the electronics industry was largely dependent on Department of Defense contracts, federal policy had a significant but shrinking role in Western urbanization and suburbanization processes over time. The early Cold War years were crucial ones in this process. For instance, in the case of California, twelve of the state's major metropolitan counties grew at their fastest rate during the Eisenhower years (1953–1961).

SEATTLE AND BOEING

Although, California, especially southern California, underwent the most dramatic defense industry-induced change during the Cold War era, other Western regions shared in the transformative impact of peacetime war spending. For instance, while Seattle shared in the pain of postwar demobilization with employment at Boeing dropping from a World War II peak of 45,000 to 15,000 at the end of the war, Boeing was successful at not only securing Department of Defense R&D contracts for military jet bombers and missiles but also developing its jets for the civilian market during the 1950s and 1960s. By 1954, one in two King County manufacturing employees was at Boeing, with employment at the company peaking in 1967 with 148,000 employees.

The dominance of Boeing in Seattle made for a dangerous over-reliance that produced the disastrous "Boeing Bust" of the late 1960s and early 1970s. A decline in the demand for its commercial jets together with the company's failure to secure key military contracts caused Boeing to lay off over 60,000 employees, or about 60 percent of its Seattle workforce. The giant billboard near the Seattle airport, which read, "Will the Last Person Leaving Seattle Turn Out the Lights?" became emblematic of the dire impact Boeings' downturn had on the entire city. Despite the billboard's sentiment, many laid-off Boeing workers did not leave Seattle, which made it easier for the company to regroup and expand when commercial jet sales and defense contracts began to boom again. During the Reagan years, Boeing became the fastest growing defense contractor with over $6 billion in government sales in 1986. By 1984, Boeing's sales to the military represented more than 70 percent of its total profits and 50 percent of its total sales.

Boeing expanded outward from the Seattle city core looking for open land, just as the aerospace industry in southern California had done, only on a much smaller scale since just one company was involved. During World War II, the Renton plant was established on the south shore of Lake Washington, at first to build Navy Sea Rangers and B-29 bombers, followed by commercial jets after the war. In 1964, Boeing unveiled the space center in Kent, Washington, even further south. During the 1970s, machine control data (MCD) was produced in the Kent plant; yet another Kent facility was built to produce air-launched cruise missiles in 1980. In 1966, work began on the Boeing Everett assembly facility, north of Seattle, which produced civilian jet planes, as well as military variants, and remains the largest building (in terms of usable space) in the world. During the 1980s, the Boeing Computer Services headquarters occupied a 90-acre site across the lake to the west in Bellevue, Washington, where work for NASA contracts, among others, was carried out. It is probably no accident that Boeing established its computer-services division in Bellevue, the exact suburb where Bill Gates had relocated his new Microsoft Company in 1979. By the same token, Boeing's long history in the Seattle region may have given Gates the concentration of white-collar, scientific, and technical labor that contributed to his company's success. By 1985, only 35 percent of employment in the greater Seattle area was dependent on military spending as the region's economy diversified into electronics, computers, and medical equipment.

COLD WAR "SUNRISE INDUSTRIES" EXPAND INTO THE INTERIOR

Although California continued to garner the largest share of defense contracts during the Cold War years, amounting to 18–22 percent of total US prime contracts from 1958 through the Reagan buildup of the early 1980s, select regions of the West away from the coast also benefited. In some cases, locales that had attracted defense spending during World War II flourished again, while other locations charted a brand new course. The fear of a possible outbreak of nuclear warfare during the early years of the Cold War made dispersing defense-related industry away from its few concentrations on the coast an idea with a menacing new urgency. As a result, most of the earliest Cold War defense-related industries in the interior Western states received their initial impetus from the perception that there was a strategic value in moving industry away from the coast. As missile technology improved and all parts of the country became potential targets, the benefit of having key defense industries located in the center of the country lessened, but in many cases the initial stimulus proved enough to make those interior Western states recipients of defense dollars throughout the Cold War and beyond.

An example of the Cold War era revival of what had earlier been a booming World War II defense industry was in Wichita, Kansas. With aviation roots back to the 1920s, Wichita was primed to benefit from a massive World War II expansion, which led the city to produce huge numbers of B-29 bombers until the end of the war, when demobilization caused the closure of the huge, government-built Plant Two. Within a few years, however, the 1948 Berlin airlift caused the Air Force to ask Boeing to reopen Plant Two to produce the new B-47 Stratojet. Despite resistance from both the Seattle-based Boeing management and other aircraft firms centered in Los Angeles—all of whom feared the trend toward "decoastalization"—Boeing moved ahead to rapidly ramp up the Wichita operations. This massive expansion was followed by even more defense spending when the Air Force selected the site for the production of B-52s in 1952. By 1957, Boeing employed 35,000 workers at the height of the B-52 program. Despite the ups and downs of the Cold War, Boeing continued to employ large numbers of workers toward the end of the period in 1985, a year when 20,000 employees labored in plants run by the Boeing Airplane Company and Boeing Computer Services. Together, these two companies were the largest employers in Kansas. In addition to Boeing, Wichita was home to other aircraft manufacturers, including Learjet, Beach, Cessna, and Gates.

Other Cold War defense industries grew up in the interior as new technologies brought certain weapons systems, products, or companies into prominence. An example is the Thiokol Chemical Corporation in Utah. The name, Thiokol, derived from the Greek words for sulfur and glue and was used to describe the synthetic rubber that the company's founders invented in 1926. The company's fortunes sagged during World War II, but some Jet Propulsion Laboratory scientists approached the company in the mid-1940s, having discovered that Thiokol's solvent-resistant polymer made excellent rocket fuel. Meanwhile, the wartime industrial dispersal policy led the government to fund construction and operation of the large Geneva Steel Works in Utah during World War II. This government spending set Utah up for a growth in manufacturing. In the 1950s, Thiokol went straight to the military with its superior rocket fuel and landed a research contract. With the onset of the Korean War, the Thiokol Corporation's solid rocket fuel began to displace the far more dangerous liquid rocket fuel. Later, the company began manufacturing solid fuel for Minuteman missiles and rocket motors for missiles and NASA's space shuttle. In the late 1950s, Sperry Rand also relocated in Utah to build missiles, while the Hercules Corporation also found Utah an attractive location for producing rocket engines.

Similar missile-related growth in defense industries can be seen in Colorado, where federal military-contract expenditures peaked in the early 1960s. Ground zero for the Colorado defense establishment was Colorado Springs, where energetic city boosters had already attracted an army air base to the town during World War II, using the same dispersal policy justification that had been used to build a steel plant in Utah hundreds of miles from any large-scale manufacturing center. Starting in 1958, Colorado Springs' role as the host to the North American Air (later Aerospace) Defense Command (NORAD) headquarters gave the city one of the most important defense operations in the country. NORAD, together with the Strategic Air Command in Omaha, Nebraska, oversaw a sophisticated communications and radar network designed to track and intercept strategic bombers or nuclear-armed missiles and commanded strings of missile silos located throughout Montana and North Dakota. Its operation center one mile deep in the heart of Cheyenne Mountain was designed to withstand a nuclear bomb blast.

The unusual concentration of military installations in Colorado Springs attracted federal contractors as early as the late 1950s. Even as the importance of NORAD to US defense strategy declined in the late 1970s, space defense functions increased. TRW had already set up a small-scale outfit called Colorado Electronics in the early 1960s, and Honeywell, United Technologies, and Ford Microelectronics came in the 1970s. With the new Reagan-era emphasis on the Strategic Defense Initiative (SDI, popularly known as Star Wars) in the early 1980s, federal money poured into Colorado Springs to start up the Consolidated Space Operation Center, the US Space Command, and the SDI National Test Bed Facility. Federal subcontractors flocked to Colorado Springs, and a Department of Defense-funded mini-boom ensued.

Texas was also an important destination for federal defense dollars, especially the region around Dallas–Fort Worth. For most of the period 1958–1982, Texas was third in the list of top ten states receiving defense-related prime contracts, amounting to between around 7 and 9 percent of the total awards. Some of the earliest relocations of aircraft firms to Texas were due to the dispersal policy in the aftermath of World War II. In 1948, the Navy persuaded East Coast-based Vought Aircraft to move its entire production facility from Connecticut to the Naval Weapons Industrial Reserve Plant in Dallas, Texas. The move, which involved 27 million pounds of equipment, company workers, and their families, was the largest industrial move up to that time. In response to evolving Cold War defense procurement needs, Vought branched out from its successful military aircraft business into aerospace and missiles. Convair, which became the Convair Division of General Dynamics in the early 1950s, built aircraft but received a contract in 1954 to build intercontinental ballistic missiles from the WDD's Air Research and Development Command. Bell Helicopter, another aircraft firm that abandoned the Northeast for the West, established its headquarters in Fort Worth and developed a variety of aircraft for the military, including the secret Bell supersonic X-1. In 1963, the Dallas–Fort Worth metropolitan area ranked fourteenth in the nation in terms of amount of federal defense R&D spending it received, after Los Angeles, San Francisco, San Diego, Seattle, Denver, Salt Lake City, and Sacramento ($124 million). By 1975–1977, Dallas–Fort Worth had advanced to eighth place, behind only three other Western locations, namely Los Angeles, San Francisco, and Seattle. The federal investment in aerospace R&D later helped make Dallas–Fort Worth an important outpost of the high-tech electronics industry.

Although the Dallas–Fort Worth area represented the most successful recipient of federal defense largesse, other Texas locations were also beneficiaries. Houston's acquisition of the Manned Spacecraft Center (MSC), later Johnson Space Center, in 1961 marked its emergence as a key concentration of aerospace corporations and space-related R&D. San Antonio, home of one of the largest concentrations of military installations in the USA, including Fort Sam Houston, the Lackland Air Force Base, and Randolph Air Force Base, has attracted its share of defense contracting. Austin boosters, including the University of Texas, succeeded in attracting Lockheed's Sunnyvale operations to the city, paving the way for future collaborations between the university and aerospace firms.

DEFENSE-RELATED ELECTRONICS AND THE INVENTION OF THE COMPUTER

Rockets, missiles, and advanced aircraft all required sophisticated avionics, communications, and guidance and control systems, which helped stimulate a large number of inventions in the electronics field. During and after World War II, the military worked to develop electronic computing machines to assist with a variety of military applications. Indeed, until sometime in the 1960s, almost all demand for computer hardware, semiconductors, and software originated with defense, energy, and space agencies.

Although the development of the first room-sized computers mainly took place on the East Coast, Western locations also played a role. The EDVAC (*Electronic Discrete Variable Automatic Computer)*, built for the US Army's Ballistics Research Laboratory by the University of Pennsylvania, had the ability to store programs, making it the first large-scale, electronic digital computer. Delivered to the Army in 1951, the EDVAC weighed 29 tons and covered 490 square feet. Another milestone was the first installation of the IBM 701 (Defense Calculator) at the AEC's Los Alamos Laboratory in New Mexico in 1953. The Livermore National Laboratory in California was also an early source of demand for high-speed scientific computers.

The invention of radar during World War II made the interception of aircraft a realistic goal, but only if radar stations were placed in dense-enough networks to give adequate notice of an attack in progress. Tracking this radar data and taking appropriate action would have overwhelmed any team of human operators. The military's answer to this problem was SAGE (Semi-Automatic Ground Environment), an automated control system designed to track and intercept enemy bombers deployed in 1959. The Air Force commissioned the computerized system from IBM, but the software that ran the system was programmed by the System Development Corporation (SDC), a RAND spin-off located in Santa Monica, California. SDC's role in providing 500,000 lines of assembly language code for SAGE made the nonprofit a candidate for "first software company."

THE INTEGRATED CIRCUIT AND THE WEST

As computer technology evolved, the West began to play a more central role in key breakthroughs. Early computers used vacuum tubes as switches and to store instructions; the EDVAC, for instance, used 6,000 vacuum tubes, and its predecessor, the ENIAC (*Electronic Numerical Integrator And Computer)* employed 18,000 vacuum tubes. Vacuum-tube technology proved cumbersome, and improvements were clearly needed. In the late 1950s, the solution to this problem—the integrated circuit—was independently invented in two Western states. Jack Kilby of Texas Instruments invented a crude integrated circuit using germanium in 1958, while in California Robert Noyce simultaneously invented an integrated circuit that used silicon as a base. While Kilby received the patent for the first integrated circuit, Noyce went on in 1960 to invent the planar process for integrated circuits, which embedded transistors and resistors on a small sliver of silicon. The later process became the basis of the computer industry and hence the name of its most famous Western location: Silicon Valley.

THE ROLE OF DEFENSE INDUSTRIES IN THE MAKING OF SILICON VALLEY

Since at least the 1980s, Silicon Valley in the Santa Clara valley of California has been viewed as the quintessential model of high-tech entrepreneurialism, firmly grounded in private sector start-ups, university–industry partnerships, and venture capital. The origins of the Silicon Valley phenomenon, however, are more complex. Scratch the surface of most historical accounts of the rise of the Silicon Valley, and the role of defense spending emerges as an underlying theme. There are three main explanations of how the Silicon Valley came to dominate the high-tech industry sector in the latter part of the twentieth century. The first and most common explanation picks up in the mid-1950s with William Shockley's return to the Santa Clara valley and the founding of Fairchild Semiconductor. Other historians have dug further into the past and point to Frederick Terman, the Dean of Stanford University's Engineering Department, as the key catalyst behind the emergence of Silicon Valley. Yet another explanation goes back even further into the early years of the twentieth century and discerns some of the Valley's defining traits in the development of the early electronics industry in the San Francisco Bay area before and after World War I. Each of these Silicon Valley origin stories highlights the importance of military procurement and defense contracts.

FROM RADIO TO VACUUM TUBES

The early years of the twentieth century were a formative time for the nascent electronics industry. On the East Coast, important innovations were quickly patented and guarded zealously by large companies that sometimes valued control over innovation. On the West Coast, there was less corporate control but also less consumer demand. Military needs, however, were an alternate source of demand for early electronics and a driver of early innovation. For instance, in the early days of radio, the greatest demand for radio communications came from ships, especially navy vessels, which could not install enormously expensive cables underwater for ship-to-ship or ship-to-shore communications in the way that land communications relied on telegraph lines at the time. In 1912, a Stanford University graduate named Cyril Elwell from the Federal Telegraph Company (FTC) based in Palo Alto approached the Navy with an alternative radio transmission technology, the Poulsen arc, which was quickly deemed to be far superior to existing methods and immediately adopted. The Stanford High Voltage Laboratory had been instrumental in perfecting the Poulsen arc and thus was an early example of Stanford University's role in assisting innovation.

Upon entry into World War I, the USA nationalized the radio industry, including the FTC. As a result, the company boomed during the war, thanks to navy orders to equip liberty ships, battleships, and shore stations around the country with radio transmitters and its role in developing the first global radio communications system. Meanwhile, the war years had seen the development of vacuum tubes as an alternative to the Poulsen arc, which left the FTC overextended at the end of the war as military orders were cancelled. It was the invention of another FTC research engineer, the three-element vacuum tube, "the audion," that would become the basis of the modern radio industry. Lee de Forest, working in Palo Alto between 1910 and 1913, enhanced his earlier invention by developing vacuum tubes that could be applied to all three stages of wireless radio communications: generation, reception, and amplification. In 1913, de Forest sold the rights to his vacuum tube design to AT&T (American Telephone & Telegraph).

In aftermath of World War I, the so-called Radio Group, composed of RCA (Radio Corporation of America), General Electric, United Fruit, Westinghouse Electric Corporation, and AT&T, formed a patent-sharing arrangement aimed at monopolizing radio technologies, especially for consumer industries. This left West Coast outfits struggling to avoid the constant threat of RCA litigation and looking for other markets. Those markets were found in military contracts and other specialized niches. In the case of the FTC, the company was able to avoid patent infringement in its move to vacuum-tube technologies because de Forest had been employed there at the time of his inventions. With navy assistance, FTC secured the contract to build a transpacific circuit for the Chinese government, but the rise of Chiang Kai-shek in the 1920s caused that project to remain incomplete. Equally important for the future of Silicon Valley were the FTC spin-offs, many of which themselves benefited from military contracts. For instance, an early spin-off was Magnovox, whose patented loudspeakers were employed on navy destroyers and battleships and its anti-noise microphones on navy flying boats patrolling for U-boats during World War I. Charles Litton, a 1928 Stanford graduate, was hired to manage FTC's in-house vacuum-tube manufacturing operation and ended by spinning off Litton Engineering Laboratories in the early 1930s. Even a young Frederick Terman interned at FTC during the early 1920s before leaving for MIT.

STANFORD UNIVERSITY, TERMAN, AND THE MILITARY

The second often-told account of Silicon Valley's origins focuses on the role of Stanford University's Engineering Department and its Dean, Frederick (Fred) Terman, during the 1930s and 1940s. Terman's father, Lewis, one of the authors of the IQ test, had moved to California in order to take up a psychology professorship at Stanford University in 1910. After young Fred returned to Palo Alto, having completed his doctorate in electrical engineering at MIT, he joined the Stanford faculty as an instructor in 1925. Over the years, his efforts to raise money for the then chronically underfunded Stanford University led to a variety of innovative overtures to industry including encouraging his graduate students to start their own companies and allowing faculty to support those efforts. Well-known beneficiaries of Terman's mentoring efforts are Bill Hewlett and David Packard, who famously set up their Hewlett-Packard Company in a garage in 1939, with Terman's encouragement. Another of Terman's innovations was the creation of the Stanford Industrial Park, officially opened in 1951 but with earlier precedents in the late 1940s. Using vacant university land, Stanford leased property to technology firms to help provide jobs for Stanford graduates.

Sometimes lost in this Stanford-centric account is the fact that the technology developed during the years before and after World War II had less to do with computer technology and more to do with military electronics. The Varian brothers, Sigurd and Russell, invented the klystron, a specialized vacuum tube used in radar equipment, in 1937. With the outbreak of World War II, the two brothers were brought to the East Coast by the Navy to work at Sperry Gyroscope Laboratories in New York in support of the war effort. Returning to California after the war, the brothers set up Varian Associates near the Silicon Valley location where Charles Litton had earlier set up his vacuum-tube factory in 1932 after the FTC moved east. Litton had also been a World War II supplier for the Navy with radio and radar tubes. Varian meanwhile had developed the airborne klystron, which was rewarded in 1948 with a sole-source contract by the Air Force. It was at this point that Varian Associates approached Fred Terman suggesting that Stanford lease the company some land near the university and thereby becoming the prototype for the Stanford Industrial Park. Between 1948 and 1958, Varian grew from six to 1,300 employees, later moving into the development of microwave tubes for the military aerospace industry. By the mid-1960s, Varian had a larger workforce than either Fairchild Semiconductor or Hewlett-Packard.

Other early Silicon Valley companies also grew out of Cold War defense procurement needs, and many benefited from Terman's efforts. For example, Hiller Aircraft Company moved its plant to Palo Alto in 1947 to make use of concrete slabs Stanford provided the company for helicopter testing. Hiller produced light military helicopters and received many R&D contracts for a variety of military aerospace applications. Ampex, known for civilian audio and videotape recorders, derived half of its sales in the 1950s from the aerospace industry, which used Ampex recorders to record missile test, instrument, and satellite data, to mention only a few of their defense-related uses. Hewlett-Packard in the 1950s also subsisted on military contracts for specialized instruments to test components for weapons used in the Korean War.

Lockheed, based in Burbank, California, came to Stanford to lease 22 acres in 1956 for its Missiles Division as well as a much larger site near Moffett Field in Sunnyvale. Lockheed won the contract for the Navy's Polaris fleet ballistic missiles, and its work on spy satellites made the company responsible for most photo-surveillance during the Cold War. By 1964, Lockheed was Silicon Valley's largest industrial employer, with 24,000 employees. One of the main contributions of Lockheed to the future of the valley was the company's development of advanced, large-scale, clean-room technology. The manufacturing of vacuum tubes, which had been going on in the region for decades, required a clean environment, but nothing compared to what the aerospace industry needed. The requirement for absolute cleanliness in the construction of missiles and satellites, where even the smallest particles could result in malfunctions in space, led to great strides in that area and provided an infrastructure for the manufacture of silicon semiconductors, which also needed clean-room technology.

The accounts of Silicon Valley origins that emphasize the role that Fred Terman played tend to gloss over the importance that the military had on Terman and his outlook toward university–industry collaboration. In part, that is because his role as Director of the Radio Research Laboratory at Harvard during World War II was a top-secret operation and not common knowledge at the time. Terman was brought to Harvard in 1942 to oversee a large effort to develop jamming devices for radar and tunable radar receivers. Later in his life he recalled the impact of his experiences during the war: "The war had made it obvious to me that science and technology are more important to national defense than masses of men. The war also showed how essential the electron was to our type of civilization."[10] Hardly had Terman been back in Stanford before the Office of Naval Research approached him with grant money for defense-related research projects, one of which resulted in a Nobel Prize for nuclear magnetic resonance. A few years later, in 1950, the Office of Naval Research came with an even more lucrative contract, which resulted in the creation of a new Applied Research Laboratory (AEL), which pursued work on classified contracts. The basic science behind the electronics research conducted under these auspices had by no means only military-related applications, and consumer-based industries benefited together with the military. However, before the 1960s, it was infusion of defense monies into Silicon Valley that spurred many early innovations.

THE VALLEY, THE AIR FORCE, AND THE SEMI-CONDUCTOR

The last of the Silicon Valley origin stories also has a military leitmotif. In this version, it is the arrival of William Shockley in 1955 to take up the directorship of the Shockley Semiconductor Laboratory run by Beckman Instruments in Mountain View, California, that sets in motion the developments that would lead to the emergence of Silicon Valley. Shockley was the co-inventor of the transistor, for which he and his partners won a Nobel Prize in Physics in 1956. He also came with significant military-related experience during World War II, for instance as Research Director at Columbia University's Anti-Submarine Warfare Operations Group. Shockley was a difficult boss, who caused most of his talented staff, the so-called "Traitorous Eight," to abandon his laboratory to set up their own shop in Palo Alto in 1957, with help from a New York investment bank and an East Coast military contractor.

In 1956, the Air Force began to encourage the move away from analog-based avionics systems to digital computers in its aircraft and missiles. This push caused the Air Force to demand that avionics firms use silicon transistors, since silicon met the more rigorous demands of airborne systems. In response to the demand for fast silicon components in the construction of jet-aircraft and missile-control systems, the new company founded by former Shockley employers, Fairchild Semiconductor, achieved several key breakthroughs in computer technology, including high-frequency transistors, the integrated circuit, and the planar process. The Air Force Minuteman project, with its stiff reliability requirements, was the most important driver of the latter two crucial innovations.

By the late 1950s, Fairchild had emerged as one of the biggest electronics component manufacturers in Silicon Valley, with military sales around $21 million in 1960. Fairchild's designs were rapidly picked up by other firms, such as Texas Instruments and Motorola, among many others. Fairchild's spectacular success, together with the amazing number of possible applications for its innovations, attracted venture capitalists to the region. Between 1960 and 1969, twenty-six silicon firms were founded in the valley, almost all of them by former Fairchild engineers and managers—the so-called "Fairchildren." In 1968, Robert Noyce and Gordon Moore founded Intel. Silicon Valley developments from the 1970s on—including the invention of the personal computer, the mouse, the graphical user interface, and the relational database—were driven more by consumer than military demand. The origins of Silicon Valley, however, cannot be understood without factoring in the role of military spending and defense procurement, but the ongoing success of the region, and other high-tech concentrations such as Phoenix's Valley of the Sun, or Austin's Little Silicon Valley has more to do with their evolution away from reliance on defense contracting. Nevertheless, the Reagan years continued to reward the Silicon Valley with defense dollars. In 1988, San Jose was the metropolitan region awarded the largest dollar amount of defense contracts per capita in the USA.

DEFENSE INDUSTRIES AND THE WEST SINCE THE 1990S

With the end of the Cold War in 1989, funding for defense purposes collapsed just as the percentage of employment in the aerospace industry to total US manufacturing employment reached its apex during the final years of the Reagan defense buildup. In 1989, aerospace employed 1.3 million workers, representing almost 7 percent of all US manufacturing employment. Six short years later, the number of aerospace workers numbered around 800,000 people, representing less than 4 percent of all manufacturing employment. Not all regions with significant defense spending were equally hard hit: California topped the list of major job losses in the aerospace, electronics, and defense industries. To some degree, the downturn in areas such as Los Angeles and Orange County represented a long-anticipated move away from dependency on defense spending toward a more diversified economy. Still, the impact on employment and on the hard-hit firms themselves was long-lasting.

Aerospace firms tended to cope with the contraction of defense dollars in several ways. Some attempted to develop markets among private companies as well as the government. Other firms tried to find consumer uses for their product lines. Probably the most widespread result was a reorganization of the industry as companies spun off divisions or merged with other companies. Some of the changes associated with this downturn included Grumman's merger with Northrup, Lockheed's merger with Martin, and Boeing's acquisition of both McDonald Douglas and North American aerospace operations from Rockwell. When the dust settled, only Lockheed Martin and Boeing remained major military contractors. Aerospace ceased to find its center in southern California.

During the Gulf War (1990–1991) and the wars in Afghanistan and Iraq since 2002, the role of defense spending in shaping the American West has continued, but in a much more muted way than during the World War II and Cold War years. On the one hand, since September 11, 2001, the USA has gone through one of the largest military buildups in the country's history. In 2004, the Department of Defense awarded $203 billion to defense contractors in all fifty states. But the distribution of these dollars has not always benefited Western states. An exception is Texas, whose share of all US defense contracts rose from 7 percent to 10 percent in 2004, largely at the expense of contracts that had gone to California and Virginia in the past. The bulk of these Texas defense dollars came in the form of a forty-year, $200 billion contract with Lockheed Martin for the Joint Strike Fighter program. Leaving aside Texas, most of the rest of the West has seen a declining share of defense spending in the twenty-first century. As the President of the Greater Colorado Springs Economic Development Corporation explained, "We don't build tanks. We don't manufacture battleships. If you are having wars, we don't manufacture a lot of 'consumables' here."[11] Nevertheless, the transformation that decades of defense spending has wrought on the Western landscape in the form of sprawling strip cities, suburbanization, concentrations of white-collar scientific labor, and high-tech consumer industries continues to define a region that might otherwise have remained an economic backwater.

NOTES

1. Ann R. Markusen, *The Rise of the Gunbelt: The Military Remapping of Industrial America* (Oxford: Oxford University Press, 1991), p. 89.
2. Gerald D. Nash, *The American West in the Twentieth Century: A Short History of an Urban Oasis* (Albuquerque, N. Mex.: University of New Mexico Press, 1984), p. 205.
3. Roger D. Launius, "End of a 40-Year War: Demobilization in the West Coast Aerospace Industry After the Cold War," *Journal of the West,* 36 (3) (1997): 85–96.
4. Dwight D. Eisenhower, Military-Industrial Complex Speech, 1961, available online at http://avalon.law.yale.edu/20th_century/eisenhower001.asp (accessed July 7, 2009).
5. Marco Casillas, "Theodore Von Karman," available online from www.csupomona.edu/~nova/scientists/articles/vonk.html (accessed April 28, 2010).
6. Ann R. Markusen, *The Rise of the Gunbelt: The Military Remapping of Industrial America* (Oxford: Oxford University Press, 1991), p. 92.
7. Jacob Neufeld, "Technology Push," available online at www.history.navy.mil/colloquia/cch9c.html (accessed April 28, 2010).
8. Fred R. Shapiro, *The Yale Book of Quotations,* New Haven, Conn.: Yale University Press, 2006, p. 841.
9. Margaret Pugh O'Mara, Cities of Knowledge: Cold War Science and the Search for the Next Silicon Valley (Princeton, NJ: Princeton University Press, 2005), p. 43.
10. Carroll W. Purcell, Technology in America: A History of Individuals and Ideas (Cambridge, Mass.: MIT Press, 1981), p. 284.
11. Thomas Hargrove, "Defense Spending Increases to States Uneven" (2005), available online at www.globalsecurity.org/org/news/2005/050224-defense-spending.htm (accessed April 28, 2010).

FURTHER READING

Abbott, Carl, *The Metropolitan Frontier: Cities in the Modern American West* (Tucson, Ariz.: University of Arizona Press, 1993).

Bauer, E. E., *Boeing: The First Century and Beyond,* 2nd edn (Issaquah, Wash.: TABA, 2006).

Bilstein, Roger E., "From Colony to Commonwealth: The Rise of the Aerospace Industry in the West," *Journal of the West,* 36 (3) (1997): 8–20.

Boyne, Walter J., "The Rise and Fall of Donald Douglas," *Air Force Magazine* (March 2006), available online at www.afa.org/magazine/march2006/0306douglas.asp.

Breihan, John R., "From Amusements to Weapons: The Glenn L. Martin Aircraft Company of California, 1910–1917," *Journal of the West,* 36 (3) (1997): 29–38.

Bright, Charles D., "The 'Seattle Air Command': A Case Study in Industrial Siting," *Journal of the West,* 36 (3) (1997): 53–59.

Bugos, Glenn E., "The Aerospace Impetus to Silicon Valley," *Journal of the West,* 36 (3) (1997): 97–104.

Butler, Anne M., and Lansing, Michael, *The American West: A Concise History* (Malden, Mass.: Blackwell, 2008).

Casillas, Marco, "Theodore Von Karman," available from www.csupomona.edu/~nova/scientists/articles/vonk.html.

Eisenhower, Dwight D., Military-Industrial Complex Speech, 1961, available from http://avalon.law.yale.edu/20th_century/eisenhower001.asp.

Etulain, Richard W., and Szasz, Ferenc Morton, *The American West in 2000: Essays in Honor of Gerald D. Nash* (Albuquerque, N. Mex.: University of New Mexico Press, 2003).

Etulain, Richard W., *The American West in the Twentieth Century: A Bibliography* (Norman, Okla.: University of Oklahoma Press, 1994).

Ewert, Eric C., "From Hinterland to Bellwether: A Century of Demographic and Economic Change in the Pacific Northwest," *Journal of the West,* 45 (1) (2006): 41–51.

Fernlund, Kevin J., *The Cold War American West, 1945–1989* (Albuquerque, N. Mex.: University of New Mexico Press, 1998).

Foster, Mark S., *Henry J. Kaiser: Builder in the Modern American West* (Austin, Tex.: University of Texas Press, 1989).

—— "Giant of the West: Henry J. Kaiser and Regional Industrialization, 1930–1950," *Business History Review,* 59 (1) (1985): 1–23.

Guillén, Reynal, "The Air Force, Missiles, and the Rise of the Los Angeles Aerospace Technopole," *Journal of the West,* 36 (3) (1997): 60–66.

Hargrove, Thomas, "Defense Spending Increases to States Uneven," Globalsecurity (2005), available online at www.globalsecurity.org/org/news/2005/050224-defense-spending.htm.

Hevly, Bruce William, and Findlay, John M. *The Atomic West* (Seattle, Wash.: University of Washington Press, 1998).

Hunner, Jon, *Inventing Los Alamos: The Growth of an Atomic Community* (Norman, Okla.: University of Oklahoma Press, 2004).

Johnson, Marilynn S., "War as Watershed: The East Bay and World War II," *Pacific Historical Review,* 63 (3) (1994): 315–331.

Kenney, Martin, *Understanding Silicon Valley: The Anatomy of an Entrepreneurial Region* (Stanford, Calif.: Stanford University Press, 2000).

Launius, Roger D., "End of a 40-Year War: Demobilization in the West Coast Aerospace Industry after the Cold War," *Journal of the West,* 36 (3) (1997): 85–96.

—— "A New Way of War: The Development of Military Aviation in the American West, 1908–1945," *Journal of the American Aviation Historical Society,* 41 (3) (1996): 220.

Lécuyer, Christophe, *Making Silicon Valley: Innovation and the Growth of High Tech, 1930–1970* (Cambridge, Mass.: MIT Press, 2006).

Lee, Chong-Moon *The Silicon Valley Edge: A Habitat for Innovation and Entrepreneurship* (Stanford, Calif. : Stanford University Press, 2000).

Lotchin, Roger W. *The Way We Really Were: The Golden State in the Second Great War* (Urbana, Ill.: University of Illinois Press, 2000).

—— "California Cities and the Hurricane of Change: World War II in the San Francisco, Los Angeles, and San Diego Metropolitan Areas," *Pacific Historical Review,* 63 (3) (1994): 393–420.

—— *Fortress California, 1910–1961: From Warfare to Welfare* (Oxford: Oxford University Press, 1992).

Markusen, Ann R., *The Rise of the Gunbelt: The Military Remapping of Industrial America* (Oxford: Oxford University Press, 1991).

Marrett, George J., *Howard Hughes: Aviator* (Annapolis, Md.: Naval Institute Press, 2004).

Milner, Clyde A., O'Connor, Carol A., and Sandweiss, Martha A., *The Oxford History of the American West* (Oxford: Oxford University Press, 1994).

Nash, Gerald D., *The Federal Landscape: An Economic History of the Twentieth-Century West* (Tucson, Ariz.: University of Arizona Press, 1999).

—— *World War II and the West: Reshaping the Economy* (Lincoln, Nebr.: University of Nebraska Press, 1990).

—— *The American West in the Twentieth Century: A Short History of an Urban Oasis* (Albuquerque, NM: Univ. of New Mexico Press, 1984).

Nash, Gerald D., and Etulain, Richard W., *Researching Western History: Topics in the Twentieth Century* (Albuquerque, N. Mex.: University of New Mexico Press, 1997).

Neufeld, Jacob, "Technology Push," (1997) available online from www.history.navy.mil/colloquia/cch9c.html.

O'Mara, Margaret Pugh, *Cities of Knowledge: Cold War Science and the Search for the Next Silicon Valley* (Princeton, NJ: Princeton University Press, 2005).

Purcell, Carroll W., *Technology in America: A History of Individuals and Ideas* (Cambridge, Mass.: MIT Press, 1981).

Rhode, Paul, "The Nash Thesis Revisited: An Economic Historian's View," *Pacific Historical Review,* 63 (3) (1994): 363–392.

Ruttan, Vernon W., *Is War Necessary for Economic Growth: Military Procurement and Technology Development* (Oxford: Oxford University Press, 2006).

Scott, Allen J., "The Aerospace-Electronics Industrial Complex of Southern California: The Formative Years, 1940–1960," *Research Policy,* 20 (5) (1991): 439–456.

Sell, T. M., *Wings of Power: Boeing and the Politics of Growth in the Northwest* (Seattle, Wash.: University of Washington Press, 2001).

Sturdevant, Rick W., and Spires, David N., "Mile-High Ventures: Highlights from Colorado Aerospace History, 1923–1997," *Journal of the West,* 36 (3): 67–77.

Tate, Michael L., *The Frontier Army in the Settlement* (1997) *of the West* (Norman, Okla.: University of Oklahoma Press, 1999).

Vander Meulen, Jacob, "World War II Aircraft Industry in the West," *Journal of the West,* 36 (3) (1997): 78–84.

Verge, Arthur C., "The Impact of the Second World War on Los Angeles," *Pacific Historical Review,* 63 (3) (1994): 289–314.

CHAPTER FIFTEEN

THE RISE OF WESTERN WOMEN TO NATIONAL STATURE IN THE EARLY TWENTIETH CENTURY

Shannon D. Smith

The twentieth century saw more change in everyday life than any previous 100 years. The industrialization and expansion of domestic and international trade that began in the nineteenth century accelerated geometrically—bringing economic and technological change that was unprecedented in human history. In 1900, automobiles, electric light bulbs, and telephones were just becoming available in the cities of the USA and had yet to be mass produced. Women were dressing much as they had throughout the 1800s with corsets defining their shape and full dresses covering their arms and legs. The ankle-length hemlines, high collars, and long sleeves of the Victorian era would be around for nearly two decades into the twentieth century. When my grandmother was born in Missouri in 1900, people traveled the West by steamship, rail, horse, and wagon, and when she moved to rural Nebraska as a young girl, she had yet to see her first car. But transportation advanced quickly throughout the West, and a decade later, when she was twenty-two, she took her first flight in an airplane when her new husband bought her a fly-over with an aerial troupe who was barnstorming through northern Nebraska. A few years later, she was surprised to read that the young pilot she rode with, Charles Lindbergh, had just completed the first solo transatlantic flight.

In the course of her ninety-one years, my grandmother not only witnessed man fly across the open sea, she also saw men walk on the moon and space shuttles make regular trips out of the earth's atmosphere. She also saw those Victorian hemlines give way to the knee-baring, convention-defying dresses of the flappers in the Roaring Twenties as they shunned traditional women's fashion and decorum. By the time mini-skirts came into vogue, she was in her sixties and still resisting wearing pants, which she did well beyond their introduction and common use. In addition to fashion, women's lives changed in immeasurable ways—from voting to employment opportunities to controlling their reproductive cycle—as a result of their gradually increasing presence in the public arena. A lot of these changes came about out of necessity. In many cases, women had to work to support their families, either their parents while they were young or their own children when they married. World wars brought women into the workforce in great numbers but when the wars were over, conservative forces in society sought to return women to their former domestically centered realm. Churches and social groups, including some women's clubs and organizations, vehemently opposed changing women's new roles, while Progressive Era reformers and suffragists challenged the status quo.

Perhaps the most astonishing change my grandmother experienced was in politics. When she turned twenty-one, the Nineteenth Amendment, granting women the right to vote, was barely a year old, and the battle to pass women suffrage in Nebraska had raged for nearly a decade. Imagine how surprised she would have been at that time if she had been told she would one day serve as the Republican Party Chairperson for her county and that she would witness a woman become governor. In fact, in 1987, Nebraska made history for not only electing the first Republican woman governor in the USA but also for hosting the first gubernatorial race ever held between two women. The pace of change in the country was spectacular in many ways, and it is easy to visualize the dramatic change in women's roles, but just how fast was this change for the women who experienced it? It wasn't until she was in her late eighties that society had changed enough for my grandmother to witness two women run against each other for governor of Nebraska. Though we can marvel at the changes in American life and women's roles that occurred in the twentieth century, change in the day-to-day life of women was not nearly as dramatic. Was my grandmother freer in her fifties than in her teens and twenties? When *did* she break from the constraints of the Victorian model of womanhood? We know the timing of change differed greatly between regions of the country as well; as a woman of the West, was her experience different from women of the East? Was this change in societal views shaped by the integration of the West into the national psyche? While the narrative of American history has been revised over the years to include women, there is no real agreement on whether Western women's roles can be explained as completely unique by virtue of their geography or whether they were nearly the same as their counterparts in the East.

One way to explore these competing narratives is to look at the story of how Western women rose to national prominence in politics and whether their experience was somehow different from women in other areas of the country. In this case, "West" is used as the territory west of the Mississippi River—nearly two-thirds of the nation's land mass—because the development of European settlement west of the Mississippi differed dramatically from the earlier settlement of the USA. However, within this huge region there are wide variations of geography, economy, and human diversity. If Iowa is considered part of the West, how different was a twentieth-century woman's life there from another Western state such as New Mexico? From a political standpoint, in 1900, eight of today's states west of the Mississippi weren't even states, and four had ratified statehood barely ten years earlier. North Dakota, South Dakota, Montana, and Washington were admitted to the Union at the end of 1889, and during the next six decades, Idaho, Wyoming, Utah, Oklahoma, New Mexico, Arizona, Alaska, and Hawaii would become states. Territorial governments and fledgling states governed half the landmass west of the Mississippi for most of the first half of the twentieth century.

In addition to the diversity of government structures and the vast range of geographic environments in which women built their lives, a defining factor of their day-to-day experience was whether they lived in rural or urban environments. In 1890, the US Census Bureau proclaimed the American frontier closed:

> Up to and including 1880 the country had a frontier of settlement, but at present the unsettled area has been so broken into by isolated bodies of settlement that there can

> hardly be said to be a frontier line. In the discussion of its extent, its westward movement, etc., it cannot, therefore, any longer have a place in the census reports.[1]

Prior to 1890, census maps defined the frontier line as a line beyond which the population was under two persons per square mile. A University of Wisconsin historian, Frederick Jackson Turner, developed this statement into his now-famous "frontier thesis" and explained it in a speech entitled "The Significance of the Frontier in American History," which he presented before the American Historical Association at the World's Columbian Exposition in Chicago in 1893. Turner proposed that the frontier that lay just west of Euro-American settlement for the first 300 years of the USA's development shaped the social character of the nation and was key to understanding the American identity. According to Turner, this frontier produced a man totally distinct from his European ancestors: coarse, strong, practical, inquisitive, and most especially *independent,* exhibiting "that buoyancy and exuberance which comes with freedom."[2]

Turner's thesis maintained that unlimited free land in the frontier offered the psychological sense of unlimited opportunity and provided a "safety valve" that protected the nation against uprisings of the impoverished. Turner's so-called "gateway of escape" theory posited that the frontier provided a place of refuge for people who could not survive in the crowded, settled Eastern lands. The free, abundant lands of the West provided a ready haven for the discontented, where one could theoretically start over unfettered by the constraints of society. Scholars now know that many factors influenced American culture besides the looming frontier, and Western historians in particular have moved beyond the idea that the West was a frontier process and now view it not as a moving line of encroaching civilization but as a vast *place* with a rich and incredibly diverse history that continues to today—well beyond the nineteenth-century "Old West" that comes to most minds when thinking about Western history.

In addition to acknowledging that Western history encompasses a place more than a frontier process, historians have recognized that the Turner thesis completely overlooked the existence of women, American Indians, and the countless other peoples who lived alongside the coarse, strong, practical, and independent Euro-American men of whom he wrote. Like Turner's thesis, the dramatic story of the frontier West that was popularized in novels, movies, and television shows was filled with Euro-American male heroes: cowboys, fur-trappers, American Indian fighters, sheriffs, and the many rugged conquerors of the wilderness. The few women who actually made it into the narrative were the unusual or noteworthy women who broke from the traditional women's role—women such as Calamity Jane, Annie Oakley, Poker Alice, Molly Brown, and Klondike Kate. Later, women were idealized as the "madonnas of the plains," the pioneer women who stuck by their men and domesticated the frontier. For the last few decades, Western women's historians have been deprogramming this narrative and have revealed the human diversity of the West, a story now peopled with men and women of a vast range of ethnicities, religions, and economic classes. The region west of the Mississippi was not a barely populated virgin territory slowly "tamed" by the westward march of Euro-American pioneer men and women; it was a place that had been settled for centuries by indigenous and Hispanic peoples.

If we delineate "West" as a place rather than the frontier-process concept, we can more readily see that the history of this huge *place* contains a vast array of cultural experiences for women. It is also necessary to delineate the era under discussion, because, even though "twentieth century" is an expedient temporal demarcation, we must keep in mind the enormous change that occurred between 1900 and 2000. Life in the American West was considerably different in the middle of the century than it was at the beginning or end. As the great women's historian Glenda Riley points out,

> the term *twentieth century* is more convenient for historians writing about the era than it was for the women living it. Women did not necessarily change their ways or ideas in 1900. Consequently, a historian of the twentieth century has to reach backward into the nineteenth century to understand more recent trends, and often has to compare patterns in the two centuries to understand twentieth-century developments.[3]

Acknowledging the nearly immeasurable range of experiences of women who lived in the enormous place we are calling the West across the century that was itself nearly immeasurable in terms of social change, historians continue to analyze their lives in search of generalizations, and there is much left to learn. To explore whether Western women rose to national prominence sooner than women in the East because of the unique nature of the region, we will first look at the predominant social norms that most women in America, including the West, were subject to at the beginning of the twentieth century. Then we will use the lens of public politics to reveal how, when, and if they changed over the next fifty years.

In 1831, two wealthy young Frenchmen, Alexis de Tocqueville and his friend Gustave de Beaumont, spent nine months exploring the USA and studying the social aspects of the nascent democracy. With the sanction of their government and money of their families in hand, they were warmly received throughout their travels from Canada to the Carolinas and west through the unsettled regions of Michigan and Wisconsin down to Tennessee and New Orleans. Throughout their journey they met with some of the most elite, prominent people of the early nineteenth century, including presidents, lawyers, businessmen, backcountry settlers, and wilderness entrepreneurs, the so-called "rugged individualists" of Turner's ever-shifting frontier region that purportedly shaped the character of the USA. Tocqueville, a truly gifted observer with a keen eye for detail, recorded hundreds of observations on America's social and political institutions and reported meticulously on the structure of government and the judicial system. The book that resulted from Tocqueville's journey, *Democracy in America,* is a classic of American history, launching a dialogue on democracy and society in the USA that remains relevant today. His observations about women's roles in America, recorded in four chapters of the second volume, were revived and analyzed in the years after World War II, and Tocqueville has since been recognized as one of the first social critics to seriously examine the situation of women in the USA.

In the course of describing American women of the era, Tocquville spent considerable time analyzing the separation of male and female roles. He proposed that the absence of a European-style aristocratic government weakened traditional patriarchal authority and afforded young American women a high degree of independence and self-confidence—these young women were much more liberated from the social constraints of their European counterparts. However, Tocqueville observed that when these women married, traditional American values and mores limited her role, "within the narrow circle of domestic interests and duties and forbids her to step beyond it."[4] This statement evolved into an interpretation of women's roles in nineteenth-century America known as the "separate spheres" theory, a set of ideas that assigned distinct and essentially opposite roles, duties, personal characteristics, and "spheres" of activity to women and men. This theory posited that women were naturally unfit for public roles in business and government activities due to their delicate constitution and higher morals. Wives, who had participated in, and even dominated, economic roles in family enterprises in the early colonial era, now left earning a living to the husband. Men were exposed to the morally corrupting public sphere in the process of earning that living, while women were devoted to the private sphere of caring for her husband, children, and home—and her virtue and morals remained protected. The Victorian ideals of the "cult of domesticity" or "cult of true womanhood" espoused that women were at the center of the private/family sphere and that she embodied the virtues of piety, purity, submission, and domesticity.[5]

At the time, this Victorian sentiment was viewed as privileging and respecting women for their duties of bearing and promoting the virtue and honor of the nation. These ideals were emphasized by ministers in sermons, physicians in health books, and in popular women's magazines. In 1860, *Godey's Lady's Book* proclaimed,

> There is the wife and mother, the centre of the family, the magnet that draws man to the domestic altar, that makes him a civilized being, a social Christian. The wife is truly the light of home, and if she will do faithfully the duties God has imposed on her condition, He will bless and sustain her.[6]

Indeed, even Tocqueville lauded this separation in his book's conclusion, saying

> For my part, I say this without hesitation: although the American woman rarely leaves her domestic sphere and in certain respects is very dependent within it, nowhere does she enjoy a higher status. And now, as I come near to the end of this book in which I have recorded so many considerable achievements of the Americans, if I am asked how we should account for the unusual prosperity and growing strength of this nation, I would reply that they must be attributed to the superiority of their women.[7]

Though this ideology was clearly aimed at Euro-American, middle- and upper-class women, and though historians are now debating the validity of the "binaric version of nineteenth-century American history" represented in the idea of the "separate spheres," this vision of the ideal woman's role in American society prevailed throughout most of the nineteenth century and well into the twentieth century.[8] In fact, on May 9, 1908, the US Senate debated and rejected a Bill to establish a "Mother's Day" national holiday partially on the grounds that motherhood was too sacred to be demeaned by a day in its honor.[9] And it would take another decade for the Senate to pass the Nineteenth Amendment granting women the ultimate public-sphere privilege: the right to vote.

Despite the fact that at the dawn of the twentieth century women activists and their male allies were advocating for—and predicting—more rights for women, in 1901 gender equality remained almost as bizarre a concept as it was in 1800. In 1900, American women were not only precluded from voting, they could not serve on juries or run for elective office, and their economic situation fundamentally depended on their marital status. The overwhelming majority of adult women were married, and their legal, economic, and social identity was subsumed into their husbands' under the traditional legal theory of marriage known as coverture. Under traditional English common law, an adult unmarried woman was considered to have the legal status of feme sole, while a married woman had the status of feme covert, literally "covered woman." A feme covert had no legal rights and obligations distinct from her husband's. Through marriage, a woman's very existence was incorporated into that of her husband, and she had very few recognized individual rights of her own. While a feme sole had the right to own property and make contracts in her own name, single women were primarily young women who were taken care of by their fathers and would come under the economic protection and "cover" of their husbands in the near future.

How did the traditional values of the "cult of domesticity" and coverture translate in the West? Did women adapt the conventions of proper female mores to their world or did they focus on conforming to the domestically based social norms of the East? Were women in the West more independent? Logic dictates that the unique tribulations of pioneer women—isolation, heavy labor, lack of medical, shopping, and educational facilities to name a few—would have an impact on the character and values of Western women and their daughters. If these women were more exposed and responsible, and, dare we say, *free* than their Eastern counterparts, did this freedom translate into different social values in the West? One way to compare the level of freedom experienced by women of the West is to examine their opportunities for education. By the twentieth century, the West had substantially more coeducational institutions of higher learning and had dominated the East in numbers of coeducational universities for three decades. One reason the West educated more women alongside men was the sheer practicality of developing institutions that would educate both rather than duplicating sex-segregated schools or classes. These new institutions could attract more students, and income, if they accepted women, plus there was great demand for teachers in these rapidly growing regions. Additionally, these new territories and states had less entrenched gender-based laws and traditions to overcome, and some thought-leaders argued that the presence of women in the classroom would have a positive influence on the rakish, rough-and-tumble atmosphere of a male-only college. In her study of women and coeducation in the American West, Andrea Radke-Moss writes that that the difference between mixed-gender education in the West and East had as much to do with the fact that "western land-grant university students and administrators viewed their own experience as unique and set themselves against the stereotypes of eastern institutions for their conservative and more rigid restrictions on social interactions."[10]

Another measure of the difference between Western and Eastern women at the turn of the twentieth century is reflected in women's participation in the political and economic arena. The movement for women's rights began in the East, but it was in the West that the first significant successes were realized. A common misperception of the drive for women's rights, particularly women's right to vote, is that it was a small, tenacious, doggedly determined set of women from the East who prevailed against the odds and persisted until men finally relented and "gave" them the vote. In reality, the fight for suffrage was a mass movement encompassing millions of American women over several generations throughout the entire nation. The movement employed highly sophisticated political strategy and organization, including some of the first organized protest movements, and developed several organizers who are today recognized as politically savvy and highly effective leaders. The movement is generally believed to have its beginning when women started participating in the anti-slavery movement in the mid-1830s. After the 1840 World Anti-Slavery Convention in London rejected the credentials of American delegate Lucretia Mott and other female American delegates, Mott and Elizabeth Cady Stanton, determined to take up the cause of women's rights. In 1848, they called a convention in Seneca Falls, New York, out of their anger with the patriarchal system of the male abolitionists—they had begun to see similarities between their own circumscribed status and that of the slaves—and this meeting energized a movement to change women's legal and political position.

These activities persisted through the 1850s, but when the Civil War broke out these activist women turned their energies to supporting the war. After the war, suffragists, assuming their patriotic war work and campaigning against slavery would be recognized, were sorely disappointed. When Congress ratified the Thirteenth, Fourteenth, and Fifteenth amendments to the Constitution that abolished slavery, conferred citizenship on African Americans, and granted the vote to African-American men, they specifically ignored women. It would be five more decades of fighting for women's rights before the Nineteenth Amendment granted full suffrage to women.

A series of events at the turn of the century revitalized the suffrage movement. Generally described as the "Progressive Era," the years from the 1890s through to the 1920s brought new life to many types of reform movements, including the women's-rights movement. This era was characterized by a loose coalition of people and organizations around the nation trying to find solutions to social and political problems. It was a turbulent period dominated by the volatile and seemingly unrelated issues of prohibition, woman suffrage, and America's growing involvement in the European conflict that would become World War I. These controversial social issues were inextricably intertwined and shaped politics and society in the West, across the nation, and around the world as they came to a head. Millions of women from all ethnic, class, and racial backgrounds became involved with public life to address severe social problems through advocating creative new reforms. As their roles in society expanded, women realized that virtually every reform they sought was regulated by law, and the legislators who passed these laws had to respond to voters. It became apparent that the social policies they were encouraging would be achievable only if women had the right to vote and so, at the dawn of the new century, the question of woman suffrage moved squarely into mainstream politics.

Interestingly, women's own acceptance and thorough indoctrination of their "cult of domesticity" cultural role was the very cause of their gradual escape from all things private and domestic into the exclusively male "public sphere" of politics. The Victorian idealization of female purity and the Darwinian conclusion that women were innately morally superior to men supported and promulgated the belief that women were the moral guardians of home. This perceived moral superiority in the home eventually translated into women's involvement in external matters dealing with social reforms. In their own minds as well as their husbands,' the mothers of the nation were expected to be the purveyors of proper religious, moral, and cultural values to their children and the inspiration for all good works in society. This opened the door for women to organize and attempt to influence community and civic affairs—women participating and hoping to be leaders in the abolition movement is one example, but women's clubs and church auxiliary organizations were the more socially accepted ways for women to learn about, and practice in, the public sphere. Management of extensive administrative organizations and budgets, interaction with political leaders and city administrators, public speeches, traveling to national conventions, and massive public-relations campaigns via newspapers and pamphlets are just some of the techniques that "club" women learned and used. These responsibilities took women far beyond their traditional submissive and domestic role but were rationalized as befitting the late-Victorian-era woman's obligation to "save" society from corruption and immorality.

This was the same strategy suffrage activists employed—expanding the "social housekeeping" concept by claiming that society would be uplifted by woman's higher moral nature and that women's vote would infuse morality into the social political system and even the economic system of the "public sphere." This gradually led to their desire for acknowledgment of their social efforts through the granting of important rights that affected women and their families. In Westering Women and the Frontier Experience, 1800–1915, Sandra L. Myres writes that American women's-rights activists were much more practical than is commonly believed, and, with a few exceptions, they were initially more interested in property rights, education, employment, divorce laws, and guardianship than they were in women's suffrage, but the movement for general rights and suffrage slowly gained momentum—particularly in the West.

> During the pre-Civil War decades much of the attention of the women's rights advocates was centered in the newer Western areas where the leaders hoped that a freer, more egalitarian society might be more willing to accord rights to the women who had helped to open and settle the country.[11]

Indeed, when the Nineteenth Amendment, granting suffrage to women, was approved by Congress in 1920, women had been voting in Wyoming for over thirty years. The next eight states to grant full suffrage to women were also Western states: Colorado (1893); Utah and Idaho (1896); Washington (1910); California (1911); and Oregon, Kansas, and Arizona (1912).

In searching for a reason why the West was first in many aspects of women's rights—particularly suffrage—it seems that there will likely never be a commonly accepted theory and that it is more likely that women's suffrage succeeded in the West for reasons as diverse as the people and places of the West itself. In 1869, the twenty-member Territorial Legislature of Wyoming approved a revolutionary measure stating: "That every woman of the age of twenty-one years, residing in this Territory, may at every election to be holden under the law thereof, cast her vote."[12] There are many theories as to why Wyoming was the first to grant women suffrage; some espouse that men granted the right to women for their own manipulative, rather than egalitarian, reasons—for instance hoping to attract women to the male-dominated territory. It has become axiomatic that the American frontier West was always more liberal than the so-called civilization of the East. It makes sense that the people who went to frontiers were the adventurous type and were amenable to new ideas, were less afraid of change, and rarely had any stake in the status quo. Women endured the same trials and tribulations as men—and then some—so the notion of the separate "sphere" didn't ring as true. It has also been proposed that with fewer women in these newly settled areas, men were more eager to please them. Perhaps all of that made men more likely to see women as similar enough to themselves to be capable of casting a ballot. Other theories have posited that Wyoming's action was a publicity stunt, a political ploy to advance partisan causes, even a panic-driven measure to counteract votes of newly enfranchised African-American men. The only thing that is clear is that, whether or not women activists pulled this off through their own efforts, there was no organized suffrage campaign, parade, debate or public display of any type other than women keeping vigil outside Governor John A. Campbell's office until he signed the Bill into law. After that, Wyoming women eagerly embraced their right to vote and would subsequently defend it against all threats.

After Wyoming, excitement for women's rights spread quickly through many parts of the West. The women's-rights activist and newspaper publisher Caroline Nichols Churchill declared, "This Shall be the Land for Women!" when Colorado became the first state to grant suffrage by popular vote in 1893.[13] The West was soon viewed as the symbol of women's enfranchisement and political equality, as the next seven states to approve women's suffrage were in the West, but it would still take two full decades for the nation, and the remaining Western states, to resolve the complicated issue of women's rights to vote in municipal, state, and federal elections. Each state in the West had such unique characters, cultural values, and laws that drawing generalizations is quite difficult—imagine how different it must have been to debate women's rights in Utah, where they were inextricably linked to the issue of polygamy, and California, where great numbers of wage-earning women of many different ethnicities banded together to finally win the popular vote in 1911.

Male voters in Oregon defeated five attempted suffrage measures between 1884 and 1912, where the leading suffrage supporter, Abigail Scott Duniway, blamed her brother, Harvey Scott, who was the editor of *The Oregonian* and a vocal opponent of suffrage, for the string of defeats. California's governmental branches demonstrated just how contentious the issue was as the Governor vetoed a suffrage Bill that had been passed by the legislature in 1893. California suffragists took the measure straight to the electorate in 1896 but the measure was defeated by a consortium led by anti-prohibitionists. Several states granted the vote—or came close—while they were still territories. In Washington, the territorial legislature defeated a women's suffrage Bill by one vote in 1854—this would have made Washington the first to grant women the right to vote. For the next thirty years women had on-again-off-again voting rights, and in 1883 the territorial legislature passed women's suffrage, the third territory after Wyoming and Utah. Several changes in the law followed over the next three years, all to clarify and reaffirm women's suffrage—all while women were voting—but in 1887 Washington's territorial supreme court overturned the law on a technicality. The next year the legislature passed suffrage but the law was once again overturned by the court. In 1889, after Washington became a state, voters defeated suffrage by a large margin, and the measure was defeated again nine years later. Finally, voters approved full women's suffrage in 1910, twenty-seven years after women had full suffrage in the early days of the territory.

Utah followed a unique path to suffrage after the territorial legislature approved a measure in 1870 to enfranchise the more than 17,000 women who had settled in the state with their families. Women in polygamist marriages were strong supporters of the measure and fought to maintain their enfranchisement as they fought the anti-polygamy laws repeatedly passed by the US Congress. When the 1887 Edmunds–Tucker Bill was passed, in addition to outlawing polygamy, it disfranchised Utah women who were outraged and labeled it an affront not only to women's rights but also to states' rights and, most importantly, to religious freedom. Mormon leaders relented to the demands of the anti-polygamy legislators, but when the state's new constitution was drawn up and approved in 1895 it included woman suffrage. After 80 percent of the state's male electorate ratified the constitution in 1896, Utah became the forty-fifth state in the Union and only the second to grant women full suffrage. Other Western states, driven by different cultural values, took a longer, slower path to suffrage. While the American Indian women of many of the tribes living in New Mexico had powerful roles within their local matriarchal societies for years, and the primarily Hispanic Euro-American women of the region enjoyed property rights which had been guaranteed under the Mexican land-grant system, the strongly patriarchal social structures of the region which became part of the USA in 1848 made the enactment of any kind of women suffrage very difficult. In fact, New Mexico's legislature was one of the last Western states to ratify the Nineteenth Amendment. At the turn of the century, suffrage activity in Texas was still very rare. The women in Texas were more like their sisters from the South than the activists of the West, and suffrage campaigns garnered little support. In 1919, a suffrage referendum failed because, in addition to woman suffrage, it called for enfranchising "aliens," a code word for African Americans and recent immigrants. Despite the lingering Southern racial and gender sentiments, Texas was one of only two Southern states to ratify the Nineteenth Amendment when the thirty-sixth state finally ratified it into law in 1920. In fact, of the eleven states that seceded, eight did not ratify the law until after 1950 and the last, Mississippi, did not ratify until March 22, 1984.

The Western states' agitation for women's rights demonstrates how the polarizing issue of women's suffrage became increasingly entangled with the equally polarizing problem of alcohol prohibition at the turn of the century. Soon these issues would end up being influenced by America's involvement in World War I, creating a complex and volatile situation that had a tremendous impact on women's roles and exacerbated the confusing and capricious nature of women's suffrage in the West. To make sense of these issues, it is useful to view these events and their impact through the life of someone who experienced them. Katherine Worley Allen became one of the nation's first women to serve in a state cabinet position and was born at the exact right time in the exact right place to experience these dramatic cultural shifts. She came to adulthood just as women were breaking from their traditional Victorian roles, and, by virtue of her having grown up in a rural, Western homestead and the eventual necessity to support her parents when she was young, she stepped out of the private sphere at an early age.

Worley was born in 1876 in Cass County, Nebraska on the homestead her grandfather, James Madison Chalfant, settled in 1856. Like the other girls she grew up with, Worley was raised with the Victorian sensibilities of the era and came of age just as women's roles began to expand. She grew up in a sod house in the heavily wooded hills near the convergence of the Platte and Missouri rivers. In 1883, at the age of seven, she started country school, walking two miles each way. When she was eleven years old, Worley moved with her parents to Elmwood, Nebraska, and completed her education in the village school. Worley described herself as being more interested in the public sphere of business than the traditional woman's place in the private sphere even as a small child, preferring to "play store" rather than "play house" with dolls. Due to her father's poor health, Worley took a job at the Elmwood newspaper as a young teenager to help support her family. She earned $1.50 a week. After three years, Worley had learned how to set type and to write brief local news and was making $3 a week.[14] Her position as primary breadwinner and the exposure to the public world of business and politics through her newspaper work clearly broke with the conventional female domestic position at that time. Like many women of the West, the necessity of working through her early years gave Worley the confidence and desire to expand beyond traditional women's roles as she matured.

Still single and supporting her parents in her early twenties, Worley became even more independent-minded. She read a pamphlet describing press-clipping bureaus and decided to start her own enterprise in 1896. She gathered together exchanges from various sources and set up business on her mother's dining-room table. Years later she described her first operation: "My entire office force consisted of myself. I wrote all letters in long hand, read all of the papers and mailed the clippings, ran to the post office for mail, and was my own general manager and office boy."[15] Worley launched her business just as the presidential campaign of 1896 was taking off. In June, at the Democratic National Convention in Chicago, Nebraska's own William Jennings Bryan delivered his now famous "Cross of Gold" speech supporting a national economic policy that would benefit farmers and the rural businesses of the plains. Bryan subsequently became the first presidential candidate from west of the Mississippi River nominated by a national party, winning both the Democrat and Populist nomination to run against the Republican, William McKinley. Contemporaries declared the campaign of 1896 as the most exciting in the USA for many years, and in Nebraska the contest was described as fierce, close, and memorable. As the campaign heated up, "political items predominated" Worley's clipping business, and her work "seemed to develop into a political bureau."[16] She never mentioned whether being a woman in this political environment was regarded as unusual by colleagues, clients, friends, or relatives. Worley's immediate success was tied to the tremendous local and national interest in the campaign. Unfortunately, her prosperity was short-lived: "all good things must come to an end, and my vision at that age not being broad enough to see the great commercial side which is now a reality in the middle west, I went out of business with the close of the campaign."[17]

For the next seven years, Worley worked in the business end of operations for several newspapers in and around Omaha. Again, there is no record that she, or anyone else, considered her work in this public economic venture to be exceptionally unusual. In 1903, she opened a press-clipping bureau for an advertising agency owned by Charles D. Thompson. The company changed hands when Thompson died in 1906, so Worley took the service she had established for Thompson and formed her own business, Universal Press Clipping Services, based in Omaha. She was now thirty years old and still living with and supporting her parents. Though she was leading a nontraditional life as a single woman owning a commercial enterprise, her home life reinforced her view of women's roles—after the death of her sister, Worley raised her five nieces and nephews for twelve years. She was representative of the time in her belief in the "special" qualities of woman's nature and always described her business acumen and success in terms of her domestic inclinations and skills or emphasized that her feminine duties superseded her business obligations. While she ran a thriving business during the day, Worley took great pride in pointing out, "I know what it is to darn stockings, mend torn trousers, and sew on buttons at the midnight hour."[18] Her invocation of women's domestic skills to either explain or downplay her success in business illustrates the transformation of Victorian women's roles. For years, under the banner of "Social Housekeeping," women's clubs had been increasing woman's "proper sphere" by including the community and nation as a larger "home." By the turn of the century, the rhetoric of women's clubs and the suffrage movement emphasized woman's moral duty to expand their positive influence in the domestic sphere to inspire and perform good works in society.

As her business grew, Worley became involved in civic affairs. However, her involvement was through women's clubs, not the male-dominated commercial and political organizations of the business world in which she was so successful. Although Worley described herself as "not a joiner," she took her feminine obligation to perform civic and social services seriously. During the early 1910s, she became active in several local and national women's organizations as a way to contribute to her community and "keep in touch with alert minds of keen business women."[19] By 1917, she was Chairman of the Civics Committee for the Omaha Woman's Club and a delegate to the national meeting of the General Federation of Women's Clubs in New York City. She was responsible for investigating poor families and women her organization might be able to assist. This work took her on tours of inspection through the "so called Slum section of the city, to gather first-hand information regarding the Housing condition."[20] She would then work with various city and state departments, such as the Welfare Board, to improve the lot of individuals—especially children—she had encountered.[21] As Chairman of the Civics Committee, Worley ran a highly publicized "Clean-Up Campaign" for the city of Omaha in 1917. She called upon and wrote to business, religious, and political leaders throughout the city enlisting their support to clean up vacant lots and public places. Her rhetoric continued to reflect women's proper roles as she promoted the campaign to the male leaders she approached. She pointed out that as her committee's "Home Year" project they wanted to clean up their "home" of Omaha to make "healthier and happier citizens and more beautiful surroundings."[22]

Worley delivered speeches, wrote newspaper articles, and campaigned tirelessly for her causes. She was particularly interested in "Americanization" efforts designed to turn immigrant minorities into model citizens. This probably evolved from her exposure to the dreadful conditions many of the immigrants of Omaha were living in at the time. Between 1910 and 1920, Omaha's population increased by over one-third to nearly 200,000 residents, nearly of third of whom were reportedly of foreign birth. These immigrants were clustered in enclaves such as "Little Italy," "Little Poland," and "Little Bohemia" around the Omaha business district and stockyards.[23] Suspicion and anxiety over the growing immigrant population was fueled by prohibitionists, who associated the problems of drinking with the traditions of the foreign-born, as well as the outbreak of war in Europe. Well-intentioned reformers embraced "Americanization" as a way to solve mounting social tensions. Business, political, and religious organizations joined together to agitate for the indoctrination of immigrants with American social values. Women's groups, already active in social work with the poor and underprivileged, were the logical arms of the physical effort to Americanize immigrants into model Christian middle-class citizens. Worley was the Director of Americanization for the Omaha Woman's Club during World War I. This was a logical progression from her position as Chairman of the Civics Committee where her social work and the contacts she made managing the Clean-Up Campaign—which emphasized cleaning up immigrant neighborhoods—gave her the knowledge of the communities she was charged with "converting." Worley organized night schools to teach English, spoke at foreign-language churches and community groups, and was on the national committee that formed the resolutions and policy for Americanization work that was adopted by the General Federation of Women's Clubs.

Worley's role in the Omaha Woman's Club and the Nebraska Federation of Women's Clubs gave her a tremendous amount of public exposure and demonstrated her organizational, managerial, and business experience. Throughout this time, her speeches, letters, and campaigns continued to use the rhetoric of woman's special qualities in the domestic sphere as justification for her public works, and she always based her success on her knowledge of motherhood and household management. This was the same strategy suffrage activists employed—expanding the "social housekeeping" concept by claiming that society would be uplifted by woman's higher moral nature and that women's vote would infuse morality into the social and political system. However, Worley did not openly agitate for woman suffrage—she seems to have preferred to demonstrate woman's capability in business and politics rather than lobby for the right to prove it.

Worley's first forty years witnessed extraordinary social, cultural, and technical changes. The Victrola phonograph, "radio music boxes," and long distance telephone—all introduced after 1915—were commonplace by 1920. In 1918, less than a decade after the Wright brothers announced their first manned flight, Nebraska Aircraft Corporation in Lincoln was making airplanes and conducting flying lessons—Charles Lindbergh would train there in early 1922 and eventually barnstorm through Nebraska and give my grandmother her first plane ride. The first automobile reportedly rolled down the streets of Lincoln in 1902. By 1910, there were 11,339 motor vehicles registered with Nebraska's Secretary of State, and by 1920 there would be more than 211,750. Yet, in 1919, there were only 5.4 miles of paved road in the entire state, a small stretch of West "O" Street between Lincoln and the small town of Emerald.[24] Change came swiftly, but old and new existed side by side for many years. As one of the last states to be settled, these changes must have seemed even more dramatic in Nebraska. As Nebraskans' everyday lives were transformed by the new technologies of the 1910s, their cultural and social values underwent Progressive Era transitions. The three intertwined and highly controversial social issues of prohibition, woman suffrage, and America's growing involvement in the European conflict shaped the two short tumultuous years of young Governor Sam McKelvie's first term and were the key motivators of his momentous decision to travel from Lincoln to Omaha on July 1, 1920, to visit Universal Press Clipping Services on the eighth floor of the Omaha National Bank Building to ask Worley to assume a senior office in his administration.

McKelvie came in person because his political future hinged on placing an unimpeachable individual in this high-profile position, but it was not going to an easy recruitment or appointment. Accepting the nomination would attract a great deal of attention—especially from the mounting opposition to McKelvie's reelection—and the constitutionally defined office would require tremendous personal sacrifice from Worley. After a recent string of articles in Democrat-backed newspapers across the state accused McKelvie's Republican administration of furloughing violent criminals from the state penitentiary, the battle lines in his reelection campaign were clear. The Democrats were going to focus on penitentiary "mismanagement" in their attempt to unseat the governor in the coming November elections.[25] In the middle of this political battle, one of his appointees to the Board of Commissioners of State Institutions, also known as the Board of Control, resigned. The Board of Control was responsible for management and control of all present and future state penal, reformatory, and charitable institutions of the state—including the controversial state penitentiary system. In the past year, the Board had been investigated by the state legislature for misappropriation and poor management, and this was McKelvie's third attempt to find someone to fill the remaining three years of a resigning commissioner's six year term.

As the governor rode the elevator to the eighth floor, he was confident he had at last found the right person for the job. Worley had impeccable credentials, proven management and administrative skills, and no political enemies. But he knew he had an uphill battle to persuade her to accept the position because, if Worley accepted the appointment and was confirmed by the legislature, she would automatically become chairman of the commission in less than a year, making her the first woman in the USA to attain that level of authority. Yet Worley had not even been allowed to vote in the election that put McKelvie in office eighteen months earlier. In fact, as the politician and the businesswoman sat in her office on that hot summer day, Worley was still ineligible to vote for the governor in the coming election. The decades-long fight for woman suffrage was still raging at the national level as individual states debated ratifying the recently passed Nineteenth Amendment. Worley voiced many concerns. If women were not enfranchised, was she eligible to serve? Was Nebraska, and the nation, ready to accept a woman in this type of highly visible public position? Was she really qualified for the job? What if the challenge was too great and she failed? Worley's appointment came at the exact point in time when Nebraska was finally coming to terms with the idea of woman suffrage, and the pitched battle surrounding suffrage in Nebraska—and the nation—gave Worley pause when contemplating whether to accept the Governor's plea.

Nebraska's contest for woman suffrage dates back to 1856, when the House of the Nebraska Territorial Legislature passed a woman suffrage Bill by fourteen to eleven votes, but the legislature's session, which was limited to forty days, adjourned before they could take a vote. Suffrage activist Amelia Bloomer later wrote that Nebraska's governor assured her the Bill would have passed had there been enough time. If it had passed, Nebraska women would have been the first to be enfranchised in America. As it turned out, Nebraska women would wait for sixty more years before male legislators would come this close to granting them full suffrage. The 1866 State Constitution explicitly limited electoral eligibility to men, but the next year the legislature passed a statute allowing women to vote on school-district issues. In 1875, the wording of this law was changed to exclude married women from voting on school issues but was changed again six years later to include any property owner or parent in the district—reflecting the struggle men and women had when debating women's roles in the West as they developed laws and constitutions and shaped the characters of their states.[26] For the next four decades, Nebraska was a hotbed of woman suffrage activity. Susan B. Anthony and many famous suffragists made appearances throughout the state as they began to focus on municipal suffrage as a step to national suffrage. Nebraska suffragists introduced municipal suffrage Bills in nearly every legislative session between 1887 and 1914. Each Bill was met with vigorous opposition from various interest groups, but the suffrage movement's tie to prohibition was the major factor that kept the vote just out of Nebraska women's grasp. For years, Nebraska's powerful German-American Alliance lobbied aggressively and successfully against the conjoined suffrage and the temperance movements, claiming suffrage was only intended to saddle the "yoke of prohibition" on their necks.[27]

By the 1910s, momentum for woman suffrage began to grow. In 1914, an initiative petition succeeded in placing a women's-suffrage amendment on the general election ballot and it lost by less than 10,000 votes. In previous elections, suffrage was defeated by a margin of four to one. Then, in late 1916, a general election in Nebraska approved prohibition, and in early 1917 the nation underwent a major shift in war sentiment when Germany torpedoed several Allied passenger and merchant ships with Americans on board. Having lost the prohibition battle and facing a wave of anti-German sentiment throughout the state, the politicians representing the German-American Alliance had bigger fish to fry when the ubiquitous municipal suffrage Bill was introduced in the 1917 legislative session. To prevent the attempted repeal of the Mockett Law authorizing foreign language instruction in public schools, the German-American lawmakers allegedly made a deal with suffrage activists. They agreed to support municipal suffrage in return for protection of the Mockett Law.[28]

The suffrage Bill passed, but soon the issue became even more contentious and polarizing. Governor Keith Neville signed the municipal suffrage Bill on April 21, 1917, fifteen days after Congress declared war on Germany and nine days before prohibition would go into effect. While the war effort supplanted the suffrage issue in other parts of the country, in Nebraska the large and politically influential German-American and anti-prohibition forces were backed into a corner and quickly retaliated against the suffrage movement. A few days after Nebraska's municipal suffrage Bill was passed, against the backdrop of America entering the war and saloons being shut down, anti-suffrage forces circulated a petition seeking to annul the municipal-suffrage law. In a matter of weeks they secured the required number of signatures that suspended the new law until it could be submitted to a vote of the people. Suffragists suspected the validity of the 30,000 signatures that had been gathered and quickly challenged the petition in court. This launched a highly publicized and protracted legal contest that attracted a tremendous amount of national attention. The fight for suffrage in Nebraska dominated the news as the case, *Barkley* v. *Pool,* worked its way through the Nebraska courts for the next fifteen months. Suffragists spent months checking more than 18,000 signatures in Omaha that had been gathered under the direction of anti-suffrage leader Mrs. L. B. Crofoot, whose husband was president of the anti-prohibition "Prosperity League." The investigation uncovered hundreds of signatures that were obtained in pool halls and cigar parlors under the pretense that the petition was to "bring back beer." Even more galling to the suffragists was the large number of legal signatures by men who were not even citizens—Nebraska's constitution allowed males who declared their intention to become citizens to sign initiative or referendum petitions.[29]

Clearly, the war between suffragists and anti-prohibitionists—backed by powerful foreign-born interest groups such as the German-American Alliance—was not over. The *Woman Citizen,* the national weekly of the woman's suffrage movement, wrote, "no battle for suffrage was ever fought harder than this legal battle in Nebraska."[30] McKelvie's term as governor began in January 1919 and encompassed the last dramatic confrontations on woman suffrage in Nebraska and around the nation. It was a period of thrilling victories followed by unimaginable defeats for both sides of the struggle. Indeed, until the final outcome—decided by a single vote thousands of miles from Lincoln—no one knew when, or how, women would ever win suffrage in Nebraska. The 1918 election that swept McKelvie into office took place against the backdrop of the close of World War I and the *Barkley* v. *Pool* suffrage battle. McKelvie defeated the incumbent, Keith Neville, a Democrat who had been elected just two years earlier. (At that time Nebraska governors served two-year terms with no term limits.)[31] With Nebraska's municipal suffrage law of 1917 suspended pending the outcome of *Barkley* v. *Pool,* Nebraska women were not able to vote in the November 1918 election that put McKelvie in office over Neville, but it is doubtful women would have made much of a difference as both men had supported suffrage. Though the election occurred on November 5, 1918—the very day the Allies conditionally accepted Woodrow Wilson's peace proposal and just six days before the official armistice—voters appear to have been influenced more by the domestic and local issues of prohibition, woman suffrage, and government reform than the war.

Neville, at thirty-three, was slightly younger than McKelvie, and though they were from opposing parties, both men were progressive in their social platforms. Their youth reflected a growing desire for social change on the part of the state's electorate. McKelvie and the Republican legislature elected with him won by large margins and used their perceived mandate to enact the largest body of new law ever created in one session in Nebraska—more than 1,200 pages. Addison E. Sheldon wrote, "[The new laws] epitomize and express the determinations of Nebraska public opinion upon some of the major questions agitated in Nebraska during the previous fifty years."[32] In addition to the new "Civil Administrative Code" reorganizing state government, this prolific 1919 legislative session ratified national prohibition; provided for a state constitutional convention to revise the constitution; levied a tax to erect a new state capitol; and declared English the official language of the state, forbidding teaching any subject in any other language.[33] This state legislature also actively supported woman suffrage, and Nebraska women had reason to be hopeful right from the start of McKelvie's term.

On January 17, 1919, ten days after McKelvie and his Republican legislature took office, the Nebraska legislature unanimously issued a joint resolution memorializing the US Senate to pass and submit to the states "without delay" the pending Woman Suffrage Federal Amendment. In this resolution, both houses called upon Nebraska's Senator, Gilbert M. Hitchcock, to vote in favor of the amendment.[34] Then, on January 24, a judge invalidated more than 4,600 signatures of Omaha's anti suffrage petitions in the *Barkley* v. *Pool* case and concluded that the required number of signatures had not been gathered. The well-publicized ruling documented the illegal tactics of the anti-suffrage movement and generated support for suffrage across the state. As a result, on January 27, State Senator C. Petrus Peterson of Lincoln introduced a joint resolution proposing an amendment to the Nebraska State Constitution to grant women full suffrage.[35]

Nebraska suffragists' celebration was short lived, however. On February 14, the US Senate defeated the proposed Nineteenth Amendment by a margin of one vote. Despite January's resolution from the Nebraska legislature and a personal visit from a delegation of Nebraska women urging him to "express the sentiments of his home state by favoring the bill," Hitchcock, an anti-prohibition Democrat from Omaha with strong German-American Alliance connections, voted against the Bill.[36] But Nebraska suffragists quickly recovered from this loss and were given much more hope in March. First, the State Attorney General ruled that women were eligible to vote in upcoming municipal elections based on the judge's ruling in *Barkley* v. *Pool*. Then, on March 24, the state legislature called for a Constitutional Convention in November: woman suffrage would almost certainly be one of the amendments to the state constitution proposed at this meeting.[37] Reflecting suffragists' confidence, on March 28 the leader of the statewide suffrage movement, Mrs. W. E. Barkley, requested that the state legislature withdraw Peterson's pending Bill to add suffrage to the State Constitution because the upcoming Constitutional Convention would submit the issue directly to the voters and avoid the possibility of another referendum ploy by anti-suffragists.

On April 15, 1919, in the middle of this intense state and national controversy over suffrage, Governor McKelvie took a very bold step. He nominated Mrs. A. G. Peterson of Aurora for an appointment to the Board of Commissioners of State Institutions, assuming the term of a commissioner who was resigning effective July 1. This was the first time a woman had ever been nominated to a position at this level. In his official message to the legislature, McKelvie went to great lengths to describe Peterson's qualifications, including her service as State President of the Federation of Women's Clubs and a director of the General Federation of Women's Clubs. McKelvie also pointed to Mrs. Peterson's business skills as a director of the Aurora National Bank and her work in her husband's mercantile business. The legislature immediately established a committee to investigate Mrs. Peterson's appointment and requested the Attorney General to "send to the Senate a written opinion upon the question of the right of Mrs. A. G. Peterson from a legal standpoint to hold a position on the Board of Control of this State."[38] The existing municipal suffrage law explicitly prohibited women from voting for constitutionally defined officers, and although the Commissioners of the Board of Control were appointed, not elected, they were constitutional officers and the legislature was unsure if there was a legal precedent that could support Mrs. Peterson's appointment.

The legislature adjourned two days later without taking action on Mrs. Peterson's appointment, but statutes provided that if the legislature was unable to confirm the position then the appointee could serve until the legislature reconvened. McKelvie and suffrage supporters were likely confident that the question of the legality of Peterson's appointment would be moot, thinking Nebraska women would surely have either state or nationally granted suffrage before the next legislative session. National suffrage would be up for another vote in the US Congress in a few months, and if that did not pass the State Constitutional Convention in November would certainly grant woman suffrage in Nebraska. In addition, Peterson was well known throughout the state for her leadership in many women's patriotic and charitable organizations and there would be little objection to her appointment. Unfortunately, by the time Peterson's term began—July 1, 1919—poor health prevented her from serving, and McKelvie was soon looking to fill the position for the third time.

A few weeks after the legislature adjourned, on April 28, 1919, a diehard contingent of anti-suffrage forces appealed *Barkley* v. *Pool* to the Nebraska supreme court. They claimed that the plaintiffs—the suffragists represented by Mrs. Barkley—as women were ineligible to vote and therefore had no right to sue because the suit pertained to a political, as opposed to a civil, right.[39] Nebraska suffragists were forced, once again, to take their case to court. But their fortune soon rebounded when, on June 4, 1919, the US Senate finally adopted the Nineteenth Amendment, sending it to the legislatures of the states for ratification—although Nebraska's Senator Hitchcock was one of the few to vote against the amendment. The Nebraska supreme court upheld the lower-court ruling in *Barkley* v. *Pool* on June 28, 1919, and a little over a month later, Nebraska's legislature, in a special session called by McKelvie, voted unanimously to become the fourteenth state to ratify the amendment on August 2, 1919. Thirty-six of the forty-eight states were required to ratify the Nineteenth Amendment for it to be added to the Constitution. Suffrage leaders expected quick ratification in twenty-eight states where women had some form of suffrage, but many of the remaining states had powerful anti-suffrage legislators who would put up a strong opposition. Like Nebraska, most of the legislatures were not in session, forcing suffragists to lobby governors to call special sessions. But suffragists grew more optimistic with each passing week as state legislatures convened and approved the amendment.

On November 4, 1919, a special election was held in Nebraska to elect delegates to the Constitutional Convention. Women were not able to vote for the delegates because Nebraska's municipal suffrage law only allowed them to vote in city and county elections and for presidential electors. Furthermore, national suffrage was not yet ratified by the required number of states. By the end of 1919, twenty-two states had ratified—and only two had rejected—the Nineteenth Amendment, and the momentum continued during the first three months of 1920. On March 22, Washington became the thirty-fifth ratified state, leaving only one more state needed to meet the required three-fourths to amend the US Constitution. Three days later, on March 25, Nebraska's Constitutional Convention adjourned after proposing a list of forty-one amendments for the state's voters to individually approve in a special general election scheduled for September 21. Women's suffrage was one of the amendments. Until the special election in September confirmed full suffrage for women in statewide elections, or the Nineteenth Amendment was ratified by one more state, Nebraska women were governed by the 1917 municipal suffrage law. They voted in the primary elections in April and attended party conventions in May, but women were excluded from voting for nominees for governor and other state offices.[40] Was the fight for suffrage nearly over? Suffragists were cautiously optimistic, but the outcome at the state and national level was by no means certain. In Nebraska, suffragists were at the mercy of male voters at the state's constitutional-amendment special election coming up in September, and the anti-suffrage lobby in the state was using prohibition—a highly emotional issue for many voters—against the suffrage movement. Nebraskans followed the Nineteenth Amendment ratification process with rapt attention: if just one more state ratified before September, women would automatically be able to vote in all state and national elections and women would be able to vote to specifically include woman suffrage in the State Constitution. Their future depended on the national suffrage campaign winning one more state, reaching what had become the suffrage battle cry: the "Perfect Thirty-Six."

Many of the state legislatures that had passed the amendment had engaged in heated debates with very close votes. Opponents still espoused the rhetoric that politics and voting would destroy woman's special moral place in society, but many anti-suffragist politicians, particularly in the South, also fought against ratification on the grounds that the USA was overstepping the bounds of states' rights. Even Nebraska's Senator Hitchcock had used the state's-rights argument when he wrote to the Nebraska legislature defending his vote against the Nineteenth Amendment.[41] When the Washington state legislature became the thirty-fifth state to ratify in March, only five Southern states had rejected the amendment and supporters finally began to believe they were going to win the long, long campaign for suffrage. But then the ratification progress stalled. The week after Washington affirmed, Mississippi became the sixth state to reject. Of the states remaining, Delaware seemed the best bet: it had a progressive Republican governor who was in favor of woman suffrage and a Republican general assembly. However, after weeks of intense debate, national scrutiny, presidential pressure, and unremitting lobbying, the Delaware representatives voted to adjourn without a decision on June 2, 1920. This was the suffragists' darkest hour. Of the six states yet to act, three were in the Deep South and adamantly opposed to woman suffrage (these three—North Carolina, Louisiana, and Florida—would not ratify for fifty more years!) and two (Connecticut and Vermont) had governors who opposed suffrage and refused to call special legislative sessions. Tennessee had no legislative session scheduled, and, with strong political opposition to suffrage and an election coming up, the governor was reluctant to call a special session.

On July 1, 1920, the day Governor McKelvie met with Katherine Fay Worley, Louisiana rejected the amendment—in spite of a desperate push by national suffrage activists. It was becoming clear that women would be shut out of the November 1920 national election, and suffragists were even more dismayed at the possibility that they would not reach the "Perfect Thirty-Six," and the law would die. As Worley and McKelvie discussed her nomination, the chances were great that women would not win national suffrage. Nebraska women had to count on the male electorate to approve statewide woman suffrage in the special constitutional election in September or they would have to accept only the municipal suffrage they won in the two-year *Barkley* v. *Pool* battle. McKelvie was up for re-election in November, and it was not at all clear if women would be able to vote for him. By placing Worley in a visible, high-level position in his administration, McKelvie was telling voters across the state that he not only believed women had the right to vote but that he had confidence in their capability to serve in powerful senior positions in government as well. But his decision was not without risk: there were still many who felt that women were not ready for this level of responsibility. In the previous six months, Iowa's governor had twice attempted to appoint women to his state board of control, but Iowa's legislature had refused to confirm the appointments.[42]

Woman suffrage was the burning issue throughout the long, hot summer of 1920. Worley understood that if she accepted this position both factions of the women's movement would scrutinize her performance. McKelvie's first female appointee, Mrs. A. G. Peterson, had been too frail to bear the pressure, making it even more critical to place a strong woman who would succeed in this position. Worley knew all eyes would be on her and that every action she took would be judged more critically than her male counterparts. Her performance could provide ammunition for pro- and anti-suffragists at a point when the issue was more volatile, emotional, and controversial than ever. McKelvie and Worley also had to face the very real possibility that national suffrage would not pass and that an all-male Nebraska electorate would not approve statewide suffrage. In that case, the legality of a woman serving in a constitutional office would be an issue when the legislature met to approve her appointment. Worley wrote that she and McKelvie discussed the challenges that faced her, and she "realized the personal sacrifice entailed, and the feeling was that it was too great a one to make."[43] She then thought of the many times she had lectured on patriotism during her Americanization campaign. She had admonished her listeners that "a person has no right to accept all of the splendid educational and cultural advantages which our state and government extends to us without being willing to give something of ourselves in service in return if the opportunity presents."[44] Worley decided that she must demonstrate her loyalty and patriotism by accepting the position. A few weeks later, she relocated to a small apartment in Lincoln and began to serve her term on the Board of Control. Women had chipped away at the strict social code that defined their station in life for decades. Now Worley felt the full weight of her role to deliver the final blow to the Victorian values she grew up with, and she later wrote that she had wondered if she had what it would take to be "a credit to the women of Nebraska."[45]

During July and August, as Worley settled into her job, the final, dramatic conflict in the battle over suffrage played out in newspapers across the country. President Wilson was finally able to pressure the reluctant governor of Tennessee into calling a special legislative session to ratify the Nineteenth Amendment, but the opposition had been so strong that the governor waited until August 9—after his state's primary election—to call the session. The stage was set for the last, and nastiest, battle in the war for suffrage. Activists from around the nation descended upon Nashville. In a legendary battle that has become known as "The War of the Roses," legislators displayed their position on suffrage by wearing roses in their lapels—red roses for anti-suffrage supporters, yellow roses for suffragists. Tennessee's senate quickly passed the amendment, and all eyes turned to the state's house of representatives. When the house called its first vote on August 18, the roses indicated that the amendment would be defeated 47 to 49, but the first roll call brought a representative over to the suffragists' side and the vote was deadlocked at 48 to 48. A second roll call repeated the deadlock. If the amendment did not pass on the third call, it was considered defeated, so anti-suffrage legislators quickly called for a third vote. Author Cheryl Hiers describes the drama of this final vote:

> With wilted collars and frayed nerves, the legislators squared off for the third roll call. A blatant red rose on his breast, Harry Burn—the youngest member of the legislature—suddenly broke the deadlock. Despite his red rose, he voted in favor of the bill and the house erupted into pandemonium. With his "yea," Burn had delivered universal suffrage to all American women. The outraged opponents to the Bill began chasing Representative Burn around the room. In order to escape the angry mob, Burn climbed out one of the third-floor windows of the Capitol. Making his way along a ledge, he was able to save himself by hiding in the Capitol attic.[46]

The next day, the twenty-four-year-old legislator explained that just before the final vote he had received a telegram from his mother in East Tennessee urging him to do the right thing and to vote for the amendment. Burn said, "I know that a mother's advice is always safest for her boy to follow, and my mother wanted me to vote for ratification."[47] For several more days, Tennessee anti-suffrage legislators attempted to stall the ratification through parliamentary procedures, but Tennessee's governor signed the Bill on August 24, and on August 26, 1920, the Nineteenth Amendment became national law.[48] The seventy-five-year-old struggle for national women's suffrage was finally over.

On September 21, 1920, a little over 19,000 Nebraska women voted in the special constitutional general election. Separate ballot boxes were used for women's votes as there were legal questions raised whether women could vote on constitutional amendments. Women voted for Amendment Number 19, women's suffrage, by a margin of 18,012 to 954. The men also supported suffrage, but with a smaller margin of 47,471 to 14,462.[49] At the November 2, 1920, general election, Nebraska women voted for all offices. There were over 156,000 more votes cast in 1920 than in 1918, nearly doubling the 222,000 votes of the previous election.[50] With the exception of McKelvie, who was re-elected in a very tight race with his Democratic opponent, Republicans swept the election with easy victories.

Worley settled into her job and was quickly recognized as a tireless worker and tremendous asset to the Board of Control. The Board was in charge of appointing the superintendents, purchasing all supplies, personally inspecting and approving admissions for the fourteen institutions under its control. Every two years, the Board gathered statistics and published the status of the institutions in a public report.[51] Worley would later write McKelvie about a few of her "variety of experiences" during her many tours of inspection. She described climbing "twenty feet down in a 'dug' well on a rickety ladder to inspect some machinery," and being "hoisted to the top of a three story building on an outside elevator-hoist used to carry up loads of material."[52] She wrote that she was glad she was "somewhat practical and at least know [sic] a hayrake from a disc, and that a gasket is not an article of food."[53] On July 1, 1921, just a year after she started, Worley became Chairman of the Board of Control. Because of the Board of Control's huge budgetary responsibility and extensive span of control, the statutes governing the makeup of the board were quite explicit. No two members could be from the same political party or from the same congressional district. Each member was required to take the same oath of office as all constitutional officers and to give a $25,000 bond "for the faithful discharge of his duties."[54]

Worley became chairman less than a year after women finally won national suffrage. She was the first woman in the USA to hold such a powerful role in government administration. Her responsibilities caught the attention of the state and the nation. Newspaper articles around the country proclaimed: "Little Woman in Big Place. . .But is Highly Competent," "Woman Head of Central Board. . .Handles Vast Sums," "Woman Presides Over Nebraska's State Institutions—Directs Expenditure of $5,000,000 Annually—Gives Sound Advice." Much was made of her small, "95 pound" stature and her powerful new position. Worley's years in the newspaper industry made her politically savvy and sophisticated in the art of public relations. Rather than focusing on women being as able as men, she would translate the unique qualities of women into abilities for serving the public. In every interview, Worley attributed her success to her feminine experience in the domestic world. In one article titled "Control Board Is State 'Housekeeper,'" Worley defined her work as "a big job of housekeeping" saying:

> The average housewife in a small way faces much the same problems that we face as a board. The housewife has to buy supplies for the home, and she shops for the best bargains in groceries, clothing, fuel and other necessities—so do we.
>
> The housewife has her problems of discipline and of business management—so do we.
>
> She has to deal with naughty children quite often—we have to deal with those that the laws and courts have said are bad men and women and naughty boys and girls.
>
> The mother also has to face the sorrows of life, too often she has to care for the afflicted, the sick and the maimed—so do we.[55]

Although she tried to downplay the significance of her gender, Worley was always conscious that she was representing the capability of all women. When she became chairman she wrote, "It was now necessary for me to 'make good' for the sake of American womanhood, as well as for that of my own state. Woman's advent into public affairs was as yet so new that we must prove ourselves worthy of confidence and trust."[56] She claimed that she prayed every day to be helpful to the governor who selected her and for the strength to be "a credit to the women of Nebraska."[57] Worley was very successful in this regard and paved the way for women in Nebraska government.

After her term expired, Worley returned to her business in Omaha. On June 14, 1925, at the age of forty-nine, Worley married businessman Charles E. Allen of Cozad, Nebraska.[58] He had been a state senator during Worley's last year on the Board of Control. The Allens split their time between Cozad and Omaha until his death in 1962. Worley ran Universal Press Clipping services until the late 1950s when she sold it to Jim Murphy, who continues to run the business under the name of Universal Information Services, Inc. Worley maintained a relationship with the company for many years, and Murphy described her as uncommonly energetic, even well into her eighties. Murphy said Worley was "a very proper lady," always exquisitely attired in gloves and hat, and was recognized wherever she went in downtown Omaha.[59] Worley died in Omaha on March 12, 1971, at the age of ninety-four.[60] From the sod house of her birth to the high rises of Omaha in the 1960s, she lived to see Nebraska undergo almost unimaginable changes. Worley had confidence that women would continue to expand their roles in business and political positions around Nebraska and the nation. In 1924 she wrote, "It is the splendid and fine young business women, well trained in executive matters and clear thinking who will carve for women an enviable place in the world's affairs."[61] And, by the end of the 1920s, it seemed as if women were surely gaining ground in public affairs and business. However, Worley did not live to see women assume a significant role in Nebraska politics. With the exception of two appointments to fill vacancies during the 1950s, women were not elected to statewide or national offices in Nebraska until the mid-1970s. Indeed, the decades after Worley left public service saw a sharp decline in women's-rights activity. Even though the suffrage movement in the Western USA dramatically expanded women's rights at home, at work, and in the community, it appears that achieving suffrage brought about the end of agitation for women's rights.

Once the Nineteenth Amendment was ratified and put in place, Euro-American women around the nation could vote, but many women were still disenfranchised, including American Indian women who were yet to be officially declared citizens of the USA and African-American women who, like African-American men in the South, were kept from the voting booth by literacy laws and poll taxes. Male legislators and politicians did, however, respond to the new female constituency, and several federal and state laws were enacted. The Sheppard–Towner Maternity and Infancy Protection Act of 1921 provided matching federal funding to states to develop programs to support poor pregnant women, mothers, and children. Not only was this a model for future social-welfare programs, this was the first legislation passed to specifically address the female electorate's demands. The Act was repealed eight years later, primarily because politicians quickly discovered that newly enfranchised women did not vote in a bloc, and many did not vote at all. In fact, one of the first electoral studies found only one-third of eligible women voted in the Chicago mayoral election in 1923, compared to nearly two-thirds of men.[62] After such a long hard road, why did women not vote?

Some scholars theorize that the social values that defined Victorian-era women and that were used to win public support for the right to vote—the idea that woman's moral purity and superiority would improve the morality of American politics on the whole—were also the cause of women's diminishing influence in government and politics after winning suffrage. These notions about women's proper sphere reinforced the idea that politics and government were no place for women. They could vote, but the rough-and-tumble world of politics was thought to be no place for a lady. Nationwide public-opinion polls conducted in the 1930s, when only a few women had served in government positions, found that 60 percent did not want more women in politics. This aversion was even stronger on the subject of women holding elected office.[63] Another reason for women's lack of participation in government was that they had fewer routes to office than men. The few law schools that accepted women placed quotas on their admissions, and hardly any women were involved in leading the large-scale business enterprises that would connect them with the power to run. Much is made of the first women who served in public office being from the West, but most of these first political women came to office through the death or ineligibility of their husbands. In fact, examination of the first women in national offices reveals very little difference in numbers between women in the West and their contemporaries in the East.

The first woman governor, Nellie Tayloe Ross of Wyoming, was selected in 1925 in a special election to succeed her deceased husband. Though women had been voting in statewide elections in Wyoming for almost four decades, Ross's selection had more to do with political party maneuvering and her own pragmatism than vindication of women's suffrage. Ross's husband, William, a Democrat, had run for office several times and lost in the Republican-dominated state until he finally won the gubernatorial election in 1922 because he appealed to many Republicans who were turned off by the Teapot Dome scandal in President Warren G. Harding's administration. Halfway into his term, William died as a result of complications from an appendectomy. State law mandated a special election to find a replacement, and the Democratic Party, in an obvious attempt to leverage voter sympathy, asked Nellie to run on the afternoon of her husband's funeral. Ross had been a housewife, raising their three young children and, like most women of the era, showed her support of social issues through participation in the Cheyenne Women's Club. Though she had not previously demonstrated personal interest in public service, she accepted the party's nomination for two reasons: she needed a job to support her family, and she wanted her husband's programs to continue. Ross, who was mourning the death of her husband, chose to forgo active campaigning. Instead, friends posted political advertisements, and she wrote two open letters stating her intentions. She easily won the race on November 4, 1924, and on January 5, 1925, Nellie Tayloe Ross became the first female governor in the history of the USA.[64]

Fifteen days later, a second woman, Miriam "Ma" Ferguson, was inaugurated as

Governor of Texas after she was elected as a stand-in for her husband, an impeached former governor who was constitutionally barred from running. Ferguson's campaign slogan was "Two governors for the price of one." It took until 1967 for the next woman to serve as a governor when Lurleen Wallace of Alabama campaigned on the slogan, "Let George do it," to replace her husband who was constitutionally prohibited from seeking another term. A decade later, a woman was finally elected in her own right when Ella Grasso was elected Governor of Connecticut from 1975 to 1980.[65] In 1986, Democrat Helen Boosalis ran against Republican Kay Orr for Governor of Nebraska in the nation's first gubernatorial election featuring two women. That momentous election did not portend a dramatic increase in women governors, nor women politicians from the West. Of the seventeen women governors who served in the twentieth century, eleven were from states west of the Mississippi. Of the eleven female governors who were elected in their own right during the century, six were from the West.[66]

Just as Western women blazed a trail to national prominence as state administrators but didn't necessarily predominate during the subsequent years, a few Western women were pioneers in the US Congress. In 1916, four years before the Twentieth Amendment granted women universal suffrage, Jeanette Rankin, a suffrage activist in Montana, decided to run for the US House of Representatives. After Rankin spent a decade working in women's and children's homes in Washington state, she became involved in the suffrage movement out of the belief that women would support social legislation more than men. She became a tireless leader in the women's suffrage movement in Montana, and her renowned persuasive speaking skills and experience in grassroots activism won women the right to vote in that state in 1914 and made her an instant celebrity. Her new notoriety paved the way for her run for election to the US House of Representatives in 1916.

Rankin, who became famous for voting against US entry into World War I, made it clear that war was not her primary issue for running for office saying, "I've said a good many times that I never ran for Congress; I ran for women's suffrage," and she later said of her campaign, "My speeches weren't so very different from the speeches I made in the suffrage campaign."[67] Despite her avid stance on women's rights, when her opponents in the congressional race claimed "A woman's place is in the home," she would respond that the way women could protect the home was for them to have a voice in government. As a suffragette, she spoke in front of the Montana legislature in 1911, saying "It is beautiful and right that a mother should nurse her child through typhoid fever, but it is also beautiful and right that she should have a voice in regulating the milk supply from which typhoid resulted."[68]

As the first woman in Congress, and serving before national women's suffrage had been enacted, Rankin was certainly in the national limelight, and her story seems to support the idea of women in the West having more power and influence. But of the forty women who served in the US House of Representatives before 1950, only eight were from west of the Mississippi. The US Senate was an even harder club for women to break into. Only twenty-five women served in the US Senate in the twentieth century, and just over half—thirteen—were from Western states. Just eleven of the twenty-five first entered the Senate through regular elections, and of these only three were from the West. The political history of women in the twentieth century incorporates a bold narrative of early successes in Western states and territories that is easy to portray as a story of Western exceptionalism. There are many fascinating stories that can be argued could have only happened in the culturally unique atmosphere of the West—even though the West is a mighty big place and women's experiences were as diverse within the West as they were different from women of the East. Perhaps the greater lesson that can be gleaned from the story of Katherine Worley is that, despite the rapid changes in technology and society, women's roles didn't change quite as dramatically. Despite her acknowledgment that she was unique as a woman in such a prominent role, Worley always described her significance in terms that reinforced her female role.

My grandmother lived from 1900 to 1991, and though much changed during her nine decades, it would be hard to identify any one decade that marked the penultimate shift in women's prominence on the political or national stage. Women had served in most political offices at the state and national level, and a few had even run for president by 1990—but their numbers were still quite small, and the idea of a woman president seemed as remote in the early 1990s as it did when women won suffrage. Future historians will likely identify the real shift in women's political prominence as beginning in the late twentieth century and carrying forward into the twenty-first. It will be a story of Hillary Clinton, Madeline Albright, Condoleezza Rice, and many other women who changed the narrative from that of uniqueness or prominence of women in high public offices to one of utter normalcy.

NOTES

1. Frederick Jackson Turner, *The Frontier in American History* (New York: Holt, Rinehart & Winston, 1921), p. 1.
2. Turner, *The Frontier in American History,* p. 37.
3. Glenda Riley, "Twentieth-Century Western Women: Research Issues and Possibilities," in Gerald D. Nash and Richard W. Etulain (eds.), *Researching Western History: Topics in the Twentieth Century* (Albuquerque, N. Mex.: University of New Mexico Press, 1997), pp. 123–124.
4. Alexis de Tocqueville, *Democracy in America,* vol. II, trans. Henry Reeve (New York: D. Appleton, 1901), p. 688.
5. The concept of separate spheres and the Victorian ideals known as the "cult of true womanhood" is described in Barbara Welter, "The Cult of True Womanhood: 1820–1860," *American Quarterly,* 18 (2, Part 1) (1996): 151–174.
6. "Editors Table," *Godey's Lady's Book,* Philadelphia, Pa., January 1860.
7. Tocqueville, *Democracy in America,* p. 702.
8. Cathy N. Davidson, "Preface: No More Separate Spheres!" *American Literature,* 70 (3) (1998): 443–463.
9. Bill Kauffman, "They Voted against Mother's Day," *American Enterprise,* April 1, 2002, p. 50.
10. Andrea G. Radke-Moss, *Bright Epoch: Women and Coeducation in the American West* (Lincoln, Nebr.: University of Nebraska Press, 2008), p. 7.
11. Sandra L. Myres, *Westering Women and the Frontier Experience, 1800–1915* (Albuquerque, N. Mex.: University of New Mexico Press, ca. 1982), p. 216.
12. *General Laws, Memorials and Resolutions of the Territory of Wyoming, Passed at the First Session of the Legislative Assembly, convened at Cheyenne, October 12th, 1869* (Cheyenne, 1870; Wyo 1 1869). Law Library of Congress.
13. Caroline Nichols Churchill, *The Queen Bee,* November 29, 1893, p. 1. The *Queen Bee* was a weekly paper published and edited by Churchill that superseded her previous Colorado Antelope weekly on in July 1882. Although Colorado newspapers reported that Churchill continued publication of the *Queen Bee* until shortly before her death in 1926, surviving issues exist only through September 4, 1895. Carolyn Stefanco, "Networking on the Frontier: The Colorado Women's Suffrage Movement, 1876–1893," in Susan Armitage and Elizabeth Jameson (eds.), *The Women's West* (Norman, Okla.: University of Oklahoma Press, 1987), p. 275.

14. Worley's childhood was described in a biographical newspaper article by Lynnelle Greer, "Individual Effort and Determination to Succeed Guiding Rule of First Woman Chairman of the Board of Control," *The Lincoln Sunday Star,* June 19, 1921.
15. Greer, "Individual Effort and Determination."
16. Greer, "Individual Effort and Determination."
17. Greer, "Individual Effort and Determination."
18. Greer, "Individual Effort and Determination."
19. Greer, "Individual Effort and Determination."
20. Katherine F. Worley, "Report of Civics Committee, Omaha Women's Club," December 11, 1916, Katherine Fay Worley Allen Papers, MS174, Nebraska State Historical Society (NSHS), Lincoln, Nebraska.
21. Worley, "Report of Civics Committee, Omaha Women's Club."
22. Katherine F. Worley, Omaha, Nebraska, Form Letter to Religious Leader, Katherine Fay Worley Allen Papers, MS174, NSHS.
23. Michael L. Lawson, "Omaha, A City in Ferment: Summer of 1919," *Nebraska History,* 58 (3) (1977): 395–417.
24. George E. Koster, *A Story of Highway Development in Nebraska* (Lincoln, Nebr.: Department of Roads, 1997), pp. 22–26.
25. Addison Erwin Sheldon, *Nebraska: The Land and the People* (Chicago: Lewis Publishing, 1931), p. 958.
26. Ann L. Wiegman Wilhite, "Sixty-Five Years till Victory: A History of Woman Suffrage in Nebraska," *Nebraska History,* 49 (2) (1968): 149–164.
27. Wilhite, "Sixty-Five Years till Victory," p. 160.
28. James E. Potter, "Barkley vs. Pool: Woman Suffrage and the Nebraska Referendum Law," *Nebraska History,* 96 (1) (1988): 11–18.
29. Potter, "Barkley vs. Pool," pp. 14–15.
30. "Nebraska Suffragists Win," *Woman Citizen,* IV (July 19, 1919), p. 168.
31. Nebraska's first state constitution, adopted in 1866, established two-year terms for governor. In 1962, Nebraska voters approved a constitutional amendment specifying a four-year term beginning in 1966. In the 1966 general election, voters limited governors to two consecutive terms. *Nebraska Blue Book* (Lincoln, Nebr.: Nebraska Legislative Reference Bureau, 2000–2001), p. 419.
32. Sheldon, *Nebraska: The Land and the People,* 1: 971–972.
33. Sheldon, "The True Story of Nebraska," p. xx.
34. *Laws, Resolutions, and Memorials Passed by the Legislature of the State of Nebraska at the Thirty-Seventh Session* (Lincoln, Nebr.: Jacob North & Co., 1919), p. 65.
35. *Senate Journal of the Legislature of the State of Nebraska, Thirty-Seventh Session* (Lincoln, Nebr.: Kline Publishing Company, 1919), p. 151.
36. Sheldon, *Nebraska: The Land and the People,* p. 958.
37. Sheldon, *Nebraska: The Land and the People,* pp. 961–964; *Laws, Resolutions, and Memorials of the Thirty-Seventh Session,* pp. 872–879.
38. *Senate Journal of the Legislature of the State of Nebraska, Thirty-Seventh Session,* pp. 1225–1226, 1233.
39. Potter, "Barkley vs. Pool," p. 16.
40. Sheldon, *Nebraska: The Land and the People,* pp. 964–969; Grace M. Wheeler, *Handbook for Nebraska Citizens* (Lincoln, Nebr.: Grace M. Wheeler, 1919), p. 24.
41. *Senate Journal of the Legislature of the State of Nebraska, Thirty-Seventh Session,* pp. 135–136.
42. "Omaha Woman Who Helps State of Nebraska Spend $4,000,000 Annually," *Omaha World Herald,* October 2, 1921.
43. Katherine F. Worley, "Foreword," February 1924, Katherine Fay Worley Allen Papers, MS174, NSHS.
44. Worley, "Foreword."
45. Worley, "Foreword."

46. Cheryl Hiers, "The Nineteenth Amendment and the War of the Roses," available online at www.blueshoenashville.com/suffragehistory.html (accessed December 13, 2008).
47. Carol Lynn Yellin, "Countdown in Tennessee, 1920," *American Heritage,* 30 (1) (1978): 12–23, 26–35.
48. Yellin, "Countdown in Tennessee, 1920."
49. Addison E. Sheldon (ed.), *The Nebraska Blue Book and Historical Register, 1920* (Lincoln, Nebr.: Nebraska Legislative Reference Bureau, 1920), p. 122; Sheldon, *Nebraska: The Land and the People,* pp. 961–962.
50. Sheldon, *Nebraska: The Land and the People,* pp. 954, 970.
51. Sheldon, *The Nebraska Blue Book and Historical Register, 1920,* pp. 171–172.
52. Katherine F. Worley, Lincoln, Nebraska, June 25, 1923, to Mr. S. R. McKelvie, Lincoln, Nebraska, Katherine Fay Worley Allen Papers, MS174, NSHS.
53. Worley to McKelvie.
54. *Compiled Statutes of the State of Nebraska, 1922* (Columbia, Mo.: E. W. Stephens; Lincoln, Nebr.: State of Nebraska, 1922), p. 2142.
55. "Control Board is State 'Housekeeper,' Miss Katherine Worley, Only Woman Chairman of State Board of Control, Tells of Her Work in Caring for Penal Inmates," *The Omaha Daily News Magazine,* Sunday, September 11, 1921.
56. Worley, "Foreword."
57. Worley, "Foreword."
58. "Worley, Allen Wedding," *Omaha Sunday Bee,* June 14, 1925.
59. Conversation with Mr. Jim Murphy, Universal Information Services, Inc., Omaha, Nebraska, March 14, 2002.
60. "Mrs. Allen's Rite Tuesday; She Was '1st,'" *Omaha World-Herald,* March 14, 1971.
61. Katherine F. Worley, "Foreword," February, 1924, Katherine Fay Worley Allen Papers, MS174, NSHS. Worley also realized the historical significance of her appointment, and in 1924 she created a scrapbook of clippings and reminiscences that was eventually donated to the Nebraska State Historical Society.
62. Jeffrey D. Schultz, and Laura van Assendelft (eds.), *Encyclopedia of Women in American Politics* (Phoenix, Ariz.: Oryx, 1999), p. 210.
63. Schultz and Assendelft, *Encyclopedia of Women in American Politics,* p. xviii.
64. Teva J. Scheer, *Governor Lady: The Life and Times of Nellie Tayloe Ross* (Columbia, Mo.: University of Missouri Press, 2005), pp. 67–71.
65. Susan J. Carroll, "Women in State Government: Historical Overview and Current Trends," in *The Book of the States* (Lexington, Ky.: Council of State Governments, 2005), p. 1, available online at www.cawp.rutgers.edu/fast_facts/levels_of_office/StateLeg-Research.php (accessed December 13, 2008.)
66. Center for American Women in Politics, "History of Women Governors," available online at www.cawp.rutgers.edu/fast_facts/levels_of_office/documents/govhistory.pdf (accessed December 13, 2008).
67. John C. Board, "The Lady from Montana," *Montana: The Magazine of Western History,* 17 (3) (1967): 2–17.
68. Board, "The Lady from Montana."

CHAPTER SIXTEEN

AMERICAN POLITICS MOVE WEST

Gordon Morris Bakken

From the time of Thomas Jefferson, Americans and American politicians looked west for the future of the nation. First, with the promise of new fertile land for settlement in the eighteenth century and then in the explorations of Lewis and Clark in the early nineteenth century, Americans looked west for a wide variety of reasons. American politicians followed suit. By the mid-nineteenth century, "manifest destiny" was firmly entrenched in American political rhetoric, and the explorations of Zebulon M. Pike (1806–1807) and Stephen Long (1819–1820) further popularized the West. The settlement of Oregon put the "Oregon Question" into national politics; the Mormon ordeal of moving to the Salt Lake valley put polygamy out of newspaper headlines until the Republican Party put it at the top of the political list with African slavery; and migration into Texas focused national attention on revolution and America's manifest destiny.

Americans also wanted to move information on goods west. As early as 1848, the postal service contracted for a New York to San Francisco monthly mail service. With the business explosion of the 1849 Gold Rush in California, business leaders complained of the monthly service and the stale information. An 1857 Appropriations Bill authorized the surveying of routes to the West. Surveying teams rode for Fort Kearney, South Pass, El Paso, and Fort Yuma in search of a fast track to the West. The Post Office Appropriations Act of 1857 opened the mail service to bids, and Postmaster General Aaron Brown of Tennessee chose the John Butterfield–William Fargo bid over a St. Louis, Fort Smith, El Paso, and Fort Yuma route. The Butterfield Overland Mail was born and made the 2,800-mile trip in two weeks. Their success encouraged others, such as the Stockton Express and the San Antonio Express, to enter the service. Russell, Majors, and Waddell got into the Western freighting business in 1855 with 300 wagons and were using 3,500 three years later. Russell's Pony Express started taking mail west in April 1860 ten days faster than the stagecoach, but the Pacific Telegraph and Overland Telegraph Company set its first pole in July 1861 and connected on October 24, 1861. Information now travelled at the speed of the telegrapher's hand across America. Business and political information shot over the wires rapidly. Both the private and public sectors saw the potential of the West.

While the mail service improved, the movement of bulk product from west to east and manufactured goods from east to west remained mostly by sea and the few watercourses such as the Missouri River. Railroads reached the Mississippi River by the 1850s.

The Missouri Pacific Railroad stretched from St. Louis to Kansas City. The Chicago and Northwestern Railroad was in Clinton, Iowa, by 1855, the same year that the Rock Island Railroad bridged the Mississippi River. In 1856, the Republican Party platform called for a transcontinental railroad. The party also was for "Free soil, free speech, and Fremont." Importantly, the party had chosen an icon of the American West, John Charles Fremont. He was an explorer who popularized the West, a man who involved himself in the Bear Flag Revolt in California, and one of California's first US senators. Fremont was a symbol of the West as the path marker of South Pass, the Oregon Trial, and the Great Basin of Utah. Fremont lost to James Buchanan, but the successful third party kept the dream of a transcontinental railroad alive.

The Republican Party was successful in the next election, and, with Abraham Lincoln in the White House and Republicans in control of Congress, the transcontinental railroad and other Western legislation won easy victory despite the demands of the Civil War. The 1862 Bill put the Central Pacific and Union Pacific in motion building across the West. Completed in 1869, the transcontinental railroad was soon linked up with the Southern Pacific and later the Texas and Pacific. Congress granted a charter to the Northern Pacific in 1864, and in 1869 Jay Cooke took a gamble building the line. Congress funded much of this railroad building with public lands and bonds. Congress continued to spend land with the Homestead Act of 1862, and subsequent land laws expanding the amount of land a person could acquire from the public domain. Whether the railroad or the equipped farmer headed west, nineteenth-century thinking believed in putting land into the hands of entrepreneurial Americans to increase the productivity of Western lands. The product of the land was shipped to American cities and international markets, increasing the capital stock of the nation.

Along the way, politicians found that subsidies were needed. The Great Plains and the Rocky Mountain West were characterized by aridity. Farmers and ranchers quickly found that water was the key to productivity. State incentives and American technology aided the process. Daniel Halladay's windmill went into large-scale manufacture by the US Wind Engine and Pump Company in the 1860s. Towns, railroads, and farmers put the system into widespread use from 1873 to 1880. But a windmill could lift only so much water. Individual irrigation plans were limited by capital. The Carey Act of 1893 authorized a state irrigation project, but only when the federal government under the Reclamation Act of 1902 put national dollars into regional projects did the West have enough water for agriculture. So it seemed to some in early twentieth-century America.

Great Plains and Rocky Mountain farmers and ranchers steadily increased the size of their holdings. Railroads brought machinery and hauled away bulk product. Ranchers drove cattle to railheads during the period of the range cattle industry but steadily moved to more intensive ranching methods, improved breeding, and improved bovine medicine. Farmers, particularly in Dakota and eastern Montana, turned to dry land farming. Hardy W. Campbell established experimental farms in coordination with railroads and demonstrated that deep plowing, frequent cultivation after every rain, and specific soil types and development produced commercial crop harvests. Coupled with Congress passing the Kincaid Act offering 640-acre tracts in Nebraska and in 1909 the Enlarged Homestead Act offering 320 acres to settlers, settlers moved onto the Plains. In 1907, the Trans-Missouri Dry Farming Congress started the promotion of dry farming on the Plains. Later titled the International Dry-Farming Congress, its meeting in Lethbridge, Canada, drew 10,000 people. Regardless of the enthusiasm of the participants, some doubted the bonanzas anticipated. In 1906, Ellery Channing Chilcott, a South Dakota experiment station scientist, sounded a word of caution arguing that more research was needed, particularly on drought-resistant plants. Regardless, railroads selling off their lands born of federal subsidies, and dry-farming promotions lured many to the land. The weather did not cooperate. Drought in western South Dakota, 1910–1913, drove many off the land. Eastern Montana experienced similar devastating drought on the advent of World War I. Yet, throughout the twentieth century, the Dakota and Montana produced wheat in abundance.

Although Congress was responsive with some legislation in support of Western agricultural entrepreneurs, farmers in antebellum America started organizing to protect their interests, frequently through politics. In the Old Northwest, farmers' clubs organized on the local level for broad purposes. State agricultural societies formed quickly in this period and their officers wielded great political influence. Some societies in collaboration with state agricultural colleges formed interest groups that lobbied state legislatures. Their lobbying work was influential in the passage of the Homestead Act and the Morrill Land Grant Act for state agricultural and mechanical colleges. From these local farmer's clubs, Oliver Hudson Kelly formed the Patrons of Husbandry or the Grange. Convinced that education and organization were necessary, Kelly started with the handful of Grange chapters in the 1860s, but the early 1870s witnessed an explosion of membership. With the prosperity of the Civil War gone with peace, farmers faced problems with transportation, credit availability, agricultural technology and the costs associated with mechanization, growing commercialization, monopoly capitalism, the growth of the crop lien system, and falling commodity prices coupled with railroad and grain elevator rate increases. Farmers focused their discontent on the railroads. In particular, farmers were discontent with marketing institutions, the consolidation of the railroads in the early 1870s, the processing industries such as the flour mills, and the variability of markets. They put political pressure on Congress, and the 1874 Windom Committee produced the results of studies of sea routes and Mississippi River alternatives to railroads. The report recommended state or federal ownership on one or more railroads and the improvement of waterways. Congress did nothing.

Although the Grange was unsuccessful at the federal level, they did pass an Illinois statute in 1871 establishing a commission with the authority to ascertain whether railroads and warehouses were complying with state law. By the time the Illinois statute arrived before the US Supreme Court in 1877, five states had such laws. In *Munn* v. *Illinois* (1877), the high court held that when private property was devoted to a public use, it was subject to state regulation. The constitutional stage was set for regulation.

The Grange also set the stage for political action. In California, the Grange persisted particularly with prosperous valley farmers. Isaac Friedlander, the grain king of California, used the Grange to fight shippers. John Bidwell, California Grange President, was a large landholder, fought the cattlemen, and achieved legislation requiring cattlemen to fence their livestock, preached agricultural diversification particularly into wine production, moved into peaches as a specialty crop, saw that food processing was critical to success, founding a flour mill and fruit-processing plant, and supported the Chinese despite death threats. He ran for governor in 1875 on the Grange ticket, campaigned for a normal school in Chico, ran for governor in 1890, and president on the Prohibition Party ticket in 1892. Bidwell recognized Indian rights and sought racial accommodation as early as 1850 and protection for Mechoopda Indians on Rancho Chico from Butte's Indian hunters. This successful agricultural entrepreneur mixed advocacy and politics in California for decades.

Some farmers hurt by the Panic of 1873 thought that the 1875 Resumption Act would lead to economic disaster. Farmers who acquired debt wanted to pay back creditors in depreciated greenback rather than specie. The Resumption Act put America on the gold standard in 1879, making greenback of equal value to gold. The Greenback Party of 1874 started with this currency premise and in 1876 put Peter Cooper forward as their candidate for President of the USA. Unfortunately, many farmers worried more about grain grading or the "brass kettle" than they did about currency value. In 1878, anti-resumption candidates won several congressional elections, but when resumption went into effect on January 1, 1879, nothing happened. The party expanded its platform in 1880 to include an income tax and female suffrage. James P. Weaver garnered only 300,000 votes and the party faded away.

Lampassas County, Texas, farmers were more worried about horse thieves and the price of cotton in 1877 than resumption. They formed the National Farmers' Alliance and Industrial Union. In 1878, the club developed a ritual and organized other clubs. By 1885, the clubs were statewide with a lecturer, S. O. Daws, to educate and uplift farmers. The Alliance deployed multiple strategies to lift the burdens of debt and retail prices. First, they worked at the county level to arrange trade agreements, but the cash terms required by merchants disadvantaged debt-ridden farmers. More successful was a second strand of policy. The Alliance opened Alliance-owned cooperative stores and sold below retail to Alliance members. The Alliance also opened cooperative flour, cottonseed, and corn mills, but cash-strapped farmers could not take advantage of them. The Alliance leadership, after the "strictly business" members bolted the 1886 convention, quickly saw political action coupled with a central, statewide exchange focused on cotton marketing as a better strategy under the leadership of Charles William Macune. The Alliance system called for each county alliance to bulk its members' cotton and to send a sample bale to the Dallas exchange headquarters of direct sale to buyers. The exchange plan cut out middlemen and theoretically increased producer profits. The system ran into a jute-bagging monopoly and faltered.

In 1889 Macune shifted Alliance strategy to monetary reform based on the perception of a money-supply problem. The Alliance adopted a "subtreasury plan" for the Congress to aprove with federal warehouses, loans to farmers at low interest, and the printing of money to a volume needed for cash-strapped farmers. The Texas Alliance supported the candidacy of James Stephen Hogg for governor and helped him win. The only problem was that Hogg's Democratic Party did not support the subtreasury plan. The Alliance members soon become supporters of the People's Party.

While Macune's Alliance was struggling and the Dallas Exchange closed in 1889, the Northern Alliance held a convention in St. Louis with its Southern Alliance brethren. Started by Milton George, Editor of the *Western Rural* of Chicago that promoted cheap transportation, the Northern Alliance had grown rapidly, grounded in local farmer's clubs and an elaborate network of farm weeklies and Alliance newspapers. The 1889 St. Louis platform called for the federal government to make more free land available, the cancellation of railroad land grants, the repeal of the National Bank Act to flood the country with cheap money, the liquidation of the national debt, an income tax, a tax on mortgages, lower property taxes, improvements in education, government ownership of the railroads, federal subsidies for the merchant marine, more federal money for canal and irrigation projects, support for labor unions, a national Australian ballot, diversification of agriculture, and support of candidates who were sympathetic to Alliance goals. Alliance political action yielded electoral victories in Kansas and Nebraska. In Nebraska, Alliance politicians and officials formed the "People's Party of Nebraska." The Populist Party was born.

In the Far West, farmers and ranchers were far more concerned with the economic and political dominance of corporations they called "beasts." As David Berman has demonstrated, these anti-corporation reformers wanted to shift a greater proportion of the tax burden to large business, to create effective rate and service regulation of corporations, to adopt protective labor legislation, and to reform the political system via the initiative, referendum, and recall to limit the reach of "the beast." Their greatest victory came in 1892 in Colorado with the election of David H. Waite to the governor's seat. The Colorado Populist Party blended ideology and pragmatism, calling for free silver, increased regulation of railroad passenger and freight rates, the eight-hour day, and cutting state officials' salaries in the interest of ending graft. The party also adopted state specific planks such as the banning of the future sale of coal lands, state-owned coalmines for the benefit of workers, and female suffrage. Fused with the Democrats they swept state offices and won both seats in the US House of Representatives as well as twelve seats in the state senate and twenty-seven in the state house.

Yet the People's Party was not yet a party in a traditional sense of the term. It was the aggregation of farmers who saw monopoly and the gold standard as monsters crushing their aspirations. It also was talented orators such as Mary Elizabeth Lease who told farmers they should "raise less corn and more *Hell!*" "Sockless" Jerry Simpson, Ignatius Donnelly, and James B. Weaver joined Lease in giving voice to agrarian grievances. In 1890, the farmers, small-town merchants, and ranchers marched to the polls and changed the face of American politics. Alliance candidates won control of the lower house of the Kansas legislature, five went to the House of Representatives, and William A. Peffer won election to the US Senate. Nebraska witnessed independents winning both legislative houses and a seat in Congress. Southern Alliance victories were similarly substantial.

Alliance leaders met in 1891 in Cincinnati to hammer out a party structure but failed. In St. Louis in February 1892 they tried again, and, after speeches by Ignatius Donnelly, Mary "Hell Raising" Lease, "Sockless" Jerry Simpson, and Terrence Powderly of the Knights of Labor, they established a committee of fifteen to organize a convention to nominate candidates for office and adopted resolutions. The committee of fifteen looked west to Omaha, Nebraska, and set July 4, 1892, for the nominating convention of the People's Party. Thirteen hundred delegates gathered in bustling Omaha and adopted a platform of demands. The national government was to authorize the free and unlimited coinage of silver, increase the money supply to $50 per person, adopt a graduate income tax, establish postal savings banks, nationalize railroads and telegraph companies, abolish alien land holdings, and reclaim all railroad lands in excess of their needs. Some delegates unsuccessfully proposed federal limits on hours of labor and laws forbidding the admission of undesirable alien others. Other delegates wanted the direct election of US senators, a single term for the President of the USA, and the adoption of the initiative and referendum. This Omaha Platform, the delegates thought, would lift the West out of its economic mire and usher in an era of prosperity.

The election returns were not as expected. Populist presidential candidate James B. Weaver of Iowa garnered only 1 million votes, carrying only Kansas, Nevada, Colorado, and Idaho. He also earned an electoral vote from North Dakota and Oregon. The party was clearly regional, and its center was west of the Missouri, not the Midwest. Yet its strength put Western issues on the Republican and Democratic Party tables. The Republicans responded in 1890 with the Sherman Silver Purchase Act that authorized the Treasury Department to purchase 54 million ounces of silver each year and to coin silver dollars.

Silver producers were paid for their metal with legal tender treasury certificates payable in gold or silver. The Treasury Department paid in gold, and the inflation the Populists had hope for was not realized. Rather, the gold standard was secure.

Bungling Republicans in and out of Congress and the fact that gold on deposit fell below $100 million precipitated the Panic of 1893. The fear that the value of the dollar was thereby jeopardized caused massive layoffs and runs on banks and curtailed wholesale spending. For farmers, the panic meant that commodity prices declined but their debt service hung over them. Across the West, farmers and politicians demanded the unlimited coinage of silver. Inflation was the debtor's answer to the cost-price squeeze. President Cleveland and the Democratic Party leadership did not listen. Rather, a special session of Congress repealed the Sherman Silver Purchase Act to the great distress of the West. This divided the country's politicians on sectional lines, with Western politicians trumpeting the cause of free silver and the East staying rock-ribbed for the gold standard.

Mountain West Populists supported the free coinage of silver but held anti-monopoly, working conditions, women's rights, direct legislation, and Chinese exclusion as significant public-policy values. California led the West in its racist bias against the Chinese, but other mountain states and territories favored limiting Chinese immigration. As Robert W. Larson demonstrated, Mountain West populism was far more complex and nuanced. Some supported a protective tariff for wool producers in New Mexico and Montana. Populists across the region favored the regulation of railroad freight rates and limitations on the railroad monopoly. Unionization in the mining states of Montana and Colorado was a positive good for Populists as part of a means of controlling monopoly power. Agricultural Populists wanted governmental intervention to protect workers from unjust corporate personnel policies. Further, they advocated public ownership of irrigation facilities. Land and cattle monopolies drew the attention of other Populists, particularly in Wyoming and New Mexico. Populism in Montana, Colorado, and Utah had a distinctly urban flavor. There Populists combined pronounced anti-monopoly with worker's rights advocacy. As Greg Hall has argued, agricultural workers in Washington and the Great Plains were drawn to the Industrial Workers of the World. These "Harvest Wobblies" were briefly successful in obtaining wage and working-condition concessions from corporate farmers, but in 1919 California and Washington corporate farmers combined with government to destroy the Agricultural Workers' Industrial Union. In Oregon, Washington, and Idaho, loggers and miners faced low wages, long hours, and hazardous working conditions. As Jeffrey A. Johnson has demonstrated, the new economic climate created strained relationships between labor and industry because of dangerous working conditions, deflated hopes of advancement, rapid social and economic change, and racial tensions in the workforce. Corporate control of communities exacerbated these conditions. The movement had greater staying power on the Great Plains, but by 1924, it too fell apart.

The Populists decided to delay their 1896 national convention until the Democratic and Republic conventions had concluded, hopefully for free silver. Rather, the Republicans adopted a gold plank, nominated William McKinley of Ohio, and caused thirty-four Western delegates to walk out of the convention led by Senator Henry M. Teller of Colorado. The Democrats adopted a "free and unlimited coinage of both gold and silver at the present legal ration of sixteen to one" plank and nominated the spellbinding orator from Nebraska, William Jennings Bryan.[1] Bryan told the delegates that the Republican Party "shall not crucify Mankind upon a cross of gold."[2] The Populists' delegates, meeting in July, decided to support Bryan and adopted a platform including free silver. They also demanded a graduated income tax, postal savings banks, government ownership of railroads, immediate foreclosure on defaulting railroad obligations and government purchase of those railroads, government ownership of telegraph companies, reclamation of railroad lands in excess of their needs, the exemption of mineral lands from railroad lands purchased by settlers and miners, free homes for settlers on the public lands, direct legislation via the initiative and referendum, direct election of the president, vice-president and US senators, home rule in the territories and the District of Columbia, public salaries set at the price of labor and products, public-works jobs in times of great industrial depressions, legislation limiting labor injunctions, pensions for disabled Union soldiers, and legislation to make elections open to all, subject of a free and fair ballot and an honest counting of ballots.

The campaign of 1896 was bitter and polarizing in the West. The monetary issue pushed some merchants and businessmen who supported Granger Laws into the Republican camp. Farmers and ranchers moved toward the Democratic Party. Bryan carried the West except North Dakota, California, and Oregon. McKinley won, and the gold standard was the law of the land in 1900. Yet the 1896 platform slowly became part of the American economic system. Gold flowed in, and money supply steadily increased. Rain fell in 1897, and crop yields enabled debtor agriculturalists to pay their debts. President Theodore Roosevelt gave the farmers railroad regulation. William Howard Taft gave them postal savings banks. Woodrow Wilson helped the currency supply with the Federal Reserve System, albeit areas of North Dakota were so short on currency and coins that merchants had to use trade tokens well into the first half of the twentieth century. The initiative and referendum was part of progressive reform in the early twentieth century and in common use by 1912.

By the turn of the century, radicalism was deeply rooted in the Rocky Mountain states. The Western Federation of Miners and the Socialist Party of America were the standard-bearers of the Populist and socialist agendas. Socialist candidates won elections in twentieth-century Colorado, Montana, and Idaho. Moving further into radicalism, the Industrial Workers of the World came to dominate Western mining disputes. Despite the radical left in the region, the more moderate message of municipal socialism produced political victories. Butte, Montana's Lewis Duncan won election to mayor in 1911 along with eight other Socialist Party members. They won reelection in 1913. As David R. Berman demonstrates, these Western Socialist Party officials adopted tempered reforms appealing to a broad-based multi-ethnic electorate. They continued the campaign for basic social reform and rights for labor. The 1901 Socialist Party platform called for a cooperative commonwealth with collective ownership of the means of production, transportation, communication, and public utilities. Legislation was to reduce hours and increase wages. They demanded state and national health, employment, and old-age insurance. The platform advocated free public education as well as state aid for books, clothing, and food. Politically, they wanted equality for men and women along with the initiative, referendum, and recall.

Despite the progressive politics emerging in some states and in national politics, the labor movement in the West had bitter consequences. Coal miners struck in 1884–1885, 1894, 1903–1904, and 1913–1914 for the eight-hour day, fair weights measuring their individual coal production, better ventilation in the mines, a 20-percent increase in tonnage rates, twice-monthly pay, and nondiscrimination against union workers. The tensions at the Rockefeller-owned Colorado Fuel and Iron Company coalfield erupted into a shooting war in 1914 called the Ludlow Massacre. A "Ten Day's War" followed, culminating in the intervention of federal troops with at least seventy-five dead. As Thomas Andrews has demonstrated, this was part of a fifty-year struggle between capitalists and mineworkers in the framework of migration, mine work characterized by craft pride, independence, solidarity, and masculinity, and labor struggle. Labor's battle for union recognition continued in regional politics with incremental and episodic gains in legislation and politics.

Many of the Populist demands found fruition in state and federal progressive laws. Women's suffrage won in the West, one state at a time. Wyoming was the first to grant women the right to vote in 1869. Utah followed in 1870, but Republicans worked to withdraw the right as part of their campaign against polygamy. Women would regain the franchise in the 1895 state constitution. Colorado extended the vote to women in 1893, Idaho in 1896, Washington in 1910, California in 1911 after a bitter loss in 1896, Oregon and Arizona in 1912, and Montana and Nevada in 1914. As Jeffrey A. Johnson has demonstrated, many of these Populist policy goals were appropriated by Progressive Party politicians. In 1912 in the Northwest, Progressivisim was at its high tide with reforms mirroring socialist goals. Legislation creating workmen's compensation, the eight-hour day, and suffrage were Populist Party planks now in the hands of Progressives. The West was clearly ahead of the nation on the suffrage issue.

In Arizona territory, Populist William "Buckey" O'Neill ushered a suffrage Bill through the legislature in 1897 that gave taxpaying women the right to vote in municipal elections. The law was overturned in 1899 by the territorial supreme court. Buckey O'Neill died in 1898 in the Spanish–American War with Roosevelt's Rough Riders. Women won the vote in 1912 with 68 percent of the voters supporting the amendment to the state constitution. Women quickly moved from voting booth to public service. C. Louise Boehringer was the first woman elected to public office in 1913, served in legislature and organized the Arizona Federation of Business and Professional Women's Clubs (BPWC). The women's clubs were the driving force behind reform and women's rights. The BPWC endorsed candidates and made the difference when voters went to the polls. For example, between 1912 and 1970, 300 California women ran for state or national office and only seventeen won. In the same period, sixty-one women served in the Arizona legislature thanks to endorsements from the BPWC. In 1914, Rachel Berry, a Mormon woman from St. Johns, won election to the Arizona House of Representatives. Rosa McKay, a Cochise County House member, introduced a minimum wage for women bill and secured its victory in 1917. In 1926, Anastasia "Ana" Collins Frohmiller won election as state auditor and served until 1950. She was the first woman in the nation to hold such an office. In 1933, Isabella Selmes Greenway, a longtime Democratic Party organizer, won a special primary election and 74 percent of the vote in the general election to a seat in the US House of Representatives. Nellie Trent Bush gained a seat in the state legislature and represented Yuma County for seven sessions. In 1924, she was appointed the first female US Commissioner in Arizona. Lorna Lockwood similarly started in the legislature and won election to county judge in 1950. A decade later she joined the Arizona supreme court. After working in the attorney general's office, Sandra Day O'Connor won appointment to a seat in the Arizona state senate in 1970. In 1979, the governor appointed her to the Arizona court of appeals, and President Ronald Reagan elevated her to the US Supreme Court in 1981.

In 1999, Sandra Day O'Connor joined five other Arizona women at a most significant event. In the 1998 election, women captured five Arizona state government positions and 37 percent of the seats in the state legislature. They were Lisa Graham-Keegan, Superintendent of Public Instruction; Attorney General Janet Napolitano; Governor Jane Hull; Treasurer Carol Springer; and Secretary of State Betsey Bayless. Napolitano moved on to the governor's office in 2002 and now serves in the Obama Administration. In 2006, Gabrielle

Giffords was the third woman sent by the people of Arizona to the US House of Representatives. The Populist beginnings for women truly bore rich fruit.

Populists also supported direct government and the Progressives would pass initiate, referendum, and recall in many states in the early twentieth century. For example, in North Dakota, Norwegians organized the Nonpartisan League in 1915, and in 1916 when the Republican Party did not support their demands, they used direct government. Under the leadership of Arthur C. Townley, they used the direct primary system to elect candidates who supported their goals. They elected their candidate for governor, gained control of the lower house, and placed a strong minority in the state senate. In 1917, the legislature put their goals into law with a grain-grading system, guaranteed bank deposits, and a nine-hour day for women. Legislators also enacted regulations for the railroads to prevent rate discrimination and authorized female suffrage.

Progressives also responded to Western demands for monetary reform. The Wilson administration made great strides with the Federal Reserve Act of 1913, the Warehouse Act of 1916, the Farm Loan Banks Act of 1916, and the Federal Intermediate Credits Bank Act of 1923. Unfortunately, the federal intermediate credits banks were a dismal failure. Indecisive and nonaggressive management combined with clumsy administrative structure resulted in a policy of no active solicitation of loans, particularly crop loans. Rather, the bulk of the loans went to farmers' cooperative marketing associations secured by commodities stored in federally licensed warehouses. Although these Grange institutions benefited from the loans, the farmers in need of short-term credit were ill served.

Populists demanded railroad regulation, and, at the federal level, Congress enacted the Hepburn Act of 1906, the Mann–Elkins Act of 1910, and the Transportation Act of 1920. Progressives provided some anti-monopoly measures such as the Clayton Anti-Trust Act of 1914 and the Federal Trade Commission Act of 1914. The Clayton Act also declared labor unions not to be combinations in restraint of trade taking a major legal argument away from corporate attorneys seeking to cripple unions.

Parallel with these gains in protective legislation, the struggle for women's suffrage and participation in government started bearing fruit. Backed by the California Populist Party, women's suffrage lost in 1896. In 1904, the Los Angeles Women Suffrage League organized women's club statewide in its campaign. In 1911, due to work of the federation of women's clubs and the Progressive Party victory of 1910, a female suffrage measure won by a narrow margin. Women quickly moved into appointive and elective office in the state.

The Progressive victory of 1910 brought Hiram Johnson to the governor's office and a good deal of procedural change in government. The Progressive objective was to eliminate political parties and thereby reduce the influence of special interests. The Progressives in the legislature quickly changed the ground rules. Legislation required all county and municipal candidates to run on a nonpartisan ticket. Their ballots were devoid of party affiliation. The civil-service system was expanded. Merit was emphasized in public service. To put merit before the voting public, a ballot-reform measure required that the qualifications of candidates be put before the voters. Another law prohibited parties from officially supporting candidates in primary elections. In 1913, "cross filing" allowed candidates to run on all party tickets. Two years later, the legislature extended nonpartisan ballot requirement to statewide elections. All of this procedural change tended to help incumbent Republicans and virtually destroyed the Democratic Party. Special interests gained strength because political parties no longer provided central public-policy direction to candidates.

As Jackson K. Putnam has persuasively argued, California progressives focused on pragmatism, activism, and moderation. They rejected ideology as simplistic in a complex world. They thought special-interest ideas were too narrow, and they turned to social-science research, independent fact gathering, and expert opinion to inform legislation. They were twenty years ahead of the New Deal approach to policy formation. Their activism focused on legislative performance churning out a mass of constituent-driven legislation. Doing something in Sacramento was both symbolic and real to progressives. In moderation, pro gressives sought compromise with lobbyists and legislators developing long-term symbiotic relationships to forge public-policy legislation that worked. Compromise was to be favored over the nineteenth-century practices of bribes and corruption in government.

Hiram Johnson advanced to the US Senate and handed the reigns of power to Governor William D. Stephens in 1917. On his watch, the legislature passed the Criminal Syndicalism Act of 1919 and plunged the state into the Red Scare era. Stephens supported prohibition, another program to save citizens from themselves and to foster middle-class Euro-American morality. The governor also supported an anti-Japanese initiative to bar landownership by these peoples. Yet Stephens signed a Bill prohibiting racial discrimination in public places, a measure providing for vocational rehabilitation training and wages and hours regulations. Harking back to the Johnson administration, Stephens appointed pro-regulation Progressives to the Public Utilities Commission over the objections of the railroads and Pacific Gas and Electric Company. What emerged during his term was a Progressive ideology. Progressives came to understand that not all businesses were always benign and needed regulation. Even more certainly, natural monopolies needed regulation and strong regulatory oversight. Regardless of government's best efforts, even the best socioeconomic system produced victims. But working with business government could enhance prosperity. Government needed to look more like a business, and, like any good business, government needed more money to provide more services and regulation. The first Stephens budget for 1917–1919 doubled expenditure, setting the modern pattern, and revenue enhancement came through higher taxes such as the King Bill of 1921 that increased corporate taxes 35 percent.

In 1922, Stephens failed to receive the nomination of the Republican Party, and a new style of politics entered the governor's mansion. Friend W. Richardson used ideology to defeat a pragmatist and worked to dismantle Progressivism with budget cuts and executive orders. In 1924, State Senator Herbert Jones and Lieutenant Governor C. C. Young organized the Progressive Voters League to win control of the legislature and increase social services. In 1926, Young won election as governor of California.

Young's term as governor, 1927–1930, demonstrated what government could do in the name of Progressivism. California state government expanded spending for highways, water projects, forestry, social services for people with disabilities, the needy, and the blind, the nation's first comprehensive old-age-pension program, agricultural pest-control stations at the borders, the founding of state water-conservation districts, a state park system, a comprehensive survey of water resources and hydroelectric power sites, law-code reform, and a comprehensive budget system to take control of state finances. The budget ballooned from $191 to $244 million through "tax revision" increasing revenues via tax increases on banks and corporations. In 1929, the state had a $30 million treasury surplus which was retained as a "rainy day fund." Conservatives wanted the surplus returned to tax payers, and, when surplus dollars emerged, California revisited this dilemma from time to time.

In 1930, the nation was recoiling from the stock-market crash of 1929 and heading into depression. In California, San Francisco mayor James Rolph won the governorship with the slogan "Smile with Sunny Jim." The budget went up, not down. The "rainy day" fund dried up in the process. Rolph and the legislature set up youth camps for conservation work ahead of the New Deal's Civilian Conservation Corps. For the period 1933–1935, the Depression resulted in a $24 million budget cut, increasing the eligibility age for old-age pensions from seventy to seventy-five, and the passage of a sales-tax bill. Rolph vetoed an income-tax bill but did not live to see the repercussions. A heart attack took him in June 1934. Lieutenant Governor Frank Merriam took over and brought modern politics to the statehouse.

Merriam set the stage for the no-party politics that dominated the state well into the late twentieth century. He hired professional managers, used the media for his message, developed attack advertisements, did not make personal attacks in public, and advocated pragmatism in public statements. In concert with the legislature, Merriam increase the budget to $400 million. He increased taxes. In 1935, taxes went up. Sales taxes and taxes on banks, insurance companies, corporations, liquor, and inheritances went up. California had its first income tax. The Republican Party was the tax-and-spend party.

The 1938 election brought change to California. Culbert Olson, an ideologue, won election to the governor's seat on the slogan "Bring the New Deal to California." Earl Warren used the Progressive electoral system to marginalize political parties by cross-filing on the Republican, Democratic, and Progressive tickets for attorney general. He won on all three ballots. Olson promised to pardon Tom Mooney, languishing in prison for the 1916 Preparedness Day bombing, and, when elected, he had Stanley Mosk, his pardons secretary, prepare the papers for signature. He accomplished little else because he was in constant conflict with the legislature. In 1943, the people sent Warren to the governor's mansion as Fortress California geared up for war production.

Warren found California awash with tax revenues generated by the war economy. For the first time, California experienced cuts of income, sales, bank, and corporate taxes. Warren and the legislature spent more on education, corrections, road construction, public health services, state unemployment compensation benefits for medical and hospitalization, and the newly created Department of Mental Hygiene. Public money created fact-finding panels, citizen advisory groups, and other progressive think tanks. Warren pushed for the Central Valley Project to increase irrigation and hydroelectric power. He fought for but did not obtain a fair employment practices commission. Warren proposed a system of universal health care for all Californians, lobbied the Assembly Committee to bring the measure to the floor, and lost on a 39–38 vote. Even Earl Warren had limits, but when he left the state in 1953 to become Chief Justice of the US Supreme Court, he left a budget surplus. He also left behind an anti-Communist hysteria managed by State Senator Jack Tenney and Richard Nixon, whose 1946 campaign against Helen Gahagan Douglas included the sobriquet "Pink Lady," linking her with the suspected Communist leanings of Hollywood movie personalities. To Warren's credit, he refused to institute loyalty checks on state employees, supported Hollywood's best people, and called the professor's loyalty oath unconstitutional.

Lieutenant Governor Goodwin Knight kept the Republican leadership liberal and spending public money at record rates. Knight opposed right-to-work legislation, signed a $2 billion budget that spent more on higher education, old-age pensions, unemployment and disability compensation, and created new programs. Crime-control programs garnered public money, as did childcare and mental-health programs. The Republican Party stayed a tax-and-spend party.

In 1958, conservative Republicans lead by Richard Nixon and US Senator William Knowland split the party when the Democratic Party was finally getting organized. The California Democratic Council of 1954 gave the party organization. Led by Elizabeth Snyder, the first women state chairperson, the California Democratic Council energized women and the middle class. Edmund G. "Pat" Brown won the governor's race. Clair Engle defeated Knowland for the US Senate seat, and the people sent Stanley Mosk to the Attorney General's office.

Brown worked with the legislature to increase taxes and balloon the budget from $2.2 billion in 1959 to $5 billion in 1966. Money flowed into higher education, road construction, worker's compensation, old-age pension, and other social programs including limited health care. The 1959 session of the legislature was a productive as the 1911 session when Hiram Johnson's progressive package of Bills passed and arrived on his desk for signature. The Burns–Porter Water Bonds Act of 1959 authorized the issuance of bonds to fund the State Water Project to bring still more irrigation water to California's agricultural sector. The Fisher Act of 1959 required public-school teachers to have an academic major. The master plan for California higher education came with the Donahoe Higher Education Act of 1960, specifying the missions of the state's community colleges, state university, and the University of California. Brown also won the Fair Employment Practice Commission from the 1959 session. A key provision of the law was barring racial discrimination in hiring workers. The Rumford Fair Housing Act followed in 1963, barring race as a factor in renting or purchasing housing. As time would tell, it was easier to raise taxes in California than to end race-based practices and racism.

The Democratic Party splintered in 1966, allowing Ronald Reagan, an ideologue, to win the governor's race with lower taxes and small government rhetoric. In 1967, Reagan and the legislature raised taxes, especially on banks, corporations, and the wealthy. They increased the budget by half a billion dollars. Reagan failed to obtain the repeal of the Rumford Fair Housing Act, opposed the damming of the middle fork of the Feather River for more water and electricity, and signed the Beilenson Bill in 1967, ushering in abortion on demand in California. Reagan avoided the details of government, used a kitchen cabinet that produced briefing papers on a single page, maintained a conservative image with the media, and teamed up with Jesse Unruh, the Speaker of the Assembly, to stop property-tax reduction under the Watson Amendment to avoid state tax increases. While governor, Reagan increased taxes and doubled the state budget to $10 billion and continued to utter anti-tax conservative slogans. Although ideological in rhetoric, Reagan was a pragmatist when it came to tax-and-spend politics. Inheriting a deficit from Pat Brown, Reagan worked with state senator George Deukmejian to craft a budget. Publicly, he announced a hiring freeze and a purchasing freeze on state vehicles. Working the Bill through the Senate, it emerged as a $1 billion tax increase, quadruple the size of Brown's 1959 tax increase. Reagan's cooperation with Unruh and Deukmejian enabled the largest state tax increase in history. When Reagan left office, the "watch my lips" governor had engineered an increase in corporate taxes for 5.5 percent to 9 percent. Taxes on banks rose from 9.5 percent to 13 percent under the "no new taxes" governor. Personal income-tax top bracket ballooned from 7 percent to 11 percent. Reagan was the master of the media and one of the most able tax-and-spend governors of California.

In 1974, the Republican Party was in disarray, and the people put Democrats in charge. Edmund G. "Jerry" Brown, Jr., was governor with a Democratically controlled legislature and a Democrat in every constitutional office except Attorney General. George Deukmejian won the office of Attorney General and kept the Republican message alive in Sacramento. Jerry Brown's appointments went to liberals and minorities. Mario Obledo was the first Mexican American in the governor's cabinet. Cruz Reynoso was the first Mexican American appointed to the state Supreme Court. Wiley Manuel, the first African American, and Rose Bird, the first woman, on the Supreme Court, joined Reynoso. In all, Brown appointed about 1,600 women to state offices. In 1975, the legislature created the Agricultural Labor Relations Board enabling Cesar Chavez to unionize farm labor. The solons also created the California Conservation Corps, giving youth an option to work in public service for a better environment. Brown vetoed a death-penalty measure in 1977 and the legislature quickly overrode the veto. Brown pronounced a "new spirit" of government in the "era of limits."

In 1978, Proposition 13 passed rolling back property taxes and imposing a two-thirds majority rule on the legislature for any budget or tax increase. It was the culmination of a twenty-year campaign to contain the legislative and executive orgy of taxing and spending. Brown moved to the right, and the state recoiled in the wake of a property-tax shortfall.

Proposition 13 was a California initiative constitutional amendment approved in June 1978 that started an American anti-government tax revolt. The ballot measure set real-estate property value for tax purposes at 1975–1976 market value, limited real-estate taxes to 1 percent of that value, limited tax increases to 2 percent per year for continuing owners, provided for a full reassessed value base for new owners, required a two-thirds vote for legislative revenue increases, and made any local government tax increase dependent upon a two-thirds approval of the local voters.

Howard Jarvis and Paul Gann led this anti-government crusade. Howard Jarvis was the chairman of the United Taxpayers Organization with an unsuccessful political record including a failed non-qualifying property-tax reform initiative in 1972. Paul Gann had a political career dating back to the 1950s. Active in the Republican Party, Gann organized the People's Advocate, Inc. in 1974, focused upon crime issues. In 1977 they met and agreed to join forces to battle against rising property taxes. What they drafted with the early help of Los Angeles Assessor Philip Watson was Proposition 13, popularly known at the Jarvis–Gann Initiative. Working almost independently in the campaign with grassroots and county organizations, they gathered 1, 263,000 signatures for the ballot measure. More than 1 million of those signatures or about 81 percent were deemed valid; the first time that over 1 million signatures had ever been solicited for a ballot measure. Their main fundraising group "Yes on 13" collected $1.5 million, with just under half of the contributions less than $50. The Jarvis–Gann campaign shook California's politics at its core.

The legislature, part of a tax-and-spend history regardless of party power, responded with the Property Tax Relief Act of 1978, and Proposition 8 designed to give 30 percent property-tax cuts compared with Proposition 13's 50 percent cuts. This put a tax-cut measure approved in Sacramento on the ballot as an alternative to Jarvis–Gann.

Jarvis–Gann forces launched an effective public campaign. They accused government of a political snow job. In particular, they pointed out that the 1 percent tax rate would provide more than enough to run government and encourage government to cut wasteful spending such as "phony sick leave grants," fat pension plans, and union-inflated prevailing wages for state workers. Further, the media loved Howard Jarvis, who gave them quotable sound bites and memorable flourishes. Jarvis was the Ronald Reagan of the media with a fixation on ten-word phrases.

State Democrats, the PTA, and labor unions opposed Proposition 13. They paraded horrors such as closed libraries, crippled schools, abandoned paramedic services, and massive layoffs. They were on the defensive and saddled with a projected $3.6 billion budget surplus in Sacramento.

On election day, 1978 voters turned out in near record numbers and they passed Proposition 13 with 65 percent of a 69 percent turnout. The turnout was the highest since 1916, even more remarkable in an off-year election. The voters crushed Proposition 8 by 53 percent to 47 percent. What was most striking was the breadth of support for Proposition 13. Jarvis–Gann had touched a political nerve that set off a national phenomenon. In the West, this was the beginning of the sagebrush revolution.

In 1982, the state moved into what Jackson K. Putnam termed the era of ill feeling. George Deukmejian was governor, and he preached fiscal restrain and increased taxes. The "Duke" worked with the legislature to impose user fees on California. In Republican parlance, a fee was not a tax. So, community-college students now paid fees for their education. Building on his tough-on-crime reputation as Attorney General, rewriting the death-penalty initiative, sponsoring the "Use a Gun, Go to Prison" statute and supporting a prison-building program, Duke took on the Supreme Court. In 1986, he worked with H. L. Richardson and William Roberts, his former campaign chairman, to unseat Chief Justice Rose Bird, Associate Justice Joseph Grodin, and Associate Justice Cruz Reynoso. Deukmejian was savvy enough to allow Stanley Mosk run unopposed. The judicial election campaign was successful, funded by the agribusiness interests who remembered Bird's actions in favor of farm workers. The governor appointed his former law partner Malcolm Lucas to the Chief Justice position.

The 1988 legislative session was the beginning of legislative deadlock as a political art form with revenue shortfalls and budget deadlines. When Pete Wilson entered the governor's mansion in 1991, he faced the worst budget shortfall in the state's history until the 2009 session. State tax revenues fell with the Berlin Wall as the aerospace industry scaled back. His first budget was a negotiated tax-increase and spending-cut masterpiece of personal and media persuasion. Wilson personally lined up Republican votes to increase taxes and cut welfare checks and state employee salaries. He shifted mental- and public-health facilities to county control and off the state tab. As revenues subsequently grew with the resurgence of the economy, Republican Wilson signed larger and larger budgets. He also sponsored the three-strike law putting three-time losers in prison for life. Wilson got behind the anti-affirmative action drive on Proposition 209.

Proposition 209, a California initiative constitutional amendment, was a November 1996 ballot measure to end affirmative action in state programs. This measure was part of a national debate in the 1990s and a Republican Party drive to end all affirmative-action programs in California.

Officially the "Prohibition Against Discrimination or Preferential Treatment by State and Other Public Entities" measure, it was the handiwork of Ward Connerly, a conservative African-American businessman on the University of California Board of Regents, Governor Pete Wilson, and Attorney General Dan Lungren. Preferential admissions practices at the University of California were annual targets for critics of the system, particularly Connerly. The primary beneficiaries of affirmative action in California had been women and African Americans. The obvious losers in the University of California, Berkeley, admissions process had been Asian Americans with demonstrably higher test scores.

The public campaign was well funded. Supporters raised $5 million of mostly Republican money, and opponents filled their chest with a similar amount from Democrats, unions, and feminists. The battle lines between conservatives and liberals was clear.

Although big business did not play a significant role in the campaign, national politics played a major role. This was a presidential-election year, and Bob Dole needed California. He came to California and spoke in favor of Proposition 209, a reversal of his prior support for affirmative action. Bill Clinton supporters and liberals stoutly called for the preservation of affirmative action as part of a civil-rights ideology. The national rhetoric had a distinctly California twang with Protestants decisively in favor, Roman Catholics in the middle, and Jews slightly on the other side. African-American and Hispanic voters decisively opposed Proposition 209, but 55 percent of the voters put it into law, at least for a day.

One day after the election, a federal civil-rights suit on behalf of female and minority contractors, labor unions, and students found its way into court. Governor Wilson moved quickly to petition the federal judge to stay proceeding to allow the state courts to rule on the proposition. The court denied Governor Wilson and issued a temporary restraining order against the Governor and Attorney General because of the strong probability that the initiative was unconstitutional. The proceedings in federal district court did not go Wilson's way, and, by the end of 1996, the court made the restraining order permanent. The Clinton administration joined the suit against Wilson and Proposition 209, but in April 1997 the Ninth Circuit Court of Appeals overturned the trial judge, ruling Proposition 209 to be constitutional. In November 1997, the US Supreme Court let stand the Circuit Court of Appeals ruling.

Proposition 187, a California initiative statute, was also a November ballot measure in the spirit of Proposition 13 designed to save the state $5 billion per year by reducing public services for illegal immigrants. The measure denied public social, health, and education services to illegal immigrants. It required state and local agencies to report suspected illegal aliens to state and federal authorities. The measure declared that the manufacture, sale, or use of false citizenship or residency documents was a felony.

Popularly known as the "Save our State" initiative, Proposition 187 raised serious constitutional issues regarding illegal-alien access to public education and federalism questions in terms of the regulation of immigration. Regardless, Governor Pete Wilson and other Republican leaders, including Harold Ezell, President Reagan's Western regional director of the Immigration and Naturalization Service, joined in support. Democratic and liberal leaders, Cardinal Roger Mahoney of Los Angeles, Los Angeles County Sheriff Sherman Block and the League of Women Voters opposed the measure. Opponents outspent supporters three to one.

Republican, moderate Euro-American, and African-American voters passed the measure with 59 percent of the vote. Democratic, liberal, and Hispanic voters overwhelmingly voted against the measure.

Proposition 187 was quickly in the courts. In 1996, Federal District Judge Mariana R. Pfaelzer ruled that the denial of services to illegal immigrants was unconstitutional. In 1998, she made her injunction permanent, grounding her decision on the federal government's exclusive authority to legislate on immigration.

This measure created deep ill will in the Hispanic community, and many immigrants responded to its threats by becoming citizens.

Proposition 187 designed to drive illegal immigrants out of the state, and Proposition 209 produced an extremely strong coefficient of correlation between the votes later manifested in the votes for Proposition 227 to end bilingual education in California's public schools. Research has demonstrated that increased wealth, increased education, and urban residency made one most likely to support such sociocultural measures. The old Progressive politics designed to minimize the impact of political parties and make direct popular government possible came home to roost in urbanizing Eden.

While Progressivism was twisting California from left to right, national politics witnessed a presidential campaign between two Westerners: Lyndon Johnson and Barry Goldwater. Johnson inherited the White House because of the assassination of President John F. Kennedy in Dallas, Texas. The Texan faced Barry Goldwater of Arizona in the 1964 election. Johnson's "Daisy" television advertisement frightened Americans into the Johnson camp. The images were of a little girl picking daisies and an atomic-bomb blast. The message was: Vote Johnson or risk world war with Goldwater. Johnson won and pushed the country deeper into the Vietnam War. Yet Goldwater's conservative message to the Republican Party stuck. Goldwater opposed the New Deal and its legacy of big government, labor-union advocacy, and the welfare state. He thought the executive branch had too much power and the military needed modernization.

Domestic opposition to Johnson's war in Vietnam drove him from seeking another term in office. Richard Nixon, a Californian, won in 1968, with a secret plan to end the war and to restore conservative values in government. Nixon's Watergate scandal, in term, ousted him from office. In 1980, Ronald Reagan, another Californian, entered the White House, telling the nation what he had largely told California voters: small government, lower taxes, and a strong military were the keys to American peace and security. Reagan cut taxes his first term and raised them thereafter, still telling the American people that he was cutting their taxes. Texas inhabitants of the oval office included George H. W. Bush in the twentieth century and his son, George W. Bush in the twenty-first century. In 2008, John McCain, who entered the US Senate taking Barry Goldwater's seat, ran an unsuccessful campaign for the presidency. Despite the loss, the influence of the West in Washington is clearly visible in the Congress.

In 2009, the Speaker of the House is Nancy Pelosi of San Francisco, California, and US Senator Harry Reid of Nevada is Senate Majority Leader. California's delegation holds many of the key positions in Congress. Representative Henry A. Waxman of Beverly Hills is the Chairman of the House Energy and Commerce Committee. Senator Barbara Boxer chairs the Senate Environment and Public Works Committee. Representative Howard L. Berman of Valley Village chairs the House Foreign Affairs Committee. George Miller of Martinez is Chair of the House Education and Labor Committee. Chula Vista's Robert Filner chairs the House Veterans Affairs Committee. In Democratic Party leadership positions are Representative Barbara Lee of Oakland heading the Congressional Black Caucus. Representative Xavier Becerra of Los Angeles is vice chairman of the House Democratic Caucus. Zoe Lofgren of San Jose is chairwoman of the California House Democratic delegation. Senator Dianne Feinstein sits on the Senate Appropriations Committee and Jerry Lewis of Redlands is the senior Republican on the House Appropriations Committee. Americans voted for change in 2008. In 2009, many Western politicians are in powerful positions in Congress to mold change for the nation.

Perhaps the most important political agenda emerging from the West was the demand for sustainability as part of a larger environmental movement. Andrew Kirk argues that particularly in the West the counterculture entrepreneurs who skipped the protest movement and the polls were the ones who made the most lasting contribution to the politics of the last decades of the twentieth century. Further, these libertarian-leaning environmentalists offered an alternative to the Progressive model of environmentalism that preserved and protected land rather than revolutionizing behavior. They focused on modifying consumption to improve the quality of life. Included in this phalanx for change were Yvon Chouinard for Patagonia, David Brower of the Sierra Club, Stewart Brand of the *Whole Earth Catalog,* Buckminster Fuller of the geodesic dome, and Steve Jobs and Steve Wozniak of Apple Computer. These entrepreneurs perfected cause-based and liberation marketing systems directly connected to the counterculture, enabling them to create powerful corporations, find new market niches, and reshape the American economy. Even more certainly, Mo Siegel, founder of Celestial Seasonings Corporation, founded a multi-billion-dollar natural-foods industry in the hip capitalist world of 1970s Boulder, Colorado. As Amy L. Scott points out, Siegel and other Boulderites articulated a visionary spirit of place. It was a micropolitan urbanism based on environmental preservation and sustainability. They celebrated tolerance and difference. They created economic systems that offered products for health and sustainability.

Politically, the Green Party's Ralph Nader articulated the different ways of seeing government's relationship with people and the environment. Environmentalism in the West went mainstream via conservation, preservation, and recreation. People came to realize that protecting the environment was a personal choice influencing their quality of life. Whether voting for environmental causes, donating to environmental organizations, or simply choosing to eat organic foods, Westerners made those choices. People walking into Serena Rundberg's Nova Café on Main Street in Bozeman, Montana, immediately see a chalkboard displaying her commitment to locally produced organic foods. Dairy products come from farms in the Gallatin Valley, goats cheese from Melvin and Sue Brown's Amaltheia Organic Dairy in Belgrade, Montana; chai from Tipu's Tiger Chai in Ronan, Montana; mushrooms from Garden City Fungi, Missoula, Montana; beef from Twin Bridges, Montana; and bison from Steve and Daryl Kroon's Montana Bison Ranch, Manhattan, Montana. What started in the West continues to delight the dining public. Whether politics or palate, the Nova Café serves up the best of the West in conversation or healthy taste.

NOTES

1. Gretchen Ritter, *Goldbugs and Greenbacks* (Cambridge: Cambridge University Press, 1997), pp. 169–70.
2. Ritter, *Goldbugs and Greenbacks,* p. 21.

FURTHER READING

Allswang, John M., *The Initiative and Referendum in California, 1898–1998* (Palo Alto, Calif.: Stanford University Press, 2000).

Andrews, Thomas G., *Killing for Coal: America's Deadliest Labor War* (Cambridge, Mass.: Harvard University Press, 2008).

Argersinger, Peter H., *The Limits of Agrarian Radicalism: Western Populism and American Politics* (Lawrence, Kans.: University Press of Kansas, 1995).

—— "The Most Picturesque Drama: The Kansas Senatorial Election of 1891," *Kansas Historical Quarterly,* 38 (spring) (1972): 43.

—— *Populism and Politics: William Alfred Peffer and the Peoples Party* (Lexington, Ky.: University of Kentucky Press, 1974).

—— "Road to a Republican Waterloo: The Farmer's Alliance and the Election of 1890 in Kansas," *Kansas Historical Quarterly*, 33 (winter) (1967): 443.
—— "No Rights on this Floor: Third Parties and the Institutionalization of Congress [Populism]," *Journal of Interdisciplinary History*, 22 (spring) (1992): 655–690.
—— "Pentecostal Politics in Kansas: Religion, the Farmers' Alliance, and the Gospel of Populism," *Kansas Quarterly*, 1 (fall) (1969): 24.
—— "Populists in Power: Public Policy and Legislative Behavior," *Journal of Interdisciplinary History*, 18 (1987): 81–105.
—— "Ideology and Behavior: Legislative Politics and Populism," *Agricultural History*, 58 (January) (1984): 43–58.
Barnes, Donna A., *Farmers in Rebellion: The Rise and Fall of the Southern Farmers Alliance and People's Party in Texas* (Austin, Tex.: University of Texas Press, 1984).
Barton, D. Scott, "Party Switching and Kansas Populism," *The Historian*, 52 (May) (1990): 453–467.
Berman, David R., *Radicalism in the Mountain West, 1890–1920: Socialists, Populists, Miners, and Wobblies* (Boulder, Col.: University Press of Colorado, 2007).
Bicha, Karel D., "Jerry Simpson: Populist without Principle," *Journal of American History*, 54 (September 1967): 291–306.
Billington, Monroe, "Susanna Madora Salter: First Woman Mayor," *Kansas Historical Quarterly*, 21 (fall) (1954): 173.
Brodhead, Michael J., *Persevering Populist: The Life of Frank Doster* (Reno, Nev.: University of Nevada Press, 1969).
Chavez, Lydia, *The Color Bind: California's Battle to End Affirmative Action* (Berkeley, Calif.: University of California Press, 1998).
Goodwyn, Lawrence, *Democratic Promise: The Populist Movement in America* (Oxford: Oxford University Press, 1976).
Hall, Greg, *Harvest Wobblies: The Industrial Workers of the World and Agricultural Laborers in the American West, 1915–1930* (Corvallis, Oreg.: Oregon State University Press, 2001).
Johnson, Jeffrey A., *"They Are All Red Out There": Socialist Politics in the Pacific Northwest, 1895–1925* (Norman, Okla.: University of Oklahoma Press, 2008).
Hurt, R. Douglas, "Agricultural Politics in the Twentieth-Century American West," in Jeff Roche (ed.), *The Political Culture of the New West* (Lawrence, Kans.: University Press of Kansas, 2008), pp. 51–73.
Kirk, Andrew G., "Free Minds and Free Markets: Counterculture Libertarians, Natural Capitalists, and an Alternative Vision of Western Political Authenticity," in Jeff Roche (ed.), *The Political Culture of the New West* (Lawrence, Kans.: University Press of Kansas, 2008), pp. 281–309.
Larson, Robert W., *Populism in the Mountain West* (Albuquerque, N. Mex.: University of New Mexico Press, 1986).
Osselaer, Heidi J., *Winning their Place: Arizona Women in Politics, 1883–1950* (Tucson, Ariz.: The University of Arizona Press, 2009).
Perlstein, Rick, *Before the Storm: Barry Goldwater and the Unmaking of the American Consensus* (New York: Hill & Wang, 2001).
Putnam, Jackson K., *Modern California Politics*, 4th ed. (San Francisco, Calif.: Boyd & Fraser, 1996).
Rarick, Ethan, *California Rising: The Life and Times of Pat Brown* (Berkeley, Calif.: University of California Press, 2005).
Roche, Jeff (ed.), *The Political Culture of the New West* (Lawrence, Kans.: University Press of Kansas, 2008).
Schrag, Peter, *California: America's High-Stakes Experiment* (Berkeley, Calif.: University of California Press, 2006).
—— *Paradise Lost: California's Experience, America's Future* (New York: The New Press, 1998).
Scott, Amy L., "Remaking Urban in the American West: Urban Environmentalism, Lifestyle Politics, and Hip Capitalism in Boulder, Colorado," in Jeff Roche (ed.), *The Political Culture of the New West* (Lawrence, Kans.: University Press of Kansas, 2008), pp. 251–280.

CHAPTER SEVENTEEN

GOVERNMENT AND LAW IN THE AMERICAN WEST

Aaron Steven Wilson

The history of Western political and legal institutions is a study in American imperialism.[1] Territorial governments provided temporary legal and political structures during the occupation and conquest of a foreign region. Statehood completed this process by irrevocably connecting the region to the USA and supplying a permanent form of government based on republican principles. Although current trends in Western history emphasize urbanization, gender, sexuality, culture, and ethnicity in studying the American imperial experience, it is necessary to pause and re-examine the role of government and legal institutions in this process. Throughout the West, American expansion has received ample attention from historians. While this article does not seek to add any new findings to the overall history of the Americans' longest-running imperialist venture, it will introduce the reader to the origin and evolution of Western legal and governmental institutions involved in westward expansion.

To begin, one must define the American West. Currently, historians generally agree that the region begins at the Mississippi and Missouri rivers, extends to the Pacific Ocean, and includes Alaska and Hawaii.[2] This article, however, must go beyond these political borders. In its treatment of the development of governmental and legal institutions in the American West, it will examine all lands west of the Appalachian Mountains. Even American territories lying in the Pacific Ocean will receive attention here. To study the territorial history, Western legal history, or Western state constitutional history, historians cannot have a myopic view that blinds them to crucial developments in other parts of the nation and world. One must examine the origins of territorial and state governments and their composition both in the West and beyond. After all, events in the East shaped the government and legal institutions in the West, which, in turn, affected Americans' approach to overseas territories. As such, the article examines not only the Eastern origins of territorial policy but also territorial and state policies in the Pacific Ocean.

Beginning with the foundations of the territorial system in American history, one must look east across the Mississippi River to developments along the eastern seaboard, in the Ohio River Valley, and in the South. Following their independence from Great Britain, the confederated states desperately needed money. They had fought an eight-year engagement against one of the most powerful European nations in Western civilization. Having failed miserably to raise domestically the money and supplies necessary to feed, clothe, and arm the Continental Army and various militia units, the Second Continental Congress relied heavily on loans from the government of King Louis XVI of France. Following the war, the American economy also suffered from loss of trade with Great Britain. Once willing to trade openly with the colonies, the Empire considered the independent states with the same cold, protectionist eye it viewed other nations. Save for irreplaceable commercial items such as tobacco, indigo, and cotton, American ships, barrels, and other products became subject to mercantilist policies. Set adrift in international currents, Americans had to rebuild a postwar economy and to establish new trading routes. Needless to say, the challenges were rather difficult. After all, the interest accrued by the national debt stood at over $10 million alone, and, to make matters worse, America had a trade deficit! It is only understandable that between 1785 and 1787 the US government lacked the necessary capital to continue repaying its erstwhile ally, France, for the massive loans extended to the young nation during the war.[3]

Although the incredibly flawed Articles of Confederation may not list among the most notable achievements of a generation that produced the Constitution of the United States of America, the Declaration of Independence, *Common Sense,* and the Virginia Declaration of Rights, their response to the national debt, ailing economy, and expansionist impulse was masterful. British mercantilist policies had left the American economy with a limited industrial base. Burdened by laws restricting the production of everything from iron to fur hats, American faced the task of building an industrial base in a colonial economy. Needless to say, their resources were somewhat limited. While the Americans may not have had enough foundries, shipyards, and textile mills to match the economic might of Great Britain, they did have one thing in abundance: land. Since the first English settlers disembarked from the *Susan Constant, Discovery,* and *Godspeed,* land was an invaluable commodity to Americans. It acted as a source of personal wealth as well as a marketable item with which speculators potentially could realize vast profits. Congressional leaders recognized this in the postwar years. Expansion and settlement were in the American blood. Americans and European immigrant looked west and hungered for the lands available beyond the Appalachian Mountains. All Congress needed to do was to provide a guide for governing, surveying, and selling Western lands to generate revenue from this overwhelming desire.

Those involved in devising a territorial process were most concerned with maintaining unity between new states and the thirteen original states. Unlike conservationist and preservationist views of the wilderness as a national treasure, Americans during the colonial era viewed undeveloped lands as a region dominated by savagery. It was a wasteland that settlers must conquer and subdue. Instead of the Edenic vision of a place where mankind existed in a state of nature, it threatened to turn the civilized European into a barbarian.[4] This perception of the wilderness as a threat civilization did not end with the conclusion of the colonial era. American statesmen feared that unchecked expansion into Western lands posed a direct threat to unity. If left to their own devices and urges, settlers could revert to a state of nature in which men worked only to feed their immediate wants. Instead of John Locke's vision of a state of nature, Americans viewed it as "the lawless reign of anarchy and vice" that would erode the moral fabric and national bonds of pioneer settlements.[5]

From 1783 to 1787, leading statesmen wrestled with the challenge of creating a national policy for westward expansion that would prevent Americans in the West from reverting to a primitive state of existence or severing their union with the USA. In response to this challenge, congressmen passed the Ordinance of 1784 and Northwest Ordinance of 1787.

The key, they determined, to orderly Western expansion would be one of gradual, directed settlement. They wanted to devise a policy that transformed Western lands into schools for republican ideology. As American pioneers developed communities in the West, federal policy would provide self-government in incremental stages aimed at connecting Western economies and governments to the older states along the eastern seaboard.[6]

Creating a process of settlement and governance in Western territories did not come overnight. Just as the founding generation experimented with state and national forms of government, congressmen relied on trial and error to craft a plan for settlement. The first attempt to develop a policy for Western lands resulted in the Ordinance of 1784. The creation of Thomas Jefferson, it was the first legislative attempt to order settlement in the American West. The Ordinance called for the division of the extensive Northwest territory into multiple states. While only five states ultimately would emerge from the region, Jefferson's plan called for Congress to carve six states out of the lands north of the Ohio River. These he imaginatively named, A, B, C, D, E, and F.[7]

According to the Ordinance of 1784, territorial government passed through four stages. In its earliest phase, government was autocratic, directed by a representative of the Continental Congress. Once enough settlers established themselves in the region, "free males of full age" meeting in assembly would establish a more democratic territorial government.[8] The Ordinance permitted them to create governments based upon existing state constitutions. This provision intended to give them a template for proper republican government. Once the population reached 20,000, the inhabitants could request admission into the Union. Congressional approval signified the fourth and final stage of the territorial process as the new state entered into the Union on an equal footing with the older states.[9]

The process outlined by Jefferson had two important characteristics. First, it was colonial in nature, inspired by the process experienced in British North America. Along the eastern seaboard, colonial governments slowly evolved into the government, council, assembly form that was standard in the eighteenth century. One can look at the evolution of Virginian government—the one most influential on Jefferson—as an example. The original charter of the Virginia Company of London called for a president and a council of seven to oversee local government at Jamestown. The charter of 1609 replaced this system with a governor and council. It was not until the Great Charter of 1618 that the House of Burgesses appeared in colonial Virginia. By that time, the original settlers had overcome the starving time by developing self-sustaining agriculture, had established tobacco as the profitable cash crop, and had hived out smaller settlements from Jamestown. From its humble origins in the early seventeenth century, the colonial government, consisting of a governor and council working in tandem with the colonial assembly, did not emerge until the imperial reshuffling during the late seventeenth century and early eighteenth century. A stabilized colonial society and growing imperial framework had made the older, more arbitrary form of government unmanageable, calling for greater autonomy at the local level.[10] Just as the original colonies required a period of tutelage before assuming greater control over their affairs, congressmen viewed self-government as a product of time and proper instruction, not government fiat.

Second, the gradual process and wording of the Ordinance of 1784 ensured the loyalty of the new states. The idea that republics must remain small political entities remained a cornerstone of American thought during this period. As James Madison's *Federalist Paper* Number 10 reveals, many leading political minds of the time doubted that a republic could preserve internal cohesion as its borders expanded. Unlike a monarchical empire, factions and distance would erode the bonds of unity between new and old provinces. Throughout his writing and in the Constitutional Convention, Madison faced the monumental task of refuting this age-old axiom. He argued that larger republics ensured liberty by preventing factions and unworthy men from controlling government. The larger the republic, the more disparate and numerous the parties represented in Congress.[11] While Madison himself did not go so far as to advance this position, one could say that expanding into new territories and adding new states was a way of further preventing tyranny. Jefferson's Ordinance of 1784 also attempted to quell any such fears toward expansion by clearly stating that new states would remain permanently tied to the American government. According to this legislation,

> the preceding articles shall be formed into a charter of compact [...] and shall stand as the fundamental constitutions between the thirteen original states, and each of the several states now newly described, unalterable [...] but by the joint consent of the USA in Congress assembled, and of the particular States within which such alteration is proposed to be made.[12]

In other words, new territories and their subsequent states entered into a compact that was binding and permanent.[13]

Initially, it was thought that the Ordinance of 1784 ended the debate over the process of directing Western settlement. By 1785, congressmen had moved beyond this question and began addressing the issue of selling and surveying the lands acquired from the British at the Treaty of Paris in 1783. With the Articles of Confederation prohibiting the imposition of taxes on the thirteen states by Congress, statesmen needed a way of raising revenue consistent with the current frame of government. Given the precarious nature of the American economy and the gargantuan war debts held by Congress, this was becoming a pressing problem.

In passing the Land Ordinance of 1785, politicians devised an ingenuous means of raising revenue through the orderly surveying and selling of the national domain. In accordance with the ordinance, federal surveyors divided the Western territory into townships of 36 square miles. Townships were further subdivided into 1-square-mile sections numbered 1–36 with lots reserved for Continental Army veterans' service to their country in a time of need. Since it could not levy taxes, land was the only means by which Congress could remit their debts to former soldiers. Congress also provided a means to ensure civic virtue in the West by supporting education. Since education was necessary for a civilized, republican state, the Land Ordinance reserved Lot 16 in each township to support a school.[14]

With the Ordinance of 1784 and Land Ordinance of 1785 in place, it seemed that the national government had all the prerequisite legislation for their orderly, gradual settlement of the West, at least until James Madison returned from a tour of the region in 1785. In order to settle and sell the land, Congress had to survey it first. This was a rather slow process given the detailed orders of Land Ordinance and the sheer size of the national domain. Returning from his trip, Madison reported that the process as envisioned by the Ordinance of 1784 was not likely to produce any visible results in the near future. According to Jack Ericson Eblen's research on this subject, Madison concluded that the "size of the proposed state was too small and the population requisite for statehood too large."[15] It would take an inordinate amount of time to populate the region, and, even then, their economies would be limited by the small boundaries called for in the ordinance. In short,

Congress may have produced a rough draft for Western settlement, but too many details were left unresolved for proper settlement.[16]

For Americans, 1787 was a very big year. A group of fifty-five delegates met in Philadelphia to draft a new federal government for the nation that has remained for over 200 years. Meanwhile, Congress devised new legislation for the Northwest territory that provided a template for American expansion that lasted into the twentieth century. Looking at these two events, one can honestly say that the federal tradition of representative democracy and aggressive American imperialism were born almost simultaneously in the USA. Over the course of the next 200 years, they would mature in tandem, both having an equally significant impact on the American history, government, and law. But the similarities stop there. Whereas its citizens willingly and readily embraced their traditions of representative government as enshrined in the Constitution, the imperial heritage born of the Northwest Ordinance remains one that Americans rarely, and only reluctantly, acknowledge and embrace.

The Northwest Ordinance is a work of legislative art. It is purely imperial as it created a process by which settlers and legislators transformed land wrested from American Indians, and later Mexicans, into states. It did so by grafting the British imperial system of colonial governance onto institutions used in popular government, thus incorporating the early American experience with westward expansion. Simultaneously, it created a method of automatic decolonization as its authors incorporated mechanisms that fully incorporated territories into the American government once they reached a certain milestone. The only drawback was its failure to provide for full civil rights of all citizens living in new states. If one adopts Frederick Cooper's definition of an empire as "a political unit that is large, expansionist (or with memories of an expansionist past), and which reproduces differentiation and inequality among the people it incorporates," then the process of expansion as provided by the Northwest Ordinance, American Indian policy, and American ethnic law created a imperial republic with a federal model.[17] As Americans expanded into new territories, they utilized the modified British imperial model to recreate their democratic institutions in the West. While the legislation provided for the incorporation of these new entities into the American empire-state, it automatically decolonized the regions for its Euro-American inhabitants. Meanwhile, voting requirements, anti-non-European ethnic legislation, and American Indian policy excluded groups deemed "un-American." It was only through the civil-rights legislation of the mid- to late twentieth century that African Americans, Chinese Americans, Japanese Americans, Mexican Americans, and a myriad of other ethnic groups received full incorporation (not to forget women in the Nineteenth Amendment). Today, the debate over the exclusion or inclusion of American Indian groups remains unresolved.

Howard Lamar was quite correct when he called the Northwest Ordinance the "Constitution of the West" as it provided Euro-American settlers with a working frame of government, complete with articles acting as a bill of rights.[18] The legislation streamlined the four-stage process of the Ordinance of 1784 into three simple stages, each marked by very specific population milestones. During the first stage, Congress organized the Northwest territory into a district. Although the framers called for one district to govern the entire region, they authorized the subdivision of the territory into two districts. Government during the district stage followed the model of British colonial government during the early 1600s. A governor held most of the power. He was the commander-in-chief of militia forces, appointed the magistrates, held power over the judiciary, and conducted American-

Indian relations. Recognizing the demands of the office, Congress provided for a secretary to keep records of legislation and other official proceedings. Finally, the ordinance provided the district with three judges who held office for tenure of good behavior.[19]

Together, the governor and three judges formed the legislative branch of government just as the governor and councilors did during the earliest stages of settlement during the colonial period. Collectively, they could pass any law necessary to maintain civil government and peace, so long as they took the legislation from statutes enacted by other states. Congress undoubtedly included this stipulation as part of their policy of using the territorial stage as an educational period in the development of the state. Unfortunately, there arose issues over this clause. According to the Northwest Ordinance,

> [t]he governor and judges, or a majority of them, shall adopt and publish in the district such laws of the original States, criminal and civil, as may be necessary and best suited to the circumstances of the district, and report them to Congress from time to time: which laws shall be in force in the district until the organization of the General Assembly therein, unless disapproved of by Congress; but afterwards the Legislature shall have authority to alter them as they shall think fit.[20]

Governors and judges faced challenges in adhering to this clause. Eastern legislation did not always prove adequate in dealing with economic, legal, or social needs arising from conditions in newly acquired territories.[21] When devising legislation for their territories, Western officials often had to adjust legal ordinances borrowed from the states to fit the needs of the settlers, a practice that raised legal questions. To what extent did they have to borrow from Eastern law? Did Congress intend for them to copy older statutes, or did the Northwest Ordinance simply demand that they use them as models for Western law? This would not be the last time that Westerners had to address the incompatibility of Eastern legal tradition with the new legal culture developing in lands of imperial expansion.[22]

As stated earlier, the adopted, modified British colonial model was intended by the framers of the Northwest Ordinance to serve as a temporary government. They designed the territory as a training ground for republicanism. Once the population of the territory reached "five thousand free male inhabitants, of full age," they could petition the governor to call a general assembly.[23] Unlike the Turnerian claims of the frontier as a bastion of American democracy, political participation in the general assembly required a degree of wealth. The Constitution of the West required settlers to own at least 200 acres of land to serve in the assembly. With land prices sitting at a minimum of $1 per acre, the democratic right to office cost at least $200. Suffrage, however, was somewhat cheaper since settlers only needed 50 acres to cast a vote.[24]

The more recognizable governor–council–assembly form of colonial government appeared in the second stage. The governor, council, and house of representatives made up the general assembly, just as the governor, council, and assembly made up a miniature parliament in British North America. The general assembly assumed the responsibility of electing local officials at this point. Also, it replaced the governor and territorial judges as the source of law.[25] Yet, one must not assume that all authoritarianism disappeared from territorial government at this stage. The Northwest Ordinance very clearly reaffirmed the governor's role in territorial politics by giving him the power to "convene, prorogue, and dissolve the General Assembly," as well as making all legislation subject to his approval.[26]

Finally, when drafting the Northwest Ordinance, its framers remembered to include something conspicuously absent from the constitution produced in Philadelphia: a bill of rights. Bills of rights were commonplace in state constitutions, so it is understandable that the congressional drafters of this law would remember to include one.[27] This section of the ordinance protected the very basic rights of settlers. They had freedom of religion. The articles prohibited cruel and unusual punishment. Residents had the right to trial by jury, habeas corpus, and to post bail (a feature in and of itself that rejects the myth of the West as a region devoid of American legal institutions). Also, the legislation recognized American Indians' rights to their land in word, if not in deed. Finally, the Northwest Ordinance recognized the right of all citizens to an education.[28]

The Northwest Ordinance established the pattern for westward settlement and government, but the process was not stagnant. Instead, territorial politics evolved as new generations of pioneers encountered conditions and needs specific to their region. The transformation of the territorial system is an aspect of American expansion that has captured the attention of leading scholars in the field. In reviewing literature on this topic, it is not uncommon to encounter a section addressing the evolution of territorial government and policy throughout the nineteenth and twentieth centuries. For example, one of the leading historians in this field, Howard Lamar, saw four distinct evolutions in territorial policy. The first stage ran from roughly 1789–1819, a time during which international events, trade, and indigenous relations dominated the territorial process. The second stage extended to the Civil War. Democratization and standardization characterized this period of territorial history as the district stage disappeared and organic acts assumed a uniform appearance. The third stage extended to the turn of the century. This period witnessed the unbridled introduction of the spoils system into territorial life. It was a time of further standardization of Western government during which

> continuing administrations continued to standardize the rules governing every facet of territorial administration. Every federal appointee had to make reports to his superiors in Washington, secure permission for leaves of absence, and make his public expenditures justifiable to an ever-suspicious treasury comptroller. After 1869, territorial assemblies were told that they could meet only biennially and for only sixty days at a time.[29]

The final stage lasted until the admission of Arizona and New Mexico. The characteristic feature of this stage was the equal treatment of the final two territories with overseas territories acquired from Spain in 1898. This was a feature caused by congressmen's concerns over the significant presence of Mexican-American residents in the regions.[30]

Jack Ericson Eblen, on the other hand, envisioned a territorial evolution that reflected the larger shift in empire. He perceived the history of American expansion in four phases or empires: 1787–1848, 1848–1890s, 1890s–1920s, 1920s and beyond. The first empire served as the training ground for later American empires. Here, expansion was not driven by capitalistic impulses. By the second empire, the industrial machinery of America pushed the borders westward into American-Indian territories. One key characteristic shared by both of these phases of empire was their geographical dimensions. While the first and second empires witnessed American expansion across the North American continent, the third empire was an "Oceanic Empire."[31] The first and second empires may have provided the template for government in these areas, but the process of imperialism had other significant differences. Finally, Eblen vaguely defined the last American empire as consisting of the post-1920s era characterized by the USA as a creditor nation to other major powers.[32]

While Eblen's theory is interesting, territorial history does not fit into neat disparate empires. Instead, it is more productive to view American expansion as the history of a single empire divided into phases—or evolutions as Lamar suggests—that often overlap. For example, the territory of Hawaii was overthrown in 1893 during Lamar's third phase and Eblen's second and third phases. However, American penetration into Hawaiian politics and society began in 1820, and Hawaiians experienced imperial control from 1898 to 1959 that was definitely the product of a standardized system. Like their counterparts in the continental territories, governors were appointed by the president, had restrictions placed on their power by a territorial organic act, had to reside in the region, were subject to partisan politics emanating from Washington, DC, and even had to submit annual reports to Congress describing economic, social, and political conditions in the islands. Residents of Guam and the Trust Territories of the Pacific Islands (TTPI), on the other hand, were conquered in the late 1800s and remained under martial law.

Lamar's views on territorial transformations and Eblen's problematic concept of multiple empires can help historians better understand the nature of territorial government and law, as well as compartmentalize the changes in American expansionist policy. This article accepts the work produced by Lamar, Eblen, and Max Farrand on the evolution of territorial government in the 1800s; however, it rejects separating noncontiguous integrated territories and colonial holding from the process. Instead, historians must view American imperial policy as the result of three stages of evolution that span from late eighteenth century to the twentieth century. First, the period from 1787 to 1836 witnessed early American experiments in territorial administration as politicians worked to create a workable imperial governmental process from the blueprint offered by the Ordinance. Second, territorial administration and incorporation witnessed a new phase from 1836 to 1898. During this time period, experiences in the continental territories witnessed the evolutions described by Howard Lamar that produced the characteristic statehood process. Congressmen organized territories using the model Wisconsin Organic Act of 1836. Once the territory reached a sufficient population or met extra requirements—such as the prohibition of polygamy in Utah—they formed state constitutions and entered the Union. Finally, 1898 inaugurated the most recent phase. In this year, the US government won overseas territories from Spain and began to address questions of governance for its two noncontiguous territories, Hawaii and Alaska. During this phase, territorial government witnessed another important transformation. While Congress considered Alaska and Hawaii integral territories and applied the basic territorial model in a slightly modified form, it adapted the methods developed over the past 111 years to govern colonial insular territories. This process has not yet ended since the USA still maintains control over the Virgin Islands, Puerto Rico, and Guam, and continues to exercise influence in the politics of the former TTPI.

Territorial historians agree that the move from piecemeal policies to a standardized approach characterized the period from 1787 to 1836. The biggest problems facing territorial governments during this phase emerged from the chaotic district stage. Criticism arose toward this method of government from a variety of sources. Governors Arthur St. Clair and William Blount offered their insights on administration in the Northwest and Southwest. The size of the regions complicated efforts to govern them effectively, and citizens disliked the limitations placed on political participation as the governor held extensive powers over law and legislation in the region.[33] Furthermore, the district stage witnessed a move away from the Enlightenment ideologies of government that the Revolutionary War generation fought to uphold. Charles Secondat, Baron de Montesquieu, first articulated this principle in *The Spirit of Laws* when he stated that

> there is no liberty, if the judiciary power be not separated from the legislative and executive. Were it joined with the legislative, the life and liberty of the subject would be exposed to arbitrary control, since the judge would then be the legislator. Were it joined to the executive power, the judge might behave with violence and corruption.[34]

The framers resolved this issue in American government by separating the judicial, executive, and legislative powers among the three branches of government. In the territories, the principle of separation of powers was not upheld with equal vigor. Aside from the territorial governor's extensive powers over politics in the region, he and the judges sat on the council and devised law. This colonial model hardly represented the principles enshrined in the Constitution. While one may argue that the justices oversaw the judicial processes and even prevented Winthrop Sargent from overstepping his bounds as acting governor during St. Clair's many absences, territorial residents' dismay at the form of district stage government is understandable.[35]

Although the restrictions on legislating in the territory hampered the district government and conflict emerged over its form, the district government provided valuable experience for Congress and territorial administrators in the transition toward a standardized system for westward expansion, as well as innovations in local government and law. For example, Arthur St. Clair left his mark on territorial legal institutions with his 1795 legislation adopting English common law for the territories. He quite creatively did so by adopting a Virginian statute passed in 1776 that incorporated the legal tradition established since the reign of King James VI and I, thus opening the Northwest to an invaluable source of law by which to govern its population.[36]

Congress also reformed local government in its attempts to systematize the legal and political structures of the West. These developments made the county as opposed to the township the prominent administrative district for civil and criminal matters. Moving away from the broadly defined roles of the territorial magistrates and justices, a streamlined form of county government emerged as the dominant local institution by the close of 1836. Instead of the three justices trying to administer law while assisting the governor in borrowing Eastern statutes, St. Clair oversaw the expansion of the court system to include a myriad of lesser courts, thus dividing the responsibilities of criminal, civil, and estate law among a cadre of justices. This process steadily continued until its culmination in Illinois when the legal institutions were changed to include a superior court, appellate courts, and lesser courts. The process was completed in 1816 when Congress gave the general assembly control over the structure of the judicial branch within its borders.[37]

This process offers students of American expansion with yet another insight into the formation of its unique imperial model of governance. During the colonial era, historians have recognized the distinctions between the regional forms of local governance. In New England, expansion and local government relied on the township. As opposed to organizing new settlements into counties, the Puritan government gave lands to townships for distribution to its settlers according to need and social rank. Law required the maintenance of schools, and the meetinghouse served as the primary forum for community debate. Moreover, local constables provided the invaluable services of distributing warrants and ensuring order, while selectmen fulfilled legislative functions. In contrast, the county was the primary administrative unit in the Southern colonies. For example, in Virginia, the court acted as the principle political unit for the county. Oligarchies of planters dominated political affairs at the county and colonial level, often dominating the office of sheriff.[38]

The American model for the West combined these two systems. Under the Land Ordinance of 1785, the township served as the principle model for land surveying complete with a lot reserved for educational purposes. Yet purchasing sections within the township was an individual affair, not a communal one. Urban centers would serve as an important element in Western development; however, it would not necessarily dominate life and regional politics as it did in colonial New England. Instead, the county acted as the primary administrative unit. Territorial officials organized tax assessors, courts, sheriffs, clerks, and other legal offices at the county level. Also, research has shown that local oligarchies as opposed to town meetings dominated local politics during the first phase of territorial history providing a situation akin to that in the southern colonies.[39] While the local oligarch would not continue as a primary political force through the second stage, he was later replaced by the large corporate presence from railroad, mining, mercantile, and petroleum interests during the rapid industrialization of the post-Civil War era.

In addition to the lasting mark left by residents and administrators, congressmen made important structural changes to territorial policy during this period. Historians agree that the first real move toward standardization and democratization of the process began with the organization of Orleans territory in 1804. This congressional act made created a government for the French, American, Spanish, African, and Creole residents in which political power rested in the hands of the governor and council of thirteen. Together, these two bodies formulated law; however, the governor could not force his will on the council as it had to approve or disapprove of legislation. Moreover, Congress provided a separate, organized judicial system complete with a superior court and appellation process. Another important innovation was in congressional support. Unlike the earlier district governments that had to generate their own revenue, the federal government gave monetary support to the region. Finally, the Orleans territory moved through the statehood process with relative speed. Congress organized it in 1804 and awarded second-stage government the following year. With an established population and the important and growing port of New Orleans within its borders, development of this territory was unlike that of the others. Whereas residents in the Northwest territory, Missouri territory, and Southwest territory had to reproduce American civilization in a region with limited exposure to Western industrial and urban structures, Orleans territory had already been exposed to French and Spanish colonial cultures. The French established Orleans in 1718. Although the French failed to plant a vibrant colonial economy during the reign of Louis XV, the presence of American settlements in the Trans-Appalachian West gave new impetus for settlement in the region as New Orleans became the main entrepôt for agricultural products. Its growing population and economic activities called for its quick absorption into the Union as of 1812.[40]

Following the organic act for Orleans territory, congressional territorial policies staggered toward a more democratic and standardized process. In areas such as Florida, this process moved backward when Congress gave the president temporary power over the region until elected representatives created a formal government. During the interim period between acquisition in 1819 and formal organization in 1822, authority rested in the hands of General Andrew Jackson.[41] While developments such as these suggest that territorial policy moved away from a more democratic form, other congressional policies propelled territorial government down another path. Under the provisions of the Alabama Organic

Act, residents in the region began their political experience in the second stage as opposed to the district stage, thus setting a precedent for the Wisconsin Organic Act of 1836. Meanwhile, American statesmen increased political participation in other Western areas by easing suffrage restrictions and opening public offices to non-landholders. The right to participate in the government and elections now required only residency and taxpayer status. Moreover, the territorial delegate increased in prestige during this period. Now elected by the public, this public official replaced the governor as the voice of Western residents in Congress. Although denied the right to vote in Congress, he could lobby official congressmen for monetary and political support, suggest legislation, and urge incorporation into the Union. Throughout the second stage of territorial history, delegates continued to build on these early developments. Finally, the federal government shifted from the unicameral form of government to a bicameral house during this period, thus replicating the national model in the West.[42] This act left a lasting legacy as state constitutional conventions would often maintain this form of representative government, until reformers of the Progressive and New Deal eras agitated for unicameral houses as a cost-efficient method of government that would be more responsive to public needs and less prone to corporate interference.

The Wisconsin Organic Act of 1836 marked the beginning of systematized territorial government for the second phase. Between 1836 and 1912, the national government used this act as a template for territorial governments until they ran into the murky problems of governing the noncontiguous, incorporated territories of Hawaii and Alaska and the overseas colonies of Puerto Rico, Guam, Midway, the Virgin Islands, the TTPI, American Samoa, and the Philippine Islands. At that point, Congress adapted the model to account for populations consisting primarily of indigenous and non-European populations that the nation had no interest in absorbing into the body politic.[43]

On April 20, 1836, Congress formally adopted the Wisconsin Organic Act, thereby initiating a new era in territorial government. In this legislation, legislators preserved the governor, council, and assembly; however, they altered these institutions to reflect the transformations and experiences from the first phase of territorial government. The governor remained as the chief executive of the region. He was the commander-in-chief, approved or disapproved legislation, oversaw American Indian affairs in compliance with federal directives, and held the power of pardon or reprieve; however, he no longer held the power to dominate the legislative process or to prorogue the assembly as he did during the district stage. The secretary assisted the governor in daily affairs. He also maintained records of the legislative and executive actions and transmitted them to the president and house of representatives, thus showing a more direct federal interest in territorial affairs.[44]

Legislative power now rested in a bicameral legislative assembly, consisting of a council and house of representatives democratically elected by "free white male citizens" residing in the county elected their representative and councilors.[45] Together, these two branches formulated law for the territory. Absent from the legislation was the onerous clause in the Northwest Ordinance restricting its powers of legislation to the adoption of Eastern state statutes. However, Congress did restrict the legislative assembly in that it could not tax federal properties or institutions, interfere in federal lands, and had to pass uniform tax codes. Finally, interference in the affairs of the legislative assembly was limited to a gubernatorial and a congressional veto.[46]

In terms of the territorial legal system, the Wisconsin Organic Act provided a rationalized judiciary. It stipulated that "the Judicial power of the said Territory shall be vested in a supreme court, district courts, probate courts, and in justices of the peace."[47] Each supreme court justice was assigned to one of three district courts that oversaw cases of civil, criminal, and chancery cases. Recognizing the strains placed on a court system administered by only three justices in a large region, the federal government authorized justices of the peace to decide on cases involving less than $50 worth of property. To correct another flaw in first-phase organic acts, the provisions of the Wisconsin Organic Act provided for a system of checks and balances within the territory. Although the governor nominated the justices, they retained their powers during good behavior. Furthermore, supreme-court justices and justices of the peace did not have legislative duties as their duties involved only judicial affairs.[48]

From the Wisconsin Organic Act sprang all others until the Insular Cases and congressional colonial policies changed the American approach to its overseas policies. Examining the organic acts establishing governments in the continental contiguous territories, it is quite evident that congressmen used it as their template. Territories west of the Mississippi River generally began with a government consisting of a governor, council, and house of representatives. On average, the bicameral legislative assemblies started thirteen councilors and twenty-six representatives. Granted, some territories such as Oregon began with fewer legislators. In the case of Oregon, it began with only nine councilors and eighteen representatives. Later, the territorial legislature could increase this number with congressional approval, as in the case of Colorado territory which had its councilors and representatives increased to thirteen and twenty-six, respectively, bringing it up to par with other regions.[49]

The legacy of the first phase of territorial reorganization is seen clearly in terms of suffrage. Free, Euro-American males over twenty-one years of age elected the territorial delegate, councilors, and representatives; however, the governor, justices, justices of the peace, clerks, and sheriffs remained appointed posts, thus effectively removing the judicial system from popular government.[50] The Treaty of Guadalupe Hidalgo and the American Civil War, however, slightly complicated the expansion of suffrage in the territories dominated by Euro-Americans endowed with ethnic prejudices imported from the East. The Treaty of Guadalupe Hidalgo of 1848 allowed Mexicans the option of remaining in their homelands or staying citizens of their republic by fleeing south. For those choosing to stay, the treaty was very precise. According to Article IX of said treaty,

> Mexicans who, in the territories aforesaid, shall not preserve the character of citizens of the Mexican republic, conformably with what is stipulated in the preceding article, shall be incorporated into the Union of the United States, and be admitted at the proper time (to be judged of by the Congress of the United States) to the enjoyment of all the rights of citizens of the United States, according to the principles of the constitution; and in the meantime shall be maintained and protected in the free enjoyment of their liberty and property, and secured in the free exercise of their religion without restriction.[51]

Since Mexican Americans composed a significant portion of the population in Southwestern territories, congressmen had to account for their rights when drafting organic acts for newly acquired territories. The Colorado Organic Act provides a solid example for the method by which Congress addressed this concern. In this document, Mexican Americans had the right to vote in the first election. However, Congress also recognized ethnic prejudices prevalent in the USA at the time. Immediately after extending them suffrage, Congress included another clause stating that "the qualifications of voters and of holding office at all subsequent elections shall be such as shall be prescribed by the legislative assembly," thus giving Western settlers a method of later excluding Mexican Americans from the democratic process.[52] Likewise, the Fourteenth Amendment and the Fifteenth Amendment to the Constitution extended the right to vote to African Americans. In the Oklahoma Organic Act, Congress incorporated the language of the Fifteenth Amendment by including a clause stating that "no denial of the elective franchise or of holding office to a citizen on account of race, color, or previous condition of servitude" was permitted in the territory.[53]

There were exceptions to the Wisconsin template. Some of the modifications were minor and were intended to meet unique regional needs. For example, the Oregon Organic Act recognized missionary land holdings within American Indian lands. Responding to the propaganda of men such as Hall Jackson Kelley, missionaries traveled to the Oregon Country intent on spreading Christianity to indigenous peoples. Individuals such as Marcus and Narcissa Whitman traveled to the Northwest and established missions where they ministered to the American Indian inhabitants. With the British cession of the southern portion of Oregon Country, their land claims required congressional recognition just as other lands needed to be surveyed. Other sections of the Oregon Organic Act added to these modifications by prohibiting dams along salmon runs, appropriating $5,000 for the construction of a library for territorial legislators, jurists, and bureaucrats, and setting aside $15,000 for the construction and maintenance of buoys and lighthouses.[54]

The Oklahoma Organic Act arguably contained the greatest modifications to the territorial model since the Wisconsin Organic Act. Passed in 1890, this legislation was part of larger legal and political developments in national Western policy. Starting in 1871, Congress initiated a massive assault on the remnants of American Indian sovereignty. That year, the federal government formally declared an end to treaty-making with indigenous nations. No longer would the federal government treat with them on an equal footing. Instead, Congress claimed plenary power over American Indians living within the borders of the USA, a claim recognized by the Supreme Court in *Lone Wolf* v. *Hitchcock* (1903). In the past, the federal government had to negotiate change with American Indians. As of 1871, Congress would dictate assimilation to them. Sixteen years later, the next blow came with the General Allotment Act. Seeking to end tribal culture, the Act allocated 160 acres to the heads of families and 80 acres to single males from reservation lands. Forcing American Indians to adopt Western farming practices served a threefold purpose. First, it would force American Indians into national market. Second, it hurried assimilation by severing the communal link with their nation. Then, it opened lands in Indian Territory to Euro-American settlement. Finally, statesmen and humanitarians turned their attention to American Indian legal practices. They ordered American Indian agents to organize police forces on the reservations. Afterwards, they began transformed this nascent legal structure into one similar to that of the USA. Policemen became judges. Traditional practices such as polygamy were subjected to American cultural and legal practices. The Major Crimes Act of 1885 advanced the process by making American Indians subject to federal action for crimes such as rape and murder. The Curtis Act of 1889 completed the process by extending US law into remaining reservation lands.[55]

To account for these changes, Congress organized a traditional territorial government for the new American settlers entering Oklahoma and also provided for the remaining legal structures and land holdings of American Indian nations. For the newcomers, they created the basic governor–council–assembly government. All citizens over the age of twenty-one had the right to vote, except for persons in the Army and Navy stationed in the area. As stated previously, Oklahoman legislators had to comply with the Fourteenth and Fifteenth amendments. The only major restrictions placed on immigrants pertained to land. The legislation recognized lands reserved for railroads, schools, missionaries, and American Indians. For American Indians, the Organic Act contained provisions for their legal system. It established three judicial districts for American Indians courts located at Mukogee, South McAlister, and Ardmore. Indigenous courts retained jurisdiction for criminal and civil cases involving members of the same tribe. However, Congress gave territorial courts jurisdiction over cases involving members of different nations, or between non-indigenous persons and American Indians. Finally, the Act empowered US marshals to enter reservations and make arrests.[56]

When discussing the form, function, structures, and origins of territorial government during the second phase, one cannot ignore the two major exceptions to the rule: California and Texas. American settlers residing in the provinces of Tejas and Coahuila severed their allegiance and the region from Mexico through revolution. With the establishment of the first 300 families by Stephen F. Austin in 1821, the American presence had expanded, eventually replacing the original communities as the dominant group. According to the original agreement between Austin and the Mexican government, settlers swore allegiance to Mexico and accepted Catholicism. While Mexico City restricted their population from areas bordering the USA, they permitted Texians to bring their slaves and gave them generous land grants. By the 1830s, the relationship turned sour. The Colonization Law prohibited the further importation of slaves and threatened to curtail settlers' political and legal rights. The American settlers not only imported their families and slaves into the region, they also brought their sense of cultural and political superiority. Following the arrest of Stephen F. Austin—a moderating voice in the region—and the attempts by Antonio Lopez de Santa Anna to consolidate Centralist power over the area, rebellion erupted in Tejas. On June 29, 1835, Texians forced the retreat of Mexican forces at Anahuac. Texas declared its independence on March 1, 1836, and the USA recognized it only two days later. On March 6, the death of 187 militiamen at the Alamo in San Antonio gave the rebellion a rallying point. The war concluded with the surrender of Santa Anna at San Jacinto River on April 21. From 1836 to 1845, Texians negotiated with the US government for annexation.[57]

As with the Texians, Californians rebelled against the Republic of Mexico, officially declaring their independence on June 15, 1846. For almost twenty years, small streams of Americans settlers traveled west overland to settle in the Edenic California. Travel literature, journals, and reports from Americans in the region had painted a glorious picture of the region. It was a fertile land ready for agriculture. Herds of wild cattle roamed the countryside waiting for enterprising ranchers to claim them. British and American ships regularly called on Californian ports creating a ready market for hide and tallow, agricultural produce, and finished goods. Even the US government recognized the potential profits of California. Presidents Andrew Jackson and John Tyler made half-hearted attempts to negotiate with the Mexican republic; however, it was President James K. Polk who took the aggressive measures necessary to conquer it. At the outbreak of the Mexican–American War in 1846, President Polk and Senator Thomas Hart Benton dispatched orders to John C. Frémont who was exploring the region to assist in severing the union between California and the Republic of Mexico. When a revolution against the government under General José Castro erupted on June 10, Frémont acted with deliberation. Using the momentum generated by the Mountain Men and Bear Flaggers, he extended US government control over the region which the Treaty of Guadalupe Hidalgo recognized two years later.[58]

After annexation, Congress acted speedily to admit Texas and California into the Union as states. The Texas question lingered on American minds for nine years, because it threatened the balance between slave and free states. President William Henry Harrison's successor, John Tyler, finally resolved the issue through a political sleight of hand. Having alienated the Whigs to the point that they ejected him from the party, Harrison realized that he could not raise the necessary votes to annex Texas. Instead, "His Accidency" took a different route by annexing Texas with a joint resolution on March 1, 1845. Nine months and twenty-eight days later, Texas entered the Union as a state.[59] Although Californians did not undergo nine years of independence before annexation, they endured something Texians did not: military rule. Established as a precedent for temporary government prior to organization in Arkansas territory, California—like New Mexico and Florida—remained under the control of the US Army for two years. Unlike New Mexico and Florida, California skipped both the first and second stages of territorial government. Although its population had not reached 60,000, Congress admitted it to the Union as part of the Compromise of 1850. With the California Gold Rush and the continuous trains of overland emigrants heading west, California was too valuable to await settlement.[60]

Having examined the formation of territorial governments and their basic organization, it is now necessary to inspect closer territorial governmental and legal institutions and officials. The territorial governor held an important position in the West, but it was also a thankless, and sometimes onerous one. Given little to no instructions prior to their appointment—or even after—governors were expected to administer the region in compliance with national policies while maintaining internal peace and security. Although the federal government relieved him of the position of supervisor of Indian affairs in the early 1870s, the governor still had a number of other important responsibilities that included guaranteeing law and order, promoting settlement, and administering territorial bureaucracy. To assist him, the federal government provided a secretary, and little else. His salary remained dismal throughout the three phases of territorial government. During the nineteenth century, the Departments of State and the Interior reluctantly offered any direction on Western policy. To make matters worse, they received little financial aid to fulfill governmental duties.[61]

Over the years, the governor became one of the most vilified territorial administrators to both residents and historians. Often labeled inept carpetbaggers who won their positions by political connections, the men who filled this position received a rather poor reputation. Granted, some governors were terribly unqualified for the position, such as Governor Andrew Reeder of Kansas who held his post for less than a year in conflict-torn Kansas.[62] Scholarship, however, has proven this assumption wrong. While the British and French revamped their colonial bureaus during the twentieth century as a means of modernizing the exploitation of resources and humans in their colonies, the USA never developed a formal apparatus for governing imperial and colonial holdings.[63] Instead, the executive and legislative branches relied on what Eblen termed "an informal colonial service."[64] Governors often had previous exposure to political posts, and many were familiar with Western territories. In addition to holding previous political posts, these individuals often moved between Western posts, serving as a governor in one area before assuming the position of justice or delegate in another. Regardless of their pedigree, popular perception of the governor as the carpetbagger politician came from two sources. First, federal policies did not require governors to be residents of the territories. Without a sufficient population from which to choose an adequate administrator, presidents needed flexibility in appointing someone to represent the national government in the West. Limiting his choices to residents of the territory would have greatly diminished the pool of acceptable candidates. Second, territorial governors became the target of residents' ire toward minimal government involvement in the region. The failure of the federal government to provide directions or funding to the territorial administration understandably irritated Westerners, and the closest target for their anger was the governor.[65]

Turning to the territorial judicial system, popular perceptions of the American West portray it as a region not only of tough individuals but also of lawlessness and vigilante justice. A good example of this prevailing stereotypical depiction of harsh Western justice is *Unforgiven* (1992), starring Clint Eastwood, Gene Hackman, and Morgan Freeman. Although an improvement over the 1950s and 1960s depiction of rough-and-ready marshals dueling with villains at high noon, *Unforgiven* contains remnants of the standard version of individual justice on the frontier. Eastwood plays William Munny, a run-down ex-gunfighter, who rides off to Big Whiskey with his friend Ned. They intend to collect a $1,000 bounty by killing two cowboys who mutilated a local prostitute. Of course, the town prostitutes would not have resorted to hiring mercenaries had Sheriff Little Bill Daggett meted out proper justice on the perpetrators of the crime instead of fining them for destruction of property to compensate their employer Skinny Dubois. In attempt to maintain order in his city, their nemesis, Little Bill, resorts to beating and whipping any would-be assassins arriving in town. After he kills Ned as an example to other murderers, William Munny must dispense a more traditional form of justice by becoming incredibly intoxicated, walking into the bar, and shooting Sheriff Daggett and his men. Before riding away into the stormy night, Munny ensures that proper justice will prevail in Little Whiskey by threatening to "kill all you bastards" if they ever mutilate another "whore."[66]

Vigilante justice, blazing gunfights, slightly crooked sheriffs, and harsh punishments are all classical depictions of justice in the West, but recent scholarship on Western legal history has proven these stereotypes wrong. Granted, vigilantism did occur, and lawmen sometimes found themselves using force to stop criminals. After all, the West was no utopia. It was a violent region when one accounts for attacks on ethnicities, wars against American Indians, conflict during labor strikes, and violence as a result of typical crimes. But this violence was not the result of a *lack* of law. Violence had very specific causes, and it took place within territories possessing an established judiciary. In comparison with the violence that erupted as a result of white racism in the South following the Civil War, one could argue that vigilante violence was relatively limited. In his assessment of vigilante violence in the West, Richard White noted that "there were 210 vigilante movements" from 1849 to 1902 resulting in 527 deaths.[67] By comparison, the Freemen's Bureau recorded 1,000 murders of African Americans by Euro-Americans from 1865 to 1868 alone.[68] This comparison is neither intended to excuse vigilante justice nor offered as an apologia to Western violence. It is simply offered to give evidence that communities did not rely solely on group or individual violence for social order and to place Western vigilantism in the broader national context. Instead of the normal method of enforcing order, Western communities generally turned to vigilantism when traditional law-enforcing agencies proved inadequate, to relieve social tension, or as the result of ethnic prejudice (such as the Californian anti-Chinese activities in late 1800s).[69]

Instead of a region characterized by lawlessness or rough individual justice, one must view the West as a region in which westward migrants replaced indigenous and Spanish legal systems to allow for the orderly exploitation of natural resources and establishment of

American settlements. As William G. Robbins's study, *Colony and Empire: The Capitalist Transformation of the American West,* reveals, industrialization and the market economy were driving factors in the settlement of Western lands. Foreign and domestic investors funded the mining ventures, cattle ranches, and other institutions hoping for a meaningful return.[70] In order for this to succeed, however, settlers, laborers, and entrepreneurs required a judicial system that protected the social order, private property, and investments.

There is an ongoing debate over the nature of law in the American West. Some historians contend that Americans simply transported Eastern legal institutions and traditions to the West. Others argue that environmental determinism transformed English common law into a completely new and unique Western legal culture. Modern legal historians, however, have taken a more moderate stance between these two extremes. Kermit Hall refutes arguments made by previous Western historians that Western legal history either imitated Eastern patterns or were bound to environmental factors. As an example, he noted that state constitutions often included innovative and progressive elements, such as banning child labor in Coloradan and Idahoan mines.[71] Likewise, Gordon Bakken has provided evidence "that Rocky Mountain lawmakers adopted English common law, but abrogated it where it did not apply to their frontier environment."[72] In his study on Rocky Mountain contract law, Bakken suggested adopting Earl Pomeroy's approach to Western law by examining juridical "continuity" and "innovation" in the context of "legal heritage."[73]

Regarding the continuities between Eastern and Western legal cultures, John Phillip Reid produced groundbreaking work in *Law for the Elephant.* In this seminal work, Reid proved that migrants traveling on the Overland Trail may have entered a region void of American law prior to the organization of territorial government, but this does not mean that they themselves abandoned the law on their departure. According to Reid, "to say there was no law on the overland trail, after all, may be to say nothing more than there were no lawyers from whom advice could be obtained and no courts to resolve disputes."[74]

Although courts, lawyers, sheriffs, and marshals were not always present to punish criminals, to administer justices, or to adjudicate cases, Western migrants carried with them their legal and cultural habits. Entering a region in which American judicial systems and social mores had not yet fully penetrated, migrants still maintained their adherence to Victorian culture and American legal practices. In other words, they carried their civilization with them. Also, Westerners traveled with their families. The wagon train became a temporary and mobile community preserved through cooperation, sharing, and law mindedness in which the social stigmas attached to criminality and immorality remained a very real presence.[75] As noted in *Law for the Elephant,* "rules shaping conduct on the trail had their roots in laws remembered from the youths of emigrants."[76]

Commercial activity during the crossing gives further evidence that American legal tradition in terms of property rights and law-mindedness traveled with overland travelers as they entered unconquered Western lands.

> A question that we seek to answer is whether legal behavior on the trail was the same as law east of the Missouri river. Here we encounter evidence indicating that in one respect it was: when a person owned property another wished to use, the owner of the property set the terms.[77]

A hybrid monetary-barter economy thrived on the Overland Trail. In addition to their belongings and necessary supplies, migrants carried money with them on the trip. Individuals brought extra provisions to trade and sell. Ferries operated by Americans and American Indians catered to the needs of westward travelers.[78] Keeping with the most cherished of American legal practices, parties entered into contracts when renting or lending property. Contracts came in the form of verbal agreements for families and individuals to share resources. Promissory notes provided another form of exchange for individuals with limited resources. Direct purchases of oxen, wagon, and goods were recognized as an exchange of title between parties.[79] Recognition of private property and rules of exchange were not limited to Western migrants. Mountain men and soldiers also engaged in commerce on the Overland Trail. Moreover, exchanges between these men and settlers used both cash and barter. While barter remained a major method of property transfer, cash exchanges were not rare. Even trappers carried cash with them and purchased goods, livestock, food, whiskey, and other products from travelers in hopes of later selling them at a higher price.[80]

To finance the crossing, some individuals entered into legal partnerships in which they recognized concurrent, private, and mess property rights. When entering partnerships, individuals understood that the agreement was contractual, yet without the formalities of the East. "It is revealing that one of the few extant partnership contracts negotiated on the frontier and reduced to writing was never signed."[81] In other words, Western migrants relied on verbal contracts as opposed to written ones using the formal legal terminology present in nineteenth-century America. Western migrants even made adjustments to established legal practices to meet new challenges. While traditional partnerships were present, Western migrants also recognized "mess partnerships" in which individuals lent property or labor to a group without losing formal private interest in it, thus allowing individuals to transport resources more efficiently and to guarantee continued ownership while sharing their beneficial uses with fellow migrants.[82]

When Congress organized a territory, it established an official judicial system for residents. As stated earlier, it contained three justices at the top, serving as the supreme court. Three district courts divided the workload beneath the supreme court. At the bottom, justices of the peace and probate courts oversaw minor cases involving property and intestate issues. While a working source on paper, the territorial judiciary faced many challenges, some owing to Western conditions, others to the quality of the man holding the office. According to Earl S. Pomeroy, "the judicial system was one of the weakest parts of the territorial institutions."[83] One of the major challenges facing Western legal institutions was the size of the territory and population serviced by a small group of legal bureaucrats. With travel time between settlements becoming a factor in their abilities to oversee court cases, judges often found themselves facing backed-up dockets. For plaintiffs and defendants, this could become quite frustrating as the wheels of justice slowly turned. Inferior justices added to the frustration. Not only did unqualified justices and political hacks frustrate lawyers seeking a fair trial, but their removal also added to the burden of competent judges. Since territorial legislators had little power over inept jurists, their only remedy was to create new departments in sparsely populated regions. Afterwards, they shipped them to these regions, thus effectively removing the individuals from the territorial court system. Of course, this meant shifting the docket load to other already overburdened districts. For a competent, overworked justice, this meant an increased workload for an already dismally paying position.[84]

This is not to imply that all territorial justices were corrupt or incompetent. Kermit Hall's research shows that a significant portion of Western jurists had formal legal training. His study on the subject of corrupt and incompetent territorial judges revealed that 65.5 percent had college training, placing them in a comparable position to the number of educated statesmen and jurists in the East. He recognized that there was a decrease in educated jurists between 1829 and 1897. Since this period is most often studied by Western historians, political hacks appointed by congressmen and presidents have become synonymous with territorial justice. However, Hall's research, in conjunction with other Western legal historians, suggests that the incompetent territorial judge was the exception rather than the rule.[85] Considering the low pay and heavy workload of justices in the West, one should not expect an abundance of justices with glowing portfolios and legal reputations as the region was not a magnet for the best and brightest of American jurists.[86]

In addition to competent, trained justices, territories had the infrastructure necessary to enforce law. At the local level, sheriffs provided important services such as issuing warrants, preventing public disturbances, and catching criminals. Contrary to popular perceptions, Wild Bill Hickok and Wyatt Earp did not represent the typical Western law officer, and gunfights at the OK Corral were not commonplace. Instead, law enforcers were usually ordinary citizens who accepted the position and learned with experience. Granted, there were lawmen who became noted figures in Western legal history; however, most held their position for a limited time with little fanfare. When they engaged with criminals, officers were cognizant of legal restraints on their actions, preferring to settle conflict with minimal violence. In most cases, they dealt with criminal elements with fists, physical strength, and warrants as opposed to guns. Of course, catching criminals was one thing; holding them was another story. With limited funds to construct prisons and jails, criminals had ample opportunities to escape prior to large state expenditures on an adequate prison system.[87]

Although federal officials used Eastern legal tradition as a template for Western judicial systems, the region developed its own legal culture in response to economic and environmental conditions. One of the most commonly addressed innovations in Western law pertained to water. Whereas English common law governed water rights in Eastern states, ensuring each landowner's riparian rights threatened to impede economic progress west of the Mississippi River where mining operations required owners to drastically redirect the flow of rivers and streams. Over time, most Western states moved from a policy of enforcing the policy of reasonable use to a policy of prior appropriation in which the first person to use a water source could redirect as much as needed for natural or economic purposes.[88]

The mining industry required other innovations in Western legal culture. Beginning under territorial government and continuing through statehood, Western judiciaries adopted a more adaptable form of contract law to meet the needs of an industry in which property readily changed hands and partnerships fluctuated. When mining camps first sprouted during a gold or silver rush, miners collectively entered into agreements to arbitrate and govern property within the community. Needing a way to protect property and water interests in the absence of a direct judicial presence, these informal structures offered the legal atmosphere necessary to engage in capitalist enterprise. The desire to protect private property and business with the law did not disintegrate with the formation of territorial and state governments. On the contrary, legislators and justices tried to promote and protect industry by giving the judicial system flexibility. Instead of rigid contract laws, partnerships could be formed and dissolved more readily. Laws governing property exchange evolved in such a way as to protect private owners' interests by guaranteeing the seller his price at the time of the original agreement in the event that his commodity experience rapid devaluation prior to the actual exchange.[89] Contrary to the stereotypes offered by popular memory that the mining industry relied on rough frontier violence and justice, Western judiciaries developed sophisticated, rational mining laws that reflected the capitalist world economy emerging during the late nineteenth century and early twentieth century.

Another piece of evidence linking Western legal culture in the late nineteenth century and early twentieth century is the willingness of territorial, state, and federal governments to use police forces and militias to counter labor movements in mining regions. Western strikes at places such as Ludlow, Colorado, were not unique to the American experience. Since the first Industrial Revolution of the late 1700s to the mid-1800s, Western industries increasingly relied on surplus labor, low wages, government control, and mechanization to increase profits and productivity. As this process intensified, Western European countries experienced an upsurge in anarchist, social, and labor movements. To counter them, countries such as Germany relied on repressive government policies and police action mixed with reform measures to prevent labor uprisings or political threats to the conservative social order. While the USA never developed a viable socialist, Communist, or anarchist party to threaten the capitalist-oriented Democratic and Republican parties, labor organizations became increasingly organized. Labor strikes such as that of Ludlow, Colorado, threatened to slow down the extraction of mineral resources necessary to fuel Eastern factories and industrialists' bank accounts. A standard response of the US government, as well as Western state and territorial governments was to mobilize the militia, private security agencies, and conservative citizens to protect industrial property and productivity as opposed to providing for laborers' general welfare. This determination to subjugate workers' rights to industrial interests clearly shows the connection between Western political and judicial institutions to the larger world economy emerging during the nineteenth century and early twentieth century.[90]

While Western legal culture has a record of positively adapting to the economic conditions in a region of US imperialism, it also had its low moments. Ethnic violence and conflict was a characteristic of the West. To subdue American Indian nations, Americans relied on military aggression when treaty negotiations failed. Although guaranteed civil, political, and religions equality in the Treaty of Guadalupe Hidalgo, Mexican Americans lost land and social status following 1848. Businessmen willingly imported Chinese, Japanese, and Filipino laborers, until they became a perceived threat to Euro-American economic dominance. Then, vigilante squads and riots targeted their communities for terrorist attacks. The question is, what was the role of the legal system in this atmosphere? Euro-Americans used the legal system to place ethnic minorities at a disadvantage. Here, scholars can clearly see the politics of difference at work as the Western judiciary subjected non-European ethnicities to a legal system reminiscent of the Jim Crow South. For example, Western courts refused to accept Chinese testimonies and dying declarations as evidence. In order for a witness to testify, he or she had to take an oath based on American Christians' moral code and belief in divine judgment. Since non-Christian Chinese religion did not include a belief in Heaven, Hell, or Divine Judgment, their testimony was considered invalid. This practice limited their ability to seek justice against Euro-American perpetrators in a court of law.[91]

The Spanish–American War of 1898 marked the beginning of the final phase in territorial history. Although the USA had acquired Alaska and rebels had overthrown the Hawaiian monarchy prior to the outbreak of hostilities, the conflict between America and Spain was the start of an aggressive overseas US policy. Within two years, the federal government annexed Hawaii, American Samoa, Puerto Rico, Guam, and Wake Island, and installed a colonial government in the Philippines. Public response to this new move, however, was mixed. Imperialists accepted the idea of claiming overseas colonies for coaling stations, markets, and raw materials. On the other hand, anti-imperialists—a broad, loose coalition of Northern humanitarians and Southern racists, as well as various politicians and intellectuals—disliked the idea of colonization. For some, it was simply inimical to the democratic ideals dating back to the American Revolutionary War. For others, the concept of governing lands in which non-European and non-American peoples composed the majority brought a shudder to anti-imperialists intent on preserving the continued ethnic domination of the country by Euro-Americans.

Complications of governance and the legality of colonialism required the American political and legal system to revisit older patterns of imperial government. The last time this had occurred was in 1854 with the Kansas–Nebraska Act, an Act which proved disastrous at best. That year, Senator Stephen A. Douglas successfully maneuvered legislation through Congress organizing the territories of Kansas and Nebraska. At the time, slavery still divided the country. To overcome potential opposition, Douglas incorporated the idea of popular sovereignty into the legislation that empowered the residents to accept or reject slavery. The result was the formation of two governments at Topeka and Lecompton created by two parties of free state and proslavery men, respectively. The division soon boiled over into regional civil war that inflamed national feelings over slavery, ultimately contributing to the national bloodbath that erupted in 1861.[92]

By the late nineteenth century, the federal government had a tested system in place for the incorporation of new territories into the Union, but the close of the century brought a very clear shift in governmental models following the Spanish–American War. As a result of this brief conflict, the federal government won its first overseas colonies in Puerto Rico, Guam, and the Philippines. The McKinley administration also took the opportunity to annex the Hawaiian Islands in the process. For empires such as Great Britain and France, the addition of additional territories would not have posed major ideological or administrative questions since they already had an extensive experience with overseas colonies. The US government, however, did not. The acquisition of these lands challenged the territorial model that evolved between 1787 and 1912. Its imperial system was intended to prepare continental settlements composed of Euro-American settlers for statehood. When confronted with territories separated from the USA physically or populated by non-Western Europeans, the model suddenly appeared inadequate.

The first test to the American system came when the US government tried to tax goods traded between Puerto Rico and the USA. In a series of Supreme Court cases, known as the Insular Cases, decided in the early 1900s, constitutional questions emerging from colonization were settled, opening the way for the creation of a purely colonial system of territorial administration. Of all the Insular Cases, *Downes* v. *Bidwell* (1901) was perhaps the most important in defining the relationship between the territories and the US government. Lacking any serious intention of admitting Puerto Rico, Guam, and the Philippines to the Union—and uncertain over the destinies of Hawaii and Alaska—Congress began taxing Puerto Rican imports as they would foreign goods. The major constitutional question in *Downes* v. *Bidwell* was whether or not constitutional rules for taxes extended to the territories. Writing the opinion of the court, Justice Henry Billings Brown declared that the Constitution did not apply to the territories automatically. Only an Act of Congress could extend constitutional protections to these regions. Although the inhabitants possessed certain natural rights such as trial by jury and freedom of religion, they were guaranteed little else. The decision of the court basically gave Congress wide latitude in determining the form of government to be granted to the colonies and rejected the proposition that the promise of statehood extended to all territorial acquisitions. From this point on, two new terms emerged in reference to lands occupied by the US incorporated territories were those granted more traditional territorial governments with the possibility of statehood. On the other hand, unincorporated territories may or may not have a traditional territorial government and were not guaranteed the possibility of statehood.[93]

With the constitutional questions settled, Congress could now implement colonial governments in its new territories. Since Congress determined which aspects of the Constitution applied to overseas, unincorporated territories, the need to create a system of government that reflected the republican traditions established in 1787 no longer applied to unincorporated territories. As such, Congress placed a presidentially appointed American governor over a colonial government consisting of American administrators and indigenous representatives in Puerto Rico and the Philippines. Since American lawmakers considered the indigenous population incapable of self-government yet publicly denounced European colonialism throughout the rest of the world, this method of control gave the appearance of a benevolent USA preparing the islands for self-government without officially tying them to decolonization. More importantly for the industrializing USA, it opened the islands for commerce and resource extraction on the one hand, while providing the US Navy with coaling stations and defensive outposts on the other. Whereas Filipino assistance in the course of World War II moved the USA to grant it independence in 1946, Puerto Rico still remains a commonwealth within the American sphere. Of course, independence did not necessarily mean the end of American interference in Filipino affairs, though it did give them autonomy.[94]

Guam and American Samoa, on the other hand, remained under the most basic of colonial governments: military rule. Executive orders 180-A, 10077, 1900 gave the US Navy complete control over civilians and military personnel in the islands. The TTPI briefly experienced full military control; however, UN oversight forced naval administrators to accept some indigenous participation. Since international commissions regularly visited the islands on fact-finding tours, the federal government needed to show the development of democratic institutions, health facilities, jobs, and schools the TTPI. Although islanders' loyalty in World War II, public pressure from the continent, and UN oversight pushed the federal government toward reconsidering military domination of the islands, judicial, political, and civilian services remained in the hands of naval administration without any clear recourse for indigenous persons to criticize or appeal. When compared to the democratic governments established by the Wisconsin Organic Act of 1836 for Western continental, and later noncontiguous, governments, historians' claims that the Western government was colonial is thrown into doubt.

Unlike other noncontiguous territories, Alaska and Hawaii received more traditional forms of territorial government. The Hawaiian Organic Act provided the standard political system headed by a governor assisted by a secretary, treasurer, attorney general, and marshal. Congress, however, did try to rectify the most onerous aspect of the executive branch by requiring Hawaiian governors to be citizens of the islands. The legislature maintained the traditional bicameral structure, and a chief justice sat with two lesser justices on the higher court. Aside from provisions governing imports, exports, wharves, and fisheries, the Organic Act was similar in tone and form to other territories, except for the Section 101.

Hawaiian sugar planters began importing Chinese laborers to supplement Native Hawaiians on the plantations starting in the mid-1800s. With the anti-Chinese hysteria present in the West during the 1880s and 1890s, federal legislators included this section to extend the Chinese Exclusion Act to the islands by prohibiting Chinese laborers from moving to the continental territories from Hawaii.[95]

The Alaskan territorial government also had unique characteristics. The first Organic Act of 1884 brought a return of the district stage to the American West. Given the scant population in the noncontiguous region and the increase in mining-town populations, it was clear that some form of government was necessary; however, the Alaskan population, composed of small pockets of Euro-American and indigenous peoples, did not warrant a second-stage government. Also, national legislators did not provide a territorial delegate to Alaska until 1906. Following a population increase and growing demands for political autonomy, Congress finally consented to establishing a government with an assembly and other traditional offices in the Second Organic Act passed in 1912. Whereas continental territories associated with the second phase began with a government based on the Wisconsin model, Alaska slowly received a second-stage government through a series of congressional legislation. Last but not least, Alaska received ample federal funding following 1912. Alaskan mineral required the construction of more railroads which the US government willingly subsidized. During the turmoil of the interwar years of 1918–1941, federal expenditures increased again in an effort to construct military bases and fortifications in the strategically valuable region.[96]

Territorial status officially ended when the US Congress passed legislation admitting the region into the Union as a state. This was the moment for which most territorial residents lived. It signaled an end to federal tutelage, as well as government by appointed officials. On average, judges, governors, and other bureaucrats tried to foster positive relations with territorial residents. Aside from the basic considerations territorial appointees had to take into account when living and working among strangers, they were also dependent on the legislature for income supplements. On average, a territorial government only received $2,600 from 1875 to 1880, a sum that hardly met living expenses. Recognizing the limitations these men faced at earning supplemental income, legislatures were willing to give deserving governors extra money to compensate for their dismal salaries.[97] Of course, this was not always the case. Territorial officials sometimes clashed bitterly with settlers as in the case of Utah. In 1855, the territorial legislature attempted to give probate courts jurisdiction over civil and criminal cases, therefore bypassing the federal judiciary. Successful at first, the plan soon failed as an economic downturn and poor harvests lured Mormons to jury duty as a way to earn money. By 1856, the situation deteriorated into another struggle over the judiciary that became mixed with religious revivalism. In the end, the struggle rekindled tensions between territorial residents and federal officials.[98]

By the time Euro-Americans began replacing American-Indian and Mexican societies and governments in the lands west of the Mississippi River with American political institutions, the process for statehood had long since been established. Once a territory obtained a population of 60,000 residents or showed a capability for self-government, Congress passed an enabling Act, and a constitutional convention framed the fundamental laws of the new state. Having completed these steps, national lawmakers formally voted on admitting the state, thus ending the territorial phase for that region.

In some states, internal politics, ethnicity, religious beliefs, or location delayed statehood. For Utah, Mormon religious practices involving polygamy complicated relations between territorial inhabitants and the federal government until the eve of statehood. American viewed polygamy as an affront to the dominant Protestant national culture. Initially, the struggle to enforce orthodoxy on the Church of Jesus Christ of Latter-Day Saints centered on territorial government in Utah territory. Federal officials wanted Mormons to accept federal control over the region; however, residents preferred self-government based on their religious views. The judicial system quickly became the battle ground between Mormons and territorial officials. Justices dispatched to the region attempted to enforce US policy through the legal system, only to have the community rebuff their attempts by relying on locally controlled probate courts to settle disputes. Between 1857 and 1858, tensions briefly flared in the Mormon War, the worst phase of which was the Mountain Meadows Massacre. In the end, Mormons willingly conceded to some US demands in hopes of achieving home rule through statehood. They turned over suspects involved in the Mountain Meadows Massacre, dissolved the Nauvoo Legion, and banned polygamy in 1890, eight years after the passage of the Edmunds Act outlawing polygamy in the USA and its territories. This final concession ended the last major point of contention between the Mormon community and the US Congress, paving the way to statehood six years later.[99]

Statehood for territories such as New Mexico and Arizona was delayed because of ethnic issues. Following the conquest in 1848, many Mexican residents elected to remain in their homelands rather than retreat south of the border. While they chose to remain under the control of the USA, they worked to preserve their culture, religion, and language. By the turn of the century, citizens of Arizona and New Mexico were ready for statehood; however, congressmen disliked the notion of fully incorporating persons of non-Western European descent. Aside from holding prejudices against their ethnicity, language, and Catholicism, federal officials worried that admitting these people as full citizens of the USA would have greater political repercussions. Having won the Philippines and Puerto Rico from Spain, accepting New Mexican and Arizonan statehood threatened to set a precedent whereby American colonies may later demand the right for inclusion.[100]

Statehood for Hawaii stalled over a number of issues including Communism, ethnicity, and noncontiguity. Regarding race, Americans began questioning the logic of annexing a territory known to contain large segments of Asian laborers. By the mid-1900s, the Asian population had outgrown the Euro-Americans and Native Hawaiians. Moreover, they grew increasingly involved in labor movements in the islands. Initially, Japanese, Filipino, and Chinese laborers engaged in labor stoppages for members of their own communities; however, the International Longshoremen and Warehousemen's Union opened membership to them by the 1930s.[101]

While the large presence of Asian Americans and immigrant laborers gave the perception of islands dominated by non-Euro-Americans, the Massie case fueled the continental public's imagination of potential racial violence toward Euro-Americans. The Massie Case involved the rape of Thalia Massie in 1931. She accused Joseph Kahahawai, Henry Chang, and Benny Ahakuelo of the crimes. Following a mistrial, Lieutenant Thomas H. Massie, her husband, decided to deliver his own vigilante justice to the young men. He and some fellow sailors abducted and killed Kahahawai with the help of Mrs. Granville Fortescue, Mrs. Massie's mother. In the following trial of Massie, Fortescue, and their co-conspirators, the courts found them guilty of second-degree murder. Shortly thereafter, Governor Lawrence M. Judd commuted their sentences, and they left the islands. The damage, however, had been done. National newspapers ran headlines reporting the supposed outbreak of racial violence in the islands leading Congress to consider revoking elements of Hawaiian home rule. Even the US Navy expressed outrage by threatening to boycott the harbors if something was not done. Following the Massie and Major–Palakiko cases, the Hawaiian territorial legislature would not only confront congressional opposition to statehood based on noncontiguity, but they would also have the confront negative stereotypes of islands as a region politically dominated by foreigners hostile to American government and culture.[102]

After World War II, the issue of Communism would combine with race in Hawaii to act as an obstacle to statehood, while the battle against the USSR stimulated the national desire to add Alaska to the Union. Hawaiian labor became the focal point of a national and territorial conspiracy theory. Although some Hawaiian union members such as Ichiro Izuka publicly admitted to Communist activities, business leaders and conservatives used the red specter as a tool to crush the International Longshoremen and Warehousemen's Union. In the process, they succeeded in imprisoning prominent union members; however, they also gave Hawaii the image of being a bastion of Communist activity. Now, territorial officials needed to combat this reason of opposition to statehood in addition to noncontiguity and ethnicity. Meanwhile, the strategic value of Alaska skyrocketed as the territory became the perfect staging ground for the air attacks on Russia should the Cold War turn hot.[103]

Once a state overcame any prerequisites or challenges in their bid for statehood, framing a constitution was the only major task left remaining. Territorial residents may have complained, at times bitterly, about Eastern economic domination, carpetbagger governors, and federal negligence leading students of this history to view them as disadvantaged for entering the Union last. However, territories did benefit from their place as younger siblings to the older states. When they sat down to draft a state constitution, they had an advantage Massachusetts, Rhode Island, Virginia, Pennsylvania, and other Revolutionary War states did not. Territorial delegates to constitutional conventions had the combined wisdom and knowledge—not to forget the actual documents and records—from other constitutional conventions. As Christian D. Fritz noted in his article on Western state constitutions,

> A close examination of how western delegates went about their task of creating frames of government dispels the notion that they did so in isolation. The identification of concerns that were peculiar to the West—aridity, mineral extraction, and Mormonism, for example—have obscured the extent to which the western conventions were part of a broader, dynamic process of constitutional understanding. Moreover, western delegates were aware not only of the latest innovations in constitution making, but had access to all the existing state constitutions—including those of the "older" states—as models.[104]

For some conventions, this meant borrowing almost verbatim from existing constitutions.[105]

Territories often had previous experience writing constitutions. Community and political leaders in Utah and Alaska believed that presenting Congress with a well-framed document would hurry the statehood process and suppress opposition by clearly demonstrating working knowledge of republican government.[106] Of course, the plan did not work for either of them. Polygamy automatically ended any possibility for Utah. On the other hand, support existed for fully admitting Alaska, yet political contests between Republicans and Democrats and its noncontiguous territorial status worked against the Alaskan statehood campaign. Linking their case for statehood with Hawaii later helped Alaskans overcome congressional opposition.

Hawaiian residents also had extensive knowledge in drafting constitutions. In fact, Native Hawaiian nobles and monarchs had drafted four constitutions with the assistance of American missionaries prior to the overthrow of the monarchy by American and American Hawaiian citizens in 1893. These frames of government created sophisticated constitutional monarchies with power emanating from the monarch yet separated between judicial, legislative, and executive branches. They included guarantees of certain basic rights, such as freedom of religion and trial by jury. Moreover, they were products of Native Hawaiians and not systems of government thrust on them by a foreign power. From the first constitution drafted in 1840 until the fall of the monarchy, their government operated efficiently, forging international treaties with Great Britain and France while preserving Hawaiian independence from aggressive European empires.[107]

The directions of Congress set forth in enabling Acts required each constitutional convention to frame a government based on republican principles. The actual form adopted, however, was left to the delegates. Overall, every Western convention remained faithful to the established three-branch system with a judiciary, executive, and legislature. Whereas the constitutional conventions of the Revolutionary War era often had to address the question of whether or not to adopt a unicameral house—as in the case of Pennsylvania—or a bicameral one, all Western states initially accepted the traditional bicameral house for the state legislature. This is not to say that the unicameral house was not considered during the constitutional conventions. Delegates for the Washington convention considered trying a unicameral house until it was deemed a change too "drastic" for adoption.[108] Nor did all states maintain a unicameral house. In 1934, Nebraskans responding to the challenges of the Great Depression and the legacy of Progressivism passed an amendment using the initiative that changed the state legislature to a unicameral body. They believed this switch offered two benefits. First, it would cut state expenditures by reducing the cost of government. Second, a unicameral house was seen as more responsive to the will of the people, thereby eliminating the influence private, wealthy interests exercised over legislators.[109]

Although they commonly adopted governments with a bicameral house, governor, and judiciary, delegates changed government institutions to fit the unique experience and needs of their territory. For example, Rocky Mountain states usually split the executive department between a governor and departmental heads thus distributing power in this particular branch in response to their experiences with federally appointed territorial governors.[110] Alaska, on the other hand, emerged from the territorial phase with the view that only a strong executive could provide the necessary leadership for development. To this end, they gave the governor control over all aspects of the executive branch, the right to appoint cabinet members and judges, and a strong veto.[111]

Texas, however, went in the opposite direction. During Reconstruction, the Lone Star State was made into a military district following the drafting of the 1866 Constitution that allowed Confederate officers to participate in government while refusing political and civil rights to African Americans. The 1869 Constitution included a governor with effective powers over the executive branch, a bicameral house, and a judicial system. It also ensured civil and political rights for African Americans, as well as funding for a school system. Unfortunately, it was quickly replaced in 1876 after radical Republicans, moderates, and African Americans lost control of the government to traditional Democrats. Reacting to the hated policies and governors of Radical Reconstruction, the 1876 Constitution targeted the executive branch for reduction. The governor lost most of his powers. Aside from the line item veto and personal influence, he or she has little control over legislation or administration in modern Texas politics. In essence, the governor was turned into a figurehead of republican government. Meanwhile, the lieutenant governor received broader powers. He or she sets the agenda in the senate and can use his or her powers to promote or destroy legislation.[112]

The judicial systems of Western territories also changed with statehood. No longer forced to accept the territorial structure of three appointed justices overseeing district courts, delegates designed more sophisticated court systems. In Western states, a supreme court served as the highest judicial body in the state. Below the supreme court, delegates created appellate, probate, district, and county courts all designed to provide the people with an efficient, responsive legal system. While they could not correct all the flaws experienced during the territorial era, framers of state constitutions did try to develop a system that was not constantly backlogged or completely removed from the democratic process. This does not mean that all states used the same template for their judicial systems. Legal institutions and processes have their own unique flavor. For example, Alaskan judges are not elected by the people. Instead the Alaska Judicial Council consisting of three attorneys and three non-attorneys submit a list of candidates to the governor who then appoints them to the bench. While citizens must vote on retaining a judge, they have no direct say over who fills the bench.[113] Also, in the state of Texas, the role of the supreme court is split between a court of appeals for civil cases and a court of criminal appeals, clearly a divergence from the standard judicial model prevalent in the USA.[114] Nor did all states successfully resolve the issue of backlogged court dockets. States with limited resources often place great pressure on their courts to keep up with heavy case loads; however, the costs of doing so can exceed available tax revenues.

The Populist and Progressive traditions strongly affected Western state constitutional conventions. Home to a large agrarian population, citizens faced high transportation and granary rates that dipped into family incomes. Railroad companies were also notorious for controlling state legislatures and vast tracts of land awarded to them for constructing rail networks. As a result, most Western states embraced the referendum, initiative, and recall as a means of returning democracy to government. The referendum permitted citizens to vote on laws passed by the legislature. If a majority of voters considered the Bill inappropriate, corrupt, or unwise, they simply voted to reject it. Likewise, recall gave the masses more control over politicians. Inept or corrupt legislators, local officials, and other politicians stood to lose their positions if enough voters demanded their removal from office, a tool designed to prevent representatives from becoming paid agents of large mining, industry, banking, or railway interests. Finally, the initiative was a process whereby voters could begin the legislative process by signing a petition and later voting on a Bill that originated with the people.

States that accepted these democratic innovations often incorporated them directly into the body of their constitutions. For example, states such as New Mexico and Oklahoma gave voters broad power over the legislative process by placing few restrictions over the use of the referendum, while Idaho recognized limitations. Passed in 1912 and 1911, respectively, the initiative and referendum provided the people with an avenue of refusing or promoting legislation; however, the state courts recognized the supremacy of the state legislature when drafting law. Kansas, on the other hand, used the referendum as a means of checking some of the powers of the state legislature. While the initiative and recall were granted through a series of legislation stretching from 1909 to 1927, constitutional convention delegates wrote the referendum directly into the fundamental laws. In order to amend the constitution, to pass non-defense-related expenditures, or to draft banking laws, the constitution required the state legislature to submit the legislation to the people for approval.[115]

The movement toward popular participation in the legislative process was not successful in all Western states and territories. For instance, Texas has remained adamantly tied to representative government over direct democracy. The Texas supreme court rejected proposals to include the initiative in 1914 on the grounds that direct democracy was counter to representative government established by the state constitution. While the state legislature occasionally conducts polls to gain an understanding of voters' attitudes toward a piece of legislation, this aspect of the Populist movement and Progressive Era of the late nineteenth century and early twentieth century has yet to take hold in the Lone Star State.[116]

The frequency by which voters use these powers also changes from state to state. Alaskan voters only acted on twelve initiatives by 1985, while Oklahomans voted on 102 initiatives and referendums between 1970 and 1988.[117] Meanwhile, California voted on 173 initiative and referendum proposals and accepted fifty-three between 1911 and 1968.[118] Regardless, some of their propositions have managed to capture national attention, such as Proposition 8, which banned gay marriages in 2008.

In terms of taxes in the Western political tradition, state governments must consider the public needs and economic growth. This issue is not new. With limited economic bases requiring fresh capital to expand, Western states have hesitated to place heavy taxes on businesses for fear of repelling potential employers and investors. When framing their governments, Rocky Mountain States faced the political conundrum of accepting farmers' demands that mining companies pay their share of the state budget while limiting taxes on the primary employer for their state. To resolve this issue, states tried to find a balance by devising innovative legislation that placed moratoriums on mining taxes or only took revenue on proceeds and not property and equipment.[119]

Other states have avoided finding a balance between business and the public. In Texas, private-interest groups lobby hard to prevent taxes from threatening private property. Although the oil boom of the 1980s generated great revenues and the industry remains a dominant source of income, the sales tax continues to remain the primary source of income for the state. While taxation policies in Rocky Mountain state constitutional conventions pitted farming interests against mining interests, Texas revenue measures are divisive issues between liberal and conservative groups with the former trying to increase them and the latter rejecting such proposals as bad for business.[120]

States with limited revenue sources must rely on sales taxes and federal expenditures to meet budgetary needs. Although Texas and California have enjoyed the presence of highly profitable oil and mining companies within their borders, both states also had viable agricultural and livestock industries as well as high-tech and, in the case of California, entertainment sectors. The diversity in their economies offers additional outlets for revenue. States such as Arizona, on the other hand, have the difficulty of generating funds from fewer sources. To date, the Arizona state legislature must depend primarily on sales taxes, property taxes, and federal funds to meet budgetary requirements. Moreover, social debates over increased taxes have highly politicized attempts to generate additional funds for education and other programs.[121]

While initiative, referendum, recall, and taxation were constitutional innovations emerging from the experiences and politics of Western territories, states in the region must also face other modern challenges. Home to many state and national parks, the West is often associated with the wilderness. It is not surprising, therefore, that Western states have developed environmental agencies and policies designed to meet environmental concerns. The development of an environmental consciousness in the West can be traced back to the turn of the century when preservationists and conservationists first appeared on the national scene. Men such as Gifford Pinchot and John Muir recognized the danger to unrestrained use of Western lands. While Muir firmly believed in leaving natural preserves untouched and unblemished, Gifford Pinchot used his position as head of the Forest Service to manage the consumption of natural resources in Western lands. The belief in a vanishing frontier gave urgency to preserving undeveloped areas of the national domain, yet this desire clashed with industrial, commercial, and social needs. Mineral deposits, timber, water, livestock, and grain were necessary to support not only newly formed states but also national progress in industrialization and international economic competition. As a result of these conflicting interests, the modern debate emerged over whether to preserve, conserve, or open federal lands, national parks, and rivers to commercial interests.

Environmental activism continued to grow as a viable social and political presence in the USA through the course of the twentieth century. Rachel Carson's *Silent Spring* gave the movement extra momentum by describing the effects of releasing chemicals into an ecosystem. By scientifically proving that pesticides did not disappear but remained in rivers, lakes, groundwater, and soil, she illustrated the need for government and social action to limit humans' impact on the ecosystem. The environmental movement that emerged from the 1960s and 1970s has placed pressure on Western states to monitor the use of natural resources. States such as Arizona, Washington, and Oregon have established state agencies designed to manage water resources, limit air pollution, protect wildlife, monitor natural resources, and preserve designated areas. Although government officials do not always embrace the most liberal reforms, there is evidence that Western states seek to establish a balance between eco-friendly legislation and promoting industry and commerce. For example, Californians used the initiative in 1990 to pass a constitutional amendment banning the use of gill and trammel nets in commercial fishing. Recognizing the financial hardships private businesses faced during the transition to ecologically safer nets, the amendment authorized the legislature to compensate fishermen for lost revenue.[122]

While the initiative, referendum, and recall were products of the Progressive Era and the twentieth-century environmental movement pressured state governments into adopting boards, committees, departments, and laws to protect Western ecosystems, the Great Depression and New Deal changed the relationship between the individual and the state. Prior to the 1930s, liberalism did not support intervention by the state in the economy and citizens' lives. It was an ideology that promoted minimal government interference in commerce and industry, supported participation in politics by responsible individuals, and demanded the guarantee of basic personal liberties such as freedom of the press.

The Great Depression, however, changed Westerners' views of liberalism. Following the catastrophic failure of the global economy in the fall of 1929, Western nations struggled to find a solution to rapid deflation and high unemployment. Nations such as Germany and Italy found their answer in fascism, a political ideology that rejected the products of the Enlightenment and French Revolution. Although fascism assumed different forms, they found a common answer to the failing economy by allowing a small elite control industry and finance, as well as relying on militarism. The USA, Great Britain, France, and other democratic nations, on the other hand, could not support such programs. Instead, they increased the role of the state in commerce, industry, and private life, and, in doing so, completely changed the meaning of liberalism.[123]

> [By] 1945, the ideology of American liberalism looked strikingly different. The critique of modern capitalism that had been so important in the early 1930s (and, indeed, for several decades before that) was largely gone, or at least so attenuated as to be of little more than rhetorical significance. In its place was a set of liberal ideas essentially reconciled to the existing structure of the economy and committed to using the state to compensate for capitalism's inevitable flaws—a philosophy that signaled, implicitly at least, a resolution of some of the most divisive political controversies of the industrial era.[124]

Following the election of 1932, classical liberalism, which had dominated American economic theory throughout the nineteenth century, faded from political policy. The USA instead turned to state intervention in the economy, similar to the Five Year Plans of Soviet Russia. Using legislation such as the Social Security Act and Civilian Conservation Corps, President Franklin Delano Roosevelt's New Deal coalition successfully transformed government into an entity that ensures a minimum standard of living for its citizens by intervening in the economy and in private lives when traditional capitalist structures fail.[125]

Western states were neither immune to the Great Depression nor exempt from the shift to New Deal liberalism. As Neil Maher illustrated in his article "'Crazy Quilt Farming on Round Land': The Great Depression, the Soil Conservation Service, and the Politics of Landscape Change on the Great Plains during the New Deal Era," programs such as the Soil Conservation Service greatly affected traditional sources of social support. While small-town residents once looked toward City Hall, churches, and local organizations for support during trying economic times, government agents' actions in teaching farmers modern agricultural techniques designed to preserve the soil and to prevent erosion created a new bond between Washington, DC, and Western farmlands.[126] Meanwhile, the Civilian Conservation Service, the Works Progress Administration, the Resettlement Administration, and the Agricultural Adjustment Act injected money directly into Westerners' pockets by offering employment, housing, and agricultural subsidies to individuals facing financial ruin.

While the federal and state governments have utilized social security, unemployment benefits, public housing projects, and educational programs to offer citizens a safety net, their efforts to address the issue of healthcare still have yet to address one of the most challenging social issues. As early as the New Deal, liberal legislators have toyed with the idea of a national healthcare program to ensure each person access to medical services. Western states have established the basic infrastructure to oversee public health through the establishment of state boards, departments, and committees entrusted with the task of containing contagious diseases, regulating medical boards, and educating the public. However, Medicaid remains the primary vehicle for addressing this issue in Western states.[127] Established in 1965, it is a program intended to offer basic medical services to the disadvantaged and impoverished through a combination of state and federal funds. States have the right to participate in this program, but they must match federal funds from tax revenues. This can put quite a strain on a state budget. For example, the Colorado state legislature considered withdrawing from the program in 1992 when its costs increased by 6 percent for a total of $100 million.[128] As the cost of pharmaceuticals, treatments, and primary-care visits continue to increase, state governments can expect further drains on state coffers, unless Congress devises a comprehensive national policy to address these problems.

While Medicaid serves as the primary vehicle for healthcare access in the West, some states have taken the initiative to develop their own solutions. In 1974, the Hawaii State Legislature passed the Prepaid Health Care Act requiring employers to provide coverage for part-time workers as a way of supplementing Medicaid funds distributed by MED-QUEST.[129] Arizona joined the Medicaid program in 1974, but the refusal of the state government to allocate any funds to meet federal contributions made it an empty gesture. Instead of relying on the traditional Medicaid structure, Arizona devised an alternative program: the Arizona Health Care Cost Containment System (AHCCCS). Operating since 1981, the AHCCCS offers limited services to the people earning an income that is below 34 percent of the established federal poverty level by contracting health-maintenance organizations (HMOs) as set rates to manage their health care. To control costs, the HMOs and designated physicians limit the number and types of treatments.[130] While AHCCCS offers a more conservative example of a public healthcare program, Oregon Health Plan (OHP) is considered one of the more progressive state initiatives in the country. It was created by State Senator John Kitzhaber in 1989 using Medicaid as a model. OHP offers impoverished Oregonians access to proven treatments, drugs, vaccinations, and prenatal care. State expenditures are supplemented by contributions from employers who do not offer insurance. Although it established a promising record of accomplishment throughout the 1990s, criticism from the right, increasing costs from larger-than-expected enrollments, opposition to tax increases, and a return to federal conservatism under the George W. Bush administration forced Oregon to scale back its plans to use OHP as a means of providing uninsured working poor with medical coverage.[131]

Territorial and state governments were crucial features of westward expansion. They assisted in the orderly settlement of new lands. They provided political and legal structures that protected and promoted industry, commerce, and social stability. More still, congressional territorial policies, state constitutional convention delegates, and Western emigrants transmitted American political traditions to newly conquered regions through the territorial process and framing of fundamental laws. Needless to say, American imperialism was as much a political as a cultural process.

In addition to the internal aspects, scholars must also pay attention to events surrounding territorial and state government in the West. The process by which a region moved from a territory to a state did not suddenly appear west of the Mississippi River. It had roots spanning back to the colonial era. Moreover, it did not end at the west coast in 1890. The experiences gained from 1836 to 1912 shaped political events in noncontiguous incorporated territories and American colonies. Finally, students of Western history must always remain cognizant of the fact that statehood did not usher in an end to political and legal innovation. Today, Western state governments continue facing environmental social, economic, and political challenges that require further innovation to guide society through the twentieth century. With current battles over health care, gay marriage, alternative energy, and educational reform generating debate within society, the West will undoubtedly remain an interesting region of study for students interesting in the history of government and law in the USA.

NOTES

1. Raymond Betts once defined imperialism as "that consciously undertaken state activity in which force, intrigue, or even negotiation is employed to secure the long-range political or economic domi-

nation by the state of foreign territory or foreign peoples it wishes for some reason to control. [...] Imperialism should not be equated with war, nor should it be confused with the sudden incursion, or *razzia,* which does not lead to permanent occupation. Moreover, it should not be considered purposeless, a sort of unconscious movement outward. [...] It is generally agreed, therefore, that imperialism is a significant aspect of foreign policy." See Raymond F. Betts, *Europe Overseas: Phases of Imperialism* (New York: Basic Books, Inc., 1968), pp. 5–6. Some Western historians confuse this with colonialism, often referring to the colonial relationship between the West and East. However, European studies and modern international bodies consider a colonial relationship between two powers to include separation by an ocean or large body of water (blue-water theory) and indirect rule by the controlling nation with no intention of incorporation into the larger body politic.

2. Western historical scholarship that continues relying on the Turnerian thesis views the Appalachian Mountains as the beginning of the West and excludes Hawaii.
3. David Hawke, *The Colonial Experience* (New York: Bobbs-Merrill, 1966), pp. 661–664; and Robert Middlekauff, *The Glorious Cause: The American Revolution, 1763–1789* (Oxford: Oxford University Press, 2005), pp. 200, 616–618.
4. For an interesting account of the evolution of Europeans' perceptions of the wilderness, see Roderick Nash, *Wilderness and the American Mind* (New Haven, Conn.: Yale University Press, 1967), pp. 25–33.
5. Peter S. Onuf, *Statehood and Union: A History of the Northwest Ordinance* (Bloomington, Ind.: Indiana University Press, 1987), p. 8.
6. Onuf, *Statehood and Union,* pp. 16–17.
7. Jack Ericson Eblen, *The First and Second United States Empires: Governors and Territorial Government, 1784–1912* (Pittsburgh, Pa.: University of Pittsburgh Press, 1968), p. 21; and Ordinance of 1784, *Journals of the Continental Congress, 1774–1789,* vol. XXVI, *1784,* ed. Gaillard Hunt (Washington, DC: GPO, 1928), p. 275.
8. Ordinance of 1784, p. 276.
9. Eblen, *The First and Second United States Empires,* pp. 22–23; and Ordinance of 1784, pp. 275–278.
10. Hawke, *The Colonial Experience,* pp. 89–101, 345–350.
11. For Madison's writings on this matter, see James Madison, "No. 10," in *The Federalist Papers,* ed. Charles R. Kesler (New York: Mentor, 1999), pp. 45–52.
12. Ordinance of 1784, pp. 278–279.
13. Ordinance of 1784, pp. 276–278.
14. Land Ordinance of 1785, *Journals of the Continental Congress, 1774–1789,* vol. XXVI, *1785.* ed. John C. Fitzpatrick (Washington, DC: GPO, 1933), pp. 375–378, 381.
15. Eblen, *The First and Second United States Empires,* p. 28.
16. Eblen, *The First and Second United States Empires,* pp. 28–29.
17. Frederick Cooper, *Colonialism in Question: Theory, Knowledge, History* (Berkeley, Calif.: University of California Press, 2005), p. 27.
18. Howard R. Lamar, *The Far Southwest, 1846–1912: A Territorial History,* rev. edn (Albuquerque, N. Mex.: University of New Mexico Press, 2000), p. 6.
19. Northwest Ordinance of 1787, *Journals of the Continental Congress, 1774–1789,* vol. XXXII, *1787,* ed. Roscoe R. Hill (Washington, DC: GPO, 1936), pp. 334–337.
20. Northwest Ordinance of 1787, p. 336.
21. The inadequacy of Eastern laws in Western lands remained a defining trait throughout Western history. Whether it was water rights derived from English common law or laws governing land sales, Western states had to modify Eastern laws to fit their needs.
22. Eblen, *The First and Second United States Empires,* pp. 87–92.
23. Northwest Ordinance of 1787, p. 337.
24. Northwest Ordinance of 1787, pp. 337–338; and Land Ordinance of 1785, p. 378. See also Lamar, *The Far Southwest, 1846–1912,* pp. 6–7; and Max Farrand, *The Legislation of Congress for the Government of the Organized Territories of the United States, 1789–1895* (Buffalo, NY: William S. Hein & Co., 2000), p. 15.

25. Northwest Ordinance of 1787, pp. 338–339.
26. Northwest Ordinance of 1787, p. 339.
27. State constitutions either began with or simply included a Bill of rights stating the principles and freedoms upon which government rested. This tradition continued in Western states.
28. Northwest Ordinance of 1787, p. 340; and Lamar, *The Far Southwest,* p. 6. Lamar's work provides an in-depth look into the territorial politics of New Mexico, Utah, Arizona, and California.
29. Lamar, *The Far Southwest,* pp. 9–10.
30. Lamar, *The Far Southwest,* pp. 8–13.
31. Eblen, *The First and Second United States Empires,* p. 7.
32. Eblen, *The First and Second United States Empires,* pp. 2–15.
33. Eblen, *The First and Second United States Empires,* pp. 143–144.
34. Charles Secondat, Baron de Montesquieu, *The Spirit of Laws,* trans. Thomas Nugent (New York: Hafner, 1949), p. 152.
35. See Eblen, *The First and Second United States Empires,* pp. 80–85, for the struggle between Winthrop and the territorial justices.
36. Max Farrand, *The Legislation of Congress for the Government of the Organized Territories of the United States, 1789–1895* (Buffalo, NY: William S. Hein & Co., 2000), pp. 18–19; and Eblen, *The First and Second United States Empires,* pp. 95–96.
37. Farrand, *The Legislation of Congress,* pp. 29–30; and Eblen, *The First and Second United States Empires,* pp. 116–119.
38. David Hackett Fischer, *Albion's Seed: Four British Folkways in America* (Oxford: Oxford University Press, 1989), pp. 181–199, 398–410; and Hawke, *The Colonial Experience,* pp. 151–154, 117–118.
39. For a description of county governments and oligarchic control, see Eblen, *The First and Second United States Empires,* pp. 117–122, 130–132.
40. For the changes made by the Organic Act for Orleans territory, see Farrand, *The Legislation of Congress,* pp. 21–23; and Eblen, *The First and Second United States Empires,* pp. 145–147. For French imperial policies and expansion in Louisiana, see Carl J. Eckberg, *French Roots in Illinois Country: The Mississippi Frontier in Colonial Times* (Urbana, Ill.: University of Illinois Press, 2000), and W. J. Eccles, *France in America,* rev edn (East Lansing, Mich.: Michigan State University Press, 1990).
41. Farrand, *The Legislation of Congress,* p. 30.
42. Farrand, *The Legislation of Congress,* pp. 25–26, 36–37; and Eblen, *The First and Second United States Empires,* pp. 149, 160–166.
43. It is noted that Alaska was different that these other territories demographically. However, the sparsely populated region became involved in the imperialist and anti-imperialist debates over the effect colonialism could have on the republic. For a solid work on the debates over Alaskan statehood and territorial government in the region, see John S. Whitehead, *Completing the Union: Alaska, Hawai'i, and the Battle for Statehood* (Albuquerque, N. Mex.: University of New Mexico Press, 2004).
44. An Act Establishing the Territorial Government of Wisconsin, Chapter 54, *US Public Statutes at Large,* vol. V (Boston, Mass.: Little, Brown, & Co., 1856), pp. 11–12 (Hereafter referred to as the Wisconsin Organic Act of 1836).
45. Wisconsin Organic Act of 1836, p. 12.
46. Wisconsin Organic Act of 1836, p. 13.
47. Wisconsin Organic Act of 1836, p. 13.
48. Wisconsin Organic Act of 1836, p. 13. It is also important to note that the Wisconsin Organic Act also included a Bill of rights for territorial residents, only in condensed form. It stipulated that "the inhabitants of the said Territory shall be entitled to, and enjoy, all and singular the rights, privileges, and advantages, granted and secured to the people of the Territory of the United States northwest of the River Ohio, by the articles of the compact contained in the ordinance for the government of the said Territory, passed on the thirteenth day of July, one thousand seven hundred and eighty-seven" (Wisconsin Organic Act of 1836, p. 15).

49. See USA, An Act to Divide the Territory of Wisconsin and to Establish the Territorial Government of Iowa, Chapter 96, *US Public Statutes at Large,* vol. V (Boston, Mass.: Little, Brown, & Company, 1856), pp. 235–241 (hereafter Iowa Organic Act); USA, An Act to Establish the Territorial Government of Oregon, Chapter 177, *US Public Statutes at Large,* 9 (1862): 323–331 (hereafter Oregon Organic Act); USA, An Act to Establish the Territorial Government of Utah, Chapter 51, *US Public Statutes at Large,* 9 (1862): 453–458 (hereafter Utah Organic Act); USA, An Act to Provide a Temporary Government for the Territory of Colorado, Chapter 59, *US Statutes at Large,* 12 (1863): 172–177 (hereafter Colorado Organic Act); and USA, An Act to Provide a Temporary Government for the Territory of Dakota, and to Create the Office of Surveyor General therein, Chapter 59, *US Statutes at Large,* 12 (1863): 239–244 (hereafter Dakota Organic Act).
50. See Iowa Organic Act, p. 237; Oregon Organic Act, p. 325; Utah Organic Act, p. 454; Colorado Organic Act, p. 173; Dakota Organic Act, p. 241; USA, An Act to Provide a Temporary Government for the Territory of Oklahoma, to Create the Enlarge the Jurisdiction of the US Court in the Indian Territory, and for Other Purposes, Chapter 182, *US Statutes at Large,* 26 (1891): 84 (hereafter Oklahoma Organic Act).
51. USA, Treaty of Peace, Friendship, Limits, and Settlement with the Republic of Mexico, *US Statutes at Large,* 9 (1862): 922 (hereafter Treaty of Guadalupe Hidalgo).
52. Colorado Organic Act, p. 173. This clause is also present in the Dakota Organic Act and in modified form in the Utah Organic Act.
53. Oklahoma Organic Act, p. 84.
54. Oregon Organic Act, pp. 323, 328, 330, 331.
55. See John R. Wunder, *"Retained by the People": A History of American Indians and the Bill of Rights* (New York: Oxford University Press, 1994), pp. 27–41; and David R. Morgan, Robert E. England, and George G. Humphreys, *Oklahoma Politics and Policies: Governing the Sooner State* (Lincoln, Nebr.: University of Nebraska Press, 1991), pp. 24–25.
56. Oklahoma Organic Act, pp. 83–84, 85–88, 93–94.
57. For a general synthesis of the Texas Revolution, see Ray Allen Billington and Martin Ridge, *Westward Expansion: A History of the American Frontier,* 6th edn, abridged (Albuquerque, N. Mex.: University of New Mexico Press, 2001), pp. 125–150; and Robert V. Hine and John Mack Faragher, *The American West: A New Interpretive History* (New Haven, Conn.: Yale University Press, 2000), pp. 163–174. For a more in-depth account of the events leading to and during the revolution, see Marquis James, *The Raven* (Austin, Tex.: University of Texas Press, 1994); and James L. Haley, *Sam Houston* (Norman, Okla.: University of Oklahoma Press, 2002).
58. Billington and Ridge, *Westward Expansion,* pp. 203–214. Although a very traditional account of American expansion in the mold of Frederick Jackson Turner, Billington and Ridge's work offers an informative synthesis of the conquest of California.
59. USA, Joint Resolution for Annexing Texas to the United States, No. 8, *US Statutes at Large,* 5 (1856): 797–798; and USA, Joint Resolution for the Admission of the State of Texas into the Union, No. 1, *US Statutes at Large,* 9 (1862): 108.
60. Eblen, *The First and Second United States Empires,* pp. 152, 229; and Hine and Faragher, *The American West,* pp. 234–241.
61. Eblen, *The First and Second United States Empires,* pp. 301–314; and Earl S. Pomeroy, *The Territories and the United States, 1861–1890: Studies in Colonial Administration* (Seattle: University of Washington Press, 1969), pp. 6–27.
62. Eblen, *The First and Second United States Empires,* p. 273.
63. For information on the restructuring within the British and French Empires, see Ronald Hyam, "The British Empire in the Edwardian Era," in Judith M. Brown and William Roger Louis (eds.), *The Twentieth Century,* vol. IV of William Roger Louis (ed.), *The Oxford History of the British Empire* (Oxford: Oxford University Press, 1999), pp. 47–63; and William B. Cohen, *Rulers of Empire: The French Colonial Service in Africa* (Palo Alto, Calif.: Hoover Institute Press, 1971).
64. Eblen, *The First and Second United States Empires,* p. 280. A more accurate term for this would be

"an informal imperial service" since American territorial holdings did not fit the colonial pattern found in British, Belgian, Portuguese, Spanish, and German territories throughout Africa, Asia, and the South Pacific. See the beginning of this article for the difference between colonialism and imperialism.

65. Eblen, *The First and Second United States Empires,* pp. 271–304.
66. *Unforgiven,* directed by Clint Eastwood (1992). USA.
67. Richard White, *"It's Your Misfortune and None of My Own": A New History of the American West* (Norman, Okla.: University of Oklahoma Press, 1993), p. 333.
68. Eric Foner, *Reconstruction, America's Unfinished Revolution: 1863–1877* (London and New York: Harper & Row, ca. 1988), p. 120.
69. For a synthesis of justice, vigilantism, banditry, and ethnic violence, see White, *"It's Your Misfortune,"* pp. 328–351.
70. William Robbins, *Colony and Empire: The Capitalist Transformation of the American West* (Lawrence, Kans.: University Press of Kansas, 1994), pp. 122–125. Overall, Robbins' work is an intriguing study of forces of capitalism in Western expansion. It covers a wide range of subjects, even offering comparative analyses of these forces in the borderland regions between the USA, Canada, and Mexico.
71. Kermit L. Hall, "The 'Magic Mirror' and the Promise of Western Legal History at the Bicentennial of the Constitution," *Western Historical Quarterly,* 18 (4) (October 1987): 431–433; and Gordon Bakken, "Contract Law in the Rockies, 1850–1912," *American Journal of Legal History,* 18 (1) (1974): 33–34. See also John Phillip Reid, "The Layers of Western Legal History," in John McLaren, Hamar Foster, and Chet Orloff (eds.), *Law for the Elephant, Law for the Beaver: Essays in the Legal History of the North American West* (Pasadena, Calif.: Ninth Judicial Circuit Historical Society, 1992), pp. 23–25.
72. Bakken, "Contract Law in the Rockies," p. 34.
73. Bakken, "Contract Law in the Rockies," p. 34.
74. John Phillip Reid, *Law for the Elephant: Property and Social Behavior on the Overland Trail* (San Marino, Calif.: Huntington Library, 1980), p. 10.
75. Reid, *Law for the Elephant,* pp. 19–30.
76. Reid, *Law for the Elephant,* p. 265.
77. Reid, *Law for the Elephant,* p. 83.
78. Reid, *Law for the Elephant,* pp. 68–79.
79. Reid, *Law for the Elephant,* pp. 84–88.
80. Reid, *Law for the Elephant,* pp. 109–112.
81. Reid, *Law for the Elephant,* p. 131.
82. Reid, *Law for the Elephant,* pp. 68–76, 131–132.
83. Pomeroy, *The Territories and the United States,* p. 61.
84. Pomeroy, *The Territories and the United States,* pp. 52–55.
85. Kermit L. Hall, "Hacks and Derelicts Revisited: American Territorial Judiciary, 1789–1959," *Western Historical Quarterly,* 12 (3) (1981): 280–283. See also, John D. W. Guice, "Role of Territorial Supreme Courts," in John Porter Bloom (eds.), *The American Territorial System* (Athens, OH: Ohio University Press, 1973).
86. Pomeroy, *The Territories and the United States,* pp. 54–55.
87. Mark R. Ellis, *Law and Order in Buffalo Bill's Country: Legal Culture and Community on the Great Plains* (Lincoln, Nebr.: University of Nebraska Press, 2007), pp. 54–57. Ellis's work offers a great study on the development of legal institutions in Nebraska. It is a remarkable source for the court system, law-enforcement officers, judges, and lawyers who ensured the public welfare during its territorial and statehood phases.
88. For a traditional synthesis of this issue, see Walter Prescott Webb, *The Great Plains* (Lincoln, Nebr.: University of Nebraska Press, 1959), pp. 431–452.
89. Gordon Bakken, "Contract Law in the Rockies," pp. 38–42.

90. For European industrialization, political ideology, economic policies, and labor movements, see Robert Gildea, *Barricades and Borders: Europe 1800–1914* (Oxford: Oxford University Press, 2003), pp. 20–28, 123–130, 143–154, 219–236, 292–305. For a synthesis of Western labor conflict, see White, *"It's Your Misfortune,"* pp. 346–351. William G. Robbins's book *Colony and Empire* provides a detailed account of the conflict between labor and capital during the development of an industrialized, capitalist system in the American West.
91. John R. Wunder, "Chinese in Trouble," *Western Historical Quarterly,* 17 (1) (1986): 27, 34.
92. For Douglas's views on popular sovereignty, see Robert W. Johannsen, "Stephen A. Douglas and the Territories in the Senate," in John Porter Bloom (ed.), *The American Territorial System* (Athens, OH: Ohio University Press, 1973), pp. 84–85. For a solid account of Bleeding Kansas, popular sovereignty, and politics in the territories, I highly recommend Nicole Etcheson, *Bleeding Kansas: Contested Liberty in the Civil War Era* (Lawrence, Kans.: University Press of Kansas, 2004).
93. *Downes* v. *Bidwell,* 182 US 244 (1901), pp. 251, 257, 279, 286–287, 305–313.
94. For information on American colonial governments in Puerto Rico and the Philippines, see Efrén Rivera Ramos, *American Colonialism in Puerto Rico: The Judicial and Social Legacy* (Princeton, NJ: Marcus Wiener Publishers, 2007), pp. 55–59; Surendra Bhana, *The United States and the Development of the Puerto Rican Status Question, 1936–1938* (Lawrence, Kans.: University Press of Kansas, 1975), pp. 93–165; José A. Cabranes, *Citizenship and the American Empire: Notes on the Legislative History of the United States Citizenship of Puerto Ricans* (New Haven, Conn.: Yale University Press, 1979), pp. 1–14; H. W. Brands, *Bound to Empire: The United States and the Philippines* (Oxford: Oxford University Press, 1992), pp. 86, 88–93, 112–116; and Paul A. Kramer, *The Blood of Government: Race, Empire, the United States, and the Philippines* (Chapel Hill, NC: University of North Carolina Press, 2006), pp. 162–166. Aside from the basic system of colonial government established by the USA, these works are good resources for issues ranging from citizenship to ethnicity to self-government.
95. An Act to Provide a Government for the Territory of Hawaii, Chapter 339, *US Public Statutes at Large,* 331 (1901), pp. 141–162.
96. Gerald A. McBeath and Thomas A. Morehouse, *Alaska Politics and Government* (Lincoln, Nebr.: University of Nebraska Press, 1994), pp. 38–49; and Whitehead, *Completing the Union,* pp. 42–50.
97. Pomeroy, *The Territories and the United States,* pp. 35–45, 47–48.
98. Lamar, *The Far Southwest, 1846–1912,* pp. 288–293.
99. Lamar, *The Far Southwest,* pp. 317–318, 330–351.
100. Lamar, *The Far Southwest,* pp. 143–145, 423–427.
101. For ethnicity and labor movements during the early 1900s, see Moon-Kie Jung, "No Whites, No Asians: Race, Marxism, and Hawai'i's Preemergent Working Class," *Social Science History,* 23 (3) (1999): 357–393; and Gavan Daws, *Shoal of Time: A History of the Hawaiian Islands* (Honolulu, HI: University of Hawaii Press, 1968), pp. 314–317.
102. The Major–Palakiko case involved the murder of an elderly European American woman in 1948. The suspects were classified as "Hawaiian." The court case not only revived the memories of the Massie case within non-European communities, but also raised the question of legal equality between these ethnic groups. An examination of these two cases in Hawaii, as well as other cases in the American West involving person of European and non-European descent, suggest that Frederick Cooper's politics of difference, in which imperial citizens enjoy greater legal, social, political, and economic freedoms than ethnic and indigenous others, was fully operational in continental and insular territories. For more information on the Majors–Palakiko case, see Bernhard L. Hormann, "The Significance of the Wilder or Major–Palakiko Case, a Study in Public Opinion," *Social Process in Hawaii* 17 (1953): 1–13. For Cooper's theory on the politics of difference, see Cooper, *Colonialism in Question,* 22–24. Daws, *Shoal of Time,* pp. 317–326; Whitehead, *Completing the Union,* pp. 21–24; and Aaron Wilson, "'West of the West?': The Territory of Hawai'i, the American West, and American Colonialism in the Twentieth Century," Ph.D. dissertation (University of Chicago, 2008), pp. 201–204. The Massie case is widely written about.

103. Whitehead, *Completing the Union,* pp. 137, 166, 201–205, 220–225; and Daws, *Shoal of Time,* pp. 357–381.
104. Christian D. Fritz, "Constitution Making in the Nineteenth-Century American West," in John McLaren, Hamar Foster, and Chet Orloff (eds.), *Law for the Elephant, Law for the Beaver: Essays in the Legal History of the North American West* (Pasadena, Calif.: Ninth Judicial Circuit Historical Society, 1992), p. 295.
105. Fritz, "Constitution Making," pp. 298–301.
106. Jean Bickmore White, *Charter for Statehood: The Story of Utah's State Constitution* (Salt Lake City, UT: University of Utah Press, 1996), pp. 19–23; and Whitehead, *Completing the Union,* pp. 254–262.
107. There are a number of scholarly works on Native Hawaiian government. While Daws' *Shoals of Time* offers a good synthesis of Hawaiian government, the best sources detailing this process are Jonathan Kay Kamakawiwo'ole Osorio, *Dismembering Lāhui: A History of the Hawaiian Nation to 1887* (Honolulu, HI: University of Hawai'i Press, 2002); Ralph S. Kuykendall, *The Hawaiian Kingdom,* vol. I: *1778–1854: Foundation and Transformation* (Honolulu, HI: University of Hawai'i Press, 1968); and Ralph S. Kuykendall, *The Hawaiian Kingdom,* vol. II: *1854–1874: Twenty Critical Years* (Honolulu, HI: University of Hawai'i Press, 1966), 115–117; and Ralph S. Kuykendall, *The Hawaiian Kingdom,* vol. III, *1874–1893 The Kalakaua Dynasty* (Honolulu, HI: University of Hawai'i Press, 1967). Osorio's account is especially valuable as it gives the reader an account of these developments from the Native Hawaiian view. Kuykendall's is a well-written work as well but has come under some scrutiny for its treatment of Native Hawaiian history from traditional European and American accounts.
108. Mary W. Avery, *Government of Washington State,* rev. edn (Seattle, Wash.: University of Washington Press, 1973), pp. 19–20.
109. James C. Olson and Ronald C. Naugle, *History of Nebraska,* 3rd edn (Lincoln, Nebr.: University of Nebraska Press, 1997), pp. 327–330.
110. White, *Charter for Statehood,* p. 63; Gordon Morris Bakken, *Rocky Mountain Constitution Making, 1850–1912* (New York: Greenwood Press, 1987), p. 38.
111. Gerald A. McBeath and Thomas A. Morehouse (eds.), *Alaska State Government and Politics* (Fairbanks, AK: University of Alaska Press, 1987), pp. 38–42, 261, 269–271.
112. Beryl E. Pettus and Randall W. Bland, *Texas Government Today: Structures, Functions, Political Processes,* 3rd edn (Homewood, Ill.: Dorsey Press, 1984), pp. 200–215.
113. McBeath and Morehouse, *Alaska State Government and Politics,* pp. 326–329.
114. Pettus and Bland, *Texas Government Today,* p. 267.
115. Avery, *Government of Washington State,* pp. 248–251; Chuck Smith, *The New Mexico State Constitution: A Reference Guide* (Westport, Conn.: Greenwood Press, 1996), pp. 50–51; Donald Crowley and Florence Heffron, *The Idaho State Constitution: A Reference Guide* (Westport Conn.: Greenwood Press, 1994), pp. 74–75; and W. Drury, *The Government of Kansas* (Lawrence, Kans.: University of Kansas Press, 1961), pp. 39–40.
116. Janice C. May, *The Texas State Constitution: A Reference Guide* (Westport, Conn.: Greenwood Press, 1996), pp. 39–40, 76.
117. Morgan et al., *Oklahoma Politics & Policies,* p. 135; and McBeath and Morehouse, *Alaska State Government and Politics,* p. 35.
118. John R. Owens, Edmond Costantini, and Louis F. Weschler, *California Politics and Parties* (London: Macmillan, 1970), p. 270.
119. Bakken, *Rocky Mountain Constitution Making,* pp. 51–63.
120. Pettus and Bland, *Texas Government Today,* pp. 284–295, 383–396.
121. Dan A. Cothran, "The Arizona Budget in an Age of Taxpayer Revolt," in Zachary A. Smith (ed.), *Politics and Public Policy in Arizona,* 2nd edn (Westport, Conn.: Praeger, 1996), pp. 173–186.
122. Avery, *Government of Washington State,* p. 81; Brent S. Steel and Denise Lach, "Environmental Policy," in Rihcard A. Clucas, Mark Henkels, and Brent S. Steel (eds.), *Oregon Politics and Government:*

Progressives versus Conservative Populists (Lincoln, Nebr.: University of Nebraska Press, 2005), pp. 225–241; Joseph R. Grodin, Calvin R. Massey, and Richard B. Cunningham, *The California State Constitution: A Reference Guide* (Westport, Conn.: Greenwood Press, 1993), pp. 172–180; and Marin A. Nie and Zachary Smith, "Environmental Policy in Arizona," in Zachary A. Smith (ed.), *Politics and Public Policy in Arizona,* 2nd edn (Westport, Conn.: Praeger, 1996), pp. 238–240, 245–246.

123. Eric Hobsbawm, *The Age of Extremes: A History of the World, 1914–1991* (New York: Vintage Books, 1994), pp. 96, 114, 128–129.
124. Alan Brinkley, *The End of Reform: New Deal Liberalism in Recession and War* (New York: Vintage Books, 1996), p. 6.
125. There are a number of reliable sources to reference for information pertaining to the USA and Europe during the Great Depression and the New Deal. Robert S. McElvaine, *The Great Depression: America, 1929–1941* (New York: Three Rivers Press, 1993) and Richard Lowitt, *The New Deal in the West* (Norman, Okla.: University of Oklahoma Press, 1993) both offer solid overviews of the federal legislation passed during the New Deal. For a work on the shift in ideology, Alan Brinkley's *The End of Reform* has much to offer. Finally, Hobsbawm's *The Age of Extremes* is a good work to reference for a brief explanation of the global effects on politics and ideology during this time.
126. Neil Maher, "'Crazy Quilt Farming on Round Land': The Great Depression, the Soil Conservation Service, and the Politics of Landscape Change on the Great Plains during the New Deal Era," *Western Historical Quarterly,* 31 (3) (2000): 319–339.
127. For examples, see Richard C. Pratt with Zachary Smith, *Hawai'i Politics and Government: An American State in a Pacific World* (Lincoln, Nebr.: University of Nebraska Press, 2000), pp. 164–165; Avery, *Government of Washington State,* pp. 97, 101–102, 168; and Pettus and Bland, *Texas Government Today,* pp. 456–462.
128. Tomas E. Cronin and Robert D. Loevy, *Colorado Politics and Government: Governing the Centennial State* (Lincoln, Nebr.: University of Nebraska Press, 1993), pp. 80–81.
129. Pratt and Smith, *Hawai'i Politics and Government,* pp. 164–165.
130. Deborah Mathieu, "Health Policy in Arizona," in Zachary A. Smith (ed.), *Politics and Public Policy in Arizona,* 2nd edn (Westport, Conn.: Praeger, 1996), pp. 130–134.
131. William Lunch, "Health Policy," in Richard A. Clucas, Mark Henkels, and Brent S. Steel (eds.), *Oregon Politics and Government: Progressives versus Conservative Populists* (Lincoln, Nebr.: University of Nebraska Press, 2005), pp. 244–245, 248–254.

CHAPTER EIGHTEEN

REGION OF EXTREMES AND CONTRAST

Alicia M. Dewey

Along the central coast of California near San Simeon, La Cuesta Encantada, the palatial Hearst castle, perches high upon cliffs overlooking the Pacific Ocean. It serves as a symbol of the vision that Anglo-Americans brought to the West, a lifestyle that blended wealth-making with an attempt to incorporate and enjoy the best of Euro-American art, architecture, music, and other aspects of high culture. Built between 1919 and 1947 in a Mediterranean style with terraced gardens, marble pools, and tennis courts, the castle has fifty-six bedrooms, nineteen sitting rooms, and over 90,000 square feet of space. Its halls contain the paintings and sculptures collected by its original owner, publishing magnate William Randolph Hearst, whose father, George, purchased the land on which it stood in 1865. Hearst loved to entertain Hollywood stars, American diplomats, and millionaires on the grounds of his exclusive retreat.[1]

The world created and enjoyed by Hearst and his guests depended on the labor of thousands of people whose lives were virtually invisible to the wealthy. Many lived in squalid tenement houses and shacks without running water or electricity just a few hundred miles away from San Simeon in the city centers of Los Angeles and San Francisco. Unlike Hearst's guests, most of whom were Euro-American, this laboring class consisted of a mixture of different ethnic groups, some Italians, some Russian Jews, but many more African Americans, Mexicans and Mexican Americans, Chinese, and Japanese. Working long hours in garment and cigarette factories or doing back-breaking work in fields and orchards, often unable to earn enough to keep food on the table, Hearst's castle seemed as far away from their world as the moon.

One of the ironies of the American West is that despite its place in American mythology as a place of unlimited bounty, opportunity, and success for any rugged individual willing to move there, its growth in the nineteenth and early twentieth centuries actually spawned an economy where only a relatively small number of Americans harvested the region's rich natural resources. These successful entrepreneurs created enormous commercial wealth in a variety of mining, lumber, oil, agricultural, mercantile, and railroad enterprises. They built huge mansions on large lots in beautifully landscaped neighborhoods and gave generously to educational and cultural institutions that eventually rivaled those of the Eastern United States.[2]

These Western entrepreneurs formed a society that fueled a growing gap between rich and poor. Their need for cheap labor attracted millions of unskilled and semi-skilled workers from around the globe, many of whom settled in the mushrooming urban areas near newly created industrial districts, forming densely populated enclaves of poverty. Although the philanthropy of Western business leaders helped to create an intellectual and social infrastructure that attracted people to the region and in turn enhanced economic development, eventually leading to the growth of a diverse and broad-based middle class, it offered very little to the working poor. Only a few of the descendants of the workers who helped to build the region were eventually able to move into the middle class and participate in the fruits of Western philanthropy through educational scholarships and grants. Despite some advances in the late twentieth century, especially after the civil-rights movement began to break down barriers to the entry of ethnic minorities into higher paying jobs, as the West enters the twenty-first century, extreme differences between the lifestyles of rich and poor have not disappeared. Instead, they have become even more pronounced as the post-industrial economy continues to require low-paid workers for the service and retail sectors.

ENTERPRISE, WEALTH, AND PHILANTHROPY

American entrepreneurs established what the governments of Spain and Mexico were never able to, that is, a thriving, highly developed urban society in the region that is now the Western USA. They did this primarily through the creation of huge fortunes that in turn funded the growth of educational, cultural, and social institutions. Today, the result is that the modern American West contains important, emerging global centers of capital and culture, the primary one being the city of Los Angeles, California.

American Enterprise in the West

The Americans brought with them a system of competitive, commercial capitalism that created not only the possibility of greater wealth accumulation but also sharper disparities in the overall distribution of that wealth.[3] San Francisco and Denver were the first truly American cities in the West, created within ten years of each other by two different gold rushes, one in 1848 and the other in 1858.[4] They grew rapidly, becoming "instant cities" as thousands of people poured in to get rich. These were cities centered on making money, and they set the pattern for others throughout the West.[5] In the early years, social categories remained fluid and unstable because fortunes were often won and lost in a single day. Members of the elite, middle class, and working class rubbed shoulders on a day-to-day basis, not only in their jobs but also on the streets and in restaurants, stores, and boardinghouses.[6]

Eventually, however, a small group of people began to rise to the top "out of the plebeian entrepreneurial mass."[7] Most fortunes initially arose, directly or indirectly, from the extraction of natural resources, especially mining of gold, silver, and other minerals. George Hearst made his millions initially by investing in the Comstock Lode in Nevada in 1859 and parlayed the profits into silver, gold, and copper mines in Utah, South Dakota, Idaho, Montana, and Mexico. Most of Denver's elite, including Horace A. W. Tabor and Simon Guggenheim, profited from the lead, silver, and iron mines that crisscrossed the Colorado Rockies.[8]

Mining booms created demands for supplies and equipment that, in turn, led to booms in trade and industry. Many merchants benefited from the rush for gold in California more so than the gold diggers themselves. Among them were Collis P. Huntington, Leland Stanford, Mark Hopkins, and Charles Crocker—the "Big Four," middle-class migrants from New England and upstate New York who used their mercantile proceeds to build the transcontinental railroad in the 1860s and became multi-millionaires as a result. Others acquired their riches in industry. Peter Donahue traveled to California to find his fortune in gold but failed. Success came after he started a small iron foundry, which later became Union Iron Works. He sold it to Irving Murray Scott and another partner in 1865, and then branched out into gasworks, railroads, banking, timber, and real estate. Irving Murray Scott made enormous profits selling equipment to mine owners and railroad companies.[9]

Wealth-making branched out from Denver and San Francisco as enterprising Americans found ways to make money in other parts of the West as well. The first fortunes in Seattle, including that of the Stimson family, came from the logging industry. The city later became a supply point for the Klondike Gold Rush in 1896, providing a basis for expanding mercantile and banking operations. Edward L. Doheny launched an oil boom in southern California in the 1890s, and Henry L. Huntington, nephew of Collis Huntington, added to the inheritance from his uncle by building networks of electric urban railways throughout Los Angeles and its environs.[10] In the countryside, families such as the Kings and Kenedys of south Texas and the Boswells of California's Central Valley built massive commercial ranching and farming enterprises.[11] On the edge of the West, the cities of Houston and Dallas began to emerge in the late nineteenth and early twentieth centuries as centers for the distribution of cotton, creating the fortunes of William Marsh Rice and the Caruth family. After the discovery of oil at Spindletop in 1901, Houston became the core of the oil industry and oil services, led by people such as Joseph S. Cullinan, owner of the Texas Company, later known as Texaco. Dallas became a major retail and financial center for its vast rural hinterland.[12]

Some of the most powerful families controlled the region's newspaper and media outlets. These families have had enormous staying power. William Randolph Hearst, George Hearst's son, begged his father to allow him to run the ailing *San Francisco Examiner,* which his father had purchased a few years earlier. He became the publisher in 1887 while he was still studying at Harvard. He later built a publishing empire by acquiring newspapers across the country.[13] His rival, Michael de Young, owned the *San Francisco Chronicle.* De Young, of Dutch Jewish descent, married his daughters to the scions of wealthy Anglo-American families in order to preserve the family's wealth and to enhance its status.[14] In Los Angeles, Harrison Gray Otis made the *Los Angeles Times* a successful daily newspaper. He created the Times Mirror Company and bought the paper in 1884. When he died in 1917, his son-in-law Harry Chandler took over and later passed leadership on to his son, Norman, in 1944.[15]

Lured by opportunities to make money and largely disconnected from family and community ties, the *nouveau riche* of the American West created a society that classified people mainly based on wealth. Most of them had come from Euro-American Protestant middle- or lower-middle-class backgrounds; upper-class Easterners who invested in Western enterprises generally did not move. John D. Rockefeller, Jr., for example, had extensive mining operations in Colorado, but he continued to live in New York. This led to some degree of social flexibility and upward mobility for Euro-American ethnic groups such as Jews and Irish Catholics who faced barriers in other parts of the country and gave Western cities their reputation for openness and progressive attitudes.[16] This flexibility did not extend, however, to members of ethnic minorities, including African Americans, Asians,

with trees, acre-sized lots or larger with required setback lines, and single family homes isolated from commerce and industry located near parkland. Mansions such as the one built between 1906 and 1914 by William Wrigley, Jr., founder of Wrigley's chewing gum, lined South Orange Grove Boulevard in Pasadena (see Figure 18.1). Palos Verdes Estates and Huntington Palisades in Los Angeles illustrated these trends as did the Highlands area of Seattle, River Oaks in Houston, and Highland Park in Dallas.[35] Most importantly, these neighborhoods were almost always exclusively Euro-American. Several Westside neighborhoods in Los Angeles even contained occupancy restrictions against Jews, although one of the wealthiest, Beverly Hills, had a distinctly Jewish flavor because of its proximity to Hollywood.[36]

Western cities became divided into affluent Euro-American neighborhoods and "foreign quarters" where poor working-class Euro-Americans, including Jewish, Italian, and Irish immigrants mingled with African Americans, ethnic Mexicans, Chinese, and Japanese laborers. Wealthy Euro-Americans lived on the west side of Los Angeles, the south side of Denver, the northwest side of San Antonio, and the north side of Phoenix.[37] Even Santa Fe began to attract affluent Euro-Americans to the east side of the old plaza. Although these wealthy areas contain more diversity today than they did a century ago, they remain overwhelmingly Euro-American in terms of their demographics, and they tend to reflect Anglo-American tastes in architecture and layout.

Wealthy entrepreneurs also spent lavishly on their leisure time and social life. The Palace Hotel of San Francisco, built in 1875 with the most advanced modern luxuries of the time including air conditioning, electronic call buttons, five elevators, seven grand stairways, fifteen-foot high ceilings in every room, and elaborate woodwork, was the gathering place

Figure 18.1 Front grounds and façade of the Wrigley Mansion at 391 South Orange Grove Avenue, Pasadena, California.
Source: UCLA Charles E. Young Research Library Department of Special Collections, Los Angeles Times Photographic Archives. © Regents of the University of California, UCLA Library.

for the city's elite.[38] In the early twentieth century, desert, beach, and mountain resorts sprang up across the West, from the California coast to Tucson and Colorado Springs.[39] By the mid-to late twentieth century, affluent Texans and Californians were building second and third homes in the Colorado Rockies in order to enjoy skiing in the winter and hiking in the summer, driving up local property values and taxes, making it difficult for long-time residents to remain in their homes.

Philanthropy

In addition to spending money on their material possessions, Western entrepreneurs also funneled significant portions of their fortunes into philanthropy that benefited Western cities. It is nearly impossible to detect any sort of pattern in Western giving because people gave to a wide variety of organizations and institutions according to their own personal interests and goals. Much of it was unplanned and uncoordinated. The overall effect of Western philanthropy, however, was the creation of an educational and cultural infrastructure that promoted economic growth and intellectual development in the region. Through the farsighted leadership of a few Western businessmen and women, Western cities once considered cultural backwaters have evolved to become internationally renowned centers of the arts.

Philanthropy lagged behind capital accumulation, which meant that for most of the nineteenth century very little money was available for cultural development. Western philanthropy generally followed patterns set on the East Coast during the early nineteenth century, in part because many of the wealthiest Westerners were born and raised there. Philanthropists divided their efforts into two main areas: first, charity, or the relief of suffering and poverty through the construction of hospitals, orphanages, housing for the poor, and settlement houses; and second, arts and education.[40]

Women played crucial roles as philanthropic leaders in the West. The wives, daughters, and widows of wealthy entrepreneurs used their time and access to the fortunes created by husbands and fathers to improve their communities. Mary Crocker, Phoebe Hearst, and Jane Stanford were at the center of charitable work in San Francisco in the late nineteenth and early twentieth centuries. Through their leadership, they provided an example to others that wealth brought with it the responsibility to serve and enhance the community.[41]

Efforts to provide services to improve the lives of those who struggled in the poor areas of the cities were particularly noticeable during the Progressive era at the turn of the twentieth century. In Dallas, Texas, the Dallas Federation of Women's Clubs led by May Dickson Exall, wife of the developer of the high-end suburb of Highland Park, worked for such diverse causes as prison reform, temperance, and the establishment of orphanages, hospitals, settlement houses, and infant-welfare stations.[42] Los Angeles had a similar federation. In San Francisco, a group of Jewish philanthropists and charity workers, led by Judge Solomon Heydenfeldt and merchant Samuel Levy, started a movement to create kindergartens. Mary Crocker and Jane Stanford both contributed funds to the effort.[43] The Houston Negro Hospital, built in 1924, was the gift of J. S. Cullinan, founder of Texaco Oil Company, in memory of his son who died in World War I.[44]

One major emphasis of Western philanthropy has been higher education. The thousands of buildings in universities across the West bearing the names of some of its most widely recognized citizens testify to the ongoing close relationships among Western business leaders, higher education, and research. Business leaders have long perceived this relationship to be absolutely crucial because the universities serve as training and recruiting grounds for future entrepreneurs, managers, and employees. Some see a connection between the prestige of local educational institutions and overall income levels in a city.[45] The universities, in turn, have always relied heavily on wealthy donors to fund their programs and to build their buildings. In California, the symbiosis between business and higher education has led to the national prominence of several of its universities, including the University of California at Berkeley, the University of California at Los Angeles (UCLA), and Stanford University. Because of the close relationship, some have criticized universities as being merely extensions of modern corporations and completely beholden to the captains of industry who dominate their boards and act as their chief donors.[46]

Mining moguls were among the first benefactors of higher education in Western states. Enos Wall, organizer and director of the Utah Mining Company, gave funds to endow a fellowship in metallurgy at the University of Utah in 1907. Jesse Knight, the Mormon developer of the Humbug mine, gave over $500,000 to Brigham Young University between 1898 and 1921. Winfield S. Stratton, who discovered the Independence Mine in Cripple Creek Colorado in 1891, gave $55,000 to Colorado College to construct a building of arts and sciences. Another mine owner, Henry R. Wolcott, contributed funds to build an astronomical observatory at Colorado College. John B. Evans, the second territorial governor of Colorado, and James B. Grant, Vice President of the Omaha and Grant Smelting and Refining Company in Denver, funded the opening of the University of Denver in 1880. Simon Gannett Reed of Portland, Oregon, who made a fortune from the Bunker Hill and Sullivan mines in north Idaho's Coeur d'Alene district, gave anonymous gifts in 1893 to the predecessor of the California Institute of Technology.[47]

The University of California at Berkeley, which opened in 1869 as the result of a merger between the private California College and a new state agricultural and mechanical land-grant college, was closely tied to San Francisco's financial district and its business elite. William C. Ralston and Darius O. Mills, organizers of the Bank of California, gave generously to the university and also served on its Board of Regents. George Hearst, whose original fortune derived from the Comstock Lode in Nevada, was responsible for much of its early growth. His wife, Phoebe, took a particular interest in the school and financed buildings, endowed professorships, donated libraries, paid for scholarly research, and endowed a Department of Anthropology. William Randolph Hearst continued his mother's giving to Berkeley, offering student fellowships and funding the construction of additional edifices across the campus.[48]

Wealthy local citizens were instrumental in starting many of the other colleges and universities throughout the West. In 1884, Leland Stanford, lawyer, railroad financier, and Republican governor of California, decided to build a private university in memory of his only son, Leland, Jr., who had died of typhoid fever at age sixteen while on a tour of Europe. After Stanford died in 1894, his widow, Jane Stanford, continued to donate generously to the school, mainly to build buildings. Unlike Phoebe Hearst, she rarely endowed scholarships or professorships. Millionaires in Pasadena, California, help found and maintain the California Institute of Technology. Ozro W. Childs, former California governor; John G. Downey, an Irish Catholic pharmacist; and Isaias W. Hellman, a German-Jewish banker, provided land and funds for the establishment of the University of Southern California in 1880.[49] The Caruth family of Dallas, whose wealth originated from cotton farming, gave the land to build the campus of Southern Methodist University, and over the years the Dedman family, who reaped millions from establishing and operating golf courses and private clubs, has endowed many of its buildings and programs. Robert H. Dedman, Sr., also supported nutrition and cancer research at the University of Texas Southwestern Medical Center and established a hospital bearing his name.[50]

On the southeastern edge of the West, Rice University in Houston was the dream of William Marsh Rice, one of the wealthiest businessmen in Houston by the late nineteenth century. Rice asked Captain James A. Baker of the law firm Baker & Botts to draw up the charter for the William M. Rice Institute for the Advancement of Literature, Science, and Art. Rice appointed a Board of Trustees, designating Baker as the head. Baker remained in that position for fifty years, playing a key role in Houston's cultural development. The Rice Trust was one of the largest accumulations of capital in the entire region. Baker grew the trust by making careful loans; it had $9.8 million by 1912 when Rice Institute was established, enough money to hire a prominent, well-trained, and diverse faculty.[51] Today, Houston is also the home of the world renowned M. D. Anderson Cancer Center, established in 1941 with funds from the estate of Monroe D. Anderson, wealthy cotton broker, banker, and partner in the prominent local firm of Anderson, Clayton, & Company. Anderson's gifts made the creation of the Texas Medical Center possible.[52]

In addition to higher education, Western philanthropists have spent millions collecting works of art and creating museums to house them, financing symphonies, and building musical centers and schools. This commitment to the arts has historically been one of the primary ways in which the newly rich of the Western states have developed their identities as cultured and refined people, paving the way for social acceptance not only among fellow Western elites but also among those on the East Coast as well.[53] Because San Francisco was the home of some of the earliest large fortunes in the West, it became the West's first major cultural center. Artists and writers attracted to the vibrancy of the city and the beauty of California's rugged coastline and the Sierra Nevada Mountains moved there in significant numbers. Responding to this new presence and to criticism that the city lacked cultural sophistication, some of the business elite decided to fund the San Francisco Art Association, which opened the first art school in the West, the San Francisco School of Design, in 1874, under the direction of landscape painter Virgil Macey Williams. The school supported a number of young painters, including Albert Bierstadt, whose paintings of the Rocky Mountains became well known. Renamed the San Francisco Art Institute in 1961, today it is one of the finest schools of contemporary art in the nation.[54]

Michael de Young, publisher of the *San Francisco Chronicle,* spearheaded an effort to create the first major public art museum in the West. After the California Midwinter International Exposition in San Francisco's Golden Gate Park in the winter of 1894, de Young insisted that the Fine Arts Building constructed for the fair remain standing as San Francisco's first public art gallery. He donated part of his European art collections along with funds to pay maintenance costs. He later financed a more permanent structure, which was renamed the M. H. de Young Memorial Museum.[55] The first three decades of the twentieth century saw the growth of more San Francisco cultural institutions, including the San Francisco Symphony Orchestra, the San Francisco Opera Company, and the San Francisco Ballet.[56]

As Los Angeles began to rise in importance as a center of business in the twentieth century so did its place in the world of art and education. The earliest philanthropists of Los Angeles included Harrison Gray Otis, Otis' son-in-law Harry Chandler, and Henry E. Huntington. The Otis and Chandler families of the *Los Angeles Times* would continue to be major leaders in Los Angeles philanthropy throughout the twentieth century. Otis gave land in 1918 for the establishment of the first public art institute in Los Angeles, the Otis College of Art and Design.[57] Henry Huntington provided the basis for another important cultural institution, the Henry E. Huntington Library and Art Gallery in San Marino. Huntington had started collecting rare books and artworks in the early 1900s. He established a research library in 1919 and an art gallery in the early 1920s. When he died in 1927, he left a rich endowment. In 1919, William Andrews Clark, Jr., a copper magnate from Arizona, gave the Los Angeles Philharmonic a huge endowment to begin in 1919.[58]

Los Angeles largely remained a cultural backwater, however. It had no central plan for the development of the arts, and its few institutions reflected a very individualistic drive on the part of the individuals who gave. This all changed between 1960 and 1990. Dorothy Buffum Chandler, department-store heiress and wife of Norman Chandler, publisher of the *Los Angeles Times,* was instrumental in the transformation. Dorothy Chandler's impact on the Los Angeles cultural community began in the mid-1950s when she became the de-facto leader of the Board of Directors of the Los Angeles Philharmonic. She was motivated partly by her desire to keep the center of the city going as the result of Chandler interests located there; the family had made fortunes not only in newspapers but also in oil, ranching, television, and insurance. She led efforts to build a performing-arts center for city that would serve as the Philharmonic's new home and helped bring prominent conductor Zubin Mehta to lead the orchestra between 1962 and 1978. In course of her fundraising activities for the new Los Angeles Music Center, she broadened her campaign to include prominent Jews, much to the chagrin of many of her friends who derided them as "movie people." Her efforts proved successful, and the Dorothy Chandler Pavilion finally opened on December 6, 1964.[59]

Chandler also helped to promote Franklin D. Murphy, a man with a vision for the arts in southern California, to a position where he was able to influence and facilitate philanthropic giving to promote Los Angeles as a major cultural center of the nation. Murphy had moved to California in 1960 from Kansas to become the Chancellor of UCLA. Chandler became acquainted with Murphy while he was working to make UCLA a major national university. Murphy recognized that national political power and industrial supremacy were beginning to shift to the West, and he believed that cultural and intellectual influence must follow. He felt that Los Angeles had been far too driven by money and economic growth to the detriment of creating cultural spaces. Murphy first tried to inspire cultural change through his position at UCLA but became disillusioned with the internal political battles within the university system and the scarcity of funds. When Dorothy and her husband Norman invited him to become chairman and chief executive of the Times Mirror Company in 1968, he accepted, believing he could do much more in that position for Los Angeles. Murphy was highly regarded as an authority on the arts; he was the first person west of Pittsburgh asked to serve on the board of the National Gallery of the Art in Washington, DC. As he became known throughout Los Angeles, he served as an adviser to many of the area's millionaires and as a trustee on the boards of most of the major foundations and trusts. Through his positions on these boards as well as his relationships with wealthy entrepreneurs, Murphy was able to substantially influence the direction of giving to the arts in southern California.[60]

Murphy's first major project, begun while he was at UCLA, involved the advancement of the Los Angeles County Museum of Art (LACMA). Los Angeles had had a city museum since 1910, but it housed a hodgepodge of items in a building called the Los Angeles County Museum of History, Science, and Art in Exposition Park near the University of Southern

California. Retail giant Edward Carter met Murphy through his position as a regent of UCLA and asked him to join with him and other wealthy businessmen to discuss the formation of a new art museum. Carter negotiated with the city for the land in Hancock Park on Wilshire Boulevard near the La Brea Tar Pits and became LACMA's founding president. Both he and Murphy served as trustees. Dr. Richard F. Brown became its first director. The new building on Wilshire Boulevard opened in spring 1965, but it took a long time to gain the respect of the art world. The *New York Times* criticized Los Angeles as "culturally immature" and said that its donors were wealthy people who were only interested in advancing themselves. Over the years, however, as a member of the board, Murphy was able to blunt such criticism as he successfully negotiated for both funds and collections for LACMA, eventually helping to make it a world-class museum.[61]

Murphy convinced many of Los Angeles' wealthy to donate to LACMA. Murphy became fast friends with Howard Ahmanson shortly after he moved to Los Angeles. Ahmanson, owner of Home Savings Bank, had acquired the bulk of his enormous fortune during the huge postwar boom in housing. He had funded a number of projects in the 1960s, including the Ahmanson Theatre, which was built in 1962 as part of the Los Angeles Music Center, and the Ahmanson Building at LACMA. When he died suddenly and unexpectedly in 1968 of a massive heart attack while on a vacation in Europe, Murphy became involved in the management and distribution of his estate. Murphy helped to engineer the appointment of Harold's nephew, Robert A. Ahmanson, to the position of director of the Ahmanson Foundation, which received a large portion of the estate's assets. As adviser to Robert and as trustee of the foundation, Murphy influenced the recipients of the Ahmanson fortune. In its fifty-year history, the Ahmanson Foundation has given approximately $550 million to the Los Angeles community, including money for art acquisitions at LACMA, endowed professorships in biology and medicine at the University of Southern California, an endowment fund for the Ahmanson Theatre, neurological research at UCLA, and a building to house the History of Business and Economics at the California Museum of Science and History. Ahmanson money even helped save the Huntington Library during an especially dire period of financial struggle in the early 1990s.[62]

Murphy's influence over other wealthy Angelinos was substantial but limited. Private money continued to flow in different directions, reflecting the individualistic nature of the philanthropists. He was unable to convince Norton Simon of Hunt's Foods, Inc., and Armand Hammer, head of Occidental Petroleum, both of whom had given money to LACMA over the years, to give their private art collections permanently to the museum. To Murphy's great disappointment, in January 1988, despite a pledge to bequeath his impressionist and post-impressionist artworks to LACMA, Armand Hammer announced his intention to build the Armand Hammer Museum of Art and Cultural Center. Simon built the Norton Simon Museum in Pasadena to house his extensive collection of impressionist and old European masters' paintings. Murphy was also unable to acquire the extensive collections of J. Paul Getty but did end up playing a role in the founding of the new Getty Museum, funded by the J. Paul Getty Trust. Due to a recommendation from people close to the Trust, Murphy eventually served on its board and helped direct the design of the new Getty Museum, which opened three years after Murphy's death on the west side of Los Angeles in December 1997.[63]

Although Los Angeles philanthropists initially funded cultural institutions that primarily showcased American and European heritage, many began to recognize the need for diversity in the arts even as different racial and ethnic groups were making significant strides politically and economically. Murphy in particular contributed to cultural diversity in Los Angeles through his influential role on the board of the Ahmanson Foundation. He was interested in supporting cultural centers that emphasized the artistic achievements of particular groups, not institutions that would house small displays of a variety of groups. The Foundation gave grants to the Pacific Asia Museum, the Japanese American Cultural and Community Center, Plaza de la Raza, the California Afro-American Museum, and the Southwest Museum of the American Indian. In the late 1970s, Rabbi Uri Herscher, Dean of Hebrew Union College Jewish Institute of Religion, approached Murphy to assist in building a cultural center dedicated to exploring the intersection of Jewish heritage and American democratic ideals. The Skirball Cultural Center began in June 1981 and opened in 1996. Murphy served on its Board of Directors.[64]

Other Western cities have generally followed the pattern of San Francisco and Los Angeles in terms of cultural development. A few major business leaders typically spearhead the establishment and growth of artistic institutions, appointing friends to manage the process. Today, although most Western cities have their own art museums and musical associations, most lag behind those in California cities in terms of reputation and quality. This seems to be a direct result of a combination of factors, including the comparative scarcity of a significant concentration of wealthy philanthropists and the lack of strong leadership to promote the arts. There are a few cities, however, that are trying to change this and to promote the arts in ways that they did not in the past. Fort Worth, Texas, is the best example of one of these new rising cultural centers in the West. Once a remote cattle town, today it boasts that it houses the nation's third largest cultural district, and some of its museums have gained national and even international renown. Fort Worth was generally a more conducive environment for local artists and writers than either Dallas or Houston, mainly because of leadership in the area of the arts on the part of a few of its wealthy residents.[65]

Amon G. Carter, Sr., was Fort Worth's first major philanthropist and its leading citizen for many years. He was the publisher and president of the Fort Worth Star-Telegram, the local newspaper. He actually made his fortune by investing in oil wells. He collected Remington and Russell paintings, an interest he developed through his friendship with Will Rogers. Carter was instrumental in the creation of the cultural district in the 1930s during the Texas Frontier Centennial Celebration, including the Will Rogers Memorial Complex, a multi-purpose center that could house a variety of exhibitions and performances. Carter later stipulated in his will that he wanted to construct a museum near the Will Rogers Complex to house his collection of Western art. Six years after his death in 1955, the Amon Carter Museum, designed by architect Philip Johnson, opened. Its purpose, according to its founder, who said that while he was growing up he was "denied the advantages which go with the possession of money," was "to give to those who have not had such advantages, but who aspire to the higher and finer attributes of life, those opportunities which were denied to me."[66]

Another wealthy Texas businessman, Kay Kimbell, donated substantial sums of money to build a museum to house his collections of Renaissance art, eighteenth-century English paintings, and French and American nineteenth-century paintings. The Kimbell Art Museum, now internationally renowned, opened in 1972. Its first director, Richard F. Brown, had also been the first director of LACMA and played a key role in the Kimbell's subsequent rise to prominence. He collaborated closely with American architect Louis I. Kahn to design the building, which was located just down the hill from the Amon Carter

Museum and across the street from the Will Rogers Memorial Complex.[67] The construction of the Kimbell completed the Fort Worth Cultural District.

Probably the most influential philanthropists in Fort Worth's history are the Bass brothers, who grew up there, made their fortunes elsewhere, and began pouring money back into the city beginning in the 1980s. They have given millions to the Fort Worth Zoo and local art museums. They redeveloped the central business district, Sundance Square, restoring historic buildings and creating venues for entertainment and dining. One of those venues, the Bass Performance Hall, opened in May 1998 as the permanent home of the Fort Worth Symphony Orchestra, the Texas Ballet Theater, Fort Worth Opera, and the Van Cliburn International Piano Competition. Bass Hall already has an international reputation.[68]

Although the city of Houston has had a number of wealthy philanthropists, none was as infamous and few were as generous as Kenneth Lay, the founder and chief executive officer of Enron Corporation. Enron and Ken Lay brought new money and new recognition to Houston as a cultural center in the 1990s, which almost all evaporated overnight after the dramatic plunge of Enron's stock, its subsequent bankruptcy, and the criminal convictions of its former officers, including Lay. Lay had a vision of making Houston a "world class city, with opera, sports and other amenities, if for no other reason than to help Enron attract ambitious high achievers who did not want to live in a backwater."[69]

Ken Lay was well known for his philanthropy, giving regularly to the Houston Ballet, the Alley Theater, the Houston Opera, and the Museum of Fine Arts as well as Habitat for Humanity and the United Way.[70] Lay and Enron provide an example of the late-twentieth-century development of corporate philanthropy, where large corporations and their employees donate money to a variety of causes. In addition to its corporate giving, Enron began its own collection of world-class art that was never completed. This became a source of controversy during the bankruptcy proceedings because Enron was apparently purchasing masterpieces worth millions of dollars while its employees were daily losing value in their retirement portfolios.[71] When Enron collapsed, Houston lost a major source of revenue for the arts that would not quickly be replaced. The Lay and Enron scandal provides a good example of the downside of this close relationship between business and the arts. When businesses suffer losses or participate in scandals, the art world suffers as well.[72]

Unlike Ken Lay, who sought to use his wealth to benefit Houston, Bill Gates, the founder of Microsoft, is pursuing a new kind of global philanthropy. Gates's maternal ancestors, the Maxwells, were an old, socially prominent Seattle family who had set an example of philanthropic involvement in the community.[73] Gates is following in those footsteps but with a different twist. He and his wife established the Bill and Melinda Gates Foundation in 2000 to fund a variety of efforts to improve health and economic development around the world. Not focused on the arts at all, its goals involve lifting people out of poverty and hunger and addressing health problems around the world. Gates believes science and technology have great potential to improve lives around the world. Although he has given some money to ventures in Seattle, most of his funds go overseas.[74] Recently, Warren Buffet has said that he plans to give most of his money to the Bill and Melinda Gates Foundation, suggesting that globalism is not only affecting the scope of business but also of philanthropy.[75]

Although most Western philanthropy directed toward the arts funded institutions and collections similar to those found in New York City and Washington, DC, a few people decided to invest in the promotion of regional art and architecture. This approach was innovative rather than imitative, and it centered in the ancient city of Santa Fe, New Mexico, rather than in the new American cities where wealth was being generated. Northern New Mexico became a magnet for a community of transplanted Eastern and Midwestern artists and intellectuals who appreciated the natural beauty of the area as well as its Pueblo Indian and Spanish cultural heritage. These artists and intellectuals led the way in making the area a major cultural center. Their efforts gave rise to an artistic culture that built on, romanticized, and sometimes exploited the heritage of the region.

One of the first people to write extensively about the Southwest was Charles Fletcher Lummis, a newspaper editor from Cincinnati who took a job with the *Los Angeles Times* in 1884. Lummis helped create a Southwestern regionalism through his descriptions of the Indian and Hispanic cultures of the area. The Santa Fe Railroad used his writings to promote travel and tourism to the Southwest, especially northern New Mexico, which was very isolated from the rest of the nation.[76] In contrast to California and Colorado, there had been few reasons for many Euro-Americans to move there, and the railroad did not arrive until 1880.[77]

The efforts of a few scholars conducting research on native cultures in the 1880s also drew attention to the cultures of the Southwest. The Federal Bureau of Ethnology funded Frank Cushing's study of Zuñi Pueblo in 1879.[78] Adolph Bandelier began his work among the ruins of the Anasazi in 1880, establishing Santa Fe as his base of operations in 1885. Another archaeologist, Edgar Lee Hewett, hired Kenneth Milton Chapman, an artist from Indiana who had trained at the Chicago Institute of Art, to help conduct fieldwork on the Pajarito Plateau in 1899. Chapman ended up playing a key role in developing and managing the cultural institutions that define Santa Fe today.[79] He, along with a few other Anglo-Americans, helped to transform Santa Fe from a "dusty adobe town to a bustling cultural center and the capitol of Indian art [...] within the span of a single generation, between 1907 and 1931."[80]

Intellectuals and artists flocked to Santa Fe in the early twentieth century, intrigued with the land and people of the desert Southwest. The open skies, vast lands, and clean air contrasted sharply with the crowded and polluted conditions of New York and Chicago. Alice Corbin Henderson, the former associate editor of *Poetry Magazine,* based in Chicago, and artist Frank Applegate provided leadership among the artists, writers, and other intellectuals who came to Santa Fe, including Georgia O'Keefe, D. H. Lawrence, Carl Sandburg, and Ansel Adams.[81] Many of them also visited Taos at the invitation of Mabel Dodge Luhan, an iconoclastic woman from New York who facilitated an artistic community in the rugged mountains of northern New Mexico.[82] All of these artists were interested in one way or another in reviving some aspect of Indian or Spanish culture. Cather's novel *Death Comes for the Archbishop* is based on her travels to Santa Fe and Taos and explores the contrasts between the indigenous Mexican people of the area and the European bishop who was appointed to oversee the local parishes.[83] One of Ansel Adams' most famous photographs, "Moonrise, Hernandez," was taken near the village of Abiquiu, New Mexico, close to where Georgia O'Keefe made her home for many years.[84]

The regional emphasis inspired changes in architecture. John Gaw Meem, an architect who moved to the area hoping to cure his tuberculosis, helped redesign the buildings of Santa Fe, which by the late nineteenth century were all mainly in the Victorian style, into a romantic regional style known as Pueblo Revival. In contrast to Mission Revival, which became popular in California and revealed a Spanish-Mediterranean influence, Pueblo Revival was intended to resemble the adobe structures of the Pueblo peoples who inhabited the Rio Grande Valley of New Mexico.[85]

Chapman, Henderson, and others convinced several philanthropists to invest in cultural institutions that would showcase the indigenous and Hispanic arts of the region. Many of these philanthropists were Easterners with an interest in the cultures of the Southwest. At Chapman's urging, John D. Rockefeller funded a Laboratory of Anthropology and the Pueblo Pottery Fund in 1926, which were both later incorporated into the Indian Arts Fund.[86] Bostonian Mary Cabot Wheelwright gave money to establish the Museum of Navajo Ceremonial Art, later renamed the Wheelwright Museum of the American Indian, which exhibited works of American Indian artists such as the beautifully crafted rugs of Daisy Taugelchee, one of the most prominent Navajo weavers of the twentieth century. Alice Henderson served as the first curator of the museum. Henderson, along with Kenneth Chapman, Frank Applegate, Mary Austin, and others founded the Spanish Colonial Arts Society in 1925. Cyrus McCormack, Jr., one of the owners and managers of International Harvester Company, was its main benefactor, funding several exhibitions and the acquisition of a permanent art collection. The Museum of Spanish Colonial Art opened on Museum Hill in 1987 to exhibit the artworks that the Society had accumulated over the years.[87]

Philanthropists in other parts of the West have facilitated the development and preservation of regional art and history. Charles Fletcher Lummis organized the Southwest Society in Southern California in 1903 and convinced some of Los Angeles' wealthy elite to finance a public museum that initially housed a hodgepodge of scientific, archaeological, and artistic artifacts. By the 1920s, it decided to focus solely on the cultures of the native peoples of the Southwest and became known as the Southwest Museum of the American Indian. Gene Autry, "America's Favorite Singing Cowboy," along with his wife and another couple, Monte and Joanne Hale, founded the Gene Autry Museum of Western Heritage in Griffith Park in Los Angeles order to provide a venue for exhibits expressing the culture and history of different aspects of the American West. Today, the Autry National Center encompasses the Southwest Museum of the American Indian, the Museum of the American West, and the Institute for the Study of the American West. A similar institution, the Buffalo Bill Historical Center in Cody, Wyoming, houses several museums, including one dedicated to Plains Indian culture and another to the history of Buffalo Bill Cody, along with a research library. Its roots date back to the weeks after Cody died, in March 1917, and it had a variety of local and national benefactors.[88] In Phoenix, Dwight Heard, major investor, real-estate developer, and owner of the *Arizona Republican* newspaper, and his wife Maie, gave funds as well as their collection of Indian artifacts to build a museum dedicated to the preservation of Indian art. The Heard Museum, which was completed in 1929, is currently one of the premier institutions involved in collecting and studying the native art and cultures of the Southwest.[89]

Philanthropy directed toward the development of regional arts and culture in places such as Santa Fe and Taos as well as philanthropy directed toward universities and a more European-oriented artistic culture in modern cities like Los Angeles, San Francisco, and Denver has helped to make the West a place of culture and creativity, attracting artists, intellectuals, and other educated individuals.[90] By stimulating migration and creating educational opportunities, educational and artistic institutions contribute to the growth of an affluent middle class. Yet not everyone benefits. In some cases, the growth of artistic communities and the wealth that has accompanied that growth have adversely affected the people whose cultures are supposedly being promoted. Property values and taxes have increased, forcing poorer residents out of the cities.[91] Arts-training programs for Hispanic and Indian people have helped them preserve some of their culture but often at the expense of training for higher paying jobs.[92]

POVERTY

The philanthropy of the wealthy did not ameliorate the conditions of poverty in the West, despite well-intentioned efforts to promote the public good. In fact, many of the businesses run by the wealthy philanthropists produced poverty on a more massive scale than had previously existed because of the need for cheap labor. The low wages paid for most jobs created by Western enterprises in the nineteenth and early twentieth centuries contrasted sharply with the enormous profits reaped by some of the entrepreneurs, resulting in huge disparities in income. These extremes of wealth and poverty were especially visible in the cities where both entrepreneurs and workers lived. In the late twentieth century, although more people have shared in the affluence of Western enterprise, the need for cheap labor has never disappeared. Jobs with minimal pay, few benefits, and little opportunity for advancement continue to contribute to the problem of poverty in the post-industrial West.

Poverty in the West before World War II

Members of the Western American Indian tribes were among the very poorest people in the region by the end of the nineteenth century. They had been consigned to reservations and deprived of their lands and the freedom of movement necessary to sustain the nomadic or semi-nomadic lifestyle that most had followed. Reservation lands were often marginal and insufficient for farming and grazing livestock on a large scale. Some American Indians obtained low-skilled jobs in mining and seasonal agriculture (see Figure 18.2). Many just barely subsisted on the reservations, depending heavily on federal aid, often facing high rates of disease and nutritional deficiencies in their diets. Federal Indian policy for much of the first half of the twentieth century emphasized assimilation, resulting in the suppression of traditional native beliefs and culture, which contributed to a sense of decline and sometimes hopelessness. Many of the tribes resisted this and managed to maintain a strong cultural identity, but they still struggled with poverty and remained on the margins of Western economic development.[93]

The numbers of American Indians in the Western USA remained relatively small compared to the vast numbers of migrants moving there to work in its rapidly proliferating enterprises. These workers became the most familiar face of poverty in the West prior to World War II. They consisted of a mixture of people, including impoverished European immigrants, Irish and Italian Catholics, Russian Jews, ethnic Mexicans, Chinese and Japanese immigrants and their native-born children, as well as African Americans. Their heritage, ethnicity, and socioeconomic status contrasted sharply with the wealthy Euro-American owners of the mines, railroads, agribusinesses, and oil wells of the West who for the most part were fairly homogenous with roots in middle-class America.

For the workers, many of whom had left dire circumstances, the American West symbolized the possibility of a new kind of freedom and the promise of economic advancement. If they could not realize the dream, perhaps their children might. They were preoccupied primarily with survival, regardless of their ethnic background. Each group, however, had a somewhat different experience. Members of the Euro-American working class had the greatest chance for occupational mobility, but lack of education and resources made it difficult

Figure 18.2 South Dakota Sioux Indian men picking strawberry crop in Orange County, California, 1965.
Source: UCLA Charles E. Young Research Library Department of Special Collections, Los Angeles Times Photographic Archives. © Regents of the University of California, UCLA Library.

for many of them to climb out of poverty. Yet they still had advantages. Many belonged to labor unions, which at times were able to negotiate for better wages and working conditions. Depending on their skill level, they were able to move into higher paying industrial jobs that required a greater skill level. By contrast, ethnic Mexicans, African Americans, and Asians were completely marginalized both economically and politically for over 100 years, regardless of their wealth or family background. Most were barred from labor unions and remained in the lowest skilled, lowest paying jobs within the Western economy. Their only path to a better standard of living was through self-employment and independent enterprise; employers generally did not hire minorities for highly skilled blue- and white-collar jobs. The situation did not begin to change for them until the 1950s and 1960s.[94]

The patterns of minority and working-class settlement differed across the West. Southern California was the most racially and ethnically diverse and had the most complicated race relations. It had sizeable numbers of African Americans, Mexicans, Jews, and Asians. Moving north along the coast of California and into the Pacific Northwest, there were fewer and fewer ethnic Mexicans, but small communities of African Americans and Asians intermingled with larger Irish Catholic and Italian groups. Ethnic Mexicans constituted the largest minority population and the bulk of the working class in Arizona, New Mexico, and South Texas. Colorado's workers were mostly ethnic Mexicans, Italians, and Eastern Europeans. Most Western cities had very small African-American populations; the largest

African-American community west of the 100th meridian was in Los Angeles. Although all groups occupied a very low socioeconomic position, their status often varied from place to place, depending on the mix of groups in any given local community.[95]

The working poor usually lived in areas of the city known as "foreign districts" or "immigrant quarters" to the affluent Anglo-Americans. The "tent cities" of the gold-rush days in San Francisco and Denver gave way to more permanent working-class neighborhoods located near iron foundries, brickyards, factories, and warehouses, often in low-lying areas subject to flooding where pollution, dirt, noise, and noxious smells were pervasive.

Similar neighborhoods emerged in other Western cities as more and more workers moved in to work on railroads, in agriculture, or in developing industry. These areas were attractive to workers because of cheap rents, proximity to places of employment, and the presence of others who shared their language and culture. The central business districts rapidly became overcrowded with the influx of people in the early decades of the twentieth century. Working-class neighborhoods spread into blocks abandoned by affluent Euro-Americans who were moving in droves to the streetcar suburbs. Invariably, these inner-city areas deteriorated and suffered from the absence of basic infrastructure such as sewers, paved roads, and electrical lines.[96]

Not all of the people who lived in the "immigrant quarters" were foreigners, but the name provides insight into how affluent Anglo-Americans perceived the mixture of cultures, ethnic groups, and languages found there. Anglo-Americans equated Americanism with the clean, neatly landscaped, single-family homes of the Euro-American neighborhoods, which embodied the values of the dominant Euro-American culture. By contrast, they associated

Figure 18.3 Views of Hoover Towns in the LA area from the Los Angeles Department of Health and Human Services Collection, Vol. I, *Habitations of the Unemployed*, ca. 1930–1932.
Source: Courtesy of the Huntington Library, San Marino, California.

the densely populated foreign quarters with heathenism, poverty, crime, disease, filth, and vice. Despite the mixed nature of most working-class neighborhoods, Anglo-Americans began to associate certain blocks with particular ethnic groups. This reflected the fact that people with similar backgrounds tended to congregate together to escape the discrimination and hostility of the dominant Anglo-American society and to promote their own values and culture. At the same time, the identification of particular subdivisions as "Japan Town" or "Chinatown" or "Little Mexico" was a way for Anglo-Americans to designate and cordon off certain areas for the expression of non-Euro-American culture, implying that the rest of the city should reflect Anglo-Saxon Protestant culture and values.[97]

There was a noted absence of large enclaves of Euro-American ethnic groups in Western cities similar to those found in Chicago and New York. Members of the Euro-American working class settled throughout the city, moving to better neighborhoods as they achieved higher incomes. The most distinctive groups included the Irish, the Jews, and the Italians. Irish Catholics were the most numerous among San Francisco's foreign-born residents in 1870. They found California attractive because of the promise of greater economic mobility and less discrimination. Most had arrived by sea before the completion of the transcontinental railroad, traveling from East Coast cities such as Boston and New York. They started off settling in neighborhoods in the heart of the old city but were able to gradually move out and purchase homes. They developed extensive networks of benevolent societies dominated by the Catholic Church and became involved in the politics of the city from its earliest days. They blended into the larger community fairly rapidly.[98]

The influx of sizeable numbers of Yiddish-speaking Jews from Eastern Europe in substantial numbers in the late nineteenth and early twentieth centuries alarmed many Westerners and led to an increase in anti-Semitism. Unlike the first wave of migrants, the more affluent German Jews, these people were, for the most part, poorer and less educated, and some had radical political leanings. They often made a living by peddling fish, vegetables, fruit, clothing, shoes, and other types of dry goods or by working in garment factories. Many had fled the pogroms in Russia and Poland; they were joined by Sephardic Jews from Turkey.[99] Like other members of the working class, they originally settled in poor areas around the central business districts of Western cities. Two-thirds of the Los Angeles Jewish population lived in south central Los Angeles in 1900, primarily in and around Boyle Heights, close to downtown factories. Dallas's Little Mexico on the outskirts of downtown began as a community of Russian Jews. In early twentieth-century Denver, Eastern European Jews lived along the South Platte River, where they tried to make a living by retrieving and selling scraps and rags.[100]

Despite difficult beginnings, many working-class Jewish families were able to prosper after a generation or two. By the 1950s, a number had moved to the suburbs where they still tended to congregate with other Jews.[101] Assimilation was less of an issue for the working class than for the more affluent Jews who tried to blend in the dominant culture by celebrating Christmas, sending their children to the best private schools, steering them into "respectable occupations," and building homes that reflected Anglo-American tastes.[102]

Italians also migrated to the Western USA in large numbers between 1890 and the institution of immigration restrictions in the early 1920s. Several Western cities had neighborhoods designated as "Little Italy." Denver had a distinct Italian neighborhood near the South Platte River. Immigrants initially settled in the flood plain, moving to higher ground in the Highland area once they had saved enough to purchase one of the small, wooden-frame houses. Our Lady of Mount Carmel Church served as the center of social and cultural life in an area where few spoke any language other than Italian. Many of Denver's Italians initially accumulate some money by peddling on foot, which required virtually no capital, and then they later upgraded to a cart or horse-drawn wagon. Some eventually bought small truck farms.[103]

Like Denver, San Francisco had a distinctive Italian community, which by 1920 was the sixth largest in the USA. Most of San Francisco's Italians congregated in the North Beach area, where Italian immigrants speaking the same language and sharing the same traditions could find a refuge from the larger Anglo-American community. The area had the lowest rents in the city before 1940. In 1912, the average rent in North Beach was $31 per month compared to $56 a month elsewhere. The majority of these Italian immigrants were initially young males who arrived alone, intending to work, save money, and return to Italy. Fewer than one in ten actually farmed or fished using skills learned in their homeland. Over 40 percent worked in domestic and personal services, and almost 50 percent worked in trade and manufacturing. Most served as unskilled laborers, partly because of discrimination but also because of their own illiteracy, rural habits, lack of industrial skills, and transience. The creation of Italian banks was the most significant step in the growth of the Italian-American economy in San Francisco. Italians had historically experienced difficulty borrowing from American banks because they were viewed as unreliable. Beginning in the 1890s, the development of banks owned and operated by Italians such as James Fugazi, Andrea Sbarboro, and Peter Giannini that were willing to lend to Italian-born businesspeople and consumers made it possible for the community to prosper financially.[104]

The Italians in both San Francisco and Denver had achieved a level of material success by the 1930s that was higher than that of most other immigrant groups at the time. Forty-seven percent of Italians in San Francisco owned their own homes by 1930. Hundreds of Italians became naturalized citizens between 1941 and 1943 in order to prove their patriotism during World War II, more than in the previous decade. Naturalization accelerated the movement of Italians into mainstream Euro-American America. Many of the children of Little Italy's residents left North Beach after the war, blending into the larger Euro-American community, and Americanization accelerated.[105] After World War II, ethnic Mexicans replaced Denver's Italian population in what had been known as Little Italy.[106]

Ethnic Mexicans have had one of the most complex and varied experiences in the cities of the West. Although many tried to identify themselves as just another Euro-American ethnic group, they were unsuccessful. In general, they occupied a sort of intermediate position within the racial hierarchy, higher than that of African Americans, American Indians, Chinese, and Japanese, but lower than that all of the other Euro-American groups. Their experience was different from that of other groups because their presence predated the Anglo-American conquest. Some descendants of Spanish *vecinos* and Mexican land-grant families experienced a downward slide into poverty and working-class status joining first- and second-generation Mexican immigrants in *barrio* neighborhoods, which continually expanded with the almost constant influx of people from Mexico.[107]

The effect of the American conquest of northern Mexico in 1848 and the subsequent introduction of capitalism was that the Mexican elite lost most of their economic and political power over time to the Anglo-Americans. Although there were instances where Anglo-Americans converted to Catholicism and learned Spanish in order to gain economic and political power, for the most part, once Anglo-Americans arrived they displaced the ruling elite and rejected their culture and religion as inferior, reviving aspects of it in romanticized form only when their power and position had become entrenched.

A few members of the Mexican elite assimilated, primarily through intermarriage with the new Anglo-Americans business elite, but many others experienced downward social and economic mobility. Land loss was common because of new, higher property taxes, the decline of ranching as a viable business, the costs of defending land titles, and occasionally downright fraud.[108]

By the late nineteenth century, newcomers from central and southern Mexico were moving north seeking seasonal labor and higher wages in the modernizing economy of the American Southwest, swamping the native Spanish-speaking population. Most of the immigrants were poor peasants from villages in central Mexico, displaced by land consolidation and, later, the 1910 Revolution. These newcomers reinforced Anglo-American stereotypes of Mexicans as dirty, superstitious, and unassimilable. Anglo-Americans rarely drew distinctions between the newcomers and descendants of the earliest settlers. Wealthier ethnic Mexicans sought to adapt by emphasizing their Spanish heritage, calling themselves "Spanish American" or "Hispano" to avoid affiliation with the Mexican immigrants.[109]

The eventual outcome of the process of displacement and migration was that the ethnic Mexican population of the Southwest gradually became marginalized at or near the bottom of the occupational ladder. "Barrioization" occurred, a term coined by scholar Albert Camarillo to describe how land loss, the decline of the pastoral economy, and racial prejudice led to segregated neighborhoods, usually around or near the central plaza in cities built during the Spanish colonial period. These areas became overcrowded as immigrants from Mexico settled there to be near others who spoke their language and shared their culture. A distinct ethnic Mexican neighborhood developed north of the Los Angeles Plaza in the mid-nineteenth century. Anglo-Americans who settled south of the Plaza called the northern area "Sonoratown" to identify it as a foreign, Spanish-speaking enclave. Similar *barrios* emerged elsewhere, including in Santa Barbara and San Diego on the West Coast, and in San Antonio, Texas.[110]

Living standards in rural Hispanic communities, such as the remote villages in the mountains of northern New Mexico, also experienced substantial declines. Villagers lost access to communal lands and thus the ability to sustain themselves by herding sheep and planting vegetables. Their residents turned to wage labor, resulting in a pattern of seasonal migration to work in the mines and commercial beet fields of Colorado. Some families moved to Denver and lived in areas once inhabited by Italians, Poles, and Russians, especially along the South Platte or in Adams County. Joined by immigrants from Mexico, they lived in the worst housing in the city, often one-room shacks with no electricity or running water.[111]

The growth of industry in Western cities coupled with economic disruption and instability in Mexico led to substantial increases in the ethnic Mexican population across the Southwest in the 1910s and 1920s. The growth of commercial agriculture in South Texas and California was also a driving force.[112] Migrant workers often moved into urban neighborhoods, resulting in overcrowding, forcing ethnic Mexicans away from central plazas. New *barrios* formed in areas and in cities without a Hispanic past. Restrictive covenants and economics determined the new settlement patterns. In Los Angeles, ethnic Mexicans moved onto the flat areas near a heavily industrialized portion of the Los Angeles River, a place that flooded frequently. Mexican workers generally lived in whatever they could find, including old boxcars or wood shacks. Many lived in *cholos* or house courts, consisting of three to thirty small houses on a common lot measuring about 40 by 170 feet with a shared

Figure 18.4 Boy playing in Hoyo Mara Eastside slum in Los Angeles, California, 1963. Source: UCLA Charles E. Young Research Library Department of Special Collections, Los Angeles Times Photographic Archives. Copyright © Regents of the University of California, UCLA Library.

yard, toilet, and washing facilities. They suffered from high rates of infant mortality and tuberculosis.[113] Eventually, those who had better jobs moved across the river into East Los Angeles where there were more business opportunities for small businesspeople and nicer, relatively inexpensive housing (see Figure 18.4).[114]

In Dallas, Texas, ethnic Mexicans settled northwest of downtown in what became known as "Little Mexico," in some cases renting shacks still owned by Russian Jews. The middle class and working class lived side by side along and near the northern stretch of McKinney Avenue. Small businessmen established tortilla factories and stores that served the laborers. The WPA Guide for Dallas written in the 1930s described it as "a close-packed mass of flimsy, tumbled-down frame shanties and 'shotgun' houses threaded by narrow, twisting, unpaved streets, muddy or dusty according to the weather," but noted that it was "one of the few slum areas in Dallas possessing individual character."[115] Houston and Phoenix also had large Mexican populations by 1920.[116]

Despite their problems, the *barrio* neighborhoods served as places where recently arrived immigrants could experience a more gradual adjustment to American life. There they could speak Spanish and interact with others of their own cultural background. The *barrio* was also a refuge from a sometimes hostile Anglo-American culture. In many cities across the Southwest, ethnic Mexicans experienced both actual and de-facto segregation in schools, public places, and restaurants. The *barrios* grew larger than any Asian or Euro-American ethnic neighborhoods in Western cities, which remained small because of distance and immigration restrictions, especially after 1924. The proximity of Mexico and the continuous migration helped create places in the Southwestern USA where the Spanish language and Mexican customs flourished. By 1930, Los Angeles had the largest Mexican community in the USA.[117] The Depression, however, decimated that community and others as jobs disappeared. The government deported roughly 600,000 Mexicans during the decade of the 1930s.[118]

Like ethnic Mexicans, African Americans also found themselves at the bottom of the socioeconomic hierarchy, but life for them was somewhat better in the West than in the rural South. In the late nineteenth century, a relatively small number moved to escape the growing oppressiveness of the Jim Crow system, believing they would find greater freedom and opportunity. This migration remained small but steady, even during World War I when thousands moved to Chicago, Detroit, Pittsburgh, and Cleveland to work in the wartime factories. Some settled in rural communities on the Great Plains, but most traveled to the cities where they established small but thriving communities. Those that migrated were usually not the very poorest African Americans, and many had middle-class aspirations.[119]

A sizeable number of African Americans in the Far West, including up to a quarter of those living in Los Angeles by the 1920s, had moved there from Texas.[120] The influx of Mexican immigrants into Houston, Dallas, and the surrounding countryside by the early twentieth century complicated race relations in east and central Texas, where there was already a large number of freedmen and their descendants. African Americans and Mexicans competed for jobs, and, in many cases, Mexicans had the advantage because African Americans were unquestionably at the bottom of the racial hierarchy, one of the legacies of slavery.[121] In the cities, African Americans worked in the very lowest paying jobs in domestic service and unskilled labor. They lived in the poorest parts of town, often in narrow, one- or two-room shotgun houses crowded along dirt roads. They attended segregated schools and faced restrictions in public places, hotels, restaurants, and movie theaters. Even working-class neighborhoods became increasingly segregated in the early twentieth century, enforced by custom, economics, deed restrictions, and occasionally by municipal ordinances. Despite poverty and segregation, Texas African Americans built thriving communities, such as Frenchtown in Houston's Fifth Ward and the State-Thomas District in Dallas. There was a small but influential African-American middle class consisting of skilled workers, independent African-American businessmen, teachers, ministers, and newspaper editors, who provided much needed services to the community as a whole. They usually settled on the edges of poor African-American neighborhoods because deed restrictions prevented them from moving to less crowded, cleaner Euro-American neighborhoods. Members of this group were the most likely to have enough capital to move west.[122]

San Antonio was a major destination for African Americans migrating from rural Texas as well as from Dallas and Houston. Kenneth Mason has argued that San Antonio represents the confluence of African-American Southern and Western experiences. As one of the oldest cities in the state with roots deep in the Spanish and Mexican past, it had a sizeable ethnic Mexican population. It served as the gateway to the Southwest and to Mexico and thus drew thousands of Mexican immigrants. African Americans benefited from the fact that they were not the largest minority population in the city, although their numbers were sizeable, totaling approximately 20,550 by 1880. They had greater access to public facilities and arguably more autonomy than in either Houston or Dallas. They lived in less segregated neighborhoods on the east side near working-class Euro-Americans. There are no records of mobs lynching African Americans in San Antonio as there were in east Texas. However, they still attended segregated schools, faced Jim Crow ordinances, and had difficulty improving their socioeconomic status.[123]

In cities west of San Antonio, African Americans initially experienced more freedom than they had in Texas.[124] A combination of circumstances actually made life better for most African Americans in cities such as Seattle, Denver, Los Angeles, and San Francisco prior to World War II. First, the African-American population in these places was relatively small. This meant that the problems of scant, overcrowded housing were not as severe. Los Angeles had the largest African-American community, totaling 15,500 by 1920, but it comprised a fairly small percentage of the overall population. African Americans were generally able to find decent, affordable housing in the neighborhoods south of downtown Los Angeles and in more rural areas like Watts as well as along streets on the east side. Second, especially along the West Coast, African Americans were not always at the bottom of racial hierarchy. Asians, particularly the Chinese and the Japanese, usually occupied that position, which moderated discrimination toward African Americans somewhat. Asians also substantially outnumbered African Americans in Los Angeles, San Francisco, and Seattle. Finally, African Americans had higher incomes, at least in some Western cities, than they did in the South, but they still lagged behind Euro-Americans and faced barriers to employment in higher skilled jobs.[125]

The West was attractive to African Americans more because of its promise of opportunity than the actual reality they experienced. This was especially true in California. San Francisco had a reputation as a progressive, liberal city. It had few segregation laws, and the San Francisco Board of Education voted to integrate its schools in 1875. The entire state of California outlawed segregated schools in 1880. Los Angeles also lacked an entrenched, rigid Jim Crow system. African Americans who moved there could vote, and jobs were plentiful.[126]

African Americans did not escape discrimination in the West, however, which exacerbated problems of poverty, especially as the twentieth century progressed (see Figure 18.5). Euro-Americans in Los Angeles and San Francisco still viewed African Americans as inferior. Anti-African-American sentiment increased as the numbers of African-American migrants grew and as more and more Euro-American Southerners moved to California. African Americans struggled for equal rights in education, housing, employment, and public accommodations.

When African Americans tried to move to Euro-American suburbs around Los Angeles or to the "Westside," they faced restrictive covenants and block agreements. African-American professionals who succeeded in purchasing homes in Euro-American areas often paid inflated prices, and sometimes faced vandalism and death threats. Los Angeles became a center of challenges to restrictive covenants, which remained legal until the Supreme Court declared them unconstitutional in *Shelly* v. *Kraemer* in 1948. African Americans also experienced social and recreational segregation. They were excluded from many clubs, hotels, restaurants, beaches, and theaters. They had little success in changing the situation, mainly because they had virtually no political power, partly because of their position within the racial hierarchy and partly because their numbers were small compared to the overall population. Before 1940, in San Francisco, they never represented more than about 1 percent of the population, and in Los Angeles, they constituted about 3.5 percent of the population.[127] They faced similar obstacles in other Western cities. In 1926, Phoenix opened the only segregated high school in all of Arizona to serve its small but growing African-American population, the Phoenix Union Colored High School.[128] African Americans sat in segregated areas in local theaters in Denver, swam separately from Euro-Americans at limited times in Curtis Park, and attended segregated schools.[129]

Figure 18.5 Abandoned apartments in Watts, Los Angeles, California.
Source: UCLA Charles E. Young Research Library Department of Special Collections, Los Angeles Times Photographic Archives. Copyright © Regents of the University of California, UCLA Library.

In all Western cities they held a lowly position on the socioeconomic ladder. A few became entrepreneurs, struggling to keep afloat with limited access to credit, but most worked in low-wage jobs similar to those they had in Texas as maids, janitors, and porters.[130] There were some successes, however, especially in Los Angeles. The city hired more African-American police officers and firemen than most other cities. By 1920, there was a very small group of African-American professionals and an increase in skilled laborers. African Americans purchased homes and invested in real estate, providing the foundation for further accumulation of capital. They claimed the Central Avenue district for themselves in the early twentieth century and set up a flourishing community of African-American churches, businesses, nightclubs, and even a "Colored YMCA." It was in part the contradiction between the promise of the West and its reality that led Western African-American residents to begin to push for civil rights in the early twentieth century. Their desire to move out of poverty into the middle class was a driving force.[131]

Of all of the minority groups in the West, Asians probably experienced the most severe discrimination before World War II. Unlike other members of the working class, Asians who immigrated to the West had no hope of ever becoming American citizens. Even the status of their American-born children was constantly questioned. Moreover, they faced immigration restrictions and prohibitions against owning land. Socially, they were completely ostracized. Their religious beliefs, their appearance and dress, and their habits seemed very peculiar to Euro-Americans. Many Euro-Americans considered them to be heathens, dishonest, and prone to vice and criminal behavior. This combination of legal restrictions and social attitudes kept Western Asian communities in a very tenuous position. Few experienced any kind of economic stability. The only wage-laboring jobs they could find involved unskilled, manual labor, and they faced intense competition from ethnic Mexicans, who eventually became more favored by employers.[132]

Analysis of the Asian experience in the West is complicated by their ethnic diversity. Before 1950, the Chinese constituted the largest group of Asians, followed by the Japanese, and then the Filipinos. Koreans, Vietnamese, and Cambodians did not arrive in large numbers until after the Immigration Act of 1965 eliminated restrictions on immigration from non-European countries. The Chinese were the first to arrive. Although there is some evidence of their presence in California before 1848, the majority came during and after the California Gold Rush. Spurred by problems in China, including famines, floods, droughts, poverty, and the corruption of the Manchu dynasty, many saw "Gold Mountain" as a place to achieve the prosperity that had eluded them in their own country. San Francisco's Chinatown emerged in the late 1840s and 1850s when these Chinese immigrants set up temporary homes along fisherman's wharf where they landed.[133]

The Chinese faced intense discrimination from the very beginning, targeted directly at their ability to make a living. In the Sacramento Valley, they were harassed by both Euro-Americans and Mexicans and had difficulty establishing individual mining claims because of the foreign miners' tax. Some left to work on the railroads while still others drifted back to the city where some set up small proprietorships such as restaurants and laundries. By 1880, about 30 percent of the Chinese in San Francisco operated laundries, which took little capital to open. The Board of Supervisors of San Francisco, however, passed a series of taxes and ordinances designed to inhibit the growth of these Chinese laundries. The Chinese also took jobs in light manufacturing such as cigar-making, shoes, and woolens where they worked in dim light in poorly ventilated rooms. Wealthy business owners preferred Chinese labor because they could pay them less than Euro-American workers, which had the effect of lowering wages overall and creating resentment that frequently led to violence. Boycotts against Chinese-made goods drove them out of manufacturing. Riots against Chinese mine and railroad workers occurred in California, Oregon, Nevada, Colorado, and Washington. The most famous episode took place in Rock Springs, Wyoming, where Euro-American miners killed twenty-eight Chinese men, wounded at least fifteen others, and drove hundreds out of town, confiscating their property. In Seattle, Tacoma, and Eureka, angry mobs forcefully evicted Chinese people from their homes.[134]

After the Chinese Exclusion Law passed in 1882, Chinese laborers faced deportation. The Chinese community had no political power or voice, and thus politicians did little to help them. Moreover, there was no public outcry on their behalf. Chinese people had to produce on demand documentation showing their right to be in America, and they suffered from ongoing immigration raids. They lived a shadowy existence, constantly anxious about their situation. The exclusion laws reinforced the social, political, and economic segregation of the Chinese and led to increased return migration. Some created false identities to enter and remain in the USA, which alienated them even from themselves. Congress revived the Chinese Exclusion Laws year after year until they were repealed on December 17, 1943 during World War II.[135]

Many Chinese chose to live in Chinatowns in cities along the west coast to escape the hostility of the wider Euro-American world (see Figure 18.6). These Chinatowns provided respite from discrimination and racial prejudice and functioned as a separate space where the Chinese language was freely spoken and Chinese traditions were observed. Chinese businesses and restaurants adorned with bells, flowered lanterns, and painted balconies, windows filled with jade and ivory carvings, and signs covered with Chinese characters

Figure 18.6 Chinatown, San Francisco, California, ca. 1910. Source: Courtesy of the Huntington Library, San Marino, California.

lined the streets. Moreover, rents were low, and housing was located near manufacturing plants where the Chinese worked.[136]

For over a century the Chinese population in the USA was overwhelmingly male, and most of these males came with the idea that they would stay only temporarily. This influenced the nature and development of Chinese neighborhoods. They were filled with boarding houses and tenement buildings rather than small houses, geared to serve a transient population. Many of the young men who lived in them frequented the prostitution houses and opium dens near their residences. Between 1848 and 1854, only about sixteen out of 45,000 Chinese immigrants were women. In 1890, the population of San Francisco's Chinatown was 90 percent male. Despite an influx of Chinese women in the 1930s and 1940s, Chinatown was still 78 percent male in 1978.[137] Families tended to be patriarchal and to favor boys; gender roles only gradually began to change in the late twentieth century.[138]

The streets and buildings of the West's Chinatowns were crowded, filthy, and disease-ridden, containing blocks of dilapidated apartments, open sewers, and unpaved streets. They lacked basic infrastructure and city services such as garbage collection. Rats scampered through rickety wooden buildings. San Francisco's Chinatown, the largest in the country, consistently suffered from substandard housing, unemployment, overcrowding, crime, and disease throughout the twentieth century and remains a low-income area today.

A study in the 1930s revealed that out of 3,830 housing units, 3,000 lacked heat and only 447 met acceptable standards. The tuberculosis rate was three times that of the rest of the city.[139]

In many Western towns, the area where the Chinese lived also became the local vice district, largely by design. The Board of Supervisors in San Francisco purposely located prostitution houses, opium dens, and gambling saloons in Chinatown to keep them away from "respectable neighborhoods."[140] Los Angeles' Chinatown, on the eastern edge of the Plaza, was the center of prostitution, gambling, and drugs for many years.[141]

Like the Chinese, the Japanese and Filipinos who immigrated to the USA worked in manual labor, especially farm work and domestic service. Many of the Japanese in particular found ways to climb the agricultural ladder by first leasing and later purchasing their own land. Fearing increased competition from these Japanese truck farmers, several Western states passed alien land laws to prohibit foreigners from purchasing real estate.[142] The alien land laws forced most Japanese and other Asians into the cities. Along the coast, a few operated small fishing fleets. In Los Angeles, Japanese started leasing Jewish stores and displacing African-American residents from the Five Points area down to Azusa Street in the 1910s. By 1920, Little Tokyo was a center of Japanese culture, but the neighborhood still had many non-Japanese residents.[143] Seattle also had a fairly sizeable entrepreneurial Japanese community, which created its own business organizations such as the Japanese Chamber of Commerce and the Japanese Hotel Association.[144] Japanese businesses were decimated, however, during World War II. Japanese communities along the west coast disappeared overnight after Franklin Delano Roosevelt issued Executive Order 9066. Thousands were shipped to inland camps in the California desert and in Idaho. Little Tokyo in Los Angeles turned into a African-American neighborhood called Bronzeville, and Chinese truck farmers took over the empty stalls left by the Japanese at the City Market.[145] After the war, Japanese Americans returned to find all of their property confiscated.

The Depression of the 1930s devastated all minority working-class communities across the West. With jobs disappearing, minorities were usually the first to go. Temporary settlements sprang up on the outskirts of cities where the unemployed scavenged in garbage dumps to piece together scraps to create shelters (see Figure 18.7). A huge migration to California by poor Euro-Americans displaced by the new agricultural policies of the Roosevelt administration and by the dust storms swirling across the Great Plains resulted in job losses for large numbers of ethnic Mexicans, Asians, and African Americans in fields and factories. Although these "Okies" and "Arkies" faced intense discrimination at first, they were able to assimilate and eventually move into the middle class by changing their accents and customs associated with their rural roots. They experienced no barriers of race or ethnicity, language, or religion. Like other Euro-Americans before them, those who were willing could overcome class and regional barriers.[146]

Poverty in the West before World War II was largely a function of low wages and no political voice. People lacked the funds to improve the often deplorable conditions of their homes and neighborhoods on their own and had limited ability to influence mayors and city councils to provide the services they needed. Moreover, city funds were often limited in many places because taxes were low or nonexistent. Municipal governments frequently paid attention only when conditions in the neighborhoods appeared to threaten the city as a whole. For example, in Los Angeles in 1924, for several weeks the city government forcibly quarantined the "Mexican district" near downtown after several people died of bubonic plague, and the disease began to spread. In addition to spreading

Figure 18.7 Views of Hoover Towns in the LA area from the Los Angeles Department of Health and Human Services Collection, Vol. I, *Habitations of the Unemployed*, ca. 1930–1932.
Source: Courtesy of the Huntington Library, San Marino, California.

poison to kill rats carrying the disease, municipal workers and volunteers burned clothing, furniture, and hundreds of houses without compensation, leaving families homeless and destitute.[147]

By the 1920s, the business elite increasingly became concerned that large areas of dilapidated shacks, which seemed to be expanding daily with the influx of workers, would adversely impact the economic growth and overall attractiveness of their cities. During the New Deal era, with the availability of federal funds to build low-cost public housing, city governments began to make plans to clear slum areas and replace them with new multi-family projects.[148] World War II delayed many of these projects, and most were not completed until the 1950s. In most cases, they did little to actually solve the problem of overcrowding because World War II brought an even larger influx of people to work in a variety of industries associated with the war effort across the West.

Civil Rights, Poverty, and Progress in the Postwar Period

World War II and the subsequent Cold War between the USA and Russia boosted industrialization substantially in the American West, created new opportunities for scientific research, and launched a new period of prosperity. The federal government partnership with Western entrepreneurs and universities to mobilize for the war effort expanded fortunes and created thousands of white-collar, skilled manufacturing, and unskilled laboring jobs, attracting new migrants to urban centers in the West. The African American and ethnic Mexican populations surged, leading to even more crowded neighborhoods and strained resources. Middle- and upper-middle-class Euro-Americans also moved; the city of Albuquerque, New Mexico had more Ph.D.s per capita than any other city in the nation. California's role as a leading research center after the war contributed to the boom in high tech at the end of the twentieth century.[149]

The struggle for civil rights, which had really always been present in the West, intensified during this period; members of ethnic minorities who sacrificed for their country saw the inequities of a nation that required much of them but returned little. A few began to make substantial strides economically, socially, and politically in the postwar period, benefiting from more open attitudes and changed laws that allowed them to participate in the prosperity created by federal investment and wealthy entrepreneurs. Greater awareness of minority rights steered the fruits of philanthropy in the direction of minority students who profited from scholarships and revised curriculum. Prosperity and progress for some, however, highlighted the growing gap between the middle class and the poor within groups of ethnic minorities.

In the postwar period, class divisions between rich and poor, middle class and lower class increased. Most Euro-American ethnic groups, including Jews, Italians, and Greeks, had successfully blended into the Euro-American mainstream by mid-century and moved into higher-paying jobs. Euro-Americans continued to resist the integration of their neighborhoods even after the Supreme Court nullified restrictive covenants but more readily accepted the presence of middle-class Asians and ethnic Mexicans than African Americans. Affluent African Americans slowly began to move into the suburbs as well—however, with more difficulty. For example, in the 1950s, African Americans with steady blue-collar jobs began to move out of Watts and South Central Los Angeles and to purchase new tract homes in Compton, a predominantly Euro-American working-class suburb. Resident Euro-Americans reacted strongly and sometimes violently, but most ended up moving elsewhere. In 1964, Proposition 14, supported by 65 percent of California voters and 70 percent of Los Angeles voters, revealed the extent of Euro-American opposition to racially integrated neighborhoods. It removed all restrictions on the rights of property owners to decide whether or not to lease or sell their property to a given individual. Proposition 14 probably contributed to the Watts Riot in 1965, and although the state supreme court declared it unconstitutional in 1966, tensions remained high.[150] Still, segregation gradually became a function primarily of occupation and income level.

The out-migration of more successful members of minority groups may have contributed to a sense of hopelessness for many inhabitants of the West's inner-city cores.[151] Although middle-class African Americans, Asians, and Latinos were advancing in politics, education, income, and occupational levels, those left behind became more and more isolated socially, culturally, economically, and politically. Unemployment rates rose as factories laid off workers during the phase of deindustrialization. Multiracial neighborhoods disappeared. For the first time, places such as San Francisco and Seattle had large African-American ghettos similar to those found in Dallas and Houston. South Central Los Angeles became almost exclusively African American, while East Los Angeles was almost exclusively ethnic Mexican. Mexicans and Mexican Americans moved into Denver's Little Italy after World War II. Hunters Point in San Francisco, once the residence of a diverse group of shipyard workers, was 95 percent African American by 1960.[152]

Public housing projects, designed to ease the crowding of inner-city neighborhoods and replace substandard residences, contributed to the emergence of these mono-racial neighborhoods. In some cases, such as the Ping Yuen Housing Project in San Francisco and Little Mexico in Dallas, they were built specifically for a particular group. In other cases, they simply reflected the racial demographics of the areas in which they were located, usually in or near downtown cores and away from suburbs. Dallas built some of its most extensive public housing in West Dallas, one of its poorest neighborhoods filled with dilapidated shacks, garbage dumps, and roads filled with dust or mud, depending on the weather.[153]

The "projects" temporarily improved slum areas and provided clean, new residences, but deterioration and decline frequently set in fairly rapidly.[154]

Redlining contributed to the erosion of the tax base in inner-city areas, leaving little funds to repair and maintain infrastructure. Businesses moved out. Crime, high unemployment rates, police brutality, poor sanitation, and overcrowding plagued inner-city neighborhoods. Watts had unemployment rates as high as 30 percent.[155] Conditions in San Francisco's Chinatown in the 1970s were probably even worse in some ways than they had been in the nineteenth century. Eighty-eight percent of Denver's ethnic Mexican population lived in substandard housing in the 1950s.[156] Increasingly, in the minds of those on the outside, these neighborhoods became associated with the jobless and chronically underemployed or those working in the most menial jobs.[157]

Inner-city residents also faced the threat of neighborhood destruction. Cities tried to combat urban blight through government-funded slum-clearance programs. Condemnation of poor, ethnic neighborhoods increased as cities built freeways and public facilities. It was usually far more cost-effective, as well as in line with the public policy of eliminating nuisance areas, to use the power of eminent domain in poor areas. In Dallas, Central Expressway divided and eventually destroyed the thriving African-American State-Thomas district, and the Woodall Rodgers freeway decimated Little Mexico.[158] Networks of freeways in East Los Angeles as well as zoning and commercial development downtown adversely affected Mexican-American *barrios*. One of saddest examples of neighborhood destruction was the condemnation of Chávez Ravine, a poor, working-class area consisting of small frame houses tucked away in a small canyon near a brickyard and tortilla factory. In the 1940s, the City of Los Angeles decided the homes were substandard, lacking running water or electricity and determined to build public housing units in their place, with total disregard for the feelings of the people who lived there. The city never built the public housing but instead signed a deal for the construction of Dodger Stadium, which opened in 1962.[159]

The problems of poverty, including unemployment and underemployment, low wages, substandard housing, and marginal schools, contributed substantially to the modern civil-rights movement. Economic subordination within the society was the result, many believed, of social segregation and institutional barriers to advancement.[160] One way to address these problems was by influencing government to intervene, beginning on the local level. All ethnic minorities suffered from differing degrees and types of discrimination that exacerbated poverty, but they did not always join together in their efforts to improve their situations. Tensions between African Americans and Mexican Americans were especially high in many Western cities, in part because more affluent Mexican Americans saw themselves as a Euro-American ethnic group and did not want to be identified with African Americans. By the same token, African Americans felt they should occupy a superior station to Asians, whom they viewed as foreigners. These tensions did not disappear by the end of the twentieth century but actually increased as groups fought over how to allocate resources.[161]

African Americans throughout the West demonstrated as early as the 1930s and 1940s for the desegregation of schools and other public places. For example, Eleanor and Lincoln Ragsdale, African-American entrepreneurs in Phoenix, led a series of successful efforts to integrate schools and neighborhoods in the city. As a result, Phoenix schools desegregated one year before the 1954 Supreme Court decision of *Brown* v. *Board of Education,* which mandated desegregation throughout the nation.[162] School desegregation was a key front in the effort to eliminate poverty because it was poor, substandard schools that contributed to African-American children's difficulty in acquiring the skills necessary for higher-paying jobs.

African Americans were the most successful of all minority groups in the political arena, at least initially. They served in the state legislature of California beginning in the 1940s.[163] A coalition made up primarily of liberal Jews and African Americans elected Thomas Bradley as the first African-American mayor of Los Angeles in 1973.[164] Willie Brown, originally from East Texas, served in the California state legislature and eventually as the only African-American mayor of San Francisco. Denver had two African-American legislators in 1951 and an African-American lieutenant governor by the mid-1970s.[165] Dallas began electing African Americans to the city council in the 1970s, and Ron Kirk, lawyer in a major Dallas law firm, became its first African-American mayor in 1995.[166] Changes in the method of electing council members facilitated the presence of members of minorities in municipal governance. Western cities changed from at-large elections of mayors and city-council members to district election systems, creating the potential for greater representation of minority communities, but the system did not always work to the benefit of everyone equally.[167]

As the result of the civil-rights movement, the African-American experience in the West has become increasingly diverse. Some African Americans made substantial advances after the 1960s, moving into white-collar jobs, purchasing homes, educating their children in good schools, and sending them to college. Income levels increased faster for African-American men than for any other group between 1969 and 1989 although they still lagged behind those of Euro-Americans. Conditions have worsened, however, for African Americans remaining in the inner cities. Crime, drugs, unemployment, and bad schools persist as almost intractable problems. Furthermore, the rapid growth of the ethnic Mexican population and their increased participation in politics has threatened the formerly dominant role of African Americans as leaders in minority politics in many cities.[168]

The first movements for greater equality for ethnic Mexicans in the West were generally more cautious than those of the African-American community. Middle-class Mexican Americans thought of themselves as another Euro-American ethnic group and did not want to identify in any way with African Americans. Groups such as the League for United Latin American Citizens and the American GI Forum, both of which originated in South Texas, emphasized citizenship, education, and cooperation with the Euro-American community. At the same time, they advocated the preservation and cultivation of a distinct cultural identity.[169]

By the 1960s, among people frustrated with the slow pace of change, the more radical and confrontational Chicano/a movement began to emerge in South Texas, Colorado, New Mexico, and California. César Chávez organized grape pickers into the United Farm Workers and led a strike against growers in the San Joaquin valley of central California in 1965. José Gutiérrez helped form the third party La Raza Unida in Crystal City, Texas, in 1970, to represent the interests of Chicanos and Chicanas. La Raza Unida gained support and members in California and Colorado, especially from the urban *barrios*. Another organization popular among Chicano and Chicana youth was the Crusade for Justice, founded in Denver in 1966 by Rodolfo "Corky" Gonzalez, the child of sugar-beet workers. Although it pursued a number of issues, one of its major goals was to promote a curriculum relevant to the culture of Mexican Americans in order to help them better understand and appreciate their ethnic identity and heritage. Reies López Tijerina championed the use of violence to protest the loss of Hispanic village lands in northern New Mexico to national parks and private developers over the course of the previous century. The Chicano/a Movement led to a flowering of artistic and musical expression among ethnic Mexicans who found new inspiration to resist assimilation into Euro-American culture.[170]

Mexican Americans have made some advances in the political arena, especially in Texas. In San Antonio, Texas, Mexican Americans began serving on the city council in the 1950s. Residents of south- and east-side *barrios* elected Henry B. González to the city council in 1953, the state senate in 1956, and the US Congress in 1962, where he served until 1998 and was succeeded by his son, Charles. Gus García served on the San Antonio School Board in the 1950s. In 1981, San Antonio elected the first Mexican-American mayor of a large American city, Henry B. Cisneros, who also served as Secretary of Housing and Urban Development under President Bill Clinton until scandal forced him from office in 1997.[171] Despite the large community of ethnic Mexicans in Los Angeles, their political influence lagged behind that of African Americans, creating a rivalry between the two groups. Edward Roybal was the first ethnic Mexican elected to the city council of Los Angeles in 1949 by a multiracial coalition. In 1962, he became the congressional representative for the Twenty-Fifth District, which included East Los Angeles as well as Boyle Heights. He served in the US Congress for thirty years. When he left Los Angeles, however, no other Mexican Americans served on the city council for another twenty years; redistricting generally favored African Americans over Mexican Americans.[172]

As a result of the various civil-rights efforts in the twentieth century, some Mexican Americans have made substantial gains. A growing number are moving into white-collar and professional jobs and are purchasing houses in once exclusively Euro-American suburbs. Mexican Americans are represented on city councils, school boards, and in state legislatures throughout the West. They serve in the US Congress. New Mexico had a governor of ethnic Mexican heritage between 2002 and 2008, Bill Richardson, who left to serve in President Barak Obama's cabinet. The curriculum in schools from the elementary level to the graduate level in universities is beginning to include the Spanish, Mexican, and Indian heritage of the nation.

Ethnic Mexicans are not a monolithic group, however. The material and political progress of a few has not affected the overall plight of the many, which has been exacerbated by ongoing waves of immigration from Mexico. They remain in the lowest-paying jobs and live in some of the worst housing. In the postwar period especially, the illegal status of many ethnic Mexican workers has made them more vulnerable to workplace exploitation than other groups. Employers often pay them well below the minimum wage. Poverty, language barriers, and lack of citizenship among a significant percentage of ethnic Mexicans has contributed to an intense uphill struggle to gain political influence. Most attend poorly financed schools and struggle for access to medical care. They continue to suffer from a dwindling tax base as more and more businesses move to the suburbs of the large cities.[173]

Some Asians experienced substantial gains in the postwar period because they were finally allowed to become American citizens. Emphasizing education as the primary path to upward mobility, many successfully moved into higher paying jobs and purchased homes in the suburbs, giving rise to the image of the "model minority."[174] Those who remained in the inner cities were usually poor and increasingly elderly.[175] The Immigration Act of 1965, which repealed virtually all of the earlier exclusionary provisions and quotas on immigration from Asia, had a significant impact on the Asian population in the West, contributing to class divisions. Refugees from Korea, Laos, Vietnam, and Cambodia poured in to Los

Angeles and other Western cities. In 1970, about 57 percent of Asians were native-born American citizens whereas by 1990 that proportion had dropped to 30 percent. A substantial number of the immigrants were poor and illiterate, lived in substandard housing, and worked in low-paying service-sector jobs.[176] Thousands of Vietnamese refugees settled in Westminster, Orange County, California, in an area known as "Little Saigon." Koreans flocked to Orange County and the San Fernando valley, moving away from Koreatown in Los Angeles as they progressed economically. Houston was another major entry point for Vietnamese and Cambodian refugees. It has the largest Southeast Asian population outside of the Los Angeles area and remains an attractive destination, mainly because of its low cost of living, mild climate, and growing ethnic Asian enclaves.[177]

Different groups of Asian Americans have participated in civil-rights movements, but their efforts have been less publicly visible. A few began to enter local and regional politics in California in the 1970s. California elected its first female Chinese American secretary of state, March Fong Eu, in 1974. Asian Americans still have very little political representation on the national level, however, especially because most remain concentrated on the west coast.[178] President George W. Bush appointed Norman Mineta, a Japanese American from Santa Clara, California, Secretary of Transportation from 2001 until 2006, the only Democrat to serve in his cabinet. The most successful political effort led by Asian Americans was the redress movement. In 1988, Congress formally apologized to Japanese Americans for the internments of World War II.[179]

Finally, the original inhabitants of the Western USA also experienced some of the benefits of civil rights, although they continue to lag behind most other groups in terms of education and income. The federal policy of Termination in the 1950s brought hundreds of Indians to live and work in Western cities. High unemployment rates and poor living conditions became common experiences. Urban living fostered cooperation among members of many different tribes, especially in Los Angeles and San Francisco, leading to the development of a network of social and political organizations designed to address some of the difficult issues Indians faced, including poverty. Out of these networks emerged a pan-Indian identity and an active civil-rights movement, the American Indian Movement (AIM), which emphasized self-determination, Indian sovereignty, and the preservation of tribal culture. The AIM marked the beginning of a period in which Indians would begin to see progress in their attempts to regain lands and political rights as well as in areas of improving education, literacy, and income levels.[180] Much Indian activism centered on strengthening the sovereignty of Indian tribes. Some Indian politicians, such as Ben Nighthorse Campbell of Colorado, successfully moved into mainstream American politics where they could contribute to this effort. Campbell, a member of the Northern Cheyenne Tribe, served as a Democratic representative from Colorado in the US Congress from 1987 until 1992. He became Colorado's US Senator in 1993 and served until 2005. He switched to the Republican Party in 1995.[181]

Advances in civil rights have not erased poverty and its problems, nor have they eliminated the gap between rich and poor. On the positive side, more non-Euro-Americans in the West are participating in the prosperity created by enterprise and enjoying the flowering of high culture in Western cities. Yet each successive generation deals with a new set of issues generated by the paradox of capitalism, its propensity to produce extremes in both wealth and poverty. "Gentrification," the latest form of urban renewal, is increasingly displacing poor inner-city residents. Partnerships between private developers and city governments have arisen as part of an overall effort to revitalize central business districts.

They have also attempted to recreate "walking cities" as an antidote to urban sprawl, the destruction of wild lands, and excessive automobile pollution. Portland, Oregon, is one of the best examples of this approach to land use. The city has created an urban growth boundary limiting development beyond a certain point. Yet as old homes are razed to build condominiums and retail centers and as young professionals convert modest shacks into comfortable cottages, property values and hence tax assessments increase far beyond the capacity of many long-standing inhabitants.

Christmas lights adorned the last remaining houses of Dallas's Little Mexico in December 2007. They are difficult to find, surrounded as they are by luxury high-rise condominiums, office buildings, and the American Airlines Center. Jesse Salinas, who moved into his house sixty years before when it was on a dirt road on the edge of town said he could no longer afford the property taxes. "Everything is gone, used to be I knew all the friends and neighbors around here, not no more, I don't know them no more [...] Little Mexico, it was, but Harwood is the owner of all this property. [...] I don't want to spend another Christmas here."[182] People such as Salinas have to find alternative housing, usually in decaying apartment buildings in some of the older suburban areas. The cycles of wealth and poverty continue with each successive generation in the American West.

NOTES

1. Gray Brechin, *Imperial San Francisco: Urban Power, Earthly Ruin* (Berkeley, Calif.: University of California Press, 1999), pp. 200–202.
2. Gunther Barth, *Instant Cities: Urbanization and the Rise of San Francisco and Denver* (Oxford: Oxford University Press, 1975), pp. x–xxiii.
3. Before the arrival of the Americans, American Indians were the region's most numerous inhabitants, and the European powers simply maintained isolated outposts. The Spaniards had the most dominant presence, but they never succeeded in establishing any major cities as they had in New Spain. Spanish *pueblos* such as Santa Fe, Los Angeles, San Diego, and San Antonio remained small and failed to develop rich cultural and intellectual institutions comparable to those in Mexico City. The largest city in New Spain was Santa Fe, which at its height had about 6,000 settlers. The northern frontier suffered from chronic shortages of funds, the dominating role of the military, hostile American Indians, and an inability to attract many settlers. Although there were differences in wealth and social class, a sense of egalitarianism prevailed. Such wealth that existed was tied up largely in land grants. Cattle ranching and the hide-and-tallow trade never produced much in the way of commercial capital. The major social dividing line was between Indians and Spaniards. This state of things prevailed during the short-lived Mexican period but changed rapidly once the Americans arrived. John W. Reps, *The Forgotten Frontier: Urban Planning in the American West before 1890* (Columbia, Miss.: University of Missouri Press, 1981), pp. 5–24, 39–51, 80–81, 101; David J. Weber, *The Spanish Frontier in North America* (New Haven, Conn.: Yale University Press, 1992), pp. 322, 333–334.
4. Barth, *Instant Cities,* pp. 59–60, 227. There is extensive scholarship on the California Gold Rush. Two recent examples include Malcolm J. Rohrbaugh, *Days of Gold: The California Gold Rush and the American Nation* (Berkeley, Calif.: University of California Press, 1997) and Susan Lee Johnson, *Roaring Camp: The Social World of the California Gold Rush* (New York: W. W. Norton & Company, 2000). For an edited primary-source account, see J. S. Holliday, *The World Rushed In: The California Gold Rush Experience* (Norman, Okla.: University of Oklahoma Press, 2002). Elliott West, *The Contested Plains: Indians, Goldseekers, and the Rush to Colorado* (Lawrence, Kans.: University Press of Kansas, 1998) discusses the impact of the 1858 Gold Rush on the Plains environment and the American Indians who lived there. Wealth-seeking played a key role in the settlement of the American West. Reps, *The Forgotten Frontier,* p. 131.

5. Salt Lake City was also an American city, but it does not follow the typical pattern because it was established by a persecuted religious group led by Brigham Young: the Mormons. It was a closed, "temple city," that combined both ecclesiastical and civil affairs. It was the center of both migration and trade for the Mormons who moved west. They lived there, learned to farm in the arid desert and then moved out from the city. Established in 1847, the oldest American city in the West, it was highly planned. Reps, *The Forgotten Frontier,* p. 131; Barth, *Instant Cities,* pp. 59–60.
6. Barbara Berglund, *Making San Francisco American: Cultural Frontiers in the Urban West, 1846–1906* (Lawrence, Kans.: University Press of Kansas, 2007), pp. 4–5, 16–31.
7. Philip J. Ethington, *The Public City: The Political Construction of Urban Life in San Francisco, 1850–1900* (Cambridge: Cambridge University Press, 1994), p. 53.
8. John H. Davis, *The Guggenheims: An American Epic* (New York: William Morrow & Co., 1978), p. 245; Stephen J. Leonard and Thomas J. Noel, *Denver: Mining Camp to Metropolis* (Niwot, Col.: University Press of Colorado, 1990), pp. 45–51.
9. Brechin, *Imperial San Francisco,* pp. 126–130. There are numerous works on the building of the transcontinental railroad and the "Big Four." The seminal study is Oscar Lewis, *The Big Four: The Story of Huntington, Stanford, Hopkins, and Crocker, and of the Building of the Central Pacific* (New York: A. A. Knopf, 1938).
10. William B. Friedricks, *Henry E. Huntington and the Creation of Southern California* (Columbus, OH: Ohio State University Press, 1992), p. 49. Two biographies that discuss Doheny's impact on the oil industry include Margaret L. Davis, *Dark Side of Fortune: Triumph and Scandal in the Life of Oil Tycoon Edward L. Doheny* (Berkeley, Calif.: University of California Press, 1998) and Dan La Botz, *Edward L. Doheny: Petroleum, Power, and Politics in the United States and Mexico* (New York: Praeger, 1991).
11. Don Graham, *Kings of Texas: The 150-Year Saga of an American Ranching Empire* (Hoboken, NJ: John Wiley & Sons, Inc., 2003) and Mark Arax and Rick Wartzman, *J. G. Boswell and the Making of an American Empire* (New York: Public Affairs, 2005).
12. Gerald Nash, *The American West in the Twentieth Century: A Short History of an Urban Oasis* (Albuquerque, N. Mex.: University of New Mexico Press, 1977), p. 34; Patricia Everidge Hill, *Dallas: The Making of a Modern City* (Austin, Tex.: University of Texas Press, 1996), pp. xxiv–xxv.
13. Brechin, *Imperial San Francisco,* p. 177. A popular biography of William Randolph Hearst is W. A. Swanberg, *Citizen Hearst: A Biography of William Randolph Hearst* (New York: Scribner, 1961).
14. Brechin, *Imperial San Francisco,* pp. 126–146.
15. Kevin Starr, *Inventing the Dream: California through the Progressive Era* (Oxford: Oxford University Press, 1985), pp. 72–75; Margaret Leslie Davis, *Culture Broker: Franklin D. Murphy and the Transformation of Los Angeles* (Berkeley, Calif.: University of California Press, 2007), p. 38.
16. Barth, *Instant Cities,* pp. 197–199, 227; Ethington, *The Public City,* pp. 53–54.
17. Arnoldo De Léon, *Racial Frontiers: Africans, Chinese, and Mexicans in Western America, 1848–1890* (Albuquerque, N. Mex.: University of New Mexico Press, 2002), pp. 48–52.
18. William Issel and Robert W. Cherny, *San Francisco, 1865–1932: Politics, Power, and Urban Development* (Berkeley, Calif.: University of California Press, 1986), pp. 33, 110–111, 139–150. Additionally, silver kings James Fair, James Flood, John Mackay, and William O'Brien of San Francisco, who made their fortunes in the 1860s and 1870s, were all Irish Catholic. Kevin Starr, *California: A History* (New York: Modern Library Chronicles, 2007), p. 121.
19. Dino Cinel, *From Italy to San Francisco: The Immigrant Experience* (Palo Alto, Calif.: Stanford University Press, 1982), pp. 236–239.
20. Leonard and Noel, *Denver,* pp. 77–78.
21. Fred Rosenbaum, *Cosmopolitans: A Social and Cultural History of the Jews of the San Francisco Bay Area* (Berkeley, Calif.: University of California Press, 2009) tells the story of both the Jewish elite and working class of San Francisco. Jacob Furth, a Jew who had immigrated from Bohemia in the nineteenth century, became successful in Seattle by 1882 when he became the president of Puget Sound National Bank. Richard C. Berner, *Seattle, 1900–1920: From Boomtown, Urban Turbulence,*

to Restoration (Seattle, Wash.: Charles Press, 1991), pp. 69–72. Brothers Philip, Isaac, and Alexander Sanger started a substantial wholesale dry-goods business in Dallas in the nineteenth century and established branches all over Texas. Stanley Marcus and Al Neiman began a luxury clothing store in Dallas in the early twentieth century that is now a national retailer. Texas wholesalers Adolphe, Mayer, and Solomon Halff became ranchers in West Texas, buying up thousands of acres of land. Natalie Ornish, *Pioneer Jewish Texans: Their Impact on Texas and American History for Four Hundred Years, 1590–1900* (Dallas, Tex.: Texas Heritage Press, 1989), pp. 73–76; Leon Harris, *Merchant Princes: An Intimate History of Jewish Families Who Built the Great Department Stores* (New York: Harper & Row, 1979), pp. 201, 207. Herbert and Mortimer Fleishhacker were banking executives in San Francisco; they financed the transmission of hydroelectric power to the city in the 1920s and 1930s. Kevin Starr, *The Dream Endures: California Enters the 1940s* (Oxford: Oxford University Press, 1997), p. 127.

22. A recent biography tells his story: Frances Dinkelspiel, *Towers of Gold: How One Jewish Immigrant Named Isaias Hellman Created California* (New York: St. Martin's Press, 2008).
23. Neal Gabler, *An Empire of Their Own: How the Jews Invented Hollywood* (New York: Anchor Books, 1989) discusses the Jewish origins of most of Hollywood's early directors and the impact of their background on the development of the film industry. For a survey of Hollywood's early history, see Starr, *Inventing the Dream,* pp. 283–339.
24. Michael Phillips, *White Metropolis: Race, Ethnicity, and Religion in Dallas, 1841–2001* (Austin, Tex.: University of Texas Press, 2006), p. 142; Brechin, *Imperial San Francisco,* pp. 126–146; *Starr, The Dream Endures,* p. 125; Raphael J. Sonnenschein, *Politics in Black and White: Race and Power in Los Angeles* (Princeton, NJ: Princeton University Press, 1993), pp. 24–25. Jews had a small political presence in Western cities from the nineteenth century on, but they had little influence. Jews had served on Los Angeles' city council for nearly the entire nineteenth century. They had a minimal but regular presence on San Francisco Board of Supervisors from 1900 forward. Dallas's wards elected five Jewish aldermen between 1875 and 1905. Except for some working-class labor activism, for the most part Jews accommodated to the existing power structure because of their tenuous relationship with the Anglo-American elite and their own economic status.
25. Mark S. Foster, *Henry J. Kaiser: Builder in the American West* (Austin, Tex.: University of Texas Press, 1989), pp. 211–233.
26. For an overview of these men and their companies, see Steve Weinberg, *Armand Hammer: The Untold Story* (Boston, Mass.: Little, Brown & Co., 1989) and Loren Fox, *Enron: The Rise and Fall* (Hoboken, NJ: John Wiley & Sons, Inc., 2003).
27. James Wallace and Jim Erikson, *Hard Drive: Bill Gates and the Making of the Microsoft Empire* (New York: HarperBusiness, 1993), p. 7.
28. Sonnenschein, *Politics in Black and White,* pp. 13–14; Hill, *Dallas,* pp. 159–160; Bradford Luckingham, *Phoenix: The History of a Southwestern Metropolis* (Tucson, Ariz.: University of Arizona Press, 1989), p. 3; Issel and Cherny, *San Francisco,* pp. 17, 34; Leonard and Noel, *Denver,* pp. 131–133.
29. Berner, *Seattle,* pp. 62–63.
30. Leonard and Noel, *Denver,* pp. 46–47.
31. William David Estrada, *The Los Angeles Plaza: Sacred and Contested Space* (Austin, Tex.: University of Texas Press, 2008), pp. 44–45; Reps, *The Forgotten Frontier,* p. 89–107.
32. Reps, *The Forgotten Frontier,* p. 89–107.
33. Robert M. Fogelson, *The Fragmented Metropolis: Los Angeles, 1850–1930* (Cambridge, Mass.: Harvard University Press, 1967), pp. 89–94, 105–139; Nash, *The American West in the Twentieth Century,* pp. 221–222.
34. Leonard and Noel, *Denver,* pp. 55–58; Roger Sale, *Seattle: Past to Present* (Seattle, Wash.: University of Washington Press, 1976), pp. 68–80; Sister Agatha, *The History of Houston Heights, 1891–1918* (Houston, Tex.: Premier Printing Company, 1956); A. C. Greene, *Dallas, USA* (Austin, Tex.: Texas Monthly Press, 1984), pp. 171–173.
35. Fogelson, *The Fragmented Metropolis,* pp. 157–163; Sale, *Seattle,* pp. 80–81; Phillips, *White Metropo-*

lis, p. 64; Don Riddle, *River Oaks: A Pictorial Presentation of Houston's Residential Park* (Houston, Tex.: River Oaks Corporation, 1929).

36. John H. M. Laslett, "Historical Perspectives: Immigration and the Rise of a Distinctive Urban Region, 1900–1970," in Roger Waldinger and Bozorghmehr Mehdi Waldinger (eds.), *Ethnic Los Angeles* (New York: Russell Sage Foundation, 1996), pp. 39–75; p. 51; Starr, *The Dream Endures,* pp. 270–271; Fogelson, *The Fragmented Metropolis,* p. 145.
37. Mark Wild, *Street Meetings: Multiethnic Neighborhoods in Early Twentieth Century Los Angeles* (Berkeley, Calif.: University of California Press, 2005), pp. 9–14; Leonard and Noel, *Denver,* pp. 277–283: Bradford Luckingham, *Minorities in Phoenix: A Profile of Mexican American, Chinese American, and African American Communities, 1860–1992* (Tucson, Ariz.: University of Arizona Press, 1994), pp. 17–18; Rodolfo Rosales, *The Illusion of Inclusion: The Untold Political Story of San Antonio* (Austin, Tex.: University of Texas Press, 2000), pp. 12–13.
38. Berglund, *Making San Francisco American,* pp. 41–47.
39. Starr, *The Dream Endures,* pp. 10–27; Nash, *The American West in the Twentieth Century,* pp. 38–41.
40. Robert F. Dalzell, Jr., *Enterprising Elite: The Boston Associates and the World They Made* (New York: W. W. Norton & Co., 1987), pp. 113–163. Boston was one of the first cities in which philanthropy occurred on a large scale. The wealthy Boston families who had made their fortunes in mercantile activities, the textile industry, banking, insurance, and real estate, began to invest some of their profits in the community at large. Their early efforts, prior to 1845, emphasized social cohesion and harmony within the community. Thus, institutions that helped the poor were paramount because they would help relieve the potential conflict that might come from a growing class of people whose basic needs were not met. The wealthy had a vested interest in stability. Increasingly, however, by the mid-nineteenth century and beyond, Boston's elite gave to educational institutions and the arts, places that benefited their own social class. They maintained their control through serving on boards and as trustees of large endowments. It was apparent that Western businessmen, many of whom had recently migrated from the East, followed the example of Boston's elite, men such as Nathan Appleton or Frances Cabot Lowell, in deciding when and how to give. Andrew Carnegie was also a model for giving. Richard H. Peterson, "The Spirit of Giving: The Educational Philanthropy of Western Mining Leaders, 1870–1900," *The Pacific Historical Review,* 53 (3) (1984): 309–336.
41. Issel and Cherny, *San Francisco,* p. 106; Judith Robinson, *The Hearsts: An American Dynasty* (Newark, Del.: University of Delaware Press, 1991), p. 266.
42. Elizabeth York Enstam, *Women and the Creation of Urban Life: Dallas, Texas, 1843–1920* (College Station, Tex.: Texas A & M University Press, 1998), pp. 96–97.
43. Issel and Cherny, *San Francisco,* p. 106.
44. Marguerite Johnson, *Houston: The Unknown City, 1836–1946* (College Station, Tex.: Texas A & M University Press, 1991), pp. 243–251.
45. Peterson, "The Spirit of Giving," pp. 326–328. Richard W. Fisher, president and chief executive of the Federal Reserve Bank in Dallas, recently expressed concern over Texas's relative lack of prestigious universities and what that might mean for the future of Dallas's economy. "[But] it worries me terribly that the University of Texas at Austin ranks [at or] below six of the University of California campuses in terms of excellence." Brendan M. Case, "Dallas Fed Chief Richard W. Fisher Speaks His Mind in Q & A," *Dallas Morning News,* August 17, 2008, Sec. D, p. 1.
46. Brechin, *Imperial San Francisco,* pp. 303–307.
47. Peterson, "The Spirit of Giving," pp. 315–322.
48. Brechin, *Imperial San Francisco,* pp. 280–283; Robinson, *The Hearsts,* pp. 267–211.
49. Brechin, *Imperial San Francisco,* pp. 283–284.
50. A local history of the Caruth family of Dallas is found in Barot Sanders, *Caruth—Dallas' Landed Gentry* (Irving, Tex.: Sanders Press, 1988); for information on the Dedman family, see "Robert H. Dedman, Sr.: SMU Benefactor Was a Champion of Giving," *Dallas Morning News,* August 21, 2002.

51. Kenneth Lipartito and Joseph Pratt, *Baker & Botts and the Development of Modern Houston* (Austin, Tex.: University of Texas Press, 1991), pp. 47–60.
52. David G. McComb, *Houston: The Bayou City* (Austin, Tex.: University of Texas Press, 1969), p. 241; Johnson, *Houston,* pp. 308–309.
53. Leonard and Noel, *Denver,* pp. 79–80; Brechin, *Imperial San Francisco,* p. 185.
54. Peterson, "The Spirit of Giving," p. 323; Nash, *The American West in the Twentieth Century*), pp. 55, 110–111.
55. Issel and Cherny, *San Francisco,* pp. 111–112; Berglund, *Making San Francisco American,* pp. 173–175; Brechin, *Imperial San Francisco,* pp. 184–186.
56. Nash, *The American West in the Twentieth Century,* p. 127; Issel and Cherny, *San Francisco,* p. 114. San Francisco architecture changed in the late nineteenth century from wooden Victorian structures to more seemingly permanent ones of brick and cement. Arthur Page Brown was the architect involved in the transformation. Starr, *Inventing the Dream,* pp. 176–198.
57. Nash, *The American West in the Twentieth Century,* p. 122.
58. Nash, *The American West in the Twentieth Century,* pp. 120–122.
59. Nash, *The American West in the Twentieth Century,* p. 38.
60. Davis, *Culture Broker,* pp. xi–xiv, 63, 83, 99–111.
61. Davis, *Culture Broker,* pp. 29–30, 80–82.
62. Davis, *Culture Broker,* pp. 110–111, 181–191.
63. Davis, *Culture Broker,* pp. 210–211, 274–287, 307–317, 332–334.
64. Davis, *Culture Broker,* pp. 357–359, 362.
65. Greene, *Dallas, USA.,* p. 247. Texas philanthropy did not really take off until the 1920s, after the discovery of oil and the subsequent creation of new fortunes. Most giving was channeled into research, education, museums, and community parks. Mary L. Kelley, *The Foundations of Texan Philanthropy* (College Station, Tex.: Texas A. & M, Press, 2004), pp. 80–85.
66. "An American Legacy," available online at www.cartermuseum.org/about (accessed September 30, 2008). Kelley, *The Foundations of Texan Philanthropy,* pp. 58–69.
67. "Director's Message," available online at www.kimbellart.org (accessed September 30, 2008).
68. Richard F. Selcer, *Fort Worth: A Texas Original!* (Austin, Tex.: Texas State Historical Association, 2004), pp. 106–109. After lagging behind Fort Worth for many years, Dallas has recently made great strides in developing a world-class cultural center, which is in the process of completion. David Dillon and Scott Cantrell, "Dallas Arts District at 30: What's Next?" *Dallas Morning News Guide Live,* June 8, 2008, p. 3.
69. Michael Hedges and Mike Tolson, "Ken Lay, 1942–2006: Lay's Criminal Record May Vanish with His Death," *The Houston Chronicle,* July 6, 2006, sec. A, p. 1.
70. Fox, *Enron,* p. 191.
71. Bill Murphy, "The Fall of Enron: Enron's Art Hopes Just a Pipe Dream: Buying Spree Now Drawing Fire," *The Houston Chronicle,* July 29, 2002, sec. A, p. 1.
72. "The Enron Effect: Ripples in Houston Could Become Dangerous Waves," *The Houston Chronicle,* February 17, 2002, sec. A., p. 1.
73. Wallace and Erickson, *Hard Drive,* pp. 8–15.
74. See www.gatesfoundation.org (accessed October 3, 2008).
75. Warren Buffett, "10 Ways to Get Rich," *Parade Magazine,* September 7, 2008.
76. Chris Wilson, *The Myth of Santa Fe: Creating a Modern Regional Tradition* (Albuquerque, N. Mex.: University of New Mexico Press, 1997), p. 88–89. Lummis established the Southwest Society in 1896, which was the Western branch of the Archaeological Institute of America. In 1907, he helped found the Southwest Museum, which later became the Southwest Museum of the American Indian, the oldest museum in Los Angeles, for the purpose of housing Indian artifacts and providing a place for Western artists to show their work. He also led an effort to rebuild the old Spanish missions of California. Nash, *The American West in the Twentieth Century,* p. 55.

77. Nash, *The American West in the Twentieth Century,* p. 55; Marc Simmons, *New Mexico: An Interpretive History* (Albuquerque, N. Mex.: University of New Mexico Press, 1977), pp. 134, 156–164.
78. Wilson, *The Myth of Santa Fe,* pp. 88–89.
79. Marit K. Munson, *Kenneth Chapman's Santa Fe: Artists and Archaeologists, 1907–1931* (Santa Fe, N. Mex.: School for Advanced Research Press, 2007), pp. 4–10, 32–33, 77, 94, 119–129.
80. Munson, *Kenneth Chapman's Santa Fe,* p. 1.
81. Nash, *The American West in the Twentieth Century,* pp. 120–121.
82. Lois Palkin Rudnick, *Mabel Dodge Luhan: New Woman, New Worlds* (Albuquerque, N. Mex.: University of New Mexico Press, 1984), pp. 145–190. For Mabel's own version of her experience in moving to Taos and creating a new life there, see Mabel Dodge Luhan, *Edge of Taos Desert: An Escape to Reality* (Albuquerque, N. Mex.: University of New Mexico Press, 1937). Mabel Dodge Luhan arrived in Taos more than a decade after artists Bert Phillips and Ernest Blumenschein from New York discovered it by accident on their way to Mexico. Phillips and Blumenschein decided it would be an excellent place for artists to work with its landscapes, light, and American Indian and Hispanic cultures. They convinced Irving Couse to come in 1902, and others soon followed, forming the Taos Society of Artists in 1915. Members of the Taos Society of Artists sold many of their works to the Southern Pacific Railroad, which used them in brochures written by Charles Lummis.
83. Willa Cather, *Death Comes for the Archbishop* (New York: Vintage Classics, 1927).
84. Lesley Poling-Kempes, *Valley of Shining Stone: The Story of Abiquiu* (Tucson, Ariz.: University of Arizona Press, 1997), pp. 192–194.
85. Wilson, *The Myth of Santa Fe,* pp. 274–281. Charles Montgomery, *The Spanish Redemption: Heritage, Power, and Loss on New Mexico's Upper Rio Grande* (Berkeley, Calif.: University of California Press, 2002), pp. 89–97.
86. Munson, *Kenneth Chapman's Santa Fe,* pp. 4–10, 32–33, 77, 94, 119–129.
87. Montgomery, *The Spanish Redemption,* pp. 172–173.
88. Nash, *The American West in the Twentieth Century,* p. 55; "About the Autry National Center," available online at http://theautry.org/about-us/what-is-the-autry (accessed May 13, 2010); "History of the Buffalo Bill Historical Center," available online at www.bbhc.org/news/BBHChistory_01.cfm (accessed May 13, 2010).
89. Luckingham, *Phoenix,* pp. 90–92.
90. Nash, *The American West in the Twentieth Century,* p. 257. For example, the Santa Fe Opera, founded in 1954, was funded by John Crosby, an Easterner.
91. Wilson, *The Myth of Santa Fe,* pp. 269–273.
92. Sarah Deutsch, *No Separate Refuge: Culture, Class, and Gender on an Anglo-Hispanic Frontier in the American Southwest, 1880–1940* (Oxford: Oxford University Press, 1987), pp. 179–183.
93. Richard White, *"It's Your Misfortune and None of My Own": A New History of the American West* (Norman, Okla.: University of Oklahoma Press, 1991), pp. 111–112, 237. For a complete, brief history of American Indians in the twentieth century, see Peter Iverson, *"We Are Still Here": American Indians in the Twentieth Century* (Wheeling, Ill.: Harlan Davidson, Inc., 1998). Other works dealing with particular Western tribes include Peter Iverson, *Diné: A History of the Navajo* (Albuquerque, N. Mex.: University of New Mexico Press, 2002); Frederick E. Hoxie, *Parading through History: The Making of the Crow Nation* (Cambridge: Cambridge University Press, 1995).
94. White, *"It's Your Misfortune,"* 282.
95. Tomás Almaguer, *Racial Fault Lines: The Origins of White Supremacy in California* (Berkeley, Calif.: University of California Press, 1994), pp. 45–53, 158–159; Neil Foley, *The White Scourge: Mexicans, Blacks, and Poor Whites in Texas Cotton Culture* (Berkeley, Calif.: University of California Press, 1997), pp. 1–16.
96. Fogelson, *The Fragmented Metropolis,* pp. 145–147; Issel and Cherny, *San Francisco,* p. 58; Leonard and Noel, *Denver,* pp. 53–56; Phillips, *White Metropolis,* pp. 63–65; Sale, *Seattle,* pp. 55–61, 80–81; Wild, *Street Meetings,* pp. 1–14.

97. William Deverell, *Whitewashed Adobe: The Rise of Los Angeles and the Remaking of its Mexican Past* (Berkeley, Calif.: University of California Press, 2004), p. 43.
98. R. A. Burchell, *The San Francisco Irish, 1848–1880* (Berkeley, Calif.: University of California Press, 1980), pp. 34–47, 60–70, 94–127; Issel and Cherny, *San Francisco, 1865–1932,* pp. 130–134; William A. Bullough, *The Blind Boss and His City: Christopher Augustine Buckley and Nineteenth Century San Francisco* (Berkeley, Calif.: University of California Press, 1979) discusses the life and career of Irish political boss Christopher Buckley and his role in the history of San Francisco.
99. Ricardo Romo, *East Los Angeles: The History of a Barrio* (Austin, Tex.: University of Texas Press, 1983), pp. 64–65; Berner, *Seattle,* pp. 69–72; Elaine H. Maas, *The Jews of Houston: An Ethnographic Study* (New York: AMS Press, 1989). They never represented a sizeable percentage of the overall population. San Francisco's community numbered about 40,000 in 1940, approximately 6 percent of the overall population. Starr, *The Dream Endures,* p. 125. In 1870, there were about 330 Jews in Los Angeles, roughly 6 percent of the total population. By 1946, their numbers had reached 168,000, but they still constituted less than 10 percent of the city's populace. Sonnenschein, *Politics in Black and White,* pp. 47–51. As more and more realtors from the area advertised to Jews living in New York City, the Jewish population mushroomed from 2,500 to 20,000 by 1920. The homes along Brooklyn and Soto Streets became occupied almost exclusively by Jews. Wild, *Street Meetings,* pp. 25–30; Laslett, "Historical Perspectives," pp. 41, 52. For a study of Jews in Portland, Oregon, see William Toll, *The Making of an Ethnic Middle Class: Portland Jewry over Four Generations* (Albany, NY: State University of New York Press, 1982).
100. Leonard and Noel, *Denver,* pp. 194–195.
101. Richard C. Berner, *Seattle, 1921–1940: From Boom to Bust* (Seattle, Wash.: Charles Press, 1992), pp. 205–213.
102. Gabler, *An Empire of Their Own,* pp. 237–245.
103. Leonard and Noel, *Denver,* pp. 183–185.
104. Cinel, *From Italy to San Francisco,* pp. 101–114, 234–235.
105. Cinel, *From Italy to San Francisco,* pp. 196–197, 249–250. Italians in San Francisco were accused of anti-Americanism because of ties to Fascism and Mussolini. The California legislature launched an investigation into San Francisco's Little Italy because of the perception that they were engaging in anti-American activities; people viewed them as having developed a self-contained and almost self-sufficient enclave with only superficial ties to the USA.
106. "The Italians of Denver," Exhibition at the Colorado Historical Society, April 20–July 6, 2008. For a discussion of how Eastern and Southern Europeans, including Jews, Italians, Greeks, and others, eventually became part of the Euro-American mainstream, see Matthew Frye Jacobson, *Whiteness of a Different Color: European Immigrants and the Alchemy of Race* (Cambridge, Mass.: Harvard University Press, 1998).
107. Almaguer, *Racial Fault Lines,* pp. 47–48.
108. This happened in California, Texas, and New Mexico, the centers of Spanish-Mexican settlement in the Southwest. Montgomery, *The Spanish Redemption,* pp. 20–53; Armando C. Alonzo, *Tejano Legacy: Rancheros and Settlers in South Texas* (Albuquerque, N. Mex.: University of New Mexico Press, 1998), pp. 256–258; David Montejano, *Anglos and Mexicans in the Making of Texas, 1836–1936* (Austin, Tex.: University of Texas Press, 1987), pp. 50–74; Leonard Pitt and Ramon A. Gutíerrez, *Decline of the Californios: A Social History of the Spanish-Speaking Californias, 1846–1890* (Berkeley, Calif.: University of California Press, 1999).
109. George I. Sánchez, *Becoming Mexican-American: Ethnicity, Culture, and Identity in Chicano Los Angeles, 1900–1945* (Oxford: University of Oxford Press, 1993), p. 87–107; Arnoldo De León, *They Called Them Greasers: Attitudes toward Mexicans in Texas, 1821–1900* (Austin, Tex.: University of Texas Press, 1983); Montgomery, *The Spanish Redemption,* pp. 54–88. See also David G. Gutiérrez, *Walls and Mirrors: Mexican Americans, Mexican Immigrants, and the Politics of Ethnicity* (Berkeley, Calif.: University of California Press, 1995).
110. Albert Camarillo, *Chicanos in a Changing Society: From Mexican Pueblos to American Barrios in*

Santa Barbara and Southern California (Cambridge, Mass.: Harvard University Press, 1979), pp. 53–78.

111. Deutsch, *No Separate Refuge,* pp. 168–180.
112. Jorge Iber and Arnoldo De Léon, *Hispanics in the American West* (Santa Barbara, Calif.: ABC-CLIO, Inc., 2006), pp. 151–205.
113. Deverell, *Whitewashed Adobe,* pp. 40–44, 172–206; Romo, *East Los Angeles,* pp. 68–73. Another important study of the nineteenth-century Los Angeles *barrio* is Richard Griswold del Castillo, *The Los Angeles Barrio, 1850–1890: A Social History* (Berkeley, Calif.: University of California Press, 1970).
114. Deverell, *Whitewashed Adobe,* pp. 40–44, 98–108, 172–206; Romo, *East Los Angeles,* pp. 68–73; Estrada, *The Los Angeles Plaza,* p. 105.
115. Maxine Holmes and Gerald D. Saxon, *WPA Dallas Guide and Its History, Written and Compiled from 1936 to 1942 by the Workers of the Writers' Program of the Works Progress Administration* (Denton, Tex.: University of North Texas Press, 1992), p. 306.
116. As ethnic Mexicans moved to Houston to work on the railroads, they settled in the central city near places of employment, frequently inhabiting homes formerly owned by middle- and upper-class Euro-Americans that had been turned into rooming houses. Houston's first Mexican *barrio* took shape in the second ward east of downtown between 1910 and 1920, where Jews had once been the majority. Immigrants built *jacales* along the bayou made of any material they could find, including tin and sheet iron. In the 1920s, lower-middle-class ethnic Mexicans who owned small businesses eventually moved to the Magnolia Park area where they set up Spanish theatres, drug stores, gas stations, bakeries, and restaurants. Arnoldo De Léon, *Ethnicity in the Sunbelt: Mexican Americans in Houston* (College Station, Tex.: Texas A. & M. University Press, 2001), pp. 11–25. Fort Worth also developed a Mexican *barrio* neighborhood around 1900 when a couple of meatpacking plants were built by Swift & Company and Armour Packing Company in south Fort Worth. Texas Steel also built a plant. Mexicans flooded in to work in these places. Carlos E. Cuéllar, *Stories from the Barrio: A History of Mexican Fort Worth* (Fort Worth, Tex.: Texas Christian University Press, 2003), pp. 11–16. For an overview of ethnic Mexicans in Phoenix, see Luckingham, *Minorities in Phoenix,* p. 53.
117. Romo, *East Los Angeles,* pp. 12–13, 61. For a discussion of Mexican immigrant adaptation to the USA, see Sánchez, *Becoming Mexican-American.*
118. Rodolfo Acuña, *Occupied America: A History of Chicanos,* 4th edn (New York: Longman, 2000), pp. 216–225.
119. Douglas Flamming, *Bound for Freedom: Black Los Angeles in Jim Crow America* (Berkeley, Calif.: University of California Press, 2005), pp. 35–59.
120. Flamming, *Bound for Freedom,* pp. 49–50. Most of the Texans migrated from the cities of Houston, Dallas, and San Antonio.
121. Foley, *The White Scourge,* pp. 1–16.
122. Cary D. Wintz, "Blacks," in Fred. R. Von Der Mehden (ed.), *The Ethnic Groups of Houston* (Houston, Tex.: Rice University Studies, 1984), pp. 16–18; Kenneth Mason, *African Americans and Race Relations in San Antonio, Texas, 1867–1937* (New York: Garland Publishing, Inc., 1998), pp. xiv–xv; James M. SoRelle, "Race Relations in 'Heavenly Houston,' 1919–1945," in Howard Beeth and Cary D. Wintz (eds.), *Black Dixie: Afro-Texan History and Culture in Houston* (College Station, Tex.: Texas A. & M. University Press, 1992), pp. 9–40; Harry Robinson, Jr., Marc Sanders, and Ruthe Winegarten, *The Lives and Times of Black Dallas Women* (Austin, Tex.: Eakin Press, 2002). Before 1910, Houston did not have distinct African-American areas. The African-American populations of Dallas and Houston mushroomed in the early twentieth century. In Dallas after 1910, some Euro-American families became concerned about the rapid increase of the African-American population and wanted to take steps to prevent African-American families from spilling over into their neighborhoods. The African-American population nearly doubled in the following decade, growing from 9,035 to 18,024 by 1920. Restrictive covenants began to appear in warranty deeds to prevent sales to

African Americans in Euro-American neighborhoods. Still, some Euro-Americans continued to sell to African Americans and the neighbors were generally powerless. They petitioned the city to enforce segregation in residential areas. In response, the city passed a series of ordinances, beginning in 1916, labeling various blocks as "colored," "white," and "mixed." Although the US Supreme Court invalidated zoning laws requiring racial segregation in urban neighborhoods in the 1917 case of *Buchanan* v. *Warley,* the ordinances remained in place for a few more years. Additionally, the city shifted from a ward system to an at-large commission system of government to reduce African-American political power. Terry Anne Schulte-Scott, Marsha A. Prior, and Melissa M. Green, *From Freedmantown to Roseland Homes: A Pioneering Community in Dallas, Texas* for the US Army Corps of Engineers, Fort Worth District and US Department of Housing and Urban Development, Texas Antiquities Permit No. 2374, Misc. Report of Investigations Number 252 (April, 2005), pp. 42–67; Phillips, *White Metropolis,* p. 65; Hill, *Dallas,* p. 35.

123. Mason, *African Americans and Race Relations,* pp. xiv–xv.
124. Flamming, *Bound for Freedom,* p. 197.
125. Albert S. Broussard, *Black San Francisco: The Struggle for Racial Equality in the West, 1900–1954* (Lawrence, Kans.: University Press of Kansas, 1993), p. 5; Quintard Taylor, "Blacks and Asians in a White City: Japanese Americans and African Americans in Seattle," *Western Historical Quarterly,* 22 (4) (November, 1991): 401–429.
126. Broussard, *Black San Francisco,* pp. 6–7, 15–20; Flamming, *Bound for Freedom,* pp. 60–91.
127. Flamming, *Bound for Freedom,* pp. 60–91.
128. Luckingham, *Minorities in Phoenix,* p. 143.
129. Leonard and Noel, *Denver,* pp. 193–194.
130. Leonard and Noel, *Denver,* pp. 194–195; Flamming, *Bound for Freedom,* pp. 73–78; Broussard, *Black San Francisco,* p. 14. In a study of Seattle, Quintard Taylor noted that African Americans were far less successful in pursuing entrepreneurial ventures than the Japanese population. Taylor, "Blacks and Asians in a White City," p. 413.
131. Flamming, *Bound for Freedom,* pp. 8–9, 92–93.
132. Almaguer, *Racial Fault Lines,* pp. 158–160, 204; Sucheng Chan, *Asian Americans: An Interpretive History* (Boston, Mass.: Twayne Publishers, 1991), pp. 25–42; Erika Lee, *At America's Gates: Chinese Immigration during the Exclusion Era, 1882–1943* (Chapel Hill, NC: University of North Carolina Press, 2003), p. 25. The Chinese people who immigrated to America were unable to become naturalized citizens. Since most of the immigrants were men who were either single or had left wives and children back in China, the number of Chinese children born in the USA, and thus able to automatically become American citizens, was relatively small. The 1790 naturalization law had limited citizenship to "free white persons." After the Civil War, Congress modified the law to extend rights to people of African descent in order to effectuate the provisions of the Fourteenth Amendment. Neither of these laws expressly provided for the naturalization of Asian immigrants, and the courts consistently interpreted the laws as preventing Asians from becoming citizens. It was not until 1943 that the Chinese were allowed to become citizens. They were also restricted from landownership by alien land laws, which were particularly stringent in California, where most Asians lived. These were not repealed until 1952. Nash, *The American West in the Twentieth Century,* p. 218.
133. Yong Chen, *Chinese San Francisco, 1850–1943: A Trans-Pacific Community* (Palo Alto, Calif.: Stanford University Press, 2000), pp. 37–41, 49–69; Chan, *Asian Americans,* pp. 6–8.
134. Chan, *Asian-Americans,* pp. 32–33; Chalsa M. Loo, *Chinatown: Most Time, Hard Time* (New York: Praeger, 1987), pp. 37–38, 43–45.
135. Lee, *At America's Gates,* pp. 221–243.
136. Chen, *Chinese San Francisco,* pp. 70–95; Loo, *Chinatown,* pp. 19–21. Some of the most beautiful pictures of early twentieth-century Chinatown include Genthe's photographs, but they present a misleading picture of the community because they give the impression that there were more families and children than there were, and they were largely taken on ceremonial days when the Chinese were wearing traditional dress, something they did not do every day. John Kuo Wei Tchen and Arnold

Genthe, *Genthe's Photographs of San Francisco's Old Chinatown* (New York: Dover Publications, 1984).

137. Issel and Cherny, *San Francisco,* p. 71; Chen, *Chinese San Francisco,* pp. 55–57; Loo, *Chinatown,* pp. 19–21, 41–45. It is important to recognize, however, that there were families in Chinatown and that the presence of children, though small, was important to the development of the community. Children growing up in Chinatown were often pioneers in the struggle to combat anti-Chinese exclusionary practices while establishing their own Chinese-American identity. Wendy Rouse Jorae, *The Children of Chinatown: Growing Up Chinese in San Francisco, 1850–1920* (Chapel Hill, NC: University of North Carolina Press, 2009), pp. 1, 228.
138. Judy Yung, *Unbound Feet: A Social History of Chinese Women in San Francisco* (Berkeley, Calif.: University of California Press, 1995), pp. 285–287.
139. Loo, *Chinatown,* p. 48.
140. Loo, *Chinatown,* pp. 19–21, 41–45; Issel and Cherny, *San Francisco,* p. 71.
141. Wild, *Street Meetings,* pp. 59, 203. Apart from those in San Francisco and Los Angeles, there were no other Chinatowns in the West of any size. For a short time, Seattle had a very small one located along Washington and Main between Sixth and Seventh Streets, but anti-Chinese harassment and violence during the winter of 1885 essentially eliminated its Chinese population. Berner, *Seattle,* pp. 64–66. Chinese crowded along Sixteenth Street and Wazee Street in Denver. Leonard and Noel, *Denver,* p. 65. There were also small Chinatowns in Tucson and Phoenix, Arizona. Luckingham, *Minorities in Phoenix,* pp. 83–93. A few Chinese settled in Dallas and Houston but not enough to create a distinctive neighborhood. Edward C. M. Chen and Fred R. Von Der Mehden, "Chinese," in *The Ethnic Groups of Houston* (Houston, Tex.: Rice University Studies, 1984), pp. 63–79; pp. 68–69.
142. Chan, *Asian Americans,* p. 47.
143. Issel and Cherny, *San Francisco,* p. 66; Wild, *Street Meetings,* pp. 24–25.
144. Berner, *Seattle,* pp. 67–69.
145. Wild, *Street Meetings,* p. 201; Sale, *Seattle,* p. 177.
146. James N. Gregory, *American Exodus: The Dust Bowl Migration and Okie Culture in California* (Oxford: Oxford University Press, 1989), pp. 61–62, 118–123.
147. Deverell, *Whitewashed Adobe,* pp. 185–195.
148. Robert B. Fairbanks, *For the City as a Whole: Planning, Politics, and the Public Interest in Dallas, 1900–1965* (Columbus, OH: Ohio State University Press, 1998), pp. 4–6, 136, 147, 161–163.
149. White, *"It's Your Misfortune,"* pp. 496–519.
150. Josh Sides, *LA City Limits: African American Los Angeles from the Great Depression to the Present* (Berkeley, Calif.: University of California Press, 2003), pp. 108–130; Lawrence B. De Graaf, "African American Suburbanization in California, 1960–1990," in Lawrence B. de Graaf, Kevin Mulroy, and Quintard Taylor (eds.), *Seeking El Dorado: African Americans in California* (Seattle, Wash.: University of Washington Press, 2001), pp. 405–437.
151. Laslett, "Historical Perspectives," pp. 61–66.
152. Arthur E. Hippler, *Hunter's Point: A Black Ghetto* (New York: Basic Books, Inc., 1974), pp. 14–15.
153. Loo, *Chinatown,* pp. 52–53; Fairbanks, *For the City as a Whole,* pp. 161–163, 196–197.
154. Sides, *LA City Limits,* pp. 112–120.
155. Sides, *LA City Limits,* pp. 112–120.
156. Leonard and Noel, *Denver,* p. 369.
157. Sides, *LA City Limits,* pp. 115–120.
158. Phillips, *White Metropolis,* pp. 169–172; Darwin Payne, *Big D: Triumphs and Troubles of an American Supercity in the 20th Century* (Dallas, Tex.: Three Forks Press, 1994), pp. 341–346.
159. Wild, *Street Meetings,* pp. 203–207; Don Normark, *Chávez Ravine: A Los Angeles Story* (San Francisco, Calif. Chronicle Books, 1999).
160. White, *"It's Your Misfortune,"* p. 577.
161. Quintard Taylor, *The Forging of a Black Community: Seattle's Central District from 1870 through the*

Civil Rights Era (Seattle, Wash.: University of Washington Press, 1994), pp. 6–7; Wild, *Street Meetings,* p. 207; Sides, *LA City Limits,* pp. 203–205; Payne, *Big D,* pp. 401–414.

162. Matthew C. Whitaker, *Race Work: The Rise of Civil Rights in the Urban West* (Lincoln, Nebr.: University of Nebraska Press, 2005), pp. 4–5, 94–123.
163. Sides, *LA City Limits,* pp. 131–168; Broussard, *Black San Francisco,* p. 97. African Americans in Denver demonstrated unsuccessfully against the segregated swimming beach at Washington Park in 1932. Leonard and Noel, *Denver,* p. 366. The Los Angeles The *National Association for the Advancement of Colored People* filed court cases challenging restrictive covenants and segregated swimming pools in the 1920s. African-American Angelinos heartily supported A. Phillip Randolph's Double V Campaign to wipe out fascism overseas and racism at home. They consistently agitated for racial equality, long before Congress passed the landmark Civil Rights Act of 1964 and the Voting Rights Act of 1965, and their influence played a key role in the passage of those two Acts. Flamming, *Bound for Freedom,* pp. 211–225, 372–374.
164. Sonnenschein, *Politics in Black and White,* pp. 99–114.
165. Leonard and Noel, *Denver,* p. 381.
166. Payne, *Big D,* p. 172.
167. Fairbanks, *For the City as a Whole,* pp. 211, 249; Rosales, *The Illusion of Inclusion,* pp. 142–144; Sonnenschein, *Politics in Black and White,* pp. 26–27.
168. Sides, *LA City Limits,* p. 204.
169. De Léon, *Ethnicity in the Sunbelt,* pp. 127–130; Richard A. García, *Rise of the Mexican American Middle Class: San Antonio, 1929–1941* (College Station, Tex.: Texas A. & M. University Press, 1991). For a study of early Mexican American civil-rights movements in California, which occurred mainly through union organizing, see Douglas Monroy, *Rebirth: Mexican Los Angeles from the Great Migration to the Great Depression* (Berkeley, Calif.: University of California Press, 1999), pp. 209–256.
170. Acuña, *Occupied America,* pp. 351–355, 369–373. For further study, see Richard Griswold del Castillo, *César Chavez: A Triumph of Spirit* (Norman, Okla.: University of Oklahoma Press, 1995); Ernesto B. Vigil, *The Crusade for Justice: Chicano Militancy and the Government's War on Dissent* (Madison, Wisc.: University of Wisconsin Press, 1999).
171. Rosales, *The Illusion of Inclusion,* pp. 1–13, 45–70.
172. Acuña, *Occupied America,* pp. 43, 139.
173. Iber and De Léon, *Hispanics in the American West,* pp. 334–342.
174. Chen, *Asian Americans,* pp. 167–169.
175. Loo, *Chinatown,* pp. 49–51, 61–63, 81.
176. Lucie Cheng and Philip Q. Yang, "Asians: The 'Model Minority' Deconstructed," in Roger Waldinger and Bozorghmehr Mehdi Waldinger (eds.), *Ethnic Los Angeles* (New York: Russell Sage Foundation, 1996), pp. 305–44; pp. 316–319.
177. Fred Von Der Mehden, "Indochinese" and "Japanese" in *The Ethnic Groups of Houston* (Houston, Tex.: Rice University Studies, 1984), pp. 88–97, 104; My-Thuan Tran, "Flocking from SoCal to Houston," *The Los Angeles Times,* December 21, 2007.
178. Chen, *Asian Americans,* pp. 171–175; Some of the Asian civil-rights groups include the Chinese for Affirmative Action founded in San Francisco in 1969 and Asian Americans for Civil Rights and Equality based in Sacramento.
179. For an analysis of the redress movement, see Yasuko I. Takezawa, *Breaking the Silence: Redress Japanese American Ethnicity* (Ithaca, NY: Cornell University Press, 1995).
180. Troy R. Johnson, *The Occupation of Alcatraz Island: Indian Self-Determination and the Rise of Indian Activism* (Urbana, Ill.: University of Illinois Press, 1996), pp. 6–14, 217–221; Iverson, *"We Are Still Here,"* pp. 149–174.
181. Iverson, *"We Are Still Here,"* p. 196.
182. "New Development Demolishes Dallas' Little Mexico," *The Dallas Morning News,* December 21, 2007.

CHAPTER NINETEEN

AMERICAN INDIANS IN THE TWENTIETH CENTURY

Lisa E. Emmerich

INTRODUCTION: THE TEARS OF THE SEVEN GENERATIONS

On September 8, 2000, Assistant Secretary of the Interior for Indian Affairs Kevin Gover stunned participants at a ceremony commemorating the 175th anniversary of the Bureau of Indian Affairs (BIA) with an impassioned and remorseful acknowledgment of the "burden" of the agency's history. Gover, a Pawnee Indian appointed during the Clinton administration, tearfully identified himself as the spokesman of an organization that had "committed acts so terrible" their cumulative effect was the near destruction of Indian people and their communities. His litany of ruinous policies included the allotment of tribal lands, the suppression of traditional languages, the wholesale disruption of tribal social networks, and the persistent refusal to defend tribal interests. The inventory of the BIA's failures, while painfully familiar to those American Indians in attendance, must have been stunning for the non-Indians in the audience.

Lest any of those listening to these unprecedented remarks assume that the BIA's "bad old days" were out of mind because they were out of sight, the Assistant Secretary's observations recalled that the "wars of assimilation" waged in the nineteenth century had their own more recent counterparts. Gover's recitation of fiascos alluded to twentieth-century initiatives such as reorganization, termination, relocation, and self-determination. One hundred and seventy-five years of misguided and ethnocentric policies, he alleged, had poisoned federal–Indian relationships.[1]

Though Gover later pointed out that his emotionally charged statement was intended to illuminate the "legacy of inhumanity and racism," he also acknowledged that the fifteen-minute American Indian history primer cum apology may have resonated more with the non-Indian members of the audience than the tribal representatives. The American Indians present at the ceremony needed no reminders of the tortured relationship linking BIA policy failures to current conditions in Indian country. Even the most stable and prosperous Native nations today struggle with the remnants of past federal initiatives intended to "save the Indian." And in many less fortunate tribal communities, political marginalization, endemic poverty, and the social upheaval coincident with healthcare crises ranging from alcoholism to rampant diabetes continue to be everyday realities. The irony of the situation—it took an Indian to apologize to other Indians for past and ongoing actions taken against Indians by an agency of the federal government now staffed mostly by Indians—must have been inescapable for the Native men and women there.[2]

In retrospect, two things are notable about Gover's remarks. First, a Native man held sufficient rank within Department of the Interior to make his repudiation of federal policies newsworthy on both the domestic and international scene. The Assistant Secretary seized the bully pulpit afforded by his status to illuminate some of the most appalling episodes in American Indian history and call his own agency to account for its past bad acts toward tribal peoples. Second, he spoke to an audience that included tribal representatives from around the USA. In a roundabout fashion, Gover's tally sheet of losses obliquely paid tribute to American Indian persistence in the face of nearly insurmountable odds. Their very presence at the ceremony testified to the incredible resolve that American Indians historically have shown in withstanding threats to their survival. Simply put, absent that remarkable tenacity, no apology would have been necessary because there would have been no Indians left to apologize to in 2000.

Destruction and renewal, surrender and persistence, loss and recovery: for more than 400 years, the history of the Native peoples of the USA has unfolded in the spaces bounded by those opposites. This is especially the case for the most recent period of American Indian history. During the last decades of the nineteenth century, Americans armed themselves with guns, with legislation, with bibles, with romanticized ideas about the "rescue of a race," and with federal approval to obliterate tribalism and annihilate any substantive tribal resistance. A cast of characters that included federal officials, miners, missionaries, ranchers, railroad executives, military men, professional hunters, adventurers, corrupt politicians, farmers, and town boosters took up the challenge on behalf of the nation. They discounted the terrible price American Indians paid as necessary for national development, the unfortunate but wholly reasonable precursor to assimilation.

American Indians faced repeated challenges to sovereignty, to resource rights, to economic development, to social cohesion, to population stability, and to cultural continuity after 1900. Modern rhetoric may have changed the language of the old ethnocentric and ethnocidal policies but it certainly did not alter their intent. Against a national backdrop of world wars, economic booms and busts, domestic upheavals and international crises, American Indians survived and continually pushed back against renewed attempts to solve "the problem" that they, and their cultures, represented. Their response to federal acculturation plans was direct and (deceptively) simple: preserve the core elements of indigenous identity and keep hope for the future alive. Sooner or later, through human agency or the forces of the universe, American Indians would resume their lives as American Indians.

Tribal commonalities and shared concerns formed the heart of much of the twentieth-century American-Indian resistance to the ongoing federal assimilation agenda. Cooperative indigenism, however, did not guarantee any collective perspectives. The cultural and political diversity inherent within the boundaries of Indian country frequently complicated attempts to ensure tribalism's survival. With no monolithic outlook or universalized approach, American Indians came at the challenges they faced from a multiplicity of directions and perspectives.

It is that very array of attitudes and experiences, of similarities and differences, of successes and failures that makes the twentieth-century history of American Indians so compelling. Individuals and their communities experienced this period as one of disjuncture and continuity, of near-devastation and powerful resurgence. Theirs is a powerfully moving account of pain and persistence that unfolds across generations and across the USA. Sovereignty, education, health care, resource management, and economic development each have their own parts in this narrative. Crafted by many, the American-Indian narrative of the twentieth century incorporates all of the disparate elements that typify anything created by humans, a little at a time. And like all stories from Indian country, it is one best told from the beginning.

"I NEVER FORSOOK THE BLANKET": 1890–1932

In 1879, a young Lakota boy traveled east from the Dakota territory on a terrifying journey that took him into a strange new world. Plenty Kill, the son of Lakota leader Standing Bear, was a member of the first class of students recruited by Richard Pratt for his experimental Indian boarding school at Carlisle, Pennsylvania. Once there and renamed Luther Standing Bear, he was immersed in the federally sponsored assimilation process. In a memoir published in 1933, Standing Bear recalled how the American version of "civilization" denigrated his culture, his language, his religion, and his very sense of self. He struggled with this message at Carlisle and for the rest of his life. It was a private and protracted battle, he wrote, to "retain my individuality and my life as a Lakota" by holding as much as possible to the old ways.[3] Short hair and "citizen's dress" may have changed his outward appearance but, as Standing Bear proudly wrote, "I never forsook the blanket."[4]

Luther Standing Bear's world was one shared by American Indians throughout the USA in the last years of the nineteenth century. The "civilization" juggernaut emerging after the Civil War demanded the surrender of all that defined individuals and communities as Native. According to the American legal system, tribes were "domestic dependent nations" and as such occupied the subordinate position in their relationship with the national government. Federal officials therefore did not hesitate when they instituted assimilation policies and programs intended to eradicate the very "blanket" lifestyle that Luther Standing Bear refused to surrender. But tribalism, much to the surprise and chagrin of American officials and so-called "friends of the Indian," proved itself exceptionally difficult to eradicate.

Native communities across the country struggled, each in their own way and on their own terms, to maintain identity and cohesion in the face of this onslaught that bridged the nineteenth and the twentieth centuries. Coping strategies took some tribal communities down the "white man's road" toward Americanization and others along the path of active military resistance. Some American Indians found ways to resist the pressure to de-Indianize with religious movements such as the Ghost Dance, the Indian Shaker Church, and the peyote religion. Others, such as the Ponca tribal leader Standing Bear, used the American court system to promote Indian tribal and human rights. These decisions, sometimes made after careful deliberation and sometimes made hastily at the end of a bayonet, had a real human cost and cumulative effects on individuals and communities.

Steadily, misery accumulated in Indian country in these years like windblown snow piling up in a blizzard. Any attempt to ascertain just how hard life had become for American Indians by the last years of the nineteenth century is simplified, tragically, by a glance at population statistics. In 1870, the US Census Bureau reported a total American Indian population of 313,712. This number dropped to 306,543 in 1880. In 1890, the same year that the American frontier officially "closed," census records indicated that American Indians numbered fewer than 250,000. Often referred to today as the nadir population figure, this was the lowest number ever recorded for tribal people. For two decades, Native population hovered fairly close to that number. Only after 1910 did a slow but steady demographic rebound begin.[5]

Field matrons, BIA physicians, and other reservation-based personnel compiled empirical evidence indicating that rising morbidity and mortality rates could be linked to both familiar culprits as well as new threats. Epidemic diseases such as measles, influenza, and diphtheria continued to ravage tribal members of all ages just as they had from the onset of European contact. Tuberculosis, diabetes, and alcohol-related illnesses were newer threats proving themselves just as dangerous. And though not lethal, the ever-increasing incidence of trachoma robbed otherwise able-bodied Indians of their vision and opportunities to better their life circumstances.

Facing the dawn of the twentieth century, American policymakers long obsessed with making Indianness vanish suddenly found themselves confronted by Indians who *were* vanishing. What had gone wrong? Were they simply unable to adapt to modernity and therefore doomed to extinction? Or was the process of adaptation itself proving fatal? The Indian school system, Richard Pratt's invention from the 1870s, *should* have been equipping Native children and youth with the skills necessary for survival in the post-tribal world. The agency farmer and field matron programs that began in the 1880s and 1890s *should* have been assisting Indian adults in adopting their own new "civilized" and gender-specific duties. Allotment, one of the primary assimilation engines from the 1880s, *should* have been acculturating American Indians to private property, nuclear family structure, and life unfettered by the bonds of tribal social networks. Missionaries *should* have been bringing Indians into the Christian fold, thus guaranteeing their salvation while encouraging their further separation from the land base. All of these programs and all the best efforts of those working in them *should* have been enough to save the American Indians from themselves and the weight of their cultures. But they were not.

The appalling state of life in Indian country as the rest of the USA entered the twentieth century engendered a wide range of reactions in and outside the federal government. With extinction apparently in the offing, some capitalized on the general public's sudden nostalgia for things Indian. Buffalo Bill Cody and Edward Curtis spoke to differing American perceptions of the so-called "vanishing race," one through carnival spectacle and the other through lyrical, sepia-toned photographic images. In the early years of the twentieth century, Cody continued to tour nationally and internationally with his "Wild West" extravaganzas. Artfully staged attacks on wagon trains and daily performances of Custer's Last Stand thrilled those hoping for a glimpse at "authentic" American Indians. Spectators could encounter Indians from the safety of their seats for a small price.

Visual arts offered viewers a different kind of excursion through the history of Native America. Edward Curtis began his photographic record of American Indians in the late 1890s. By 1906, his images of tribal life and the "disappearing race" won him a lucrative book contract from J. P. Morgan. The North American Indian, a twenty-volume series of pictures accompanied by ethnographic materials published over a twenty-year period, shaped national perceptions of Native life. Acting on a similar impulse to document the "last" Indians, the Bureau of Printing and Engraving commissioned James Earle Fraser to memorialize two of the nation's most endangered species—bison and American Indians—on the Buffalo Head Nickel in 1913. This beautiful coin, in circulation until 1938, made the linkage between the last of the Indians and the last of the frontier permanent for many Americans.

Others who also envisioned a continuing decline for tribal peoples and their communities set about trying to rescue what they could. The Bureau of American Ethnology, an offshoot of the Smithsonian Institution, promoted salvage anthropology as a means of conserving the remnants of dying cultures. A veritable anthropological "who's who" worked under the Bureau of American Ethnology's auspices to gather songs, oral histories, cultural materials, and ceremonial information from a wide range of tribes. Francis Densmore, Alfred Kroeber, Alice Cunningham Fletcher, Franz Boas, James Mooney, Francis La Flesche, and Washington Matthews were among those scholars who worked for decades to preserve information they believed otherwise was destined for oblivion.

Kroeber's efforts at cultural recovery work took an especially poignant turn in 1911, when a mysterious middle-aged Indian man appeared in Oroville, California. Soon known only as "Ishi," he was apparently the sole survivor of the Yahi Indian tribe of northern California. From 1911 until his death in 1916 from tuberculosis, Ishi worked as an informant for Kroeber and lived in the Lowie Museum at the University of California, Berkeley. Surrounded by the inanimate detritus of other cultures, he flaked obsidian points for visitors who flocked to see the human artifact one newspaper labeled "man primeval."

Salvage anthropology, commemorative coins, and so-called human relics notwithstanding, direct federal responses to the declining American Indian population evolved more slowly. While BIA officials carefully tracked the number of children matriculating at Indian boarding schools, the amount of acreage tribal farmers tilled, and the cost of provisions for reservation communities, the agency collected no statistical information about American Indians until the early twentieth century. The first comprehensive study of sanitary conditions in the Indian school system and on reservations was not conducted until 1903. Budgetary constraints meant that an exclusive, federally funded healthcare system for American Indians with physicians, nurses, public-health experts, and hospitals was years in the offing.

Facing a rapidly shrinking American Indian population, the BIA turned to the emerging national public-health community for information and assistance in solving this version of the "Indian problem." Programs such as the "Swat the Fly" (1912) and "Save the Babies" (1912–1918) campaigns were early attempts to reverse the decline in Indian population through education about household cleanliness and scientific childrearing techniques. Emphasizing adherence to Americanized models of domesticity and gender relationships, both initiatives reinforced the notion that tribal women shouldered all the responsibility for the welfare and, indeed, survival of their families. Neither venture made any genuine attempt to address the systemic poverty defining the conditions on most reservations. What good they accomplished was curtailed by federal cuts to the Indian-affairs budget that were coincident to the American entry into World War I.

Much like the actual survival of Native people, the continued existence of Indian autonomy seemed to hang in the balance during the early twentieth century. What hadn't been undercut by nineteenth-century policies such as the Major Crimes Act (1885), the Dawes Act (1887), and the Curtis Act (1898) came under attack again after 1900. Land policy remained a hotly contested arena. Congress continued in the early twentieth century to look for ways to speed up the allotment process, particularly in highly desirable areas such as those still under tribal jurisdiction in Indian territory (Oklahoma). Representatives from the Kiowa, Comanche, and Apache nations protested a unilateral congressional decision to ignore the explicit provisions of previous treaties and to seize tribal land. They brought suit against the US government in a case that ultimately went to the Supreme

Court. The verdict rendered in 1903 in *Lone Wolf* v. *Hitchcock* was stunning to these, and other, American Indian communities. The court held against the tribes and, by extension, the principles underlying treaties. It affirmed congressional plenary power and that body's right to abrogate any treaty provided it acted with "perfect good faith" toward the Indians.[6] The Supreme Court's decision was a severe blow to tribal sovereignty.

The passage of the Burke Act in 1906 was a product of the political elation accompanying the verdict in this pivotal case. Congressional leaders, anxious to capitalize on their plenary powers, authorized the Secretary of the Interior to clear the way for additional sales of land by granting fee-simple title to any "competent" Native allottee. They and commissioners of Indian Affairs who were more interested in supporting Western economic development than sovereignty engineered the sale and leasing of hundreds of thousands of acres of once tribally held lands over the next thirty years.

Roughly simultaneous with this decision to seriously challenge tribal autonomy and erode reservation land holdings, the Supreme Court also took up two cases dealing with Native access and control of vital natural resources. Indians fared somewhat better in this context. In 1905, the justices rendered a pro-treaty decision in *US* v. *Winans.*[7] Yakima Indians in Washington State brought suit after non-Indian private-property holders denied them fishing rights in their "usual and accustomed places." The Supreme Court upheld the language of the 1855 treaty and the principle of "reserved rights." American Indians, the court concluded, retained any and all rights, powers, practices, and privileges not explicitly surrendered in treaty negotiations. Thus, they could continue to fish in their "usual and accustomed places," even if those were located off the reservation.

In the 1908 case *Winters* v. *US,* dealing with the equally precious resource of water, the Supreme Court amplified this argument.[8] Assiniboine and Gros Ventres residents of the Fort Belknap Reservation in Montana had their claims to water from the reservation's Milk River sustained under the doctrine of reserved rights. The court held that Native agreement to treaties requiring their adoption of agriculture clearly, if not overtly, substantiated their right to Milk River water. Denying the citizens and state of Montana the use of prior appropriation principles to usurp tribal water rights, *Winters* v. *US* affirmed both the supremacy of treaty law and American Indian rights. Unfortunately, the Indian victories in neither Winans nor Winters had little immediate or substantive impact on the emerging conflicts over tribal resources. These cases did, however, establish important legal precedents that would be employed to defend Native resource rights later in the twentieth century.

Perhaps as a response to omnipresent tribal concerns, perhaps as a manifestation of an emerging pan-Indian consciousness, and perhaps as a reflection of the national spirit of progressivism, Native national and local political organizations began to develop during the early twentieth century. By the 1920s, scattered through Indian Country, a number of distinctly different groups had coalesced. On Columbus Day, 1911, in Columbus, Ohio, leading American Indian activists such as Charles Eastman, MD (Santee Sioux), Carlos Montezuma, MD (Yavapai), Laura Cornelius (Oneida), and Henry Standing Bear (Lakota Sioux) joined with others to found the Society of American Indians (SAI). SAI founders wanted to introduce mainstream America to articulate, well-educated Natives who had their own ideas about the solution of the so-called "Indian problem." The SAI was an unabashed supporter of assimilation, and its members believed Native communities needed to adapt to changing circumstances. Unlike other reformers, however, they also believed that a strong sense of cultural heritage and Indian identity worked in favor of, and not against, this goal.

Other organizations sought to encourage recognition of and respect for Indian civil and sovereign rights during this period. Far to the north, in the territory of Alaska, the creation of a segregated educational system separating most indigenous children from their Euro-American counterparts in the early twentieth century underscored the marginalization of the Native people there. Recognizing the need for unified action, Tlingit Indians from the southeastern part of the territory joined with members of other Native communities in Sitka in 1912 to form the Alaska Native Brotherhood (ANB) and Alaska Native Sisterhood (ANS). Early ANB and ANS leaders such as Eli Paul (Tlingit) and Elizabeth Petratrovich (Tlingit), graduates of missionary schools, insured that their organizations based their activism on the promotion of assimilation rather than the strict defense of indigenous cultures. The agendas for the ANB and ANS reflected this emphasis on acculturation in their attention to citizenship rights and equal educational opportunities for Alaskan Natives while endorsing the abolition of traditional ways of life.[9]

While internal dissent over issues such as the spread of peyotism and the role of the BIA in tribal life quickly disrupted the SAI and rendered it impotent by the early 1920s, the ANB and the ANS continued successfully to promote Alaskan Native agendas. These two organizations were joined in this locally focused work by the Northwest Federation of American Indians (1914), the Mission Indian Federation (1919), the All-Pueblo Council (1922), and the Navajo Tribal Council (1922). These groups actively promoted tribal sovereignty in disputes over fishing rights, reservation home rule, the control of valuable reservation resources, and the restoration of illegally taken lands.

As the men and women of these early Indian political organizations looked for ways to improve the status of their fellow American Indians, events outside the USA were creating the context for substantial change. America's entry into World War I in 1917 would have profound effects on Indian country. Chief among these was the Selective Service Act of 1917 that established the process of drafting American men for military service. Though most were not citizens, American Indian men were expected to register and, if called, serve in the US Army. This amorphous political space occupied by Natives resulted in a wide range of reactions to the draft. Support for military service varied among the elites of the Indian progressive movement. In the *American Indian Magazine,* Gertrude Bonnin (Yankton Sioux) implored Native men to "stand by the flag...it is your flag."[10]

Others were not so sure. Carlos Montezuma used his newsletter *Wassaja* to criticize efforts to draft Indians, wondering how those legally identified as federal dependents had suddenly become competent enough to go to war. He argued that military service in the absence of citizenship "is another wrong perpetrated upon the Indian."[11] Men from tribal communities as disparate as the Goshute of Utah, the Pamunkey of Virginia, and the Six Nations of New York shared his perspective and resisted the draft.

Thousands of other young American Indian men did decide that the "First Americans" should participate in the American war effort. About half of the approximately 12,000 who served during World War I were volunteers. Like Robert Spott from the Yurok reservation, who won the *Croix de Guerre,* they did not regard patriotism as solely the province of citizens and proudly defined themselves as Americans. Serving in the US Navy and in the Allied Expeditionary Force, they compiled enviable records that were touted at home as proof of Native assimilation into American culture.

Ironically, among the best-known Indian heroes of World War I were those from the celebrated 36th "Panther" Division recruited largely from Oklahoma. Choctaw Indians who still retained their traditional language skills served as the pioneering "code talkers" during the Argonne Forest campaign in 1918. By the armistice, Osage, Comanche, Cheyenne, and Lakota men had joined them, using languages long considered a deterrent to the de-Indianization process in the service of the country that legally still was not theirs.

As the strains of the jauntily heroic "Over There" and more risqué "Mademoiselle from Armentieres" faded after November 1918, many Natives began to reassess their status in American society. On the positive side, participation in the "Great War" resulted in increased visibility and public approval for American Indians. Those who did not engage in active service still supported the cause by purchasing $25 million of war bonds, the equivalent of $75 for every Native person. Tribal women showed their commitment by volunteering in record numbers for the Red Cross. Native farmers joined in the drive to maximize agricultural productivity to feed both the USA and war-torn Europe. All in all, many American Indians across the USA believed that the war finally gave them a chance to publicly and proudly affirm their American identity.

To respond to and acknowledge these efforts, Congress in 1919 extended citizenship to all those Indians who had served as soldiers and sailors. The mechanics of the legislation were cumbersome and minimized the number who eventually applied for citizenship under its auspices. Nonetheless, Indian reform advocates of all stripes hailed the law as a potent acknowledgment of Native importance. Five years later, congressional action finally extended citizenship to all American Indians residing within the USA. The Snyder Act of 1924 was, for many, the long overdue recognition of their obvious status. Others were not so sanguine about its implications. Some Natives regarded the Snyder Act as irrelevant and potentially damaging to people who were already citizens of sovereign nations.

This ambivalence over citizenship hinted at some of the significant undercurrents in American Indian communities. Hidden behind the triumphant red, white, and blue bunting that bedecked postwar America was a Native population in social, economic, and political crisis. The hemorrhaging of tribal lands increased dramatically during the war. The BIA pushed Indians to open individual allotments and communally held properties for leasing by non-Indians, ostensibly to address the need for increased agricultural productivity. The results were dramatic. Euro-Americans farmed approximately six times more tribal land than Indians by 1918. While Native men fought "over there," the federal government accommodated and facilitated lucrative land grabs by non-Indians "over here."

Federal appropriations and BIA personnel were other casualties of the war effort. Spending on health care and education contracted and with it so did the ranks of the Indian Service staff who worked most closely with reservation populations. Medical employees left for military service or to work outside Native communities. Soaring rates of tuberculosis, trachoma, epidemics, and infant mortality were the consequences. As conditions deteriorated, American Indian communities found even the most rudimentary Western-style health care hard to secure. The personal and public health implications of these budget cuts had a devastating and prolonged impact on tribes struggling to rebound from the nadir population point.

While far less dramatic, the decline of the Indian school system was no less significant. Budgetary reductions deemed necessary to support the war effort robbed Native students and their communities of educational opportunities. Cuts in faculty and staff salaries, maintenance, and operating funds hit the chronically under-funded system hard. Hardly models of effectiveness or efficiency, the BIA schools were often the only option available for reservation children. And dwindling funds meant that those children who lived near non-Indian communities might be mainstreamed into hostile public-school systems. Few were prepared for the academic and social upheaval accompanying this haphazard integration. Indian education, once regarded central to the success of the assimilation campaign, languished during and after World War I.

Increasingly in the early 1920s, then, it seemed as if the postwar era would prove a difficult period for American Indians. This also was true for those living outside the continental USA. During that decade, Alaskan Natives furiously defended their civil rights in a territorial political system more than ready to accord them second-class status. Politics took a different turn in Hawaii. In 1921, Congress passed the Hawaiian Homes Commission Act. It was intended to establish an excusive land base on which Native Hawaiians could promote economic development. The law may have been envisioned as a stepping stone to community stability, but many Americans—including members of some of those families who had engineered the kingdom's annexation—believed it threatened their property rights and privileges. Native Hawaiians placed their names on ever-lengthening rosters and waited to receive their allocation of land; today, some of their descendants are still waiting.

Waiting was not the strategy employed by members of the New Mexico Pueblo communities when, in 1922, congressional action challenged their tribally held lands. The Bursum Bill required the Indians of the state's pueblos to prove their ownership rights. If they could not, non-Indians would receive title to the land they claimed under the principle of squatter's rights. The SAI worked with the All-Pueblo Council, the venerable Indian Rights Association (IRA) and the newly organized American Indian Defense Association (AIDA) to increase public awareness and sympathy for the Pueblo Indians' plight and defeat the Bill. Hoping to counter this support, the BIA began to spread rumors about the alleged immorality of traditional Pueblo religious practices. Though the disinformation campaign failed, it did result in the decision to issue BIA Circular 1665. This extended a ban on ceremonies, dances, and rituals conducted in *any* American Indian community. Apparently, American citizenship was no guarantee for American Indian citizens that their religions and their ceremonies would be protected from outside interference or suppression.

American Indians recognized that the Bursum Bill and Circular 1665 both highlighted their precarious political status in the late 1920s. Pervasive destitution, public-health problems bordering on the catastrophic and cultural disintegration drove home the same messages about their very survival. But most non-Indian Americans, including those in policymaking positions, did not realize how close to the edge Native communities really were. Or, worse, they saw the problems but ascribed to the familiar ethnocentric stereotype that Indians were quite content to live in poverty and squalor. In response, the Board of Indian Commissioners urged the Department of the Interior in 1926 to conduct a focused study of the conditions in which American Indians lived. The results could be used to counter ethnocentric assumptions and to provide valuable data for those advocating Indian reform measures. Researcher Lewis Meriam worked for two years with Indian adviser Henry Roe Cloud (Winnebago) to survey Indian country for the Brookings Institute. The results of his work became a substantial document entitled The Problem of Indian Administration, later known simply as the "Meriam Report."

Over 872 pages of meticulous of research in the Meriam Report unfold a story that still makes for heartrending reading. In measured, dispassionate prose, the author described for his readers the myriad problems facing American Indian communities. He described the ongoing and genuine "suffering and discontent" common among those living in Indian country. Every category of evaluation—sanitation, health, housing, economic status—indicated that federal policies, mismanagement, corruption, and ineptitude were responsible for their appalling conditions. American Indians had been the victims, rather than the beneficiaries, of each attempt to accomplish and speed assimilation. Their journey along the much-vaunted "white man's way" had by 1928 turned out to be a one-way trip down a dead-end street.

If anyone in the BIA hoped that Meriam's research would vindicate nearly a century of assimilation efforts, they were sorely disappointed and publicly humiliated by the results. The release of the report unleashed more fierce criticism against the BIA, the Department of the Interior, Congress, and every federal official who had anything to do with Indian Affairs. This time, detractors were armed with the Meriam Report's documentation. Leaders in the IRA, the General Federation of Women's Clubs, the National Council of American Indians (1926), and the AIDA all expressed their outrage and determination to see American Indians and their communities redeemed. In a pamphlet entitled "Our Indian Citizens: Their Crisis," the AIDA served notice to the American public and particularly to the 1928 presidential candidates that the problems plaguing Indian country demanded immediate action.

There's no question that the Meriam Report generated positive political momentum for creating innovative solutions to the historic but newly rediscovered "Indian problem." Initially, Republican President Herbert Hoover won widespread approval with his appointment of Charles Rhoads, the former president of the IRA, as his commissioner of Indian Affairs. He took up the call for change and began working to mend the federal Indian policy system. Rhoads reached out to tribal leaders and looked for new ways to build positive connections between them and the federal government. And he selected W. Carson Ryan to head the Indian school system and embraced his strategy for improving and "re-Indianizing" the educational process. But once the economy began to sour, American Indians and their concerns took a back seat to the ever-worsening national calamity.

When did the Great Depression arrive in Indian country? For most American Indians, there was no single moment that heralded the collapse of tribal economies. As the Meriam Report had demonstrated so vividly, most tribal people lived far beyond the economic margins. By the eve of the 1932 presidential election, conditions for American Indians had become desperate. State and charitable agencies, their resources stretched to the breaking point, could not even begin to address the needs extant in Indian country. Reservation per-capita incomes ranged from $135 among the Navajo to less than $28 for the Catawba of South Carolina in the first years of the economic crisis. Half of all American Indians had no land on which they could raise food for themselves or their families. In 1931, one year after Congress authorized more than $1 million in emergency aid to reservations, the BIA asked the US Army and the American Red Cross to distribute food among starving residents in tribal communities. Circumstances in the early 1930s were no less dire than between 1890 and 1910, when American Indian survival seemed to hang in the balance. And, as in that earlier dark period, the solutions seemed equally elusive.

"SON, YOU ARE AN INDIAN": 1932–1969

In November 1969, a small group of American Indian activists claimed Alcatraz Island, site of the former federal penitentiary in San Francisco harbor, on behalf of all American Indians. Don Patterson, a Tonkawa Indian raised in Oklahoma, was one of them. Speaking about the event thirty years later, Patterson noted that publicity about the takeover brought people previously reluctant to claim, or ignorant of, their tribal identities to act publicly in support of the occupation. In particular he remembered one young man who arrived on the island and reported that he had only learned of his Comanche heritage a few days earlier. Watching televised news coverage with his mother, he had expressed a desire to help if only he could claim some Native identity. Patterson could still recall the young man's account of his mother's response: "Son, you are an Indian." His mother was a Native relocatee to the Bay Area who had, until then, never discussed her ancestry. Moved by his interest and perhaps by her own sympathy for the activists, she chose that moment to reclaim her Comanche identity for herself and her son.

Patterson's amusing and poignant story of one young man's discovery of his tribal heritage speaks to far more than the revelation of a family's hidden history in 1969. From the 1930s through the 1960s, American Indians experienced three decades of dramatic and profound change during which their identities became politicized in previously unanticipated ways. Economic and political chaos again ushered the world into war, this one an unprecedented orgy of violence that owed its origins partly to ideological, ethnic, racial, and religious hatreds. Americans furiously debated the nature of citizenship at home after World War II while aggressively countering the twin perceived threats of Communism and decolonization abroad. Against this backdrop, American Indians explored and tested the inherent contradictions and surprising continuities of late-twentieth-century indigeneity.

When Richard Oakes (Mohawk), Adam Fortunate Eagle (Red Lake Chippewa), Grace Thorpe (Sac and Fox/Potowatamie), Dennis Hastings (Omaha), and other Indian activists landed at Alcatraz on November 9, 1969, they did so under the legal authority of the 1868 Treaty of Fort Laramie. Then more than a century old, this document ended the Powder River War, a bloody conflict that had pitted an unsuccessful US Army against warriors led by the Lakota leader Red Cloud. The treaty ceded abandoned American installations along the Bozeman Trail in Montana to the Lakota and Arapaho victors of the conflict. One hundred years later, Indians used specific provisions of the treaty that ended an "old" Indian war to legitimize their claims in what many saw as the first battle of the "new" Indian war.

While the Fort Laramie treaty had the starring role in the 1969 activism, other historic Indian policies helped shape the actions and rhetoric of the American Indian protesters. Some of the impetus to exercise sovereign rights in such a public and muscular fashion can be traced back to one document with a more modern provenance: the Indian Reorganization Act. Termination policy and federal relocation efforts also figured into the Alcatraz equation. By implementing those controversial policies in the 1950s and 1960s, the BIA quite unintentionally set in motion a chain of events that would result in a predominantly urban Native group seizing a derelict federal installation to protest threats to tribal culture and sovereignty. Much of what defines Indian country politics today can be directly linked to the period book-ended by the Indian "New Deal" of the 1930s and the 1960s surge of Indian activism. This period was the crucible in which modern Indian identity was formed.

In some ways, the American Indian protest activities that transfixed the USA during the 1960s were directly connected to the terrible economic crisis of the 1930s. Franklin D. Roosevelt's election to the presidency in 1934 was largely attributable to the Great Depression that gripped the nation and the world. Americans embraced his call for a "New Deal" and his inaugural promises to help all members of American society. And for once, those promises included American Indians.

Franklin D. Roosevelt's plans for Indian country were made abundantly clear soon after he took office. Harold Ickes' appointment as Secretary of the Interior was the first tangible signal that real change was on the way to Indian country. Ickes was an AIDA co-founder and a long-standing BIA critic who, with Roosevelt's designation, moved into an unparalleled position to make a difference. Soon after this encouraging selection, Roosevelt signed off on the nomination of John Collier for Commissioner of Indian Affairs. If Ickes' presence at the Department of the Interior echoed the call for change, Collier's assumption of leadership at the BIA constituted a full-throated cry for revolution.

Collier's first order of business as BIA chief was securing the extension of New Deal relief efforts into Indian country. The Indian Arts and Crafts Board was a unique federal attempt to promote the training of tribal artists and the marketing their work to the public. National programs such as the Public Works Administration, the Works Progress Administration, the Resettlement Administration, and the Civil Works Administration eventually employed thousands of American Indians whose wages infused crippled tribal economies with much-needed cash. Indian workers in these programs laid sewer lines, dug wells, reseeded acreage, built irrigation systems, and constructed roads that improved reservation conditions.

The Civilian Conservation Corps–Indian Division (CCC-ID) arguably had the most impact on Indian country of all the federal relief programs. Within its first six months of operation, the CCC-ID hired more than 14,000 Indians who worked on reservation conservation and reclamation projects across the USA. By 1941, that number would grow to more than 80,000 with many employed in supervisory and administrative jobs within the program. Along with their regular work, CCC-ID personnel could enroll in evening classes to acquire more training in English and useful skills such as drafting. Their salaries, like those of the other Indian relief workers, went home to help families and communities. As Vine Deloria, Jr., later wrote, the CCC-ID helped the Lakota climb "from absolute deprivation to mere poverty."[12]

Immediate economic relief to Indian families and communities was wonderful, but Collier's ambitious long-term goal was the reformation of the general framework of American Indian policy. To that end he collaborated with Senator Burton K. Wheeler (Montana) and Congressman Edgar Howard (Nebraska) to draft legislation designed to finally bring Indians into the policymaking process as equal participants rather than subordinates. This ultimately became the Indian Reorganization Act and was passed in 1934.

Collier's original ideas for reforming federal Indian policy included the creation of a national Indian court, support for traditional cultures, tribal control over BIA budget expenditures, promotion of tribal self-governance, and the extension of tribal control over allotted lands. Many of these did not make it into the Indian Reorganization Act. What survived the vetting and legislative processes embodied something of Collier's desire to transfer power to tribal governments and to undo the worst excesses of the assimilation era. The Indian Reorganization Act halted the policy of allotment, allowed for the consolidation of allotments into communally held lands, promoted the organization of tribal governments along corporate lines, and increased federal funding for Indian education in critically important areas such as resource management and public health. Its provisions promised to empower Native nations and to chart a new direction for federal–Indian interaction.

Reaction to the Indian Reorganization Act and the reform agenda it incorporated varied greatly in Indian country. Some Native individuals based their critique of the Indian Reorganization Act on Collier's apparent desire to unwrite the history of the preceding half-century by undoing assimilation programs. Navajo leader J. C. Morgan argued that the legislation encouraged federal paternalism and pushed his tribe into greater dependency. Cahuilla Indian Rupert Costo of California castigated Collier's methods and called his policies the "Indian Raw Deal." Seneca Alice Lee Jemison complained that the Indian Reorganization Act's emphasis on the recovery of Indian languages and the recreation of a traditional land base required Natives to live in an overly idealized state that bordered on Communism.

Ambivalence about the Indian Reorganization Act extended beyond individuals to entire communities. Many Indian tribes argued that treaty law, not congressional legislation, was the basis of their relationship with the USA and should stay so. The Oneida of New York demanded that the Treaty of Canandaigua, not the Indian Reorganization Act, govern its interaction with the federal government. From their perspective, the "old deal" struck in 1794 remained intact and applicable. No "new deal" was required, they believed, if the government lived up to its treaty-based promises.

Other American Indians believed that Collier's plans were in their best interests and that the tribes would truly be getting something new and important if they adopted the Indian Reorganization Act. Joe Hillaire of the Northwest Federation of American Indians characterized Collier's reform agenda as a hopeful sign. Antonio Luhan from Taos Pueblo lobbied other Indians for the BIA and encouraged their support for the law. He told them that Collier's Indian Reorganization Act was a protective barrier sheltering them from the excesses of the assimilationists still seeking to destroy tribal culture. Lakota Alfred DuBray believed that the advantages associated with the Act, especially the agricultural and educational loans, really did offer their tribe a "new deal." And Tuscarora anthropologist J. N. B. Hewitt argued that the new governments could end the corruption and tyranny that plagued some Native communities.

Buoyed by his supporters and undeterred by his critics, Collier pushed the Indian Reorganization Act's governmental reorganization through in Indian communities across the USA. Within two years of the law's passage, 181 tribes voted to accept its provisions and seventy-seven rejected the Indian Reorganization Act. These hotly contested referenda sometimes featured the use of political techniques that advocates as well as detractors found unsettling. The tenets of the centralized, constitutionally based system at the heart of the Indian Reorganization Act's governmental renovation scheme were not easily translated in communities where consensus traditionally dominated political discourse. Did those who chose to sit out the vote on the new tribal governments actually vote yes, no, or neither? According to Collier and the BIA, non-votes could be counted as yes votes, and were. Given this creative tabulation strategy, it is not surprising that many Indians believed the political process had been rigged in favor of a positive outcome.

Sixty years later, the report card for the Indian "New Deal" presents a mixed record. On the plus side are some notable accomplishments. Relief work clearly benefited many tribal communities. Collier's aggressive promotion of increased funding for the existing Indian schools did actually enhance the quality of the educational experience for hundreds of students. The BIA's increased attention to tribal health care resulted in an expanded medical and public health staff offering an increased array of services to a larger segment of Indian country residents. Unthreatened by allotment and fraudulent land sales, more Indians owned more land and more tribes controlled more property and resources. Native communities tapped into federal funding sources and strengthened their economies. And, of course, the long delayed recognition of the legitimacy of tribal cultures acknowledged their centrality to the Native world.

On the other side of the Indian "New Deal" report card is a fairly substantial list of deficits. Tribal communities could not define their own models of government because Collier's "one size fits all" bureaucratic framework was not flexible enough to accommodate them. The fact that the Indian Reorganization Act governmental model was not of tribal origin was particularly galling to Indians. The process of establishing and adopting a tribal constitution privileged some American Indians—especially those who were comfortably literate in English—over others. Certainly, the heavy-handed treatment of the voting process raised eyebrows, hackles, and ultimately concerns about the reality of self-government. The omission of Oklahoma Indians and Alaskan Natives from the original Indian Reorganization Act put them at a disadvantage that had to be rectified with the passage of additional legislation. His controversial Navajo stock-reduction program created long-lasting resentments. Finally, Collier's abrasive personality and inability to countenance disagreement did more to sow doubt and encourage cynicism than it did to persuade potential Indian allies that they could trust him, his programs, and his agency.

Of course, the Indian "New Deal" was no more than rhetoric and legislation without American Indian acceptance and participation. Native nations, not Collier, ultimately determined its success or the failure. Some tribes, including the Klamath, the Navajo, and the Crow, feared a loss of sovereignty and said no to the Indian Reorganization Act. Others such as the Stockbridge–Munsees, the Poncas, and the Seminoles said yes because they saw the chance to enhance their autonomy. American Indians spent the rest of the 1930s sorting out how to best take advantage of more direct connections with the federal government and the chance to reestablish a coherent tribal land base. The Puyallups, reduced to pauperism on a 60-acre reservation by allotment, used the Indian Reorganization Act to regroup themselves socially and politically. The tribes decided whether reorganization and corporate structure enhanced or detracted from historic Native identities. Disagreements between "New Dealers" and "Old Dealers" abounded on reservations across the country where differences of opinion morphed into disputes over which side was more authentically Indian. And American Indians did all of this against the backdrop of a continuing national economic emergency and the rising international political crisis.

As tensions in the Pacific and Europe escalated in the late 1930s and early 1940s, American Indians and their communities assessed the prospect of war from a number of different viewpoints. The peacetime draft, instated in October 1940, once again brought questions of identity and sovereignty to the fore. Some, like men from the Tohono O'odham reservation in Arizona and the Six Nations in New York, believed that their distinct legal status under treaty law precluded such demands by the federal government. Most Native draftees, though, defined military service as a way to protect their families and communities from potential harm. Regardless of their tribal affiliation, they saw no real contradiction between being an Indian defending sovereignty and being an American serving "Uncle Sam."

American Indian men and women went to war in a variety of ways. On some reservations, as many as 25 percent of the able-bodied residents left home to participate in war-related work. Of the total American Indian population, perhaps as many as 25 percent of the women lent their efforts to the cause. As civilians, both men and women took jobs in airplane factories, shipyards, munitions factories, and other production sites for war materiel. They worked in mines, in sawmills, as truck drivers, and as farmers. Their salaries fueled continuing economic recovery in countless communities. And their increased visibility enhanced the larger national reputation of American Indians.

Of course, those Indian men and women who joined the armed services had the greatest visibility. During World War II, some 25,000 served in the American military. Native women who enlisted in organizations such as the Women's Army Corps and the Women's Air Force comprised 800 of that number. Fighting in all theatres of war, Indian soldiers distinguished themselves. Smiling up from the pages of newspapers and magazines, their uniformed images reminded the general public of American Indian patriotism. The BIA regularly publicized the exploits of men such as Ernest Childers (US Army, Creek), Ira Hayes (US Marines, Pima), Henry Nolatubby (US Navy, Choctaw), and William R. Fredenberge (US Army Air Corps, Menominee) as exemplars of Native manhood. By the spring of 1945, the agency's honor's list for American Indian service personnel included seventy-one Air Medals, fifty-one Silver Stars, forty-seven Bronze Stars, thirty-four Distinguished Flying Crosses, and two Congressional Medals of Honor.[13]

The Navajo "Code Talkers" earned the greatest notoriety as American Indian participants in World War II. Like their Native counterparts in World War I, the Navajo used their linguistic skills to assist with military communications in the field. Unlike other tribal groups such as the Oneida who used their languages to convey general military information, the Navajo employed a unique code designed for them. The US Marines established the Navajo code program, trained over 400 soldiers by war's end, and dispersed them throughout the Pacific theatre. In a battle zone where communications were dangerously permeable, the Navajo (and their language) were a tremendous asset to the American military effort.

World War II demanded home-front sacrifices of all kinds, and American Indians were not exempted from them. Any and all governmental expenditures not deemed essential were redirected to the defense budget. By 1942, spending cuts of 15 percent were in place, and the federal government had actually moved the entire BIA infrastructure to Chicago to open up more space for war workers in Washington, DC. For the duration of the war the BIA functioned as an absentee bureaucracy with a considerably reduced base budget.

Financial shortfalls and the complications associated with the relocation of this agency were not the only problems facing Collier and the BIA during the war years. Support for the Indian Reorganization Act had begun to erode even before 1941. One of the Bill's sponsors, Senator Burton K. Wheeler, started to advocate for its repeal as early as 1937. Collier spent considerable time during the late 1930s and early 1940s trying to keep up the momentum for his reorganization plans. But by 1945 he and his Indian policy had been left behind by changing times and changing perspectives. He resigned from the BIA that year, unsure if any of his reform plans would survive in postwar Indian country.

As it turned out, Collier's concerns were well founded. Senator Wheeler's change of heart about the Indian Reorganization Act manifested itself in repeated calls during the war for the reduction of BIA services to tribal communities. He also began to discuss ending, or terminating, federal guardianship over American Indians. The idea garnered increasing interest as the war effort whittled down the BIA and reduced its effectiveness. Opposition to the strategy came mainly from the Association of American Indian Affairs, a group whose roots dated back to the fight over the Bursum Bill. They argued successfully at first in favor of sustaining the federal commitment to tribal people. Collier's departure, however, cleared the political and ideological space necessary for termination to resurface as federal policymakers considered the place of Indians in postwar America.

As the long war wound down to an end, many Americans believed that the following era would be a time of uncertainty. Would the wartime economy's robust strength continue in peacetime? How would the demobilization of millions of service personnel affect the nation? What role would America play in the postwar world? What would happen to the terrifying new source of power revealed so dramatically at Hiroshima and Nagasaki? American Indians had similar concerns and asked those same questions. But they also had a list of their own, specific to Indian country. How would the end of the war affect federal policies? What would happen to tribal economies and tribal governments? Could the war's rhetoric of freedom and self-determination expand to include both Native civil and sovereign rights? And what did it mean to be an American Indian in the postwar world?

In November 1944, representatives of more than fifty Native tribes assembled in Denver, Colorado, to address some of those questions. Their goal was to create an organization that would allow Indians to set their own agendas and develop tribal solutions to tribal problems. Their deliberations resulted in the formation of the National Congress of American Indians (NCAI). The NCAI immediately and self-consciously set itself apart from its predecessor, the SAI, by limiting membership exclusively to Natives and focusing on the "everyday" civil and sovereign rights issues important in most tribal communities. NCAI leadership and members came from widely different communities. There were women and men, full bloods and mixed bloods, old and young, reservation and non-reservation dwellers, the unschooled and the graduates of the Indian school system. The organization did not intend to replace local groups such as the All-Pueblo Council or the ANB. Charter NCAI members instead wanted to a national political platform for the promotion and protection of Indian interests. Little did these women and men know how quickly federal policies would test their resolve.

In a history replete with ironies, one of the greatest is that the path to termination began with John Collier's reform agenda. The Indian Reorganization Act contained provisions for tribes to sue the federal government for lands illegally taken and reservation funding that never arrived. In 1946, a variety of congressional supporters joined forces to enact the Indian Claims Commission Act. The Act, which was supported by the NCAI, created a road map for tribes to bring suit against the USA. American Indians could file their claims in a five-year window between 1946 and 1951. No claims filed after 1951 would be considered. Wildly optimistic policymakers expected the claim assessment process would conclude in 1956; the Indian Claims Commission (ICC) required multiple congressional extensions and finally completed its work by 1978! Tribes had to prove aboriginal title to the lands they claimed and would only receive cash payouts (from which federal expenditures on their behalf could be debited). The ICC could not return any land to any tribe. Indian nations filed 850 suits with the ICC; only 484 were adjudicated. Of that number, tribal claimants received 285 favorable decisions. The ICC distributed awards totaling $657 million. This seemingly enormous sum actually was the equivalent of $1,000 for every American Indian in the USA in the 1970s.

The adjudication of historic Indian land claims occurred in a political environment much different from the 1930s. As the ICC received Indian suits, a conservative Cold War-era Congress moved ahead with the second step in an emerging three-part strategy to get the federal government "out of the Indian business" forever. Neo-assimilationists who likened tribal communities to "communistic" social organizations decided that a legislative approach offered the best and most certain solution to the "Indian problem." The Indian "independence movement" supporters such as E. Y. Berry, Dillon S. Myer, Arthur Watkins, and Glenn Emmons planned to use termination to cut all federal ties to the tribes and free them from wardship. Compensation would be redefined to constitute phase one of the process; termination was phase two.

Congress enacted termination through two companion laws passed in 1953: HCR 108 and PL 280. HCR 108 severed relationships with selected tribes, transforming politically distinct American Indian nations into ethnic or racially defined groups with no special claim on the federal government. No de-Indianization plan ever devised before by Congress could compare in result to HCR 108. Its sole purpose was to extinguish sovereignty. Tribes located in California, New York, Texas, and Florida were to be terminated, as were the Flatheads (Montana), the Klamaths (Oregon), the Menominees (Wisconsin), the Potowatamies (Kansas and Nebraska), and the Turtle Mountain Chippewas (North Dakota).

The companion piece, PL 280, reassigned civil and criminal jurisdiction over the Indians identified in HCR 108 to those respective states. With that one law, Congress undid more than a century of legal precedent, robbing Natives and autonomous tribal communities of their legal rights. Arthur Watkins, a US senator for Utah and ardent advocate of termination, later floridly compared PL 280 and HCR 108 to the Emancipation Proclamation. With termination, Watkins said he saw in fiery words about the heads of Indians the phrase, "These People Shall Be Free."[14]

It's highly unlikely that American Indians thought of Abraham Lincoln as they faced termination and the imminent loss of their sovereignty. Indeed, it's probable that the policy—which had originally been called "liquidation"—conjured up more recent and far less pleasant historical associations for them. Some tribes did favor the policy initially as they hoped for an end to federal paternalism. But during implementation individuals and communities protested what they saw as heavy-handed attempts to erase their very existence. The latter was especially true if their reservation lands boasted valuable resources. Then Native communities discovered to their dismay that compensation had become termination's handmaiden: to get the cash award, some tribes had to agree to termination.

The third and final phase of the postwar assimilation campaign was relocation. Urban living was nothing new for Indians. From residence in colonial towns to more recent experiences linked to war work, American cities almost always had tribal residents. In 1951, Commissioner of Indian Affairs Dillon S. Myer institutionalized federal support for Indian urbanization by establishing the Bureau of Placement and Relocation. Over the next decade, an estimated 30,000 young men and women from reservations across the USA were given bus tickets to cities such as Cleveland, Denver, Los Angeles, and Phoenix. The great American urban melting pot would take care of the rest.

Relocation's goal was to immerse those Natives most open to change in environments where they could discover all of the benefits associated with city life. The relocatees were told they might find good jobs, better housing, better educational opportunities, and greater access to mainstream America. They received some support from federal Indian centers that were opened in selected cities. Some did find a better lifestyle and stayed to create urban Indian enclaves in places like Minneapolis–St. Paul, Oakland, and Chicago. Others found employment opportunities but never fully surrendered ties with their reservation homes. More wound up marooned in cities where discrimination on the basis of race and ethnicity in housing, education, and employment remained legally accepted practice in the 1950s and early 1960s.

Termination and its associated policies infuriated American Indians who went public with their outrage at the audacity and breathtaking rapidity of the postwar assault on tribalism. Blackfeet elder Earl Old Person asked federal policymakers "how can we plan our future when the Indian Bureau threatens to wipe us out as a race?"[15] Long-time Seneca activist Alice Lee Jemison reminded a senate committee that "no Indian asked for the introduction of these bills [...] we have kept our shares of the treaties, and we are here to ask that you keep yours."[16] The NCAI galvanized non-Indian public attention by noting that termination "hit lists" were compiled without tribal input. Moreover, the organization explained that tribes had no protection from over-zealous BIA personnel anxious to "liberate" them. Rather than a solution to the historic "Indian problem," federal officials belatedly realized that their actions set the stage for an entirely new set of problems with Indians.

Termination directly affected 109 American Indian tribes and constituted a political turning point for nearly all American Indians. The threat of termination kindled the flames of postwar activism, and Indian country at that moment was nothing if not a fuel-rich environment. From the mid-1950s to the mid-1960s, American Indian determination to defend civil and sovereign rights expressed itself in lobbying, court cases, and direct engagement. Sometimes this took the form of independent individual protest; sometimes it manifested itself in the coordinated efforts of national organizations. Either way, American Indians began a new period of public activism in which they, speaking as both fellow citizens and as citizens of sovereign nations, addressed the larger non-Indian society.

For Alaska Natives and Native Hawaiians, such as the Indians in the continental USA, the changes begun during World War II continued unabated during the decades after it. Non-American Indians flocked to these wildly different territories during and after the war, resulting in statehood for both Alaska and Hawaii in 1959. The postwar population boom in both placed increased pressure on Alaska Natives and Native Hawaiians to assimilate into mainstream culture and to forgo any concerted efforts to secure those rights unique to their indigenous status. Access and title to land, educational access, and civil rights became their priorities and demanded their constant attention.

Natives in the "lower 48" during the postwar period increasingly tested the boundaries of American citizenship and the limits of their civil rights. Arizona, New Mexico, Utah, Washington, and Idaho continued to deny them the right to vote after World War II because the states challenged their legal status. Protesting their exclusion from the American political process, Indian men and women began to press for relief in the state and federal court systems. In 1948, in *Harrison* v. *Laveen,* the Arizona Supreme Court ruled that residence on a reservation did not automatically disenfranchise prospective Native voters.[17] Utah was the last state to extend full suffrage in 1956.[18] New Mexican legal resistance to Indian suffrage continued intermittently until 1962. It took the passage of the 1965 Voting Rights Act to reassure many American Indians that their franchise rights had some level of federal protection.

Along with civil rights, the courts remained the principal arena for Indians to contest their treaty rights. In 1963, the US Supreme Court ruled on a long-standing dispute between Arizona and California concerning water from the Colorado River. Those two states, and the others in the Colorado drainage area, were constantly fighting over usage rights. The court addressed these issues and examined the status of tribal water rights for reservations along the lower Colorado. The court's decision resurrected the reserved rights doctrine first articulated in *Winters* v. *US* and defended Indian claims to the Colorado River.[19] They, not the states, received first right to the water. And they received water allocations sufficient for all practicably irrigable acreage, not just what was then under cultivation.

Treaty rights also emerged as a substantial arena for activism outside the courts during this same period. Borrowing a strategy from the African-American civil-rights movement, members of Indian communities in the Pacific Northwest staged a series of "fish-ins" to defend rights acknowledged in treaties dating to the 1850s. Natives courted arrest and acquired publicity by fishing in rivers throughout the region without licenses or using traditional technologies such as gill nets. Celebrities such as Marlon Brando, Jane Fonda, Dick Gregory, and Buffy Sainte-Marie (Cree) cheered on the protesters from the shorelines and sometimes accompanied them, braving violence at the hands of game wardens, police, and non-Indian fishers anxious to maintain the status quo. Largely symbolic, the "fish-ins" alerted the American public to the reality of tribal sovereignty and the emerging cadre of young Indian activists.

The young men and women advocating direct action like the "fish-ins" were members of a new generation of Natives who were frustrated with the pace and the policies of older Indian leaders and organizations. Meeting first in June 1961 at a gathering of American Indians in Chicago, budding activists such as Clyde Warrior (Ponca), Herbert Blatchford (Navajo), Melvin Thom (Paiute), and Shirley Hill Witt (Akwesasne Mohawk) reconvened that August in Gallup, New Mexico, to found the National Indian Youth Council (NIYC). The NIYC represented young Indians who were disenchanted with tribal leaders, the IRA tribal governments, and the NCAI for its cautious tactics. Labeling some tribal leaders as "Uncle Tomahawks" and accusing them of being "apple Indians—red on the outside, white on the inside," they served notice to the larger Native community that the old pan-Indian political system had to expand to accommodate new leaders, philosophies, and strategies.

Anyone in the early 1960s wondering what philosophy would guide the new Indian activists had only to glance at the eloquent 1961 "Declaration of Indian Purpose" to learn they wanted the free exercise of cultural and spiritual rights, the protection of trust and treaty lands, civil-rights protection, and the clear recognition of tribal rights to self-government. This ambitious statement clearly signaled the importance of both civil and sovereign rights to this new group of leaders. Similarly, the early "fish-ins" quickly indicated to others how the younger Native generation planned to advocate for those rights.

In fact, Melvin Thom encouraged NIYC members to embrace direct action, arguing that "it is time to raise some hell."[20] Raising hell could mean occupying Alcatraz for the first time in 1964 and again in 1969. Raising hell could mean publicly and persistently questioning tribal governments. Raising hell could mean calling out Congress and the BIA for their actions. Raising hell could mean embracing pan-Indianism while maintaining familiar tribal traditions. Raising hell ultimately did mean reminding other Indians and everyone else involved in Indian affairs that, as Vine Deloria, Jr. (Standing Rock Lakota) had said, "The days of Sitting Bull and Geronimo and Crazy Horse may be gone, but there *is* an Indian Revolt and the 20th Century warriors are deadly serious."[21]

"KEEPING THE NAME ALIVE": 1969–2000

One of those early and serious young warriors was a woman named Wilma Mankiller. In 1987, she became Principal Chief of the Cherokee Nation. She was the first Cherokee woman elected to hold that political position. Mankiller, who was born in 1945 in Oklahoma and relocated to San Francisco with her family by the BIA in 1956, became involved in Indian activism when she was in her twenties. In 1969, she was one of the "Indians of All Tribes" occupying Alcatraz. Her experiences there determined her political future. Mankiller returned to Oklahoma with her children in 1977 and went to work for the Cherokee Nation. Six years later, she was elected Deputy Chief of the Cherokee in the administration of Principal Chief Ross Swimmer. She became the acting chief for the tribe when Swimmer left to head the BIA in 1985. Two years after she won on her own right and did so again in 1991 with 83 percent of the vote.

Despite the Cherokee's historic support for women in leadership roles, the post-removal power structure that had become ensconced in Oklahoma was decidedly male-dominated and oriented. One of Principal Chief Mankiller's first priorities was to address the barriers limiting women's participation in modern Cherokee politics. She later joked that her name helped counter resistance; she sometimes told those who opposed her that it had been "well earned." Knowing that the name Mankiller had been carried down to her over the generations by Cherokee men and women who constantly fought to protect their identity helped, she once wrote, to keep her resolve strong. To honor those before her, her job was to "keep the name alive."[22]

Keeping their names, their families, their communities, their cultures, their religions, and their sovereignty alive continued to be the priority for American Indians during the last decades of the twentieth century. As usual, this required individuals and tribal communities to grapple with the role of the federal government in their lives. Lingering doubts and profoundly deep cynicism colored government-to-government relationships even when the USA seemed ready to support American Indian interests. Pan-Indianism remained a powerful and sustaining political force in Indian country during this period. But some of those same doubts and cynicism spilled over into intratribal and intertribal conflicts. On those occasions, whether in the 1960s or the 1990s, differing Native agendas motivated divergent Native responses.

Melvin Thom's challenge to American Indian to "raise some hell" in the 1960s did not go unanswered or unchallenged. Throughout the 1960s and 1970s, young Native men and women undertook a series of direct actions that drew public attention to civil and sovereign rights concerns. America, they believed, needed to be educated to the realities of Indian life and the rights of Indian nations. Increasingly, these activists associated themselves with an organization that had been founded in Minneapolis–St. Paul but was gaining in influence nationally, the American Indian Movement (AIM). More so than groups such as Indians of All Tribes or Women of All Red Nations, AIM would become the national symbol of the "Red Power" movement.

The AIM's emergence coincided with a period of significant social upheaval in the USA. Between 1968 and 1970, Clyde and Vernon Bellecourt (Chippewa), Dennis Banks (Chippewa), John Trudell (Santee Sioux), and Russell Means (Oglala Lakota) became the AIM's first leaders. They incorporated ideas and programs from other protests group into theirs. They ran street patrols like those of the Black Panthers that helped to lessen incidences of urban police brutality against Indians. They created the Little Red School House to address the racist treatment of American Indians in the American educational system. They reached out to reservation dwellers and looked for connections to tribal traditions. And, above all, they relished opportunities to employ political theatre.

Charismatic public figures, the "shades and braids" members of AIM, took their activism to the streets in ways that other Natives criticized as "un-Indian." In 1970, AIM made national headlines by capturing the *Mayflower II* as a protest against the observance of Thanksgiving. The next year, they brought attention to the long-standing controversy over the Black Hills when more than twenty Natives camped out atop Mount Rushmore. National media attention brought the debate over sovereignty into American homes via the evening news programs. These activities and the attention they garnered also attracted the interest of the Federal Bureau of Investigation (FBI). It soon decided that the group was a threat and began to engage in infiltration and surveillance efforts.

AIM's use of direct confrontation and its predisposition for delivering withering critiques about existing tribal and federal power structures did little to endear it to members of the NCAI, the newly formed National Tribal Chairman's Association, or even its closest predecessor, the NIYC. They were accused of being federal puppets, of moving too slowly and too cautiously, and of trading sovereignty for federal funds. NIYC members who once accused tribal leaders of being "apple Indians" now found themselves the target of that same epithet. Those active in the more established Indian-rights groups dismissed the AIM as an organization that was too busy looking for publicity and too enamored with its own militant image to do American Indians any good. Generational discord, differences in leadership styles, and competing visions for the future of Indian country created an enormous political divide among the groups.

Despite the criticism of more established groups, the AIM's membership expanded and, by 1972, included American Indians from across the country. Anxious to capitalize on these growing ties, the organization decided to launch a car caravan protest tour that would start on the west coast in October and end in Washington, DC. To clarify the AIM's views, Hank Adams (Assiniboine and Sioux) drafted a visionary statement for the future of Native America. Simply named the Twenty Points, it called for the resumption of treaty making, the restoration of tribal lands, the abolition of the BIA, immediate relief for treaty violations, greater attention to Indian health and educational issues, and the protection of religious freedom. AIM members would carry the Twenty Points with them on their cross-country journey and deliver it to, ideally, President Richard Nixon. Like African Americans who a decade earlier had taken the national spotlight with their March on Washington, AIM leaders hoped to bring national attention to the troubles in Indian country with their Trail of Broken Treaties.

The Trail of Broken Treaties did bring AIM plenty of attention but not exactly in the way it had originally intended. Imagine the 300 or 400 Indian participants converging on the BIA offices on November 3, 1972. They are looking for an opportunity to raise grievances, looking for someone in charge to accept the Twenty Points, and looking for some help in finding beds for the night. They are ignored and dismissed by federal officials. When police threatened to remove the group by force, they proceeded to take over the building (the BIA is housed in the Department of the Interior) and declared it the "Native American Embassy." During the six days of the occupation, the AIM was transformed by the press and the participants into a violent and radical insurgent group. The FBI began a surveillance and infiltration program. Somewhere in the melee, Hank Adams' carefully crafted call for Indian rights got lost.

The AIM's reputation as violent and radical was further solidified three months after the Trail of Broken Treaties occupation when Dennis Banks and Russell Means led a group of Lakota supporters to seize and hold the historic hamlet of Wounded Knee on the Pine Ridge (South Dakota) reservation. They did so to protest the corrupt political administration of then-tribal chair Richard Wilson. From February 27 to May 8, 1973, AIM members and supporters from around the country held the town and faced off against US Army units, National Guard detachments, and assorted federal law enforcement officials. The seventy-one-day siege attracted international news coverage, the support of groups such as the Vietnam Veterans against the War, and the involvement of celebrities such as Marlon Brando. It also increased the distance between the AIM and other Natives equally concerned about sovereignty and civil rights but favoring more moderate tactics.

The final act of the AIM drama played out two years later when a shoot-out erupted on the Pine Ridge reservation in June 1975. An AIM member and two FBI agents were killed. The federal government accused the AIM of staging an ambush and prosecuted four AIM leaders. One of the four, Leonard Peltier, was convicted and sentenced to two life terms in prison in a trial that many observers characterized as little more than a witch hunt. Banks, Means, and other members of the AIM leadership were also prosecuted for their roles in various confrontations with police. These lengthy trials, their cost, and the negative publicity associated with them effectively cancelled out the AIM's legitimacy for many American Indians.

The true extent of AIM's political influence and its effectiveness were often the subjects of heated debate across Indian country during the organization's heyday. Critics believed AIM members were more interested in "playing" Indian, especially for the press, than getting down to the difficult work of "being" Indian. Some of them, men and women who preferred a less confrontational and more collaborative approach to solving tribal problems, had been working behind the scenes while AIM had been "raising hell." These political moderates became more visible as infighting and protracted litigation weakened the Red Power movement. This increased stature positioned them to become the first power brokers of the self-determination era.

Beginning with John F. Kennedy and his "New Trail" for American Indians, every American president has pronounced himself a supporter of self-determination. So, too, have virtually all federal officials involved with any aspect of Indian affairs. But what is it that they are supporting? Indian self-determination has had as many different definitions as there are tribes. At the national level since the 1960s, self-determination has represented a new recognition of and respect for sovereignty. It has opened up new opportunities for Indians to influence the framing and implementation of policies affecting them. Self-determination has meant that American Indians could use federal programs to take on reservation problems such as poverty, minimal health care, and poor resource management. Ultimately, self-determination has been characterized as the remedy for the years of neglect, and worse, that American Indians had experienced.

It's no accident that self-determination became a federal policy watchword at the same time that groups such as the NIYC and AIM and individuals such as Vine Deloria, Jr., were drawing unwanted attention to the BIA's disastrous record. Politicians in Congress and the White House all understood that efforts enhancing self-determination would go a long way toward diffusing the criticism leveled by Native militants. It would also, conveniently, help to empower more moderate political factions in reservation governments. The NCAI and tribal leaders from around the country recognized this too and kept up the pressure. The specter of Indian political activism gave them enough leverage to demand inclusion in the political process and real solutions for their problems.

Politicians in Congress and the White House worked to insure that support for Indian self-determination was woven into federal policy whenever possible during the 1960s and 1970s. Kennedy appointee Philleo Nash, head of the BIA, included fifty-six reservations and four Alaskan Native regions in early federal economic-development programs. Lyndon Baines Johnson's War on Poverty and Great Society campaigns were a boon to tribal communities seeking help with housing, food assistance, economic development, rural electrification, and job training. Richard Nixon followed in his predecessors' footsteps, authorizing the expansion of Indian employment opportunities in the BIA and additional economic development funding. He also encouraged the return of the sacred Blue Lake to the Taos Pueblo people, signing the legislation in the company of traditional Taos religious leaders.

Congress devised a variety of methods for promoting Indian self-determination. Between 1969 and 1979, it enacted a series of laws on behalf of American Indians. In 1968, it passed the controversial Indian Civil Rights Act. Intended to rein in the power of tribal governments by affirming the applicability of constitutional protections to Indian people living in Indian country, the law also created the context for suits that challenged tribal authority and sovereignty. Congress enacted the Alaska Native Claims Settlement Act in 1971, transferring millions of acres of land and mineral rights to the Native population. These Native-owned corporations authorized by the legislation would become the federally recognized governing bodies for Alaska Natives. Federal legislators effectively repudiated termination with the passage of the Menominee Restoration Act in 1973. Indian education received a tremendous boost with the Tribally Controlled Community Colleges Act in 1978. Tribes could encourage post-secondary education by creating colleges that would offer reservation students a chance to earn college degrees. This veritable legislative downpour went a long way to dampen down fires lit by termination and the Red Power movement.

Three particular laws stand out from this decade. Supporters heralded the 1975 Indian Self-Determination and Educational Assistance Act as the late twentieth century version of the Indian Reorganization Act. Tribal people and their governments became the arbiters of community goals and corresponding programmatic needs. They, not the BIA, would determine which services were needed. Tribes might opt to continue with the standard BIA services or to contract with other providers for health, education, and social-welfare initiatives. In 1978, Congress enacted the Indian Child Welfare Act (ICWA) and the American Indian Religious Freedom Act (AIRFA). The ICWA sought to reverse the destructive, federally sanctioned policy of fostering and adopting Indian children in non-Indian homes. It vested tribal governments with the power to intervene in adoptions to reclaim children and acknowledged the centrality of the traditionally defined family unit in Indian life. The AIRFA took on the problem of spiritual freedom in Indian country and affirmed the rights of Natives to express and exercise their traditional religions. In particular, AIRFA attempted to safeguard the Native American Church, promote the rights of Indian prisoners to ceremonial activities while incarcerated and protect access to sacred sites. Though lacking an enforcement clause, the AIRFA extended explicit protection over American Indian religious freedom. It was a far cry from the 1920s and Circular 1665.

Of course, American Indians understood that these attempts to formalize self-determination in the 1960s and 1970s were yet another attempt to "solve the Indian problem." Some prominent Native leaders criticized federal efforts as "self-administration" and argued that transferring programs to the tribes was a new way to duck old fiduciary responsibilities. Recalling the old AIM rallying cry, "If you want to be sovereign, you have to act sovereign," they argued that real self-determination was possible only when tribes determined and acted upon their own agendas. The rhetoric of Vine Deloria's call for "Red Power" was outdated by the 1980s but the message itself—tribal people exercising power over their own lives—was more relevant than ever.

American Indians "being" and "acting" sovereign at the end of the twentieth century pursued a variety of strategies. One was legal action. Tribes and Native individuals used the courts to address civil and sovereign rights questions. New organizations formed during this period also promoted autonomy through their involvement with law, natural resources, health, education, and cultural rights. New leaders, many of whom were working in the federal government and in academia, emerged and raised their voices to support these causes. Native women and men with a wide range of political perspectives joined the remaining "shades and braids" activists in pushing the boundaries of self-determination.

American Indians made unprecedented efforts between the 1970s and 2000 to define, defend, and expand the parameters of their legal rights. By 1970, Dinebeiina Nahiilna Be

Agaditahe (*Attorneys Who Work for the Revitalization of the People*) was providing free legal services to Navajo and Hopi tribal members. The Native American Rights Fund, founded in 1970, quickly became a leading advocate for tribal sovereign and civil rights. Native attorneys such as Kirke Kickingbird (Kiowa) and John and Walter Echo Hawk (Pawnee) helped make courtrooms the new battlefields for Indians in cases pertaining to land, identity, water, fishing, hunting, gathering rights, tribal authority over Indians and non-Indians, and religious freedom.

Over thirty years, the list of cases with national significance for Indian Country grew extensive. Some, like *Morton* v. *Mancari* (Indian identity, 1974), *Santa Clara Pueblo* v. *Martinez* (definition of tribal membership, 1978), *Mississippi Band of Choctaw Indians* v. *Holyfield* (tribal and parental rights under the ICWA, 1989), and *Rice* v. *Cayetano* (political rights of Native Hawaiians, 2000) spoke to issues of tribal authority and Indian identity. Two cases, *US* v. *Washington* (tribal fishing rights in the Pacific Northwest, 1974) and *US* v. *Dion* (hunting rights and the Endangered Species Act, 1986) examined treaty fishing and hunting rights. *Passamaquoddy Tribe* v. *Morton* (tribal land rights, 1975) and *US* v. *Sioux* (compensation for illegal land seizures, 1980) focused on Native land rights. The courts took up the subject of tribal jurisdiction in *Oliphant* v. *Suquamish* (tribal jurisdiction over non-Indians, 1978), *Brendale* v. *Confederated Tribes* (tribal jurisdiction over non-Indian fee lands, 1989), and *Alaska* v. *Native Village of Venetie* (tribal powers of taxation, 1998). Religious freedom was addressed in *Lyng* v. *Northwest Indian Cemetery Protective Association* (First Amendment and AIRFA protections for sacred land, 1988) and *Oregon* v. *Smith* (First Amendment and AIRFA protections for sacramental peyote users, 1988). While *Morton* v. *Mancari, US* v. *Washington, Passamaquoddy Tribe* v. *Morton, Santa Clara Pueblo* v. *Martinez, US* v. *Sioux,* and *Mississippi Band of Choctaw Indians* v. *Holyfield* were all decided favorably for American Indians, the majority of legal decisions after 1970 were not. The dominant trend, especially with the Supreme Court, has been to circumscribe tribal sovereignty and treaty rights while sharply limiting tribal powers within Indian country.

As tribal legal representatives fought to keep sovereignty and civil rights alive for American Indians, other activists worked just to keep American Indians alive. From the end of World War I until the 1950s, the BIA maintained supervision over Indian health. The Indian Health Service (IHS), a branch of the US Public Health Service, assumed responsibility for Native health care in 1955. At that time, the public-health issues facing tribal communities and urban ethnic enclaves were mainly associated with infectious diseases, unintentional injury, pregnancy-related morbidity and mortality, and the mixed bag of illnesses associated with chronic poverty. Despite the transfer of services to the IHS, change was a long time coming to the Native healthcare system.

In 1976, Congress passed the Indian Health Care Improvement Act. This legislation acknowledged the then-dreadful state of the IHS and mandated a required level of funding for the agency. Such specific attention had not been paid to the Indian healthcare system since 1921. Congressional action at least attempted to address structural issues in tribal health care. But nothing could be done to encourage American Indians to trust IHS medical care or its practitioners. This issue became especially apparent in many communities after Constance Redbird Pinkerton-Uri, MD, revealed surprisingly high rates of sterilization for Native women of childbearing age during the early 1970s. Women of All Red Nations and tribal governments were particularly concerned with her findings.

By the 1990s, the IHS had made significant steps in addressing the most desperate Indian health problems and in developing better relations with its service population. Infant mortality rates declined by more than 60 percent, tuberculosis cases were down more than 80 percent, maternal mortality had dropped by 75 percent, and deaths by unintentional injuries had dropped by 50 percent.[23] This feat, however, was accomplished on a shoestring. Studies concluded during the 1990s indicated that the IHS budget typically ran 40 percent below its recommended funding level. These shortfalls were especially troubling given the shift in Native needs away from infectious disease prevention and toward chronic disease and behaviorally related health problems. The list of the most pressing Indian healthcare problems by the end of the decade included diabetes, hypertension, cardiovascular disease, obesity, alcoholism, fetal alcohol affect, fetal alcohol syndrome, liver disease, and suicide.

Care and attention to the needs of living American Indians was mirrored during this era by the decision of many tribes to extend a similar degree of concern to ancestral human remains and sacred objects no longer under their guardianship. In the 1980s, the Pawnee, Zuni, and Iroquois Indians all demanded the repatriation of ceremonial artifacts, sacred objects, and human remains that had been taken from burial sites and religious shrines. American Indians faced off against academics who believed that such practices would cost them research opportunities. Reuben Snake, Jr. (Winnebago) and Michael Haney (Seminole/Lakota) were influential spokesmen for tribal communities. Fortunately, most policymakers and the general public could understand and empathize with Native concerns for the proper disposition of their ancestors and their sacred objects. In 1989, Congress passed the Native American Graves Protection and Repatriation Act that directed all institutions receiving federal funding to begin the process of returning human remains and associated materials. It was a significant step in acknowledging and addressing Indian human rights and religious freedom issues.

As American Indians took on legal threats to sovereignty, public-health concerns, and threats to religious freedom, their work emphasized the importance of education for the continued survival of tribal communities. The historic (and highly imperfect) Indian school system was one of the first targets for reform during the self-determination era. Tribal educators involved in K–12 instruction founded the National Indian Education Association in 1969 to promote their concerns at the federal and local levels. They wanted culturally relevant instruction, more accuracy in instructional materials developed for use in BIA and public-school systems, greater opportunities for community educational outreach, and the chance to harness the power of education. Leaders such as David Risling (Hoopa) pushed the state school systems to accommodate the needs of Indian students. At the same time, tribal communities such as the Navajo and the Mohawk developed schools such as the Rough Rock Demonstration School and the Akwesasne Freedom School to bring their own educational values to Indian education.

Post-secondary education also became a priority for tribal leaders. Legislation such as the Tribally Controlled Community Colleges Act (1978) gave more than twenty reservations the funds needed to create colleges. Most of these owe their existence to the persistence of their founders. Janine Pease (Crow), President of Little Big Horn Community College, started her career in Indian higher education by sweeping classroom floors to save money. The American Indian Higher Education Consortium, organized in 1972, helped her and others to organize curricula, recruit faculty, develop expansion plans, and survive the predictably unpredictable changes in federal funding that have remained a constant in Indian affairs.

Overall, gains have been made in Indian education since the start of the self-determination era. In K–12 education, Indian communities have taken the lead in encouraging the BIA and local school districts to attend to the unique needs of tribal students. More than thirty tribal colleges annually serve a post-secondary population of over 30,000 part and full-time students. But problems still abound. Drop-out statistics remain far higher for Native students than for any other American population except Hispanics. The percentage of tribal young adults (aged sixteen to twenty-four) who did not complete their education was 15 percent, compared with 9.9 percent for the USA as a whole. American Indian children are disproportionately placed in special-education programs and programs for the learning disabled. They have among the lowest national percentage of access to computers. A study conducted in the early 1990s revealed that realizing any meaningful change in the persistence rates of Native students will take more time, more creativity, and more money.[24]

The fluctuating state of tribal economies was perhaps the greatest commonality among all American Indian nations during the self-determination era. Congressional appropriations for Indian country varied greatly depending on which party and which philosophy of government held sway in Washington, DC. Monies available in the 1960s and 1970s dried up in the Reagan Revolution of the 1980s and then reappeared again in the 1990s. Tribal communities looked for ways to enhance their economic stability against a backdrop of uncertain federal appropriations. Some looked to their unique resource bases as a way to create and maintain wealth. Others had to develop more inventive ways to solve their financial problems.

In the aftermath of the allotment and termination eras, American Indian nations looked increasingly to their resources as a way of generating funds to support tribal programs. Representatives of more than twenty tribes with substantial energy resources formed the Council of Energy Resource Tribes (CERT) in 1975. Peter McDonald (Navajo), La Donna Harris (Comanche), and Charles Lohah (Osage) brought together members of Native nations uncomfortably familiar with watching their resources make others wealthy. Leases negotiated by the federal government typically paid tribes 1 to 2 percent of the market value of coal, oil, and natural gas. CERT's policy was to push for contract revisions whenever possible to benefit the Native resource owners. The organization also encouraged energy-rich Indian communities to become independent producers and to increase their earnings. It was this philosophy that led McDonald as CERT's chair to meet with representatives of the Organization of the Petroleum Exporting Countries, the international energy cartel associated with the oil embargoes of the 1970s.

Even tribes without energy resources could appreciate CERT's desire to bring a new degree of autonomy to the process of resource management. The problems with the group's aggressive advocacy revealed themselves when others in Indian country began to critique CERT for encouraging the same kind of environmental and cultural carelessness that was typical of the non-Native energy corporations. The Navajo decision to allow the Peabody Coal Company to mine Black Mesa, a site of great spiritual importance to its tribal members as well as the Hopi, was interpreted by some as an example of CERT's preference for capitalism over culture. Elders from both communities criticized the "un-Indian" social values underlying the decision as well as its environmental impact on the Four Corners region.

American-Indian communities without energy resources to exploit sought other ways to create financial stability. In the late 1970s, tribes such as the Stockbridge-Munsees, the Seminole, the Pequots, and the Cabazon Band of Mission Indians took advantage of federal economic development programs and opened the first reservation bingo "palaces." By the early 1980s, tribes across the country were discovering how much money could be made through gaming. Successful bingo operations led them to look toward more lucrative gambling activities such as house-banked card games and electronic gaming machines. High-stakes gaming was clearly the fastest way for Indian nations to escape poverty and marginalization.

The tribes faced off with the states as they became more involved in gaming. Florida and California were nervous about these new Indian gaming ventures; they wanted to be able to supervise and tax the businesses and their profits. Florida attempted to rein in Seminole gaming by using PL 280 to assert jurisdiction over tribal lands. In a stunning decision, the federal appeals court decided in a 1981 landmark case that Florida had no legal standing to curtail tribal gaming because elsewhere in the state, non-Indian gaming operations were *regulated* but not *prohibited*.[25] PL 280 could not be used to circumscribe tribal rights in this context. In subsequent cases brought forward by tribal communities across the USA, the federal courts consistently defended the rights of Indian nations to operate bingo games free from state interference.

Six years later, the Cabazon Band of Mission Indians went to the Supreme Court to defend their gaming rights against interference by the state of California. The Supreme Court's 1987 decision in *State of California* v. *Cabazon Band of Mission Indians* upped the ante in the Indian gaming controversy considerably. California, like Florida in *Butterworth,* cited regulatory concerns and criminal jurisdiction as justification for invoking PL 280 and shutting down the Cabazon gaming operations. The tribe, in the wake of litigation dating from *Butterworth,* argued that the state had no right to interfere in Indian gaming if the state already had existing forms of legal gambling. The Supreme Court found in favor of the Cabazon Band, thereby paving the way for the rapid spread of Indian gaming throughout the country.

Congress, seeing the tidal wave on the horizon, hurriedly passed the Indian Gaming Regulatory Act in 1988. This legislation attempted to mediate the tense tripartite relationship linking the states, the gaming tribes, and the federal government. The Act established a three-tier classification for gaming in Indian country that ranged from traditional social and hand games (Class I), to bingo and lottos (Class II), to the most sophisticated high-stakes activities that included house-banked card games and slot and electronic card games (Class III). The tribe alone regulates Class I gaming; the tribe and its state partner may collaborate on the regulation of Class II activities. Class III gaming, the most lucrative and potentially problematic, required the negotiation of a tribe–state compact that would specify and define the terms under which the highest stakes gambling could occur.[26] The Indian Gaming Regulatory Act also created the National Indian Gaming Commission to provide regulatory oversight and advice to states and tribes.

By the end of the twentieth century, Indian gaming had grown into a multi-billion-dollar industry employing thousands across the country and bringing much-needed funds into Native communities. But for every tribe like the Mashantucket Pequot, the Mohegan, and the Seminole, who have prospered dramatically, there are others for whom gaming has done little economically or politically. Tribes such as the Oglala Lakota, who are located far from major transportation routes, have casinos that struggle to stay afloat. At the Elem Colony of Pomo Indians near Clear Lake, California, conflicts over gaming helped precipitate sporadic gun battles between members in 1995. If the long-term impact of gaming on Indian country remains to be seen, in the short run successful tribes have been able to parlay their winnings into better reservation living conditions and increased political power beyond Indian country.

CONCLUSION: "ONE HAMBURGER AT A TIME"—NATIVE AMERICA AFTER 2000

In late 2006, commuters and tourists in the heart of New York City confronted a sight that likely surprised even the most blasé urbanite. On December 5, a small group of Seminole Indians stood atop a balcony overlooking the Hard Rock Café marquee in Times Square.[27] Waving to passers-by, they sported the brightly colored ribbon shirts and skirts that typify "traditional" Seminole dress. The group was in New York to announce the tribe's purchase of Hard Rock International, the chain of hotels, restaurants, and casinos known around the globe for its embrace of the iconography of rock and roll music and American pop culture. Shortly before their appearance on the marquee, the Seminole closed the $965 million deal with the Hard Rock's parent company, the Rank Group of Britain. Applauding his tribe's momentous transaction, Max Osceola, Jr., recalled Manhattan's supposed "sale" by the Canarsie Indians to sixteenth-century Dutch traders. He declared that the Seminoles' objective was to reclaim the island, "one burger at a time."

The Seminole's promise to redeem Manhattan is an expression of self-confidence, sovereignty, and cultural persistence that is very much a product of the twentieth century. Today, there are more than 4.1 million American Indian people and more than 560 tribes in the USA. These men and women, and the communities to which they belong, are the descendants of those who teetered on the brink of oblivion in 1900. Their survival and resurgence is the story of the past 100 years of American Indian history. Their determination to maintain their identities as American Indians, preserve their traditional cultures, and protect their tribal communities will be the story of the twenty-first century—especially if, like the Seminole, they also plan on redeeming America one hamburger at a time.

NOTES

1. Kevin Gover, "Special Feature: Remarks at the Ceremony Acknowledging the 175th Anniversary of the Establishment of the Bureau of Indian Affairs," *American Indian Law Review*, 25 (1) (2000/2001): 161–163.
2. Christopher Buck, "Never Again: Kevin Grover's Apology for the Bureau of Indian Affairs," *Wicazo Sa Review*, 21 (1) (2006); p. 106.
3. Luther Standing Bear, *Land of the Spotted Eagle* (Lincoln, Nebr.: University of Nebraska Press, 1978), p. 236.
4. Luther Standing Bear, *Land of the Spotted Eagle*, pp. 229–237.
5. The actual nadir population number for American Indians is contested. US Census documentation records the American Indian population dropping to 248,000 in 1890, declining to 237,000 in 1900, and rebounding to 266,000 in 1910. The Annual Reports of the Commissioners of Indian Affairs for the 1890–1910 period estimate that the Indian population was 228,000 in 1890, 250,000 in 1900, and 279,000 in 1910. See Russell Thornton "Aboriginal North American Population and Rates of Decline, ca. A.D. 1500–1900," *Current Anthropology*, 38 (2) (1997): 310–315 and James P. Collins, "Native Americans in the Census, 1860–1890," *Prologue*, 38 (2) (2006): 54–59.
6. *Lone Wolf* v. *Hitchcock*, Supreme Court of the USA, 1903 (187 US 553, 23 S. Ct. 216, 47 L. Ed. 299).
7. *US* v. *Winans*, Supreme Court of the USA, 1905 (198 US 371, 25 S. Ct. 662, 49 L. Ed. 1089).
8. *Winters* v. *US*, Supreme Court of the USA, 1908 (207 US 564, 28 S. Ct. 207, 52 L. Ed. 340).
9. Terrence M. Cole, "Jim Crow in Alaska: The Passage of the Alaska Equal Rights Act of 1945," *Western Historical Quarterly*, 23 (4) (1992): 429–449.
10. Michael L. Tate, "From Scout to Doughboy: The National Debate over Integrating American Indians into the Military, 1891–1918," *Western Historical Quarterly*, 17 (4) (1986): 437.

11. Carlos Montezuma, "Drafting Indians and Justice," *Wassaja*, 2 (7) (October 1917): 3.
12. Vine Deloria, Jr., "This Country Was a Lot Better Off When the Indians Were Running It," in Alvin Josephy, Jr. (ed.), *Red Power: The American Indian's Fight for Freedom* (New York: McGraw-Hill, 1971), p. 239.
13. "Indians in the War," Office of Indian Affairs, US Department of the Interior (Haskell Okla.: Haskell Indian School, 1945).
14. Arthur Watkins, "Termination of Federal Supervision: The Removal of Restrictions over Indian Property and Person," cited in John Wunder, *"Retained by the People": A History of American Indians and the Bill of Rights* (Oxford: Oxford University Press, 1994), p. 101.
15. Earl Old Person quoted in James J. Rawls, *"Chief Red Fox is Dead": A History of Native Americans since 1945* (Belmont, Calif.: Wadsworth Group, 2001), p. 42.
16. "Statement of Alice Lee Jemison" quoted in R. David Edmunds, Frederick E. Hoxie, and Neal Salisbury, *The People: A History of Native America* (Boston, Mass.: Houghton-Mifflin, 2007), p. 413.
17. *Harrison* v. *Laveen*, 67 Ariz. 337, 196 P. 2d. 456 (1948).
18. *Allen* v. *Merrell*, 6 Utah 2d. 32 (1956) and *Montoya v. Bolack, 70 N. M. 196 372 P. 2d. 387* (1962).
19. *Arizona* v. *California*, Supreme Court of the USA, 1963 (373 US 546, 83 S. Ct. 1468, 10 L. Ed. 542).
20. R. David Edmunds, Frederick E. Hoxie, and Neal Salsibury, *The People: A History of Native America* (New York: Houghton-Mifflin, 2007), p. 416.
21. Vine Deloria, Jr., "Indians in Revolt, 1970," in The Council on Interracial Books for Children (ed.), *Chronicles of American Indian Protest* (Greenwich Conn.: Fawcett Premier Books, 1971), p. 296.
22. Wilma Mankiller, "Old Names in Charge," quoted in Peter Nabokov (ed.), *Native American Testimony* (New York: Penguin Press, 1999), p. 453.
23. Harvard Project on America Indian Economic Development, *The State of Native Nations: Conditions under US Policies of Self-Determination* (Oxford: Oxford University Press, 2008), pp. 219–235.
24. Harvard Project on America Indian Economic Development, *The State of Native Nations*, pp. 199–219.
25. *Seminole Tribe of Florida* v. *Butterworth*, 658 F. 2nd 310, 311 (5th Cir. Unit B 1981). See especially William E. Horowitz, "The Scope of Gaming under the Indian Gaming Regulatory Act of 1988 after *Rumsey v. Wilson:* White Buffalo or Brown Cow?" *Yeshiva University Cardozo Arts & Entertainment Law Journal*, 153 (14) (1994), for an excellent discussion of PL 280, termination, and their ramifications for Indian gaming; and N. Bruce Duthu, "Crow Dog and Oliphant Fistfight at the Tribal Casino: Political Power, Storytelling, and Games of Chance," *Arizona State Law Journal*, 171 (29) (1997) for discussions of *Bryan* v. *Itasca County* (423 US 373, 1976) and its relationship to this debate.
26. Indian Gaming Regulatory Act, 25 USCA, Section 2701–2721.
27. Michael J. de la Merced, "Florida's Seminole Tribe Buys Hard Rock Cafes and Casinos," *The New York Times*, December 8, 2006.

FURTHER READING

Adams, David Wallace, *Education for Extinction: American Indians and the Boarding School Experience, 1875–1928* (Lawrence, Kans.: University of Kansas Press, 1995).

Ambler, Marjane, *Breaking the Iron Bonds: Indian Control of Energy Development* (Lawrence, Kans.: University of Kansas Press, 1990).

Bernstein, Alison, *American Indians and World War II: Toward a New Era in Indian Affairs* (Norman, Okla.: University of Oklahoma Press, 1990).

Britten, Thomas A., *American Indians in World War I: At War and At Home* (Albuquerque, N. Mex.: University of New Mexico Press, 1997).

Burton, Lloyd, *American Indian Water Rights and the Limits of the Law* (Lawrence, Kans.: University of Kansas Press, 1999).

Cornell, Stephen, *The Return of the Native: American Indian Political Resurgence* (Oxford: Oxford University Press, 1988).

Deloria, Vine, Jr., and Clifford Lytle, *The Nations Within: The Past and Future of American Indian Sovereignty* (New York: Pantheon, 1984).
Deloria, Vine, Jr., and David E. Wilkins. *Tribes, Treaties, and Constitutional Tribulations* (Austin, Tex.: University of Texas Press, 1999).
Calloway, Colin, *First Peoples: A Documentary Survey of American Indian History,* 3rd edn (Boston, Mass.: Bedford/St. Martin's Press, 2008).
Council on Interracial Books for Children (ed.), *Chronicles of American Indian Protest* (Greenwich, Conn.: Fawcett Premier Books, 1971).
Cowger, Thomas, *The National Congress of American Indians: The Founding Years* (Lincoln, Nebr.: University of Nebraska Press, 1999).
Edmunds, R. David (ed.), *The New Warriors: Native American Leaders since 1900* (Lincoln, Nebr.: University of Nebraska Press, 2001).
Edmunds, R. David, Hoxie, Frederick E., and Salisbury, Neal, *The People: A History of Native America* (Boston, Mass.: Houghton-Mifflin, 2007).
Fixico, Donald L., *Termination and Relocation: Federal Indian Policy, 1945–1960* (Albuquerque, N. Mex.: University of New Mexico Press, 1986).
——*The Urban Indian Experience in America* (Albuquerque, N. Mex.: University of New Mexico Press, 2000).
Getches, David H., Wilkinson, Charles F., and Williams, Robert A. Jr., *Federal Indian Law,* 5th edn (St. Paul, Minn.: West Group Publishers, 2004).
Harvard Project on America Indian Economic Development, *The State of Native Nations: Conditions under US Policies of Self-Determination* (Oxford: Oxford University Press, 2008).
Holm, Tom, *The Great Confusion in Indian Affairs: Native Americans and Whites in the Progressive Era* (Austin, Tex.: University of Texas Press, 2005).
Hoxie, Frederick E., *A Final Promise: The Campaign to Assimilate the Indians, 1880–1920,* 2nd edn (Lincoln, Nebr.: University of Nebraska Press, 2001).
Josephy, Alvin (ed.), *Red Power: The American Indian's Fight for Freedom* (New York: McGraw-Hill, 1971).
Kelly, Lawrence C., *The Assault on Assimilation: John Collier and the Origins of Indian Policy Reform* (Albuquerque, N. Mex.: University of New Mexico Press, 1983).
McNickle, D'Arcy, *Native American Tribalism: Indian Survivals and Renewals* (Oxford: Oxford University Press, 1973).
Nabokov, Peter (ed.), *Native American Testimony: A Chronicle of Indian–White Relations from Prophecy to the Present, 1492–1992,* rev. edn (New York: Viking Press, 1999).
Nagel, Joanne, *American Indian Ethnic Renewal: Red Power and the Resurgence of Identity and Culture* (Oxford: Oxford University Press, 1996).
Philp, Kenneth, *Termination Revisited: American Indians on the Trail to Self-Determination, 1933–1953* (Lincoln, Nebr.: University of Nebraska Press, 1999).
Rawls, James J., *Chief Red Fox Is Dead: A History of Native Americans since 1945* (Fort Worth, Tex.: Harcourt Brace, 1996).
Smith, Paul Chaat and Warrior, Robert Allen, *Like a Hurricane: The Indian Movement from Alcatraz to Wounded Knee* (New York: New Press, 1999).
Trask, Haunani-Kay, *From a Native Daughter: Colonialism and Sovereignty in Hawai'i* (Monroe, ME: Common Courage Press, 1990).
Wilkins, David, *American Indian Politics and the American Indian Political System* (New York: Rowman & Littlefield, 2002).
Wilkins, David E. and K. Tsianina Lomawaima, *Uneven Ground: American Indian Sovereignty and Federal Law* (Norman, Okla.: University of Oklahoma Press, 2001).
Wilkinson, Charles F., *Blood Struggle: The Rise of Modern Indian Nations* (New York: W. W. Norton, 2005).
Wunder, John R., *"Retained by the People": A History of American Indians and the Bill of Rights* (Oxford: Oxford University Press, 1994).

CHAPTER TWENTY

LEADERSHIP OR DOMINANCE

The Role of the Western US Economy in the World

Neil Greenwood

In a 2008 article David Kennedy posed the question if the West could lead the nation to a better future.[1] This question recognizes the role for the West in the national and global economies. The economy of the American West today is the product of a transformation that occurred during the latter half of the twentieth century. The components of this transformation include federal government spending, population growth, and the continuing development of the manufacturing and recreational sectors in the regional economy. This transformation has made the Western USA a leader in American and, to a lesser extent, global economic development in the early twenty-first century.

The economy of the modern American West is marked by dualities. The key duality is geographic. The geography of the West is defined by expansive metropolitan areas surrounded by huge rural hinterlands, particularly in the Intermountain West. While the economies of modern Western cities are based upon service industries, manufacturing, and growth, their hinterlands still draw their wealth largely from agriculture and the extraction of natural resources.

As a region, the West is a historical construct and a product of geography. However, in analyzing the economic development of the West, including its economic primacy nationally and internationally, the West can be seen as two regions, with economic development and growth varying by state. One of the two regions includes the Great Plains and the Intermountain West. The other consists of the periphery going down the Pacific coast and eastward across Arizona and New Mexico into Texas. Alaska and Hawaii, while nominally part of the American West as extensions of US power into the Pacific, are detached from the two geographically contiguous regions. With respect to the variations among these regions population growth has paralleled economic development.

For minority groups, Western economic development reflected national trends in terms of both incidental and purposeful discrimination. Such discrimination is reflected in the divisions among neighborhoods and communities within Western, as well as Eastern, cities. In terms of breaking down barriers to equality of opportunity across ethnic, gender, racial, and religious lines, progress since the Second World War has been uneven across the region. The racial discrimination that African Americans have faced in Texas and, to a lesser extent, Oklahoma has been akin to that in Southern states to the East through Virginia. In various metropolitan areas along the Pacific coast Asian Americans have made significant progress in breaking down historic housing, occupational, and social barriers that date back to the middle of the nineteenth century. Hispanic Americans have experienced traditional forms of discrimination in the West, particularly the Southwest, as informed by historic forces of alienation if not animosity generated by events going back to the Mexican War. Those communities and areas of the West that have moved toward equality of opportunity are leaders not only in the USA but also, with globalization, around the world.

Carol O'Connor characterized the West as "a region of cities."[2] Along with reflecting one of the biases in the historiography of the American West, this characterization may epitomize one of the most important characteristics in understanding the economic development of the modern trans-Mississippi West. While the harvest of natural resources is still a big part of the West's economy, it is the region's cities that are the engine of this economy. These cities are the focal points of manufacturing, population growth, and regional, national, and international trade. The competition among them constitutes one of the most important dynamics in the West.

THE FEDERAL PRESENCE

The foundations of the modern West are rooted in the nineteenth century when the harvest of natural resources dominated, if not defined, Western economic development. At the time, cities were the focal points for the economic activity that grew out of the exploitation of these natural resources. In his essay "Wage Earners and Wealth Makers," Carlos Schwantes characterizes Western cities in 1900 as "urban outposts juxtaposed with vast and lightly populated hinterlands."[3] During the nineteenth century, the region's economy was controlled by external demand for its agricultural products and natural resources and by capital in the East and Europe.

During the twentieth century the federal government was the driving engine in the economic development of the West. Federal support for Western economic development occurred in several waves of spending prompted by regional, national, and international events. These waves of federal spending in the West differed from federal subsidies of Western economic development during the nineteenth century in that the twentieth-century spending was more proactive with respect to the encouragement of development. Federal subsidization of Western development during the nineteenth century consisted largely of land grants to railroads, settlers, and speculators, carried out in response to the demands of those parties involved in the settlement of the West. This approach by the federal government to the development of the West during the nineteenth century reflected the external control of the process through capital investment and the disposition of federal lands. Twentieth-century federal spending marked the progress of the West's maturation from junior to full partner with the East in the nation.

The first wave accompanied the Progressive Era reforms by the federal government during the early twentieth century. In the nation as a whole, Progressive reforms sought to address the dramatic changes that came with immigration, industrialization, the rise of big business, and urbanization. Progressive reforms specifically tailored for the West reflected a new role for the federal government of facilitating development, as distinct from the federal government's role in the nineteenth century of opening the region to settlers. Possibly the most important congressional Act that came with this wave was the National Reclamation Act, also known as the Newlands Act, passed in 1902, which created the Bureau of Reclamation. Congress passed this act with the enthusiastic support of President Theodore

Roosevelt. This act ushered in a new era of federally sponsored projects that epitomized this new responsibility for Washington. The Bureau of Reclamation presided over water-diversion projects throughout the West, principally in the form of dams, part of what Donald Worster characterizes as the federalization of irrigation development.[4] Irrigation became a priority for the federal government. As William Cronon points out in his essay "Landscapes of Abundance and Scarcity," reclamation took on a new meaning associated with the development of the West through irrigation. Along with the Bureau of Reclamation, Congress also established a Committee on Irrigation and Reclamation of Arid Lands. During the twentieth century the Bureau of Reclamation and the Army Corps of Engineers became the primary government agencies in Western dam construction, in a complex relationship with each other, alternately marked by competition and cooperation.

One such project was the Roosevelt Dam in Arizona. Built in a mountain gorge on the Salt River about 65 miles northeast of Phoenix, it fueled the initial growth of that desert city during the first half of the twentieth century by ensuring a stable water supply for the irrigation of the growing farmland in the Salt River Valley. Local leaders established the Salt River Valley Water Users Association in 1903 to promote the construction of a dam on the Salt River financed through the Newlands Act. Completed in 1911, its contribution to Phoenix's growth can be seen in the population figures for 1900–1920. Phoenix grew from 5,544 in 1900 to 29,053 in 1920. Projects such as Roosevelt Dam grew out of efforts by community leaders and organizations, such as the Salt River Valley Water Users Association, to assert control over the new subsidies that the federal government offered for the development of the West, a hallmark of this development during the twentieth century. Other Western water projects completed through the Newlands Act include Elephant Butte Dam 125 on the Rio Grande which provided a source of reservoir water for agricultural irrigation in and around El Paso and the Owens Valley Project designed to provide a stable source of water for the growth of Los Angeles.[5] The Owens Valley Project reflected the growing struggle for water among rival groups, communities, and states in the West.

The second wave came in response to the Great Depression with the Reconstruction Finance Corporation and Franklin Delano Roosevelt's New Deal. The Reconstruction Finance Corporation, established in 1932 as a lender of last resort to private companies and state and local governments, provided the financing for the bond issued to raise funds for the construction of the Colorado Aqueduct, a water project designed to tap the Colorado River for the West's growing water needs during the first half of the twentieth century. With the initiation of the New Deal, the West benefited from both national New Deal programs, along with other regions of the USA, and regional projects, which provided jobs and federal funds for irrigation and flood control.

One of the most significant projects of the New Deal era was the Hoover Dam. Originally known as Boulder Dam, it was built on the Colorado River near Henderson, Nevada, at the southern tip of the state. Behind this dam lay Lake Mead, which provided water and electricity for various areas of the West and recreational opportunities that would transform southern Nevada and northwestern Arizona into a cluster of entertainment areas during the third quarter of the twentieth century. Hoover Dam was one of the most visible manifestations of Franklin D. Roosevelt's solicitation of Western political support, a manifestation itself of the region's growing political importance, and of US government efforts during the twentieth century to control Western rivers. By the end of 1936 the Bureau of Reclamation had nineteen other major dams under construction in the West, a number of which were inspired by the success of Hoover Dam.[6]

A number of New Deal Acts and programs also significantly aided and/or otherwise stabilized a number of natural-resource industries in the West, including the 1933 Agricultural Adjustment Act, the National Recovery Administration, the Rural Electrification Administration, the 1934 Silver Purchase Act, the Soil Conservation Service, and the 1934 Taylor Grazing Act.[7] Instead of being at the mercy of external demand and outside capital, these Western resource industries could count on the federal government for regulation if not protection. While the Tennessee Valley Authority did not involve any Western states, it did inspire efforts to establish similar government corporations dedicated to harnessing and administering a number of Western rivers, the most notable of which were the Columbia and Missouri rivers. Ultimately these efforts failed. Arguably the Missouri Valley Authority, which would have controlled the Missouri River and its tributaries in the northern half of the Great Plains, came the closest to fruition, but its supporters failed to overcome opposition from the Army Corps of Engineers and the Bureau of Reclamation, not to mention ideological opposition who saw such a public corporation as a statist or socialist threat to the nation's market economy.[8]

Along with the rest of the nation the West benefited from the new role that the New Deal established for the federal government of guaranteeing the public welfare. Social security, as one of the linchpins of the New Deal, provided an incomplete national pension system. The National Labor Relations Act helped Western as well as Eastern unions. The Civilian Conservation Corps, the Public Works Administration, and the Works Progress Administration all provided opportunities for temporary employment.

The third wave of federal spending came with the US war effort in World War II. Fighting on two sets of fronts against the Germans and the Japanese made both the Atlantic and the Pacific coasts staging areas for US military operations. Unlike World War I, the West was fully integrated into the US war effort in World War II. As part of the US war effort in World War II, the national government spent money in the West on new military facilities, scientific research projects, and technological development programs. In the construction of these facilities in the West the federal government took advantage of the region's open spaces and rivers. The federal government dispersed military bases and production facilities across the nation to provide better security against enemy attack.

New military bases were the most tangible manifestations of federal spending in the West during World War II. The federal government selected sites for these bases on the basis of strategic, geographic, transportation, and political considerations. These considerations dovetailed with the ongoing competition among Western cities within that region. The geographic distribution of these bases across the West took the shape of a "U" that went down the Pacific coast across southern Arizona and New Mexico and up the Great Plains and the front range of the Rockies. Fewer bases were constructed or expanded in the Intermountain West states of Idaho, Montana, Nevada, Utah, and Wyoming. Not surprisingly, Los Angeles, San Diego, San Francisco, and Seattle hosted the most extensive naval facilities on the Pacific coast. Army and Army Air Force facilities were spread throughout the West.

California was the home of many of the new and expanded wartime facilities. This process began before World War II with the transfer of part of the US Navy from the Atlantic coast to the Pacific coast because of the growing importance of the Pacific Ocean to the country. More specifically, the US Navy was reacting to a growing threat from Japan. More visible was the transfer of the Pacific Fleet to Pearl Harbor in May 1940. With this transfer Hawaii became the center of the US military capability in the Pacific. The construction of military facilities in Hawaii contributed to the solidification of its strategic and, to a lesser extent, economic integration with the forty-eight contiguous states. Location dictated that Alaska remain on the fringe of the US war effort against Japan.

For some Western cities, World War II was a crucible. Possibly the best known example was Las Vegas. A small railroad junction at the beginning of World War II, Las Vegas was selected as the site for several military facilities. In 1941, the federal government established the Las Vegas Aerial Gunnery School 8 miles northeast of Las Vegas. The next year, this school for training in the use of aircraft weaponry became Nellis Air Force Base.[9] That same year of 1941, the federal government planned the construction of a magnesium plant in Henderson, southeast of Las Vegas. The construction of factories for the production of military goods was part of the government's strategy to spread wartime military production across the continental USA. The advantages of this site reflect the factors that dictated the selection process. The Las Vegas area was accessible to raw materials. In addition, the recently completed Hoover Dam provided both water and electricity. Finally, the plant would be built near a transportation route to the Pacific coast. The construction and operation of this wartime plant transformed Henderson into a "boomtown." During the war more than 12,000 workers came to Henderson to build and then operate the plant.[10]

Other Western cities came to depend on the federal government as an economic foundation during World War II. Biggs Field and Fort Bliss punctuated the growth of military employment in and around El Paso, Texas.[11] El Paso came to rely on both for the local economy.

Farther north in the Intermountain West the federal government established ten new military bases in Utah during the war. These bases, along with other forms of military spending during the war, contributed to the development of other major sectors of the Utah's economy beyond its prewar dependence on mining.[12] In Colorado, Denver and Colorado Springs hosted a number of Army, Army Air Corps, and Navy facilities.[13]

Along with new bases, World War II brought the first large-scale aircraft, ship, and steel manufacturing facilities to the West. Southern California, running from San Diego to Los Angeles, became one of the major aircraft-production areas for the war effort, producing about a third of the nation's total output.[14] Los Angeles and San Francisco became major shipbuilding centers. Fontana, outside of Los Angeles, became the home in 1942 of one of the two largest steel factories west of the Mississippi River.[15]

The US war effort in the West during World War II also gave birth to the region's modern research and development capabilities. The Office of Scientific Research and Development (OSRD), created by the federal government in 1941, spearheaded the establishment of research and development facilities throughout the nation. The OSRD distributed money through contracts to businesses and universities. Western facilities that benefited from the federal largesse, doled out by the OSRD among other federal agencies, included the Jet Propulsion Laboratory at the California Institute of Technology, the Scripps Institute of Oceanography in La Jolla, California, and the Boeing Corporation in Seattle.[16]

The single largest federal government project during World War II which affected both the West and the nation as a whole was the Manhattan Project. As part of the War Department's policy of decentralizing US military facilities in the war effort the Manhattan Project included a number of sites spread throughout the West, as well as the East, including Berkeley and Inyokern, California; Uravan, Colorado; Alamogordo and Los Alamos, New Mexico; Monticello and Wendover, Utah; and Hanford, Washington. Manhattan Project managers selected these sites because of their proximity to natural resources or their relative geographic isolation. For example, Hanford was selected as the site for a plutonium production reactor facility because of its proximity to the Columbia River. The federal government already had constructed dams on the Columbia, including Bonneville Dam in 1937 and Grand Coulee Dam in 1941, which generated a surplus of electricity used at Hanford to run the nuclear reactors and manufacturing facilities involved in the production of plutonium. The Los Alamos facilities completed portions of the technical development of the first bombs. The first nuclear detonation occurred at the remote Trinity site near Alamogordo, New Mexico on July 16, 1945.[17] This project, involving both research and development, left as its legacy a number of facilities devoted to either or both, as well as trained and experienced communities of scientists and engineers.

In addition to the military bases constructed in the West as part of the war effort against Japan, the federal government also constructed ten wartime internment camps and built or used a number of related facilities for assembling, processing, and detaining approximately 110,000 Japanese Americans and Japanese immigrants, about three-quarters of the detainees being Japanese Americans. Both in the short-term and the long-term these camps had a much greater impact on the detainees than the communities from which they were expelled, most being along the Pacific coast, or the communities near the internment camps. These detainees lost or were forced to sell most of their real estate and personal property.

World War II fueled tremendous population growth in the West, in the form of soldiers, sailors, airmen, and defense-industry workers who migrated to the region. This wartime migration was one of a number of growth spurts that have punctuated the development of the West. This population boom was driven by the construction of new military bases, the expansion of extant military bases, and the growth of manufacturing for the war effort, particularly aircraft production and shipbuilding, as part of the overall war effort against Japan.

As during other periods of the twentieth century, population growth rates varied among the different areas of the West. Nearly half the people who moved to the West during the 1940s settled in the Pacific Coast states, with California experiencing the most dramatic population growth between 1940 and 1950, increasing from 6,907,000 to 10,586,000.[18] Select cities punctuated this growth. San Diego grew from 203,341 in 1940 to 334,387 in 1950 as a result of the greatly increased naval presence and new aircraft-production facilities. Seattle's population increased from 368,302 in 1940 to 467,591 in 1950. Based on the wartime migration, the West continued to attract new residents during the latter half of the twentieth century.

The fourth wave was propelled by Cold War military spending and population growth in the West. In a number of ways, this Cold War spending and population growth continued or resumed after a brief respite in the late 1940s following the spending and growth generated by World War II. This Cold War defense spending in the West continued to be a foundation of the region's economic development after the World War II.

One of the defining components of this wave was federal military spending in the form of Western military bases and defense contracts for corporations in the West. Much of this Cold War federal military spending was part of the US government's global strategy of containing Communist expansion. This strategy was a continuation of the US strategy during World War II of fighting two separate war efforts at both ends of Eurasia. Having faced and defeated Germany and Japan, US leaders after World War II embraced a new international and interventionist orientation, looking both eastward across the Atlantic Ocean and westward across the Pacific Ocean.

During the Cold War the federal government invested a number of extant Western military bases with new responsibilities and constructed a number of new bases. The federal government built new bases and expanded existing bases in the West on the same bases used to determine base locations and new tasks as had been used during World War II—decentralization of the nation's military facilities and satisfying the demands of Western congressmen and senators for bases in their districts and states. Military personnel stationed at these bases added another source of federal revenue to local economies that was measurable if not significant.

In the Intermountain West, Colorado received the largest share of military installations. Colorado Springs received one of the crown jewels of the federal government's newly independent Air Force: the Air Force Academy. Construction of the Air Force Academy in Colorado provided geographic balance with and distance from the Army's academy in West Point, New York, and the Navy's academy in Annapolis, Maryland. West of Denver, the federal government constructed the Rocky Flats Nuclear Weapons Center. Between the 1949 Soviet detonation of a nuclear bomb and the construction of intercontinental ballistic missiles some in the federal government even considered making Denver an alternate capital in case of a Soviet nuclear strike on Washington because of the "mile high" city's location in the Intermountain West.[19]

As during World War II, Cold War military spending provided a major economic boost to a number of Western cities. In and around El Paso, Fort Bliss, which had been a training center for antiaircraft artillery units, became the US Air Defense Center in 1957, and Biggs Field became Biggs Air Force Base, as part of the newly independent Air Force.[20] Consequently, the military became a major employer in the El Paso area.

Military installations in Alaska and Hawaii were part of the US projection of power across the Pacific to contain first Soviet and later Chinese and Korean aggression against US allies in East Asia. Pearl Harbor, among other military installations, continued to be one of the linchpins of the US military presence in the Pacific. Accordingly, these installations have been a measurable part of Alaska's and Hawaii's economies.

The Korean and the Vietnam Wars punctuated Western-oriented military spending during the Cold War.[21] The Pacific states, including Hawaii and Alaska, housed forward military bases used for staging US military operations in Korea and Southeast Asia and for maintaining US military bases in Okinawa and the Philippines. After a lull during the late 1940s, San Diego experienced renewed growth during the 1950s, driven in part by the Korean War.[22] The US war efforts in Korea and Vietnam also reinforced the demand for military aircraft and naval vessels produced in Western manufacturing facilities.

Another consequence for the West of the Cold War in East Asia has been the migrations provoked by various events of the Cold War. The first such migration consisted of Chinese Nationalists fleeing China in the wake of the 1949 establishment of the People's Republic of China. From 1880 to 1940 the recorded number of people in the USA of Chinese extraction varied from 85,000 to 110,000. From 1950 to 2000 that number increased progressively from 150,005 in 1950 to 2,432,585 in 2000. As a proportion of the total US population, those from China, Hong Kong, and Taiwan, or of Chinese extraction increased from 0.10 percent to 0.86 percent during that same period.[23] Consequently, ethnic Chinese communities grew throughout the West. The growing number of Chinese Americans has solidified ties between the Western USA and East Asia.

The end of the Vietnam War in 1975 provoked a similar migration of Vietnamese to the USA. As a result, the number of people in the USA from Vietnam or of Vietnamese extraction increased from 245,025 in 1980 to 1,122, 528 in 2000. The largest numbers of these immigrants and Vietnamese Americans live in California and Texas.[24]

During the Cold War, cities throughout the West experienced economic and population booms driven by federal military spending. Denver led the way in the Intermountain West if not the entire West. In 1966, construction of the headquarters of the North American Aerospace Defense Command was completed in Cheyenne Mountain, southwest of Colorado Springs. With the construction of military facilities around Denver a number of aerospace and electronics companies established operations in the Denver area, including Dow Chemical, Hewlett-Packard, Honeywell, IBM, Johns-Manville, and Martin-Marietta (at the time Martin Company).[25] Outside of Ogden, Utah Hill Air Force Base, which the federal government had established during the late 1930s, became the key missile facility in the West and attracted other related military installations and defense industries.[26]

Correspondingly, cuts in defense and related government spending adversely affected local and area economies in the West, as well as the East. In 1959, San Diego endured the economic pain when Convair laid off 21,000 workers when its jet program was discontinued. The end of the Apollo program caused more widespread economic pain in aerospace centers, including Florida, Houston, and southern California. Once the USA had won the "race" with the Soviet Union to the moon and manned moon landings became commonplace, even routine, to Americans, many saw no need to continue what they considered an expensive program.

In terms of infrastructure, the most visible manifestation of Cold War spending in the West by the federal government was the interstate highway system. Created by Congress in 1956 with passage of the Interstate and Defense Highway Act, this highway system was designed to be a military ground transportation network. The interstate highways that served as linear corridors between Western cities along east–west and north–south corridors were built along natural transportation routes and reflected the local geographic features. In this way, these highways reinforced historic relationships among the Western cities they connected.[27] This system became the foundation of the nation's transportation infrastructure after World War II, particularly in the vast expanses between metropolitan areas in the West.

The interstate highway system accelerated, or *exacerbated,* Western urban growth after World War II. Various interstate highways contributed to the growth of a number of urban corridors. These highways include Interstate 5 in California, Oregon, and Washington; Interstate 15 in Utah, Idaho, and Montana; Interstate 25 in New Mexico, Colorado, and Wyoming; Interstate 35 in Texas; Interstate 10 in Arizona, New Mexico, and Texas; Interstate 80 in California; Interstate 90 in Washington and Montana; and Interstate 94 in North Dakota. A number of select Western cities which punctuate these urban corridors are identified in Table 20.1.

The interstate highways contributed to urban population growth and territorial expansion. As an example, in Texas the construction of Interstate 10 and Interstate 410 through and around San Antonio encouraged the building of housing and shopping malls progressively northward of the city during the latter half of the twentieth century. Interstate 410 is typical of a number of loop highways built around cities as part of the interstate highway system. This northward expansion epitomizes one of the challenges Western cities faced after World War II of determining whether to allocate resources to the renovation of existing, and often neglected if not rundown, neighborhoods within these cities or new middle- and upper-class subdivisions on the fringes which promised increased tax revenues. In a number of cities, the decision-making process laid bare racial, ethnic, geographic, and economic and class divisions.[28]

Table 20.1 Intrastate interstate highway connections among Western cities

Interstate highway	*State*	*Cities*	*Description*
I-5	California	Sacramento, Stockton	North-south agricultural valley corridor
I-5	Oregon	Eugene, Salem, Portland	North-south corridor along the Willamette River
I-5	Washington	Olympia, Tacoma, and Seattle	North-south corridor along Puget Sound and its tributaries
I-15	Utah	Provo, Salt Lake City, and Ogden	Metropolitan cluster centered on Salt Lake City
I-15	Idaho	Pocatello and Idaho Falls	Small-scale urban cluster
I-15	Montana	Butte, Helena, Great Falls	North-south corridor
I-25	Colorado	Pueblo, Colorado Springs, Denver, Fort Collins	North-south corridor along the Front Range of the Rockies
I-25	New Mexico	Albuquerque, Santa Fe	North-south urban cluster
I-25	Wyoming	Casper, Cheyenne	North-south small-scale urban corridor
I-35	Texas	San Antonio, Austin, Dallas-Fort Worth	North-south corridor connecting metropolitan clusters
I-10	Arizona	Phoenix, Tucson	East-west corridor
I-10	New Mexico	Las Cruces; El Paso, Texas	Urban cluster
I-10	Texas	Houston, San Antonio	East-west corridor
I-80	California	Oakland, San Francisco	Urban cluster
I-90	Montana	Billings, Bozeman, Butte, Missoula	East-west corridor
I-90	Washington	Seattle, Spokane	East-west corridor
I-94	North Dakota	Bismarck, Fargo	East-west corridor

The construction of interstate highway loops, crescents, and spurs around, out from and through Western cities epitomized one of the signature characteristics of these cities: room to grow. The decision to build a crescent, loop, or a spur in any given Western, or, for that matter, any given US, city was dictated by the potential for and directions of expansion and the terrain of the city. As examples, highway loops were built around Dallas and Fort Worth individually and together. These two cities are located in the relatively flat lands of northern Texas, in the southern end of the Great Plains. By way of contrast, a crescent highway was built around Denver covering all but the northwest quadrant of the metropolitan area.

Interstate highways also fed the growing recreation and tourism industries in the Western USA, as well as the eastern half of the country. These highways allowed people to visit tourist sites across the vast expanses of the West on their own terms, at least within the context of the mid- and late-twentieth-century American automobile culture. These highways also propelled the development of a number of travel industries dedicated to providing auto tourists with the food, lodging, and souvenirs when and where they wanted them.

In 1978, President Carter proposed the construction of the MX missile system, possibly the most elaborate Western military project during the Cold War. As part of this system, intercontinental ballistic missiles would be placed on trains to make them more difficult to target for destruction by Soviet missiles. The goal of this system was to increase to deterrent value of the US nuclear force. This project generated opposition from groups in the region where the railroad was to be constructed. These groups feared increased vulnerability to a Soviet nuclear strike. They also resented the proposed placement of this system in their region, particularly by a president from the Southeast.

One part of this Cold War federal spending wave consisted of more dams built in the West after World War II. A variety of different Western constituencies embraced the construction of individual multipurpose dams over the creation of systems of such dams as the Tennessee Valley Authority. These individual dams epitomize the desire of many Westerners during the twentieth century for federal dollars without federal supervision.

NATURAL RESOURCES

While the harvesting of natural resources continues to be an important part of the West's economy, the exploitation of these resources has not contributed significantly to the West's economic dominance and preeminence. The exploitation of natural resources hearkens back to the nineteenth century when the West's economy was driven by external demand for its resources and dependent on capital from eastern US and European investors. One significant difference between the nineteenth-century West and the modern West is the new regulatory role played by the federal government in the natural resource economy. This is a role that has been maturing slowly since the early twentieth century.

The fortunes and prominence of the lumber industry within the West further illustrate the limited importance of the exploitation of natural resources in the Western economy. Lumber is a renewable resource managed by the federal and the state governments. Like other natural resources, lumber is harvested in select areas of the West, including the Pacific Northwest and the Intermountain West.

The most tumultuous category of natural resources in this sector of the West's economy is fossil fuels. During the last third of the nineteenth century, prices for these commodities have fluctuated dramatically. The growing national and global demand for energy has increased interest in all of the fossil-fuel resources in the West. They epitomize the volatile nature of the old resource-driven economy of the West before World War II. However, regardless of how much the demand for fossil fuels grows, revenue from their exploitation is by the very nature of these resources finite. Western states have led the nation in the production of fossil fuels during the first decade of the twenty-first century.

The rising price of gasoline in conjunction with increasing US reliance on foreign oil has intensified demands for drilling in Alaska and off the coasts of the USA. From 2002 to 2007, Texas, Alaska, and California have led the nation in drilling oil. A 1.5-million-acre area of coastal plain along the Arctic Ocean in the Arctic National Wildlife Refuge in Alaska is possibly the most famous unexploited source of petroleum in the Western USA. By one estimate, this tract may hold 10.4 billion barrels of crude oil. While the exploitation of these deposits would certainly provide work for participating oil companies in the short run, that being anywhere from ten to thirty years, the oil revenue would not provide long-term economic stability for Alaska. On the contrary, the exploitation of these deposits would possibly, if not probably, result in a boomtown situation with all the economic, demographic, and political instability that accompanies such situations.

Rising oil prices have also sparked interest in oil shale, a form of sedimentary rock containing a petroleum-like substance called kerogen that can be heated into liquid form. The world's largest deposits of oil shale are located in the Green River formation in Colorado, Utah, and Wyoming. The amount of recoverable oil in these deposits may be several times the proven oil reserves in Saudi Arabia. However, the process of recovering oil from oil shale requires a great deal of water and energy. That the Bush administration and members of the 110th Congress considered the exploitation of oil shale in this formation as part of the solution to the nation's energy problem indicates the desperation Americans were feeling about gas prices, an appropriate provocation for another Western boomtown industry.

Western coal deposits offer the promise of profits over a longer timeframe for the companies involved, if the people of the USA decide to accept the expenses involved in the employment of this energy source. Western coal production surpassed Appalachian coal production in 1998. Through 2007, the West has led the nation in coal production, and Wyoming led the West.[29] Over half the nation's demonstrated reserves west of the Mississippi are in the Intermountain West, with Wyoming ranked third and Montana first in 2005.[30] In 2006, Wyoming accounted for 38.4 percent of the nation's total coal production that year.

One "clean" fossil fuel is natural gas. US natural-gas reserves have provoked conversations about rebuilding our energy infrastructure in order to use it more widely. A number of Western states have led the nation in natural-gas production. During the early twenty-first century, Texas led the nation in the number of producing gas wells. Other Western states among the top ten states in this category during this period have been Colorado, Kansas, New Mexico, Oklahoma, and Wyoming.

Uranium mining has been more problematic. Although the industry benefited from demand generated by the production of nuclear weapons and the construction of nuclear power plants during the Cold War, that demand has dropped since the end of the Cold War and the continuing concerns about nuclear energy, including the disposal of nuclear waste. As with fossil fuels, the growing demand for energy has overshadowed concerns about the use of uranium as an energy source. The states of Wyoming, New Mexico, and Utah lead the pack in potential for future uranium production with all the traditional economic benefits that accompany a mining boom.[31] As with other economic endeavors involving the exploitation of natural resources in the modern West, uranium-mining operations will require federal government approval, reflecting the Washington's regulatory role.

Conversely, the West does not stand out in the number of nuclear power plants. This is due to a number of factors, including the relatively greater reliance on rivers for water than in the East. The question of where to store nuclear waste has prompted the politically contentious proposal to build a storage facility in Yucca Mountain in Nevada, which epitomizes the regional stereotype of the West as sparsely populated, even empty. The proposal has provoked a great deal of political controversy growing in part out of long-held feelings in the West that their section is again being discriminated against in favor of the East. Opponents of the Yucca Mountain site complain that political considerations rather than scientific considerations influenced the decision, as demonstrated by the congressional vote to focus on Yucca Mountain as a candidate for this facility before the completion of studies concerning other sites.[32] In particular, opponents stress that the site is approximately 90 miles from Las Vegas, a major entertainment and residential center, despite its isolation from other major cities.

Reliance on profits from fossil fuels and uranium has shaped the economic development of areas and states within the West. A number of Western state governments have configured their revenue systems to take advantage of if not become reliant on such profits. As an example, Wyoming has traditionally relied on the income from farming, cattle ranching, and mining, most recently coal. Wyoming does not have individual or corporate income taxes, and its single largest source of taxes in 1998 was sales tax. Similarly, neither Alaska, Nevada, South Dakota, Texas, nor Washington has levied individual income taxes frequently or at all, and Nevada, Texas, and Washington have largely done without corporate income taxes. Conversely, in 1998, Florida was the only state east of the Mississippi River that did not levy an individual income tax, while all of the Eastern states levied corporate income taxes.[33] The Alaska state government may have gone the farthest in its dependence on income from the exploitation of natural resources with the policy of providing state residents with annual dividend payments from this income. This reliance or dependence has resulted on unpredictable if not unstable revenue streams and spending patterns among these states.

Three sources of energy in the West that promise to transcend the unstable character of Western energy resource economies are geothermal, solar, and wind energy. Unlike coal, oil, and oil shale, which are finite, geothermal, solar, and wind, as theoretically infinite sources of energy, can provide long-term and stable revenue, with the support of consumers and local, state, and the federal governments, including permanent state and federal tax credits. Climate makes various areas of the West particularly profitable for solar panels and windmills with continuing advances in the technology. Away from the nation's coastlines, the Great Plains and the northeastern sector of the Intermountain West holds the greatest potential for wind power, while the desert southwest enjoys the greatest amount of sunshine in the USA. By one estimate, wind power could provide a fifth of the demand for electricity in the USA, with the right investment in infrastructure.

The West has led the nation in the production of electricity through these three renewable sources. In 2001, California led the states in the generation of electricity through geothermal, solar, and wind sources. During that year, California was one of only four Western states in the country that produced over 100 million kilowatt hours, the others being Hawaii, Nevada, and Utah. In the generation of electricity from the sun that year, only Arizona and California produced over 100,000 kilowatt hours. In the generation of electricity from wind that year, while more states produced more than 100,000 kilowatt hours than had been the case with either geothermal or solar power, the majority of those states were west of the Mississippi River.[34]

In addition to fossil fuels, the West produces various non-fuel minerals. In 2007, the top five states in non-fuel mineral production were in the West. All of the major copper-, gold-, molybdenum-, platinum-, and silver-mining operations are west of the Mississippi River. Western states host a number of the leading mining operations of copper, gold, and iron, most of them located in the Intermountain West. In particular, in the early twenty-first century, most of these leading copper operations are in Arizona, and most of the leading gold mines are in Nevada.[35]

Of course, harnessing the West's natural resources has provoked numerous disputes over federal ownership and control of Western lands. In terms of the percentage of land owned by the federal government, the top ten states were in the West in 2004, as listed in Table 20.2. Nevada has been both a symbol and a center of these disputes. One of the most famous, or infamous, of these disputes was the so-called "Sagebrush Rebellion." This "rebellion" converged with a Conservative resurgence punctuated by Ronald Reagan's elec-

Table 20.2 Percentage of federally owned land by state, 2004

Nevada	84.5
Alaska	69.1
Utah	57.4
Oregon	53.1
Idaho	50.2
Arizona	48.1
California	45.3
Wyoming	42.3
New Mexico	41.8
Colorado	36.6

Source: U.S. Census Bureau. 2008 Statistical Abstract. National Data Bank.

tion as president in 1980. Members of the Alaska Independence Party sought control over the state's natural resources without federal government restrictions as a sovereign nation, while still enjoying the military protection and diplomatic support of the USA.

AGRICULTURE

Farming continues to play a significant role in the economies of the Western USA. As has been the case since the late nineteenth century, the fortunes of farmers in the West has been subject to external demands for their products, transportation infrastructure, and the availability of water. A new factor that has been growing in importance during the twentieth century has been increasing demand within the West. With the growth of the population of the West the region has become a market for Western farmers comparable to markets in the East and abroad.

As with other areas of the economy, California and Texas have led the other states in agricultural exports during the early twentieth century. The values of agricultural exports from California and Texas from 2003 through 2007 are listed in Table 20.3. Along with California and Texas, Washington, Iowa, Kansas, Nebraska, and North Dakota have represented the West in the value of exports among the top ten states during this period. California's agriculture is diverse in various parts of the state, not necessarily playing to the state's strengths but counting on the dream of available water.

As with other areas of life in the West, agriculture is influenced if not defined by the relative scarcity of water in the region. This is reflected in the types of crops grown in the West. In a number of areas of the West farmers have not only ignored but have defied this

Table 20.3 Value of agricultural exports from California and Texas in millions of dollars

Year	*U.S.*	*California*	*Texas*
2003	56,014.0	7,423.4(13.25%)	3,330.3(5.95%)
2004	62,408.8	8,713.6(13.96%)	3,875.8(6.21%)
2005	62,516.3	9,465.0(15.14%)	3,621.5(5.79%)
2006	68,592.7	10,608.6(15.47%)	3,624.5(5.28%)
2007	81,947.1	11,301.7(13.79%)	5,198.6(6.34%)

Source: Economic Research Service. U.S. Department of Agriculture.

scarcity through huge irrigation and water-storage projects. One example is the Central Arizona Project designed to harness and control the state's limited water resources.

MANUFACTURING

The development of manufacturing in the West has largely been a recent phenomenon, most of it occurring since 1940. Western manufacturing reflects the underlying differences in the overall economic development of the various areas of the West. As with other sectors of the economy, East as well as West, manufacturing is not distributed evenly through out the region. Most of the manufacturing is located in the Pacific coast–Arizona–Texas crescent. In 2004, California and Texas, the anchors of the Western economy, led the fifty states in value-added manufacturing of all varieties of products.[36]

In the development of manufacturing, the West has benefited from several regional characteristics, including the open space, climate, and informal lifestyle of the West. Another regional characteristic shared by the Southeast, as part of the so-called Sunbelt, is an attitude by local and state business and political leaders that welcomes new businesses. This attitude is marked by support for "right to work" laws, low taxes, less government regulations, and various incentives to attract new businesses, including tax breaks, new infrastructure, and land grants. Since World War II, Western and Sunbelt states have offered the opportunity to build new facilities in a new manufacturing landscape free of many of the impediments of the older manufacturing landscapes in the so-called "Frostbelt" states of the Midwest, the Middle Atlantic, and New England.

Local government and business leaders have led the drive to attract new manufacturing in the West, with the strong support of their state governments. During the latter half of the twentieth century, community leaders in a number of Western cities worked hard to attract companies in industries that rely on the region's intangible characteristics rather than specific tangible natural resources, notably aerospace, defense, and electronics. Phoenix boosters, for example, touted the city's low humidity to attract electronics manufacturers. These industries met Americans' growing demands for consumer goods and the country's military needs in the Cold War. The efforts have driven a competition among Western cities, as well as cities across the nation, for new manufacturing facilities. Western cities, on the other hand, have not made the same efforts to attract the signature manufacturing industries of the "Frostbelt," such as steel.

The Cold War and the Space Race propelled the crystallization of the "Space Crescent," marked by the cities of Houston, southern California, and Seattle, which led in aerospace development and manufacturing. The aerospace communities of this "Space Crescent" were Western "boomtowns" that owed their prosperity to federal spending on the development and production of new military aircraft and on the Mercury, Gemini, and Apollo programs.[37] With the conclusion of the Apollo program in the 1970s, a number of the communities declined just as Western mining towns, such as Bodie, California, declined in the late nineteenth century.

After World War II, Phoenix led the cities of the Southwest between Los Angeles and Dallas in efforts to attract new manufacturers to the city, particularly in aircraft and electronics. These efforts began with the successful campaign during the late 1940s to convince Motorola to build a research and development center in Phoenix. During the 1950s, AiResearch, General Electric, Goodyear Aircraft, Kaiser Aircraft and Electronics, and Sperry Rand followed Motorola in the construction of new facilities in Phoenix. Phoenix business leaders sought businesses that would employ highly trained and well-paid workers in the related defense and electronics industries.[38] In these efforts, Phoenix was part of the larger endeavor of Arizona to turn away from its economy of the nineteenth and early twentieth centuries dominated by mining and farming toward activities that could sustain themselves in the long term.

Boeing epitomizes the development of aircraft manufacturing in the West after World War II. Boeing has attempted to tap into the post-World War II demand for civilian and military jet aircraft. The fortunes of this company have waxed and waned during the latter half of the twentieth century with the fortunes of this industry, including its own success in developing successful aircraft. Delays in the development and production of the 787 epitomize the volatility of the aerospace industry.

Boeing has been an important part of Seattle's economy, providing a number of high-paying jobs. Seattle leaders, like the leaders of a number of Western cities, including Phoenix, have sought to build a local economy based on a wide range of sectors, including technologically advanced manufacturing industries such as aerospace.

The development of manufacturing in the West after World War II, particularly in the aerospace and aircraft, defense, and electronics industries constitutes the linchpin in the West's move away from economic dependence on extractive industries.[39] This development has been shaped by the conscious efforts of business and political leaders first and foremost at the local level to take advantage of the absence of an established large-scale manufacturing base, in contrast with the Northeast and the Midwest, to respond to the military and consumer demands of Cold War America. In this respect, it is not the West that has been a leader but Western cities as national and, to a lesser extent, global competitors.

TRADE AND THE WEST

The involvement of the West in the development of the nation's international trade since World War II is reflective of the West's economic development in that it has varied among the different parts of the region. Since 1999, Texas and California led the fifty states in the value of exports. However, during that period, Washington has been the only Western state among the top ten states of the Union in the value of their exports.[40]

During the twentieth century, the importance of the West in the national trade network has increased, particularly with the growing trade with East Asia, Canada, and Mexico. Canada, China, and Mexico were the USA's top trading partners in 2007. Trade with China and Mexico in particular reflect the growing importance of the West in trade, as a consumer of imports, a producer of exports, and a conduit to the East. By the end of the twentieth century, Americans had realized a dream held since shortly after independence of trade with China, although not in the way imagined.

The North American Free Trade Agreement (NAFTA) has been a significant force in the current economic development of the West. Implemented in 1994, this treaty has significantly increased trade across the West with Canada and Mexico. One manifestation of the trade generated between Mexico and the USA has been the maquiladora, factories in Mexico that assemble parts sent south from US manufacturers into finished products which are then shipped back to the USA. Maquiladora take advantage of the lower labor costs for labor-intensive portions of manufacturing in Mexico and inexpensive transportation between the two countries, particularly through California. Not surprisingly, the largest numbers of these plants are in the Mexican states that border the USA. Among these border states, Baja California has the largest number of maquiladora, among related types of manufacturing facilities.[41]

NAFTA has also increased the flow of exports from individual Western states to Mexico and, to a lesser extent, Canada. In 2003, Texas and California were first and second respectively among the fifty states in the value of exports shipped to Mexico. In fact, the value of exports from those two states was over half the value of all US exports to Mexico that year. Canada purchased a smaller portion of exports from Western states in 2003. California and Texas still led among Western states in exports to Canada, but the value of their exports was only about 13 percent of the US total.[42] This difference is due to the regional orientations of Canada's and Mexico's economies with the Midwest and the Southwest, respectively. The US automakers' subsidiaries in Canada account for much of the ties between Canada and the Midwest.

Of course, NAFTA has raised issues such lower labor costs, unsafe working conditions, and industrial pollution in Mexico, particularly as they affect the USA. Labor activists, including union leaders and environmentalists, prompted by concerns about these issues, have been and continue to be among the strongest opponents of NAFTA.

In terms of proportionate increases, NAFTA has not benefited Western states more than Eastern states in their export trade to Canada and Mexico. Well over half the states enjoyed proportionate increases of more than 50 percent in their export trade to these two countries. Overall, NAFTA has facilitated the integration of the US economy into an overarching North American economy, along with the economies of Canada and Mexico. This integration has resulted in the further internationalization of the Western regional economy, particularly with Mexico because of geographic proximity and historic ties.

As a conduit, the West experienced a greater increase in traffic with Mexico than with the USA. The principle routes between California and Baja California pass between San Diego and Tijuana and Calexico and Mexicali. The principle routes between Canada and Mexico through the eastern half of the Great Plains are Interstate 35 and Interstate 29. Interstate 35's southern terminus is Laredo in south-central Texas on the Rio Grande border with Mexico. Going northward through Texas, Oklahoma, Kansas, Missouri, Iowa, and Minnesota, it ends in Duluth, Minnesota, on Lake Superior. Interstate 29 juts off from Interstate 35 in Kansas City, Missouri, proceeding through Kansas, Nebraska, and the Dakotas, ending on the Canadian border, about 50 miles south of Winnipeg. In terms of border crossings by train or truck, Texas was either first or second, behind Michigan, in 2007.[43]

In a development related to the growth of trade with Mexico, a number of Western cities have become member cities of international metropolitan clusters straddling the US-Mexican border. The largest of these cities are San Diego, Tijuana, El Paso, and Ciudad Juarez. From 1996 to 2007, El Paso experienced a steady increase in overall border crossings from Mexico, although there were spot decreases in individual categories of crossings.[44] There are a number of smaller pairs of US and Mexican cities in and across Arizona and Texas. The growth in the populations of these pairs and in the volume of US–Mexican trade through them reflects the growing importance of trade between the two countries and the globalization of the USA's southwest periphery.

One measure of the growing role of the West in US foreign trade is the trade through the nation's maritime ports. Between 1980 and 2005, west-coast ports exceeded both Gulf and Atlantic coast ports in their respect growth rates of containerized exports and imports.[45] This trend reflects the country's growing trade with Asia. The Pacific-coast ports of Los Angeles, Long Beach, Oakland, and Seattle/Tacoma, along with Houston, are among the nation's busiest. These ports not coincidentally are along the West's urban crescent that runs down the Pacific coast and eastward to the Texas gulf coast. The Ports of Long Beach and Los Angeles in particular have led the rest of the nation's ports in container traffic from 2001 to 2006.[46]

With the economic maturation of the West during the twentieth century, domestic trade within the West and between the West and other regions of the country has grown dramatically. This growth has been facilitated by the construction and enlargement of airports, highways, ports, and railroads, as components of the national transportation infrastructure. Because the USA is one of the largest nations in the world in area, the importance of this domestic trade should not be underestimated. In a way, the USA has fifty state economies. Slightly more than two centuries before NAFTA, the Constitution established what would become one of the wealthiest free trade zones in the world.

TOURISM

During the twentieth century, areas and cities in the West took full advantage of the maturation of the tourist industry made possible by the mass production of affordable automobiles, the country's burgeoning middle class, and the inclusion of leisure time in the work year. Western tourist sites are oriented around the natural and human history of the West, leisure activities, scenery, and climate. In the minds of tourists, what distinguishes sites in the West from those east of the Mississippi River are the regional histories behind them and the open spaces that separate them. These sites constitute a category of natural resources.

One of the most visible categories of Western tourist attractions consists of the scenic wonders in the National Park Service. These parks, including Death Valley, the Grand Canyon, Grand Teton, Lake Mead, Mount Rushmore, Yellowstone, Yosemite, and Zion embody some characteristics of the natural or historic West, and they enjoy among the highest annual visitation rates among in the park system.

The West, like the East, as well as other parts of the world, also has developed the tourist potential of the region's historic sites. In this endeavor, the West has led through the preservation work of the National Park Service, the various state park services, and privately owned historic attractions. The historic themes of these sites have revolved around exploration of the region, conflict between colonizers and indigenous peoples, and boom-time settlements based on the exploitation of natural resources, particularly mining.

The most economically vulnerable leisure activity in the West has been skiing. Despite developing technology to make artificial snow, ski resorts are dependent on snowfall. Western resorts have to compete with resorts in the East and, to a lesser extent, those outside the USA, even North America. Western resorts have the advantages of higher mountains than Eastern resorts and proximity to Western population centers.

A unique tourist site in the West has been Las Vegas. The builders of the first casinos in Las Vegas after World War II exploited two characteristics of southern Nevada: undeveloped and inexpensive land and proximity to other Western centers with resources. The city's increasingly sophisticated entertainment complexes have figuratively if not literally made something out of nothing through the creation of imaginary landscapes on the *tabula rasa* that is the southern Nevada desert. Not having much in the way of conventional natural resources, the Las Vegas entertainment industry made proximity to Los Angeles a resource to be exploited.

Ironically, one of the signature tourist attractions of the West, Disneyland, does not have a traditionally Western theme. It consists of a series of unrelated imaginary landscapes grounded in fantasy and American history. Disneyland is a manifestation of post-World War II urbanization. By building the Magic Kingdom in Anaheim, California, Walt Disney was banking on the continued population growth of southern California, not to mention its appeal to vacationers already attracted by the weather, beaches, and other tourist sites of the area. The fact that Disneyworld in Orlando, Florida, eclipsed Disneyland confirms the phenomenon of post-World War II Sunbelt urbanization, overlapping and reinforcing as it does Western urbanization.

Many of the West's tourist attractions have become integrated into the nation's automobile infrastructure through the course of the twentieth century. During the last century, Americans came to rely increasingly on the mobility that the car could provide to visit Western tourist destinations. Death Valley, the Grand Canyon, Grand Teton, Lake Mead, Mount Rushmore, Yellowstone, Yosemite, and Zion all have enjoyed a steady increase in annual visitation since the early twentieth century. This increase illustrates how the automobile and leisure travel are integrated into the national consumer culture. Interestingly, all of these parks enjoyed their greatest proportionate but not absolute increases either in 1945 or 1946 when Americans began indulging their pent-up desires to take vacations after four years of war and over a decade of depression. These indulgences were part of the phenomenon of Americans spending with a vengeance during the latter half of the 1940s.

Among Western tourist destinations, Alaska and Hawaii are the most exotic. Commercial passenger-jet service has revolutionized the tourist industry in both states. The Hawaiian islands have embodied all the traditional attractions of such destinations, including climate, scenery, and a relaxed lifestyle. However, as a Pacific archipelago, Hawaii adds a new dimension to the Western USA. Hawaii has been able to market its unique island geography along with its indigenous culture to tourists in the USA, as well as around the rest of the Pacific Rim. The growth of Hawaii's tourist industry reflects the globalization of tourism grounded as it is in the growth of middle classes around the Pacific. Alaska's tourist industry has also successfully marketed its natural features, including even the weather, albeit on a smaller scale.

While tourism is not one of the foundational industries in the modern Western economy, its role is measurable. Tourism will be an important sector of the West's economy for the foreseeable future. More importantly, Western tourism is part of a global trend in the marketing of tourist sites, which takes advantage of the growing global middle classes, the crystallization of a global leisure culture patronized by those middle classes, the development of a global transportation infrastructure, and the spread of cultural and historical literacy that transcends traditional national borders. Accordingly, the West's tourist industry, along with the tourist industries of other regions of the world, is dependent on continuing middle-class prosperity, particularly of urban and suburban dwellers in search of something new or unusual.

THE TWENTIETH-CENTURY GROWTH OF WESTERN CITIES

One of the key manifestations of the West's economic primacy if not dominance during the latter half of the twentieth and into the early twentieth-first centuries has been the explosive growth of Western cities and their surrounding suburbs since World War II. This growth has defined the modern American West. This explosive urban growth in the West was driven by a massive migration to the West that began during World War II. Migration to the West during the latter half of the twentieth century dovetails and overlaps with a similar migration from the so-called "Frostbelt" to the so-called "Sunbelt," along with military demobilization and an aberrant and unexpected prosperity.[47] This growth and development in the aftermath of the war was part of a larger trend in Western growth and development. It was these metropolitan areas that have been the driving force in the development of the West since World War II.

The roots of urban and suburban growth in the modern American West go back to the early nineteenth century. Since the nineteenth century, cities have punctuated Western economic development both literally and figuratively.[48] They have served as economic hubs for trade, transportation, and the limited manufacturing in the region prior to World War II. With the exception of the development of new manufacturing industries, not much has changed in the urban West since the Second World War.

Western cities, among US cities as a whole, have grown both in population and land area since World War II. The growth of a number of Western cities, in both geographic area and population, has been dramatic during select decades in the latter half of the twentieth century. The availability of space in southern Arizona attracted migrants during the third quarter of the twentieth century, an attraction common to many growing Western cities. Phoenix and Tucson are the anchors of a linear metropolitan cluster extending along I-10. Phoenix is one of the most striking examples of this dramatic growth. Between 1940 and 1960 the city expanded from 9.6 square miles and 65,414 to 187.4 square miles and 439,170, most of that expansion occurring during the late 1950s. Over the following twenty years, Phoenix grew to 324.1 square miles and 789,704.[49] During the twentieth century, Maricopa County, which includes Phoenix, went from a population density of a little more than two persons per square mile to over 220 people per square mile.[50] In May 1956, the Phoenix city government had made the decision pursue the annexation of surrounding areas aggressively and proactively in order to avoid being hemmed in by emerging suburbs that incorporated themselves, a fate that had occurred to many Eastern cities.[51] Phoenix boosters successfully sold the city as a land of new opportunities that epitomized the American dream as conceived during the latter half of the twentieth century.

A number of factors have contributed to Western urban and suburban growth since the mid-twentieth century. Along with economic development, Western metropolitan growth has been fed by certain features of the West that have attracted migrants, including climate, scenery, recreational opportunities, and living space. These features constitute the appeal that has been one of the defining characteristics of Western cities since the early twentieth century. During the nineteenth century, migrants to Western cities only expected economic opportunities from their new homes. Since the early twentieth century, many newcomers to Western cities have expected their new homes to be nice places to live.

Advances in climate-control technology constitute a vital factor in the growth of Southwestern cities during the latter half of the twentieth century. Without doubt the single most important, if not revolutionary, technological development in climate control has been air conditioning. Air conditioning has been one of the foundations of the population growth experienced by cities across Texas, New Mexico, and Arizona, as well as other cities in states to the east, north, and west. Introduced in the 1920s, air conditioning became an increasingly common feature across the South and the West after World War II. Mass production of residential air-conditioning units began in the 1950s.[52]

The dramatic if not revolutionary impact of air conditioning can be seen most clearly in southern Arizona's metropolitan cluster of Phoenix and Tucson. Air conditioning was a prerequisite for Phoenix growth from 65,414 in 1940 to 1,321,045 in 2000. Mesa, an incorporated suburb of Phoenix, joined Phoenix and Tucson in the 2000 census as the three most populous cities in Arizona and among the seventy-five most populous cities in the USA. While the arrival of new corporations and the establishment of Arizona State University and the University of Arizona drove the economic development and growth of these two cities and their surrounding metropolitan areas, air conditioning made each city livable, *or bearable,* from the spring through the fall.[53] The mild winter climate was one of a number of amenities that attracted migrants, including retirees, from other parts of the country.

The proliferation of and the growing dependence on air conditioning reflect one of the characteristics of the modern American West: a collective defiance of the ecological realities of that region, particularly as manifest in professional sports venues. Air conditioning made it possible for Phoenix to acquire its own franchises in the National Football League (NFL) and the National League. The Cardinals moved from St. Louis to Phoenix in 1988, and the Diamondbacks came to Phoenix in 1998. In 1994, the Phoenix Cardinals became known as the Arizona Cardinals, an appropriate change considering how many Arizonans live in the greater Phoenix metropolitan area.

One of the most visible symbols of this denial by professional sports is Chase Field in Phoenix. The official website of the Arizona Diamondbacks, the home team of this stadium, brags about the scale of the air-conditioning plant, which consists of an 8,000-ton cooling system "that is the equivalent of air conditioning for more than 2,500 typical Arizona homes."[54] This system, designed to cool the seating areas of the stadium, was meant to decrease the temperature in the cooled areas by 30 degrees in three hours.[55] As manifest in Chase Field, among other engineering efforts to combat nature, the West leads the nation and is among a number of other regions around the world that lead in this environmental defiance.

Another technological development that has made moving west a less lonely experience has been commercial air-passenger services. The introduction of jets such as the 707, the 727, and the DC-8, has made it possible to travel across the country in one day. Commercial jet passenger service also strengthened ties between the contiguous states of the trans-Mississippi West and Alaska and Hawaii. Unlike westward-bound migrants of the nineteenth century, those Americans moving westward after World War II have not cut themselves off from visits with friends and relatives back east.

The growing number of air travelers, along with growing trade with Asia, Canada, and Mexico, also has considerably increased the size of Western airports, particularly in the crescent running from Seattle down the Pacific coast, eastward across to Phoenix, and then through to Dallas–Fort Worth and Houston, along with the Intermountain West cities of Denver and Salt Lake City. In 2005, six of the nation's top ten airports in terms of passengers were located in this Western arc. On the other hand, extending the list of the nation's busiest airports to the top forty in 2005, fifteen were in the West.[56] The overall dominance of the Pacific coast–Southwest crescent in air traffic also is reflected in the distribution of large air-traffic hubs. These hubs include the Seattle, San Francisco, Los Angeles, Phoenix, Dallas–Fort Worth, and Houston International Airports. The three large air-traffic hubs in the Interior West are at the Denver, Las Vegas, and Salt Lake City International Airports.[57] These airports have become hubs for a number of major air carriers. As examples, the Dallas–Fort Worth and Houston International Airports are important centers for American and Continental Airlines, respectively, while Salt Lake City International Airport is a hub for Delta Airlines.

Because of the land, money, and other resources required for airport construction, rival cities in many areas of the West have had to cooperate for their common good. The Dallas–Fort Worth International Airport is one of the most visible products of such cooperation. Although proponents of such a joint airport raised this idea as early as 1927, planning for this project didn't begin until 1962 when the Civil Aeronautics Board initiated an investigation of this proposal. It took the post-World War II growth of both cities and the realization that only the area between them provided suitable locations for such an airport to overcome the rivalry between boosters of Dallas Love Field and the Greater Fort Worth International Airport boosters. The airport opened officially in 1973.[58] In 2005, Dallas–Fort Worth ranked third, behind Atlanta and Chicago, in passengers enplaned.[59]

Western urban growth has fed the boomtown industry of home construction. Among the four recognized regions in the US census (the Northeast, the Midwest, the South, and the West) the West has been in second place, behind the South, in new privately owned single-family houses from 1980 to 2005.[60] From 1994 to 2006, the median sales prices for existing single-family houses sold during this period was the highest in the West.

While the constructed homes were meant to be permanent, the construction communities that built them resemble the old boomtowns of the nineteenth-century West, virtually if not physically. The most significant difference is that the nineteenth-century boomtowns became ghost towns, leaving only their physical remnants, while the construction communities that have been developing subdivisions since the mid-twentieth century are truly nomadic, leaving nothing but unused construction materials and empty drink cans and bottles.

One of the paradoxes of the modern urban West is that because of the disproportionately greater growth experienced by a number of Western cities among US cities as a whole since the end of World War II the Western states in which these cities are located have shared in that disproportionate growth among all the states. A number of these states at the same time have large sparsely inhabited areas. Among the ten states with the greatest urban population density in the 2000 census, six were in the West. In the category of urban land proportion in the 2000 census, the twelve states with the lowest proportion of urban land area were in the West, located throughout the region. Nevada is the most extreme example of this trend. Ranked fourth in urban density in the 2000 census, Nevada was forty-sixth in percentage of urban land in the state. Although Las Vegas was the thirty-second most populous city in the US in 2000, aside from Carson City and Reno, the Silver State has no other sizeable cities. This paradox reflects the dual nature of the modern West, a geographically large, rural, and sparsely inhabited region punctuated on densely populated metropolitan areas. This paradox is addressed in Table 20.4, which concerns urban density and land percentage in select Western states.

More than any other state, Nevada epitomizes the model of urbanization as being the engine of population growth during the latter half of the twentieth century. Las Vegas, one of the state's few urban areas and its dominant population center, has grown from 8,422 in 1940 to 478,434 in 2000, becoming the thirty-second most populous city in the country. Between 1990 and 2000, the population of the Las Vegas metropolitan area nearly doubled from 852,737 to an astounding 1,563,282. Over the course of the twentieth century, Clark County, in which Las Vegas is located, went from a population density of just under half a person per square mile to over ninety people per square mile.[61] During the late twentieth century, Las Vegas embodied the old boom-time mentality of the West in house construction. The post-World War II growth of Las Vegas has driven Nevada's dramatic growth. From 1950 to 2000, Nevada has experienced one of the highest rates of proportionate growth among the fifty states.

Table 20.4 Urban density and land percentage

State	*Urban density rank*	*Urban land % rank*
California	3	22
Nevada	4	46
Hawaii	5	21
Colorado	7	39
Utah	8	43
Arizona	10	37

The Silver State also reflects more than any other state the limits imposed by water on population growth in the West. Population for the state, particularly Las Vegas, has far outstripped the growth of water supplies. As a result of drawing on local groundwater resources during the latter half of the twentieth century, land in downtown Las Vegas, the Strip, and north Las Vegas has subsided five to ten feet, exacerbating the problem of growing earth fissures.[62]

Water scarcity has shaped the economic development and politics of the Southwest, and, to a lesser extent, the entire West since at least the late nineteenth century. With the growth in demand across the Southwest, the competition for water among competing states, cities, and groups of users has intensified. At the forefront of these competitions is that between farmers and cities. Because city residents use less water than farmers per capita, cities have had the advantage in this competition. The greater amount of water used by farmers than urban consumers indicates that the trend of farms giving way to urban, or suburban, sprawl will continue.

The main watershed in the Southwest since the early twentieth century has been the Colorado River. The key agreement on the allotment of water from the Colorado was the 1922 Colorado River Commission Compact which sought to resolve disputes over the distribution of the river's water among seven participating states. Possibly the greatest dispute that the compact was meant to settle was between the upper-basin states of Colorado, New Mexico, Utah, and Wyoming and the lower-basin states of Arizona, California, and Nevada. The dispute reflected one of the defining divisions in the West between the faster-developing periphery of the Pacific coast and the states on the Mexican border and the slower-developing Intermountain West. The primary flaw in this compact is the optimistic allotments agreed to by the signatories. Consequently, between increasing demand and periodic droughts, these states, along with cities, user groups, Indian tribes, and Mexico, have competed with each other for their shares in the shadow of growing scarcity. As Norris Hundley explains in his seminal study of the Colorado River compact, the compact has seen increased intervention by Congress and the US Supreme Court instead of ending such intervention as intended. In 1990, Arizona, California, and Nevada consumed their legal allotments of water from the Colorado River.[63] Overuse of the Colorado River has had a number of consequences, including increased salinity.

During the latter half of the twentieth century, local, state, and federal-government officials have largely reacted to this dramatic growth in demand, being unable or unwilling to get much ahead of this problem. This population growth in each of the Western states from 1940 to 2000 is listed by decade in Table 20.5. Las Vegas residents, among the citizens of a number of other fast-growing Southwestern cities, have had to employ a variety of dif-

Table 20.5 Resident population percentage change by state, 1940–2000

State	*1940–50*	*1950–60*	*1960–70*	*1970–80*	*1980–90*	*1990–2000*
Pacific Basin States	47.75	51.2	27.15	29.65	25.9	11.65
Alaska	77.4	75.8	32.8	33.8	36.9	14.0
Hawaii	18.1	26.6	21.5	25.5	14.9	9.3
Desert Southwest States	39.1	56.65	21.4	40.85	25.55	30.05
Arizona	50.1	73.7	36.0	53.5	34.8	40.0
New Mexico	28.1	39.6	6.8	28.2	16.3	20.1
Pacific Coast States	43.3	28.23	21.57	21.9	17.13	18.43
California	53.3	48.5	27.0	18.6	25.7	13.8
Oregon	39.6	16.3	18.2	25.9	7.9	20.4
Washington	37.0	19.9	19.5	21.2	17.8	21.1
Inter-Mountain West States	23.33	30.17	21.07	38.28	14.82	29.47
Colorado	18.0	32.4	25.8	30.9	14.0	30.6
Idaho	12.1	13.3	6.8	32.5	6.7	28.5
Montana	5.6	14.2	2.9	13.3	1.6	12.9
Nevada	45.2	78.2	71.3	63.8	50.1	66.3
Utah	25.2	29.3	18.9	37.9	17.9	29.6
Wyoming	15.9	13.6	0.7	41.3	–3.4	8.9
Northern Plains States	1.125	6.8	0.925	5.125	1	6.475
Kansas	5.8	14.3	3.1	5.2	4.8	8.5
Nebraska	0.7	6.5	5.1	5.8	0.5	8.4
North Dakota	–3.5	2.1	-2.3	5.7	–2.1	0.5
South Dakota	1.5	4.3	-2.2	3.8	0.8	8.5
Southern Plains States	7.9	14.25	13.4	22.65	11.7	16.25
Oklahoma	-4.4	4.3	9.9	18.2	4.0	9.7
Texas	20.2	24.2	16.9	27.1	19.4	22.8
U.S. Average	14.5	19	13.3	11.5	9.8	13.2

Source: U.S Census Bureau. Statistical Abstract of the U.S.

ferent means for conserving and recycling water, foreshadowing the employment of such approaches in other areas of the West.

THE EMERGENCE OF WESTERN CITIES AS LEADERS

During the twentieth century, the West not only became increasingly urban, but it has also has led the East in urban development. In 1900, of the seventy-five most populous cities in the USA, only eight were located west of the Mississippi River. In 2000, of the seventy-five most populous cities in the USA, thirty-seven were in the West. More significantly, six of the ten most populous cities in 2000 were in the periphery of the West along the Pacific coast and the Mexican border. This growth in the population of the urban West has been part of two overlapping shifts in population in the USA since World War II from the Northeast and the Midwest to the South and the West.

The West also holds a number of the nation's fastest-growing cities. Some of these fast-growing Western cities are reminiscent of Western boomtowns with respect to the pace of their growth. Houston has risen from the forty-fifth most populous US city in 1920, to the twenty-first in 1940, to the seventh in 1960, to the fifth in 1980, to the fourth in 2000. Phoenix has gone from twenty-ninth in 1960 to ninth in 1980, to sixth in 2000. Los Angeles made its dramatic leap earlier than Houston and Phoenix, from thirty-sixth in 1900 to tenth in 1920, ultimately rising to second in 2000. Another mark of the dominance of these cities in urban America is the rankings of their international airports, among the busiest airports in the USA. In terms of passengers emplaned in 2005, the Los Angeles, Phoenix, and Houston International Airports ranked fourth, seventh, and eighth among US airports.[64]

The leading cities of the West, which are also included among the nation's leading cities, lie on the arc that runs along the Pacific coast, across Arizona, and into Texas. The cities and urban areas that anchor this arc include Seattle, San Francisco and San Jose, Los Angeles and Orange County, San Diego, Phoenix, Dallas and Fort Worth, and Houston. The population growth for Dallas, Houston, Los Angeles, and Phoenix, from 1940 to 2000 can be seen in Table 20.6. These cities lead the nation as well as the West in professional sports.

The wide open spaces of the West have placed Western cities on the leading edge of the dominant phenomenon of urban growth during the latter half of the twentieth century: suburban geographic expansion. Since World War II, suburbanization has been at the cutting edge of Western as well as national growth. Many of the hinterlands to which Carlos Schwantes referred have filled with suburbs, one of the hallmarks of Western and US demographic development during the latter half of the twentieth century.

This sprawl has taken two forms. In a number of metropolitan areas such as Dallas and Fort Worth, this growth has proceeded outward from a central core. In other metropolitan areas consisting of a number of centers, such as southern California, growth proceeds outward from all of the cores. Los Angeles has led the nation in this second form of low-density urban sprawl. In 1930, Los Angeles led the nation's cities of comparable size in the percentage of residents who lived in single-family houses. Since World War II, other cities in the West, as well as across the Sunbelt, have followed this trend. Another manifestation of Los Angeles' leadership in low-density urban sprawl has been the diffusion of shopping across the expanding metropolitan areas, another trend followed by many other US cities.[65]

The foundation of urban sprawl has been the automobile, and Western cities have again been on the leading edge of the nation's growing reliance, even dependence, on the car

Table 20.6 Population growth of select Western cities during the latter half of the twentieth century

	1940	*1960*	*1980*	*2000*
Dallas	294,734	679,684	904,078	1,188,580
Houston	384,514	938,219	1,595,138	1,953,631
Los Angeles	1,504,277	2,479,015	2,966,850	3,694,820
Phoenix	65,414	439,170	789,749	1,321,045

during the latter half of the twentieth century. The personal freedom to travel wherever and whenever one chose fit in well with the freedom of open spaces and available land associated with the American West. The geographic expansion of Western cities, and, to a lesser degree, US cities overall, has been predicated on automobiles as the primary form of transportation. Los Angeles has led the nation's cities in its residents' reliance on cars during most of the twentieth century. As early as 1930, Los Angeles County had more cars per resident than any other city in the world. By the late 1930s, county residents used cars for 80 percent of all their local traveling.[66]

One aspect of urban development in which Western cities have lagged behind Eastern cities, and are thus being led rather than leading, is public transportation. In 2005, California and Hawaii were the only Western states in which workers matched or exceeded the national average in the percentage of workers who use public transportation to commute to work.[67] During the latter half of the twentieth century, many Western cities relied primarily if not exclusively on buses for public transportation.

The failure of most Western cities to develop adequate public transportation has come at a cost for the residents of these cities in terms of air pollution and traffic congestion. The dedication of Westerners to the automobile culture has slowed efforts to solve these problems. During the latter half of the twentieth century, many Westerners made a choice. However, fatigue with traffic congestion, the increasing cost of gasoline and growing concerns about the environmental impact of internal combustion engines have prompted a number of Western cities to construct rail and light-rail systems. It was not until 1990 that the Blue Line of the Los Angeles Metro Rail system began operation. The Blue Line is the first of five light-rail and rapid-transit lines running across Los Angeles County as part of this system. Los Angeles trails Eastern cities such as Atlanta and Washington, DC. Certainly, no Western city and few Eastern cities have a subway system as comprehensive as that in New York City. In the construction of public transportation, Western US cities are in the pack among other cities around the world.

One of the defining characteristics of the urban domination of the West has been suburban growth at the expense of agricultural areas. The proportionately greater amount of water required to irrigate farm lands in the dry West accelerated the transformation of farms into residential housing developments. As one example, by 1980 Tucson had purchased over 10,000 acres of irrigated farm land in order to put the land to other uses that required less water.[68]

During the latter half of the twentieth century, a number of Western cities also have become leading cities in the nation. Three such leading Western cities illustrate this development. Based on its location near the Gulf of Mexico, its proximity to Texas oil fields, its oil-refining industry, and its mild winter climate, Houston has grown to become not only the Houston most populous city in the country but also one of the centers of oil production. During the 1940s and 1950s, business and political leaders in Phoenix focused on attracting manufacturers, particularly in electronics, to their city toward the end of making Phoenix the leading metropolis in the Southwest east of Los Angeles.[69]

During the early and mid-twentieth century, Los Angeles became a national cultural leader through the movie industry that developed there. The uncommon combination of diverse topography and a semi-arid climate made Los Angeles County an ideal location for the early film industry during the early twentieth century, requiring as it did sunny weather because of the inadequacies of artificial lighting. One of the pioneers of the movie industry, D. W. Griffith, took advantage of this sunshine when he came to Los Angeles in the winter of 1910. In his essay, "Improbable Los Angeles," David Clark explains that the Los Angeles movie industry "popularized" the Southern California style of dress across the nation and made Los Angeles one of the centers of gravity in US culture for the radio, television, and recording industries. The Los Angeles movie and television industries constitute one of the foundations of the national culture that crystallized in the USA during the twentieth century.

In their emergence as major cities in the nation as well as the West, Houston, Los Angeles, and Phoenix have surpassed nearby competitors. Until the early twentieth century, Galveston vied with Houston for the honor of being the dominant city along Texas's gulf coast. Because of the damage done to Galveston by a hurricane in 1900, Houston outpaced it in growth after that. By 2000, both Houston and Dallas had surpassed San Antonio as the Lone Star State's most populous cities. This urban competition is certainly not unique to the West. It occurs on many different levels throughout the world.

The locations of Western cities illustrate the trend in national and Western urbanization toward the development of extended metropolitan areas, metropolitan clusters, and megalopolises. A number of these exist throughout the West, reflecting the population distributions throughout the region, particularly the domination of the region by the Pacific coastal and Mexican border peripheries. Anchoring either end of an east–west axis along the southern periphery of the West are two sets of metropolitan clusters and megalopolises that are two of the region's most important economic hubs in terms of manufacturing, trade, and transportation. At the western end of this axis is the developing megalopolis running from San Diego to San Francisco. This megalopolis consists of two metropolitan clusters in southern California, a third in north-central California, and a line of cities in the Central Valley between the northern and the southern clusters. This megalopolis has made California the most populous state in the Union.

The largest of these four metropolitan clusters is the southern California coastal cluster of Los Angeles and Orange Counties, which during the second half of the twentieth century expanded inland to include Riverside and San Bernadino Counties and northward into Ventura County. The other metropolitan cluster in southern California centers on San Diego, the state's second most populous city. The third cluster in north-central California centers on the cities of Oakland, San Francisco, and San Jose. The Central Valley agriculturally oriented urban corridor is the smallest in population and the longest in linear distance among the four in California. It is marked by three metropolitan centers: Bakersfield, a little over 100 miles north of Los Angeles, through Fresno, a little more than 100 miles north of Bakersfield, to Sacramento, northeast of San Francisco. This cluster also functions as a transportation corridor between northern and southern California, as part of the longer transportation corridor marked by Interstate 5 through the Pacific coast states.

This megalopolis reflects the economic diversity of California, as well as the mosaic nature of the West, composed as it is of distinct and diverse regions. In terms of modern urban culture, California's megalopolis is more than the sum of its component metropolitan areas. This California megalopolis is also one of a number of harbingers throughout the West, the nation, and the world of the growing importance if not primacy of metropolitan clusters.

At the other end of this axis is the metropolitan corridor along Interstate 35, punctuated by four of Texas's six most populous cities in 2006: Austin, Dallas, Fort Worth, and San Antonio. Houston, Texas's most populous city, and the fourth most populous city in the

USA in 2006, is at the southern end of Interstate 45, the north–south interstate that runs parallel to Interstate 35 in the eastern half of Texas. The economy of this metropolitan corridor is dominated by oil-refining, chemical production, manufacturing, transportation, and trade with Mexico.[70] El Paso, Texas's sixth most populous city, and located at the western end of the state, is at the northern end of Mexico's principle transportation through the middle of northern Mexico.

California and Texas have dominated the economic development and growth in the West as well as the nation as a whole. These states are the anchors of the periphery of states running down the Pacific coast and eastward along the Mexican border that dominates the West and is a leading section of the nation and a leading region of the world. By land area, Texas and California are the second and third largest states. After World War II, California and Texas became the first and second most populous states.

One measure of the economic primacy of California and Texas in the region as well as the nation is energy consumption. In the commercial, residential, and transportation sectors of the economy in 2005, California and Texas were first or second in energy consumption. In the industrial sector that year, Texas was first, Louisiana second, and California third. In total energy consumption that year, Texas was first and California second. However, Western states on the whole lag behind Eastern states in total energy consumption.[71] Another measure of the economic dominance of the Pacific coast–Southwest crescent is the gross domestic products of the anchor states of the crescent. In 2005, California and Texas were first and second in this economic measure.

During the latter half of the twentieth century, one of the most visible signs of the growth of Western cities and their new leadership role among the nation's and the world's cities has been the relocation of existing professional sports teams from Eastern to Western cities and the creation of new teams for Western cities. Prior to World War II, the trans-Mississippi West was a hinterland for professional sports. Because of the distances from each other and cities in the East and because of lower population densities, Western cities could attract only minor league teams in professional sports during the first half of the twentieth century. With the dramatic population growth of select Western cities and the development of transcontinental jet passenger service, teams could travel quickly, if not easily, across the nation.

The selection of Western cities as hosts for major league professional sports teams has been determined by a number of factors, including population, fan base, and the ability of local leaders to raise money for stadium construction. Most of these cities are located on the arc along the Pacific coast, across southern Arizona, into Texas. This arc reveals a number of characteristics about the modern urban West, as well as the selection processes of Major League Baseball (MLB), the National Basketball Association (NBA), the NFL, and Major League Soccer (MLS). In the Intermountain West, Denver and Salt Lake City, as the major urban centers, have the only fan bases large enough to attract major league sports teams.

In 1957, the Dodgers led the way, moving from Brooklyn to Los Angeles. The intensifying competition among cities across the country is epitomized by the Dodgers' move to Los Angeles. The failure of New York City leaders to agree on a site for a new stadium for the Brooklyn Dodgers that would be more accessible for the growing number of suburban fans provided an opportunity that Los Angeles leaders exploited. They proposed a site for a new stadium at or near the junction of several "freeways."[72] This competition for manufacturing complexes and sporting teams and events has been particularly sharp and pronounced.

In the wake of the Dodgers' move, both extant and new MLB franchises have come to Western cities, mostly along the populated crescent running down the Pacific coast from Seattle through the California coastal metropolises, eastward across Arizona to Phoenix then to Dallas and Houston, Texas. The one Western MLB team outside of this crescent is the Denver Rockies, Denver being the hub city of the Rockies.

Similarly, Western cities have also attracted new or extant NBA franchises. Twelve of the NBA's thirty teams are located west of the Mississippi River. These twelve franchises are located on the same crescent along the Pacific coast and to the east in Phoenix and Texas. California has four teams, Texas three, and Oregon and Washington each have one, in Portland and Seattle, respectively. Within the Intermountain West, both Denver and Salt Lake City have NBA franchises. At least in the NBA, Plains cities may have finally made a breakthrough with the announcement in September 2008 that Oklahoma City would become the home of a new franchise: the Thunder.

NFL franchises are distributed in the same geographic pattern in the West. Out of thirty-two teams in the NFL, nine are located in the West. Three are in California, interestingly Los Angeles not included. Two are in Texas: Dallas and Houston. The other four are in Denver, Kansas City, Phoenix, and Seattle. The Kansas City NFL franchise, like its MLB counterpart, is the only host city not in the Western coastal or border periphery or the Intermountain West. Of course, professional football differs from professional baseball and basketball in that football games occur only once a week, making a single football game more important for the NFL than a single baseball or basketball game is for MLB or the NBA.

Because MLS was founded in late 1993, well after the crystallization of regional parity between the East and the West, the fourteen (sixteen by 2010) teams are evenly divided between the two sections. The host cities of the Western Conference teams lie along the same arc from the Pacific coast to Texas occupied by the major league teams of baseball, basketball, and football. Like these other sports leagues, MLS selected cities for the Western Conference that are major population centers and regional centers or hubs. In addition, MLS has selected cities on the basis of their ability to accommodate international audiences and to forge ties with soccer teams and competitions abroad. Jorge Vergara, the owner of Chivas USA, located in Carson, south of Los Angeles, has attempted to capitalize on the global popularity of soccer by appealing to both English- and Spanish-speaking fans and by highlighting his team's connections abroad. Along with Chivas USA, he owns professional teams in China and Costa Rica and fourteen soccer academies in Arizona, California, and Texas. He plans on adding teams in Europe and India to these holdings.[73]

Not surprisingly, the National Hockey League (NHL) has a smaller presence in the Western USA than MLB, the NFL, the NBA, and MLS. Of the thirty NHL teams, six are located in the Western USA. The locations of five of these six teams have been dictated by the size of the potential fan base, not proximity to winter sports venues. These five teams call the cities of Anaheim, Dallas, Los Angeles, Phoenix, and San Jose home, along the densely populated peripheral arc of the West. These teams have tapped into the popularity of street hockey. The sixth, Denver, is one of the few major cities in the Intermountain West with a winter climate appropriate to ice hockey. Anaheim, Los Angeles, and San Jose are part of the California megalopolis. The acquisition of a NHL franchise by these cities in the Western USA reflects not only their integration into the professional hockey subculture of American sports but also their pursuit of status as world-class sports cities able to compete with other major cities around the world in the arena of professional spectator sports.

The Anaheim Ducks of the NHL reflects a number of trends in American spectator sports during the late twentieth century, including increasingly sophisticated marketing operations and the placement of professional spectator sport venues in large-scale recreational and shopping complexes. The Ducks was founded by the Walt Disney Company in 1993. Located in north Orange County, south of Los Angeles country, Anaheim is part of the huge southern California megalopolis. The Disney Company's selection of Anaheim as the Ducks' home harnessed and added to Anaheim's entertainment and recreational activities. The Ducks' home stadium, originally Arrowhead Pond, now known as the Honda Center, is located on an axis, east of Disneyland and on the other side of the Orange Freeway from Anaheim Stadium, home of the Angels. Located next to the Orange Freeway, the Ducks' home stadium is able to draw fans from the huge metropolitan cluster around Los Angeles and Orange County. The Ducks are part of a trend in cities around the world of clustering entertainment and recreational venues in a concentrated area accessible by major highways. However, fans have had to deal with the accompanying trend of traffic congestion that is one of the defining characteristics of metropolitan sprawl in the USA today. No doubt this congestion, a feature of life in the densely populated periphery of the West, along with other factors such as cost, has discouraged many would-be spectators from attending Ducks games.

A number of other Western, as well as Eastern, cities have built stadiums near other entertainment and recreational venues. In Phoenix, Chase Field, home of the Diamondbacks, is located one block from the US Airways Center, home of the Phoenix Suns. This was a purposeful effort to make this area of Phoenix the sports center of the city and the greater metropolitan area, if not, more distantly, of the state.[74] In Denver, Coors Field, home of the Denver Rockies, is within walking distance of the Pepsi Center, which hosts the city's Arena Football League, NBA, MLS, Major League Lacrosse, and NHL franchises. In Seattle, Qwest Field, home of the Seahawks, and Safeco Field, home of the Mariners, are within a mile of each other in the southern end of the central city.

Owners of professional sports stadiums, in the East as well as in the West, are also attempting to attract events beyond the identified sports competitions to these stadiums. Dave Checketts, owner of the Real Salt Lake of MLS, hopes that the new team's recently completed stadium, Rio Tinto Stadium, will bring new kinds of events to the Salt Lake City area. In order to attract people, the stadium boasts the same kinds of luxury seats and suites commonly found in professional sports stadiums across the USA. The stadium's designers also built the stadium in Sandy, Utah, reportedly fifteen minutes by car from downtown and with a view of both the Oquirrh and the Wasatch Mountains, two of the area's signature scenic sites.[75] This stadium is part of the city's efforts to become a national if not international center for culture, entertainment, and sports, an aspiration common to many cities in the Western USA, the country as a whole, and across the world.

The construction of these sports venues also reflected the new roles played by local and, to a lesser extent, state governments in financing these projects. Across the country, sports stadiums have been financed by private capital and local, state, and federal government grants, loans, and loan guarantees. Local governments played a direct part in raising money for construction primarily through the dedication of part of a local sales tax.

Not surprisingly, fewer major-league sports teams are located in the Intermountain West, including the states of Colorado, Idaho, Montana, Nevada, Utah, and Wyoming, than in the more densely populated coastal and border peripheries. Denver and Salt Lake City reflect this generalization. Being the largest cities in this region, they are the only ones to host any major-league franchises. In this respect, each city is a hub for a less densely populated hinterland. This role has generated revenue for the entertainment and leisure sectors of each city's economy.

Alaska and Hawaii confirm these generalizations about the spread of professional sports teams to the West. Because of their distance from the other forty-eight states and their relatively small sports markets, both states are still not represented in any of these leagues. However, continued population growth in each state might alter that situation. Already, University of Hawaii sports teams travel to contiguous states for games. With the growing number of college sporting contests which are televised, Hawaiian sports will become increasingly integrated into the national sports culture and economy.

The metamorphosis of what had been regional professional sports networks in the first half of the twentieth century into national and transcontinental networks during the second half of the twentieth century epitomized the integration of Western cities into the national urban spectator sports subculture of modern American culture, a key indicator of their maturation as US cities. These new Western sports teams have also become part of a global spectator sports culture defined by huge marketing operations, both national and international, and divided into different classes of spectators based largely on disposable income.

The West also has played a leading role in the Olympics. In addition to the location of the national Olympic training center being located in Boulder, Colorado, a number of Western cities hosted various Olympics during the twentieth century, attesting to the economic, organization, and political maturity of the region. Los Angeles has hosted two of the four Summer Olympics held in the USA since their modern inception in 1896. Squaw Valley and Salt Lake City have hosted two of the four Winter Olympics held in the USA since they began in 1924.

THE FUTURE

So what is the economic future of the West? The answer to that question lies more with the cities of the region than the region as a whole, particularly those cities in the coastal-border periphery. Certainly, part of the West's economic future lies in the continued hopes for the profitable exploitation of the region's non-renewable and renewable natural resources. The growing global demand for energy, punctuated by the dramatic growth of consumption in China and India, has renewed efforts to harvest various fossil-fuel deposits in the West. However, revenue from these deposits is dependent upon external factors and reflects the continuing exploitive colonial character of the natural resources of the Western economy. Many involved in these exploitative economic activities accept the unstable nature of their industries as defined by "boom-and-bust" cycles. Continued military spending also will play a role in the economy of the West with the further development of US economic and strategic interests in the Pacific and East Asia.

While the West provides an economic, geographic, and historic context for the region's cities, the dominance of those cities in the region and beyond transcends any limited regional identity. As with the rest of the nation and the rest of the world, Western cities are and will be the driving forces of the region's economy. At a time when the primacy of nations has subsided somewhat in the international economy and the political arena, cities have at least partially filled the void in leadership. In this increasingly important role, Western cities, among other cities around the world, have become homes for multinational corporations, venues for international sports competitions, and centers of the global urban middle classes. Los Angeles, San Diego, and San Francisco, acting through the California state legislature, can influence the product lines of auto manufacturers and, in this way, may even challenge the authority, even the relevance, of the federal government. Regions and cities of the West also drive the national and international consumer cultures.

Of course, the West faces a number of economic challenges, the greatest being the growing demand for limited, even scarce, water resources. The challenge presented by the growing scarcity of water in the West is exacerbated by the struggle for this resource among competing states, cities, and groups. This challenge, however, presents the West with the opportunity to develop new solutions to problems growing out of its limited water resources and in this way lead other areas of the world facing these problems.

Of all the West's major cities, Las Vegas may be at the leading edge of problems growing out of the increasing scarcity of water in the region. It faces the possibility of major economic dislocations if not disruptions if steps are not taken to address problems growing out the overuse of local and regional water resources. Ultimately, if Westerners want to avoid such economic dislocations and disruptions throughout the region they will have to stop ignoring if not defying the region's limited water resources.

The economic dominance of the West is the product of a number of intersecting trends and events during the twentieth century. With the population growth of the region, the nation's growing economic and strategic presence in the Pacific Ocean and Asia, and the integration of the region into the national culture and economy, the federal government dramatically increased its expenditures on the region.

Economically, the West has not dominated the rest of the nation. Rather, it has been in the vanguard of change and development in the USA since World War II. The driving force of this vanguard has been the states of California and Texas as the anchors of the coastal and border periphery of the West. Leading this vanguard has been the cities of these states. As they have throughout the history of the USA beyond the Mississippi, cities continue to be on the leading edge.

NOTES

1. David Kennedy, "Can the West Lead Us to a Better Place?," *Stanford Magazine,* 37 (3) (2008): 46–53; p. 46.
2. Carol A. O'Connor, "A Region of Cities," in Clyde A. Milner II, Carol A. O'Connor, and Martha A. Sandweiss (eds.), *The Oxford History of the American West* (Oxford: Oxford University Press, 1994), pp. 535–563; p. 535.
3. Carlos A. Schwantes, "Wage Earners and Wealth Makers," in Clyde A. Milner II, Carol A. O'Connor, and Martha A. Sandweiss (eds.), *The Oxford History of the American West* (Oxford: Oxford University Press, 1994), pp. 431–467; p. 432.
4. Donald Worster, *Rivers of Empire: Water, Aridity, and the Growth of the American West* (New York: Pantheon Books 1985).
5. David L. Clark, "Improbable Los Angeles," in Richard M. Bernard and Bradley R. Rice (eds.), *Sunbelt Cities: Politics and Growth since World War II* (Austin, Tex.: University of Texas Press, 1983), pp. 268–308; p. 272; Bradford Luckingham, *The Urban Southwest: A Profile History of Albuquerque, El Paso, Phoenix, and Tucson* (El Paso, Tex.: Texas Western Press, 1982), pp. 40–43.
6. Michael Malone and Richard Etulain, *The American West: A Twentieth-Century History* (Lincoln, Nebr.: University of Nebraska Press, 1989), p. 97.
7. Malone and Etulain, *The American West,* pp. 94–99.

8. Malone and Etulain, *The American West,* p. 228.
9. Peter Wiley and Robert Gottlieb, *Empires in the Sun: The Rise of the New American West* (Tucson, Ariz.: University of Arizona Press, 1985), p. 209.
10. Wiley and Gottlieb, *Empires in the Sun,* p. 210.
11. Luckingham, *The Urban Southwest,* pp. 75–76.
12. Wiley and Gottlieb, *Empires in the Sun,* p. 145.
13. Gerald Nash, *The American West Transformed: The Impact of the Second World War* (Bloomington, Ind.: Indiana University Press, 1985), p. 83.
14. Clark, "Improbable Los Angeles," p. 283.
15. Nash, *The American West Transformed,* pp. 25–26.
16. Nash, *The American West Transformed,* pp. 153–156.
17. William Cronon, "Landscapes of Abundance and Scarcity," in Clyde A. Milner II, Carol A. O'Connor, and Martha A. Sandweiss (eds.), *The Oxford History of the American West* (Oxford: Oxford University Press, 1994), pp. 603–637; pp. 624–625.
18. Nash, *The American West Transformed,* p. 38.
19. Malone and Etulain, *The American West,* p. 224.
20. Luckingham, *The Urban Southwest,* pp. 75–76.
21. Although there are no studies on the impact of the Cold War, the Korean War, or the Vietnam War on the West specifically, the impact of each is treated in a number of other articles and books. In addition, the development of the military-industrial complex in the West has been addressed in a number of works, including Roger W. Lotchin, *Fortress California, 1910–1961: From Warfare to Welfare* (Oxford: Oxford University Press, 1992) and Stuart W. Leslie, *The Cold War and American Science: The Military-Industrial-Academic Complex at MIT and Stanford* (New York: Columbia University Press, 1993).
22. Anthony W. Corso, "San Diego: The Anti-City," in Richard M. Bernard and Bradley R. Rice (eds.), *Sunbelt Cities: Politics and Growth since World War II* (Austin, Tex.: University of Texas Press, 1983), pp. 328–345; p. 331.
23. US Census, Population: 1790–1990; US Census, 200, "Foreign Born Population by Sex and Race: 1850 to 1970," *Historical Statistics of the US, Colonial Times to 1970* (Washington, DC: US Government Printing Office, 1975), p. 14; "Persons Obtaining Legal Permanent Resident Status by Country of Birth: 1981–2008," US Census Bureau, *Statistical Abstract of the United States: 2010* (Washington, DC: US Government Printing Office, 2010), p. 48.
24. US Census, Population: 1790–1990; US Census, 2000, "Persons Obtaining Legal Permanent Resident Status by Country of Birth: 1981–2008," US Census Bureau, *Statistical Abstract of the United States: 2010* (Washington, DC: US Government Printing Office, 2010), p. 48.
25. Wiley and Gottlieb, *Empires in the Sun,* p. 123.
26. Wiley and Gottlieb, *Empires in the Sun,* p. 145.
27. There is still work to be done on the ways in which interstate highways have influenced the development of Western cities through which they were built and on the connections that have developed among the cities as a result of these cities.
28. David R. Johnson. "San Antonio: The Vicissitudes of Boosterism," in Richard M. Bernard and Bradley R. Rice (eds.), *Sunbelt Cities: Politics and Growth since World War II* (Austin, Tex.: University of Texas Press, 1983), pp. 235–252; p. 241.
29. Fred Freme, "US Coal Supply and Demand: 2007 Review," US Energy Information Administration: Coal, available online at www.eia.doe.gov/fuelcoal.html (accessed July 2, 2008), p. 5.
30. "Coal-Bearing Areas of the US" (map) National Mining Association; Table 885 "Demonstrated Coal Reserves by Major Producing State: 2004 and 2005," US Census Bureau, *2008 Statistical Abstract: The National Data Book,* available online at www.census.gov/compendia/statab/2008/cats/natural_resources/coal.html (accessed August 2008).
31. James Finch, "US Uranium Production Could Reach 25 Million Pounds by 2016," available online at www.stockinterview.com/News/10232006/Update-Miller.html (accessed September 12, 2008).

32. "'If Not Yucca Mountain, Then What?': an Alternative Plan for Managing Highly Radioactive Waste in the United States," available online at www.ieer.org/fctsheet/yuccaalt.html (accessed September 8, 2008).
33. "State Governments: Revenue by State: 1998" (Table 512), "State and Local Government Finances and Employment," *Statistical Abstract of the United States* (Washington, DC: Government Printing Office, 2000), pp. 297–319; p. 317.
34. Energy Information Administration, "Renewable Electric Power Sector Net Generation by Source by State, 2001," available online at www.eia.doe.gov/cneaf/solar.renewables/page/rea_data/tablec1.html (accessed October 24, 2008).
35. National Mining Association, "Major Metals Operations in the US" (map).
36. "Statistics for All Manufacturing by State: 2004," Table 1, *Annual Survey of Manufacturers.* US Census Bureau, US Department of Commerce, Economics and Statistics Administration, available online at www.fastfacts.census.gov/servlet/IBQTable?_bm=y&-ds_name=AM0431AS101 (accessed October 15, 2008).
37. Clark, "Improbable Los Angeles," pp. 284–285.
38. Luckingham, *The Urban Southwest,* pp. 80–83.
39. Malone and Etulain, *The American West,* p. 219.
40. "State Merchandise Export Totals to the World, 2001–2005," Office of Trade and Economic Analysis, International Trade Administration, US Department of Commerce available online at www.trade.gov/td/industry/otea/state/2005_year_end_totals_01_05.html (accessed September 15, 2008).
41. Federal Reserve Bank of Dallas, *Southwest Economy,* 3(May/June 2007), available online at www.dallasfed.org/research/swe/2007/swe0703d.cfm (accessed October 6, 2008).
42. "State Merchandise Export Totals to Canada and Mexico, 1993–2003," US Department of Commerce, "NAFTA: A State Export Perspective, 1993–2003" (Office of Trade and Economic Analysis, July 2004), pp. 23–24, available online at www.ita.doc.gov/TD/Industry/OTEA/nafta/NAFTA_states.pdf (accessed September 18, 2008).
43. US Border Crossings/Entries by State, 2007, US Department of Transportation, Research and Innovative Technology Administration, Bureau of Transportation Statistics.
44. "US Border Crossings/Entries through El Paso, 1995–2007," US Department of Transportation, Research and Innovative Technology Administration, Bureau of Transportation Statistics.
45. "Growth of US Maritime Containerized Exports and Imports by Coastal Port Region: 1980–2005," US Department of Transportation, Research and Innovative Technology Administration, Bureau of Transportation Statistics.
46. "US Water Transportation Statistical Snapshot," US Department of Transportation Maritime Administration.
47. For a discussion of the origin of the term "Sunbelt" see Richard M. Bernard and Bradley R. Rice, "Introduction," in Richard M. Bernard and Bradley R. Rice (eds.), *Sunbelt Cities: Politics and Growth since World War II* (Austin, Tex.: University of Texas Press, 1983), pp. 1–26; pp. 2–3.
48. Richard White, *"It's Your Misfortune and None of My Own": A History of the American West* (Norman, Okla.: University of Oklahoma Press, 1991), p. 541.
49. Luckingham, *The Urban Southwest,* pp. 87, 107.
50. Peter Schweitzer with Yew Yuan, "Population Density Data by County: An Interactive Database," *Impact of Climate Change and Land Use in the Southwestern United States, Human Impacts on the Landscape,* US Department of the Interior, US Geological Survey. Available online at geochange.er.usgs.gov/sw/resources/population (accessed October 1, 2008).
51. Bradford Luckingham, "Phoenix: The Desert Metropolis," in Richard M. Bernard and Bradley R. Rice (eds.), *Sunbelt Cities: Politics and Growth since World War II* (Austin, Tex.: University of Texas Press, 1983), pp. 309–27; p. 317.
52. Luckingham, "Phoenix: The Desert Metropolis," p. 85.
53. Luckingham, "Phoenix: The Desert Metropolis," p. 83.

54. "Chase Field: A Unique Baseball Experience," available online at http://arizona.diamondbacks.mlb.com/ari/ballpark/chase_info.jsp (accessed September 23, 2008).
55. "The Arizona Diamondbacks," available online at http://arizona.diamondbacks.mlb.com/ari/ballpark/chase_info.jsp (accessed September 23, 2008).
56. Table 1042: "Top 40 Airports in 2005: Passengers Enplaned, 1995 and 2005," US Census Bureau, *2008 Statistical Abstract: The National Data Book,* available online at www.census.gov/compendia/statab/2008/cats/transportation.html (accessed September 20, 2008).
57. "Air Traffic Hubs 2008," US Department of Transportation, Research and Innovative Technology Administration, Bureau of Transportation Statistics.
58. Martin Melosi, "Dallas–Fort Worth: Marketing the Metroplex," in Richard M. Bernard and Bradley R. Rice (eds.), *Sunbelt Cities: Politics and Growth since World War II* (Austin, Tex.: University of Texas Press, 1983), pp. 162–195; pp. 187–188.
59. Table 1042: "Top 40 Airports in 2005: Passengers Enplaned, 1995 and 2005," US Census Bureau, *2008 Statistical Abstract: The National Data Book* available online at www.census.gov/compendia/statab/2008/cats/transportation.html (accessed September 20, 2008).
60. Table 939: "New Privately Owned One-Family Houses Sold by Region and Type of Financing, 1980 to 2006, and by Sales-Price Group, 2006," US Census Bureau, *2008 Statistical Abstract: The National Data Book,* available online at www.census.gov/compendia/statab/2008/cats/construction_housing.html (accessed September 25, 2008).
61. Schweitzer and Yuan, "Population Density Data by County."
62. William Acevedo, Leonard Gaydos, Janet Tilley, Carol Mladinich, Janis Buchanan, Steve Blauer, Kelley Kruger, and Jamie Schubert, "Urban Land Use Change in the Las Vegas Valley," *Impact of Climate Change and Land Use in the Southwestern United States, Human Impacts on the Landscape,* US Department of the Interior, US Geological Survey. December 2003, available online at geochange.er.usgs.gov/sw/changes/anthropogenic/population/las_vegas (accessed October 2, 2008).
63. Jon Unruh and Diana Liverman, "Changing Water Use and Demand in the Southwest," *Impact of Climate Change and Land Use in the Southwestern United States, Human Impacts on the Landscape,* US Department of the Interior, US Geological Survey. December 2003, available online at geochange.er.usgs.gov/sw/impacts/society/water_demand (accessed October 3, 2008).
64. Table 1042: "Top 40 Airports in 2005: Passengers Enplaned, 1995 and 2005," US Census Bureau, *2008 Statistical Abstract: The National Data Book,* available online at www.census.gov/compendia/statab/2008/cats/transportation.html (accessed September 20, 2008).
65. Clark, "Improbable Los Angeles," p. 272.
66. Clark, "Improbable Los Angeles," p. 272.
67. Table 1068: "Commuting to Work by State: 2005," US Census Bureau, 2008, *Statistical Abstract: The National Data Book* available online at www.census.gov/compendia/statab/2008/cats/transportation.html (accessed September 26, 2008).
68. Luckingham, *The Urban Southwest,* p. 99.
69. Luckingham, *The Urban Southwest,* p. 96.
70. "Gross Domestic Product by State: Sorted by 2000–2005 GDP Per Capita Growth," Bureau of Economic Analysis, US Census Bureau, available online at http:/www.economics-charts.com/gdp/GDP-Per-Capita-by-gpc-growth.html accessed October 1, 2008).
71. "Energy Consumption by Sector, Ranked by State, 2005," Energy Information Administration, *State Energy Consumption Estimates, 1960 through 2005,* p. 13, available online at www.eia.doe.gov/states/_seds.html (accessed August 1, 2008).
72. O'Connor, "A Region of Cities," p. 535.
73. Chris Hawley, "Owner Takes Chivas worldwide," *USA Today,* September 17, 2008, p. 6C.
74. "Chase Field: A Unique Baseball Experience."
75. Beau Dure, "Soccer" column, *USA Today,* October 8, 2008, p. 2C; "Rio Tinto Stadium: Stadium Facts," available online at www.riotintostadium.com/stadium_facts.php (accessed October 8, 2008).

INDEX

For Product Safety Concerns and Information please contact our EU
representative GPSR@taylorandfrancis.com
Taylor & Francis Verlag GmbH, Kaufingerstraße 24, 80331 München, Germany

www.ingramcontent.com/pod-product-compliance
Ingram Content Group UK Ltd.
Pitfield, Milton Keynes, MK11 3LW, UK
UKHW050755180926
475701UK00007B/1060

9781032402604